T0344963

Introduction to Quantum Optics
From the Semi-classical Approach to Quantized Light

Covering a number of important subjects in quantum optics, this textbook is an excellent introduction for advanced undergraduate and beginning graduate students, familiarizing readers with the basic concepts and formalism as well as the most recent advances.

The first part of the textbook covers the semi-classical approach where matter is quantized, but light is not. It describes significant phenomena in quantum optics, including the principles of lasers. The second part is devoted to the full quantum description of light and its interaction with matter, covering topics such as spontaneous emission, and classical and non-classical states of light. An overview of photon entanglement and applications to quantum information is also given. In the third part, nonlinear optics and laser cooling of atoms are presented, where the use of both approaches allows for a comprehensive description. Each chapter describes basic concepts in detail, and more specific concepts and phenomena are presented in 'complements'.

Gilbert Grynberg was a CNRS Senior Scientist at the Laboratoire Kastler Brossel at the Université Pierre et Marie Curie Paris 6, and a Professor at the Ecole Polytechnique. He was a pioneer in many domains, including atomic spectroscopy, nonlinear optics and laser-cooled atoms in optical lattices.

Alain Aspect is a CNRS Senior Scientist and Professor at the Institut d'Optique and the Ecole Polytechnique. A pioneer of the field of quantum entanglement, his research covers quantum optics, laser cooling of atoms, atom optics, Bose–Einstein condensates, atom lasers and quantum atom optics. He was awarded the 2010 Wolf Prize in Physics.

Claude Fabre is a Professor in the Laboratoire Kastler Brossel at the Université Pierre et Marie Curie Paris 6, and a senior member of the Institut Universitaire de France. His fields of research are quantum optics, atomic and laser physics, both experimentally and theoretically.

Introduction to Quantum Optics

From the Semi-classical Approach to Quantized Light

GILBERT GRYNBERG
Ecole Normale Supérieure, Paris
Ecole Polytechnique

ALAIN ASPECT
Institut d'Optique and Ecole Polytechnique,
Palaisean

CLAUDE FABRE
Université Pierre et Marie Curie and Ecole Normale
Supérieure, Paris

With a Foreword by Claude Cohen-Tannoudji

CAMBRIDGE
UNIVERSITY PRESS

University Printing House, Cambridge CB2 8BS, United Kingdom

One Liberty Plaza, 20th Floor, New York, NY 10006, USA

477 Williamstown Road, Port Melbourne, VIC 3207, Australia

4843/24, 2nd Floor, Ansari Road, Daryaganj, Delhi - 110002, India

79 Anson Road, #06-04/06, Singapore 079906

Cambridge University Press is part of the University of Cambridge.

It furthers the University's mission by disseminating knowledge in the pursuit of education, learning and research at the highest international levels of excellence.

www.cambridge.org
Information on this title: www.cambridge.org/9780521551120

Original edition: *Introduction aux Lasers et á l'Optique*, Ellipses 1997

Published with the assistance of the French Ministry of Culture–National Book Centre

Ouvrage publié avec le concours du Ministère français chargé de la culture – Centre National du Livre

First published 2010

A catalogue record for this publication is available from the British Library

ISBN 978-0-521-55112-0 Hardback

Contents

Contents

Foreword

Atomic, molecular and optical physics is a field which, during the last few decades, has known spectacular developments in various directions, like nonlinear optics, laser cooling and trapping, quantum degenerate gases, quantum information. Atom–photon interactions play an essential role in these developments. This book presents an introduction to *quantum optics* which, I am sure, will provide an invaluable help to the students, researchers and engineers who are beginning to work in these fields and who want to become familiar with the basic concepts underlying electromagnetic interactions.

Most books dealing with these subjects follow either a semi-classical approach, where the field is treated as a classical field interacting with quantum particles, or a full quantum approach where both systems are quantized. The first approach is often oversimplified and fails to describe correctly new situations that can now be investigated with the development of sophisticated experimental techniques. The second approach is often too difficult for beginners and lacks simple physical pictures, very useful for an initial understanding of a physical phenomenon. The advantage of this book is that it gives both approaches, starting with the first, illustrated by several simple examples, and introducing progressively the second, clearly showing why it is essential for the understanding of certain phenomena. The authors also clearly demonstrate, in the case of non-linear optics and laser cooling, how advantageous it may be to combine both approaches in the analysis of an experimental situation and how one can get from each point of view useful, complementary physical insights. I believe that this challenge to present and to illustrate both approaches in a single book has been taken up successfully. Whatever their ultimate interests, the readers of this work will be exposed to an important example of a broad and vibrant field of research and they will better understand the intellectual enrichment and the technical developments which result from it.

To write a book on such a broad topic, the authors must obviously possess wide knowledge of the field, they must have thought long and hard about the basic concepts and about the different levels of complexity with which one can approach the topics. They must also have a deep and concrete knowledge about experimental and technical details and the many problems which daily confront a laboratory researcher. Having worked extensively with them, I know the authors of this work fulfil these requirements. I have the highest admiration for their enthusiasm, their scientific rigour, their ability to give simple and precise physical explanations, and their quest to illuminate clearly the difficult points of the subject without oversimplification. Each of them has made many original contributions to the development of this important field of physics, and they and their younger collaborators for this book work at the cutting edge of modern quantum optics. In reading the book, I am therefore not surprised to find their many fine qualities reflected in the text. The general organisation of the main chapters and complementary sections allows reading on many different levels. When the authors discuss a new physical problem, they begin the analysis with the simplest possible model. A large variety of experiments and applications are presented with clear diagrams and explanations and with constant attention to highlighting the guiding principles, the orders of magnitude and the problems which remain open.

This work will allow a broad audience an easier access to a field of science which continues to see spectacular developments. I believe that science is not simply a matter of exploring new horizons. One must also make the new knowledge readily available and we have in this book, a beautiful example of such a pedagogical effort. I would like finally to evoke the memory of Gilbert Grynberg who participated with Alain Aspect and Claude Fabre in the writing of a preliminary, much less developed, French version of this book and who passed away in 2003. Gilbert was an outstanding physicist, a fine person, and had an exceptional talent for explaining in the clearest possible way the most difficult questions. I think that the present book is the best possible tribute to be paid to him.

Claude Cohen-Tannoudji
Paris, September 2009

Preface

Since its invention in 1960, the laser has revolutionized both the study of optics and our understanding of the nature of light, prompting the emergence of a new field, *quantum optics*. Actually, it took decades until the words quantum optics took their current precise meaning, referring to phenomena which can be understood only by quantizing the electromagnetic field describing light. Surprisingly enough, such quantum optics phenomena hardly existed at the time that the laser was invented, and almost all optics effects could be fully understood by describing light as a classical electromagnetic field; the laser was no exception. As a matter of fact, to understand how a laser works, it suffices to use the *semi-classical description of matter–light interaction*, where the laser amplifying medium, made of atoms, molecules, ions or semi-conductors, is given a quantum mechanical treatment, but light itself is described by classical electromagnetic waves.

The first part of our book is devoted to presentation of the semi-classical approach and its use in describing various optical phenomena. It includes an elementary exposition of the physics of lasers, and some applications of this ubiquitous device. After recalling in **Chapter 1** some basic results of the quantum mechanical description of interaction induced transitions between the atomic energy levels, we use these results in **Chapter 2** to show how the interaction of a quantized atom with a classical electromagnetic wave leads to absorption or stimulated emission, and to derive the process of laser amplification that happens when a wave propagates in an inverted medium. **Chapter 3** gives an elementary exposition of the physics of laser sources and of the properties of laser light.

Although the quantum theory of light existed since its development by Dirac in the early 1930s, quantum optics theory in its modern sense started when Roy Glauber showed, in the early 1960s, how to apply it to classical optics devices such as the Michelson stellar interferometer or the Hanbury Brown and Twiss intensity interferometer. At that time it could have appeared to be an academic exercise without consequence, since the only known phenomenon that demanded quantization of light was spontaneous emission, and it was not clear whether quantum theory was at all useful for describing light freely propagating far from the source. Actually, Glauber developed a clear quantum formalism to describe optics phenomena, and introduced the important notion of quasi-classical states of light, a theoretical tool that allowed physicists to understand why all available sources of light, including lasers, delivered light whose properties could be totally understood in the framework of the semi-classical approach. But in doing so, he paved the way for the discovery of new phenomena which can be understood only if light is considered as a quantum system. It became possible to build sources delivering single photon wave packets, pairs of entangled photons, squeezed beams of light…

The second part of our book is devoted to the presentation of the quantum theory of light and its interaction with matter, and its use in describing many phenomena of modern quantum optics. We show in **Chapter 4** how it is possible to write the dynamical equations of a classical electromagnetic field, i.e. Maxwell equations, in a form allowing us to use the canonical quantization procedure to quantize the electromagnetic field, and obtain the notion of a photon. We then use our results, in **Chapter 5**, to describe some fully quantum effects observed in experiments with *single photons*, *squeezed light or pairs of entangled photons*. It is remarkable that many of these experiments, whose first goal was to demonstrate the highly counter-intuitive, non-classical properties of new types of light states, turned out to stimulate the emergence of a new field, *quantum information*, where one uses such properties to implement new ways of processing and transmitting data. In **Chapter 6**, we show how to use the quantum optics formalism to describe the interaction between light and atoms. We will then revisit in this new framework the phenomena of absorption and stimulated emission, already studied in Chapter 2. Moreover, we will now be able to give a *consistent treatment of spontaneous emission*.

Having introduced the full quantum optics formalism and reviewed some remarkable phenomena that could not have been discovered without such a formalism, we would not like to leave the reader with the impression that he/she can now forget the semi-classical approach. Both approaches are definitely useful. On the one hand, there is no reason to use the, usually more involved, fully quantum analysis, when the situation does not demand it. After all, nobody would use quantum mechanics to describe the motion of planets. Similarly, no experimentalist studying fusion plasmas with intense lasers would start using the quantum formalism of light. What is important then is to be able to recognize when the full quantum theory is necessary, and when one can content oneself with the semi-classical model. To help the reader to develop their intuition about this point, we present, in the third part of this book, two topics, non-linear optics in **Chapter 7**, laser cooling and trapping of atoms in **Chapter 8**, where it is convenient to 'juggle' between the two approaches, each being better adapted to one or the other particular phenomenon. As 'the cherry on the cake', we will give in Chapter 8 an elementary presentation of atomic Bose–Einstein condensates, and emphasize the analogy between such a system, where all atoms are described by the same matter wave, and a laser beam where all photons are described by the same mode of the electromagnetic field. When we started to write the first French version of this book, we had never dreamt of being able to finish it with a presentation on *atom lasers*.

This book is composed of **chapters**, in which we present the fundamental concepts and some applications to important quantum optics phenomena, and of **complements**, which present supplementary illustrations or applications of the theory presented in the main chapter. The choice of these examples is, of course, somewhat arbitrary. We present them as a snap-shot of the current state of a field which is rapidly evolving. Complements of another type are intended to give some supplementary details about a derivation or about concepts presented in the chapter.

The prerequisite for using this book is to have followed an elementary course on both electromagnetism (Maxwell's equations) and quantum mechanics (Schrödinger formulation in the Dirac formalism of bras and kets, with application to the harmonic oscillator). The book is then self-consistent, and can be used for an advanced undergraduate, or for a first graduate course on quantum optics. Although we do not make use of the most advanced tools studied at graduate school, we make all efforts to provide the reader with solid derivations of the main results obtained in the chapters. For example, to quantize electromagnetic waves, first in free space, and then in interaction with charges, we do not use the Lagrangian formalism, but we introduce enough elements of the Hamiltonian formalism to be able to apply the canonical quantization rules. We are thus able to provide the reader with a solid derivation of the basic quantum optics formalism rather than bringing it in abruptly. On the other hand, when we want to present in a Complement a particularly important and interesting phenomenon, we do not hesitate to ask the reader to admit a result which results from more advanced courses.

We have done our best to merge the French teaching tradition of logical and deductive exposition with the more pragmatic approach that we use as researchers, and as advisors to Ph.D. and Masters students. We have taught the content of this book for many years to advanced undergraduate or beginning graduate students, and this text represents the results of our various teaching experiences.

Acknowledgements

In this book, we refer to a number of textbooks in which general elementary results of quantum mechanics are established, in particular the book by Jean-Louis Basdevant and Jean Dalibard,[1] which we indicate by the short-hand notation 'BD', and the one by Claude Cohen-Tannoudji, Bernard Diu and Franck Laloë,[2] which we denote by 'CDL'. On the more advanced side, we sometimes refer to more rigorous demonstrations, or to more advanced developments, that can be found in the two books written by Claude Cohen-Tannoudji, Jacques Dupont-Roc and Gilbert Grynberg, to which we refer under the short-hand notations 'CDG I' and 'CDG II', respectively.[3,4]

It is not possible to mention all those who have contributed to or influenced this work. We would first like to acknowledge, however, our principal inspiration, Claude Cohen-Tannoudji, whose lectures at the Collège de France we have had the good fortune to be able to follow for three decades. At the other end of the spectrum, we also owe a lot to our students at Ecole Polytechnique, Ecole Normale Supérieure, Institut d'Optique Graduate School, Université Pierre et Marie Curie, as well as the many graduate students we have advised towards Masters or Ph.D. work. By their sharp questioning, never content with a vague answer, they have forced us to improve our lectures year upon year. We cannot cite all of the colleagues with whom we have taught, and from whom we have borrowed many ideas and materials, but we cannot omit to mention the names of Manuel Joffre, Emmanuel Rosencher, Philippe Grangier, Michel Brune, Jean-François Roch, François Hache, David Guéry-Odelin, Jean-Louis Oudar, Hubert Flocard, Jean Dalibard, Jean-Louis Basdevant. In addition, Philippe Grangier was kind enough to write Complement 5E on quantum information.

Martine Maguer, Dominique Toustou, and all the team of Véronique Pellouin at the Centre Polymedia of Ecole Polytechnique have done an impressive and professional job in preparing the manuscript with its figures. We would like also to thank the Centre National du Livre, of the French Ministry of Culture, for its important financial support in the translation of our French text.

[1] J.-L. Basdevant and J. Dalibard, *Quantum Mechanics*, Springer (2002).

[2] C. Cohen-Tannoudji, B. Diu and F. Laloë, *Quantum Mechanics*, Wiley (1977).

[3] C. Cohen-Tannoudji, J. Dupont-Roc and G. Grynberg, *Photons and Atoms – Introduction to quantum electrodynamics*, Wiley (1989).

[4] C. Cohen-Tannoudji, J. Dupont-Roc and G. Grynberg, *Atom-photon Interactions: Basic processes and applications*, Wiley (1992).

Special acknowledgement

This book has three authors, who wrote the original French textbook on which it is based.[5] Sadly, as we had just started to prepare the English version, **Gilbert Grynberg** passed away, and for several years we were discouraged and not able to carry on working on the English version. Eventually, we realized that the best demonstration of all that we owe to our former friend and colleague was to resume this project. But we realized then that almost a decade after writing the French version, quantum optics had evolved tremendously, and we had also personally evolved in the ways in which we understood and taught the subject. The original French book, therefore, had not only to be translated but also widely revised and updated. In this long-term enterprise, we have been fortunate to have fantastic help from our younger colleagues (and former students) **Fabien Bretenaker** and **Antoine Browaeys**. For the past three years they have devoted innumerable hours to helping us complete the revised version, and without their help this would not have been possible. There is not a single chapter that has not been strongly influenced by their thorough criticisms, their strong suggestions, and their contributions to the rewriting of the text, not to speak of the double checking of equations. Moreover, they bring to this book the point of view of a new generation of physicists who have been taught quantum optics in its modern sense, in contrast to we who have seen it developing while we were already engaged in research. For their priceless contribution, we can only express to Fabien Bretenaker and Antoine Browaeys our immense gratitude. Gilbert would have been happy to have such wonderful collaborators.

Alain Aspect and Claude Fabre,
Palaiseau, Paris, July 2009.

[5] Gilbert Grynberg, Alain Aspect, Claude Fabre, *Introduction aux lasers et à l'Optique Quantique*, cours de l'Ecole Polytechnique, Ellipses, Paris (1997).

PART I

SEMI-CLASSICAL DESCRIPTION OF MATTER–LIGHT INTERACTION

1 The evolution of interacting quantum systems

In this work we shall study the interaction of matter and light. In so doing we shall rely heavily on the description of such processes provided by quantum mechanics. This appears on a number of levels: firstly, a *quantum description of matter* is indispensable if one wants to understand *on the microscopic scale* the different kinds of interaction processes that can occur. Secondly, a *quantum description of light* often turns out to be useful, sometimes necessary, to better understand these processes. We shall study phenomena such as spontaneous emission, which can only be properly treated by a theory taking into account the quantum nature of both light and matter.

In the following chapters we shall address, amongst others, the following question: 'given an atom prepared at a given time in a particular state and subjected from this time onwards to electromagnetic radiation, what is the state of the atom and radiation at any later moment in time?' In order to be able to answer this question it will be necessary for us to know how to calculate the evolution of a quantum system in a small number of typical situations. These methods we shall demonstrate in the first chapter.

The evolution of the coupled atom–light system depends on the temporal dependence of the applied light field, which could, for example, be applied from a given moment and thereafter remain unchanged in intensity, or, perhaps, be appreciable only for a finite period of time (pulsed excitation). We shall see that the nature of the evolution depends also on the *structure of the energy spectrum* of the system considered, whether it is describable by a set of *discrete levels* well separated in energy, or by a *continuum*.

This chapter starts with a brief reminder of some elementary results of quantum mechanics (Section 1.1). In the following section we demonstrate the use of a perturbative method to calculate the probability of the transition of a quantum state from a given initial to a given final state under the influence of an interaction. Finally in Section 1.3 we shall study the case in which the initial state is coupled to a very large number of closely spaced energy levels (we speak of a *quasi-continuum* of states). We derive an important result for the transition probability, known as *Fermi's golden rule*. Finally, in the conclusion we discuss the different regimes of temporal evolution that can be obtained.

This chapter is rounded off by two *complements*. Complement 1A outlines a simple model which enables us to understand the transition between the two limiting situations of Sections 1.2 (two discrete coupled levels) and 1.3 (a discrete level coupled to a continuum). Complement 1B addresses the situation where a quantum system is interacting with a random perturbation, whose frequency spectrum is broad (broadband excitation). In that case, one finds a transition probability as in Section 1.3.

1.1 Review of some elementary results of quantum mechanics

We start by recalling some important results relating to a quantum system described by a Hamiltonian $\hat{H}_0$ independent of time.[1] We designate by $|n\rangle$ and E_n the eigenstates and eigenenergies of $\hat{H}_0$. Suppose at time $t = 0$ the system is in the most general state:

$$|\psi(0)\rangle = \sum_n \gamma_n |n\rangle . \tag{1.1}$$

Using the Schrödinger equation,[2] one can show that the system is found at a later time in the state:

$$|\psi(t)\rangle = \sum_n \gamma_n \mathrm{e}^{-\mathrm{i}E_n t/\hbar} |n\rangle . \tag{1.2}$$

The probability of finding the system in the state $|\varphi\rangle$ is then

$$P_\varphi(t) = |\langle\varphi|\,\psi(t)\rangle|^2 \tag{1.3}$$

and the probability that the system has made a transition from state $|\psi(0)\rangle$ to the state $|\varphi\rangle$ between times 0 and t is therefore

$$P_{\psi(0)\to\varphi}(t) = |\langle\varphi|\,\psi(t)\rangle|^2 . \tag{1.4}$$

In particular, if the system is initially prepared in the eigenstate $|n\rangle$ of $\hat{H}_0$, it is given at any later time t by the state vector

$$|\psi(t)\rangle = \mathrm{e}^{-\mathrm{i}E_n t/\hbar} |n\rangle . \tag{1.5}$$

The probability of finding it later in a state $|m\rangle$ of $\hat{H}_0$ with $m \neq n$ is then zero:

$$P_{n\to m}(t) = |\langle m|\,\psi(t)\rangle|^2 = 0. \tag{1.6}$$

For example, the electron of an atom of hydrogen initially in the state $|n, l, m\rangle$ would remain indefinitely in this state if the atom were not coupled to the exterior environment. In practice it undergoes transitions to different levels under the effect of *exterior interactions* of various origins: collisions with ions, atoms or electrons, oscillating electromagnetic fields etc. The coupling with the quantized electromagnetic field is also responsible for spontaneous transitions between an excited and lower energy levels accompanied by the emission of a photon. This is the process of *spontaneous emission* that we shall treat in Chapter 6.

In these different examples the evolution of the system is driven by a *time-dependent* Hamiltonian, this time dependence being sinusoidal in the case of an electromagnetic field, impulsive in the case of a collision. In general, the state vector describing the system cannot be calculated *exactly* for all time. We show in the next section, however, that one can obtain an exact expression for transition probabilities in the form of a series expansion.

[1] In this work we shall distinguish the operators of quantum theory by a hat, e.g. $\hat{H}$.
[2] See CDL, § III.D.2.

A problem that is formally similar, that we shall also treat, is that in which the total Hamiltonian $\hat{H}_0 + \hat{W}$ is independent of time, but in which the system is prepared in an eigenstate of $\hat{H}_0$ and detected at a later instant, t', in another eigenstate of $\hat{H}_0$. The corresponding transition probabilities will be calculated using a similar series method since, as we shall demonstrate, this problem is mathematically identical to the case in which the coupling $\hat{W}$ is applied transiently in the interval of time between $t = 0$ and $t = t'$.

1.2 Transition between discrete levels induced by a time-dependent perturbation

1.2.1 Presentation of the problem

We consider a system described by a Hamiltonian

$$\hat{H} = \hat{H}_0 + \hat{H}_1(t). \tag{1.7}$$

$\hat{H}_0$ is independent of time, its eigenstates and eigenvalues being denoted by $|n\rangle$ and E_n:

$$\hat{H}_0 |n\rangle = E_n |n\rangle . \tag{1.8}$$

$\hat{H}_1(t)$ is an interaction term of which the matrix elements between the eigenstates of $\hat{H}_0$ are assumed small compared to the energy differences between these eigenstates, $\langle n|\hat{H}_0|m\rangle \ll |E_n - E_m|$. At this stage the time dependence of $\hat{H}_1(t)$ is left arbitrary, it could for example be constant for a finite interval of time and zero outside this interval.

The coupling $\hat{H}_1(t)$ will be capable of inducing transitions between different eigenstates of $\hat{H}_0$. Here we propose to calculate the corresponding transition probabilities $P_{n\to m}(t)$ supposing, for simplicity, that the levels are non-degenerate in energy. More general treatments can be found in standard texts on quantum mechanics.[3]

1.2.2 Examples

Before studying the mathematical development, we present two examples of physical systems well described by the model we propose to adopt. These will be of use to us later to provide a physical illustration of the results obtained.

Interaction of an atom with a classical electromagnetic field

Consider an atom[4] described by the atomic Hamiltonian $\hat{H}_0$ and which interacts with an incident classical electromagnetic wave of which the electric field at the position of the (stationary) atom is

[3] See, for example, CDL, Chapter XIII.

[4] More precisely, we consider for simplicity a one-electron atom such as hydrogen.

$$\mathbf{E}(t) = \mathbf{E}\cos(\omega t + \varphi). \tag{1.9}$$

We shall see in Chapter 2 that the interaction of the atom with the field can, to a good approximation, be written in terms of an *electric dipole* coupling:

$$\hat{H}_1(t) = -\hat{\mathbf{D}} \cdot \mathbf{E}(t), \tag{1.10}$$

where $\hat{\mathbf{D}}$ is the electric dipole of the atom,

$$\hat{\mathbf{D}} = q\hat{\mathbf{r}}. \tag{1.11}$$

Here q is the electronic charge and $\mathbf{r}$ the radius vector between the atomic nucleus and its valence electron.

Under the action of $\hat{H}_1(t)$ the electron, initially in the eigenstate $|n, l, m\rangle$ of $\hat{H}_0$, will be able to undergo transitions to other states $|n', l', m'\rangle$. If the energy of the latter state is higher than that of the former the energy necessary to excite the atom is taken from the electromagnetic field (this is *absorption*), if it is lower there is a transfer of energy from the atom to the field (this is *stimulated emission*). We shall come back to these processes and their consequences in Chapter 2.

Collision processes

We consider a stationary atom A, of which the internal energy levels are the eigenstates of a Hamiltonian, $\hat{H}_0$ and suppose that another particle B passes in the neighbourhood of A (Figure 1.1).

$\hat{V}$ is the interaction potential between the collider B and the atom A; it depends on the distance R between B and A. For atom A this interaction is represented by an operator acting in the space of states of A. Its matrix elements between those states are a function of R and tend to zero as R becomes very large. Since R varies in time, *the interaction Hamiltonian itself also depends on time.* If before the collision, when the atoms are far apart, atom A is in the state $|n\rangle$, there is the possibility that after the collision it will be found in a different state $|m\rangle$. If the energies of the initial and final states are the same, the collision is described as *elastic*, otherwise it is termed *inelastic*. This type of collision-induced transition is responsible, for example, for the excitation of atoms in a discharge lamp (a neon lamp, for example) or, as we shall see in Chapter 3, in certain kinds of laser.

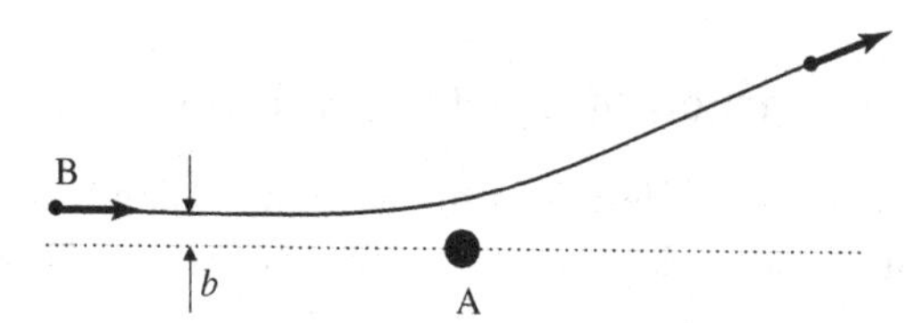

Figure 1.1 Collisional interaction between a particle B and an atom A. The distance of closest approach, b, is called the impact parameter. The interaction depends only on the R = AB distance, hence $\hat{V}(t) = \hat{V}(|\mathbf{R}(t)|)$.

1.2.3 Perturbation series expansion of the system wavefunction

The evolution of atom A is determined by solving the Schrödinger equation using the Hamiltonian of Equation (1.7). To this end we are going, in the following, to employ a method of approximate solution based on perturbation theory, which is valid provided the matrix elements of $\hat{H}_1(t)$ are small compared to those of $\hat{H}_0$.[5] In order to be able to identify the successive orders of the perturbation expansion more easily we rewrite $\hat{H}_1(t)$ in the form

$$\hat{H}_1(t) = \lambda \hat{H}_1'(t), \tag{1.12}$$

where $\hat{H}_1'(t)$ has matrix elements of the same order of magnitude as those of $\hat{H}_0$ and where λ is a real, dimensionless parameter much smaller than unity, which characterizes the relative strength of the interaction $\hat{H}_1(t)$. In the first of our two examples, λ is proportional to the amplitude of the incident electric field, in the second it is a function of the impact parameter b. In each case it will be possible to find experimental conditions in which the approximation $\lambda \ll 1$ is valid (weak electric field, large impact parameter).

The Schrödinger equation for the system is

$$\mathrm{i}\hbar \frac{\mathrm{d}}{\mathrm{d}t} |\psi(t)\rangle = \left(\hat{H}_0 + \lambda \hat{H}_1'(t)\right) |\psi(t)\rangle . \tag{1.13}$$

Expanding $|\psi(t)\rangle$ in the basis of eigenstates of $\hat{H}_0$ we get

$$|\psi(t)\rangle = \sum_n \gamma_n(t)\, \mathrm{e}^{-\mathrm{i}E_n t/\hbar} |n\rangle . \tag{1.14}$$

Here we have written the coefficient of the ket $|n\rangle$ as a product of terms $\gamma_n(t)$ and $\exp(-\mathrm{i}E_n t/\hbar)$. This separation permits us to take into account the free evolution of the system under the influence of $\hat{H}_0$ alone, since if $\hat{H}_1(t)$ is zero the $\gamma_n(t)$ are *constant* by virtue of Equation (1.2). This will facilitate later developments.

Next, we project Equation (1.13) on an eigenstate $|k\rangle$ of $\hat{H}_0$:

$$\begin{aligned} \mathrm{i}\hbar \frac{\mathrm{d}}{\mathrm{d}t} \langle k| \psi(t)\rangle &= \langle k| \hat{H}_0 |\psi(t)\rangle + \lambda \langle k| \hat{H}_1' |\psi(t)\rangle \\ &= E_k \langle k| \psi(t)\rangle + \lambda \sum_n \langle k| \hat{H}_1'(t) |n\rangle \langle n| \psi(t)\rangle, \end{aligned} \tag{1.15}$$

where we have used the *closure relation*: $\sum_n |n\rangle\langle n| = \hat{1}$.

Using expression (1.14) for $|\psi(t)\rangle$, we rewrite (1.15) in the form:

$$\begin{aligned} \left[E_k \gamma_k(t) + \mathrm{i}\hbar \frac{\mathrm{d}}{\mathrm{d}t}\gamma_k(t)\right] \mathrm{e}^{-\mathrm{i}E_k t/\hbar} &= E_k \gamma_k(t)\, \mathrm{e}^{-\mathrm{i}E_k t/\hbar} \\ &\quad + \lambda \sum_n \langle k| \hat{H}_1'(t) |n\rangle\, \gamma_n(t)\, \mathrm{e}^{-\mathrm{i}E_n t/\hbar}. \end{aligned} \tag{1.16}$$

[5] More precisely, in the basis $\{|n\rangle\}$ of the eigenstates of $\hat{H}_0$, the off-diagonal matrix elements $|\langle n|\hat{H}_1|m\rangle|$ must be small compared to the corresponding energy separations $|E_n - E_m|$.

The terms proportional to E_k in the right and left-hand sides simplify to give

$$i\hbar \frac{d}{dt}\gamma_k(t) = \lambda \sum_n \langle k| \hat{H}_1'(t) |n\rangle \, e^{i(E_k - E_n)t/\hbar} \, \gamma_n(t), \tag{1.17}$$

which is a (possibly infinite) system of differential equations. This system is *exact*, no approximations having been made thus far.

The coefficients $\gamma_k(t)$ depend on λ. Perturbation theory consists of developing $\gamma_k(t)$ as a power series in λ (which, we recall is much smaller than unity):

$$\gamma_k(t) = \gamma_k^{(0)}(t) + \lambda \gamma_k^{(1)}(t) + \lambda^2 \gamma_k^{(2)}(t) + ... \tag{1.18}$$

In substituting this series in (1.17) we can collect together terms of the same order in λ. In this way we obtain:

- to order 0

$$i\hbar \frac{d}{dt}\gamma_k^{(0)}(t) = 0; \tag{1.19}$$

- to order 1

$$i\hbar \frac{d}{dt}\gamma_k^{(1)}(t) = \sum_n \langle k| \hat{H}_1'(t) |n\rangle \, e^{i(E_k - E_n)t/\hbar} \, \gamma_n^{(0)}(t); \tag{1.20}$$

- to order r

$$i\hbar \frac{d}{dt}\gamma_k^{(r)}(t) = \sum_n \langle k| \hat{H}_1'(t) |n\rangle \, e^{i(E_k - E_n)t/\hbar} \, \gamma_n^{(r-1)}(t). \tag{1.21}$$

This system of equations can be solved *iteratively*. In fact the zeroth-order terms are already known: they are *constants determined by the initial state of the system.* On substituting these terms in (1.20) the terms of order one, $\gamma_k^{(1)}(t)$, can be found. These then lead to an expression for $\gamma_k^{(2)}(t)$, and so on. Thus it is possible, in principle, to determine successively all the terms in the expansion (1.18).

1.2.4 First-order theory

Transition probability

Suppose that at initial time t_0, the system is prepared in an eigenstate $|i\rangle$ of $\hat{H}_0$. It follows that all $\gamma_k(t_0)$ are zero except for $\gamma_i(t_0)$ which is equal to one. The solution of (1.19) is then:

$$\gamma_k^{(0)}(t) = \delta_{ki}. \tag{1.22}$$

We now consider the possibility of transitions to level $|k\rangle$ different from the initial state ($k \neq i$). Substituting the result (1.22) into (1.20) and integrating over time we find, for $\gamma_k^{(1)}$

$$\gamma_k^{(1)}(t) = \frac{1}{i\hbar} \int_{t_0}^{t} dt' \, \langle k| \hat{H}_1'(t') |i\rangle \, e^{i(E_k - E_i)t'/\hbar}. \tag{1.23}$$

The probability amplitude for finding the system in the state $|k\rangle$ at time t is, according to (1.14) and (1.18), equal to (within a phase factor)

$$\gamma_k^{(0)}(t) + \lambda\gamma_k^{(1)}(t) + ... \tag{1.24}$$

For a state $|k\rangle$ different from $|i\rangle$ the zeroth-order term is zero. From this we deduce, also using (1.23), that the amplitude for the transition $|i\rangle$ to $|k\rangle$ to first order and to within a phase factor is

$$S_{ki} = \lambda\gamma_k^{(1)}(t) = \frac{1}{\mathrm{i}\hbar}\int_{t_0}^{t} \mathrm{d}t' \langle k|\hat{H}_1(t')|i\rangle \mathrm{e}^{\mathrm{i}(E_k-E_i)t'/\hbar}, \tag{1.25}$$

since, according to (1.12), $\lambda\hat{H}'_1(t')$ is equal to $\hat{H}_1(t')$. The probability of finding the system in the state $|k\rangle$ is given by the square modulus of (1.25), that is

$$P_{i\to k} = \frac{1}{\hbar^2}\left|\int_{t_0}^{t} \mathrm{d}t' \langle k|\hat{H}_1(t')|i\rangle \mathrm{e}^{\mathrm{i}(E_k-E_i)t'/\hbar}\right|^2. \tag{1.26}$$

The formulae (1.25) and (1.26) are the important results of first-order, time-dependent perturbation theory. We shall use these in what follows. Notice, however, that this perturbative approach is only valid if

$$P_{i\to k} \ll 1, \tag{1.27}$$

that is, effectively, that the interaction Hamiltonian $\hat{H}_1$ induces only small effects in first order so that the full perturbation expansion of (1.18), which includes also the effects of higher-order terms, will converge rapidly. Condition (1.27) is in fact a *necessary* condition, but not sufficient that first-order perturbation theory can be accurately applied.

Example of a collisional process: qualitative study of the accessible energy range

In the following, we shall show that the properties of the *Fourier transform* applied to Equation (1.26) enable us to predict the range of energy over which atomic energy levels can be excited during a collision.

Suppose, for simplicity, that the interaction term $\hat{H}_1(t)$ is of the form:

$$\hat{H}_1(t) = \hat{W} f(t), \tag{1.28}$$

where $\hat{W}$ is an operator acting on atomic variables and $f(t)$ is a real function of time which tends to zero when $t \to \pm\infty$ and attains its maximum value at $t = 0$ (see Figure 1.2). We suppose that before the collision ($t_0 = -\infty$) the system is in a state $|i\rangle$. The amplitude for finding it in a state $|k\rangle$ after the collision ($t_0 = +\infty$) is

$$S_{ki} = \frac{W_{ki}}{\mathrm{i}\hbar}\int_{-\infty}^{+\infty} \mathrm{d}t\, f(t)\, \mathrm{e}^{\mathrm{i}(E_k-E_i)t/\hbar}, \tag{1.29}$$

where W_{ki} is the matrix element $W_{ki} = \langle k\,|\hat{W}|\,i\rangle$.

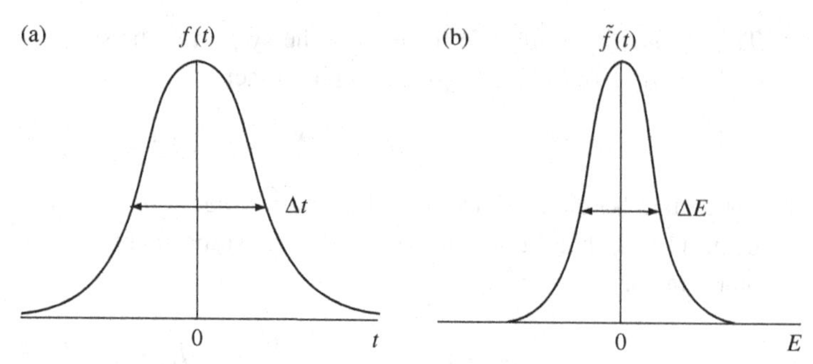

Figure 1.2 The Fourier transform of the function $f(t)$, centred on the time origin and of width Δt (a) is a function of energy centred on $E = 0$ of width $\Delta E \approx \hbar/\Delta t$ (b).

Introducing the Fourier transform, $\tilde{f}(E)$ of the function $f(t)$:

$$\tilde{f}(E) = \frac{1}{\sqrt{2\pi\hbar}} \int_{-\infty}^{+\infty} \mathrm{d}t\, f(t)\, \mathrm{e}^{\mathrm{i}Et/\hbar}, \tag{1.30}$$

we obtain the following expression for the transition probability $P_{i\to k}$,

$$P_{i\to k} = \frac{2\pi}{\hbar} |W_{ki}|^2 \left|\tilde{f}(E_k - E_i)\right|^2, \tag{1.31}$$

which depends on the value of $\tilde{f}(E)$ taken at $E = E_k - E_i$.

We now make use of a well-known property of Fourier transforms (Figure 1.2): if the width of the function $f(t)$ is Δt then that of its Fourier transform is of the order of $\hbar/\Delta t$. The expression (1.31) then shows that if the collisional interaction is of duration Δt, the energy levels for which

$$|E_k - E_i| < \frac{\hbar}{\Delta t} \tag{1.32}$$

will have a significant probability of being populated.

Consider the case of a collision in which the interaction is a decreasing function of the distance between the collision partners (see Figure 1.1). The collision duration is of the order of b/v, where b is the impact parameter and v is the relative velocity of the particles. The inequality (1.32) implies that only the states $|k\rangle$ such that

$$|E_k - E_i| \leq \frac{\hbar v}{b} \tag{1.33}$$

will be appreciably populated as a result of the collision.

Comment Formula (1.33) shows that for an impact parameter of the order of 10 nm a collision with an atom of speed 10^7 m.s^{-1} is necessary to excite a ground state hydrogen atom to its first excited state (an energy transfer of about 10 eV). This is a very large velocity associated with a kinetic energy of the order of 1 MeV, very large compared to the excitation energy of the atom. To excite a hydrogen atom with lower energy particles one must consider impact parameters of the order of a Bohr radius. For this type of 'close' collision the matrix elements of $\hat{H}_1$ are not small compared to the energy differences $E_n - E_m$, and the hypothesis of a perturbative interaction is no longer valid.

Case of a constant perturbation suddenly 'switched on'

It often arises that a system is suddenly at $t = 0$ made to interact with a perturbation $\hat{W}$ which has a constant value at all later times.[6] In this section we are going to determine the transition probabilities in first-order perturbation theory for this important situation, a result which will be of use in the remainder of this chapter.

If at time $t = 0$, the system is in the eigenstate $|i\rangle$ of $\hat{H}_0$, the amplitude for finding it in the state $|k\rangle$ at a time T may be calculated from (1.25) thus:

$$S_{ki}(T) = \frac{W_{ki}}{\mathrm{i}\hbar}\frac{\mathrm{e}^{\mathrm{i}(E_k - E_i)T/\hbar} - 1}{\mathrm{i}(E_k - E_i)/\hbar}. \tag{1.34}$$

Hence we deduce the transition probability $P_{i \to k}(T)$:

$$P_{i \to k}(T) = \frac{|W_{ki}|^2}{\hbar^2} g_T(E_k - E_i), \tag{1.35}$$

where

$$g_T(E) = \frac{\sin^2(ET/2\hbar)}{(ET/2\hbar)^2} T^2 \tag{1.36}$$

is the function shown in Figure 1.3.

The important characteristics of this function are the following:

- it has its maximum value of T^2 at $E = 0$;
- its width is of order $2\pi\hbar/T$;
- its area is proportional to T, or more precisely:[7]

$$\int_{-\infty}^{+\infty} \mathrm{d}E\, g_T(E) = 2\pi\hbar T. \tag{1.37}$$

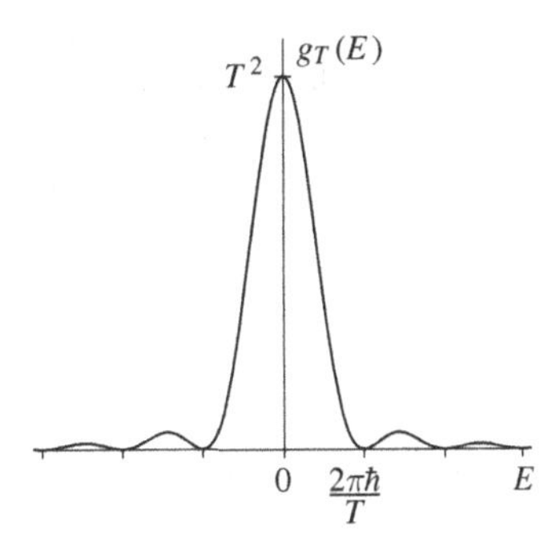

Figure 1.3 Form of the function $g_T(E) = (T \sin(ET/2\hbar)/(ET/2\hbar))^2$. Its value at $E = 0$ is T^2 and the first zeros are at $E = \pm 2\pi\hbar/T$.

[6] As we pointed out in the introduction, this calculation applies equally to the case of a Hamiltonian $\hat{H}_0 + \hat{W}$ independent of time, but when the system is prepared and subsequently detected in an eigenstate of $\hat{H}_0$.

[7] Its value is half the product of its height and of the distance between the two first zeros, as though the function g_T were triangular.

Let us write

$$\delta_T(E) = \frac{g_T(E)}{2\pi\hbar T} = \frac{2\hbar \sin^2(ET/2\hbar)}{\pi T E^2}. \tag{1.38}$$

The term $\delta_T(E)$ is a function peaked at $E = 0$ of width $2\pi\hbar/T$, of unit area. It constitutes, for sufficiently large T, an approximation to the Dirac delta-function and one can show that $\lim_{T\to+\infty}\delta_T(E) = \delta(E)$. We can therefore rewrite (1.35) in the form:

$$P_{i\to k}(T) = T\frac{2\pi}{\hbar}\,|W_{ki}|^2\,\delta_T(E_k - E_i). \tag{1.39}$$

We therefore obtain, more rigorously here, the result from above: the levels $|k\rangle$ which are *efficiently populated* are those for which the energy is such that

$$E_i - \frac{\pi\hbar}{T} < E_k < E_i + \frac{\pi\hbar}{T}. \tag{1.40}$$

The energy of the final state must therefore be the same as that of the initial state to within $2\pi\hbar/T$. Consequently, if the state $|i\rangle$ is one of a set of closely spaced levels, the halfwidth, ΔE of the energy distribution of the final states will be smaller the longer is the interaction time T. One can write

$$\Delta E.T \approx \hbar/2. \tag{1.41}$$

Finally, Figure 1.4 shows that, for a given value of E, $g_T(E)$ is an oscillatory function of T (except in the resonant case, $E = 0$). We shall return to this oscillatory behaviour in 1.2.6, where we introduce a non-perturbative treatment of the transition probability (which leads to the appearance of *Rabi oscillations*).

Comment The problem that we have just treated covers two situations which are in fact quite different from the point of view of the behaviour of the perturbation after time T: either $\hat{W}$ is switched off at time T and the system evolves no further ('top-hat'-pulsed perturbation), or $\hat{W}$ remains at a constant value (step-function perturbation) and we observe it at time T. Of course, what interests us here is the state of the system at time T, which is independent of the model chosen.

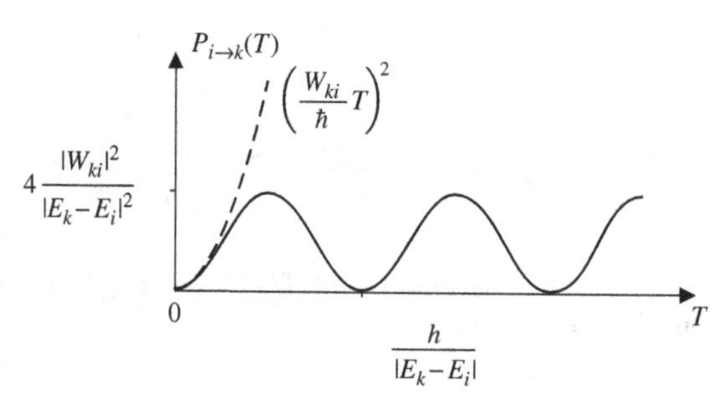

Figure 1.4 **Evolution of the transition probability between two discrete levels versus interaction time T. The dashed line represents the parabolic shape at the beginning, which is independent of $|E_k - E_i|$.**

Case of a sinusoidal perturbation

We saw in Section 1.2.2 that in the case of an interaction between an atom and electromagnetic radiation (1.10) one often has to deal with a sinusoidal perturbation, that is to say of the form:

$$\hat{H}_1(t) = \hat{W}\cos(\omega t + \varphi). \tag{1.42}$$

The probability amplitude for passing from a state $|i\rangle$ to another state $|k\rangle$, calculated according to (1.25) is then:

$$S_{ki} = -\frac{W_{ki}}{2\hbar}\left(\frac{\mathrm{e}^{\mathrm{i}(\omega_{ki}-\omega)t-\mathrm{i}\varphi} - \mathrm{e}^{\mathrm{i}(\omega_{ki}-\omega)t_0-\mathrm{i}\varphi}}{\omega_{ki}-\omega} + \frac{\mathrm{e}^{\mathrm{i}(\omega_{ki}+\omega)t+\mathrm{i}\varphi} - \mathrm{e}^{\mathrm{i}(\omega_{ki}+\omega)t_0+\mathrm{i}\varphi}}{\omega_{ki}+\omega}\right), \tag{1.43}$$

ω_{ki} being the Bohr frequency associated with the transition from $|i\rangle$ to $|k\rangle$ given by

$$E_k - E_i = \hbar\omega_{ki}. \tag{1.44}$$

The probability amplitude S_{ki} is the sum of two terms of which the denominators are respectively $\omega_{ki} - \omega$ and $\omega_{ki} + \omega$. In order to have an appreciable transition probability it is necessary to work in a domain in which one of these denominators appearing in (1.43) is small compared to the other. In the case of $\omega_{ki} > 0$, this occurs when

$$|\omega_{ki} - \omega| \ll \omega. \tag{1.45}$$

This condition, which is that for *quasi-resonant* excitation is necessary to have a non-negligible transition probability. In fact, taking the example of an atom interacting with visible radiation,[8] $W_{ki}/\hbar\omega$ rarely exceeds 10^{-6} if one uses a classical lamp, or even a continuous-wave laser source.[9] Since $\omega_{ki} + \omega$ is much larger than ω, condition (1.45) implies that the second term of (1.43), called the *anti-resonant* term is negligible compared to the former.[10] One then has a simple expression for the transition probability:

$$P_{i\to k}(T) = \frac{|W_{ki}|^2}{4\hbar^2} g_T(E_k - E_i - \hbar\omega), \tag{1.46}$$

giving

$$P_{i\to k}(T) = T\frac{|W_{ki}|^2}{4}\frac{2\pi}{\hbar}\delta_T(E_k - E_i - \hbar\omega). \tag{1.47}$$

In these relations, $g_T(E)$ and $\delta_T(E)$ are the functions introduced in the preceding paragraph (Equations (1.36) and (1.38)), and $T = t - t_0$ is the duration of the interaction.

[8] Recall that for visible radiation, the frequency $\omega/2\pi$ is of the order of 10^{14} Hz.

[9] One should note, however, that much larger values, of order one or greater, can be obtained using a mode-locked laser, which produces a train of very intense pulses of which the duration is a few tens of femtoseconds.

[10] For historical reasons the approximation obtained by neglecting the anti-resonant term is known as the *rotating-wave approximation* (it was introduced in the domain of nuclear magnetic resonance). In the optical domain the label *quasi-resonant approximation* would describe more accurately the physical situation.

In the case in which ω_{ki} is negative, the condition for quasi-resonant excitation is fulfilled when

$$|\omega_{ki} + \omega| \ll \omega. \tag{1.48}$$

It is then the first term of S_{ki} that is negligible, and one obtains the corresponding expression for $P_{i \to k}$ on replacing ω by $-\omega$ in Equation (1.46) or (1.47).

The final expressions for the transition probabilities are then analogous to those of (1.36) and (1.38) to within a factor of four if one replaces $E_k - E_i$ by $E_k - E_i \pm \hbar\omega$. The conclusions of the previous section then apply also to the case of a sinusoidal perturbation: the energy levels $|k\rangle$ which will be appreciably populated at the end of the interaction time T will be those for which the energy is such that:

$$\begin{cases} E_i + \hbar\omega - \dfrac{\pi\hbar}{T} < E_k < E_i + \hbar\omega + \dfrac{\pi\hbar}{T} \\ E_i - \hbar\omega - \dfrac{\pi\hbar}{T} < E_k < E_i - \hbar\omega + \dfrac{\pi\hbar}{T}. \end{cases} \tag{1.49}$$

For sufficiently long interaction times the only possible transitions will be those to states $|k\rangle$ for which the energy separation $E_k - E_i$ is exactly $\hbar\omega$: *the change in the atomic energy occurs by the absorption or emission of a quantum of energy* $\hbar\omega$. Thus we recover here the rule that Bohr introduced empirically in the earliest days of quantum mechanics. As we shall show in later chapters this case of atom–light interaction has a natural interpretation in terms of *photons* of energy $\hbar\omega$ absorbed or emitted during the interaction. Note that there is, however, no a-priori need to introduce this notion in order to derive Bohr's relation condition. At this level the concept of the photon is a convenience, not a necessity.

Comment The quasi-resonant approximation implies, in addition to the condition (1.45) or (1.48) that $W_{ki}/\hbar\omega$ is very much smaller than unity, which is usually the case in the optical domain. On the other hand, for high intensity lasers, or for fields having a much smaller frequency, for example in the radio-frequency domain (where $\omega/2\pi$ is of the order of 10^9 Hz), it is possible to achieve intensities for which $W_{ki}/\hbar\omega$ is in the region of one. Under these conditions it is no longer possible to neglect the anti-resonant term. Its principal effect is to displace the frequency of the centre of the resonance by a quantity of order $W_{ki}{}^2/\hbar^2\omega$. This is known as the Bloch–Siegert shift.[11]

1.2.5 Second-order calculations

In numerous situations the initial and final states $|i\rangle$ and $|k\rangle$ are not directly coupled by the interaction Hamiltonian $\hat{H}_1(t)$ (one has $\langle i|\hat{H}_1(t)|k\rangle = 0$). The transition probability (1.26) is therefore zero to first order. However, if states $|i\rangle$ and $|k\rangle$ are both coupled to other states $|j\rangle$ (that is to say $\langle i|\hat{H}_1(t)|j\rangle \neq 0$ and $\langle j|\hat{H}_1(t)|k\rangle \neq 0$) a non-zero second-order transition probability can arise. This occurrence can be described as a transition from state $|i\rangle$ to state $|k\rangle$ in which the system 'passes' via a state $|j\rangle$. Examples of such situations are found

[11] See, for example, CDG II, Complement A_{VI}.

frequently in atom–light interactions and we shall make much use of the results of this paragraph in later sections (see Figure 2.9 of Chapter 2).

In the situation in which $\langle k|\hat{H}_1(t)|i\rangle$ is zero, one must calculate the $\gamma_j^{(1)}(t)$ with j different from k, and substitute their values (given by expressions analogous to 1.23) in the expression (1.21) to calculate $\gamma_k^{(2)}(t)$. The amplitude for the transition from $|i\rangle$ to $|k\rangle$ in the interval $t = 0$ to T is then, to within a phase factor:

$$\begin{aligned} S_{ki}(T) &= \lambda^2 \gamma_k^{(2)}(T) \\ &= \frac{1}{(\mathrm{i}\hbar)^2} \int_0^T \mathrm{d}t' \int_0^{t'} \mathrm{d}t'' \sum_j \left\langle k \left| \hat{H}_1(t') \right| j \right\rangle \\ &\quad \times \langle j|\hat{H}_1(t'')|i\rangle\, \mathrm{e}^{\mathrm{i}(E_k - E_j)t'/\hbar}\, \mathrm{e}^{\mathrm{i}(E_j - E_i)t''/\hbar}. \end{aligned} \tag{1.50}$$

Note that the sum is over states $|j\rangle$ different from $|i\rangle$ and $|k\rangle$ since we have supposed that $\langle k|\hat{H}_1(t)|i\rangle = 0$.

We now consider the particular case in which $\hat{H}_1(t)$ is of the form

$$\hat{H}_1(t) = \hat{W} f(t), \tag{1.51}$$

where $\hat{W}$ is an operator and $f(t)$ is a 'switch-on' function with a characteristic time of θ (see Figure 1.5a). More precisely, we suppose that the interaction time T is very large compared to the switching-on time θ and that θ itself is very large compared to the characteristic evolution times of the free atomic system, that is $\hbar/|E_i - E_j|$ where $|j\rangle$ is one of the intermediary levels in the transition from $|i\rangle$ to $|k\rangle$:

$$T \gg \theta \gg \frac{\hbar}{|E_i - E_j|}. \tag{1.52}$$

Under these assumptions, we can give a rudimentary calculation of the transition probability $P_{i\to k}(T)$. Using the form (1.51) of the Hamiltonian, we find that the amplitude $S_{ki}(T)$, for the transition from an initial state $|i\rangle$ (at a time $t_0 < 0$ before the switching-on of the interaction) to the state $|k\rangle$ at a later time T can be put in the form:

$$S_{ki}(T) = -\frac{1}{\hbar^2} \sum_{j\neq k,i} W_{kj} W_{ji} \int_{t_0}^T \mathrm{d}t' \int_{t_0}^{t'} \mathrm{d}t''\, \mathrm{e}^{\mathrm{i}(E_k - E_j)t'/\hbar}\, \mathrm{e}^{\mathrm{i}(E_j - E_i)t''/\hbar} f(t') f(t''). \tag{1.53}$$

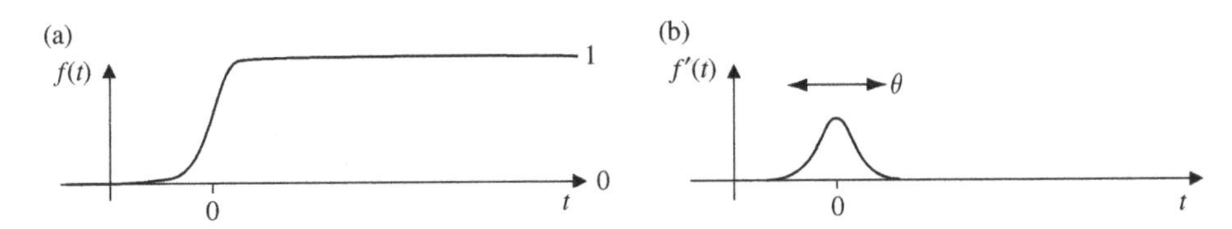

Figure 1.5 (a) Form of the 'switch-on' function $f(t)$, which varies smoothly between 0 and 1 over a time interval θ; (b) form of the derivative of $f(t)$.

Performing the integration over t'' by parts we obtain:

$$\int_{t_0}^{t'} \mathrm{d}t'' \mathrm{e}^{\mathrm{i}(E_j - E_i)t''/\hbar} f(t'') = \frac{\mathrm{e}^{\mathrm{i}(E_j - E_i)t'/\hbar}}{\mathrm{i}(E_j - E_i)/\hbar} f(t') - \int_{t_0}^{t'} \frac{\mathrm{e}^{\mathrm{i}(E_j - E_i)t''/\hbar}}{\mathrm{i}(E_j - E_i)/\hbar} f'(t'')\,\mathrm{d}t''. \tag{1.54}$$

The assumptions (1.52) about the form of $f(t)$ tell us that the maximum value of $f'(t'')$ is $1/\theta$ (see Figure 1.5b). The second term of the right-hand side of (1.54) is therefore smaller than the first by a factor of the order of $\hbar/\theta|E_j - E_i|$, which we suppose small compared to unity. Neglecting this term, (1.53) becomes:

$$S_{ki}(T) = \frac{1}{\mathrm{i}\hbar} \sum_{j \neq i} \frac{W_{kj} W_{ji}}{E_i - E_j} \int_{t_0}^{T} \mathrm{d}t' \mathrm{e}^{\mathrm{i}(E_k - E_i)t'/\hbar} \left(f(t')\right)^2, \tag{1.55}$$

$$S_{ki}(t) \approx \frac{1}{\mathrm{i}\hbar} \sum_{j \neq i} \frac{W_{kj} W_{ji}}{E_i - E_j} \int_{0}^{T} \mathrm{d}t' \mathrm{e}^{\mathrm{i}(E_k - E_i)t'/\hbar}. \tag{1.56}$$

The square of the integral appearing in (1.56) above is equal, to within terms of order θ/T to the function $g_T(E_k - E_i)$ introduced in (1.36) in the first-order calculation. We thus obtain the transition probability as

$$P_{i \to k}(T) = T \left| \sum_{j \neq k,i} \frac{\langle k|\hat{W}|j\rangle\langle j|\hat{W}|i\rangle}{E_j - E_i} \right|^2 \frac{2\pi}{\hbar} \delta_T(E_k - E_i), \tag{1.57}$$

where $\delta_T(E)$ is the function of width $2\pi/T$ introduced in (1.38). A comparison with formula (1.39) shows that the second-order result can be put in a form identical to that of the first-order result if the Hamiltonian $\hat{W}$ is replaced by an *effective Hamiltonian* $\hat{W}_{\text{eff}}$ of which the matrix element between states $|i\rangle$ and $|k\rangle$ is given by:

$$\langle k|\hat{W}_{\text{eff}}|i\rangle = \sum_{j \neq k,i} \frac{\langle k|\hat{W}|j\rangle\langle j|\hat{W}|i\rangle}{E_i - E_j}. \tag{1.58}$$

This matrix element, and hence the transition probability, $P_{i \to k}$, is significant when one or more of the intermediate levels $|j\rangle$ has an energy close to that of the initial level, $|i\rangle$.

Comment One can also show that the perturbation $\hat{H}_1(t)$ leads to shifts of the systems energy levels, by an amount corresponding to the diagonal matrix elements of the effective Hamiltonian, $\hat{W}_{\text{eff}}$ (see, for example, Section 2.3.3). The energy shift of the state $|i\rangle$ is then

$$\langle i|\hat{W}_{\text{eff}}|i\rangle = \sum_{j \neq i} \frac{\langle i|\hat{W}|j\rangle\langle j|\hat{W}|i\rangle}{E_i - E_j}. \tag{1.59}$$

In the limit of long interaction times, transitions can only occur between states of which the *shifted* energies are identical.

1.2.6 Comparison with the exact solution for a two-level system

It is interesting to compare the first-order perturbative solution of Equations (1.35–1.36) to the exact solution (in which all orders are taken into account). This exact solution can be found for the most simple quantum system: that of a two-level system on which is imposed an interaction of constant value suddenly switched on at $t = 0$.

Some useful formulae

We consider a Hamiltonian $\hat{H}_0$ defined on a two-dimensional Hilbert space, which has for eigenstates the states $|a\rangle$ and $|b\rangle$ of energies E_a and E_b, respectively. In a manner identical to that described in Section 1.2.4 we apply from time $t = 0$ an interaction $\hat{W}$ which has a constant value at subsequent times. We shall assume, for simplicity, that the diagonal matrix elements of the interaction Hamiltonian, $\hat{W}_{aa}$ and $\hat{W}_{bb}$ are zero and that the off-diagonal element $\hat{W}_{ab}$ is real.

We shall now calculate the probability, $P_{a\to b}$, of the transition to state $|b\rangle$ under the influence of the interaction, supposing that the system is initially prepared in state $|a\rangle$. To solve this problem we must determine the eigenstates $|\varphi_1\rangle$ and $|\varphi_2\rangle$ and eigenenergies, E_1 and E_2 of the total Hamiltonian $\hat{H}_0 + \hat{W}$. One can show that the eigenstates are:[12]

$$\begin{cases} |\varphi_1\rangle = \cos\theta\,|a\rangle + \sin\theta\,|b\rangle \\ |\varphi_2\rangle = -\sin\theta\,|a\rangle + \cos\theta\,|b\rangle\,, \end{cases} \tag{1.60}$$

the angle θ being given by:

$$\tan 2\theta = 2W_{ab}/(E_a - E_b). \tag{1.61}$$

The corresponding eigenenergies are then

$$\begin{cases} E_1 = \frac{1}{2}(E_a + E_b) + \frac{1}{2}\sqrt{(E_a - E_b)^2 + 4W_{ab}^2} \\ E_2 = \frac{1}{2}(E_a + E_b) - \frac{1}{2}\sqrt{(E_a - E_b)^2 + 4W_{ab}^2}. \end{cases} \tag{1.62}$$

Temporal evolution

The state of the system at time t is obtained by expressing the initial state $|a\rangle$ in terms of the eigenstates $|\varphi_1\rangle$ and $|\varphi_2\rangle$, and by multiplying the coefficients of these two states by phase factors $\exp(-\mathrm{i}E_1 t/\hbar)$ and $\exp(-\mathrm{i}E_2 t/\hbar)$. A simple calculation yields the state-vector of the system at time T,[13] and hence the transition probability $P_{a\to b}$. This is given by:

$$P_{a\to b}(T) = \frac{4W_{ab}^2}{(E_a - E_b)^2 + 4W_{ab}^2}\sin^2\left\{\sqrt{(E_a - E_b)^2 + 4W_{ab}^2}\,\frac{T}{2\hbar}\right\}. \tag{1.63}$$

[12] See CDL, Complement B_{IV}.

[13] See CDL, § IV.C.3.

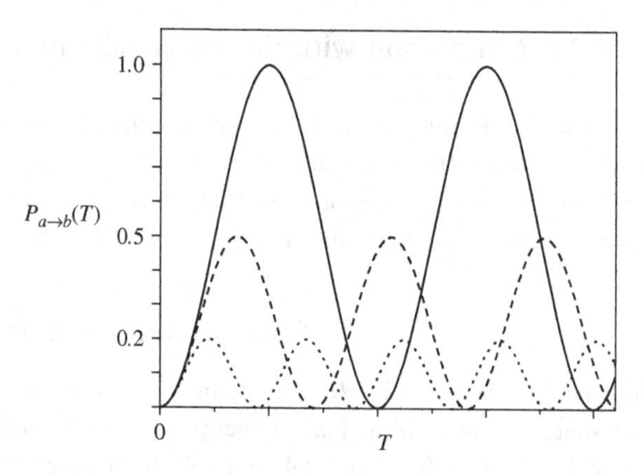

Figure 1.6 Temporal evolution of the transition probability $P_{a\to b}$ for degenerate energy levels (solid line) and non-degenerate levels (dashed and dotted lines). The corresponding values of $(E_a - E_b)/W_{ab}$ are 0, 2 and 4.

The result is an *oscillatory temporal dependence of the transition probability* known as *Rabi oscillation*. This is shown in Figure 1.6. The characteristic angular frequency of this oscillation is known as the *Rabi frequency*.

When the two eigenstates $|a\rangle$ and $|b\rangle$ have the same energy, the transition probability periodically attains the value 1 (*total* transfer of population to state $|b\rangle$), whatever the magnitude of the coupling W_{ab} between the two levels. The corresponding Rabi frequency, $2W_{ab}/\hbar$ is proportional to this coupling. If the two eigenstates $|a\rangle$ and $|b\rangle$ are *non-degenerate* in energy, $E_a \neq E_b$, the oscillation is more rapid but the transfer of population between the two states is *never complete*, however great the magnitude of W_{ab}.

Comment This oscillatory transfer of energy between the two states $|a\rangle$ and $|b\rangle$ at a frequency depending on the strength of the coupling between the two levels has many similarities to the behaviour of two coupled oscillators in classical mechanics: in the degenerate case, the energy in the system periodically alternates between the two oscillators.

Comparison with perturbation theory

We consider first of all the *short-time limit*. The transition probability (1.63) then has approximately the following value, corresponding to the small-time behaviour of the Rabi oscillation:

$$P_{a\to b}(T) \approx \frac{W_{ab}^2}{\hbar^2} T^2. \tag{1.64}$$

The time-dependence therefore obeys a T^2 *rule*, which is consistent, also in the small-time limit, with the first-order perturbation theory result (see Equations (1.35) and (1.36)). Note that this result is obtained in the limit where $\sin x \approx x$, which implies that $x \ll 1$ and therefore also that $P_{a\to b} \ll 1$.

What do we get from this comparison for longer interaction times? Consider the situation in which

$$|W_{ab}| \ll |E_a - E_b|\,, \tag{1.65}$$

which is a necessary condition for the application of perturbation theory to be valid. In this case (1.63) becomes:

$$P_{a\to b}(T) \approx \frac{W_{ab}^2}{\hbar^2} \frac{\sin^2\left[\left(\frac{E_a-E_b}{2} + \frac{W_{ab}^2}{E_a-E_b}\right)\frac{T}{\hbar}\right]}{\left(\frac{E_a-E_b}{2\hbar}\right)^2 + \frac{W_{ab}^2}{\hbar^2}}. \tag{1.66}$$

This expression is very close to expression (1.35) of first-order perturbation theory: *the maximum value of the transition probability and the frequency of the oscillation are practically identical.* Note, however, that the small difference in the frequencies leads to *instantaneous* values of the probability in total disagreement after a sufficient number of oscillations, more precisely once:

$$T > \frac{\pi\hbar\,|E_a - E_b|}{W_{ab}^2}. \tag{1.67}$$

Comment It is also possible to calculate exactly the evolution of a two-level system under the influence of a perturbation having a sinusoidal temporal dependence at a frequency ω, in the quasi-resonant case ($\omega \approx (E_b - E_a)/\hbar$). This calculation permits the generalization of the concept of Rabi oscillations, that we have introduced, and leads to analogous conclusions as far as the comparison with perturbation theory is concerned. This problem will be studied in detail in Chapter 2 (Section 2.3.2) for the particular case of the atom–light interaction.

1.3 Case of a discrete level coupled to a continuum: Fermi's golden rule

After having studied the case of the coupling of two isolated levels, we now come to consider a radically different situation in which the initial level is coupled to an ensemble of levels forming a *continuum*. We shall show that in this case the evolution of the system is radically different from what we have just considered. After giving an example of such systems, common in quantum mechanics, we shall calculate the temporal evolution of the system, the principal result being Fermi's golden rule. Complement 1A points out how one can move continuously from the situation of a coupling with a discrete state to a coupling with a continuum.

1.3.1 Example: autoionization of helium

Approximation of independent electrons

Consider the atom of helium, made of two electrons and a doubly-charged He^{2+} nucleus. If we neglect as a first approximation the electrostatic repulsion of the two electrons and effects related to the indistinguishability of the two particles, each electron is independently submitted to a Coulomb potential. The eigenstates of the system are tensor products of eigenstates of the hydrogen-like ion He^+, which are identical, apart from a change of scale due to the double charge of the nucleus, to those of atomic hydrogen. They consist of discrete bound states having an energy

$$E_n = -\frac{E_\mathrm{I}}{n^2}, \tag{1.68}$$

(the ionization energy E_I being equal to four times that of the hydrogen atom: $E_\mathrm{I} = 54.4\,\mathrm{eV}$), and of ionized states $|k\rangle$ of positive energies, in which the electron is not confined to the vicinity of the nucleus. As the energy E_k of these states is not quantized (k is a real quantity), they form a continuum. The energy of these states is represented in Figure 1.7.

Let us first consider the series of helium states that we label $(1, n)$, corresponding to one electron in state $n = 1$ and the other in any bound state of principal quantum number n. The energy of this state is the sum of the energy of the two electrons, that is

$$E_{1,n} = E_1 + E_n = -E_\mathrm{I}\left(1 + \frac{1}{n^2}\right). \tag{1.69}$$

The series of these states is represented in Figure 1.8a. Above the ground state $(1, 1)$, which has an energy $-2E_\mathrm{I}$, one has discrete bound states of energy between $-2E_\mathrm{I}$ and $-E_\mathrm{I}$, then

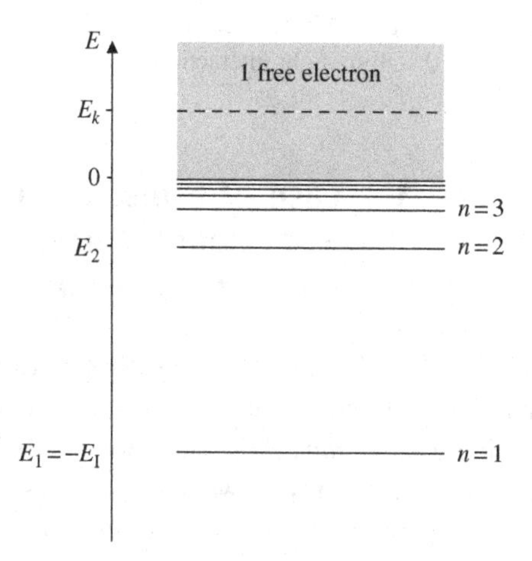

Figure 1.7 Energy levels of an electron in a Coulomb potential: they comprise bound states of quantized negative energies forming a discrete series, and ionized states of positive energy forming a continuum.

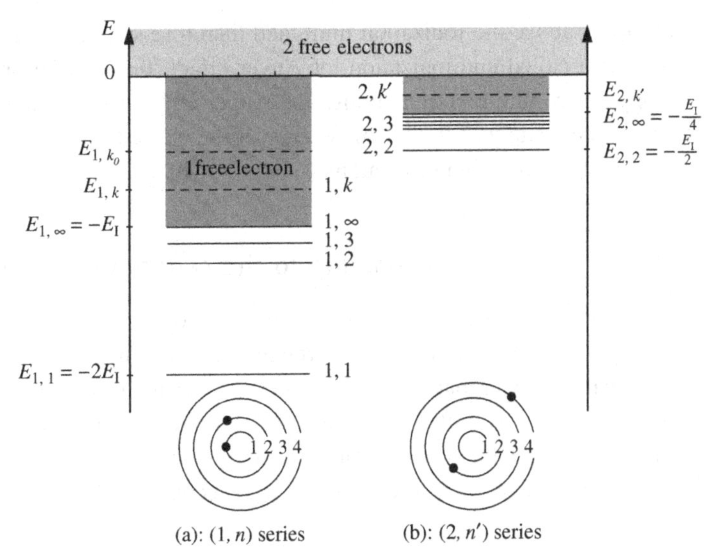

Figure 1.8 Energy levels of the helium atom at the approximation of independent electrons. (a) Levels corresponding to one electron in ground state $n = 1$ and the other in any bound or ionized state; (b) levels corresponding to one electron in the first excited state $n = 2$ and the other in any bound or ionized state.

a continuum of states $(1, k)$ above $-E_I$ describing the ion He^+ in its ground state plus a free electron.

If we now consider the series of helium states labelled $(2, n')$, corresponding to one electron in state $n = 2$ and the other in state $n' > 2$ (Figure 1.8b), one observes that the state $(2, 2)$ has an energy which is greater than $-E_I$, and lies therefore within the continuum of states $(1, k)$, which means that there is an ionized state $(1, k_0)$ having the same energy as the bound state $(2, 2)$.

Effect of electron coupling

Let us now take into account the *electrostatic coupling* between the two electrons, described by the term:

$$\hat{H}_1 = -\frac{q^2}{4\pi\varepsilon_0} \frac{1}{|\hat{\boldsymbol{r}}_1 - \hat{\boldsymbol{r}}_2|}, \tag{1.70}$$

which we shall treat in a perturbative way. In the first place, this term will modify the energies of the various states that we have previously considered. However the series $(1, n)$ and $(2, n)$ still exist, and the discrete state $(2, 2)$ still lies above the ionisation limit of the series $(1, n)$, so that there is an ionized state $(1, k_0)$ having the same energy as $(2, 2)$. Furthermore, one can show that the time-independent operator $\hat{H}_1$ has a non-zero matrix element between states $(2, 2)$ and $(1, k_0)$. There is a possibility of transition between these two states, and the excited state $(2, 2)$ can partly release its internal energy to excite one of its electrons above the ionization limit, and therefore spontaneously ionize. This is the phenomenon called autoionization, or Auger effect, that can be observed in helium and in many other multi-electron atoms. The Auger effect is of particular use in material analysis, where a study of the kinetic energy of the ejected electron reveals information on the chemical environment of an atom.

Coupling to the continuum $(1, k)$

Actually the coupling term $\hat{H}_1$ has also non-zero matrix elements between $(2, 2)$ and other states $(1, k)$ of the continuum, in particular those having an energy close to $(1, k_0)$. We have seen in this chapter that a transition is possible even if there is not an exact equality between the energies of the initial and final states. In order to treat the autoionization process one must therefore take into account the coupling between $(2, 2)$ and all the continuum states $(1, k)$. This is what will be done in the next section.

1.3.2 Discrete level coupled to a quasi-continuum: simplified model

In the example of the previous section, there is, in the final state of the system, a free electron which escapes from the atom. Clearly the emission of this electron is an irreversible process, qualitatively different from the phenomenon of Rabi oscillations that occurs when discrete levels are involved. In the following section we shall study this irreversible process more quantitatively.

The quasi-continuum

The mathematical treatment of the eigenstates of a quantum system belonging to a continuum must be performed with some care, in particular because they are not normalizable. It turns out therefore to be more practical to perform calculations for the case of an ensemble of discrete levels very close in energy, a *quasi-continuum*, and then to proceed to the limiting case of a true continuum at the end of the calculation. To this end the system of interest is thought of as being contained in a *fictitious box of large dimensions*. This changes the boundary conditions in such a manner that the positive energy levels are now bound states of the box potential and are therefore discrete. It then suffices to allow the *dimensions of the box to tend to infinity* at the end of the mathematical manipulations to obtain the continuum result.

We illustrate this procedure using the example of the last paragraph. The introduction of the fictitious box leads to adding to the Coulomb potential a potential which is zero inside the box and infinite outside it. The stationary solutions corresponding to this new potential are exclusively bound states, forming a discrete series even for positive energy values (Figure 1.9). We shall therefore be able to use here the results of Section 1.2. The essential point is that when L (a *macroscopic* quantity) is very much larger than the Bohr radius (a *microscopic* quantity), the results of measurements made on the system are independent of

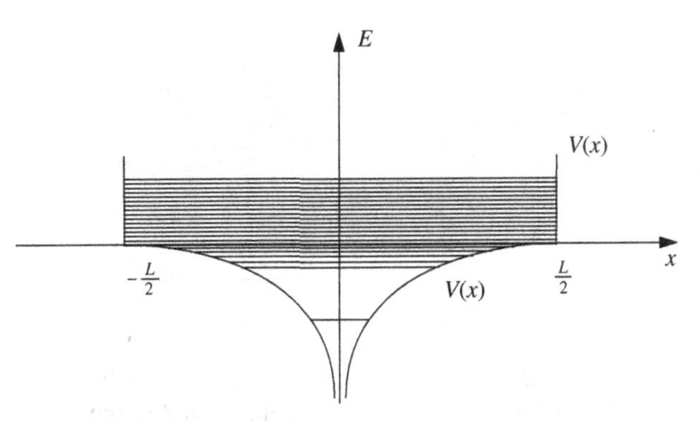

Figure 1.9 **Energy levels of an electron submitted to a Coulomb potential and confined inside a box of size *L* around the nucleus. All levels are discrete and correspond to bound states. However, the positive energy states form a *quasi-continuum* when *L* is large enough.**

L and are identical to results obtained by considering a true continuum.[14] Let us also note that the normalization of the wavefunctions imposes that their amplitude varies as $L^{-1/2}$.

We suppose, in addition, to simplify the calculations, that in the box of dimensions L, the eigenstates $|n\rangle$ differ little from those of the 'simple' infinite square well $V(x) = 0$ for $|x| < L/2$ and $V(x) = +\infty$ for $|x| > L/2$. These states have energies

$$E_k = \frac{\hbar^2\pi^2}{2mL^2}k^2, \tag{1.71}$$

where k is a positive integer. The energy separation of successive levels is then, for large k:

$$\Delta E = k\frac{\hbar^2\pi^2}{mL^2} = \frac{2\pi\hbar}{\sqrt{2m}}\frac{\sqrt{E_k}}{L}. \tag{1.72}$$

It is inversely proportional to the box size L for a given energy, and, as expected, is very small when the box size tends to infinity.

Simple model of a quasi-continuum

We return to the general problem of a discrete level coupled to a quasi-continuum which we treat first of all in a greatly simplified manner which allows us to obtain the important results whilst avoiding more complex mathematical manipulations. Consider a Hamiltonian $\hat{H}_0$, of which the eigenstates are, on the one hand, the initial level $|i\rangle$ (for which we choose the corresponding energy to be zero, $E_i = 0$) and on the other an ensemble of levels $|k\rangle$ which we suppose evenly spaced in energy by writing $E_k = k\varepsilon$ (k is an integer taking all values between $-\infty$ and $+\infty$). The energy separation of consecutive levels, ε, is presumed

[14] We have introduced the idea of a quasi-continuum here as a mathematical device to facilitate our calculations. However, examples of real quasi-continua do occur in physics. For example, the excited rotation–vibration states of a multi-atom molecule form a dense system of energy levels having the properties of a quasi-continuum.

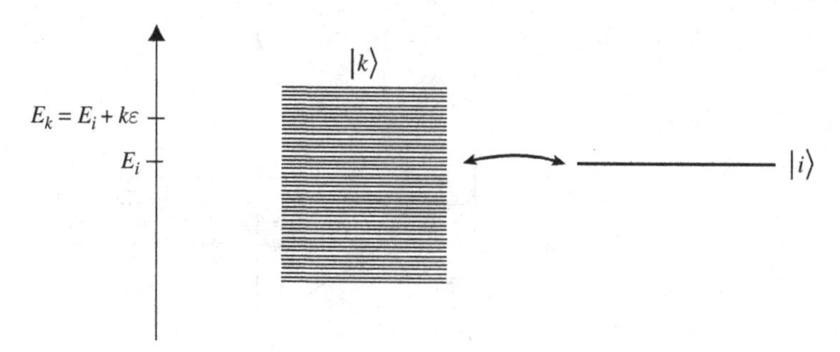

Figure 1.10 **Energy levels of a system comprising a single discrete state $|i\rangle$ coupled to an infinite set of regularly spaced states $|k\rangle$ forming a quasi-continuum.**

to be very small (we shall make a more quantitative statement of this condition later), so that the levels $|k\rangle$ constitute a quasi-continuum (see Figure 1.10).

The second simplification that we adopt is to assume that the level $|i\rangle$ is coupled to the levels $|k\rangle$ by a coupling term $\hat{W}$ of which the matrix elements are

$$\begin{aligned}\langle k|\hat{W}|i\rangle &= w \\ \langle k|\hat{W}|k'\rangle &= 0 \\ \langle i|\hat{W}|i\rangle &= 0.\end{aligned} \tag{1.73}$$

The interaction $\hat{W}$ therefore does not couple different continuum states. We suppose furthermore, for simplicity, that the matrix element w is real and independent of k.

Short-time behaviour: transition probability per unit time

We shall calculate the probability that the system, which is initially in the state $|i\rangle$, is still in this state at time T. Because the states of the continuum are mutually orthogonal, in order to find the total transition probability we have to form the sum of the transition probabilities to each of the continuum states. Therefore, the probability of still being in $|i\rangle$ is

$$P_i(T) = 1 - \sum_{k=-\infty}^{+\infty} P_{i\to k}(T), \tag{1.74}$$

where $P_{i\to k}(T)$ is the transition probability for the case of a perturbation which has a constant value in the time interval $t = 0 \to T$, of which the value is given to first order in expression (1.39):

$$P_{i\to k}(T) = T\,\frac{2\pi}{\hbar}\,w^2\,\delta_T(k\varepsilon). \tag{1.75}$$

We consider the case in which the separation of consecutive continuum levels ε is small compared to the width $2\pi\hbar/T$ of δ_T. The function $\delta_T(k\varepsilon)$ therefore varies slowly with k so that the discrete sum in Equation (1.74) can be replaced by an integral. This gives:

$$\begin{aligned} P_i(T) &= 1 - T\frac{2\pi}{\hbar}w^2 \int_{-\infty}^{+\infty} \frac{\mathrm{d}E}{\varepsilon}\,\delta_T(E) \\ &= 1 - T\frac{2\pi w^2}{\hbar\varepsilon}, \end{aligned} \tag{1.76}$$

since the total area of the function δ_T is unity. We obtain then, in the framework of the current model, an evolution of the probability of occupation of the initial level that is *linear* in T. This is in contrast to the *quadratic* evolution found for the case of two coupled discrete levels (see Equation (1.64)). Hence we can define in this case a *departure probability per unit time*, Γ where

$$\Gamma = \frac{1 - P_i(T)}{T}, \tag{1.77}$$

with

$$\Gamma = \frac{2\pi}{\hbar}w^2\frac{1}{\varepsilon}. \tag{1.78}$$

The term Γ, which has the dimensions of reciprocal time, is also known as the *transition probability per unit time*, or *transition rate*, to the quasi-continuum.

In the simple example of quasi-continuum presented above, w is proportional to the continuum wavefunction amplitude, which varies as $L^{-1/2}$, and ε varies as L^{-1}, so that the quantity w^2/ε is a real physical quantity, that does not depend on the fictitious box size L. It is precisely this property that validates the use of quasi-continuum states. It must be systematically checked at the end of all derivations using this tool.

Comment Two conditions must be satisfied to justify the use of Equations (1.77) and (1.78). They impose two independent upper limits to the interaction time T:

$$T \ll 2\pi\hbar/\varepsilon \text{ and } T \ll \Gamma^{-1} = \hbar\varepsilon/2\pi w^2.$$

The first is necessary in order that we may replace the summation by an integral in (1.74). The second ensures the validity of first-order perturbation theory ($P_i(T) \approx 1$). In addition, in realistic examples the quasi-continuum doesn't extend over an infinite domain of energy but over a finite one, of a width that we denote by Δ. It is necessary, therefore, to add another condition on T to ensure that the integral of (1.83) does indeed give unity when carried out over a domain of width Δ:

$$T \gg \hbar/\Delta.$$

These conditions are compatible when $\Gamma \ll \Delta$. Complement 1A presents a case where this condition does not hold.

Long-time behaviour: exponential decay

In order to study the evolution of the system on a timescale that could be long compared to Γ^{-1} one must use a non-perturbative method. The means of solving this type of problem was introduced in 1930 by Weisskopf and Wigner. We outline a simplified version of their treatment in the following.

The wavefunction of the system, given by (1.14) can be written in our case:

$$|\psi(t)\rangle = \gamma_i(t)\,|i\rangle + \sum_{k=-\infty}^{+\infty} \gamma_k(t)\,\mathrm{e}^{-\mathrm{i}k\varepsilon t/\hbar}\,|k\rangle\,, \tag{1.79}$$

and the corresponding Schrödinger equation is then equivalent to the following set of differential equations:

$$\begin{cases} \mathrm{i}\hbar\dfrac{\mathrm{d}}{\mathrm{d}t}\gamma_i(t) = w \displaystyle\sum_{k=-\infty}^{+\infty} \gamma_k(t)\,\mathrm{e}^{-\mathrm{i}k\varepsilon t/\hbar} \\ \mathrm{i}\hbar\dfrac{\mathrm{d}}{\mathrm{d}t}\gamma_k(t) = w\,\mathrm{e}^{\mathrm{i}k\varepsilon t/\hbar}\gamma_i(t). \end{cases} \tag{1.80}$$

The system being initially in the state $|i\rangle$, we have $\gamma_k(0) = 0$ so $\gamma_k(t)$ can be written:

$$\gamma_k(t) = \frac{w}{\mathrm{i}\hbar}\int_0^t \mathrm{d}t'\,\gamma_i(t')\mathrm{e}^{\mathrm{i}k\varepsilon t'/\hbar}. \tag{1.81}$$

Substituting this result into the first equation of (1.80) and using the expression (1.78) for Γ, we obtain the following exact equation:

$$\frac{\mathrm{d}}{\mathrm{d}t}\gamma_i(t) = -\frac{\Gamma}{2\pi\hbar}\int_0^t \mathrm{d}t'\,\gamma_i(t')\left[\sum_{k=-\infty}^{+\infty} \varepsilon\mathrm{e}^{\mathrm{i}k\varepsilon(t'-t)/\hbar}\right]. \tag{1.82}$$

We suppose still that the quasi-continuum levels are sufficiently tightly spaced such that $\varepsilon \ll \hbar/T$. The sum over k in square brackets can then be replaced by the following integral:

$$\int_{-\infty}^{+\infty} \mathrm{d}E\;\mathrm{e}^{\mathrm{i}E(t'-t)/\hbar} = 2\pi\hbar\delta(t'-t). \tag{1.83}$$

Putting $\tau = t' - t$, Equation (1.82) can be rewritten in the form:

$$\frac{\mathrm{d}}{\mathrm{d}t}\gamma_i(t) = -\Gamma\int_{-t}^{0} \mathrm{d}\tau\;\delta(\tau)\gamma_i(t+\tau). \tag{1.84}$$

The integral of Equation (1.84) is equal to $\gamma_i(t)/2$ since $\delta(t)$ is even and, for any $t > 0$

$$\frac{\mathrm{d}}{\mathrm{d}t}\gamma_i(t) = -\frac{\Gamma}{2}\gamma_i(t). \tag{1.85}$$

The solution is then straightforward and gives:

$$\gamma_i(t) = \exp(-\Gamma t/2). \tag{1.86}$$

From this the probability $P_i(t) = |\gamma_i(T)|^2$ can be deduced:

$$P_i(T) = \exp(-\Gamma T). \tag{1.87}$$

The probability of finding the system in the initially occupied state therefore *decreases exponentially* with time, tending towards zero at long times in the manner of radioactive decay. In other words (1.87) expresses the fact that the state $|i\rangle$ has a *lifetime* equal to Γ^{-1}.

Finally we note that in the limit of short times (1.87) reduces, as it should, to the perturbative result of (1.76).

The final state of the system

In contrast to the problem of the coupling of discrete levels studied previously (expression (1.63)), this system evolves to a steady state in which the probability of the initial state being occupied is zero. One might then ask what are the possible final states in the limit $t \to +\infty$? In order to answer this we need to calculate the coefficients $\gamma_k(t = +\infty)$. To achieve this we substitute the result (1.86) in Equation (1.81). The evaluation of the integral is straightforward and yields:

$$\gamma_k(t) = w \frac{1 - \mathrm{e}^{(\mathrm{i}k\varepsilon/\hbar - \Gamma/2)t}}{k\varepsilon + \mathrm{i}\hbar\Gamma/2}. \tag{1.88}$$

The probability that the system is in the state $|k\rangle$ at long times ($t \to +\infty$) is then:

$$P_k = \frac{w^2}{(k\varepsilon)^2 + \hbar^2\Gamma^2/4}. \tag{1.89}$$

Therefore, the probability $\mathrm{d}P(E)$ of finding the system in any one of the quasi-continuum states in the energy range between E and $E + \mathrm{d}E$ is:

$$\mathrm{d}P(E) = P_k \frac{\mathrm{d}E}{\varepsilon}, \tag{1.90}$$

since $\mathrm{d}E/\varepsilon$ is simply *the number of quasi-continuum states in the energy interval* $\mathrm{d}E$. Using (1.78) and (1.89) we find:

$$\frac{\mathrm{d}P}{\mathrm{d}E} = \frac{\hbar\Gamma}{2\pi} \frac{1}{E^2 + \hbar^2\Gamma^2/4}. \tag{1.91}$$

The form of the energy distribution of the possible final states is therefore a Lorentzian, centred on the energy $E_i = 0$ of the initial discrete state, of which the full width at half-maximum is equal to $\hbar\Gamma$ (see Figure 1.11).

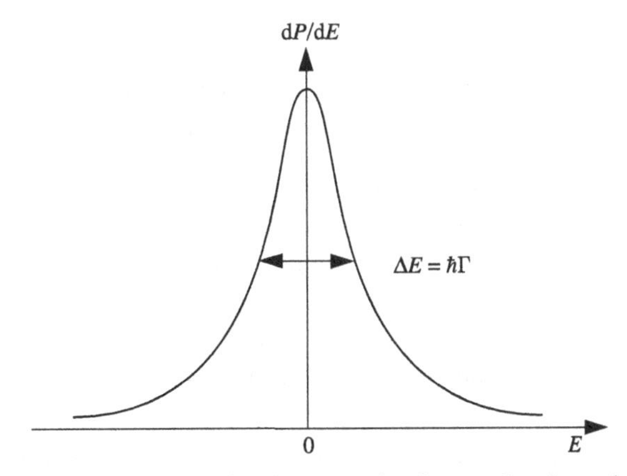

Figure 1.11 Energy distribution of the final states for the case of a discrete level coupled to a quasi-continuum by a constant perturbation. The width of this distribution is $\Delta E = \hbar\Gamma$.

One can therefore consider that the initial discrete state 'empties' into the surrounding quasi-continuum states, the range of states filled being determined by the energy uncertainty of the initial state $\hbar\Gamma$ arising from its finite lifetime Γ^{-1}.

Comment A striking characteristic of this exponential evolution (when compared with the Rabi oscillations seen before) is its *monotonic* character: the probability of occupation of the initial state decreases at all times. One could be tempted to see an irreversible behaviour here. However, the problem is rather more subtle. At any time t_0 in its evolution, the system is in a uniquely determined state $|\psi(t_0)\rangle$, which is given by the coefficients $\gamma_k(t_0)$ of Equations (1.88). Let us operate on this state with the time-reversal operator, $\hat{K}$, which is equivalent to taking the complex conjugate of its wavefunction[15] (the effect of this operator is equivalent to that of an instantaneous inversion of velocities in classical mechanics). If the system evolves from the new state $\hat{K}|\psi(t_0)\rangle$ under the influence of the same Hamiltonian it is easy to show that after time t_0 it will be found exactly in its initial state $|i\rangle$. This is a manifestation in this special case of *reversibility*, which always characterises the dynamics of systems obeying the Schrödinger equation.

However, from a practical point of view, it is much easier to prepare the system in the eigenstate $|i\rangle$ of $\hat{H}_0$ than in the state $\hat{K}|\psi(t_0)\rangle$, which requires the control of the amplitude and phase of each eigenstate of $\hat{H}_0$, and allowing the system to 'condense' back into $|i\rangle$ at a later time. The inverse of the evolution described by (1.88) is thus possible, but preparing the system in such a situation is generally very unlikely: it is therefore at this *statistical* level that irreversibility appears in this kind of process.

1.3.3 Fermi's golden rule

The case considered in Section 1.3.2 is somewhat idealized. In practice the strength of the coupling, W_{ik}, is a function of the level $|k\rangle$ and the energy separation $E_k - E_{k-1}$ of successive levels depends also on the pair considered. In the following we shall not treat the general case,[16] but shall content ourselves with stating the exact result whilst drawing attention to the similarities and differences with the simple case that we have just studied.

A more realistic quasi-continuum: the concept of a density of states

Consider then the more general case in which the energy levels are not spread regularly in energy: the energy difference $E_k - E_{k-1}$ is a function of k and furthermore the quasi-continuum energy levels are limited to a finite range of energies and no longer extend from energies of $-\infty$ to $+\infty$ (Figure 1.12). We then define a *density of states* $\rho(E)$ which is equal to the number of quasi-continuum levels in the energy range from E to $E + \mathrm{d}E$, divided by the energy width of this interval:

[15] See, for example, A. Messiah, *Quantum Mechanics*, Dover (1999), Chapter XIII C.

[16] See CDL II, Chapter XII C.

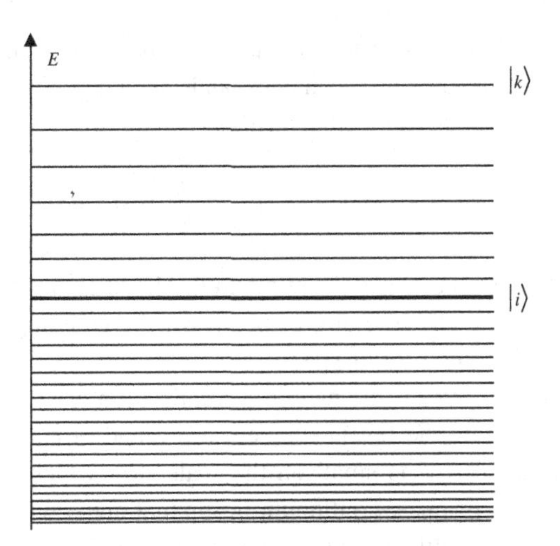

Figure 1.12 Coupling of a discrete level with a quasi-continuum in which the level spacings are a function of the energy.

$$\rho(E) = \frac{dN(E)}{dE}. \tag{1.92}$$

The value of $\rho(E)$ depends on the particular quantum system considered (in the simplified model it was equal to $1/\varepsilon$).

Finally, in a realistic model the matrix element is not necessarily constant. We suppose nevertheless that $\langle k|\hat{W}|i\rangle$ is a *slowly varying function of k*.

Fermi's golden rule for a non-degenerate continuum

Suppose the system to be initially in the level $|i\rangle$. One can generalize the results of 1.3.2 and calculate by first-order perturbation theory the probability $P_i(t)$ of finding the system in state $|i\rangle$ for times T sufficiently short. Starting from expression (1.39), we can calculate the transition probabilities to all possible final states $|k\rangle$:

$$P_i(T) = 1 - \sum_k T\frac{2\pi}{\hbar}|W_{ki}|^2\,\delta_T(E_k - E_i). \tag{1.93}$$

If W_{ik} is a function of k which varies slowly over the width $\hbar/T$ of the function δ_T, the discrete sum can be replaced by an integral by introducing the density of states $\rho(E)$:

$$P_i(T) = 1 - T\frac{2\pi}{\hbar}\int dE_k\,\rho(E_k)\,|W_{ki}|^2\,\delta_T(E_k - E_i). \tag{1.94}$$

If, in turn, $\rho(E_k)$ is slowly varying, δ_T can be approximated by a Dirac delta-function and one then finds a linear dependence on T:

$$P_i(T) = 1 - \Gamma T, \tag{1.95}$$

where the transition probability per unit time Γ is given by the following important relation often used in quantum mechanics (and called by Fermi the *golden rule of quantum physics*, now known as *Fermi's golden rule*):

$$\Gamma = \frac{2\pi}{\hbar} |W_{fi}|^2 \rho(E_f = E_i). \tag{1.96}$$

The state $|f\rangle$ is the quasi-continuum level, which has the same energy as the discrete level $|i\rangle$, and W_{fi} and $\rho(E_f)$ are, respectively, the coupling matrix element and the density of states calculated for this state.

Formula (1.96) is the natural generalization of (1.78) since we know that at the end of a time T, only the levels situated in a range of energies of width $\hbar/T$ about E_i can be populated: if $\hbar/T$ is small on the scale over which $\rho(E)$ and W_{ik} vary appreciably, the distribution of quasi-continuum levels that are coupled to the initial state coincides with that of the simplified model introduced earlier, which can therefore be used if $\rho(E)$ varies slowly in the neighbourhood of E_i.

Fermi's golden rule for a degenerate continuum

Formula (1.96) implicitly assumes that the continuum states are non-degenerate, that is to say that the specification of E is sufficient to identify a given state in the continuum. However, there are often more complex problems where we need an additional set of parameters to completely specify a continuum state. We denote these extra quantum numbers by the condensed notation β so that a unique continuum state is denoted by $|E, \beta\rangle$. In the example of autoionization introduced in Section 1.3.1, the quasi-continuum is that of a free electron enclosed in a box with large dimensions compared to those of the atom. In this case the quasi-continuum states $|E_k, \theta, \phi\rangle$ are determined by the specification of two angular coordinates, giving the direction of ejection of the electron, in addition to that of the kinetic energy E_k.

We seek an expression for the differential probability $\mathrm{d}P(T)/\mathrm{d}\beta$ of finding the system in the state $|E, \beta\rangle$ for which the parameter β has a value in the range β to $\beta + \mathrm{d}\beta$. An analogous argument to that employed above allows us to show that this differential probability increases linearly with T. We can therefore define a differential probability per unit time:

$$\frac{\mathrm{d}\Gamma}{\mathrm{d}\beta} = \frac{1}{T}\frac{\mathrm{d}P(T)}{\mathrm{d}\beta} = \frac{2\pi}{\hbar} \left|\left\langle E_f, \beta \left| \hat{W} \right| i \right\rangle\right|^2 \rho(\beta, E_f = E_i). \tag{1.97}$$

The probability that the system remains in the initial state at time T retains the linear dependence of Equation (1.95), with Γ given by:

$$\Gamma = \int \mathrm{d}\beta \, \frac{\mathrm{d}\Gamma}{\mathrm{d}\beta}. \tag{1.98}$$

This result has a straightforward interpretation: Γ is obtained by summing the probabilities of all possible transitions to distinct final states with energy E_f.

In the case of our simplified model of autoionization, formula (1.97) allows the calculation of the probability $\mathrm{d}\Gamma/\mathrm{d}\Omega$ of the electrons being ejected in a given direction 'β' $= (\theta, \phi)$. The total autoionization probability is then found by integrating over all possible directions. This gives:

$$\Gamma = \int \frac{\mathrm{d}\Gamma}{\mathrm{d}\Omega}\mathrm{d}\Omega = \int \frac{\mathrm{d}\Gamma}{\mathrm{d}\Omega}\sin\theta\,\mathrm{d}\theta\mathrm{d}\varphi. \tag{1.99}$$

An actual evaluation of this integral yields a result that is independent of the dimensions L of the fictitious box in which the atom is enclosed (see Complement 2E).

Long-time behaviour

The non-perturbative long-time solution calculated by the Wigner–Weisskopf method (Section 1.3.2) can be generalized to an arbitrary quasi-continuum. One obtains, as a result, an exponential decrease of the population in level $|i\rangle$ (formula (1.87)) with Γ now given by expressions (1.96) or (1.98). However, this formula is only valid if the quasi-continuum extends over a sufficiently large energy range. More precisely, $\hbar\Gamma$ being the 'width' of the initial level, induced by the coupling with the continuum, the decrease in population with time will be exponential when the quantity $|W_{fi}|^2\rho(E)$ changes little in the energy range of width $\hbar\Gamma$ around E_i. We shall see in Complement 1A what can arise when this condition is no longer satisfied.

Comment The energy distribution of the final states of the process is also given by a Lorentzian curve of width $\hbar\Gamma$, Γ being determined by (1.96) or (1.98). However, strictly speaking, the coupling with the quasi-continuum levels usually causes a displacement of the energy levels so that the centre of the Lorentzian is slightly displaced with respect to the position of the discrete level E_i. In the case of the simple model of Section 1.3.2 this displacement is zero because the contributions to the shifts of the levels $|k\rangle$ of which the energies are positioned symmetrically around E_i are equal and opposite.

1.3.4 Case of a sinusoidal perturbation

We can use the results of Section 1.2.4 to generalize Fermi's golden rule to the case of a sinusoidal perturbation:

$$\hat{H}_1(t) = \hat{W}\cos(\omega t + \varphi). \tag{1.100}$$

To achieve this it is sufficient to use formula (1.47) for the probability $P_{i\to k}(t)$ in place of formula (1.39). One obtains in the same manner a linear temporal variation of $P_i(T)$, allowing us to define a transition probability per unit time which now has the value:

$$\Gamma = \frac{2\pi}{\hbar}\frac{1}{4}\left[\left|W_{f'i}\right|^2\rho(E_{f'} = E_i + \hbar\omega) + \left|W_{f''i}\right|^2\rho(E_{f''} = E_i - \hbar\omega)\right], \tag{1.101}$$

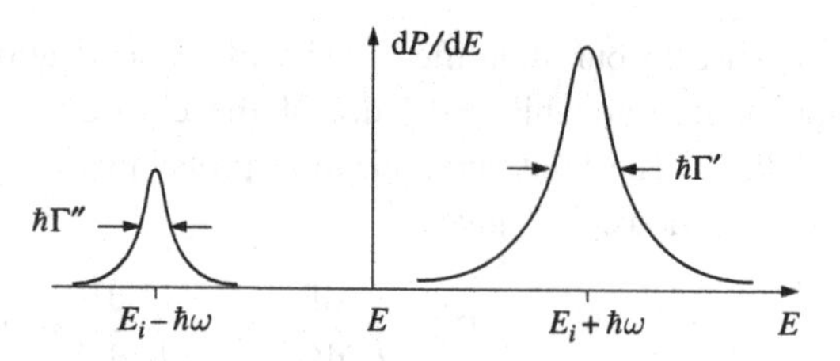

Figure 1.13 **Energy distribution of the energy of the final states at the end of the interaction between a discrete level and a quasi-continuum induced by a sinusoidal perturbation of frequency ω. The amplitudes of the two peaks are proportional to the first and second terms of relation (1.101). This result can be interpreted in terms of the emission or the absorption by the system of a quantum of energy $\hbar\omega$.**

$|f'\rangle$ and $|f''\rangle$ being, respectively, the quasi-continuum levels of energies $E_i+\hbar\omega$ and $E_i-\hbar\omega$, and $W_{f'i}$ and $W_{f''i}$ the corresponding matrix elements.[17]

At the end of the process the levels that will be populated by the sinusoidally varying coupling will be the quasi-continuum levels of which the energy is equal to $E_i \pm \hbar\omega$ to within $\hbar\Gamma$ (see Figure 1.13). As in Section 1.2.4, formula (1.101) has a natural interpretation in terms of the absorption and emission of a quantum of energy $\hbar\omega$, despite the fact that we have not yet introduced the quantization of the field that drives the transition.

Comment After studying the coupling between two discrete levels and then between a discrete level and a continuum, one could extend the study to more complicated systems for which, for example, the system initially occupies an ensemble of discrete or continuum states.

If the system starts from one continuum level to end in another continuum level (as, for example, in the case of a free electron scattered by a potential), one can still apply formula (1.97) giving the transition probability per unit time. However, if the initial state $|i\rangle$ belongs to a continuum its wavefunction, and therefore formula (1.97), depend on the dimensions of the fictitious box that encloses the system. The physically significant quantity, which is independent of the dimensions of the box, is the *differential cross-section* which is the ratio of the transition probability per unit time over the incident particle flux. We shall meet some examples of this in Chapter 6.

If the system is initially in an *incoherent* superposition (i.e. a statistical mixture) either of discrete or continuum states, it is sufficient to average the probabilities $P_{i\to k}$ (expression (1.26)) over the range of possible initial states and to sum the result over the ensemble of possible final states.

1.4 Conclusion

In this chapter we have reviewed some essential aspects of the quantum mechanical description of the dynamic behaviour of non-dissipative systems (i.e. which obey a Schrödinger equation). In the case of the simple couplings (either constant or with a

[17] If the continuum does not extend over the energy regions neighbouring $E_i + \hbar\omega$ or $E_i - \hbar\omega$, the corresponding transition probability is obviously zero, which is equivalent to setting $\rho(E_{f'})$ or $\rho(E_{f''})$ to zero (see for instance Complement 2E).

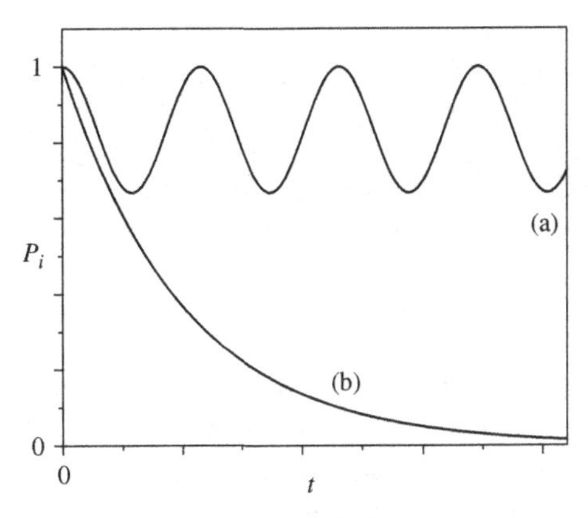

Figure 1.14 **Variation of the population of the initially occupied discrete state coupled by a constant perturbation to (a) a second discrete level, (b) a quasi-continuum.**

sinusoidal temporal dependence) that we have considered, we have met two qualitatively different modes of temporal evolution of the population, $P_i(t)$ of the initially populated state (see Figure 1.14):

(a) an evolution proportional to t^2 at short times and oscillatory at long times in the case where the initial state is coupled to an isolated final state;
(b) a probability decreasing monotonically, linearly in t at short times and exponentially at long times, in the case of a discrete initial level coupled to a tightly packed ensemble of levels forming a sufficiently broad quasi-continuum.

In the following we shall meet situations corresponding to both of these cases: for example, the coupling of a two-level atom to a monochromatic electromagnetic wave for case (a), the spontaneous decay of an isolated atom prepared in an excited state for case (b). These two cases are fundamentally different since in the former the system returns periodically to the initial state, whereas in the latter it evolves monotonically into a state different from its initial one. These two examples correspond in fact to two extreme limiting cases. In the second complement to this chapter we shall analyse the evolution of the population of a discrete state coupled to a quasi-continuum of variable width Δ. We shall show that if Δ is small compared to Γ we see behaviour analogous to mode (a) above, since the oscillatory evolution is damped when Δ is of the order of Γ, whereas we see behaviour reminiscent of mode (b) when Δ is large compared to Γ.

1A

Complement 1A A continuum of variable width

In this chapter we have met two distinct modes of temporal evolution occurring according to whether the initial state was coupled to an isolated state or a quasi-continuum. In this complement we shall study a simple model which enables us to study behaviours intermediate between these two extreme regimes. We consider the interaction of a discrete state $|i\rangle$ and a quasi-continuum of finite energy width Δ which is a parameter that can vary.

1A.1 Description of the model

We consider again the simplified model of the quasi-continuum of Section 1.3.2. This consists of an ensemble of states $|k\rangle$, equidistant in energy, $E_k = k\varepsilon$, with k taking all values between $-\infty$ and $+\infty$, coupled to a discrete state, which we take to have zero energy. Now we suppose that the matrix elements of the interaction Hamiltonian $\hat{W}$ are no longer constant but are given by:

$$\begin{aligned} \left\langle k\left|\hat{W}\right|i\right\rangle = w_k &= w\left[1+(k\varepsilon/\Delta)^2\right]^{-1/2} \\ \left\langle k\left|\hat{W}\right|k'\right\rangle &= 0 \\ \left\langle i\left|\hat{W}\right|i\right\rangle &= 0. \end{aligned} \tag{1A.1}$$

The square of the coupling matrix elements is therefore a Lorentzian function of the energy, centred on the energy of the initial discrete state. Thus, for an atom in the initial state $|i\rangle$, the quasi-continuum appears to have an effective width Δ. States for which $|E_k| > \Delta$ are therefore weakly coupled to the initial state. We suppose, in addition, that w and Δ are large compared to ε. We have chosen this Lorentzian form for the energy dependence of the square of the coupling matrix elements because, as we shall see in the following, it allows solution of the equations governing the evolution of the system for any value of the continuum width Δ.

1A.2 Temporal evolution

In order to evaluate the temporal evolution of the model system, we shall follow a procedure analogous to that of Section 1.3.2, by writing that wavefunction in the form:

$$|\psi(t)\rangle = \gamma_i(t)\,|i\rangle + \sum_{k=-\infty}^{+\infty} \gamma_k(t)\mathrm{e}^{-\mathrm{i}k\varepsilon t/\hbar}\,|k\rangle\,. \tag{1A.2}$$

The Schrödinger equation is then equivalent to the following system of differential equations:

$$\frac{\mathrm{d}}{\mathrm{d}t}\gamma_i(t) = \frac{1}{\mathrm{i}\hbar}\sum_{k=-\infty}^{+\infty} w_k\gamma_k(t)\mathrm{e}^{-\mathrm{i}k\varepsilon t/\hbar}, \tag{1A.3}$$

$$\frac{\mathrm{d}}{\mathrm{d}t}\gamma_k(t) = \frac{w_k}{\mathrm{i}\hbar}\mathrm{e}^{\mathrm{i}k\varepsilon t/\hbar}\gamma_i(t). \tag{1A.4}$$

Formal integration of the second of these equations (see 1.81) gives the following expression for $\gamma_i(t)$:

$$\frac{\mathrm{d}}{\mathrm{d}t}\gamma_i(t) = -\int_0^t \mathrm{d}t'\;\gamma_i(t')\left[\sum_{k=-\infty}^{+\infty}\frac{w_k^2}{\hbar^2}\mathrm{e}^{\mathrm{i}k\varepsilon(t'-t)/\hbar}\right], \tag{1A.5}$$

ε being very small, after having substituted (1A.1), the sum in square brackets can be replaced by the integral:

$$\int_{-\infty}^{+\infty}\frac{\mathrm{d}E}{\varepsilon}\frac{w^2}{\hbar^2}\frac{1}{1+(E/\Delta)^2}\mathrm{e}^{\mathrm{i}E(t'-t)/\hbar} = \frac{w^2}{\hbar^2}\frac{\pi\Delta}{\varepsilon}\mathrm{e}^{-\Delta|t'-t|/\hbar}. \tag{1A.6}$$

Introducing the usual notation $\Gamma = (2\pi/\hbar)(w^2/\varepsilon)$ (the transition probability per unit time given by Fermi's golden rule), we arrive at the following integro-differential equation for γ_i:

$$\frac{\mathrm{d}}{\mathrm{d}t}\gamma_i(t) = -\frac{\Gamma\Delta}{2\hbar}\int_0^t \mathrm{d}t'\;\gamma_i(t')\mathrm{e}^{\Delta(t'-t)/\hbar}, \tag{1A.7}$$

or, equivalently, by differentiating this expression with respect to t, at the second-order differential equation:

$$\frac{\mathrm{d}^2}{\mathrm{d}t^2}\gamma_i + \frac{\Delta}{\hbar}\frac{\mathrm{d}}{\mathrm{d}t}\gamma_i + \frac{\Gamma\Delta}{2\hbar}\gamma_i = 0. \tag{1A.8}$$

This differential equation may be straightforwardly solved because of our choice of the form of the energy dependence of the coupling matrix elements in Equation (1A.1).

Suppose that at $t = 0$ the state of the system is the discrete level $|i\rangle$. The initial conditions are therefore: $\gamma_i(0) = 1, \gamma_{k\neq i}(0) = 0$, from which $\mathrm{d}\gamma_i/\mathrm{d}t(0) = 0$ by virtue of (1A.3). The corresponding solution of (1A.8) can then be written, in the case of $\Delta > 2\hbar\Gamma$:

$$\gamma_i(t) = \mathrm{e}^{-\Delta t/2\hbar}\left(\cosh\frac{\Delta' t}{2\hbar} + \frac{\Delta}{\Delta'}\sinh\frac{\Delta' t}{2\hbar}\right), \tag{1A.9}$$

$$\text{with}\quad \Delta' = \Delta\sqrt{1-\frac{2\hbar\Gamma}{\Delta}}. \tag{1A.10}$$

In the case of $\Delta < 2\hbar\Gamma$, the solution is:

$$\gamma_i(t) = \mathrm{e}^{-\Delta t/2\hbar}\left(\cos\frac{\Delta'' t}{2\hbar} + \frac{\Delta}{\Delta''}\sin\frac{\Delta'' t}{2\hbar}\right), \tag{1A.11}$$

$$\text{with} \quad \Delta'' = \Delta\sqrt{\frac{2\hbar\Gamma}{\Delta} - 1}. \tag{1A.12}$$

Consider then the two extreme values for the width Δ of the continuum:
If $\hbar\Gamma/\Delta \ll 1$ (broad continuum), $\gamma_i(t)$ is given to first order in $\hbar\Gamma/\Delta$ by:

$$\gamma_i(t) = \mathrm{e}^{-\Gamma t/2}\left(1 + \frac{\hbar\Gamma}{2\Delta}\right) - \frac{\hbar\Gamma}{2\Delta}\mathrm{e}^{-\Delta t/\hbar}. \tag{1A.13}$$

For times that are long compared to the inverse frequency width of the continuum, $\hbar/\Delta$, the contribution of the second exponential is negligible and we recover the usual expression of Equation (1.86). The second term does play an important role, however, at short times, $t < \hbar/\Delta$, when it ensures that $\gamma_i(t)$ evolves quadratically and not linearly with t. In fact, close to $t = 0$ and to first order in $\hbar\Gamma/\Delta$, $\gamma_i(t)$ is given by:

$$\gamma_i(t) = 1 - \frac{\Gamma\Delta}{4\hbar}t^2. \tag{1A.14}$$

If $\hbar\Gamma/\Delta \gg 1$ (narrow continuum), the approximate solution of (1A.8) is, to lowest order in $\Delta/\hbar\Gamma$,

$$\gamma_i(t) = \mathrm{e}^{-\Delta t/2\hbar}\cos\sqrt{\frac{\Gamma\Delta}{2\hbar}}t. \tag{1A.15}$$

One obtains oscillations of angular frequency:

$$\left[\frac{\Gamma\Delta}{2\hbar}\right]^{1/2} = \frac{w}{\hbar}\left(\frac{\pi\Delta}{\varepsilon}\right)^{1/2}, \tag{1A.16}$$

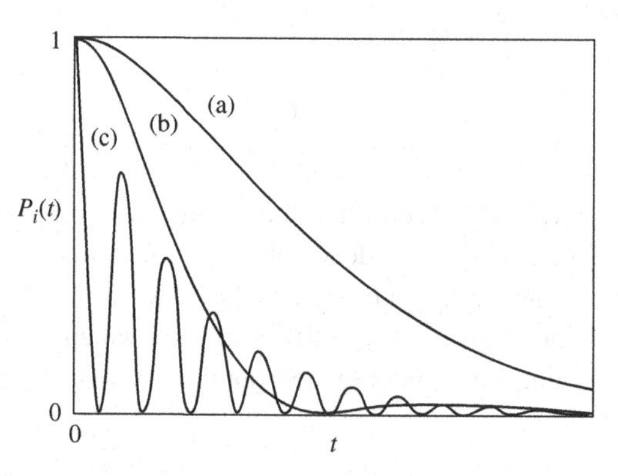

Figure 1A.1 Temporal evolution of the occupation probability of the initial state for different values of the parameter $p = \hbar\Gamma/\Delta$, corresponding to continua with increasingly narrow effective widths (Δ now being a constant) (a) $p = 0.5$, (b) $p = 2$, (c) $p = 100$. The time axis is shown from $t = 0$ to $t = 5\hbar/\Delta$.

similar to Rabi oscillations, but which are damped on a timescale of order $(\hbar/\Delta)$. At short times, when the effect of this damping is negligible, the system evolves as though the initial state were coupled to a unique final state by a coupling matrix element equal to $w(\pi\Delta/\varepsilon)^{1/2}$.

In this model the transition between the regimes in which the occupation probability of the initial state decreases monotonically or undergoes oscillations reminiscent of Rabi-flopping occurs at the value of Δ given by $\Delta = 2\hbar\Gamma$. Figure 1A.1 shows the evolution of $P_i(t) = |\gamma_i(t)|^2$ for several values of the parameter $p = \hbar\Gamma/\Delta$. We see here how the transition between the two modes of behaviour occurs.

If the precise form of $P_i(t)$ depends on the details of the particular model, its qualitative behaviour does not: in particular, quite generally *the transition from the oscillatory to the monotonically decreasing regime occurs when the transition probability per unit time calculated by Fermi's golden rule is of the order of the width of the continuum.* It should also be noted that even in the oscillatory regime, the global amplitude of the oscillations will be damped with a characteristic time of the order of $\hbar$ divided by the energy width of the continuum.

Comments

(i) Note that, in this complement, the width of the continuum is infinite because the density of states is constant and has the value $1/\varepsilon$. In fact, the important parameter is the width of the continuum, *weighted by the energy dependence of the coupling with the initial state* $\rho(E)\ |\langle k\ |\hat{W}|\ i\rangle\ |^2$, rather than the energy width of the entire band of continuum states (see CDG II, Chapter 13).

(ii) For a treatment of more complex continua see CDG II, Complement C_{III}.

(iii) The passage from an oscillating to a monotonous behaviour can have other causes than the one treated in this complement. For example, we have considered up to now only cases in which the perturbation $\hat{W}(t)$ is perfectly deterministic. However there exist cases in which the perturbation has a random temporal dependence. This is the case in the example of Section 1.2.2 concerning collisions, where the relative speed and the impact parameter have statistical variations. Complement 1B will show how such a random perturbation leads to a washing out of the Rabi oscillations. On the other hand the exponential mode of evolution will prove much less affected by the statistical averaging. We shall see an example of this in Chapter 2 (Section 2.4) when we study stimulated emission and absorption: the pumping and relaxation of the two levels involved in the transition introduce an average over Rabi oscillations commencing at different instants in time resulting in their being washed out.

Complement 1B Transition induced by a random broadband perturbation

We have seen in Section 1.2 that a transition between two discrete levels $|i\rangle$ and $|k\rangle$ is possible when the system is submitted to a perturbation which has a non-zero Fourier component close to the Bohr frequency $\omega_0 = (E_k - E_i)/\hbar$ of the transition. This is the case when the perturbation $\hat{H}_1(t)$ is sinusoidal, of the form $\hat{H}_1(t) = \hat{W}\cos(\omega t + \phi)$, and oscillates at a frequency ω close to ω_0. But in many real physical situations, the perturbation is not perfectly mastered: it is a pure sine function only for times shorter than a limit, called the *coherence time*. At longer times, the oscillating perturbation has a phase and an amplitude that vary randomly. This is, for example, the case when the perturbation is due to an incident electromagnetic wave produced by a thermal lamp, in which the thermal fluctuations induce random variations of the amplitude and phase of the wave. As the perturbation is no longer a pure sinusoidal wave, its Fourier spectrum is no longer a Dirac delta function. It has some finite width around a mean frequency. For this reason a random perturbation is also called a 'broadband' perturbation.

We shall determine in this complement the temporal evolution of the quantum system submitted to such a random perturbation. We shall see that the transition probability between two discrete levels is proportional to T for small interaction time T, and has an exponential behaviour at long times, a result which is very similar to that obtained for a transition between a discrete level and a continuum.

1B.1 Description of a random perturbation

1B.1.1 Definitions

We shall write the perturbation $\hat{H}_1(t)$ as $\hat{H}_1(t) = \hat{W}f(t)$, where $f(t)$ is a *stochastic* real function of time, the value of which can be randomly chosen according to a given statistical distribution. Such a function is also called a random process, or a stochastic process. The quantities that can be measured are only *statistical ensemble averages* over the statistical distribution (which we will denote by an overline). Different quantities characterize the random process. The simplest ones are the mean $\overline{f(t)}$, and the *autocorrelation function* $\Gamma(t,t') = \overline{f(t)f(t')}$. We shall suppose here that the mean $\overline{f(t)}$ is zero, and restrict ourselves to the case of a *stationary random process*, in which all the statistical means are invariant by a translation of the origin of time. This implies that $\overline{f(t)}$ is time independent, and that $\Gamma(t,t')$ depends only on the time difference $\tau = t - t'$. We shall write $\Gamma(t,t') = \Gamma(\tau)$. We assume in addition that the random process is ergodic, i.e. ensemble averages are equal to time averages over a large enough interval of time.[1]

Let us now define the Fourier transform of the perturbation, $\tilde{F}(\omega)$:[2]

$$\tilde{F}(\omega) = \frac{1}{\sqrt{2\pi}} \int_{-\infty}^{+\infty} \mathrm{d}t\, f(t)\, \mathrm{e}^{\mathrm{i}\omega t}. \tag{1B.1}$$

Its mean is zero, and its variance $\overline{|\tilde{F}(\omega)|^2}$ is infinite. More precisely, one can show that, defining a Fourier transform in a time window of duration T:[1]

$$\tilde{F}_T(\omega) = \frac{1}{\sqrt{2\pi}} \int_{-T/2}^{+T/2} \mathrm{d}t\, f(t)\, \mathrm{e}^{\mathrm{i}\omega t}, \tag{1B.2}$$

the quantity $\overline{|\tilde{F}_T(\omega)|^2}$ diverges linearly with T. The mean and the variance in the Fourier space cannot therefore characterize the spectral distribution of a random perturbation.

One can then define the *spectral density* (also called 'power spectrum'):

$$S(\omega) = \lim_{T \to \infty} \frac{1}{T} \overline{|\tilde{F}_T(\omega)|^2}, \tag{1B.3}$$

which is a quantity characterizing the stationary random process. It obeys the celebrated Wiener–Khintchine theorem,[1] which can be written:

$$S(\omega) = \frac{1}{2\pi} \int_{-\infty}^{+\infty} \Gamma(\tau)\, \mathrm{e}^{\mathrm{i}\omega\tau} \mathrm{d}\tau, \tag{1B.4}$$

or reciprocally:

$$\Gamma(\tau) = \int_{-\infty}^{+\infty} S(\omega)\, \mathrm{e}^{-\mathrm{i}\omega\tau} \mathrm{d}\omega. \tag{1B.5}$$

The autocorrelation function $\Gamma(\tau)$ and the spectral density are a Fourier transform of each other (within a $\sqrt{2\pi}$ factor).

The spectral density is a measurable quantity. As an example, considering thermal light from a source as a random stationary process, a spectrometer (for instance based on frequency dispersion by a prism or a grating) yields $S(\omega)$, where ω are optical frequencies. Similarly, electronic spectrum analysers give the power spectrum of an electric signal.

Note that the variance of the random process (with null average) can be expressed as:

$$\overline{|f(t)|^2} = \Gamma(0) = \int_{-\infty}^{+\infty} S(\omega)\, \mathrm{d}\omega. \tag{1B.6}$$

It can be interpreted as an incoherent sum of all spectral components $S(\omega)\,\mathrm{d}\omega$. Note however that the Wiener–Khintchine theorem remains valid even in the case where some spectral components have actually well-defined relative phases.[1] This is the case for the light emitted by a single mode laser well above threshold, which can be described with a constant amplitude, and a phase making a random walk (see Complement 3D).

[1] J. Goodman, *Statistical Optics*, Wiley (2000).

[2] Note the difference with the Fourier transform $\tilde{f}(E) = \tilde{F}(E/\hbar)/\sqrt{\hbar}$ (Equation (1.30)).

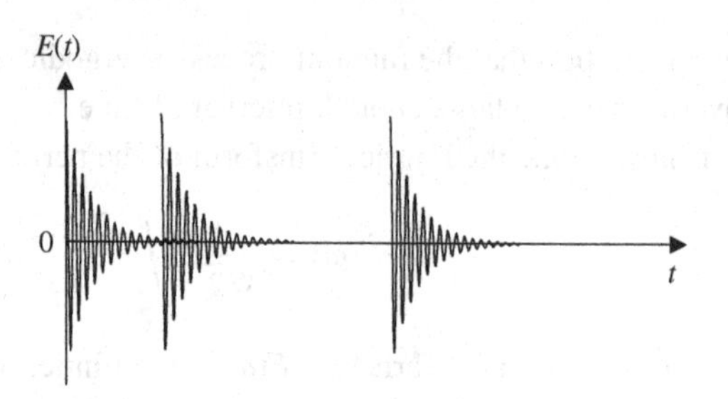

Figure 1B.1 **Random perturbation consisting of a series of damped 'wave packets' starting at random times.**

1B.1.2 Example

Let us give here a simple example of a random perturbation: we consider a succession of 'wave packets' of a monochromatic wave of frequency $\overline{\omega}$, decaying over a time T_p and starting at random times (see Figure 1B.1). The quantity T_p is in this case the correlation time of the perturbation. A wave packet starting at t_i is written as:

$$p(t) = \cos\left[\overline{\omega}(t - t_i)\right] e^{-(t-t_i)/T_p} \quad \text{for} \quad t > t_i\,; \tag{1B.7}$$

$$p(t) = 0 \quad \text{for} \quad t \leq t_i. \tag{1B.8}$$

The random function $f(t)$ is here given by:

$$f(t) = \sum_{t_i} p(t - t_i), \tag{1B.9}$$

where the starting times t_i of the pulses are randomly distributed on the time axis. We call N the mean number of such pulses during a unit time interval.

This kind of perturbation can model, for example, the interaction between an atom and the light received from an assembly of excited atoms decaying to the ground state in an independent way.

Assimilating the ensemble average to a time average over time intervals much longer than T_p (ergodicity), one can compute the correlation function:

$$\Gamma(\tau) = \frac{NT_p}{4} \cos(\overline{\omega}\tau)\, e^{-\frac{|\tau|}{T_p}}, \tag{1B.10}$$

which is a function rapidly decaying to zero for times longer than T_p. Its power spectrum is obtained by Fourier transform:

$$S(\omega) = \frac{N}{8\pi}\left[\frac{1}{(\omega - \overline{\omega})^2 + \left(\frac{1}{T_p}\right)^2} + \frac{1}{(\omega + \overline{\omega})^2 + \left(\frac{1}{T_p}\right)^2}\right]. \tag{1B.11}$$

In the positive part of the spectrum, one can neglect the contribution of the Lorentzian centred at $\omega = -\overline{\omega}$,[3] and write:

$$S(\omega) \simeq \frac{N}{8\pi} \frac{1}{(\omega - \overline{\omega})^2 + \left(\frac{1}{T_p}\right)^2}. \tag{1B.12}$$

The spectrum of the perturbation is indeed 'broadband': the spectral density $S(\omega)$ is a Lorentzian curve centred at the mean frequency $\overline{\omega}$, of width $2/T_p$ at half-maximum. The fact that the spectral width is of the order of the inverse of the correlation time of the perturbation is a very general result.

1B.2 Transition probability between discrete levels

1B.2.1 General expression

Let us first consider, as in Section 1.2, a transition between two discrete levels $|i\rangle$ and $|k\rangle$. To determine the transition probability, one first takes a given realization of the random perturbation $\hat{H}_1(t) = \hat{W} f(t)$. According to Equation (1.26), the transition probability $P_{i\to k}$ is then given by:

$$\begin{aligned} P_{i\to k} &= \frac{W_{ki}|^2}{\hbar^2} \left| \int_0^T \mathrm{d}t f(t)\,\mathrm{e}^{\mathrm{i}(E_k - E_i)t/\hbar} \right|^2 \\ &= \frac{W_{ki}|^2}{\hbar^2} \int_0^T \mathrm{d}t \int_0^T \mathrm{d}t' f(t) f(t')\,\mathrm{e}^{\mathrm{i}\omega_{ki}(t-t')}, \end{aligned} \tag{1B.13}$$

using the notation $\omega_{ki} = (E_k - E_i)/\hbar$. The measurable quantity is the transition probability $\overline{P_{i\to k}}$, averaged over the different possible values of the random perturbation:

$$\begin{aligned} \overline{P_{i\to k}} &= \frac{W_{ki}|^2}{\hbar^2} \int_0^T \mathrm{d}t \int_0^T \mathrm{d}t'\; \overline{f(t) f(t')}\,\mathrm{e}^{\mathrm{i}\omega_{ki}(t-t')} \\ &= \frac{W_{ki}|^2}{\hbar^2} \int_0^T \mathrm{d}t \int_0^T \mathrm{d}t'\; \Gamma(t-t')\,\mathrm{e}^{\mathrm{i}\omega_{ki}(t-t')}. \end{aligned} \tag{1B.14}$$

We can now replace $\Gamma(t - t')$ by its expression (1B.5) and obtain:

$$\begin{aligned} \overline{P_{i\to k}} &= \frac{W_{ki}|^2}{\hbar^2} \int \mathrm{d}\omega\, S(\omega) \int_0^T \mathrm{d}t \int_0^T \mathrm{d}t'\,\mathrm{e}^{\mathrm{i}(\omega_{ki}-\omega)(t-t')} \\ &= \frac{W_{ki}|^2}{\hbar^2} \int_{-\infty}^{+\infty} \mathrm{d}\omega\, S(\omega) \left| \frac{\mathrm{e}^{\mathrm{i}(\omega_{ki}-\omega)T} - 1}{i(\omega_{ki} - \omega)} \right|^2. \end{aligned} \tag{1B.15}$$

[3] This approximation is once again the quasi-resonant approximation, or secular approximation, which we have encountered several times in this chapter, under different forms.

We recognize the function $g_T(E_k - E_i - \hbar\omega) = 2\pi\hbar\delta_T(E_k - E_i - \hbar\omega)$ introduced in (1.36) and (1.38), so that we finally get:

$$\overline{P_{i\to k}} = 2\pi\frac{T}{\hbar}|W_{ki}|^2 \int_{-\infty}^{+\infty} d\omega\, S(\omega)\, \delta_T(E_k - E_i - \hbar\omega), \tag{1B.16}$$

an expression which is very similar to Equation (1.47) giving the transition probability for a sinusoidal perturbation, except that $\delta_T(E_k - E_i - \hbar\omega)$ is now replaced by the same function averaged over the spectral density of the perturbation.

1B.2.2 Behaviour at intermediate times

In Section 1.2.4c, we have seen that $\delta_T(E)$ is a narrow function with width $2\pi\hbar/T$. Let us now assume that the interaction time T is long enough so that $2\pi/T$ is much smaller than the characteristic variation scale of the spectral density $S(\omega)$. In the example of the random succession of short pulses of duration T_p, this condition is simply $T \gg T_p$. More generally, it is fulfilled when T is much longer than the correlation time of the perturbation. We can then replace, in the integral, δ_T by the true Dirac function and finally obtain:

$$\overline{P_{i\to k}} = 2\pi\, T\frac{|W_{ki}|^2}{\hbar^2} S(\omega_{ki}) = \Gamma\, T. \tag{1B.17}$$

The transition probability increases linearly with time, as in the case of a transition induced by a sinusoidal perturbation between a discrete level and a continuum. We can then define *a transition probability per unit time* Γ, which has an expression similar to the Fermi golden rule (Equation (1.96)), replacing the density of states $\rho(E = \hbar\omega_{ki})$ by the spectral density of the perturbation at the Bohr frequency of the transition.

1B.2.3 Behaviour at long times

Expression (1B.17) is obviously not valid at too long times, when $\overline{P_{i\to k}}$ becomes close to 1. Using the decomposition of the state vector over the two states $|i\rangle$ and $|k\rangle$, and a derivation analogous to Section 1.3.2b, one can show that the transition probability has an exponential behaviour:

$$\overline{P_{i\to k}} = 1 - e^{-\Gamma t}, \tag{1B.18}$$

with Γ defined in (1B.17). We shall retrieve such an exponential behaviour when we treat the same problem using the density matrix formalism, and end up with the Bloch equations (see Complement 2C, Section 2C.5.2).

Comment The results established here, for the case of a broadband excitation of an infinitely narrow transition, remain valid in the case of a transition with a finite linewidth, provided that the spectrum of the excitation is broader, and the transition rate Γ (1B.17) is larger, than the linewidth.

1B.3 Transition probability between a discrete level and a continuum

As in Section 1.3, we now turn to the case where the final state $|k\rangle$ is embedded in a continuum and cannot be distinguished from its neighbours. In order to calculate the probability $P_i(t)$ that the system remains in the initial state $|i\rangle$ at time T, we assume a given value of the random perturbation, and then add the effect of all the possible transitions starting from $|i\rangle$. This gives:

$$P_i(T) = 1 - \sum_k P_{i\to k} = 1 - \sum_k \frac{|W_{ki}|^2}{\hbar^2} \int_0^T \mathrm{d}t \int_0^T \mathrm{d}t' f(t) f(t')\, \mathrm{e}^{\mathrm{i}\omega_{ki}(t-t')}. \tag{1B.19}$$

Assuming that the levels $|k\rangle$ are non-degenerate, the discrete sum over k is actually an integral over the energy E of the levels, weighted by the state density $\rho(E)$ introduced in Section 1.3.3a:

$$P_i(T) = 1 - \int \mathrm{d}E \rho(E) \frac{|W_{Ei}|^2}{\hbar^2} \int_0^T \mathrm{d}t \int_0^T \mathrm{d}t' f(t) f(t')\, \mathrm{e}^{\mathrm{i}(E-E_i)(t-t')/\hbar}, \tag{1B.20}$$

where W_{Ei} is the matrix element $\langle k|\hat{W}|i\rangle$ of the perturbation between the initial state and the final state $|k\rangle$ of energy E. We then average this result over all the possibilities for the random perturbation in order to find the mean probability $\overline{P_i(T)}$:

$$\overline{P_i(T)} = 1 - \int \mathrm{d}E \rho(E) \frac{|W_{Ei}|^2}{\hbar^2} \int_0^T \mathrm{d}t \int_0^T \mathrm{d}t' \Gamma(t-t')\, \mathrm{e}^{\mathrm{i}(E-E_i)(t-t')/\hbar}. \tag{1B.21}$$

Introducing the spectral density $S(\omega)$ and using the same derivation as for Equation (1B.16), we get:

$$\overline{P_i(T)} = 1 - 2\pi \frac{T}{\hbar} \int \mathrm{d}E \int_{-\infty}^{+\infty} \mathrm{d}\omega\, \rho(E) |W_{Ei}|^2 S(\omega)\, \delta_T(E - E_i - \hbar\omega). \tag{1B.22}$$

It is then easy to calculate the probability $\overline{P_i(T)}$ for times T long enough so that the function $\delta_T(E - E_i - \hbar\omega)$ can be considered as a Dirac delta function. We obtain:

$$\overline{P_i(T)} = 1 - 2\pi \frac{T}{\hbar} \int_{-\infty}^{+\infty} \mathrm{d}\omega\, S(\omega)\, \rho(E = E_i + \hbar\omega) |W_{E=E_i+\hbar\omega,i}|^2. \tag{1B.23}$$

As in the case of a pure sinusoidal perturbation between a discrete state and a continuum, we end up with a linear variation of the probability with time, and therefore a transition probability per unit time Γ given by:

$$\Gamma = 2\pi \int_{-\infty}^{+\infty} \mathrm{d}\omega\, S(\omega)\, \rho(E = E_i + \hbar\omega) \frac{|W_{E=E_i+\hbar\omega,i}|^2}{\hbar}. \tag{1B.24}$$

This formula means that in order to calculate the transition rate, we can use the usual Fermi's golden rule for each frequency component of the broadband perturbation, and then make the integral of this quantity over all the possible frequencies weighted by the spectral density $S(\omega)$. For this calculation, we can thus consider that all the frequency components of the light give uncorrelated transition rates, that can be added independently.

15.3 Transition probability between a discrete level and a continuum

2 The semi-classical approach: atoms interacting with a classical electromagnetic field

We investigate in this chapter the general problem of the interaction of an atom (or a molecule) and a classical electromagnetic field. Its importance derives firstly from the fact that a large part of our knowledge of atoms is obtained from a study of the radiation they absorb or emit (we shall consider equally visible radiation and radio-frequency and X-rays) and, secondly, that the interaction with matter modifies the propagation of the electromagnetic field itself, notably through absorption, refraction or scattering. The interactions of atoms with light therefore encompass a vast range of physical effects that we could not hope to cover adequately in a single chapter. In this chapter we shall therefore present the fundamental features of the interaction of an atom, which will be treated quantum mechanically, with a classical electromagnetic field, that is an electromagnetic field described by real electric and magnetic vectors obeying Maxwell's equations.

A rigorous description of the atom–light interaction would have to take into account the quantum nature of light. This we shall leave to Chapter 6. It turns out, nevertheless, that many important results can be obtained from the *semi-classical* viewpoint adopted here (although it is perhaps more appropriate to refer to this treatment as 'semi-quantized'). Moreover, it can be demonstrated that the results of the approach presented here are, for the most part, equivalent to those of a fully quantum mechanical treatment if we consider the interaction of an atom with the light emitted by a laser operating far above threshold, or with light from a classical source such as a discharge or incandescent lamp, provided that one takes into account the statistical character of the electromagnetic field. Our model is equally valid for radio-frequency fields produced by standard techniques (oscillators, klystron tubes etc.).

However, we should point out that a semi-classical treatment is not able to account rigorously for a phenomenon as fundamental as spontaneous emission, which can only be derived from first principles within the formalism of a quantum theory of radiation. The same is true for some light scattering processes. It is possible, though, to introduce phenomenologically the finite lifetime of excited atomic states and, with this addition, the semi-classical formalism is capable of accounting in a simple manner for many classes of atom–light interactions, including those which are the basis of laser operation. The material covered in this chapter is therefore important, even if the explanation of some phenomena associated with spontaneous emission and the new domains of *entangled and non-classical states of the radiation field*, or *cavity quantum electrodynamics* is beyond the scope of the semi-classical approach. They will be studied in the last chapters of this book.

After having presented in the first part of this chapter a qualitative description of a number of interaction processes involving atoms and light: absorption, spontaneous and induced emission, scattering and multi-photon processes, we put into place in Section 2.2 the theoretical framework. We justify the choice of the *electric dipole Hamiltonian* to describe the interaction between an atom and a light field assumed to be a classical electromagnetic wave. We introduce also, at the end of this section, the *magnetic dipole interaction*, which is the dominant term representing the coupling of an atom with a radio-frequency field. In Section 2.3 we calculate the probability of transition between two discrete atomic states of infinite lifetimes under the effect of incident radiation. After an initial perturbative treatment, we then go on to study the important phenomenon of *Rabi oscillations*. We also encounter, in this section, *multi-photon transitions* and *light-shifts*. In Section 2.4 we reconsider quasi-resonant transitions between two levels, in introducing phenomenologically their finite lifetimes. In order to keep the treatment simple we limit ourselves to the case in which the two states have equal lifetimes. We then obtain a realistic expression for the response of an assembly of atoms subjected to an incident light wave, a response which we show to be characterizable by a dielectric susceptibility. In Section 2.5 we use this model to show how such a medium can amplify an incident wave, when suitably excited, and give rise to the possibility of *laser amplification*. This leads to the presentation of the *rate equations* in Section 2.6. These equations will be extensively used to describe lasers in Chapter 3.

This chapter is supplemented by five complements. The first of these is devoted to a completely classical model of the interaction between light and matter, introduced by Lorentz, in which the electron is elastically bound to the nucleus. In spite of its limits this model is a useful tool, widely used in the literature, and we could not ignore it. The second complement presents the important phenomenon of *optical pumping*, thanks to which it is possible to prepare an assembly of atoms in a given set of Zeeman sub-levels. This relies on the selectivity of transitions excited by polarized light (selection rules). In the third complement we present the density matrix formalism which leads to the *optical Bloch equations* which allow us to treat the general problem of the atom–light interaction for an atomic transition between states of arbitrary lifetimes. Although considered a subject for advanced courses, this subject is so important that we have done our best to present it in this book. The fourth complement describes in more detail how *atomic coherence* can be produced at will by the interaction between atoms and coherent light, and shows striking examples of the possibilities opened up by the manipulation of this coherence in two- and three-level systems. Finally, the last complement describes the *photoelectric effect* and can be treated as an exercise based on a physical phenomenon which is both interesting and also at the very root of quantum optics, having first prompted Einstein to introduce the notion of the photon.

2.1 Atom–light interaction processes

This first section is intended to introduce qualitatively the most important phenomena associated with the interaction of an atom and an electromagnetic wave. Its purpose is to familiarize the reader with phenomena which will be studied in more detail in the remainder of this book and to point out common characteristics of these effects. This part passes, therefore, beyond the scope of the remainder of the chapter and the reader shouldn't be surprised at the introduction of the concept of a photon, which will not be rigorously justified until Chapter 4, but which is fundamental to the study of modern optics.

2.1.1 Absorption

Consider an atom situated at position $\vec{\mathbf{r}}_0$ and of which the internal Hamiltonian $\hat{H}_0$ has eigenstates $|a\rangle$, $|b\rangle$, $|c\rangle, \ldots$, with corresponding energies E_a, E_b, E_c, ..., with the level $|a\rangle$ the ground (lowest energy) state. This atom is subjected to a monochromatic electromagnetic wave of which the electric field is

$$\mathbf{E}(\mathbf{r}, t) = \mathbf{E}_0 \cos(\omega t + \varphi(\mathbf{r})). \tag{2.1}$$

Under the influence of this electromagnetic wave the state of the atom can be changed to a state $|i\rangle$, of higher energy, whilst the amplitude of the electromagnetic field is reduced correspondingly (Figure 2.1). As we shall see, this type of process only becomes important if the electromagnetic wave is quasi-resonant, that is to say if its frequency is very close to an atomic Bohr frequency, for example

$$\omega \simeq \omega_{ba} = \frac{E_b - E_a}{\hbar}. \tag{2.2}$$

For ω close to ω_{ba} we can neglect the other atomic levels and use a simplified model of a *two-level atom*. In the following we restrict our discussion to such a model whenever we deal with *quasi-resonant processes*. We denote the resonance frequency, ω_{ba}, of the two-level atom by ω_0. The resonance condition suggests a natural interpretation of the atom–field energy transfer in terms of the absorption of a photon of energy $\hbar\omega$ with the simultaneous quantum jump of the atom from the ground to the excited state. This is a useful image, though one not strictly justified by the analysis of this chapter, which does not deal with a quantized model of the radiation field. As was shown in Chapter 1 (and as we shall see again in Section 2.3), this resonance condition arises from the use of perturbation theory in the case of a sinusoidally varying interaction. Strictly speaking, the concept

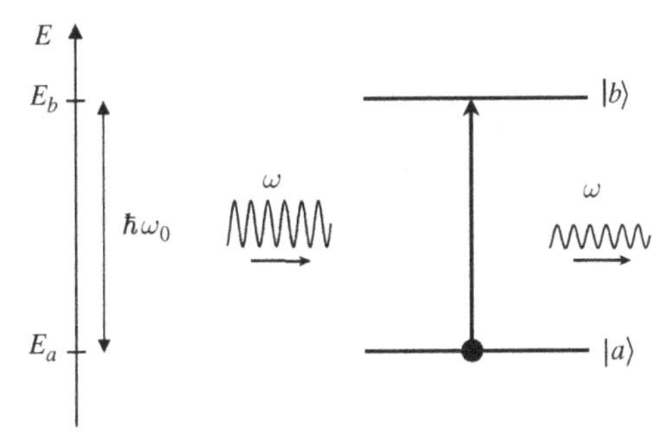

Figure 2.1 Absorption process. The excitation of the atom from its fundamental state $|a\rangle$ to an excited state $|b\rangle$ is accompanied by a decrease in the amplitude of the incident wave.

of the photon will not be justified until Chapter 4, when we shall study the quantized electromagnetic field. We shall in the meantime continue to use the concept of the photon when the image it provides is useful.

Comment

The two-level atom model might appear to be of little interest in cases where the atomic energy levels are degenerate; the resonance condition is then insufficient to enable the isolation of a unique pair of interacting levels. In fact if the incident light is polarized it will interact only with certain atomic sublevels (by the selection rules detailed in Complement 2A). The applicability of the model is therefore more widespread than it might seem at first sight.

2.1.2 Stimulated emission

We now consider how an atom in an excited state $|b\rangle$, irradiated with an electromagnetic wave of frequency ω can be de-excited to the ground state $|a\rangle$ under the influence of this wave which is consequently *amplified* (Figure 2.2). This process, known as stimulated emission, was proposed by Einstein in 1916 on theoretical grounds. Like absorption, of which it is the inverse, this process only occurs appreciably if the frequency of the electromagnetic wave is close to that of the atomic resonance, ω_0, which suggests the image of the stimulated emission of a photon of energy $\hbar\omega$ into the incident wave. To take into account the fact that the amplitude of the incident wave is consequently increased we have additionally to hypothesize that the induced radiation is emitted, not only at the same frequency as the incident field, but also with the same propagation direction and a well-defined relative phase, so that the two fields may interfere constructively.

Although for a long time the possibility of stimulated emission was considered a theoretical curiosity, because its role in traditional light sources is entirely negligible, it has lately found an important application in the amplification of light waves in lasers. As we shall see, this process constitutes the fundamental basis of laser action.

Comment

It is often considered that the primary importance of this work of Einstein's is the prediction of stimulated emission, which is certainly true from a historical point of view. On the conceptual level, and

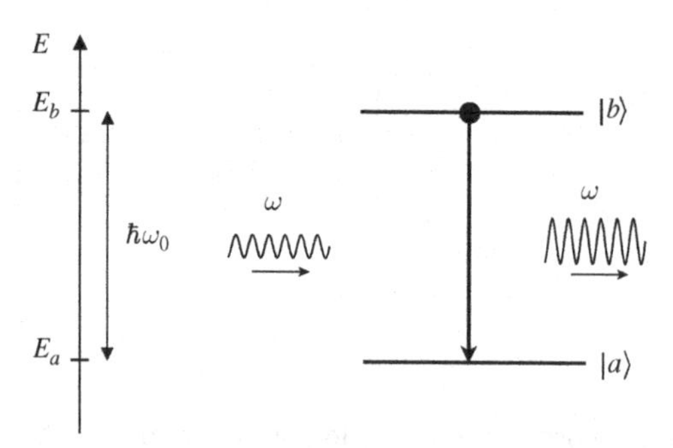

Figure 2.2 **Stimulated emission. The atom is de-excited under the influence of an incident wave, the amplitude of which is consequently increased.**

with the benefit of almost a century of hindsight, it appears, however, that stimulated absorption and emission are completely symmetric processes describable in terms of classical fields. The originality of Einstein's contribution appears then to be the demonstration of the existence of spontaneous emission and the quantitative prediction of its properties, such as its frequency variation.

2.1.3 Spontaneous emission

If the atom is initially in the excited state $|b\rangle$, it can return to the ground state $|a\rangle$ even in the absence of incident radiation. Radiation is then emitted at the frequency of the atomic resonance (the Bohr frequency), whilst the phase of the emitted field and its direction of propagation are random. This is spontaneous emission (Figure 2.3).

Here again, we can usefully employ the image of the emission of a photon. In fact this is the *only* simple model that exists, since spontaneous emission is a phenomenon fundamentally linked to the quantization of the electromagnetic field. If we present it in this chapter, it is because of its physical importance. However, in contrast to absorption and stimulated emission, spontaneous emission cannot be deduced from the semi-classical formalism presented here. Notice indeed that an isolated atom in an excited state $|b\rangle$ is in a stationary state (an eigenstate of the Hamiltonian). Elementary results of quantum mechanics then tell us that the atom should remain indefinitely in this state. It is only when one considers an extended quantum system: the atom interacting with all the modes of the quantized electromagnetic field, that the phenomenon of spontaneous emission naturally arises (see Chapter 6).

In this chapter we shall content ourselves with the knowledge that spontaneous emission exists and, where necessary, we shall account for its existence by adding in a phenomenological manner the finite radiative lifetime of a state. We shall denote by Γ_{sp} the probability per unit time of de-excitation by spontaneous emission of the state $|b\rangle$. Therefore, if the atom is in the state $|b\rangle$ at $t = 0$, the probability that it will still be in this state at a later time t is

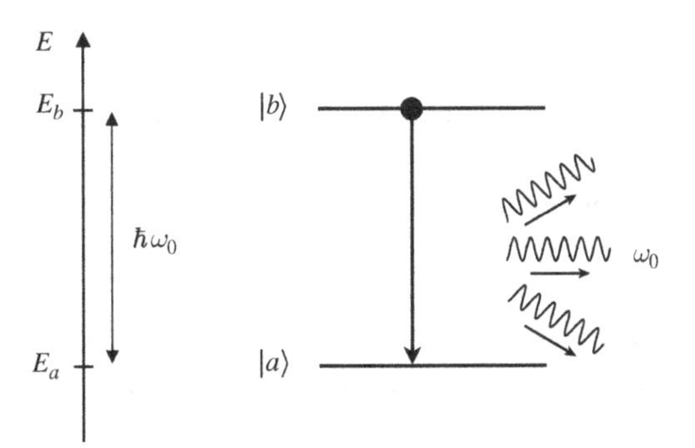

Figure 2.3 Spontaneous emission. The atom, initially in the excited state $|b\rangle$, is de-excited by the emission of a photon, even in the absence of incident radiation. This process leads to the emission of radiation in a random direction and with random phase.

$$P_b = \mathrm{e}^{-\Gamma_{\mathrm{sp}} t} = \mathrm{e}^{-t/\tau}. \tag{2.3}$$

The quantity $\tau = \Gamma_{\mathrm{sp}}^{-1}$ is the *radiative lifetime* of the excited state.

Comment We can represent the wave emitted in a given direction in terms of a damped classical wave (see Complement 2A) of amplitude

$$\mathbf{E}(\mathbf{r}, t) = \mathbf{E_0}(\mathbf{r})\,\mathrm{H}(\mathrm{t} - \mathrm{r/c}) \exp\left[-\frac{\Gamma_{\mathrm{sp}}}{2}(\mathrm{t} - \mathrm{r/c})\right] \cos\left[\omega_0(\mathrm{t} - \mathrm{r/c}) + \phi\right]. \tag{2.4}$$

Here H(u) is the Heaviside function (which is zero for u negative and unity for u positive), and ϕ is a random phase. By taking the modulus squared of the Fourier transform of this amplitude we obtain the spectrum of the emitted radiation. This has the form of a Lorentzian of width Γ_{sp} at half-maximum:

$$\rho(\omega) = \frac{\rho(\omega_0)}{1 + \frac{4(\omega - \omega_0)^2}{\Gamma_{\mathrm{sp}}^2}}. \tag{2.5}$$

The point to remember here is that the emitted radiation has a Lorentzian frequency distribution with a width equal to the reciprocal of the lifetime of the excited state:

$$\Delta\omega = \Gamma_{\mathrm{sp}} = \frac{1}{\tau}. \tag{2.6}$$

We shall establish this result more rigorously in Chapter 6.

2.1.4 Elastic scattering

Consider again an atom prepared initially in its ground state $|a\rangle$, and on which is imposed an incident electromagnetic wave of frequency ω, possibly very different from the atomic resonance frequency ω_0. In this situation a fraction of the incident light is *scattered*: the amplitude of the transmitted wave is reduced whilst the atom radiates a spherical wave of a frequency *identical to that of the incident light*. The re-radiated light has a well-defined phase with respect to the incident light (Figure 2.4).

The most straightforward interpretation of this phenomenon relies on the driving by the incident field of an atomic dipole moment, oscillating at frequency ω (see Section 2.4.3). This dipole then radiates in all directions a field of frequency ω which can be calculated by the methods of classical electrodynamics (see Complement 2A). Such a calculation accurately predicts the angular distribution and the efficiency of this scattering process.

By anticipating our fully quantum mechanical treatment of radiation, we can provide another interpretation in terms of a two-step process: the absorption of a photon of energy $\hbar\omega$ *virtually* transferring the atom into the excited state (where it remains for a very brief time of the order of $1/|\omega - \omega_0|$, compatible with the time–energy dispersion relation $\Delta E\,\Delta t \geq \hbar$) followed by the emission of a photon of energy $\hbar\omega$ in a direction different from that of the incident wave. The equality of the frequencies of the absorbed and emitted photons is then simply explained by the need to conserve energy overall.

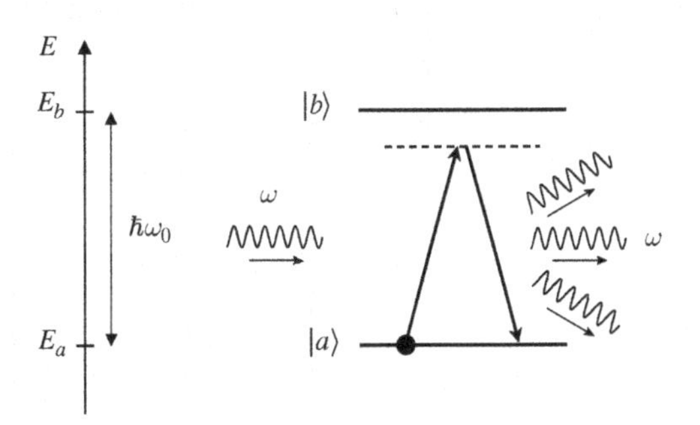

Figure 2.4 **Elastic scattering process. The amplitude of the incident wave decreases whilst light appears at the frequency of the incident wave in directions different from the propagation direction of the exciting light.**

The efficiency of elastic scattering varies strongly with the frequency of the incident light. If ω is small compared to the resonance frequency ω_0, the power scattered is proportional to ω^4 and one speaks of *Rayleigh scattering* (the scattering of solar light by atmospheric molecules, of which the electronic resonances are in the ultra-violet region of the spectrum, obeys a relation of this type, which favours the scattering of short wavelengths and thereby explains the blue colour of the sky). If, on the contrary, ω is very large compared to the resonance frequencies (X-ray scattering), the efficiency of the process is constant. This is *Thomson scattering*. At frequencies close to ω_0 the scattering efficiency is resonantly enhanced and is a Lorentzian function of the atom–light frequency difference: this is *resonant scattering* also called *resonance fluorescence*.

Comments

(i) When the only cause of broadening of the atomic transition is spontaneous emission from the excited state, the width of this resonance is the inverse of the radiative lifetime of this state, $1/\Gamma$ (compare with Section 2.1.3). This frequency dependence could prompt one to interpret the elastic scattering in terms of two uncorrelated events: the absorption of a photon taking the atom into the excited state, followed by spontaneous emission transferring the atom back into the ground state (Figure 2.3). However, such an interpretation would lead one to expect that the scattered light was of frequency width Γ and of random phase, whereas, actually, the scattered light is monochromatic and has the same frequency and phase as the incident light. Even if in quantum theory some scattering processes can be described in terms of absorption followed by spontaneous emission (Chapter 6, Figure 6.11), the emission and absorption processes invoked in elastic scattering cannot actually be treated as independent processes occurring consecutively.

(ii) It is interesting to note that here the completely classical Lorentz model presented in Complement 2A predicts correctly the behaviour of the scattering process. In particular, the scattering cross-sections for Rayleigh and Thomson scattering can be found in this manner.

2.1.5 Nonlinear processes

The effects involving absorption and stimulated emission described above appear in the first order of a perturbative treatment of the atom–light interaction. The higher-order terms

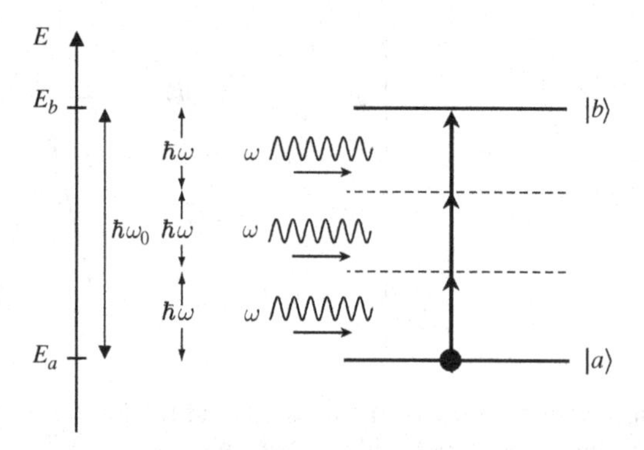

Figure 2.5 Third-order nonlinear process, resonant for $\omega = \omega_0/3$, and which can be interpreted as the simultaneous absorption of three photons of energy $\hbar\omega$.

indicate the existence of transitions resonantly enhanced when the incident field frequency is a sub-multiple of an atomic Bohr frequency:

$$\omega = \frac{\omega_0}{p}, \tag{2.7}$$

where p is an integer. These transitions, associated with nonlinear terms of degree p in the field, in the effective interaction Hamiltonian (see Section 2.3.3), can naturally be interpreted as the absorption of p photons of energy $\hbar\omega$ (Figure 2.5). These are known as *multiphoton processes.*

If the incident electromagnetic field has components at two frequencies ω_1 and ω_2, the nonlinear terms in the atom–field interaction lead to the appearance of linear combinations of the two driving frequencies in the atomic response, and a resonantly enhanced interaction occurs for

$$p_1\omega_1 + p_2\omega_2 = \omega_0, \tag{2.8}$$

where p_1 and p_2 are (possibly negative) integers. This gives rise to the very rich domain of nonlinear spectroscopy (see Complement 3G). By way of example, Figure 2.6 illustrates *stimulated Raman scattering*, which is resonant for an incident field with components at two frequencies ω_1 and ω_2 such that

$$\omega_1 - \omega_2 = \omega_0, \tag{2.9}$$

and which can be interpreted as a transition from $|a\rangle$ to $|b\rangle$, occurring through the absorption of a photon $\hbar\omega_1$ immediately followed by the stimulated emission of a photon $\hbar\omega_2$.

Comment Even if the wave at frequency ω_2 is absent, a process analogous to that of Figure 2.6 can occur: the absorption of the photon at frequency ω_1 is followed by the spontaneous emission of a photon with random phase and propagation direction at frequency ω_2. This phenomenon is known as *spontaneous Raman scattering* and, as spontaneous emission, can only be described quantitatively on the basis of a fully quantum theory of light.

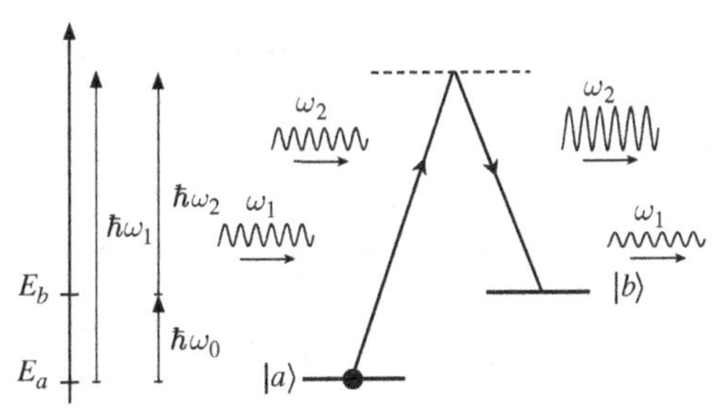

Figure 2.6 **Stimulated Raman scattering. A transition occurs from the atomic level $|a\rangle$ to the level $|b\rangle$ under the influence of incident waves at frequencies ω_1 and ω_2, provided that $\omega_1 - \omega_2 = \omega_0$.**

2.2 The interaction Hamiltonian

Now we shall put into place the formalism that enables us to describe the interaction between an electromagnetic field, treated classically, and an atom treated as a quantum object. The central element of this formalism is the Hamiltonian describing this interaction, of which the non-diagonal matrix elements between atomic states are responsible for the occurrence of transitions between those states. Contrary to what one might naively think, there is no unique form for this Hamiltonian. In fact it is a function of the electromagnetic potentials, which have a number of equivalent forms, related by *gauge transformations*. This freedom concerning the choice of gauge, and therefore of the interaction Hamiltonian, does not bring into question the unique nature of the physical predictions that result. It is therefore possible to select the gauge in which the interaction has the most convenient form for the particular interaction process under consideration. In particular, we shall show that the Göppert–Mayer transformation allows us, at the price of certain approximations that we shall detail, to employ an interaction Hamiltonian which features only the electric field of the incident waves. This is the *electric-dipole Hamiltonian* which has the form of the interaction energy between a classical electric dipole and an electric field.

In the case in which the electric-dipole Hamiltonian cannot induce a transition between two atomic levels under consideration (because the corresponding matrix element is zero), there are other interaction terms, neglected in the lowest-order approximation leading to the electric-dipole Hamiltonian, that can then come into play. The most important of these is the *magnetic-dipole* term, which is responsible in particular for radio-frequency transitions between sub-levels (for example Zeeman, fine-structure or hyperfine structure sub-levels) of a given atomic electronic level for which the electric-dipole couplings are zero. These transitions are central to radio-frequency spectroscopy, of which the most important application is currently the caesium atomic clock, the current time standard.

In this section we presume the reader to be familiar with the basic ideas of electromagnetism, which we recall briefly in the following paragraph. Occasionally we shall have need of more advanced concepts, which will be explained more fully in Chapter 4, but full comprehension of which is not necessary for the major part of this chapter.

2.2.1 Classical electrodynamics: the Maxwell–Lorentz equations

The fundamental equations of classical electrodynamics describe the evolution of a system of interacting charged particles and fields. Maxwell's equations relate the electric and magnetic fields, $\mathbf{E}(\mathbf{r},t)$ and $\mathbf{B}(\mathbf{r},t)$ to current and charge densities, $\mathbf{j}(\mathbf{r},t)$ and $\rho(\mathbf{r},t)$:

$$\nabla\cdot\mathbf{E}(\mathbf{r},t)=\frac{1}{\varepsilon_0}\rho(\mathbf{r},t) \tag{2.10}$$

$$\nabla\cdot\mathbf{B}(\mathbf{r},t)=0 \tag{2.11}$$

$$\nabla\times\mathbf{E}(\mathbf{r},t)=-\frac{\partial}{\partial t}\mathbf{B}(\mathbf{r},t) \tag{2.12}$$

$$\nabla\times\mathbf{B}(\mathbf{r},t)=\frac{1}{c^2}\frac{\partial}{\partial t}\mathbf{E}(\mathbf{r},t)+\frac{1}{\varepsilon_0 c^2}\mathbf{j}(\mathbf{r},t). \tag{2.13}$$

It is well known that Equations (2.11) and (2.12) imply the existence of vector and scalar potentials $\mathbf{A}(\mathbf{r},t)$ and $U(\mathbf{r},t)$, which characterize completely the fields. These fields can be derived uniquely from the relations,

$$\mathbf{B}(\mathbf{r},t)=\nabla\times\mathbf{A}(\mathbf{r},t) \tag{2.14}$$

and

$$\mathbf{E}(\mathbf{r},t)=-\frac{\partial}{\partial t}\mathbf{A}(\mathbf{r},t)-\nabla U(\mathbf{r},t). \tag{2.15}$$

On the other hand there are an infinity of potential pairs $\{\mathbf{A}(\mathbf{r},t),U(\mathbf{r},t)\}$ associated with the same value of the electromagnetic field $\{\mathbf{E}(\mathbf{r},t),\mathbf{B}(\mathbf{r},t)\}$. Distinct pairs of potentials are related by a *gauge transformation*:

$$\mathbf{A}'(\mathbf{r},t)=\mathbf{A}(\mathbf{r},t)+\nabla F(\mathbf{r},t) \tag{2.16}$$

and

$$U'(\mathbf{r},t)=U(\mathbf{r},t)-\frac{\partial}{\partial t}F(\mathbf{r},t), \tag{2.17}$$

$F(\mathbf{r},t)$ being an arbitrary scalar field. It is possible to do away with this arbitrariness by imposing an additional constraint, the *gauge condition* (for an example see Section 2.2.3, in which the *Coulomb gauge* is discussed).

For an assembly of point-like particles of charges q_α, localized at positions $\mathbf{r}_\alpha$ and with velocities $\mathbf{v}_\alpha$, the charge and current densities appearing in Maxwell's equations (2.10) and (2.13) can be written, with the aid of the Dirac delta-function:

$$\rho(\mathbf{r},t)=\sum_\alpha q_\alpha\delta(\mathbf{r}-\mathbf{r}_\alpha(t)) \tag{2.18}$$

and

$$\mathbf{j}(\mathbf{r},t)=\sum_\alpha q_\alpha\mathbf{v}_\alpha\delta(\mathbf{r}-\mathbf{r}_\alpha(t)). \tag{2.19}$$

On the other hand, the classical Newton–Lorentz equations describe the dynamics of the particles (of masses m_α and charges q_α) under the influence of the electric and magnetic forces exerted by the fields:

$$m_\alpha \frac{\mathrm{d}\mathbf{v}_\alpha}{\mathrm{d}t} = q_\alpha \left(\mathbf{E}(\mathbf{r}_\alpha(t), t) + \mathbf{v}_\alpha \times \mathbf{B}(\mathbf{r}_\alpha(t), t)\right). \tag{2.20}$$

The set of equations (2.10–2.13), (2.18), (2.19) and (2.20) is a closed set of coupled equations, the state of the fields depending on that of the particles and vice versa. From this set of equations can be obtained all the results of classical electrodynamics, a theory in which neither the states of the fields nor those of the particles are quantized. Our task is now to generalize these equations to the case where the states of the particles are quantized, which is indispensable to a realistic description of atoms.

2.2.2 Hamiltonian of a particle in a classical electromagnetic field

Form of the Hamiltonian

Here we describe the interaction between a classical electromagnetic field and the simplest possible atom composed of an electron in the Coulomb field of a stationary nucleus. We are therefore interested in the dynamics of a single electron, of which the motion is treated quantum mechanically and which is immersed in a classical electromagnetic field. This field is completely characterized by the potentials $\mathbf{A}(\mathbf{r}, t)$ and $U(\mathbf{r}, t)$, which include equally the Coulomb field of the nucleus and the external fields interacting with the atom.

We assume that the dynamics of this electron are determined by the Hamiltonian

$$\hat{H} = \frac{1}{2m}(\hat{\mathbf{p}} - q\mathbf{A}(\hat{\mathbf{r}}, t))^2 + q\,U(\hat{\mathbf{r}}, t), \tag{2.21}$$

where $\hat{\mathbf{r}}$ is the electron position operator and $\hat{\mathbf{p}}$ is the operator $-i\hbar\nabla$ (notice that $\mathbf{A}(\hat{\mathbf{r}}, t)$ and $U(\hat{\mathbf{r}}, t)$ are operators, being themselves functions of operators). A justification of this choice of Hamiltonian, based on a Lagrangian formalism, can be found in more advanced texts.[1] Here, we justify it by showing that this Hamiltonian leads to the classical equations of motion (2.10–2.13) and (2.20) when one considers the average values of the quantum position and momentum operators.

Velocity operator

In order to justify the use of the Hamiltonian of (2.21) we shall now establish, by using *Ehrenfest's theorem*[2] the equations that describe the evolution of the mean values of the position and velocity of the atomic electron. First of all we need to know the operator

[1] See, for example, CDG I, Chapter IV. A brief introduction of the classical Hamiltonian formalism for charged particles in an electromagnetic field is given in Complement 6A.

[2] See CDL, § III.D.

that represents the velocity when we use the Hamiltonian (2.21). The mean value of this operator $\hat{\mathbf{v}}$ must be such that

$$\langle \hat{\mathbf{v}} \rangle = \frac{\mathrm{d}}{\mathrm{d}t} \langle \hat{\mathbf{r}} \rangle . \tag{2.22}$$

Now, by Ehrenfest's theorem,

$$\frac{\mathrm{d}}{\mathrm{d}t} \langle \hat{x} \rangle = \frac{1}{\mathrm{i}\hbar} \left\langle \left[\hat{x}, \hat{H} \right] \right\rangle . \tag{2.23}$$

Since $\hat{x}$ commutes with $U(\hat{\mathbf{r}}, t)$ we obtain,

$$\left\langle \left[\hat{x}, \hat{H} \right] \right\rangle = \mathrm{i}\hbar \left\langle \frac{\hat{p}_x - qA_x(\hat{\mathbf{r}}, t)}{m} \right\rangle , \tag{2.24}$$

and from (2.23) and (2.24) we deduce that

$$\frac{\mathrm{d}}{\mathrm{d}t} \langle \hat{x} \rangle = \left\langle \frac{\hat{p}_x - qA_x(\hat{\mathbf{r}}, t)}{m} \right\rangle \tag{2.25}$$

which, following (2.22) suggests the following expression for the *velocity operator*:

$$\hat{\mathbf{v}} = \frac{\hat{\mathbf{p}} - q\mathbf{A}(\hat{\mathbf{r}}, t)}{m} . \tag{2.26}$$

Notice that generally $\hat{\mathbf{v}}$ is different from $\hat{\mathbf{p}}/m$.

Equations of motion

To find the equations of motion, we shall need to use the commutation relations of the different components of the velocity operator. Consider, for example, the commutator $[\hat{v}_x, \hat{v}_y]$. Starting from (2.26) we find:

$$\left[\hat{v}_x, \hat{v}_y \right] = -\frac{q}{m^2} \left[\hat{p}_x, A_y(\hat{\mathbf{r}}, t) \right] - \frac{q}{m^2} \left[A_x(\hat{\mathbf{r}}, t), \hat{p}_y \right] . \tag{2.27}$$

Now, whatever the function $f(\hat{\mathbf{r}})$ the commutator $[\hat{p}_x, f(\hat{\mathbf{r}})]$ is equal to $-\mathrm{i}\hbar \frac{\partial f}{\partial x}$ so that

$$\left[\hat{v}_x, \hat{v}_y \right] = \mathrm{i}\hbar \frac{q}{m^2} \left(\frac{\partial}{\partial x} A_y(\hat{\mathbf{r}}, t) - \frac{\partial}{\partial y} A_x(\hat{\mathbf{r}}, t) \right) . \tag{2.28}$$

Using the relation (2.14) between the magnetic field and the vector potential, we obtain the required relations:

$$\left[\hat{v}_x, \hat{v}_y \right] = \mathrm{i}\hbar \frac{q}{m^2} B_z(\hat{\mathbf{r}}, t), \tag{2.29}$$

$$\left[\hat{v}_y, \hat{v}_z \right] = \mathrm{i}\hbar \frac{q}{m^2} B_x(\hat{\mathbf{r}}, t), \tag{2.30}$$

$$\left[\hat{v}_z, \hat{v}_x \right] = \mathrm{i}\hbar \frac{q}{m^2} B_y(\hat{\mathbf{r}}, t). \tag{2.31}$$

Now we are in a position to calculate the evolution of the average value of the velocity operator. We apply again Ehrenfest's theorem to obtain,

$$\frac{\mathrm{d}}{\mathrm{d}t}\langle\hat{v}_x\rangle = \frac{1}{\mathrm{i}\hbar}\left\langle\left[\hat{v}_x, \hat{H}\right]\right\rangle + \left\langle\frac{\partial}{\partial t}\hat{v}_x\right\rangle. \tag{2.32}$$

Using the definition (2.26) for the velocity operator, which depends explicitly on time via $\mathbf{A}(\hat{\mathbf{r}}, t)$, we can rewrite the final term of (2.32) as

$$\frac{\partial}{\partial t}\hat{v}_x = -\frac{q}{m}\frac{\partial}{\partial t}A_x(\hat{\mathbf{r}}, t). \tag{2.33}$$

The Hamiltonian (2.21) can also be written in terms of the velocity operator:

$$\hat{H} = \frac{1}{2}m\hat{\mathbf{v}}^2 + qU(\hat{\mathbf{r}}, t), \tag{2.34}$$

so that the commutator of (2.32) becomes,

$$\left[\hat{v}_x, \hat{H}\right] = \frac{m}{2}\left[\hat{v}_x, \hat{v}_y^2\right] + \frac{m}{2}\left[\hat{v}_x, \hat{v}_z^2\right] + q\left[\hat{v}_x, U(\hat{\mathbf{r}}, t)\right]. \tag{2.35}$$

Now, we expand the first two terms of (2.35), applying the result (2.29) to (2.31) and taking care over the ordering of terms in $\hat{\mathbf{v}}$ and $\mathbf{B}(\hat{\mathbf{r}}, t)$ which do not commute. We obtain,

$$\left[\hat{v}_x, \hat{v}_y^2\right] = \hat{v}_y\left[\hat{v}_x, \hat{v}_y\right] + \left[\hat{v}_x, \hat{v}_y\right]\hat{v}_y = \mathrm{i}\hbar\frac{q}{m^2}\left(\hat{v}_y B_z(\hat{\mathbf{r}}, t) + B_z(\hat{\mathbf{r}}, t)\hat{v}_y\right) \tag{2.36}$$

and similarly,

$$\left[\hat{v}_x, \hat{v}_z^2\right] = -\mathrm{i}\hbar\frac{q}{m^2}\left(\hat{v}_z B_y(\hat{\mathbf{r}}, t) + B_y(\hat{\mathbf{r}}, t)\hat{v}_z\right). \tag{2.37}$$

The last term of (2.35) is readily simplified upon noticing that $U(\hat{\mathbf{r}}, t)$ and $\mathbf{A}(\hat{\mathbf{r}}, t)$ commute:

$$\left[\hat{v}_x, U(\hat{\mathbf{r}}, t)\right] = \frac{1}{m}\left[\hat{p}_x, U(\hat{\mathbf{r}}, t)\right] = -\frac{\mathrm{i}\hbar}{m}\frac{\partial}{\partial x}U\left(\hat{\mathbf{r}}, t\right). \tag{2.38}$$

Finally, on substituting the results (2.33) to (2.38) in (2.32) we obtain,

$$\frac{\mathrm{d}}{\mathrm{d}t}\langle\hat{v}_x\rangle = \frac{q}{2m}\left\langle\hat{v}_y\hat{B}_z - \hat{v}_z\hat{B}_y + \hat{B}_z\hat{v}_y - \hat{B}_y\hat{v}_z\right\rangle - \frac{q}{m}\left\langle\frac{\partial U(\hat{\mathbf{r}}, t)}{\partial x} + \frac{\partial A_x(\hat{\mathbf{r}}, t)}{\partial t}\right\rangle. \tag{2.39}$$

We recognize in the first term the x-component of the symmetrized operator associated with the Lorentz force ($q\mathbf{v} \times \mathbf{B}$). The second term represents the electrostatic force on the electron, being proportional to the electric field of Equation (2.15). Finally, we can represent (2.39) by the vector equation,

$$m\frac{\mathrm{d}}{\mathrm{d}t}\langle\hat{\mathbf{v}}\rangle = q\left\langle\frac{\hat{\mathbf{v}} \times \mathbf{B}(\hat{\mathbf{r}}, t) - \mathbf{B}(\hat{\mathbf{r}}, t) \times \hat{\mathbf{v}}}{2}\right\rangle + q\left\langle\mathbf{E}(\hat{\mathbf{r}}, t)\right\rangle, \tag{2.40}$$

which is the *quantum mechanical analogue of the classical Newton–Lorentz equation* (2.20). We have therefore demonstrated that the Hamiltonian we assumed in (2.21) leads to the definition (2.26) of the velocity operator, and to an equation (2.40) describing the temporal evolution of its mean value which is a natural generalization of the corresponding classical equation (2.20). Thus, although we have not presented a rigorous proof of the validity of this choice of Hamiltonian, we have shown that it is a plausible one.

2.2.3 Interaction Hamiltonian in the Coulomb gauge

The Coulomb gauge

We know that there exists an infinity of pairs of potentials $\{\mathbf{A}(\mathbf{r},t), U(\mathbf{r},t)\}$ corresponding to the same value of the electromagnetic field. Distinct pairs of potentials are related by gauge transformations (2.16–2.17). We can make use of this degree of arbitrariness by imposing an additional constraint on the potentials which amounts to the choice of a particular gauge. We shall see in the following that, amongst the many possible choices of gauge, one is particularly well adapted to problems in quantum optics. This is the Coulomb gauge which is defined by the condition,

$$\nabla \cdot \mathbf{A}(\mathbf{r},t) = 0. \tag{2.41}$$

This choice corresponds to a particular value of the vector potential, denoted by $\mathbf{A}_\perp(\mathbf{r},t)$ (the *transverse vector potential*) for reasons which will become clearer in Chapter 6. Here we give the example, which will be of use in the following, of the transverse vector potential for a plane electromagnetic wave:

$$\mathbf{E} = \mathbf{E}_0 \cos(\omega t - \mathbf{k}\cdot\mathbf{r}), \tag{2.42}$$

$$\mathbf{B} = \frac{\mathbf{k}\times\mathbf{E}_0}{\omega}\cos(\omega t - \mathbf{k}\cdot\mathbf{r}), \tag{2.43}$$

$$\mathbf{E}_0\cdot\mathbf{k} = 0. \tag{2.44}$$

The corresponding Coulomb gauge vector potential is,

$$\mathbf{A}_\perp = -\frac{\mathbf{E}_0}{\omega}\sin(\omega t - \mathbf{k}\cdot\mathbf{r}), \tag{2.45}$$

whilst the associated scalar potential is zero:

$$U(\mathbf{r},t) = 0. \tag{2.46}$$

It is easy to verify that Equation (2.41) is satisfied, and that such potentials do indeed lead to the fields (2.42–2.43) on using Equations (2.14–2.15). In particular, the electric field (2.42) of the free electromagnetic wave is,

$$\mathbf{E}(\mathbf{r},t) = -\frac{\partial}{\partial t}\mathbf{A}_\perp(\mathbf{r},t). \tag{2.47}$$

The atomic and interaction Hamiltonians

We treat the atom–field interaction then in the Coulomb gauge starting from the Hamiltonian (2.21) for the interaction between an externally applied field and an atom composed of an electron evolving under the influence of the Coulomb potential of the nucleus. An important advantage of this choice of gauge is that it permits a clear separation of the static electric field induced by the nucleus and the externally applied electromagnetic radiation. Now we consider the external field to be the plane electromagnetic wave, considered as an example above, for which the Coulomb gauge potentials are given by (2.45–2.46). As far as the nuclear electrostatic field is concerned the associated vector potential is zero, whilst the scalar potential is simply the usual Coulomb potential, $U_{\mathrm{Coul}}(\mathbf{r})$.

The Hamiltonian (2.21) can then be written, on putting $V_{\mathrm{Coul}}(\mathbf{r}) = qU_{\mathrm{Coul}}(\mathbf{r})$:

$$\hat{H} = \frac{1}{2m}(\hat{\mathbf{p}} - q\mathbf{A}_{\perp}(\hat{\mathbf{r}}, t))^2 + V_{\mathrm{Coul}}(\hat{\mathbf{r}}) \tag{2.48}$$

which, on expanding, gives,

$$\hat{H} = \frac{\hat{\mathbf{p}}^2}{2m} - \frac{q}{2m}\left(\hat{\mathbf{p}}\cdot\mathbf{A}_{\perp}(\hat{\mathbf{r}}, t) + \mathbf{A}_{\perp}(\hat{\mathbf{r}}, t)\cdot\hat{\mathbf{p}}\right) + \frac{q^2\mathbf{A}_{\perp}^2(\hat{\mathbf{r}}, t)}{2m} + V_{\mathrm{Coul}}(\hat{\mathbf{r}}). \tag{2.49}$$

The second term can be simplified, since $\mathbf{A}_{\perp}(\hat{\mathbf{r}}, t)$ satisfies (2.41), and gives,

$$\hat{\mathbf{p}}\cdot\mathbf{A}_{\perp} - \mathbf{A}_{\perp}\cdot\hat{\mathbf{p}} = \left[\hat{\mathbf{p}}, \mathbf{A}_{\perp}(\hat{\mathbf{r}}, t)\right] = -\mathrm{i}\hbar\nabla\cdot\mathbf{A}_{\perp}(\hat{\mathbf{r}}, t) = 0. \tag{2.50}$$

Finally, the Hamiltonian of the electron may be written,

$$\hat{H} = \hat{H}_0 + \hat{H}_{\mathrm{I}}, \tag{2.51}$$

with

$$\hat{H}_0 = \frac{\hat{\mathbf{p}}^2}{2m} + V_{\mathrm{Coul}}(\hat{\mathbf{r}}) \tag{2.52}$$

and

$$\hat{H}_{\mathrm{I}} = -\frac{q}{m}\hat{\mathbf{p}}\cdot\mathbf{A}_{\perp}(\hat{\mathbf{r}}, t) + \frac{q^2{\mathbf{A}_{\perp}}^2(\hat{\mathbf{r}}, t)}{2m}. \tag{2.53}$$

In the above $\hat{H}_0$ is the usual (hydrogenic) atomic Hamiltonian, that is the Hamiltonian for a charged particle moving in a Coulomb potential. The term $\hat{H}_{\mathrm{I}}$ describes the interaction between the hydrogenic atom and the applied field. It depends on the quantized variables $\hat{\mathbf{p}}$ and $\hat{\mathbf{r}}$, describing the motion of the electron, and the applied field, which is completely characterized, in the Coulomb gauge, by its vector potential $\mathbf{A}_{\perp}(\mathbf{r}, t)$.

The interaction Hamiltonian $\hat{H}_{\mathrm{I}}$ may be decomposed into two parts $\hat{H}_{\mathrm{I}1}$ and $\hat{H}_{\mathrm{I}2}$ which are, respectively, linear and quadratic functions of the vector potential:

$$\hat{H}_{\mathrm{I}1} = -\frac{q}{m}\hat{\mathbf{p}}\cdot\mathbf{A}_{\perp}(\hat{\mathbf{r}}, t), \tag{2.54}$$

$$\hat{H}_{\mathrm{I}2} = \frac{q^2\mathbf{A}_{\perp}^2(\hat{\mathbf{r}}, t)}{2m}. \tag{2.55}$$

Interaction processes that vary linearly with the applied field amplitude can be written in terms of $\hat{H}_{\mathrm{I}1}$ alone.

Long-wavelength approximation

In the atom–light interactions studied in quantum optics the wavelength, λ, of the light is usually very large compared to atomic dimensions. For example, for the hydrogen atom typical emission or absorption lines have a wavelength of 100 nm at least, whilst the atomic size is of the order of the Bohr radius, $a_0 = 0.053\,\text{nm}$. Under these conditions, the amplitude of the external field is practically constant over the spatial extent of the atom and the vector potential $\mathbf{A}_\perp(\hat{\mathbf{r}}, t)$ can be replaced by its value at the nucleus $\mathbf{A}_\perp(\mathbf{r}_0, t)$. This is the *long-wavelength approximation*.

Applying this approximation the interaction Hamiltonian (2.53) becomes:

$$\hat{H}_\text{I} = -\frac{q}{m}\hat{\mathbf{p}} \cdot \mathbf{A}_\perp(\mathbf{r}_0, t) + \frac{q^2}{2m}\mathbf{A}_\perp{}^2(\mathbf{r}_0, t). \tag{2.56}$$

This equation is considerably more simple than (2.53) since the electron position operator no longer appears in $\hat{H}_\text{I}$. Notice also that the second term of (2.56) is a *scalar* of which the matrix elements between different atomic states are *zero* and which cannot, therefore, induce transitions between two such states. Therefore, we can state the following final form for the interaction Hamiltonian in the long-wavelength approximation:

$$\hat{H}_\text{I1} = -\frac{q}{m}\hat{\mathbf{p}} \cdot \mathbf{A}_\perp(\mathbf{r}_0, t). \tag{2.57}$$

In this Hamiltonian, which is often called 'the **A.p** Hamiltonian', the externally applied field is represented by the Coulomb gauge vector potential taken at the position of the nucleus.

2.2.4 Electric dipole Hamiltonian

Göppert-Mayer gauge

We now introduce a second gauge which is often used in atomic physics, since it leads to a particularly suggestive form for the interaction Hamiltonian. This is the Göppert-Mayer gauge, which is obtained from the Coulomb gauge by the gauge transformation (2.16–2.17) for which one takes

$$F(\mathbf{r}, t) = -(\mathbf{r} - \mathbf{r}_0) \cdot \mathbf{A}_\perp(\mathbf{r}_0, t). \tag{2.58}$$

This transformation privileges the position, $\mathbf{r}_0$, of the nucleus. It leads to the Göppert-Mayer potentials:

$$\mathbf{A}'(\mathbf{r}, t) = \mathbf{A}_\perp(\mathbf{r}, t) - \mathbf{A}_\perp(\mathbf{r}_0, t), \tag{2.59}$$

$$U'(\mathbf{r}, t) = U_\text{Coul}(\mathbf{r}) + (\mathbf{r} - \mathbf{r}_0) \cdot \frac{\partial}{\partial t}\mathbf{A}_\perp(\mathbf{r}_0, t). \tag{2.60}$$

With these potentials the Hamiltonian (2.21) becomes

$$\hat{H} = \frac{1}{2m}(\hat{\mathbf{p}} - q\mathbf{A}'(\hat{\mathbf{r}}, t))^2 + V_\text{Coul}(\hat{\mathbf{r}}) + q(\hat{\mathbf{r}} - \mathbf{r}_0) \cdot \frac{\partial}{\partial t}\hat{\mathbf{A}}_\perp(\mathbf{r}_0, t). \tag{2.61}$$

Recalling that the electric field associated with the applied radiation is, according to Equation (2.47),

$$\mathbf{E}(\mathbf{r}, t) = -\frac{\partial}{\partial t}\mathbf{A}_\perp(\mathbf{r}, t) \tag{2.62}$$

and introducing the electric dipole operator of the atom,

$$\hat{\mathbf{D}} = q(\hat{\mathbf{r}} - \mathbf{r}_0), \tag{2.63}$$

the Hamiltonian (2.61) finally becomes,

$$\hat{H} = \frac{1}{2m}(\hat{\mathbf{p}} - q\mathbf{A}'(\hat{\mathbf{r}}, t))^2 + V_{\text{Coul}}(\hat{\mathbf{r}}) - \hat{\mathbf{D}} \cdot \hat{\mathbf{E}}(\mathbf{r}_0, t). \tag{2.64}$$

Long-wavelength approximation

Now we make the long-wavelength approximation which, as before, enables us to replace the potentials associated with the applied field with their values evaluated at the atomic nucleus. We therefore replace $\mathbf{A}'(\hat{\mathbf{r}}, t)$ by $\mathbf{A}'(\mathbf{r}_0, t)$ in Equation (2.64). But from (2.59),

$$\mathbf{A}'(\mathbf{r}_0, t) = 0, \tag{2.65}$$

so that (2.64) takes the form:

$$\hat{H} = \hat{H}_0 + \hat{H}'_{\text{I}} \tag{2.66}$$

where, as in (2.52),

$$\hat{H}_0 = \frac{\hat{\mathbf{p}}^2}{2m} + V_{\text{Coul}}(\hat{\mathbf{r}}), \tag{2.67}$$

which is the usual atomic Hamiltonian. As for the interaction Hamiltonian, this is,

$$\hat{H}'_{\text{I}} = -\hat{\mathbf{D}} \cdot \mathbf{E}(\mathbf{r}_0, t), \tag{2.68}$$

which is known as 'the electric dipole Hamiltonian'. This Hamiltonian has the form of the interaction energy arising from a classical electric dipole $\mathbf{D}$ localized at $\mathbf{r}_0$ in an electric field $\mathbf{E}(\mathbf{r}, t)$.

Comments

(i) We have considered here, for simplicity, a hydrogen atom. However, an electric dipole operator can be defined for any atomic species involving several electrons and the interaction Hamiltonian retains the same form.

(ii) In the following we shall need to know the numerical values of the matrix elements of the electric dipole operator corresponding to optical transitions. In principle they can be calculated from a knowledge of the eigenfunctions of the atomic states involved in the transition. For transitions between lower electronic states of the hydrogen atom a good estimate is the product of the electron's charge, $e = 1.6 \times 10^{-19}$ C and the Bohr radius, $a_0 = 0.5 \times 10^{-10}$ m, which determines the electron–proton length scale in the atom.

(iii) As shown in Complement 2B, the diagonal matrix elements of the electric dipole operator are null:

$$\langle i|\hat{\mathbf{D}}|i\rangle = 0. \tag{2.69}$$

(iv) In the case of an atom moving uniformly in a progressive wave we explicitly include the time-dependent nucleus coordinate $\mathbf{r}_0 = \mathbf{v}_{\text{at}}t$ and the electric field experienced by the atom is then,

$$\mathbf{E} = \mathbf{E}_0 \cos(\omega t - \mathbf{k} \cdot \mathbf{r}_0) = \mathbf{E}_0 \cos(\omega - \mathbf{k} \cdot \mathbf{v}_{\text{at}})t. \tag{2.70}$$

It is apparent in the above that the frequency of the field seen by the atom is displaced by the *Doppler effect* by an amount $\Delta\omega_D = -\mathbf{k}\cdot\mathbf{v}_{at}$. This phenomenon plays an important role in laser spectroscopy (see Complement 3G), and in laser cooling of atoms (Chapter 8).

Discussion

In the above, we have arrived at two distinct Hamiltonians describing the dynamics of our model system (an atom interacting with an imposed electromagnetic field). This might appear to bring into question the uniqueness of the predictions of the theory. In fact there is no problem, as long as the exact equations are employed and all calculations are carried out without the adoption of further approximations (a gauge transformation leaves the fields unchanged and physical results depend only on the fields). We can likewise state that the long-wavelength approximation leads to identical transition probabilities whether one adopts the $\mathbf{A}\cdot\mathbf{p}$ or $\mathbf{D}\cdot\mathbf{E}$ forms of the Hamiltonian.

On the other hand, when one makes approximate calculations, the results can depend subtly on the form of the Hamiltonian chosen. For example, if either the initial or final state wavefunctions are replaced by approximate expansions one can arrive at results that do depend on which Hamiltonian is employed. In this case it is found that the electric dipole Hamiltonian gives the most accurate results when the transition involves two bound levels, but the $\mathbf{A}.\mathbf{p}$ Hamiltonian gives best results if the transition involves a continuum of states.[3]

One advantage of the Göppert-Mayer gauge in cases when the long-wavelength approximation is applied is that it leads to an expression for the interaction energy involving mathematical quantities, each of which has a ready physical interpretation. For example, the electron velocity operator $\hat{\mathbf{v}}$ is then equal to the operator $\hat{\mathbf{p}}/m$ (from Equations (2.26) and (2.65)). This implies that the Hamiltonian $\hat{H}_0$ (Equation 2.67) coincides with the sum of the kinetic and the electrostatic potential energies of the electron whilst the interaction term is the usual electric dipole energy expression.

When the long-wavelength approximation is invalid it is no longer possible to use the electric dipole Hamiltonian. In that case one must either use the exact Coulomb gauge Hamiltonian or modify the gauge transformation so that higher order terms of the multipolar expansion also appear (of which the first two are the magnetic dipole and electric quadrupole interactions). These terms are particularly important when, for some reason, the electric dipole term is zero (compare Complement 2B).

2.2.5 The magnetic dipole Hamiltonian

If the electric dipole matrix element between two states vanishes, the magnetic dipole interaction can still lead to transitions between the two states. Since the electric dipole operator is odd (that is it changes sign when it undergoes a transformation equivalent to a reflection in the origin, see Complement 2B), this is the case if the two atomic states

[3] CDG1, Complement EIV, exercise 2.

under consideration have the same parity. Such a situation is encountered, for example, for two ground state hyperfine levels of an atom, which have the same principal and azimuthal quantum numbers n and l. Nevertheless, transitions between two such levels do occur, falling in the radio-frequency range. Examples include the resonance line of hydrogen at 1420 MHz (the '21 cm line', between the $F = 0$ and $F = 1$ ground state hyperfine levels) and the transition between the $F = 3$ and $F = 4$ hyperfine levels of caesium at 9192 MHz, which serves as the definition of the second.

In order to establish the interaction term leading to such transitions it is necessary to consider the next term in the multipolar expansion and also to take into account the coupling between the incident electromagnetic field and the magnetic moments of the particles of which the atom is composed.[4] One then obtains an interaction term which can readily be interpreted as the magnetic dipole coupling between the magnetic dipole moment of the atom, **M**, and the magnetic field of the incident wave, **B**:

$$\hat{H}_{\mathrm{I}}'' = -\hat{\mathbf{M}} \cdot \mathbf{B}(\mathbf{r}_0, t). \tag{2.71}$$

For the electron of a hydrogen atom, $\hat{\mathbf{M}}$ has a contribution not only from the angular momentum **L**, associated with the orbital motion, but also one owing to the electron's spin, $\hat{\mathbf{S}}$:

$$\hat{\mathbf{M}} = \frac{q}{2m}(\hat{\mathbf{L}} + 2\hat{\mathbf{S}}). \tag{2.72}$$

It is straightforward to make an estimate of the relative magnitudes of the magnetic and electric dipole couplings. For an atomic electron the matrix elements of $\hat{\mathbf{M}}$ are of the order of a Bohr magneton,

$$\mu_{\mathrm{B}} = \frac{\hbar q}{2m}, \tag{2.73}$$

q being the charge on the electron. For a travelling electromagnetic wave in free-space, B is equal to E/c. Using the approximation qa_0 for the electric dipole coupling (compare comment (ii) of Section 2.2.4) we obtain

$$\frac{\left\langle \hat{H}_{\mathrm{I}}'' \right\rangle}{\left\langle \hat{H}_{\mathrm{I}}' \right\rangle} \approx \frac{\hbar}{mca_0} = \frac{q^2}{4\pi\varepsilon_0 \hbar c} = \alpha \approx \frac{1}{137}, \tag{2.74}$$

where α is known as the fine-structure constant ($\alpha \approx 1/137$). So the magnetic dipole coupling is typically two orders of magnitude smaller than the electric dipole one. These magnetic couplings do, however, lead to numerous observable effects.

Comment To the same order in the series expansion of the Hamiltonian (2.21) appears an interaction term coupling the electric quadrupole moment of the atom with the electric field gradient of the applied field. This term also leads to transitions between states separated by an energy equal to the frequency of the field. The relative importance of the magnetic dipole and electric quadrupole terms is then

[4] CDL2, Complement A_{XIII}.

determined by the symmetries of the atomic states involved and the amplitudes of the relevant matrix elements.

2.3 Transitions between atomic levels driven by an oscillating electromagnetic field

Equipped with an interaction Hamiltonian we can now proceed to study the transitions between atomic states $|i\rangle$ and $|k\rangle$ driven by a sinusoidally oscillating electromagnetic wave. Initially, we shall consider only the case of a weak coupling and shall rely on the results of first-order perturbation theory. Then, in Section 2.3.2, we present a more exact calculation, necessary when the transition probability is no longer small compared to one, on the basis of the two-level atom model in the quasi-resonant approximation. This predicts the occurrence of *Rabi oscillations*, a physical phenomenon of considerable importance. In Section 2.3.3 we consider the questions of *multi-photon transitions* and *light-shifts*.

2.3.1 The transition probability in first-order perturbation theory

Absorption and stimulated emission

We consider an atomic system described by a Hamiltonian $\hat{H}_0$ of which the eigenstates $|n\rangle$ correspond to levels of energies E_n. At $t = 0$ the atom is in the eigenstate $|i\rangle$ of $\hat{H}_0$ and an electromagnetic field oscillating at frequency ω is imposed. We seek to know the probability of finding the atom in a different eigenstate $|k\rangle$ of $\hat{H}_0$ at a later instant, t. We shall assume that the results of Section 2.2, which concern the interaction Hamiltonian for a hydrogen atom, remain valid whatever the atom. In particular, we assume that in the long-wavelength approximation we can always use the interaction Hamiltonian,

$$\hat{H}_I'(t) = -\hat{\mathbf{D}} \cdot \mathbf{E}(\mathbf{r}_0, t), \tag{2.75}$$

where the electric field $\mathbf{E}(\mathbf{r}_0, t)$ is evaluated at the position $\mathbf{r}_0$ of the atom, and where $\hat{\mathbf{D}}$ is an atomic (electric dipole) operator having non-zero matrix elements between the eigenstates of $\hat{H}_0$ under consideration.

Writing the electric field:

$$\mathbf{E}(\mathbf{r}_0, t) = \mathbf{E}(\mathbf{r}_0)\cos(\omega t + \varphi(\mathbf{r}_0)) \tag{2.76}$$

we put the interaction Hamiltonian in the form met in Chapter 1 (Equation (1.42)):

$$\hat{H}_I(t) = \hat{W}\cos(\omega t + \varphi) \tag{2.77}$$

with

$$\hat{W} = -\hat{\mathbf{D}} \cdot \mathbf{E}(\mathbf{r}_0). \tag{2.78}$$

We can use the results obtained in Section 1.2.4 in our first-order perturbation theory calculation. We know that the transition probability is only appreciable under quasi-resonant conditions, that is to say for the case in which the energy of the final state E_k is approximately equal to $E_i \pm \hbar\omega$, the equality being satisfied to within $\Delta\omega = \pi/T$, T being the duration of the interaction (Equations (1.49)). It is then possible for the atom to pass from state $|i\rangle$ to state $|k\rangle$, provided that the frequency of the wave is close to the Bohr frequency of the transition. In the case corresponding to the first line of Equation (1.49) in which level $|k\rangle$ is higher in energy than level $|i\rangle$ there is absorption of energy from the field. In the opposite case there is induced emission to the lower lying level $|k\rangle$. In the following we shall treat these phenomena more quantitatively.

Comment As we saw above, in the case in which $\hat{\mathbf{D}}$ has zero matrix elements between the states $|i\rangle$ and $|k\rangle$, there can be a coupling between the magnetic field of the incident wave and the magnetic dipole moment of the atom. The interaction Hamiltonian can then be put in the form of (2.77) but the coupling term is then

$$\hat{W} = -\hat{\mathbf{M}} \cdot \mathbf{B}_0(\mathbf{r}_0). \tag{2.79}$$

The results obtained in the following can easily be adapted to this case.

Transition probability

In first-order perturbation theory, the probability that a transition has taken place from the state $|i\rangle$ to the state $|k\rangle$ after an interaction of duration T is given by Equation (1.46):

$$P_{i\to k}(T) = \frac{|W_{ki}|^2}{4\hbar^2} g_T(\hbar\delta), \tag{2.80}$$

in which

$$g_T(\hbar\delta) = T^2 \left(\frac{\sin(\delta T/2)}{\delta T/2}\right)^2 \tag{2.81}$$

is a sharply-peaked function of δ of height T^2 and of half-width π/T, centred on $\delta = 0$ (see Figure 1.3). The quantity

$$W_{ki} = \left\langle k \middle| \hat{W} \middle| i \right\rangle \tag{2.82}$$

is the (non-zero) matrix element for the transition and

$$\delta = \omega - \frac{|E_i - E_k|}{\hbar} \tag{2.83}$$

is the detuning of the frequency of the electromagnetic wave from that of the atomic resonance. (Taking the absolute value in (2.83) allows us to treat absorption and stimulated emission simultaneously.)

It will be useful to specify the matrix element W_{ki}. We introduce the unit vector $\boldsymbol{\varepsilon}$, parallel to the direction of the electric field $\mathbf{E}(\mathbf{r}_0)$ of the electromagnetic wave, which therefore defines its polarization:

$$\mathbf{E}(\mathbf{r}_0) = \boldsymbol{\varepsilon} E(\mathbf{r}_0). \tag{2.84}$$

The interaction Hamiltonian of (2.78) then leads to

$$W_{ki} = -\left\langle k \left| \hat{\mathbf{D}} \cdot \boldsymbol{\varepsilon} \right| i \right\rangle E(\mathbf{r}_0) = -d\, E(\mathbf{r}_0), \tag{2.85}$$

in which we have introduced the matrix element d of the component of $\hat{\mathbf{D}}$ in the direction $\boldsymbol{\varepsilon}$ of the electric field. Notice that it is always possible, by judicious choice of the relative phase of the states $|i\rangle$ and $|k\rangle$, to render this matrix element real and negative. It is then usual to write

$$W_{ki} = -d\, E(\mathbf{r}_0) = \hbar\Omega_1, \tag{2.86}$$

which defines the *Rabi frequency* Ω_1 (the justification of this name will become apparent in Section 2.3.2).

In this notation, the transition probability becomes

$$P_{i\to k}(T) = \left(\frac{\Omega_1 T}{2}\right)^2 \left(\frac{\sin(\delta T/2)}{\delta T/2}\right)^2. \tag{2.87}$$

Comments

(i) The choice of relative phases leading to Ω_1 positive has been made to simplify subsequent expressions. It is easy to generalize the following results to the more general case of Ω_1 complex. For example, in (2.87) the quantity Ω_1^2 is then replaced by $|\Omega_1|^2$.

(ii) In the case of a circular polarization, one can make use of a complex polarization vector $\boldsymbol{\varepsilon}$ (compare Complement 2B). The expression of Equation (2.76) is then replaced by

$$\mathbf{E}(\mathbf{r}_0, t) = \mathrm{Re}\left\{\boldsymbol{\varepsilon} \mathrm{E}(\mathbf{r}_0)\mathrm{e}^{-\mathrm{i}(\omega t + \varphi(\mathbf{r}_0))}\right\}. \tag{2.88}$$

The calculation then proceeds as previously in the framework of the resonant approximation. One can in particular write (2.85) and use (2.86) to define the complex Rabi frequency.

The transition probability of (2.87) is proportional to the square of the modulus of the Rabi frequency and thus to the square of the electric field amplitude. To characterize this quantity, we define the *intensity* $I(\mathbf{r}_0)$ of the electromagnetic wave at the point $\mathbf{r}_0$ to be the mean-square value of the electric field, averaged over a period of time, θ, long compared to $2\pi/\omega$:

$$I(\mathbf{r}_0) = \frac{1}{\theta}\int_t^{t+\theta} (\mathbf{E}(\mathbf{r}_0, t'))^2 \mathrm{d}t' = \frac{E^2(\mathbf{r}_0)}{2}. \tag{2.89}$$

We can then write

$$\Omega_1^2 = \frac{2d^2}{\hbar^2} I(\mathbf{r}_0). \tag{2.90}$$

Comments

(i) For the case of a uniform travelling wave ($\varphi(\mathbf{r}_0) = -\mathbf{k}\cdot\mathbf{r}_0$ and $E(\mathbf{r}_0) = E_0$ in Equation (2.88)) the *intensity* is spatially uniform and proportional to the power per unit area across a surface perpendicular to the beam direction (also known as *irradiance* in optics) which is equal to the modulus of the mean value of the Poynting vector:

$$\Pi = \frac{\varepsilon_0 c}{2} |E_0|^2 = \varepsilon_0 c I. \tag{2.91}$$

For this reason the intensity I is often expressed in units of W.m^{-2}, an inconsistency which does, however, relate the intensity to an experimentally measurable quantity. It is then understood that the numerical value refers to $\varepsilon_0 c \overline{E^2}$, where $\overline{}$ denotes a time average. If the intensity is not uniform it can be expressed in the same units provided the above convention is adopted.

(ii) Other definitions of the intensity are sometimes used. Their common feature is their proportionality to the square of the Rabi frequency. These variations in definition are not inconvenient since intensity usually appears as a fraction, the other term of which is also an intensity.

(iii) When the electromagnetic field is not monochromatic, Equation (2.89) can be used if it is averaged over a period of time θ compared to the typical period of the field, but short compared to the response time of the detectors employed. For visible light, for example, θ is taken to be 10^{-11} seconds. The resulting intensity may then be time-dependent and could oscillate, for example, at the beat note between two distinct frequency components.

(iv) It is useful to know the typical order of magnitude of Ω_1. For a laser beam of power 1 mW and of cross-section 1 mm^2, Π is 10^3 W.m^{-2}. Taking d to be of the order of qa_0 (10^{-29} C.m), one then finds $\Omega_1 \approx 10^8$ s^{-1}. On resonance the transition probability then becomes appreciable in a few nanoseconds.

Discussion

The result of (2.87) highlights several important features of absorption and stimulated emission processes, despite the fact that it is only valid if the transition probability remains small compared to unity. Far from resonance (where the detuning far exceeds the Rabi frequency), $P_{i \to k}$ is always small compared to one and has an oscillatory dependence on the interaction time. It is the perturbative limit of Rabi oscillations that will be studied in full generality in Section 2.3.2. The maximum value $(\Omega_1/\delta)^2$ of the transition probability has the resonant character around $\delta = 0$ already mentioned, but the perturbative expression cannot yield information on the Rabi oscillations for δ less than or equal to Ω_1. Note that the Rabi oscillation can be interpreted as an alternation between periods dominated by absorption and periods dominated by stimulated emission.

In the above treatment we have not taken spontaneous emission into account (Section 2.1.3) nor, more generally, the possibility of the atomic levels involved being unstable, with finite lifetimes Γ^{-1}. For transitions in the visible, the radiative lifetime (that is owing to spontaneous emission) of an excited state usually has a value, $1/\Gamma$, of between a nanosecond and a microsecond. The frequency of Rabi oscillations driven by a laser source is often considerably larger than these values of Γ and the transition probability (2.87) can become appreciable in a time which is short compared to the lifetime, and Rabi oscillations are observed. On the other hand, if the light originates from a classical source (incandescent or discharge lamp), the spectral width of the light, $\Delta\omega$, which plays a role analogous to that of Γ, is typically larger than Γ by several orders of magnitude, whilst the Rabi frequencies are smaller or at most unchanged. This *broadband excitation* situation (see Complement 1B) is not, therefore, correctly described by Equation (2.87). In contrast radio-frequency transitions can occur between levels with quasi-infinite lifetimes (atomic ground-state sub-levels, for example, or non-electronically excited molecular states). Equation (2.87) then accurately describes the situation, provided that the excitation probability remains small.

In conclusion, a perturbative calculation demonstrates the essential characteristics of the behaviour of the transition probability for an atomic system driven by an electromagnetic wave: the resonant nature of the response, its proportionality to the intensity of the incident field and the possibility of the occurrence of Rabi oscillations. Its results, however, are of limited applicability and we shall seek more general results in Section 2.3.2. We shall then develop a non-perturbative description for interactions involving two levels only.

Equivalence of the $\mathbf{A}\cdot\mathbf{p}$ and $\mathbf{D}\cdot\mathbf{E}$ interaction Hamiltonians

We consider the case of an electric dipole transition, where $\hat{\mathbf{D}}$ has a non-zero matrix element. If in place of the electric dipole Hamiltonian (2.78) we had used the **A.p** form we would have obtained an analogous result but with $\hat{W}$ replaced by

$$\hat{W}' = -\frac{q}{m}\mathbf{p}\cdot\mathbf{A}_{\perp}(\mathbf{r}_0,\mathbf{t}). \tag{2.92}$$

Here we shall show that this leads to the same transition probabilities on resonance or, equivalently, that $\hat{W}'_{ki}$ and $\hat{W}_{ki}$ have the same modulus.

First of all we recall the relation linking the electric field amplitude and the vector potential in the Coulomb gauge (compare Equation 2.45):

$$\mathbf{A}(\mathbf{r}_0,t) = \frac{\mathbf{E}(\mathbf{r}_0)}{\omega}\cos\left(\omega t + \varphi(\mathbf{r}_0) + \frac{\pi}{2}\right). \tag{2.93}$$

Additionally, we must compare $\langle k \mid \hat{\mathbf{D}} \mid i\rangle$ and $\langle k \mid \hat{\mathbf{p}} \mid i\rangle$. Consider, for example, the z-components of the two matrix elements. Starting from the expression (2.52) for the atomic Hamiltonian and using the commutation relation

$$\left[\hat{z},\hat{p}_z\right] = \mathrm{i}\hbar, \tag{2.94}$$

we obtain

$$\left[\hat{z},\hat{H}_0\right] = \mathrm{i}\hbar\frac{\hat{p}_z}{m}. \tag{2.95}$$

Projecting Equation (2.95) on $\langle k\mid$ on the left and on $|i\rangle$ on the right we find

$$\langle k\left|\hat{z}\right| i\rangle(E_i - E_k) = \frac{\mathrm{i}\hbar}{m}\langle k\left|\hat{p}_z\right| i\rangle. \tag{2.96}$$

Finally, recalling that $\hat{D}_z = q\hat{z}$ for a one-electron atom and keeping only resonant terms yields the result

$$\left|\frac{W'_{ki}}{W_{ki}}\right| = \left|\frac{\omega_{ki}}{\omega}\right|, \tag{2.97}$$

which proves the equality of the transition probabilities on resonance.

When one is no longer exactly on resonance, it might appear that the transition probabilities obtained using the alternative forms of the interaction Hamiltonian are no longer equal. In fact, the difference is of the same order as the effect of including the terms neglected in making the resonant approximation and is therefore not significant in the context of this argument. A more precise calculation reveals that the transition probabilities predicted are always identical, even off resonance.

2.3.2 Rabi oscillations between two levels

Non-perturbative solution of the equation of motion

In this section we shall calculate exactly the probability of a transition between two atomic states $|i\rangle$ and $|k\rangle$ driven by a quasi-resonant wave of angular frequency ω similar to the Bohr frequency of the transition $|E_k - E_i|/\hbar$, in the case where this probability can be appreciable compared to 1. In this quasi-resonant situation we shall make use of *the two-level atom approximation* and shall denote by $|a\rangle$ and $|b\rangle$ the lower and higher energy states. We shall set to zero the energy E_a of the lower state and denote by ω_0 the atomic Bohr frequency:

$$E_b - E_a = \hbar\omega_0. \tag{2.98}$$

The atomic Hamiltonian then has the form

$$H_0 = \hbar \begin{pmatrix} 0 & 0 \\ 0 & \omega_0 \end{pmatrix}. \tag{2.99}$$

We employ the electric dipole Hamiltonian of (2.77), and write for the matrix element W_{ba},

$$W_{ba} = -\left\langle b \left| \hat{\mathbf{D}} \cdot \mathbf{E}(\mathbf{r}_0) \right| a \right\rangle = \hbar\Omega_1, \tag{2.100}$$

(compare Equation 2.86) which allows us to define the Rabi frequency Ω_1 (recall that the arbitrary phases of $|a\rangle$ and $|b\rangle$ are chosen to render Ω_1 real and positive). The complete atomic Hamiltonian is then

$$\hat{H} = \hat{H}_0 + \hat{H}_\mathrm{I} = \hbar \begin{pmatrix} 0 & \Omega_1 \cos(\omega t + \varphi) \\ \Omega_1 \cos(\omega t + \varphi) & \omega_0 \end{pmatrix}. \tag{2.101}$$

To describe the evolution of the atom we expand its state on the basis $\{|a\rangle, |b\rangle\}$ as in Section 1.2 of Chapter 1:

$$|\psi(t)\rangle = \gamma_a(t)\,|a\rangle + \gamma_b(t)\mathrm{e}^{-\mathrm{i}\omega_0 t}\,|b\rangle\,, \tag{2.102}$$

where the exponential term $\mathrm{e}^{-\mathrm{i}\omega_0 t}$ accounts for the free evolution. The Schrödinger equation then leads to the following equations for the evolution of the amplitudes γ_a and γ_b:

$$\mathrm{i}\frac{\mathrm{d}}{\mathrm{d}t}\gamma_a = \frac{\Omega_1 \mathrm{e}^{\mathrm{i}\varphi}}{2}\mathrm{e}^{\mathrm{i}(\omega-\omega_0)t}\gamma_b + \frac{\Omega_1 \mathrm{e}^{-\mathrm{i}\varphi}}{2}\mathrm{e}^{-\mathrm{i}(\omega_0+\omega)t}\gamma_b, \tag{2.103}$$

$$\mathrm{i}\frac{\mathrm{d}}{\mathrm{d}t}\gamma_b = \frac{\Omega_1 \mathrm{e}^{-\mathrm{i}\varphi}}{2}\mathrm{e}^{-\mathrm{i}(\omega-\omega_0)t}\gamma_a + \frac{\Omega_1 \mathrm{e}^{\mathrm{i}\varphi}}{2}\mathrm{e}^{\mathrm{i}(\omega_0+\omega)t}\gamma_a. \tag{2.104}$$

These equations can be greatly simplified close to resonance. Terms in $|\omega-\omega_0|$ are then much smaller than terms in $|\omega+\omega_0|$ so that exponential terms involving $e^{i(\omega+\omega_0)t}$ oscillate very rapidly and give a negligible average contribution. We neglect these rapidly oscillating terms as we did in Section 1.2.4 for the perturbative case. We then obtain,

$$i\frac{d}{dt}\gamma_a = \frac{\Omega_1 e^{i\varphi}}{2}e^{i(\omega-\omega_0)t}\gamma_b, \tag{2.105}$$

$$i\frac{d}{dt}\gamma_b = \frac{\Omega_1 e^{-i\varphi}}{2}e^{-i(\omega-\omega_0)t}\gamma_a, \tag{2.106}$$

for the evolution of the amplitudes introduced in (2.102).

At this point we introduce the detuning from resonance $\delta = \omega - \omega_0$ (compare Equation 2.83) and make a change of variables:

$$\gamma_a = \tilde{\gamma}_a \exp\left(i\frac{\delta}{2}t\right), \tag{2.107}$$

$$\gamma_b = \tilde{\gamma}_b \exp\left(-i\frac{\delta}{2}t\right). \tag{2.108}$$

The system of equations (2.105–2.106) is then transformed into a set of coupled equations with constant coefficients:

$$i\frac{d}{dt}\tilde{\gamma}_a = \frac{\delta}{2}\tilde{\gamma}_a + \frac{\Omega_1 e^{i\varphi}}{2}\tilde{\gamma}_b, \tag{2.109}$$

$$i\frac{d}{dt}\tilde{\gamma}_b = \frac{\Omega_1 e^{-i\varphi}}{2}\tilde{\gamma}_a - \frac{\delta}{2}\tilde{\gamma}_b. \tag{2.110}$$

Such a system admits two oscillating eigen-solutions, of the form

$$\begin{bmatrix}\tilde{\gamma}_a\\ \tilde{\gamma}_b\end{bmatrix} = \begin{bmatrix}\alpha\\ \beta\end{bmatrix}\exp\left(-i\frac{\lambda}{2}t\right), \tag{2.111}$$

with λ taking one of the two values,

$$\lambda_\pm = \pm\Omega = \pm\sqrt{\Omega_1^2+\delta^2}, \tag{2.112}$$

to which correspond two solutions for the ratio $\frac{\alpha}{\beta}$:

$$\left(\frac{\alpha}{\beta}\right)_\pm = -\frac{\Omega_1 e^{i\varphi}}{\delta \mp \Omega}. \tag{2.113}$$

The quantity Ω introduced in (2.112) is known as the *generalized Rabi frequency*.

The general solution of the set of equations (2.109–2.110) is then

$$\tilde{\gamma}_a = K\frac{\Omega_1 e^{i\varphi}}{\Omega-\delta}\exp\left(-i\frac{\Omega}{2}t\right) + L\frac{\Omega_1 e^{i\varphi}}{\Omega+\delta}\exp\left(i\frac{\Omega}{2}t\right). \tag{2.114}$$

We seek the particular solution that corresponds to the initial conditions,

$$\tilde{\gamma}_a(t_0) = 1, \tag{2.115}$$

$$\tilde{\gamma}_b(t_0) = 0, \tag{2.116}$$

that is an atom in the ground state $|a\rangle$ at the instant $t = t_0$. The solution in this case is

$$\tilde{\gamma}_a(t) = \cos\frac{\Omega}{2}(t-t_0) - \mathrm{i}\frac{\delta}{\Omega}\sin\frac{\Omega}{2}(t-t_0), \tag{2.117}$$

$$\tilde{\gamma}_b(t) = -\mathrm{i}\frac{\Omega_1 \mathrm{e}^{-\mathrm{i}\varphi}}{\Omega}\sin\frac{\Omega}{2}(t-t_0). \tag{2.118}$$

Replacing Ω by its value $\sqrt{\Omega_1^2+\delta^2}$, given by (2.112), we obtain directly the probabilities of finding the atom in its ground or excited states $|a\rangle$ and $|b\rangle$. In fact, knowledge of the coefficients $\gamma_a(t)$ and $\gamma_b(t)$ allows us to derive the *mean value of any atomic operator*. We make use of this fact in Section 2.3.3 to calculate the mean value $\langle\hat{\mathbf{D}}\rangle(t)$ of the *atomic dipole* (we use this notation to express the fact that the quantum mechanical mean value $\langle\hat{\mathbf{D}}\rangle = \langle\psi(t)\,|\hat{\mathbf{D}}|\,\psi(t)\rangle$ is time-dependent, compare Equation (2.126)).

Comment The solution (2.117–2.118) of the system (2.109–2.110) is the particular solution associated with the initial condition (2.115–2.116). If the initial condition is changed the solution must be recalculated. In some cases the particular solution is very sensitive to the initial conditions, for example in the interactions of an atom with so-called Ramsey separated fields, often employed in high-resolution spectroscopy and in particular in caesium atomic clocks (see Complement 2D).

Rabi oscillations

Solution (2.118) permits us to evaluate the probability $P_{a\to b}(t_0,t)$ that an atom initially in the state $|a\rangle$ should have passed at time t into the state $|b\rangle$ under the influence of the quasi-resonant electromagnetic wave:

$$P_{a\to b}(t_0,t) = \frac{\Omega_1^2}{\Omega_1^2+\delta^2}\sin^2\frac{\Omega}{2}(t-t_0). \tag{2.119}$$

This expression is identical, apart from a change of notation, to that of (1.63) derived for the case of a time-independent coupling and not, as here, an oscillating one.

As is apparent in Figure 2.7, the transition probability oscillates in time at the generalized Rabi frequency $\Omega = \sqrt{\Omega_1^2+\delta^2}$ (Equation 2.112) between zero and a maximum value,

$$P_{a\to b}^{\mathrm{Max}} = \frac{\Omega_1^2}{\Omega_1^2+\delta^2}. \tag{2.120}$$

The maximum transition probability $P_{a\to b}^{\mathrm{Max}}$ has a resonant variation with the frequency detuning of the excitation at frequency ω from the atomic Bohr frequency ω_0 (Figure 2.8). This resonance exhibits a Lorentzian lineshape of width $2\Omega_1$ at half-maximum. Thus the width of the resonance increases in proportion to the *amplitude* of the electromagnetic wave.

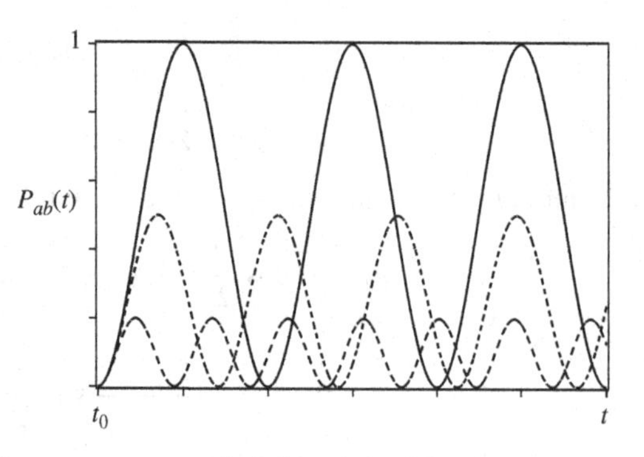

Figure 2.7 Rabi oscillations for the resonant case (solid line), and for detunings $\delta = \Omega_1$ and $\delta = 2\Omega_1$ (dashes).

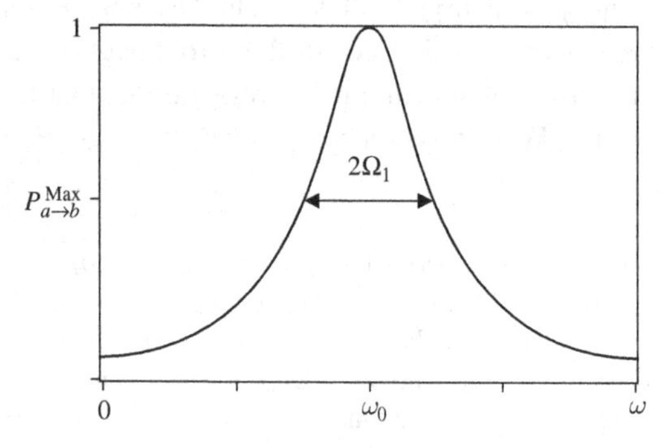

Figure 2.8 Maximum probability of finding the atom in the excited state in the course of its oscillatory evolution as a function of the frequency ω of the applied field.

Notice that exactly on resonance, the maximum value of $P^{\text{Max}}_{a\to b}$ is unity. It is then possible to transfer *the entire atomic population* from the ground state $|a\rangle$ into the excited state $|b\rangle$, provided the interaction duration is fixed at π/Ω_1. This is known as 'π-pulse excitation'.

Examples

We now consider whether Rabi oscillations are observable in the visible wavelength range. Following the discussion of Section 2.3.1 we can see that the above calculation is only valid if the interaction duration is shorter than the lifetimes of the two atomic states involved in the transition under consideration. Moreover, in order to observe Rabi oscillations the interaction duration must be at least a single oscillation period. These two conditions are compatible in the visible wavelength domain provided the Rabi frequency is at least of the order of $10^9\ \text{s}^{-1}$, which can be obtained using a laser beam of a few tens of milliwatts power focused to an area of about $1\ \text{mm}^2$. Rabi oscillations can then be observed by illuminating the atoms with a pulse of laser light, where the pulse length $(t - t_0)$ is varied over the range of a few nanoseconds, whilst observing the fluorescence produced (the intensity of which is proportional to the number of atoms in the excited state). Such an experiment is possible but presents serious technical difficulties with the result that Rabi oscillation is directly observable in the visible domain only under favourable conditions.

In the radio-frequency domain, we know from Section 2.2.5 that, by virtue of magnetic dipole coupling, transitions can be driven between ground-state hyperfine sub-levels, which have infinite lifetimes. A good example is the atomic clock transition at 9.192631770 GHz between the $F = 3$ and $F = 4$ hyperfine sub-levels of the $6^2S_{1/2}$ ground state of atomic caesium. With an intensity of a few watts per square centimetre, Rabi frequencies $\Omega_1/2\pi$ of the order of a hundred kilohertz can be induced. Techniques exist for measuring the relative populations of the sub-levels, thereby allowing the characteristics of Rabi oscillations to be studied. The use of π-pulses is now a commonplace technique for transferring an entire atomic sample to the desired sub-level.

Another important example of a system in which Rabi oscillations occur, apart from the atomic one, is that of a particle of spin 1/2 in a static magnetic field. This is another system with a pair of energy levels, of which the spacing is proportional to the applied magnetic field and which also interacts with an incident electromagnetic wave via a magnetic dipole coupling. Experiments in *nuclear magnetic resonance* often concern spin 1/2 hydrogen nuclei in a static magnetic field of the order of a few tesla for which the levels associated with the two possible spin orientations are then separated by a few hundred MHz. The precise value of the resonance frequency yields information on the *environment* of the hydrogen nuclei. This phenomenon has important applications in chemical analysis and in medical imaging.

Coherent transients and $\pi/2$ pulses

We consider again the situation described in Section 2.3.2 but suppose now that the electromagnetic field, which was switched on at $t = 0$, is suddenly switched off at time

$$t_1 = t_0 + \frac{\pi}{2\Omega_1}. \tag{2.121}$$

This is known as applying a $\pi/2$-pulse excitation. In order to simplify our calculation of the subsequent evolution of the atom we assume that the incident field is tuned exactly on resonance ($\delta = 0$). The state of the atom at time t_1 is then given by (2.117–2.118):

$$\gamma_a(t_1) = \tilde{\gamma}_a(t_1) = \frac{1}{\sqrt{2}}, \tag{2.122}$$

$$\gamma_b(t_1) = \tilde{\gamma}_b(t_1) = -\frac{\mathrm{i}}{\sqrt{2}}\mathrm{e}^{-\mathrm{i}\varphi}. \tag{2.123}$$

During the subsequent free evolution the state of the atom is described by

$$|\psi(t)\rangle = \frac{1}{\sqrt{2}}\,|a\rangle - \frac{\mathrm{i}}{\sqrt{2}}\mathrm{e}^{-\mathrm{i}(\omega_0 t+\varphi)}\,|b\rangle\,, \tag{2.124}$$

which implies that the probability of the atom being in either state does not change with time.

It should not be concluded, however, that the system does not evolve at all. This can be demonstrated by considering the evolution of the expectation of any observable for which the operator has non-zero matrix elements between the two states $|a\rangle$ and $|b\rangle$. By way of example we can take, for the case of an electric dipole transition, the operator for the component $\hat{D}_{\varepsilon} = \hat{\mathbf{D}}.\boldsymbol{\varepsilon}$ of the electric dipole moment. In the basis $\{|a\rangle, |b\rangle\}$ this can be represented by

$$\hat{D}_{\varepsilon} = \begin{pmatrix} 0 & d \\ d & 0 \end{pmatrix}. \tag{2.125}$$

The expectation value of $\hat{D}_{\varepsilon}$ for the state of (2.124) is then

$$\left\langle \hat{D}_{\varepsilon} \right\rangle (t) = -d \sin(\omega_0 t + \varphi). \tag{2.126}$$

Thus the dipole moment oscillates at the Bohr frequency ω_0. In the case of a transition in the optical domain, this oscillation is accompanied by the emission of visible light at the same frequency.

Although emitted at the same frequency ω_0 as spontaneous emission between the same two energy levels, this light has markedly different properties. These properties are related to the *coherence* of the emission. Equation (2.126) shows that the phase of the oscillations of the atomic dipole is uniquely determined with respect to that of the incident wave. An assembly of atoms all prepared by the same $\pi/2$-pulse will therefore all emit light with the same phase. This is in contrast to what occurs with spontaneous emission, when individual atoms emit light with a random phase. It is possible to observe the consequences of this coherence in experiments. These include the directionality of the emission, the appearance of phenomena related to the beating of fluorescence light with a beam coherent with the driving light. Such phenomena, known as *coherent transients*, can be observed only over a timescale which is short compared to the radiative lifetimes of the atomic states involved.[5]

Comment The radiation obviously carries away energy at the expense of atomic excitation energy and the occupation probability of the excited state correspondingly decreases in time, contrary to the prediction of (2.124). The reason for this discrepancy is that in the derivation of Equation (2.124) any interaction between the atoms and the field radiated by those atoms was neglected. In fact it is the reaction of the emitted field on the radiating atoms that causes the decay with time of $|\gamma_b(t)|$; see for instance Complement 2A for a classical treatment of this effect.

In the radio-frequency domain, and for transitions involving levels of very long lifetimes, $\pi/2$-pulses are commonly used to put a system in a freely oscillating state (at angular frequency ω_0), with a well-defined phase controlled by the exciting wave. Pulsed nuclear magnetic resonance methods make use of $\pi/2$-pulses and this technique is widely employed in atomic clocks and numerous methods involving *Ramsey fringes*. This technique will be described in Complement 2D.

[5] These phenomena have given rise to a number of beautiful experiments. See, for example, R.G. Brewer in *Frontiers of Laser Spectroscopy* (Les Houches, session XXVII) edited by R. Balian, S. Haroche and S. Liberman, North Holland, p. 341 (1977).

Comment

It should not be thought that it is necessary to generate pulses of duration exactly that of a $\pi/2$-pulse for these methods to work. The essential point is to put the atom in a linear superposition state of $|a\rangle$ and $|b\rangle$ with the two states having similar weights, so as to optimize the amplitude of the oscillations in (2.126).

2.3.3 Multiphoton transitions

Perturbative treatment

We consider the situation depicted in Figure 2.9, in which an atom, initially in the state $|i\rangle$ interacts with an electromagnetic wave of angular frequency ω, close to one half that corresponding to the energy separation between the state $|i\rangle$ and another state $|k\rangle$ ($\omega_{ki} = |E_k - E_i|/\hbar$).

Since no atomic level exists at an energy $E_i + \hbar\omega$ a calculation by first-order perturbation theory yields a very small probability for the occurrence of a transition from the initial atomic state. However, in second-order perturbation theory (Section 1.2.5 generalized to the case of a sinusoidal perturbation), one finds, in the quasi-resonant limit, a transition probability after an interaction time T of

$$P_{i\to k}(T) = T\frac{2\pi}{\hbar}\left|\frac{1}{4}\sum_{j\neq i}\frac{W_{kj}W_{ji}}{E_i - E_j + \hbar\omega}\right|^2 \delta_T(E_k - E_i - 2\hbar\omega). \tag{2.127}$$

Here again, the function $\delta_T(E)$ (centred on zero and with amplitude and half-width $T/2\pi\hbar$ and $\pi\hbar/T$, respectively) makes an appearance. It results in the resonance condition

$$E_k - E_i = 2\hbar\omega. \tag{2.128}$$

As Figure 2.9 suggests, such a process can be interpreted as a transition between the atomic levels $|i\rangle$ and $|k\rangle$ induced by the simultaneous absorption of two photons. If the atom had initially been in the higher of the two energy states rather than the lower, the atom could have made a transition from the excited to the ground state accompanied by *two-photon stimulated emission*.

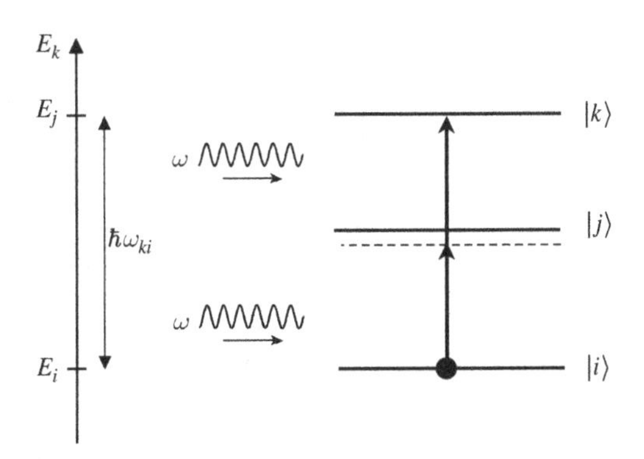

Figure 2.9 Two-photon transition from level $|i\rangle$ to level $|k\rangle$.

The two-photon absorption described here is a second-order nonlinear process. Each of the matrix elements W_{kj} and W_{ji} is proportional to the electric field amplitude of the driving wave (Equation 2.85) so that the transition probability (2.127) is *proportional to the square of the light intensity*. This result may be generalized to higher orders of perturbation theory. For example, if $\omega = |E_k - E_i|/3\hbar$, a three-photon transition can take place (compare Figure 2.5), the probability of which is proportional to the cube of the light intensity.

Comment In atomic transitions, an electric dipole matrix element is non-zero only between levels of oppostite parities (compare Section 2.2.5). As a consequence, a two-photon electric dipole transition occurs only between levels of the same parity which, in contrast, are not connected by one-photon electric dipole transitions.

Orders of magnitude

It is a straightforward task to compare the quasi-resonant two-photon transition probability of (2.127) with the one-photon transition probability of (2.80). On resonance the ratio of the two is of the order of

$$\frac{|W|^2}{(E_i - E_j + \hbar\omega)^2}, \tag{2.129}$$

equal to the square of the Rabi frequencies (assumed to be of equal values) divided by the square of the detuning from the intermediate state $|j\rangle$. Apart from the special cases in which the intermediate level is exactly equidistant in energy from the states $|i\rangle$ and $|k\rangle$, the detuning from the intermediate state is usually much larger than the Rabi frequencies of the two one-photon transitions linking, respectively, states $|i\rangle$ and $|j\rangle$ and states $|j\rangle$ and $|k\rangle$. Thus multi-photon transitions are usually much less probable than those involving only one photon. The probability of their occurring only becomes appreciable if the incident light intensity is sufficiently high, which accounts for the fact that their experimental observation usually relies on the use of tightly focused or high-power pulsed laser beams.[6]

Comment The above calculation is only valid for interaction times that are short compared to the radiative lifetimes of the initial final states $|i\rangle$ and $|k\rangle$. The frequency detuning from the intermediate state should also be large compared to its radiative width (inverse-lifetime) and also compared to the Rabi frequency $W_{ij}/\hbar$, since a more careful treatment is required if the one-photon transition linking $|i\rangle$ and $|j\rangle$ is quasi-resonant.

Effective Hamiltonian for a two-photon transition

When the intensity of the incident wave is sufficiently high, and the two-photon transition is quasi-resonant, a more exact treatment of the system is possible. To clarify the notation

[6] See G. Grynberg, B. Cagnac, and F. Biraben in *Coherent Nonlinear Optics*, edited by M. S. Feld and V. S. Letokov, Springer Verlag, Berlin, p. 111 (1980).

we denote by $|a\rangle$ and $|c\rangle$ the two states linked by the two-photon transition ($2\omega \approx \omega_0 = (E_c - E_a)/\hbar$) and by $|j\rangle$ the intermediate levels, all of which we assume to be far from the one-photon resonance. We expand $|\psi(t)\rangle$ over the basis set formed by the states $|a\rangle$, $|c\rangle$ and the set $|j\rangle$ in an analogous fashion to (2.102), introducing corresponding amplitudes $\gamma_a(t)$, $\gamma_c(t)$ and $\gamma_j(t)$. The Schrödinger equation can then be cast in a form analogous to (2.103–2.104). The $\gamma_j(t)$ can then be calculated from perturbation theory, since they are all off resonant. Supposing that the interaction is switched on slowly so that constants of integration can be eliminated (compare Section 1.2.5), we obtain

$$\begin{aligned}\gamma_j(t) = &-\frac{W_{ja}}{2(E_j - E_a - \hbar\omega)}\gamma_a(t)\,\mathrm{e}^{-\mathrm{i}(E_j - E_a - \hbar\omega)t/\hbar} \\ &-\frac{W_{ja}}{2(E_j - E_a + \hbar\omega)}\gamma_a(t)\,\mathrm{e}^{-\mathrm{i}(E_j - E_a + \hbar\omega)t/\hbar} \\ &+ \{\text{analogous terms with } a \leftrightarrow c\}.\end{aligned} \tag{2.130}$$

Substituting these terms in the equations that describe the evolution of $\gamma_a(t)$ and $\gamma_c(t)$ we find equations that represent a two-level system $\{|a\rangle, |c\rangle\}$ perturbed by an *effective interaction Hamiltonian* with non-diagonal matrix elements:[7]

$$W_{ca}^{\text{eff}} = \sum_{j \neq a,c} \frac{W_{cj}W_{ja}}{4(E_c - E_j - \hbar\omega)}\,\mathrm{e}^{\mathrm{i}(2\omega - \omega_0)}t = \frac{\hbar\Omega_1^{\text{eff}}}{2}\,\mathrm{e}^{-\mathrm{i}(2\omega - \omega_0)t} \tag{2.131}$$

and diagonal matrix elements:

$$W_{cc}^{\text{eff}} = \frac{1}{4}\left\{\sum_{j \neq c} \frac{|W_{cj}|^2}{E_c - E_j - \hbar\omega} + \sum_{j \neq c} \frac{|W_{cj}|^2}{E_c - E_j + \hbar\omega}\right\}, \tag{2.132}$$

$$W_{aa}^{\text{eff}} = \frac{1}{4}\left\{\sum_{j \neq a} \frac{|W_{aj}|^2}{E_a - E_j - \hbar\omega} + \sum_{j \neq a} \frac{|W_{aj}|^2}{E_a - E_j + \hbar\omega}\right\}. \tag{2.133}$$

Now, we just have to solve the set of coupled first-order equations, for example by following the procedure of Section 2.3.2, and so obtain

$$P_{a \to c}(t_0, t) = \left(\frac{\Omega_1^{\text{eff}}}{\Omega^{\text{eff}}}\right)^2 \sin^2\left(\frac{\Omega^{\text{eff}}}{2}(t - t_0)\right), \tag{2.134}$$

in which the generalized Rabi frequency Ω^{eff} is defined by

$$(\Omega^{\text{eff}})^2 = (\Omega_1^{\text{eff}})^2 + \left(\frac{E_c + W_{cc}^{\text{eff}} - E_a - W_{aa}^{\text{eff}}}{\hbar} - 2\omega\right)^2. \tag{2.135}$$

The result is the occurrence of Rabi oscillations between the levels $|a\rangle$ and $|c\rangle$, driven by the quasi-resonant two-photon coupling. All the remarks made in Section 2.3.2 are

[7] The term 'effective Hamiltonian' is frequently encountered in quantum mechanics texts in various contexts. Its precise meaning varies from case to case.

also directly applicable here, but in addition we point out the following. Firstly, Equation (2.134) tells us that the atom can pass entirely into state $|c\rangle$, despite the fact that the coefficients $\gamma_j(t)$ remain small compared to unity; the probability of finding the atom in one of the intermediate states is therefore negligible. This arises naturally because the two-photon transition, unlike the one-photon transitions linking the ground and intermediate states, is nearly resonant. A further novelty of (2.135) is the role of the diagonal matrix elements of the effective Hamiltonian which give rise to *light-shifts*. This important phenomenon is the object of our study in the next section.

2.3.4 Light-shifts

Formulae (2.134–2.135) show that the two-photon transition is resonant for a frequency of the incident wave,

$$2\omega = \frac{E_c + W_{cc}^{\text{eff}} - E_a - W_{aa}^{\text{eff}}}{\hbar}, \tag{2.136}$$

which can be explained by supposing that the two levels $|a\rangle$ and $|c\rangle$ are shifted in energy by W_{aa}^{eff} and W_{cc}^{eff}, respectively. From (2.131–2.132) it can be seen that these shifts are proportional to the square of the coupling, induced by the wave, of the state $|a\rangle$ (or $|c\rangle$) with the intermediate levels $|j\rangle$. They are known as *light-shifts* (or a.c. Stark shifts).[8] The light-shifts are proportional to the intensity of the incident field and inversely proportional to the detuning from the intermediate state(s). The senses of the energy displacements are as shown in Figure 2.10.

The appearance of light-shifts is not restricted to the case of two-photon transitions. They occur quite generally and we demonstrate below a further example in the case of a one photon transition between two atomic energy levels. Starting from (2.103–2.104) we can find, for the non-resonant case, a perturbative solution analogous to that presented in Section 2.3.3 above and obtain (taking $\varphi = 0$)

$$\gamma_b(t) = \frac{\Omega_1}{2}\left\{\frac{\mathrm{e}^{-\mathrm{i}(\omega-\omega_0)t}}{\omega-\omega_0} - \frac{\mathrm{e}^{\mathrm{i}(\omega+\omega_0)t}}{\omega+\omega_0}\right\}\gamma_a(t). \tag{2.137}$$

Substituting in (2.103) we obtain four terms, of which two are non-oscillating (secular terms) and which give the dominant contribution. On keeping only the largest of these, we find

$$\mathrm{i}\frac{\mathrm{d}\gamma_a}{\mathrm{d}t} = \frac{\Omega_1^2}{4}\frac{1}{\omega-\omega_0}\gamma_a(t). \tag{2.138}$$

[8] Light-shifts were predicted and observed for the first time by C. Cohen-Tannoudji. Using discharge lamps and optical pumping techniques he was able to discern a shift of only 1 Hz (C. Cohen-Tannoudji and A. Kastler, *Progress in Optics*, **5**, p. 1 (1966), edited by E. Wolf, North Holland, Amsterdam). Nowadays, with bright laser sources, light-shifts of MHz and even GHz are routinely measured.

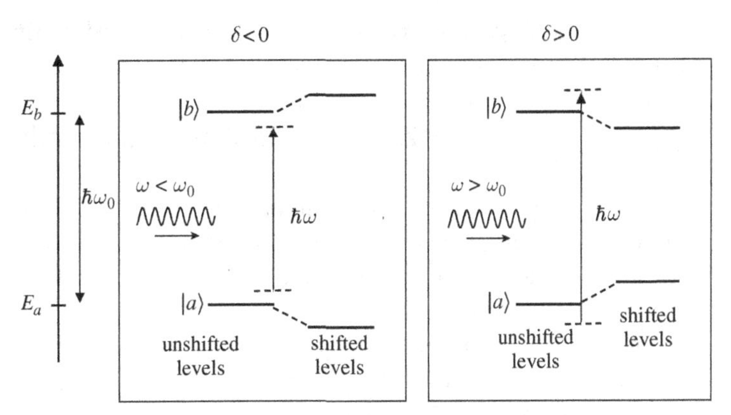

Figure 2.10 **Light-shifts in a two-level atom coupled by a laser at frequency ω for a negative detuning and for a positive detuning.**

Now, it appears that the level $|a\rangle$, originally of energy $E_a = 0$, is displaced in energy by an amount,

$$W_{aa}^{\text{eff}} = \hbar \frac{\Omega_1^2}{4\delta} = \frac{|W_{ab}|^2}{4\,(\hbar\omega - E_b + E_a)}, \tag{2.139}$$

which is of the form of (2.133) provided only the dominant term $j = b$ is kept.

The light-shift of level $|b\rangle$ is obtained in the same manner. It is

$$W_{bb}^{\text{eff}} = -\hbar \frac{\Omega_1^2}{4(\omega - \omega_0)}. \tag{2.140}$$

We find, therefore, that the levels $|a\rangle$ and $|b\rangle$ move further apart in energy when the detuning is negative and closer together when it is positive. This is shown in Figure 2.10.

Since the advent of lasers, light-shifts have played an important role in studies of atom–light interactions. Because of the high spectral brightness of these sources it is usually essential to take them into account in experiments in high-resolution spectroscopy (see Complement 3G). They are also of fundamental importance to numerous mechanisms of laser cooling and trapping of atoms (see Chapter 8).

Comments

(i) The above results, which were obtained for the case of levels of infinite lifetimes, remain valid in the case of finite lifetimes, provided the detuning is large compared to the widths of the coupled levels. Thus Equations (2.139–2.140) actually have a wide domain of applicability.

(ii) In order to measure the light-shifts of levels $|a\rangle$ and $|b\rangle$ of Figure 2.10, a second (frequency tunable) probe laser is used. This probe beam has a low intensity so that it induces negligible light-shifts of the levels under consideration. The absorption of the probe is measured as its frequency is tuned around that of a transition linking one of the levels $|a\rangle$ or $|b\rangle$ to a third level $|c\rangle$. It is then found that the position of the resonance in the probe absorption depends on the intensity and detuning of the strong laser tuned to the $|a\rangle$ to $|b\rangle$ transition in the manner predicted by the formulae of (2.139–2.140).

2.4 Absorption between levels of finite lifetimes

2.4.1 Presentation of the model

In Section 2.3 we saw how an atom (or a molecule) can undergo stimulated emission or absorption when it interacts with an incident electromagnetic wave, causing it to change its internal state. The arguments presented there were valid provided the lifetimes of the atomic states considered were very long.

In fact, very often one has to consider cases in which the finiteness of the lifetime of at least one of the atomic levels is important. For example, for a transition in the optical domain, the upper level can de-excite by spontaneous emission. The same can be true for the lower level if it is not the ground state. Apart from spontaneous emission many other processes can result in finite level lifetimes, for example, for atoms, collisions either with other atoms or with the walls of the cell containing the vapour, or, for ions in a crystalline structure, interaction with phonons. Alternatively, when the motion of an atom takes it out of the region in which it interacts with the electromagnetic wave, the effect is as though the atom itself had a finite lifetime and disappeared.

In the following, therefore, we shall take into account the finite lifetimes of the atomic levels considered. A general treatment of a two-level system interacting with an electromagnetic wave in the presence of dissipative processes requires rather more sophisticated theoretical tools than we have available in this chapter. It is necessary to use a density matrix formalism and to solve the *optical Bloch equations*. We shall present them in Complement 2C. The problem is that the dissipative processes which make levels unstable are due to a coupling between the system composed of the atom and the incident field and the external environment; they cannot, therefore, be described by the Schrödinger equation concerning the atom only, since it is only appropriate for conservative processes described by a Hamiltonian.

We describe here a simplified model which uses the Schrödinger equation and takes into account the finite lifetimes of the states of our two-level atom, with which a number of important results can be derived. This model is developed in the particular case where the two levels have the same lifetimes Γ_{D}^{-1} (Figure 2.11). Denoting by N_a and N_b the numbers of atoms in the levels $|a\rangle$ and $|b\rangle$ present in the interaction region at a given instant,[9] the rates of departure can be written

$$\left(\frac{\mathrm{d}N_a}{\mathrm{d}t}\right)_{\text{depart}} = -\Gamma_{\mathrm{D}} N_a, \tag{2.141}$$

$$\left(\frac{\mathrm{d}N_b}{\mathrm{d}t}\right)_{\text{depart}} = -\Gamma_{\mathrm{D}} N_b, \tag{2.142}$$

[9] In fact N_a and N_b should be understood as the mean values, in the statistical sense, at time t. We suppose these numbers to be sufficiently large that their relative statistical fluctuations are unimportant.

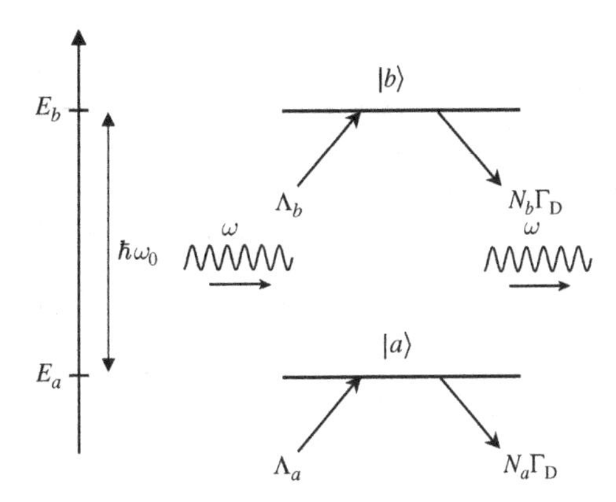

Figure 2.11 **Atomic model in which the two levels coupled by the electromagnetic wave have the same lifetime Γ_D^{-1}. The populations of the two levels are fed at rates Λ_a and Λ_b.**

which imply that the total number of atoms

$$N = N_a + N_b, \tag{2.143}$$

obeys an equation of the same form:

$$\left(\frac{dN}{dt}\right)_{depart} = -\Gamma_D N. \tag{2.144}$$

Such a situation is found when the two levels are both excited states having similar decay rates by spontaneous emission. Alternatively, this model applies to a pair of intrinsically stable atomic levels, but when an atom only spends a period of time of the order of Γ_D^{-1} in the zone of the interaction with the electromagnetic wave.

In order that a steady-state solution should exist, it is necessary for the populations of the two atomic levels to be fed, for example, by collisional excitation with charged or neutral particles – as in an electric discharge, or by optical pumping.[10] Equivalently, new atoms can enter the interaction zone to replace those that leave. The feeding rates of the two levels of our system are

$$\left(\frac{dN_a}{dt}\right)_{feed} = \Lambda_a, \tag{2.145}$$

$$\left(\frac{dN_b}{dt}\right)_{feed} = \Lambda_b, \tag{2.146}$$

which can be different.

In the *steady state* the total feeding and loss rates are equal so that the total number of atoms in the interaction region is then

$$N = \frac{\Lambda_a + \Lambda_b}{\Gamma_D}. \tag{2.147}$$

[10] See Chapter 3, Section 3.2 and Complement 2B.

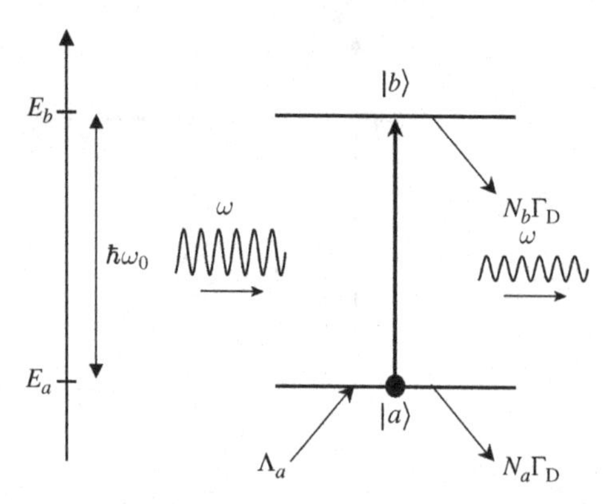

Figure 2.12 **Absorption by 2-level atoms of finite lifetime, where only the lower level $|a\rangle$ is fed. Atoms brought to the upper level $|b\rangle$ and decaying, carry away energy which is taken from the electromagnetic field.**

Notice that this is the case whatever the rate of population transfer between $|a\rangle$ and $|b\rangle$ driven by the electromagnetic wave. The simplicity of this result, which follows from (2.144), is a consequence of the equality of the lifetimes of the two levels: it is therefore specific to our model.

In this section, we shall consider the case in which Λ_b is zero, with only the population of the lower level being fed at a rate Λ_a (Figure 2.12). Under the influence of the electromagnetic wave at frequency ω some of the atoms will pass into the state $|b\rangle$ as a result of absorption. In order to describe the steady-state solution, we shall first calculate the number of atoms N_b excited to the upper state, and secondly the effect of these transitions on the electromagnetic wave, which we shall see is attenuated as it propagates.

2.4.2 Excited state population

Principle of the calculation

When only the lower level $|a\rangle$ is fed with atoms, and in the absence of an electromagnetic field, the steady-state populations are

$$N_a^{(0)} = \frac{\Lambda_a}{\Gamma_D}, \tag{2.148}$$

$$N_b^{(0)} = 0. \tag{2.149}$$

If now we apply a quasi-resonant electromagnetic field, a fraction of the atoms will pass into the excited level but, from Equation (2.147), we will always have

$$N_a + N_b = N = \frac{\Lambda_a}{\Gamma_D}. \tag{2.150}$$

For the case in which the coupling can be treated perturbatively, only a small fraction of the atoms are excited, such that N_b is always small compared to N_a, which itself remains at a value close to N.

We evaluate the probability that an atom which appears at a time t_0 in the level $|a\rangle$ will be found at a later time t in the level $|b\rangle$. This probability is the product of two terms, the first being

$$P_{a\to b}(t_0, t), \tag{2.151}$$

which is the *probability of excitation* of an atom from level $|a\rangle$ to level $|b\rangle$ by the electromagnetic field, and the second being

$$\mathrm{e}^{-\Gamma_\mathrm{D}(t-t_0)}, \tag{2.152}$$

which is the *survival probability* of an atom in either of the levels.

The total number of atoms in the level $|b\rangle$ at time t is obtained by summing the above contributions for all earlier times t_0. Knowing that $\Lambda_a \mathrm{d}t_0$ atoms appear in the level $|a\rangle$ between times t_0 and $t_0 + \mathrm{d}t_0$, we find,

$$N_b(t) = \int_{-\infty}^{t} \mathrm{d}t_0 \; \Lambda_a \; P_{a\to b}(t_0, t)\, \mathrm{e}^{-\Gamma_\mathrm{D}(t-t_0)}. \tag{2.153}$$

Perturbative solution

The probability $P_{a\to b}(t_0, t)$ was calculated earlier and is given in the perturbative case by Equations (2.80–2.81). Substituting this into (2.153) and putting $t - t_0 = T$, we obtain

$$N_b = \Lambda_a \frac{|W_{ba}|^2}{4\hbar^2} \int_0^{\infty} \mathrm{d}T \left(\frac{\sin(\omega - \omega_0)T/2}{(\omega - \omega_0)/2} \right)^2 \mathrm{e}^{-\Gamma_\mathrm{D} T}. \tag{2.154}$$

The integration can readily be performed after expanding the sine in terms of complex exponentials. It gives

$$N_b = \frac{\Lambda_a}{\Gamma_\mathrm{D}} \frac{|W_{ba}|^2}{2\hbar^2} \frac{1}{(\omega - \omega_0)^2 + \Gamma_\mathrm{D}^2}. \tag{2.155}$$

Replacing $|W_{ba}|/\hbar$ by the Rabi frequency Ω_1, and introducing the total number of atoms, N, we obtain

$$N_b = \frac{N}{2} \frac{\Omega_1^2}{(\omega - \omega_0)^2 + \Gamma_\mathrm{D}^2}. \tag{2.156}$$

The fraction of atoms transferred into the excited state is therefore proportional to the intensity of the electromagnetic wave (Ω_1^2 dependence). Furthermore, this fraction exhibits a resonant behaviour as the frequency of the wave is tuned around the Bohr frequency ω_0. It is a Lorentzian function of the detuning, the Lorentzian having a full width at half maximum of $2\Gamma_\mathrm{D}$ (see Figure 2.13).

The above calculation employed the perturbative expression for the transition probability $P_{a\to b}(t_0, t)$. It is therefore only accurate when Ω_1 is small compared either to Γ_D or to the detuning $\delta = \omega - \omega_0$. Under these conditions, the fraction of the atoms N_b/N that is transferred to the excited state also remains small.

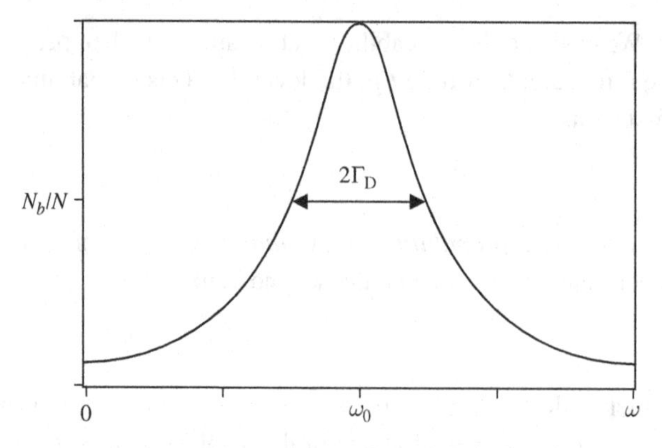

Figure 2.13 Resonant variation of the population of the excited level $|b\rangle$ as a function of the frequency of the incident field. It is described by a Lorentzian curve, with a halfwidth at half maximum of Γ_D.

Non-perturbative solution

For the non-perturbative case we employ reasoning similar to the above, but now substituting Equation (2.119) for the non-perturbative transition probability $P_{a\to b}(t_0, t)$ into Equation (2.153). Now we find

$$N_b = \Lambda_a \frac{\Omega_1^2}{\Omega^2} \int_0^\infty \mathrm{d}T \sin^2 \frac{\Omega T}{2} \mathrm{e}^{-\Gamma_D T}, \tag{2.157}$$

with (compare Equation 2.112),

$$\Omega^2 = \Omega_1^2 + (\omega - \omega_0)^2. \tag{2.158}$$

The integration is carried out as above, yielding

$$N_b = \frac{N}{2} \frac{\Omega_1^2}{(\omega - \omega_0)^2 + \Omega_1^2 + \Gamma_D^2}. \tag{2.159}$$

In the low-intensity limit ($\Omega_1^2 \ll \Gamma_D^2 + (\omega - \omega_0)^2$) this result is identical to the perturbative result (2.156): the population N_b is proportional to the intensity of the incident wave. However, at high intensity, the fraction of atoms in the excited state increases more slowly with Ω_1^2 and tends asymptotically to $1/2$. This is a manifestation of the *saturation* of the excitation. We shall return to this phenomenon at the end of Section 2.4. Additionally, formula (2.159) indicates that the width of the resonance increases with intensity, being equal to $2\sqrt{\Gamma_D^2 + \Omega_1^2}$. This effect is known as *power broadening*.

The effect of saturation appears more clearly if we introduce the saturation parameter s which, for a transition between two levels with equal lifetimes, is

$$s = \frac{\Omega_1^2}{(\omega - \omega_0)^2 + \Gamma_D^2}. \tag{2.160}$$

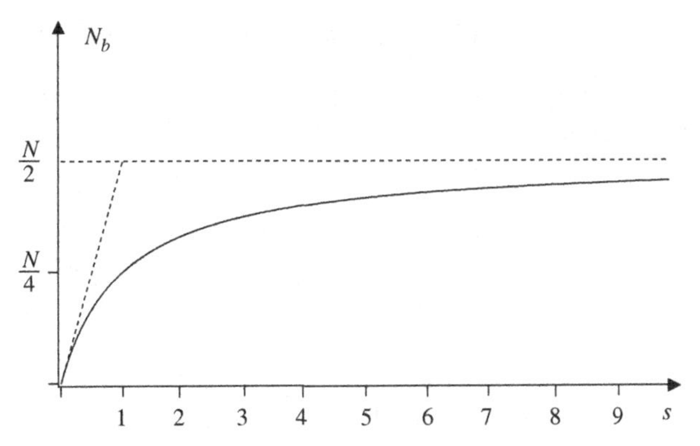

Figure 2.14 Number of excited atoms as a function of the saturation parameter *s*, proportional to the incident field intensity. It saturates at the value of half the total number of atoms.

Equation (2.159) can then be recast as

$$N_b = \frac{N}{2}\frac{s}{1+s}. \tag{2.161}$$

The function $s/(1+s)$ is approximately equal to s for $s \ll 1$ and tends asymptotically towards 1 when $s \gtrsim 1$, characterizing in a general way the phenomenon of saturation (Figure 2.14).

Using (2.90), we can express the saturation parameter as

$$s = \frac{I}{I_{\text{sat}}}\frac{1}{1+\frac{(\omega-\omega_0)^2}{\Gamma_{\text{D}}^2}}, \tag{2.162}$$

where I is the light intensity (Equation 2.90), and we have introduced the saturation intensity I_{sat}, whose value is, for the model considered here,

$$I_{\text{sat}} = \frac{\hbar^2\Gamma_{\text{D}}^2}{d^2}. \tag{2.163}$$

The saturation parameter s turns out to be a very useful quantity, since many expressions, as (2.161), remain valid for situations more complex than that of our simple model. In those cases, s assumes most often a form similar to (2.162), with the saturation intensity I_{sat} and the width $2\Gamma_{\text{D}}$ of the Lorentzian being phenomenological parameters.

2.4.3 Dielectric susceptibility

We shall now calculate the dielectric response of the assembly of atoms of Section 2.4.2 (where only the level $|a\rangle$ is fed externally) when it is subjected to an electromagnetic wave coupling the level $|a\rangle$ to the level $|b\rangle$. We find first of all an expression for the induced electric dipole and then for the susceptibility of the medium.

Mean value of the induced electric dipole

We calculate first the (quantum) mean value of the electric dipole moment on an individual atom. For an atomic state $|\psi\rangle$ we know that

$$\left\langle \hat{\mathbf{D}} \right\rangle = \left\langle \psi \left| \hat{\mathbf{D}} \right| \psi \right\rangle. \tag{2.164}$$

We calculated in Section 2.3.2 the state of an atom interacting with a field:

$$\mathbf{E}(t) = \mathbf{E}_0 \cos(\omega t + \varphi) = \boldsymbol{\varepsilon} E_0 \cos(\omega t + \varphi), \tag{2.165}$$

for times between t_0 and t, the atom starting in state $|a\rangle$ at time t_0. We found, inserting (2.117–2.118) into (2.107–2.108) and then into (2.102):

$$|\psi(t)\rangle = \gamma_a(t)\,|a\rangle + \gamma_b(t)\mathrm{e}^{-\mathrm{i}\omega_0 t}\,|b\rangle\,, \tag{2.166}$$

with

$$\gamma_a(t) = \left[\cos\frac{\Omega}{2}(t-t_0) - \mathrm{i}\frac{\delta}{\Omega}\sin\frac{\Omega}{2}(t-t_0)\right]\exp\left(\mathrm{i}\frac{\delta}{2}t\right), \tag{2.167}$$

$$\gamma_b(t) = -\mathrm{i}\left[\frac{\Omega_1}{\Omega}\sin\frac{\Omega}{2}(t-t_0)\right]\exp-\mathrm{i}\left(\frac{\delta}{2}t+\varphi\right), \tag{2.168}$$

with $\Omega = \sqrt{\Omega_1^2 + \delta^2}$. We can therefore calculate the mean value at time t of the component D_i $(i = x, y, z)$ of the single atom electric dipole, for the case of an atom appearing in state $|a\rangle$ at time t_0. This is

$$\left\langle \hat{D}_i \right\rangle (t_0, t) = d_i\, \gamma_a^*\, \gamma_b\, \mathrm{e}^{-\mathrm{i}\omega_0 t} + c.c., \tag{2.169}$$

where *c.c.* means "complex conjugate" and d_i is the matrix element $\langle a|\hat{D}_i|b\rangle$. Since this atom has a survival probability $\mathrm{e}^{-\Gamma_\mathrm{D}(t-t_0)}$, the total electric dipole moment of the ensemble of atoms can be found by a procedure similar to that employed in Section 2.4.2 above: the contributions (2.169) are summed over all t_0 after weighting them by a factor that reflects their probability of survival at time t. Introducing the volume V of the sample and denoting by $\mathbf{P}$ the *polarization* (dipole moment per unit volume) we find for its component P_i,

$$P_i V = \Lambda_a \int_{-\infty}^{t} \langle D_i \rangle (t_0, t)\, \mathrm{e}^{-\Gamma_\mathrm{D}(t-t_0)} \mathrm{d}t_0. \tag{2.170}$$

This expression can be evaluated using the above results and leads to

$$P_i = -\frac{N d_i}{V}\frac{\Omega_1}{2}\frac{(\omega_0 - \omega) + \mathrm{i}\Gamma_\mathrm{D}}{\Gamma_\mathrm{D}^2 + \Omega_1^2 + (\omega_0 - \omega)^2}\,\mathrm{e}^{-\mathrm{i}(\omega t + \varphi)} + c.c. \tag{2.171}$$

If the atomic ensemble is isotropic, the polarization $\mathbf{P}$ is necessarily parallel to the electric field of the incident wave. If d is the average matrix element of the component of the dipole along this direction we obtain finally, recalling that $\hbar\Omega_1 = -dE_0$,

$$\mathbf{P} = \frac{N}{V}\frac{d^2}{\hbar}\frac{\omega_0 - \omega + \mathrm{i}\Gamma_\mathrm{D}}{\Gamma_\mathrm{D}^2 + \Omega_1^2 + (\omega - \omega_0)^2}\frac{\mathbf{E}_0}{2}\,\mathrm{e}^{-\mathrm{i}(\omega t + \varphi)} + c.c. \tag{2.172}$$

Comment

When the atomic medium is not isotropic, the relationship between the incident field and the polarization it induces is described by a tensor and not a scalar quantity as in (2.172). The medium can then exhibit birefringence. This situation can arise, for example, if the state $|a\rangle$ is a particular Zeeman sub-level of a ground state of angular momentum different from zero, and if the feeding of the ground state populates this sub-level selectively. However, the assumption of isotropy is very often justified, in particular when the Zeeman sub-levels are equally populated and an average has to be taken. In what follows we shall, for simplicity, limit our discussion to situations in which an assumption of the isotropy of the atomic response is justified.

Dielectric susceptibility

In the last paragraph we calculated the polarization **P** induced in an atomic sample resulting from the imposition of an electric field $\mathbf{E}_0 \cos(\omega t + \varphi)$. Now, the complex dielectric susceptibility of a medium,

$$\chi = \chi' + \mathrm{i}\chi'', \tag{2.173}$$

is defined by

$$\mathbf{P} = \varepsilon_0 \chi \frac{\mathbf{E}_0}{2} \mathrm{e}^{-\mathrm{i}(\omega t + \varphi)} + c.c., \tag{2.174}$$

which is equivalent to

$$\mathbf{P} = \varepsilon_0 \left[\chi' \mathbf{E}_0 \cos(\omega t + \varphi) + \chi'' \mathbf{E}_0 \sin(\omega t + \varphi) \right]. \tag{2.175}$$

At low intensity, we can neglect the term Ω_1^2 in the denominator of Equation (2.172), thus obtaining the *linear dielectric susceptibility*

$$\chi_1 = \frac{N}{V} \frac{d^2}{\varepsilon_0 \hbar} \frac{\omega_0 - \omega + \mathrm{i}\Gamma_\mathrm{D}}{\Gamma_\mathrm{D}^2 + (\omega - \omega_0)^2}. \tag{2.176}$$

The index 1 reminds us that this is the response to first order in the electric field.

Separating the real and imaginary parts, which we shall see represent, respectively, the dispersion and absorption of the atomic vapour, we obtain

$$\chi_1' = \frac{N}{V} \frac{d^2}{\varepsilon_0 \hbar} \frac{\omega_0 - \omega}{\Gamma_\mathrm{D}^2 + (\omega - \omega_0)^2}, \tag{2.177}$$

$$\chi_1'' = \frac{N}{V} \frac{d^2}{\varepsilon_0 \hbar} \frac{\Gamma_\mathrm{D}}{\Gamma_\mathrm{D}^2 + (\omega - \omega_0)^2}. \tag{2.178}$$

Figure 2.15 shows the variations of χ_1' and χ_1'' as a function of ω.

Comment

Expression (2.176) for the linear susceptibility, obtained from our simple quantum model for the perturbative regime, has the same form as the result of a classical calculation describing the medium in which the light propagates as an assembly of elastically bound electrons (compare Complement 2A). This is the case for all quantum models, as long as one remains in the perturbative regime. Before the invention of lasers, the intensity of available light sources was sufficiently weak that one always operated in this perturbative regime. This explains the success of the elastically bound electron model in its description of the optical properties of matter.

Saturation

We now use the exact expression (Equation (2.172)) to calculate the value of the dielectric susceptibility defined by Equation (2.174). We obtain

$$\chi = \frac{N}{V}\frac{d^2}{\varepsilon_0 \hbar}\frac{\omega_0 - \omega + \mathrm{i}\Gamma_\mathrm{D}}{\Gamma_\mathrm{D}^2 + \Omega_1^2 + (\omega - \omega_0)^2}. \tag{2.179}$$

We see that χ decreases with Ω_1^2 and thus with intensity, once this reaches a sufficient level. Using the saturation parameter s (Equation (2.160)), we obtain the remarkable formula

$$\chi = \frac{\chi_1}{1+s}, \tag{2.180}$$

where χ_1 is the linear susceptibility (Equation (2.176)). *The susceptibility tends towards zero at high intensity, due to saturation of the transition.*

2.4.4 Propagation of an electromagnetic wave: absorption and dispersion

Propagation with attenuation

We know that a linear medium of dielectric susceptibility $\chi_1(\omega)$ supports the propagation of electromagnetic waves with wavevector

$$k = n\frac{\omega}{c}, \tag{2.181}$$

the refractive index being related to the susceptibility by

$$n^2 = 1 + \chi_1(\omega). \tag{2.182}$$

If the imaginary part of the susceptibility differs from zero, then the refractive index is also complex. For a dilute atomic gas the susceptibility is small compared to unity and we can write

$$n = 1 + \frac{\chi_1'}{2} + \mathrm{i}\frac{\chi_1''}{2}. \tag{2.183}$$

The complex wavevector can then be written

$$k = k' + \mathrm{i}k'', \tag{2.184}$$

with

$$k' \simeq \left(1 + \frac{\chi_1'}{2}\right)\frac{\omega}{c} \tag{2.185}$$

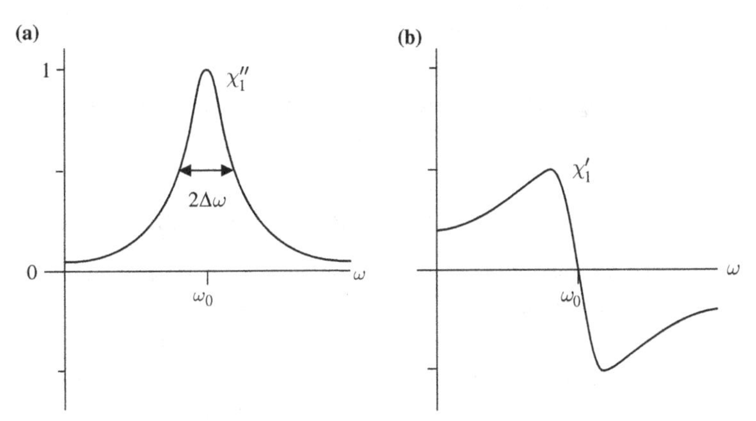

Figure 2.15 Variation of the real part χ_1' (b) and imaginary part χ_1'' (a) of the linear susceptibility of an atomic vapour as a function of field frequency ω in the vicinity of the resonant frequency ω_0. They correspond, respectively, to absorption (a) and dispersion (b) of a wave propagating in the vapour.

and

$$k'' \simeq \frac{\chi_1''}{2}\frac{\omega}{c} \simeq \frac{k'}{|k'|}\frac{\chi_1''}{2}\frac{\omega}{c}. \tag{2.186}$$

In a usual medium χ_1'' is positive and k'' has the same sign as k'.

A wave propagating in the medium along Oz with a wavevector $\mathbf{k}$ can be represented by,

$$\mathbf{E}(z,t) = \frac{\mathbf{E}_0}{2}\,\mathrm{e}^{\mathrm{i}(kz-\omega t)} + c.c. = \mathbf{E}_0\,\mathrm{e}^{-k''z}\cos(\omega t - k'z). \tag{2.187}$$

Since k' and k'' have the same sign, we see that the amplitude of the field decreases exponentially along the direction of propagation. This is *attenuated propagation*.

The *imaginary* part χ_1'' of the linear dielectric susceptibility of a medium therefore characterizes the absorption of the incident wave. Equation (2.178) shows that the linear absorption is proportional to the atomic density N/V (the number of atoms per unit volume). It also shows that this absorption manifests a resonant behaviour, having a Lorentzian dependence on the frequency (see Figure 2.15). Similar characteristics of the absorption can be obtained from the classical Lorentz model of Complement 2A (compare the comment after Equations (2.176–2.178)).

Dispersion

As Equation (2.183) shows, the real part of the susceptibility χ' is related to the real part of the refractive index, which characterizes the phase velocity of an electromagnetic wave propagating through the vapour. Expression (2.177) shows that χ' has the dispersive frequency dependence shown in Figure 2.15(b); in the wings the refractive index increases with increasing frequency (normal dispersion), whilst in the vicinity of the resonance the sense of variation is reversed (anomalous dispersion). At large detunings

$|\delta| = |\omega - \omega_0| \gg \Gamma_D$, the effects of dispersion decrease as $|\delta|^{-1}$, whilst those of absorption decrease more quickly as $|\delta|^{-2}$.

Once again, at low intensities this behaviour is exactly as predicted by the classical, elastically bound electron model.

Propagation in the saturated regime

When the saturation parameter s (Equation (2.160)) is not small compared to unity, the dielectric polarization is no longer proportional to the applied electric field and the problem of an electromagnetic wave propagating in a vapour generally becomes very complex. The problem can be simplified significantly by using the *slowly varying envelope approximation*, where the module of the complex amplitude of the electric field varies slowly on the scale of the wavelength of the light. Using this approximation, it follows that the intensity and therefore the saturation parameter s also vary slowly. We can therefore consider the latter to be constant over a slice dz which is long compared to the wavelength of the light. The results given for the linear regime can then be generalized to the high-intensity case by simply replacing χ_1 by $\chi_1/(1+s)$ (compare Equation (2.180)). The effect of saturation is therefore to reduce the absorption by a factor $1/(1+s)$ and to reduce similarly the difference between the phase velocity $c/(1+\frac{\chi'}{2})$ and the velocity of light in a vacuum, c.

Comment In the situation where the absorption and the refractive index depend on the light intensity because of saturation, the effects of nonlinear optics and nonlinear spectroscopy appear (see Complement 3G).

2.4.5 Case of a closed two-level system

Many of the results obtained from the model we have presented above can be generalized to the case where the two levels $|a\rangle$ and $|b\rangle$ have different lifetimes. An important case is that of the *closed two-level system*, where the lower level $|a\rangle$ is stable (it has an infinite lifetime) while the upper level $|b\rangle$ is unstable, but decays exclusively into the lower level (see Figure 2.16). The total population is conserved in the absence of external feeding. This is the case whenever one considers an assembly of atoms in their ground state (a gas at moderate temperature), interacting with radiation quasi-resonant for a transition coupling the ground state to an excited state which is able to decay by spontaneous emission (with a radiative lifetime Γ_{sp}^{-1}). Such a model accounts equally well for numerous optical properties of dielectric media whose lowest energy electronic resonances are at ultra-violet frequencies, when illuminated with visible radiation of much lower frequency ω. In this case, we can take the limit $\omega \ll \omega_0$ of the results of this section.[11]

To treat such a model it is necessary to solve the optical Bloch equations (see Complement 2C). The result is a susceptibility having the same form as (2.179) but with Γ_D replaced by $\Gamma_{sp}/2$ and Ω_1^2 replaced by $\Omega_1^2/2$, which gives

[11] Note that in this case it may happen that $\omega - \omega_0$ and $\omega + \omega_0$ are of similar magnitude so that the anti-resonant term cannot be neglected.

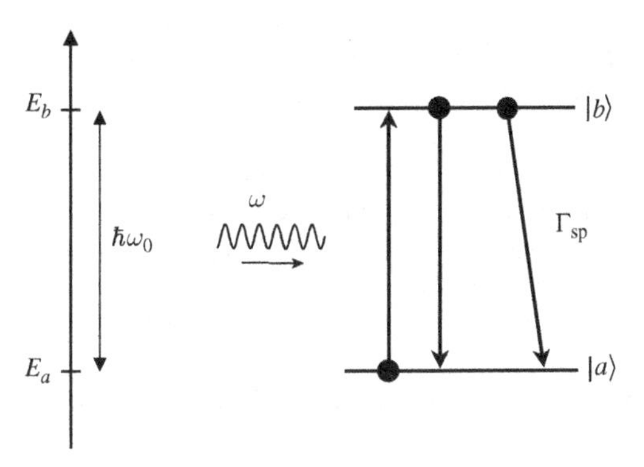

Figure 2.16 Closed two-level system. The quasi-resonant electromagnetic field (ω close to $\omega_0 = (E_b - E_a)/\hbar$) couples the two levels $|a\rangle$ and $|b\rangle$, which are not fed externally. The level $|a\rangle$ is stable and the level $|b\rangle$ decays by spontaneous emission exclusively to level $|a\rangle$.

$$\chi = \frac{N}{V}\frac{d^2}{\varepsilon_0 \hbar}\frac{\omega_0 - \omega + \mathrm{i}\frac{\Gamma_{sp}}{2}}{\frac{\Gamma_{sp}^2}{4} + \frac{\Omega_1^2}{2} + (\omega - \omega_0)^2}. \tag{2.188}$$

The similarity of this result with that of (2.179) enhances the interest of the simplified model presented in Section 2.4, of which the predictions remain generally true, provided the substitutions detailed above are made.

The similarity is even more striking when we introduce a saturation parameter s, here defined as

$$s = \frac{\Omega_1^2/2}{(\omega - \omega_0)^2 + \Gamma_{sp}^2/4}. \tag{2.189}$$

The expressions (2.161) and (2.180) then remain formally valid. As the linear susceptibility χ_1 is identical to that obtained in the elastically bound electron model, we can understand the success of this classical model in its interpretation of optical phenomena in the non-saturated regime ($s \ll 1$), which was the only observable regime before the invention of the laser.

Using expression (6.119) relating d^2 to the spontaneous emission rate, we obtain the useful formula, valid far from resonance ($|\omega - \omega_0| \gg \Gamma_{sp}$),

$$\chi_1' = \frac{3}{8\pi^2}\frac{N\lambda_0^3}{V}\frac{\Gamma_{sp}}{\omega_0 - \omega}. \tag{2.190}$$

This expression allows us to find the order of magnitude of the refractive index of gases or transparent solids. In the far off resonance limit, the detuning $|\omega - \omega_0|$ is much greater than the spontaneous emission rate Γ_{sp}, and therefore an atomic density N/V much greater than $1/\lambda_0^3$ is necessary to ensure that the susceptibility is not much less than one. Thus, far from any resonances, the refractive index (which is related to χ_1', see Section 2.4.4 above) of an atomic or molecular gas differs very little from unity: for example, $n - 1 \approx 3 \times 10^{-4}$ for atmospheric air in the visible spectral region. On the other hand, in the case of solids such as glass, or more generally dielectric media, the atomic density is much larger than

$1/\lambda_0^3$ (typically $N/V \sim 10^{28}\,\mathrm{m}^{-3}$ compared to $1/\lambda_0^3 \sim 10^{19}\,\mathrm{m}^{-3}$ for visible wavelengths), and the materials are consequently refractive as well as transparent.

Comment We shall see in Section 2.6.3 that absorption can be described in the linear regime by introducing an absorption cross-section σ_{abs}, related to the imaginary part of the susceptibility by the relation

$$\frac{\omega}{c}\chi_1'' = \frac{N}{V}\sigma_{\mathrm{abs}} \tag{2.191}$$

(here we are considering the case where only the lower level is fed, and in the linear regime $N_b \ll N_a$).

For a two-level atom, in the case where the relaxation from level $|b\rangle$ to level $|a\rangle$ is exclusively due to spontaneous emission, the absorption cross-section takes the form

$$\sigma_{\mathrm{abs}} = \frac{3\lambda_0^2}{2\pi}\frac{\Gamma_{\mathrm{sp}}^2/4}{\frac{\Gamma_{\mathrm{sp}}^2}{4} + (\omega - \omega_0)^2}, \tag{2.192}$$

where λ_0 is the wavelength on resonance ($\lambda_0 = 2\pi c/\omega_0$). Equation (2.192) shows that on resonance the effective absorption cross-section is of the order of the square of the wavelength, λ_0^2, more precisely to $3\lambda_0^2/2\pi$, a simple and useful result to remember.

2.5 Laser amplification

2.5.1 Feeding the upper level: stimulated emission

We reconsider the model presented in Section 2.4 (Figure 2.11), but now we suppose that only the population of the upper level is externally fed (Figure 2.17):

$$\Lambda_a = 0\,,\ \Lambda_b \neq 0. \tag{2.193}$$

We can repeat the calculations of Section 2.4, starting now with atoms in the upper level $|b\rangle$, to determine the steady-state populations N_a and N_b and susceptibility χ.

Population transfer by stimulated emission

In the absence of the electromagnetic field, all the atoms are in the upper level:

$$N_b^{(0)} = N = \frac{\Lambda_b}{\Gamma_{\mathrm{D}}}, \tag{2.194}$$

$$N_a^{(0)} = 0. \tag{2.195}$$

Under the influence of the incident radiation, some atoms will pass into the lower level: this is the phenomenon of *stimulated emission*. In the perturbative regime, the steady-state population in level $|a\rangle$ takes a form analogous to Equation (2.156):

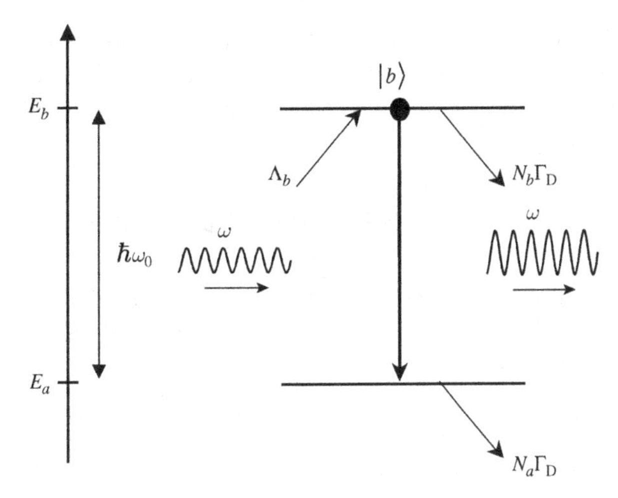

Figure 2.17 Stimulated emission for atoms with two levels of finite lifetime, when only the upper level is externally fed. Under the influence of the incident radiation, some of the atoms pass from level $|b\rangle$ to level $|a\rangle$, and the radiation is amplified.

$$N_a = \frac{N}{2} \frac{\Omega_1^2}{(\omega - \omega_0)^2 + \Gamma_D^2} = \frac{N}{2} s. \tag{2.196}$$

In the general case (not restricted to the perturbative regime), the steady-state population in level $|a\rangle$ takes the value analogous to Equation (2.161):

$$N_a = \frac{N}{2} \frac{s}{1+s}, \tag{2.197}$$

where the saturation parameter s is defined by Equation (2.160),

$$s = \frac{\Omega_1^2}{(\omega - \omega_0)^2 + \Gamma_D^2}. \tag{2.198}$$

Note that whatever the value of s, N_a always remains smaller than N_b: this is called population inversion (see Section 2.5.4).

Susceptibility

Calculation of the susceptibility gives a result analogous to that of Section 2.4.3 (see Equation (2.179)), but *with the sign inversed*:

$$\chi = -\frac{N}{V} \frac{d^2}{\varepsilon_0 \hbar} \frac{\omega_0 - \omega + \mathrm{i}\Gamma_D}{(\omega - \omega_0)^2 + \Gamma_D^2} \frac{1}{1+s}. \tag{2.199}$$

In the perturbative regime (weak saturation) we therefore have

$$\chi = -\chi_1, \tag{2.200}$$

where χ_1 is the linear susceptibility defined for atoms in the ground state (Equation (2.176)). Let us remind ourselves that the quantity χ_1 has the same form as that obtained from the classical model of the elastically bound electron and is associated with the behaviour observed most commonly (absorption, normal dispersion). We shall now see that the *change of sign has some significant consequences.*

2.5.2 Amplified propagation: laser action

We repeat the calculations of Section 2.4.4 for the case where only the upper level is externally fed. In a medium of complex linear susceptibility $\chi_1 = \chi_1' + i\chi_1''$, a monochromatic electromagnetic wave has a complex wavevector:

$$\mathbf{k} = \mathbf{e_k}(k' + ik''). \tag{2.201}$$

The unit vector $\mathbf{e_k}$ defines the direction of propagation. In the usual case, where $|\chi_1| \ll 1$ we have (see 2.186),

$$k'' \simeq \frac{k'}{|k'|}\frac{\chi_1''}{2}\frac{\omega}{c}. \tag{2.202}$$

The above expression shows the relation between the signs of k' and k''. If we consider the propagation to be along Oz, we have a field of the form (2.187),

$$\mathbf{E}(z,t) = \mathrm{Re}\left(\mathbf{E_0}e^{i(kz-\omega t)}\right) = \mathbf{E_0}\, e^{-k''z}\cos(\omega t - k'z). \tag{2.203}$$

However, since χ_1'' is negative, Equation (2.202) reveals that k'' and k' have opposite signs, that is to say, the amplitude of the field *increases* along the direction of propagation. The electromagnetic wave is therefore *amplified.* The phenomenon at the origin of this amplification is clearly stimulated emission, which is the dominant process in the case where N_b is much greater than N_a.

This amplification of an electromagnetic wave due to stimulated emission is known as LASER (Light Amplification by Stimulated Emission of Radiation). Amplification of the light intensity $I(z)$ (proportional to the square of the amplitude of the field) is characterized by a *gain per unit length* g, defined by

$$\frac{dI(z)}{dz} = gI(z). \tag{2.204}$$

In the perturbative regime considered here, Equation (2.202) gives the gain per unit length,

$$g = -2k'' = \frac{\omega}{c}\chi_1''. \tag{2.205}$$

Comment Inversion of the sign of the real part χ_1' of the susceptibility also leads to another anomalous behaviour. The refractive index $n' = 1 + \frac{\chi_1'}{2}$ decreases with increasing frequency far from resonance (transparent regions), and increases through the resonance. This anomalous dispersion is a useful experimental indicator of population inversion and thus the possibility of obtaining light amplification.

2.5.3 Generalization: pumping of both levels and saturation

We consider now the case where both levels $|a\rangle$ and $|b\rangle$ of our two-level system are pumped externally; that is atoms are fed at rates Λ_a and Λ_b, respectively, into each of the two levels. Since the pumping processes are independent and involve two different classes of atoms, we can simply add the results of the previous Sections 2.4.3 and 2.5.1 which concern pumping of either level alone. In this way, we obtain the populations of each level:

$$N_a = \frac{\Lambda_a}{\Gamma_D} + \frac{1}{2}\frac{s}{1+s}\frac{\Lambda_b - \Lambda_a}{\Gamma_D}, \tag{2.206}$$

$$N_b = \frac{\Lambda_b}{\Gamma_D} + \frac{1}{2}\frac{s}{1+s}\frac{\Lambda_a - \Lambda_b}{\Gamma_D}. \tag{2.207}$$

As previously, the steady-state populations in the absence of radiation ($s = 0$) are

$$N_a^{(0)} = \frac{\Lambda_a}{\Gamma_D}, \tag{2.208}$$

$$N_b^{(0)} = \frac{\Lambda_b}{\Gamma_D}. \tag{2.209}$$

We can therefore rewrite the equations for the populations in a more general form:

$$N_a = N_a^{(0)} + \frac{1}{2}\frac{s}{1+s}\left(N_b^{(0)} - N_a^{(0)}\right), \tag{2.210}$$

$$N_b = N_b^{(0)} + \frac{1}{2}\frac{s}{1+s}\left(N_a^{(0)} - N_b^{(0)}\right). \tag{2.211}$$

The following important relation should be noted as it will be used later,

$$N_b - N_a = \frac{N_b^{(0)} - N_a^{(0)}}{1+s}, \tag{2.212}$$

where the saturation parameter s has been defined by Equation (2.160), for our specific model. The difference between the populations decreases and tends to zero in the high-saturation limit ($s \gg 1$).

Calculation of the susceptibility is carried out in a similar fashion, by generalizing Equation (2.172). Denoting the linear susceptibility χ_1 (Equation (2.176) with $N = N_a^{(0)} + N_b^{(0)}$), we obtain

$$\chi = \frac{N_a^{(0)} - N_b^{(0)}}{N_a^{(0)} + N_b^{(0)}}\frac{\chi_1}{1+s} = \frac{N_a - N_b}{N}\chi_1, \tag{2.213}$$

where N is the total number of atoms.

2.5.4 Laser gain and population inversion

The description of light amplification given in Section 2.5.2 can be generalized to the situation of Section 2.5.3, using the slowly varying envelope approximation if the saturation parameter is not small compared with 1. From Equation (2.213), we see that the imaginary part χ'' of the susceptibility is negative for $N_b > N_a$. In this case, we shall have light amplification. The situation where a level of higher energy has a higher population than that of a lower level is manifestly *out of thermodynamic equilibrium*. For this reason, one speaks of a *population inversion*.

Following the arguments of Section 2.5.2, we define the gain per unit length,

$$g = \frac{1}{I}\frac{\mathrm{d}I}{\mathrm{d}z} = \frac{\omega}{c}\chi_1''\frac{N_b - N_a}{N} = \frac{g^{(0)}}{1+s}, \tag{2.214}$$

where we have introduced the non-saturated gain ($s \ll 1$),[12]

$$g^{(0)} = \frac{\omega}{c}\chi_1''\frac{N_b^{(0)} - N_a^{(0)}}{N_b^{(0)} + N_a^{(0)}}. \tag{2.215}$$

These formulae show clearly that laser gain is related to the population inversion. In addition, they highlight the phenomenon of *gain saturation*: the gain decreases and tends to 0 as the light intensity increases.

These results can equally be interpreted by considering the competition between stimulated emission, which permits amplification, and absorption, which opposes it. The first process is related to the population of level $|b\rangle$, the second to the population of level $|a\rangle$. We shall now develop this interpretation, first in considering the conservation of energy during these processes, then in writing the rate equations for both the radiation and the atoms.

2.6 Rate equations

2.6.1 Conservation of energy in the propagation

Absorption

We shall first consider a plane wave propagating along the positive z-axis, in a medium where atoms are fed into the state $|a\rangle$ (compare Section 2.4) and where the intensity is sufficiently weak that we restrict ourselves to the perturbative treatment. It is instructive to calculate the power dissipated by absorption in a thin slice of the medium of thickness $\mathrm{d}z$ and cross-sectional area A, perpendicular to Oz. The mean Poynting vector points in the Oz direction and has magnitude

[12] The superscript (0) indicates the non-saturated ($s \ll 1$) regime and should not be confused with the notation ω_0 where the subscript 0 denotes the resonance. To avoid such problems, we may sometimes denote the non-saturated gain at resonance as $g^{(0)}(\omega_\mathrm{M})$ where M denotes the 'maximum'.

$$\Pi = \frac{\varepsilon_0 \, c \, E_0^2 \, \mathrm{e}^{-2k''z}}{2}. \tag{2.216}$$

The power absorbed in a slice of thickness dz is, therefore,

$$-\,[\mathrm{d}\Phi]_{\mathrm{abs}} = A\,[\Pi(z) - \Pi(z+\mathrm{d}z)] = A\,\mathrm{d}z\;\varepsilon_0 c\,[E(z)]^2\,k'', \tag{2.217}$$

with

$$E(z) = E_0\,\mathrm{e}^{-k''z}. \tag{2.218}$$

Using Equation (2.186) for k'' and replacing χ_1'' by its value given in (2.178), we find

$$-\,[\mathrm{d}\Phi]_{\mathrm{abs}} = A\;\mathrm{d}z\,\frac{N}{V}\,\frac{\omega d^2\,[E(z)]^2}{2\hbar}\,\frac{\Gamma_{\mathrm{D}}}{\Gamma_{\mathrm{D}}^2 + (\omega - \omega_0)^2}. \tag{2.219}$$

This result can be simplified, using expression (2.156) for the population of the excited state $|b\rangle$. In this way we obtain

$$-\,[\mathrm{d}\Phi]_{\mathrm{abs}} = A\,\mathrm{d}z\,\frac{N_b}{V}\,\hbar\omega\,\Gamma_{\mathrm{D}}. \tag{2.220}$$

This expression can be interpreted in a straightforward way if one notices that $\Gamma_{\mathrm{D}} A\,\mathrm{d}z\,N_b/V$ is the average number of atoms in level $|b\rangle$ that leave the volume $A\mathrm{d}z$ in unit time. Since the atoms are injected in state $|a\rangle$, these departing atoms have each *absorbed an energy* $\hbar\omega$ enabling them to leave in the excited state. Equation (2.220) then simply expresses the *conservation of energy* in the interaction between the incident electromagnetic wave and the atoms.

Comment It might seem paradoxical that the transfer of atoms from level $|a\rangle$ to level $|b\rangle$ should be accompanied by the absorption of a quantity of energy $\hbar\omega$ and not of a quantity $\hbar\omega_0 = E_b - E_a$. It is easy to satisfy oneself that for each elementary process, it is a quantum of energy $\hbar\omega$ that is removed from the field – this follows naturally from the quantum model where the field is composed of photons of energy $\hbar\omega$. However, this raises doubts about energy conservation in the absorption process, where an atom is excited from $|a\rangle$ to $|b\rangle$. In fact, because the atomic excited state has a finite lifetime, it does not have a well-defined energy. More precisely, the excited state $|b\rangle$ is in fact only the intermediate state in a scattering process of which the final, emission step must also be taken into account. One can then suppose that the atom remains in the excited state only for a finite time Δt of the order of $\Gamma_b{}^{-1}$ in the case of resonant scattering, or of $|\omega - \omega_0|^{-1}$ for off-resonant scattering. There is no paradox in considering that during this time the atomic state has an energy different from E_b by $\Delta E \simeq \hbar/\Delta t$ (the Heisenberg relation). In addition, when we further consider the decay of this atom from state $|b\rangle$ to a state which we suppose to be stable, we have exact conservation of energy. For example, in the case of fluorescence, where an atom with a closed two-level system (Section 2.4.5) absorbs an incident photon and emits it in a different direction on returning to its initial state $|a\rangle$, the scattered light has a frequency ω strictly equal to that of the incident light.

Stimulated emission

We now consider in the perturbative regime the case where the atoms are injected into level $|b\rangle$ (compare Section 2.5.2). We know that this situation leads to laser amplification. As in the previous section, we calculate the power gained when the light propagates through a slice dz of the medium. We obtain

$$[\mathrm{d}\Phi]_{\mathrm{sti}} = A\,\mathrm{d}z\,\frac{N_a}{V}\,\hbar\omega\,\Gamma_{\mathrm{D}}. \tag{2.221}$$

This result can be interpreted in an analogous way. The quantity $\Gamma_{\mathrm{D}} A\mathrm{d}z\, N_a/V$ is the average number of atoms in the level $|a\rangle$ which leave the volume $A\mathrm{d}z$ in unit time. Since they were injected initially in state $|b\rangle$, each has lost an energy $\hbar\omega$ to the electromagnetic field. Equation (2.221) expresses the conservation of energy for this process.

General case

The above calculations can be generalized to the case where both levels are externally fed, and to the non-perturbative regime. Using Equation (2.213) for the susceptibility, we find the power gained by a laser beam of cross-section A propagating through a slice dz,

$$\mathrm{d}\Phi = A\,\mathrm{d}z\,\frac{\Lambda_b - \Lambda_a}{V}\,\frac{1}{2}\,\frac{1}{1+s}\,\hbar\omega. \tag{2.222}$$

Using Equation (2.211) for the population N_b, we can write

$$\Lambda_b - \Gamma_{\mathrm{D}}\,N_b = \frac{\Lambda_b - \Lambda_a}{2}\,\frac{1}{1+s}. \tag{2.223}$$

The first term of this equation is simply the number of atoms which pass from $|b\rangle$ to $|a\rangle$ per unit time, since it is the difference between the feeding rate Λ_b and the loss rate $\Gamma_{\mathrm{D}}N_b$ in state $|b\rangle$. Multiplying this number by $\hbar\omega$ we obtain the power lost by the atoms to the electromagnetic field in volume $A\,\mathrm{d}z$. Equation (2.222) therefore represents the balance of energy exchanges between the atoms and the radiation in volume $A\mathrm{d}z$.

2.6.2 Rate equations for the atoms

In the previous section we showed that attenuation or amplification of the electromagnetic wave could be understood by considering the exchanges of energy between the atoms and radiation: the energy gained by the radiation resulted from the difference in the energy gained by stimulated emission and that removed by absorption.

In this section we shall show that the values of the populations N_a and N_b obtained for the two levels $|a\rangle$ and $|b\rangle$ can be interpreted as the result of competition between the various processes: absorption and stimulated emission, internal spontaneous transfers, arrival from other levels and departure towards external levels. This is most clearly understood by writing the *rate equations*, which are simple equations summarizing the departure and arrival rates of these different processes for the populations of each level:

$$\frac{\mathrm{d}N_b}{\mathrm{d}t} = \Gamma_{\mathrm{D}}\frac{s}{2}N_a - \Gamma_{\mathrm{D}}\frac{s}{2}N_b - \Gamma_{\mathrm{D}}N_b + \Lambda_b, \tag{2.224}$$

$$\frac{\mathrm{d}N_a}{\mathrm{d}t} = -\Gamma_{\mathrm{D}}\frac{s}{2}N_a + \Gamma_{\mathrm{D}}\frac{s}{2}N_b - \Gamma_{\mathrm{D}}N_a + \Lambda_a. \tag{2.225}$$

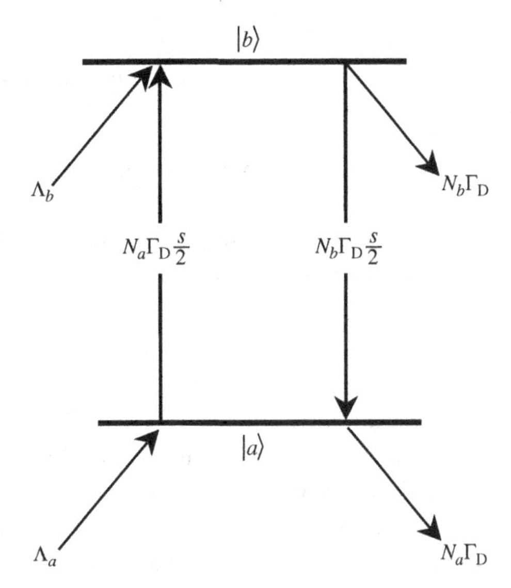

Figure 2.18 Processes included in the rate equations, describing the pumping of each level (Λ_a and Λ_b), the relaxation to external states ($N_a\Gamma_D$ and $N_b\Gamma_D$), the absorption ($N_a\Gamma_D\frac{s}{2}$) and the stimulated emission ($N_b\Gamma_D\frac{s}{2}$) of the radiation.

Einstein was, in 1917, the first to introduce in a phenomenological way these types of equations, and showed that the coefficients associated with absorption ($\Gamma_D\frac{s}{2}N_a$) and with stimulated emission ($\Gamma_D\frac{s}{2}N_b$) are necessarily equal. Actually these equations can be derived from the Bloch equations in some particular cases.[13] We shall not demonstrate this here, but instead we shall verify that they lead to results in agreement with those obtained in earlier sections. The importance of these equations is that they allow a very simple interpretation (Figure 2.18) and are easily generalized.

The structure of Equations (2.224) and (2.225) is clear. The rate of variation of each population (N_a and N_b) is the sum of the rates associated with each physical process identified in Figure 2.18. Thus we identify the pumping rates (Λ_a and Λ_b) and the relaxation rates ($-\Gamma_D N_a$ and $-\Gamma_D N_b$). In addition, the stimulated emission and absorption rates are given by

$$\left[\frac{\mathrm{d}N_a}{\mathrm{d}t}\right]_{\mathrm{abs}} = -\left[\frac{\mathrm{d}N_b}{\mathrm{d}t}\right]_{\mathrm{abs}} = -\Gamma_D\frac{s}{2}N_a \tag{2.226}$$

and

$$\left[\frac{\mathrm{d}N_b}{\mathrm{d}t}\right]_{\mathrm{sti}} = -\left[\frac{\mathrm{d}N_a}{\mathrm{d}t}\right]_{\mathrm{sti}} = -\Gamma_D\frac{s}{2}N_b. \tag{2.227}$$

[13] One can show that Equations (2.224) and (2.225) give the correct evolution of the system when the lifetime of the coherence is much smaller than the population lifetimes, which is the case in dense media, and also when the field interacting with the atom has a spectral width which is much greater than the transition linewidth. In addition, they always give the correct value for steady-state populations.

The steady-state condition is obtained by setting the left-hand side of Equations (2.224) and (2.225) to zero. Summing these two equations, we solve for the total number of atoms and find

$$N = N_a + N_b = \frac{\Lambda_a}{\Gamma_{\mathrm{D}}} + \frac{\Lambda_b}{\Gamma_{\mathrm{D}}}, \tag{2.228}$$

as was obtained in Equation (2.147). Similarly, the difference of these two equations gives the steady-state population inversion,

$$N_b - N_a = \frac{\Lambda_b - \Lambda_a}{\Gamma_{\mathrm{D}}} \frac{1}{1+s} = N \frac{\Lambda_b - \Lambda_a}{\Lambda_b + \Lambda_a} \frac{1}{1+s} = \frac{N_b^{(0)} - N_a^{(0)}}{1+s}, \tag{2.229}$$

which is identical to (2.212). Finally, we can also solve the rate equations for stationary populations N_a and N_b, and find results identical to (2.206) and (2.207). We consider that identity as a justification of the rate equations for our particular model.

Interest in the rate equations is primarily in highlighting the competition between absorption and stimulated emission terms, which is responsible for the phenomena of saturation and population inversion. Additionally, these equations are easily generalized to more complex situations: for example, unequal lifetimes of states $|a\rangle$ and $|b\rangle$, or relaxation from $|b\rangle$ to $|a\rangle$. Generally, it remains straightforward to write and solve the rate equations, whereas the optical Bloch equations rapidly become inextricable.

It should be realized, however, that the rate equations are an approximative treatment. In particular, they cannot account for phenomena in which phase (either the interacting field phase or the atomic dipole phase) plays a role. We shall see some striking examples of such phenomena in Complement 2D on atomic coherence. In particular, the rate equations are unable to give the value of the atomic dipole, and therefore of the dielectric susceptibility, which require a formalism which takes into account the phase relations between the coefficients γ_a and γ_b of Equation (2.102) describing the evolution of the atomic state. Thus in Equation (2.213) for susceptibility, which we rewrite here as

$$\chi = \frac{N_a - N_b}{V} \frac{d^2}{\varepsilon_0 \hbar} \frac{\omega_0 - \omega + \mathrm{i}\Gamma_{\mathrm{D}}}{\Gamma_{\mathrm{D}}^2 + (\omega - \omega_0)^2}, \tag{2.230}$$

the rate equations can provide the factor $N_a - N_b$, but the other terms can only be obtained a priori by a treatment capable of calculating the atomic dipole moments and taking into account their interaction with the radiation. In fact we shall now see that the rate equations are more powerful if we develop a phenomenological description of the interaction between the atoms and the photons, allowing us to write rate equations for the photons in addition to the rate equations for the atoms.

2.6.3 Atom–photon interactions. Cross-section, saturation intensity

Description of absorption and stimulated emission processes by the rate equations ((2.226–2.227) and Figure 2.18) suggests the existence of elementary processes of absorption and stimulated emission in which the atom changes state when a photon is absorbed, making the opposite transition when a photon is emitted. If this description can only be justified in a treatment where the radiation is quantized (see Chapter 6), it is of such use that we shall present this model here.

A travelling electromagnetic wave, with power per unit area Π (the Poynting vector, compare Equation (2.91)) corresponds to a flux of photons per unit area of

$$\Pi_{\text{phot}} = \frac{\Pi}{\hbar\omega}. \tag{2.231}$$

We suppose that the probability per unit time that an atom interacts with the radiation is proportional to Π_{phot}: the coefficient of proportionality, which has units of area, is called the *cross-section* of interaction, σ_{L} (the subscript L refers to a laser transition). Following Einstein, we assume that the cross-sections for absorption and stimulated emission processes are equal. If there are N_a atoms in state $|a\rangle$, the number of absorption processes per second is $N_a\sigma_{\text{L}}\Pi_{\text{phot}}$ and the rates of change of the populations due to absorption are

$$\left[\frac{\mathrm{d}N_a}{\mathrm{d}t}\right]_{\text{abs}} = -\left[\frac{\mathrm{d}N_b}{\mathrm{d}t}\right]_{\text{abs}} = -N_a\sigma_{\text{L}}\frac{\Pi}{\hbar\omega}. \tag{2.232}$$

Similarly, if there are N_b atoms in state $|b\rangle$, the rates of change of the populations due to stimulated emission are

$$\left[\frac{\mathrm{d}N_b}{\mathrm{d}t}\right]_{\text{sti}} = -\left[\frac{\mathrm{d}N_a}{\mathrm{d}t}\right]_{\text{sti}} = -N_b\sigma_{\text{L}}\frac{\Pi}{\hbar\omega}. \tag{2.233}$$

These rates of change have the same form as expressions (2.226) and (2.227). Replacing s by its expression as a function of the Poynting vector, we arrive at an expression for σ_{L}, valid for the model considered above:

$$\sigma_{\text{L}} = \frac{d^2}{\hbar\varepsilon_0 c}\frac{\Gamma_{\text{D}}\omega}{(\omega-\omega_0)^2+\Gamma_{\text{D}}^2}. \tag{2.234}$$

However, one most often treats σ_{L} as an empirical quantity to be determined experimentally. The cross-section manifests a resonant behaviour close to ω_0, so it can be specified by its value at resonance $\sigma_{\text{L}}(\omega_0)$ and the width of the resonance Γ_2. It is then written

$$\sigma_{\text{L}}(\omega) = \sigma_{\text{L}}(\omega_0)\frac{1}{1+(\frac{\omega-\omega_0}{\Gamma_2/2})^2}. \tag{2.235}$$

For the model used here, the cross-section at resonance, $\sigma_{\text{L}}(\omega_0)$, can be related to the saturation intensity (2.162) by

$$I_{\text{sat}} = \frac{\hbar\Gamma_{\text{D}}\omega}{2\varepsilon_0 c\sigma_{\text{L}}(\omega_0)}. \tag{2.236}$$

Keeping in mind that the cross-section at exact resonance is of the order of the square of the wavelength λ_0, expression (2.236) shows that a photon flux equivalent to one photon per lifetime per λ_0^2 is enough to saturate a transition. This is actually a rather low value, of the order of 1 mW/cm^2, which can be easily obtained by low-power continuous wave (c.w.)

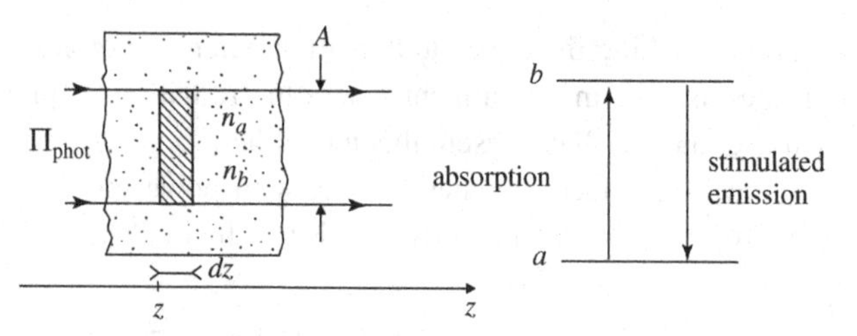

Figure 2.19 **Description of absorption and stimulated emission processes by rate equations. In the slice of thickness dz, the laser beam of cross-section A interacts with $n_a A\,dz$ atoms in state $|a\rangle$ and $n_b A\,dz$ atoms in state $|b\rangle$. For each atom in state $|a\rangle$ (respectively $|b\rangle$), the rate of absorption (respectively stimulated emission) is $\sigma_L \Pi_{phot}$, where Π_{phot} is the number of atoms per unit time and unit area. The absorption process removes photons from the laser beam, whereas the stimulated emission process adds photons to the incident laser beam.**

lasers such as semiconductor lasers.[14] In more complex systems than the toy model used here, the saturation intensity is an empirical quantity, determined experimentally.

2.6.4 Rate equations for the photons. Laser gain

Now, considering the absorption process from the point of view of the photons, each process removes a photon from the laser beam. A beam of cross-section A propagating through a slice of thickness dz (Figure 2.19) interacts with $A\mathrm{d}z N_a/V$ atoms in state $|a\rangle$, and the number of photons absorbed per unit time is

$$\left[\frac{\mathrm{d}\mathcal{N}}{\mathrm{d}t}\right]_{\mathrm{abs}} = -\mathrm{d}z\,A\frac{N_a}{V}\sigma_{\mathrm{L}}\Pi_{\mathrm{phot}}. \tag{2.237}$$

This quantity represents the change of the photon flux $A\Pi_{\mathrm{phot}}$ per unit time when the beam passes through the slice of thickness dz, and so we can write

$$\left[\mathrm{d}(A\Pi_{\mathrm{phot}})\right]_{\mathrm{abs}} = \left[\frac{\mathrm{d}\mathcal{N}}{\mathrm{d}t}\right]_{\mathrm{abs}} = -\mathrm{d}z\,A\frac{N_a}{V}\sigma_{\mathrm{L}}\Pi_{\mathrm{phot}}, \tag{2.238}$$

from which we obtain the remarkably simple equation,

$$\left[\frac{\mathrm{d}\Pi_{\mathrm{phot}}}{\mathrm{d}z}\right]_{\mathrm{abs}} = -\sigma_{\mathrm{L}}\frac{N_a}{V}\Pi_{\mathrm{phot}}. \tag{2.239}$$

We usually then rewrite this expression in terms of the density $n_a = N_a/V$ of absorbing atoms,

$$\left[\frac{\mathrm{d}\Pi_{\mathrm{phot}}}{\mathrm{d}z}\right]_{\mathrm{abs}} = -\sigma_{\mathrm{L}} n_a \Pi_{\mathrm{phot}}. \tag{2.240}$$

[14] This value is valid only at exact resonance: the main experimental requirement to saturate a transition is to be able to reach – and stay at – the exact resonance with the atomic transition, which requires good control of the laser frequency.

If we have exclusively absorbing atoms with a uniform density n_a, we have an exponential decrease in photon flux with propagation distance (the 'Beer-Lambert' law), characterized by the $(1/e)$ decay length,

$$l = (n_a \sigma_L)^{-1}. \tag{2.241}$$

In the case where one has atoms exclusively in the upper level $|b\rangle$ with a uniform density n_b, an analogous reasoning leads to

$$\left[\frac{d\Pi_{phot}}{dz}\right]_{sti} = \sigma_L \frac{N_b}{V} \Pi_{phot} = \sigma_L n_b \Pi_{phot}, \tag{2.242}$$

where $n_b = N_b/V$ is the density of atoms in state $|b\rangle$.

Despite the similarity with the reasoning applied to the absorption process, that leading to Equation (2.242) implies the hypothesis that photons emitted by stimulated emission are added to the incident laser beam with the same frequency, polarization and direction of propagation, and with a well-defined phase with respect to the incident beam. This property is the result of a completely quantum treatment of the atom–light interaction. Here we justify this result by the fact that we obtain an amplification of the electromagnetic wave identical to that given by the semi-classical treatment presented earlier in this chapter.

In the general case of a laser medium where the density of atoms in the states $|a\rangle$ and $|b\rangle$ are n_a and n_b, respectively, we have the global rate equation,

$$\frac{d\Pi_{phot}}{dz} = (n_b - n_a)\,\sigma_L \Pi_{phot}, \tag{2.243}$$

or again, multiplying by $\hbar\omega$,

$$\frac{d\Pi}{dz} = (n_b - n_a)\,\sigma_L \Pi. \tag{2.244}$$

The gain per unit length is, therefore,

$$g = \frac{1}{\Pi}\frac{d\Pi}{dz} = (n_b - n_a)\,\sigma_L. \tag{2.245}$$

This formula generalizes the formula (2.214) which we obtained using our simple model. It is a fundamental equation in the modelling of lasers.

In fact, for each laser transition, we can find experimental values for the effective cross-section at resonance $\sigma_L(\omega_0)$ and the width of the resonance Γ_2, from which we can deduce σ_L at any frequency using Equation (2.235). It is generally fairly easy to write the rate equations which allow one to calculate the population inversion $n_b - n_a$. In the steady state we generally obtain an expression of the form of Equation (2.212), where the saturation parameter can be expressed in the form of Equation (2.161), in which the saturation intensity I_{sat} is another empirical value characterizing the laser transition.

2.7 Conclusion

In this chapter we presented the semi-classical model of atom–light interaction, where the atom is quantized and the radiation is described by a classical electromagnetic field. Introducing a two-level model of the atom with equal lifetimes for the two levels, we were able to take into account relaxation processes in a simple way. In this frame, we could describe many important phenomena: absorption, stimulated emission, elastic scattering, Rabi oscillations, light-shifts, multi-photon processes, saturation, laser amplification... Complement 2C will give an introduction to the optical Bloch equations, which are a generalization of this approach to any kind of relaxation process. With these equations, one can describe the great majority of physical situations where electromagnetic waves interact with matter.

We have seen that it is also possible to use the much simpler formalism of rate equations to describe the atom–light interaction. These rate equations are constructed by simply adding the microscopic processes of absorption and stimulated emission in which an atom changes state by removing or adding a photon to the incident light beam, the latter considered as a photon flux. Rate equations give the correct behaviour of the system in many instances, but there exist situations where the phase of the electromagnetic field and the phase of the atomic dipole play a crucial role. Such situations cannot be described by the rate equations, but only by the Bloch equations. We shall see examples of these situations in Complement 2C.

To describe in a simple way the phenomena of absorption and stimulated emission, and to derive the rate equations, we have made use in a phenomenological way of the concept of the photon. In the present context of a classical electromagnetic field interacting with the atom, it is not strictly necessary, but it turns out to be convenient. The complete internal coherence of the description of light–matter interaction will be reached when we treat in a quantum way both the light and the atoms, which will be done in Chapter 6. We shall then be able to establish the conditions of validity of the present approach. We shall see that in order to model numerous phenomena in light–matter interaction, and in particular the laser, one can indeed use classical electromagnetism to describe propagation, interference and diffraction, whereas atom–light interaction can be described using the rate equations which take into account the major part of its quantum aspect. Phenomena that cannot be described by the approach sketched here belong to modern quantum optics as it has been developed over past decades. These will be presented in the second and third parts of this book.

2A Complement 2A Classical model of the atom–field interaction: the Lorentz model

This first complement is devoted to a completely classical approach of light–matter interaction which was proposed by Lorentz at the end of the nineteenth century, before the advent of quantum mechanics, but after the discovery of the electron. Lorentz' phenomenological model is based on the experimental fact that atoms have well-defined and sharp absorption lines: he assumed that atoms behaved like harmonic oscillators, in which the electrons are bound to the atomic nucleus by a restoring force which varies linearly with its displacement (from its equilibrium point close to a nucleus), and makes them oscillate at a given frequency ω_0 equal to the experimentally determined absorption frequency.

Within the frame of this model, we first calculate the electromagnetic field radiated by an oscillating electron. We show that in the absence of an externally applied force the free oscillations of the electron are damped, because electromagnetic energy is radiated at the expense of mechanical energy. We then study the characteristics of the radiation that is emitted when the oscillations are forced by the application of an external oscillatory electromagnetic field of angular frequency ω. We characterize the different regimes of this *scattering* of the incident electromagnetic wave and finally determine the polarization induced in the atomic medium by the incident electromagnetic wave.

The Lorentz model can be considered as a lowest order approximation to a description of the light–matter interaction, a better approximation being the semi-classical treatment presented in Chapter 2, and the rigorous treatment being the completely quantum mechanical model presented in Chapter 6. It might appear that the model we discuss here is not obviously well adapted to the description of an electron in an atom, which is a bound system of which the quantum mechanical character is predominant. However, as we shall show in the following, it produces results that are in good agreement with the quantum mechanical results for a two-level atom provided it is only weakly perturbed by the electromagnetic field (i.e. provided saturation effects are unimportant). We shall actually demonstrate at the end of this complement that the quantum equations reduce to the classical equations of the Lorentz model, within a single scaling factor, named the oscillator strength, at the limit of small intensities.

2A.1 Description of the model

Consider an electron of charge $q = -1.6 \times 10^{-19}$ C performing an oscillatory motion about a point O at angular frequency ω. We describe this motion by a complex amplitude $\mathcal{S}_0$ such that the position vector $\mathbf{r}$ is

$$\mathbf{r} = \boldsymbol{\mathcal{S}} + \boldsymbol{\mathcal{S}}^* = \boldsymbol{\mathcal{S}}_0\,\mathrm{e}^{-\mathrm{i}\omega t} + \boldsymbol{\mathcal{S}}_0^*\,\mathrm{e}^{\mathrm{i}\omega t}. \tag{2A.1}$$

A rectilinear oscillation for amplitude a along z and phase φ is thus described by a complex amplitude:

$$\boldsymbol{\mathcal{S}}_0 = \frac{a}{2}\,\mathrm{e}^{-\mathrm{i}\varphi}\mathbf{e}_z, \tag{2A.2}$$

whilst a circular trajectory of radius a in the xOy plane is described by

$$\boldsymbol{\mathcal{S}}_0 = \frac{a}{2}\,\mathrm{e}^{-\mathrm{i}\varphi}\boldsymbol{\varepsilon}_\pm, \tag{2A.3}$$

with

$$\boldsymbol{\varepsilon}_\pm = \mp\frac{1}{\sqrt{2}}\left(\mathbf{e}_x \pm \mathrm{i}\mathbf{e}_y\right), \tag{2A.4}$$

where $\mathbf{e}_x$, $\mathbf{e}_y$ and $\mathbf{e}_z$ are unit vectors of the axes x, y and z. We denote by $\boldsymbol{\mathcal{D}}_0$ the complex amplitude associated with

$$\mathbf{D} = \boldsymbol{\mathcal{D}} + \text{c.c.} = \boldsymbol{\mathcal{D}}_0\,\mathrm{e}^{-\mathrm{i}\omega t} + \text{c.c.}, \tag{2A.5}$$

with

$$|\boldsymbol{\mathcal{D}}_0| = \frac{d_0}{2}. \tag{2A.6}$$

The quantity $\mathbf{D}$ is the atomic dipole, the electron-forming part of a globally neutral system together with the nucleus at rest at position O. In order to simplify as far as possible the calculation of the dipole moment we adopt the following approximations:

1. *Long-wavelength approximation*: we shall only consider the case in which the charges remain localized in a volume of dimensions much smaller than the wavelength λ of the emitted radiation:

$$|\boldsymbol{\mathcal{S}}_0| \ll \lambda. \tag{2A.7}$$

   This implies that the speed of the electron, which is of the order of $\omega|\boldsymbol{\mathcal{S}}_0|$, is small compared to the speed of light c.
2. *Far-field approximation*: we consider the radiated field only at distances that are large compared to the wavelength:

$$|\mathbf{r}| \gg \lambda. \tag{2A.8}$$

In the general case the radiated field can be written as a power series in λ/r; the approximation made here is equivalent to keeping only the lowest order term. If the motion of the electron is not purely sinusoidal, the conditions given above will be assumed to be satisfied for all the Fourier components of $\boldsymbol{\mathcal{D}}(t)$.

2A.2 Electric dipole radiation

Retarded potentials

The formulation of electromagnetism that is best adapted to a description of this problem (i.e. that which leads most naturally to the introduction of simplifying approximations) is that of the retarded potentials of the Lorentz gauge.[1] We shall therefore use the following expression for the retarded vector potential:

$$\mathbf{A}_{\rm L}(\mathbf{r},t) = \frac{1}{4\pi\varepsilon_0 c^2}\int \mathrm{d}^3 r' \frac{\mathbf{j}\left(\mathbf{r}', t - |\mathbf{r}-\mathbf{r}'|/c\right)}{|\mathbf{r}-\mathbf{r}'|}. \tag{2A.9}$$

Calculation of the electric dipole radiation is simplified by the use of the approximation $|\mathbf{r}-\mathbf{r}'| \approx |\mathbf{r}| = r$, justified by the above assumptions (2A.7) and (2A.8). It remains to substitute an expression for the current density associated with the moving electron. This is given exactly by

$$\mathbf{j}(\mathbf{r}',t) = q\delta\left[\mathbf{r}' - \mathbf{r}_e(t)\right]\dot{\mathbf{r}}_e(t), \tag{2A.10}$$

since the velocity and position of the electron are $\dot{\mathbf{r}}_e$ and $\mathbf{r}_e$, respectively. Making use again of (2A.7) and (2A.8) enables (2A.10) to be approximated by a current localized at O:

$$\mathbf{j}(\mathbf{r}',t) \approx q\delta\left(\mathbf{r}'\right)\dot{\mathbf{r}}_e(t). \tag{2A.11}$$

Substituting this expression into (2A.9) we find the following expression for the retarded vector potential:

$$\mathbf{A}_{\rm L}(\mathbf{r},t) = \frac{1}{4\pi\varepsilon_0 c^2}\frac{\dot{\mathbf{D}}(t-r/c)}{r}. \tag{2A.12}$$

The Lorentz condition allows the corresponding scalar potential $U_{\rm L}$ to be obtained:

$$\frac{\partial U_{\rm L}}{\partial t} = -c^2\nabla\cdot\mathbf{A}_{\rm L}(\mathbf{r},t) = \frac{\mathbf{r}}{4\pi\varepsilon_0 c r^3}\cdot\left[c\dot{\mathbf{D}}(t-r/c) + r\ddot{\mathbf{D}}(t-r/c)\right]. \tag{2A.13}$$

Integrating with respect to time and dropping the electrostatic term, which plays no role in the following, we obtain

$$U_{\rm L} = \frac{\mathbf{r}}{4\pi\varepsilon_0 c r^2}\cdot\left[\dot{\mathbf{D}}(t-r/c) + \frac{c}{r}\mathbf{D}(t-r/c)\right] \tag{2A.14}$$

in which, according to (2A.8), the first term is dominant.

The radiation from an accelerating charge in the far-field

The magnetic field is straightforwardly obtained by taking the *curl* of the vector potential $\mathbf{A}_{\rm L}$ (Equation 2A.12) and using in addition the results

$$\nabla\left(\frac{1}{r}\right) = -\frac{\mathbf{r}}{r^3};\quad \nabla\times\dot{\mathbf{D}}(t-r/c) = -\frac{1}{rc}\mathbf{r}\times\ddot{\mathbf{D}}(t-r/c). \tag{2A.15}$$

[1] See for instance J. D. Jackson, *Classical Electrodynamics*, Wiley (1998).

We find

$$\mathbf{B}(\mathbf{r},t) = -\frac{\mathbf{r}}{4\pi\varepsilon_0 c^3 r^2} \times \left[\ddot{\mathbf{D}}(t-r/c) + \frac{c}{r}\dot{\mathbf{D}}(t-r/c)\right], \qquad (2A.16)$$

in which, according to (2A.8), the first term is dominant.

The electric field is derived from $\mathbf{E} = -\partial\mathbf{A}_L/\partial t - \nabla U_L$. Calculating its general expression is tedious. However, for large distances $r \gg \lambda$ we need keep only the terms in the lowest order of $1/r$, which gives

$$\begin{aligned}\mathbf{E}(\mathbf{r},t) &= -\frac{1}{4\pi\varepsilon_0 c^2 r}\cdot\left[\ddot{\mathbf{D}}(t-r/c) - \frac{\mathbf{r}}{r^2}\left(\mathbf{r}\cdot\ddot{\mathbf{D}}(t-r/c)\right)\right] \\ &= \frac{1}{4\pi\varepsilon_0 c^2}\frac{\mathbf{r}}{r^3}\times\left[\mathbf{r}\times\ddot{\mathbf{D}}(t-r/c)\right]. \end{aligned} \qquad (2A.17)$$

Comparing this result with that of (2A.16) we see that

$$\mathbf{B}(\mathbf{r},t) = \frac{\mathbf{r}}{rc}\times\mathbf{E}(t-r/c). \qquad (2A.18)$$

Thus the magnetic and electric fields **E** and **B** are each orthogonal to **r** and are, moreover, in phase. As Equation (2A.17) indicates, the radiated fields at large distances are proportional to $\ddot{\mathbf{D}}$, that is to the acceleration of the radiating charge, but oscillate with a time delay with respect to the oscillatory motion of the charge, of r/c, equal to the propagation time to the point of interest. In a fixed direction $\mathbf{r}/r$ the electric and magnetic fields decrease as $1/r$ and not as $1/r^2$ and $1/r^3$ as in electro- and magneto-statics. The radiated electromagnetic field can therefore be described as a non-uniform spherical wave, propagating outwards from the origin with an amplitude decreasing as $1/r$. The electromagnetic field at point M has therefore locally the same form as a plane wave propagating in the direction OM.

The Poynting vector Π at M is

$$\Pi(\mathbf{r},t) = \varepsilon_0 c^2\mathbf{E}\times\mathbf{B} = \frac{1}{16\pi^2\varepsilon_0 c^3}\frac{\mathbf{r}}{r^5}\left[\mathbf{r}\times\ddot{\mathbf{D}}(t-r/c)\right]^2. \qquad (2A.19)$$

Thus the power density of the radiated wave is proportional to $1/r^2$ and is proportional to the square of the acceleration of the charge.

Sinusoidally oscillating charge. Polarization of the emitted radiation

We return now to the case of a sinusoidally oscillating dipole moment (Equation 2A.5). Expressions for the radiated electric and magnetic fields in the complex notation of Equations (2A.1) or (2A.5) can be found from (2A.17) and (2A.18):

$$\mathcal{E}(\mathbf{r},t) = -\frac{\omega^2}{4\pi\varepsilon_0 c^2}\frac{\mathbf{r}\times(\mathbf{r}\times\mathcal{D}_0)}{r^3}\,\mathrm{e}^{-\mathrm{i}\omega(t-r/c)} \qquad (2A.20)$$

$$\mathcal{B}(\mathbf{r},t) = \frac{\omega^2}{4\pi\varepsilon_0 c^3}\frac{(\mathbf{r}\times\mathcal{D}_0)}{r^2}\,\mathrm{e}^{-\mathrm{i}\omega(t-r/c)}, \qquad (2A.21)$$

and the Poynting vector *averaged over an optical period* is then

$$\overline{\Pi(\mathbf{r},t)} = \frac{\omega^4}{8\pi^2\varepsilon_0 c^3}\frac{|\mathbf{r}\times\mathcal{D}_0|^2}{r^5}\mathbf{r}. \tag{2A.22}$$

We shall now study successively two important special cases described by this model, firstly the case of a dipole oscillating along the line Oz (π-polarized case) and secondly that of a dipole tracing out a circular trajectory in the xOy-plane (σ_+ or σ_- polarized case, depending on the sense of the rotation):

π-polarization

We have in this case $\mathcal{D}_0 = (d_0/2)\mathbf{e}_z$. Introducing the spherical polar coordinates (r, θ, ϕ) and putting $k = \omega/c$, we obtain (see Figure 2A.1):

$$\mathcal{E}_\pi(\mathbf{r},t) = -\frac{\omega^2 d_0}{4\pi\varepsilon_0 c^2}\frac{\sin\theta}{r}\,\mathrm{e}^{-\mathrm{i}(\omega t - kr)}\mathbf{e}_\theta \tag{2A.23}$$

$$\mathcal{B}_\pi(\mathbf{r},t) = -\frac{\omega^2 d_0}{4\pi\varepsilon_0 c^3}\frac{\sin\theta}{r}\,\mathrm{e}^{-\mathrm{i}(\omega t - kr)}\mathbf{e}_\phi \tag{2A.24}$$

$$\overline{\Pi(\mathbf{r},t)} = \frac{\omega^4 d_0^2}{32\pi^2\varepsilon_0 c^3}\frac{\sin^2\theta}{r^2}\mathbf{e}_r. \tag{2A.25}$$

Thus a linearly oscillating dipole emits radiation preferentially into the plane perpendicular to its vibration axis ($\theta = \pi/2$ for a dipole vibrating along z). The electric field is linearly polarized in the plane containing the dipole and the propagation direction (Figure 2A.2). An expression for the total radiated power Φ passing through a

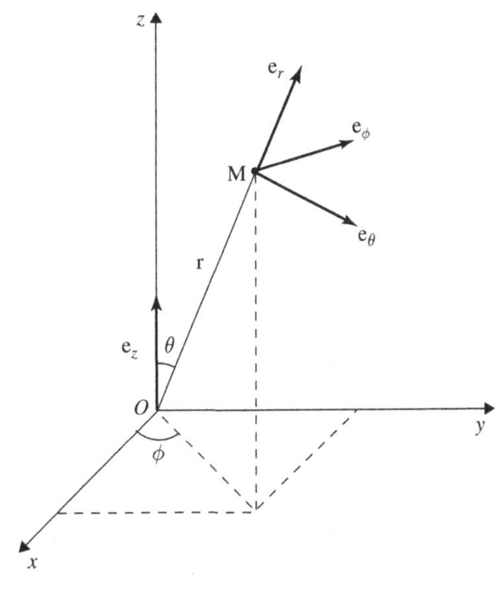

Figure 2A.1 Spherical polar coordinates.

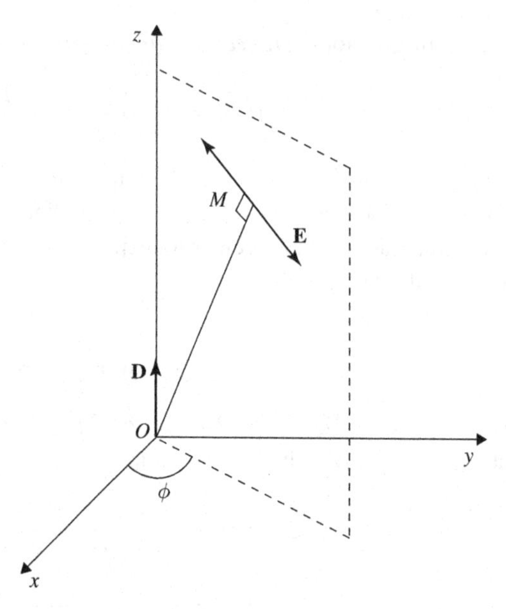

Figure 2A.2 **The radiation emitted by a dipole oscillating along *Oz*. The electric field at point *M* is linearly polarized perpendicular to *OM*, in the plane containing *Oz* and *OM*.**

closed surface enclosing the dipole can be found by integrating (2A.25). The result is independent of r:

$$\Phi = \frac{\omega^4 d_0^2}{12\pi\varepsilon_0 c^3}. \tag{2A.26}$$

σ_+-polarization

In this situation the dipole rotates in the xOy-plane in the anti-clockwise sense, as seen from the direction of positive z, and we have

$$\mathcal{D}_0 = \frac{d_0}{2}\boldsymbol{\varepsilon}_+. \tag{2A.27}$$

The polarization of the radiation emitted in an arbitrary direction OM is in general elliptical and the angular distribution of the radiation has cylindrical symmetry. However, there are directions in which it can be described more simply. Consider first of all the radiation emitted along the z-axis. One finds:

$$\mathcal{E}_{\sigma_+}(r\mathbf{e}_z, t) = \frac{\omega^2 d_0}{4\pi\varepsilon_0 c^2 r}\,\mathrm{e}^{-\mathrm{i}(\omega t - kr)}\boldsymbol{\varepsilon}_+. \tag{2A.28}$$

The electric field is circularly polarized, rotating in the same sense as the dipole. This is, of course, true also of the magnetic field $\mathbf{B}$ which is at all times perpendicular to the electric field $\mathbf{E}$. Thus the field radiated along this axis is σ_+ circularly polarized.

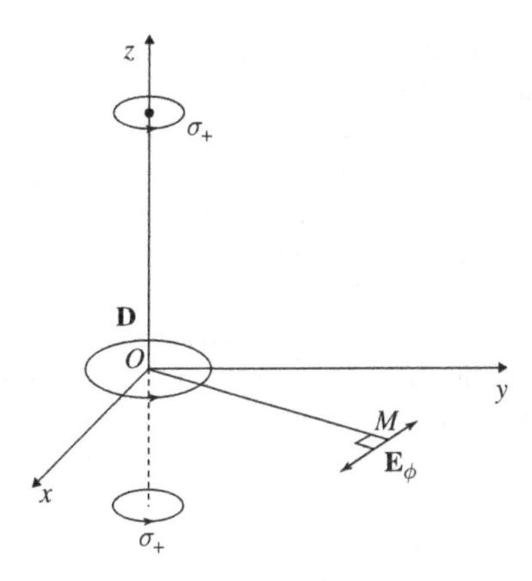

Figure 2A.3 **Radiation from a σ_+ rotating dipole. For a dipole rotating around the point O in either sense, the light emitted in the Oz direction is circularly polarized; it is linearly polarized for emission directions in the xOy plane and elliptically polarized in any other direction.**

Now we consider, for the same rotating motion of the dipole, an emission direction lying in the xOy plane and characterized by an azimuthal angle ϕ. In this case,

$$\mathbf{e}_r \times (\mathbf{e}_r \times \mathcal{D}_0) = -\frac{\mathrm{i}}{\sqrt{2}}\,\mathrm{e}^{\mathrm{i}\phi} d_0 \mathbf{e}_\phi \tag{2A.29}$$

and

$$\mathcal{E}_{\sigma+}(\mathbf{r},t) = \frac{\mathrm{i}}{\sqrt{2}} e^{\mathrm{i}\phi} \frac{\omega^2 d_0}{4\pi\varepsilon_0 c^2 r}\,\mathrm{e}^{-\mathrm{i}(\omega t - kr)}\mathbf{e}_\phi. \tag{2A.30}$$

Thus the electric field is linearly polarized along $\mathbf{e}_\phi$, perpendicular to both the propagation direction OM and the axis about which the dipole rotates. This is easily understood, since the oscillation of the dipole as seen from the emission direction that we are considering here appears to be along a line in the direction of $\mathbf{e}_\phi$.

Whilst the angular distribution of the radiation is therefore different from the case of the linearly oscillating dipole, the total radiated power, found by integrating (2A.22) over all directions, is the same (Equation 2A.26). More simply, since the fields radiated by the components of the rotating dipole along the Ox and the Oy directions are everywhere in quadrature, the total average radiated power is just the sum of their two averages, the cross-term being zero.

σ_--polarization

This situation, in which the dipole rotates in the left-handed sense, corresponds to $\mathcal{D}_0 = (d_0/2)\boldsymbol{\varepsilon}_-$.

This gives rise to the emission of σ_- circularly polarized light along the Oz axis and, as previously, to that of linearly polarized light in the xOy plane.

Comment Consider a quantized atom emitting light of angular frequency $\omega_0 = (E_b - E_a)/\hbar$ in passing from an excited state $|b\rangle$ to the ground state $|a\rangle$. The radiated field in this case can be derived from the complex value of the mean value of the atomic dipole moment $\mathbf{D}(t) = \langle\psi(t)|\hat{\mathbf{D}}|\psi(t)\rangle$, which can be substituted in expressions (2A.20) and (2A.21). Radiation distributions identical to those found in this complement are obtained. More precisely, if the transition takes place between two levels with the same magnetic quantum number with respect to Oz (a $\Delta m = 0$ transition), the radiation distribution is that found for the dipole oscillating along the z-axis i.e. the π case; if on the other hand the magnetic quantum number changes by one in the course of the transition ($\Delta m = \pm 1$), the radiation distribution is of the σ_+ or σ_- type with respect to Oz. Notice that the electric dipole matrix element is zero if $|\Delta m| > 1$ so that no electric dipole radiation is emitted in this case (see the account of the selection rules for electric dipole radiation of Complement 2B).

2A.3 Radiative damping of the elastically bound electron

The radiation reaction

As a consequence of the conservation of energy, the mechanical energy of the oscillating dipole must decrease as energy is radiated away. The mechanism for this transfer of energy is the reaction of the radiated field on the system that emits it. This interaction of the electron with the field it radiates gives rise to a damping force which acts on it, which is known as the radiation reaction.[2] This force has a component proportional to the acceleration of the electron, $\ddot{\mathbf{r}}$,[3] which can be interpreted as a radiative correction to the mass of the electron, and a term proportional to the third time derivative of the position $\dddot{\mathbf{r}}$ which can be written

$$\mathbf{F}_{\mathrm{RR}} = \frac{q_e^2}{6\pi\varepsilon_0 c^3}\dddot{\mathbf{r}} = \frac{2mr_0}{3c}\dddot{\mathbf{r}}. \tag{2A.31}$$

Here we have introduced the *classical radius of the electron*, $r_0 = q_e^2/4\pi\varepsilon_0 mc^2$, of the order of 2.8×10^{-15} m. We shall not seek to prove the expression (2A.31) here, but we shall demonstrate that it is consistent with the results established in Equation (2A.2). Let us calculate the average work done per unit time by this force on the electron when the latter oscillates with amplitude a. It is

$$\frac{\mathrm{d}W}{\mathrm{d}t} = \overline{\mathbf{F}_{\mathrm{RR}} \cdot \dot{\mathbf{r}}} = -\frac{q_e^2\omega^4 a^2}{12\pi\varepsilon_0 c^3}. \tag{2A.32}$$

Thus the electron does work at a rate equal to exactly the average radiated power (Equation (2A.26)), which shows that the force given by (2A.31) is consistent with the conservation of energy: the energy carried away by the electromagnetic radiation is taken from the electron's oscillatory motion, which is thereby damped. The damping force F_{RR} is simply the mechanical counterpart of the electromagnetic radiation.

[2] See J. D. Jackson, *Classical Electrodynamics*, Chapter 16, Wiley (1998).

[3] We adopt here the simplified notation of $\mathbf{r}$ for $\mathbf{r}_e$ since we are only interested in the position of the electron.

Radiative damping of the electron oscillations

Let us now return to the elastically bound electron. To account for the radiative damping of its motion, we must add a term representing the damping force to its equation of motion:

$$m\ddot{\mathbf{r}} = -m\omega_0^2\mathbf{r} + \frac{2mr_0}{3c}\dddot{\mathbf{r}} . \tag{2A.33}$$

We seek a solution (again employing complex notation) of the form $\boldsymbol{\mathcal{S}}_0 \exp(-\mathrm{i}\Omega t)$. The complex frequency Ω is then given by

$$\Omega^2 = \omega_0^2 - \mathrm{i}\frac{2r_0}{3c}\Omega^3. \tag{2A.34}$$

We consider only frequencies Ω such that $\Omega \ll c/r_0 \approx 10^{23}\ \mathrm{s}^{-1}$. In this case,[4] Ω is close to ω_0 and we can seek a power series solution of the form

$$\Omega = \omega_0 + \delta\Omega^{(1)} + \delta\Omega^{(2)} + ... \tag{2A.35}$$

The first-order term is then found to be

$$\delta\Omega^{(1)} = -\mathrm{i}\frac{r_0}{3c}\omega_0^2. \tag{2A.36}$$

The motion of the electron is therefore a damped oscillation:

$$\mathbf{r} = \boldsymbol{\mathcal{S}}_0 e^{-\Gamma_{\mathrm{cl}}t/2}\mathrm{e}^{-\mathrm{i}\omega_0 t} + c.c., \tag{2A.37}$$

with

$$\Gamma_{\mathrm{cl}} = \frac{2r_0}{3c}\omega_0^2 = \frac{4\pi}{3}\frac{r_0}{\lambda_0}\omega_0, \tag{2A.38}$$

in which we have introduced the wavelength $\lambda_0 = 2\pi c/\omega_0$. The radiated power is therefore likewise damped with a time constant $\Gamma_{\mathrm{cl}}^{-1}$, which is known as the *classical radiative lifetime*.

It is interesting to consider the implications of (2A.38) as far as the radiative decay of atoms is concerned. Firstly, for an atom radiating in the visible spectral region, we have $\omega_0/2\pi \approx 5 \times 10^{14}$ Hz corresponding to $\lambda_0 \approx 600$ nm, so that the radiative lifetime is $\Gamma_{\mathrm{cl}}^{-1} \approx 16$ ns, a result close to the experimental value. In the X-ray spectral region, the radiative lifetime is much shorter since Γ_{cl} varies as ω_0^2. Notice, however, that even in this domain the *quality factor* (or *Q-factor*), $\omega_0/\Gamma_{\mathrm{cl}}$, which characterizes the number of oscillations completed by the electron before the damping of its motion becomes significant, remains very much larger than one.

Comments

(i) Here again the result obtained from this classical treatment is in agreement with the result of a quantum mechanical model of a spontaneously radiating two-level atom. The quantum

[4] The domain of validity of the theory presented in this complement is, in fact, limited to frequencies much smaller than c/r_0. For example, relativistic quantum effects impose a limit $\Omega \ll mc^2/\hbar = \alpha c/r_0$ where $\alpha \approx 1/137$ is the fine-structure constant.

mechanical probability of finding the atom in the excited state decays exponentially in time, the radiative lifetime characterizing this decay being broadly in agreement with that given by (2A.38) (see Chapter 6).

(ii) When the electron traces out a circular trajectory in the xOy plane, its radiation leads to a loss of energy at the rate given by expression (2A.32), but also to a *loss of angular momentum*. The rate of change of its angular momentum $\mathrm{d}\mathbf{L}/\mathrm{d}t$ is given by

$$\frac{\mathrm{d}\mathbf{L}}{\mathrm{d}t} = \mathbf{r} \times F_{\mathrm{RR}}. \tag{2A.39}$$

The coordinates of the electron at time t being $x = a\cos\omega t/\sqrt{2}$ and $y = a\sin\omega t/\sqrt{2}$ (rotation in the positive sense), one finds from Equations (2A.31) and (2A.39) that

$$\frac{\mathrm{d}L_z}{\mathrm{d}t} = -\frac{mr_0}{3c}a^2\omega^3 = -\frac{q^2a^2\omega^3}{12\pi\varepsilon c^3}. \tag{2A.40}$$

A comparison of this result with Equation (2A.32) reveals that the radiative rate of loss of angular momentum is proportional to that of energy:

$$\frac{\mathrm{d}L_z}{\mathrm{d}t} = \frac{1}{\omega}\frac{\mathrm{d}W}{\mathrm{d}t}. \tag{2A.41}$$

Thus, the total energy and the total angular momentum radiated by the electron before its motion is completely damped are related by

$$L_z = \frac{W}{\omega}, \tag{2A.42}$$

which is obtained by integrating (2A.41) over time.

This angular momentum, lost by the rotating electron has been transferred to the circularly polarized electromagnetic field. This result, which was obtained from a purely classical argument, also turns out to be consistent with a quantum mechanical approach: in undergoing a transition from an excited state to a state of lower energy, the total energy radiated W is just that of the photon $\hbar\omega$ and, therefore, the *angular momentum of the σ_+ photon emitted has a component L_z along Oz of magnitude $\hbar$.*

The same reasoning, starting from the assumption that the electron rotates in the xOy plane in the negative sense, leads to an equality analogous to (2A.42), but with a change of sign and thus to the complementary conclusion that a σ_- photon has an angular momentum of $-\hbar$ along Oz.

We shall return in Complement 4B to the general problem of the angular momentum of light.

2A.4 Response to an external electromagnetic wave

Suppose that the electric field of the incident electromagnetic wave is $\mathbf{E}(t) = \mathbf{E}_0\cos\omega t$. In an electromagnetic wave the effect of the magnetic field is of order v/c times that of the electric field and can therefore be neglected. The motion of the electron can then be found by solving the differential equation:

$$m\ddot{\mathbf{r}} = -m\omega_0^2\mathbf{r} + \frac{2mr_0}{3c}\dddot{\mathbf{r}} + q\mathbf{E}_0\cos\omega t. \tag{2A.43}$$

This equation has a forced solution oscillating at the driving field frequency ω, which is, in complex notation (with $\mathcal{E}_0 = \mathbf{E}_0/2$),

$$\mathcal{S}_0 = \frac{q\mathcal{E}_0}{m} \frac{1}{\omega_0^2 - \omega^2 - \mathrm{i}\Gamma_{\mathrm{cl}}\omega^3/\omega_0^2}. \tag{2A.44}$$

This solution can be used to calculate the field radiated by the atom in all directions, i.e. the scattering of the incident radiation, but also to determine the induced atomic dipole, and therefore the atomic susceptibility.

Rayleigh, Thomson and resonant scattering

From (2A.44) one can deduce the scattered power Φ_{d} with the help of Equation (2A.26):

$$\Phi_{\mathrm{d}} = \frac{q^4\mathbf{E}_0^2}{12\pi\varepsilon_0 c^3 m^2} \frac{\omega^4}{(\omega_0^2 - \omega^2)^2 + \Gamma_{\mathrm{cl}}^2\omega^6/\omega_0^4}. \tag{2A.45}$$

The scattering cross-section $\sigma(\omega)$ is defined as the ratio of the scattered power and the incident energy flux:

$$\Phi_{\mathrm{d}} = \frac{\mathrm{d}\Phi_{\mathrm{i}}}{\mathrm{d}S}\sigma(\omega), \tag{2A.46}$$

where $\mathrm{d}\Phi_{\mathrm{i}}/\mathrm{d}S$ is the incident flux, or, more precisely, the average incident power per unit surface area:

$$\frac{\mathrm{d}\Phi_{\mathrm{i}}}{\mathrm{d}S} = \Pi_{\mathrm{i}} = \frac{1}{2}\varepsilon_0 c\mathbf{E}_0^2 = 2\varepsilon_0 c\mathcal{E}_0 \cdot \mathcal{E}_0^*. \tag{2A.47}$$

From the above we deduce the general expression of the scattering cross-section:

$$\sigma(\omega) = \frac{8\pi r_0^2}{3} \frac{\omega^4}{(\omega_0^2 - \omega^2)^2 + \Gamma_{\mathrm{cl}}^2\omega^6/\omega_0^4}. \tag{2A.48}$$

We shall now discuss this expression in different ranges of the driving field frequency.

Low frequency limit: Rayleigh scattering

Consider first of all the limit $\omega \ll \omega_0$. The cross-section is then equal to

$$\sigma(\omega) = \frac{8\pi r_0^2}{3}\left(\frac{\omega}{\omega_0}\right)^4. \tag{2A.49}$$

It increases rapidly with the increasing frequency of the incident radiation. It is this limit that is concerned with the scattering of solar radiation by the atmosphere, the resonance transitions of which fall in the ultra-violet spectral region, and which are responsible for the blue colour of the sky.[5]

[5] In fact the violet component of the visible spectrum is scattered even more strongly than the blue component, but the decreasing sensitivity of the human eye at short wavelengths on the one hand, and the decreasing solar emission in the same spectral region on the other, result in a sky that is apparently blue and not violet.

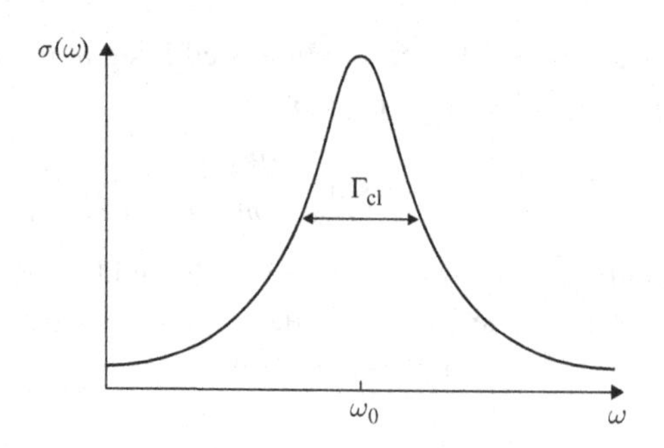

Figure 2A.4 **Frequency dependence of the cross-section for resonant scattering. The cross-section has a Lorentzian frequency dependence, of full width at half maximum Γ_{cl}.**

High frequency limit: Thomson scattering

Consider now the case $\omega \ll \omega_0$. In this limit the cross-section tends to a constant value given by:[6]

$$\sigma(\omega) = \frac{8\pi r_0^2}{3} \approx 6.5 \times 10^{-29}\ \mathrm{m}^2. \tag{2A.50}$$

This is the regime that concerns the scattering of X-rays by matter, and which played an important role in the early years of the twentieth century in enabling the number of electrons in certain atomic species to be deduced from measurements of their absorption of X-rays.

Resonant scattering

Finally, we consider the situation in which the frequency of the incident light is very close to the resonant frequency of the scattering dipole, $\omega \approx \omega_0$ (Figure 2A.4). On keeping only the terms of lowest order in (2A.48), using the fact that Γ_{cl} is much smaller than ω_0 and applying relation (2A.38) we obtain the remarkable formula:

$$\sigma(\omega) = \frac{3\lambda_0^2}{2\pi}\frac{\Gamma_{cl}^2/4}{(\omega_0-\omega)^2+\Gamma_{cl}^2/4}. \tag{2A.51}$$

The cross-section is therefore resonantly enhanced at frequencies close to ω_0, the resonance having a Lorentzian frequency dependence of full width at half height Γ_{cl}. Notice that the cross-section at exact resonance, $3\lambda_0^2/2\pi$ is of the order of the square of the resonant wavelength. This cross-section is very much larger than that found in the Thomson scattering regime, by eighteen orders of magnitude in the visible spectral region. The resonance lineshape is very sharp indeed!

[6] Notice that the validity of this model is limited to the situation $\Gamma_{cl}\omega/\omega_0^2 \ll 1$, so that the second term in the denominator can be neglected ($\omega_0^2/\Gamma_{cl} \approx 10^{23}\ \mathrm{s}^{-1}$ in the optical domain).

As we shall see in Chapter 6, the results found above are in good agreement with those predicted by quantum mechanics. The results in the Rayleigh and Thomson scattering regimes are identical with the quantum mechanical ones, whilst in the case of resonant scattering it is sufficient to replace the classical linewidth, Γ_{cl}, by the natural width Γ_{sp} of the transition concerned, the value on resonance, $3\lambda_0^2/2\pi$ being unchanged.

Atomic susceptibility

From Equation (2A.44) we can derive an expression for the dielectric susceptibility of a medium composed of 'classical atoms' of the kind described above by the elastically bound electron model. If the number density of the atoms is N/V, the polarization induced in the medium is, in the complex notation,

$$\mathcal{P} = \frac{N}{V}\frac{q^2\mathcal{E}_0 e^{-i\omega t}}{m}\frac{1}{\omega_0^2 - \omega^2 - i\Gamma_{cl}\omega^3/\omega_0^2}. \tag{2A.52}$$

Using the definition (2.174) for the susceptibility in the complex notation we find

$$\chi(\omega) = \frac{N}{V}\frac{q^2}{m\varepsilon_0}\frac{1}{\omega_0^2 - \omega^2 - i\Gamma_{cl}\omega^3/\omega_0^2}, \tag{2A.53}$$

so that the real and imaginary parts of the susceptibility are

$$\chi'(\omega) = \frac{N}{V}\frac{q^2}{m\varepsilon_0}\frac{\omega_0^2 - \omega^2}{(\omega_0^2 - \omega^2)^2 + \Gamma_{cl}^2\omega^6/\omega_0^4} \tag{2A.54}$$

$$\chi''(\omega) = \frac{N}{V}\frac{q^2}{m\varepsilon_0}\frac{\Gamma_{cl}\omega^3/\omega_0^2}{(\omega_0^2 - \omega^2)^2 + \Gamma_{cl}^2\omega^6/\omega_0^4}. \tag{2A.55}$$

Not far from the resonance, that is when $\Gamma_{cl} \ll |\omega - \omega_0| \ll \omega_0$, the real part of the susceptibility takes the simple form

$$\chi'(\omega) = \frac{N}{V}\frac{q^2}{2m\omega_0\varepsilon_0}\frac{1}{\omega_0 - \omega} = \frac{\chi'_q}{f_{ab}}, \tag{2A.56}$$

where χ'_q is the real part of the susceptibility obtained from quantum theory in the quasi-resonant approximation (Equation 2.177), and where f_{ab} is a dimensionless multiplicative factor known as the *oscillator strength* of the transition:

$$f_{ab} = \frac{2m\omega_0 z_{ab}^2}{\hbar}. \tag{2A.57}$$

In addition, using the expressions of r_0 and Γ_{cl}, the imaginary part of the susceptibility can be re-expressed in the form

$$\chi''(\omega) = \frac{N}{V}\frac{c}{\omega}\sigma(\omega), \tag{2A.58}$$

in which the cross-section $\sigma(\omega)$ is given by expression (2A.48). Equation (2A.58) shows the close relationship that exists between the absorption coefficient for the incident wave (which is proportional to χ'') and the scattering cross-section $\sigma(\omega)$. This result is actually much more generally applicable than the model used here to derive it;[7] it is known as the *optical theorem*.

2A.5 Relationship between the classical atomic model and the quantum mechanical two-level atom

In this section we shall study the relationship that exists between the predictions of the classical model outlined above and those of the quantum mechanical model of a *two-level atom*. We assume, as usual, that the energy separation of the ground and excited states $|a\rangle$ and $|b\rangle$ of the quantum mechanical atom is $E_b - E_a = \hbar\omega_0$, and we employ the semi-classical formalism introduced in Chapter 2. In the electric dipole approximation the Hamiltonian of the system is given by

$$\hat{H} = \hat{H}_0 - q\hat{\mathbf{r}} \cdot \mathbf{E}_0 \cos \omega t, \tag{2A.59}$$

where $\hat{H}_0$ is the atomic Hamiltonian. Choosing the zero of energy to coincide with the energy of the ground state and restricting the interaction Hamiltonian to the sub-space spanned by the two states $|a\rangle$ and $|b\rangle$, we can re-express the total system Hamiltonian in the form

$$\hat{H} = \hbar\omega_0 \begin{bmatrix} 0 & 0 \\ 0 & 1 \end{bmatrix} - qz_{\mathrm{ab}} E_0 \cos \omega t \begin{bmatrix} 0 & 1 \\ 1 & 0 \end{bmatrix}, \tag{2A.60}$$

where we have assumed, in addition, that $\mathbf{E}_0$ is parallel to Oz and that $z_{\mathrm{ab}} = \langle a|\hat{z}|b\rangle$ is real. This Hamiltonian, like any 2×2 matrix, can be rewritten in terms of the unit 2×2 matrix and the Pauli matrices $\hat{\sigma}_x, \hat{\sigma}_y$ and $\hat{\sigma}_z$,[8]

$$\hat{\sigma}_x = \begin{bmatrix} 0 & 1 \\ 1 & 0 \end{bmatrix} \quad \hat{\sigma}_y = \begin{bmatrix} 0 & -\mathrm{i} \\ \mathrm{i} & 0 \end{bmatrix} \quad \hat{\sigma}_z = \begin{bmatrix} 1 & 0 \\ 0 & -1 \end{bmatrix}, \tag{2A.61}$$

satisfying the commutation relation

$$\left[\hat{\sigma}_x, \hat{\sigma}_y\right] = 2\mathrm{i}\hat{\sigma}_z \tag{2A.62}$$

and its cyclic permutations. The Hamiltonian $\hat{H}$ can be written as

$$\hat{H} = \frac{\hbar\omega_0}{2}(\hat{1} - \hat{\sigma}_z) - qz_{\mathrm{ab}}\hat{\sigma}_x E_0 \cos \omega t. \tag{2A.63}$$

In order to make a comparison with the classical model outlined above, we need to determine the equation that governs the *mean value of the position* $\mathbf{r}$ of the atomic electron, or, more precisely, of its z-coordinate. To this end we apply Ehrenfest's theorem:

$$\frac{\mathrm{d}}{\mathrm{d}t}\langle\hat{z}\rangle = \frac{1}{\mathrm{i}\hbar}\langle\left[\hat{z}, \hat{H}\right]\rangle = \frac{z_{\mathrm{ab}}}{\mathrm{i}\hbar}\langle\left[\hat{\sigma}_x, \hat{H}\right]\rangle = z_{\mathrm{ab}}\omega_0\langle\hat{\sigma}_y\rangle, \tag{2A.64}$$

[7] See, for example, J. D. Jackson, *Classical Electrodynamics*, Wiley (1998).
[8] See, for example, CDL, Complement A_{IV}.

from which we deduce the following expression for the second time derivative of $\langle\hat{z}\rangle$,

$$m\frac{\mathrm{d}^2}{\mathrm{d}t^2}\langle\hat{z}\rangle = mz_{\mathrm{ab}}\omega_0\frac{\mathrm{d}}{\mathrm{d}t}\langle\hat{\sigma}_y\rangle = \frac{mz_{\mathrm{ab}}\omega_0}{\mathrm{i}\hbar}\langle\left[\hat{\sigma}_y,\hat{H}\right]\rangle. \tag{2A.65}$$

Given the form (2A.63) of $\hat{H}$, the commutator has two non-zero terms so that

$$m\frac{\mathrm{d}^2}{\mathrm{d}t^2}\langle\hat{z}\rangle = -m\omega_0^2\langle\hat{z}\rangle + \frac{2mz_{\mathrm{ab}}^2\omega_0}{\hbar}\langle\hat{\sigma}_z\rangle qE_0\cos\omega t. \tag{2A.66}$$

Finally, we introduce into this expression the oscillator strength of the transition f_{ab} (given by relation (2A.57)) and the classical force acting on the electron, $F_{\mathrm{cl}} = qE_0\cos\omega t$, and obtain

$$m\frac{\mathrm{d}^2}{\mathrm{d}t^2}\langle\hat{z}\rangle = -m\omega_0^2\langle\hat{z}\rangle + f_{\mathrm{ab}}\langle\hat{\sigma}_z\rangle F_{\mathrm{cl}}. \tag{2A.67}$$

This expresses Newton's third law as applied to the motion of an elastically bound atomic electron, with the distinction that the z-coordinate appearing in the equivalent classical expression is replaced here by its expectation value. In addition, the force exerted on the electron by the incident electromagnetic wave is modified by two factors, the first being the oscillator strength, a quantity which depends on the nature of the transition considered. As far as the resonance transitions of the alkali metals are concerned, for example, this oscillator strength is very close to unity. The second factor $\langle\hat{\sigma}_z\rangle$, is proportional to $P_a - P_b$ where P_a and P_b are, respectively, the probabilities of finding the atom in its ground or in its excited state. In the perturbative non-saturating case considered here, P_b is small so that $\langle\hat{\sigma}_z\rangle \approx 1$.

Thus, apart from the correction related to the oscillator strength, the quantum theory of the two-level atom that we have considered has reproduced the equation that served as our starting point in the discussion of the classical model. It is therefore evident that the results of the classical model are accurate in the perturbative regime. If the system is strongly excited, however, it fails since the $\langle\hat{\sigma}_z\rangle$ term falls to zero if the populations in the two atomic levels are equalized (saturation), and even changes sign if a population inversion is induced. This is, in particular, the case in amplifying media and lasers, which are not correctly described by the Lorentz model.

2B

Complement 2B Selection rules for electric dipole transitions. Applications to resonance fluorescence and optical pumping

2B.1 Selection rules and the polarization of light

2B.1.1 Forbidden electric dipole transitions

As we saw in Chapter 1, radiation can only induce a transition between two levels $|a\rangle$ and $|b\rangle$, to first order in perturbation theory, if the interaction Hamiltonian $\hat{H}_{\mathrm{I}}$ has a non-zero matrix element between these two states:

$$\langle b|\hat{H}_{\mathrm{I}}|a\rangle \neq 0. \tag{2B.1}$$

When this matrix element is zero it is often for fundamental reasons of symmetry. For example, two states of the same parity are not coupled by the electric dipole Hamiltonian, which is itself odd. This can straightforwardly be verified for a one-electron atom for which the electric dipole Hamiltonian can be written

$$\hat{H}_{\mathrm{I}} = -\hat{\mathbf{D}}.E(\mathbf{0}, t) = -q\hat{\mathbf{r}}.E(\mathbf{0}, t), \tag{2B.2}$$

where $\hat{\mathbf{r}}$ is the electron position operator. If ψ_a and ψ_b have the same parity then the function $\psi_b^*(\mathbf{r})\psi_a(\mathbf{r})$ is even, so that

$$\langle b|\hat{H}_{\mathrm{I}}|a\rangle = -\langle b|\hat{\mathbf{D}}.E(\mathbf{0}, t)|a\rangle = -q\mathbf{E}(\mathbf{0}, t).\int \mathrm{d}^3r\ \mathbf{r}\ \psi_b^*(\mathbf{r})\ \psi_a(\mathbf{r}) = 0, \tag{2B.3}$$

since we have integrated an odd function over all space. We say that a transition between two levels of the same parity is *forbidden* at least as far as the electric dipole coupling is concerned. This is an example of a *selection rule.*

Comments

(i) The operator $\hat{\mathbf{p}}$ is also odd since each of the components of the gradient operator satisfies

$$\frac{\partial}{\partial x} = -\frac{\partial}{\partial(-x)}. \tag{2B.4}$$

As might be expected, a transition that is forbidden for the electric dipole coupling is also forbidden for the **A.p** coupling.

(ii) It should be remembered that in the case that the electric dipole coupling between two states is zero, it is not necessarily true that transitions between them, leading to the absorption or emission of light, are impossible. The electric dipole interaction term (and the equivalent term **A.p** is simply the dominant term of a series expansion of the interaction Hamiltonian (Section 2.2.4).

Very often a transition that is electric dipole forbidden is allowed by the magnetic dipole (Section 2.2.5) or electric quadrupole couplings or can occur by a multi-photon process (compare Sections 2.1.5 and 2.3.3). However, the corresponding matrix elements are smaller by several orders of magnitude and so we can neglect these processes whenever the electric dipole transition is permitted.

(iii) If the nucleus is at position $\mathbf{r}_0$ and not at the origin, the reasoning above remains valid. Replacing $\mathbf{r}$ by $\mathbf{r} - \mathbf{r}_0$ in (2B.3), Equation (2B.4) then becomes

$$\frac{\partial}{\partial x} = \frac{\partial}{\partial(x - x_0)} = -\frac{\partial}{\partial(x_0 - x)}. \tag{2B.5}$$

We shall meet other selection rules in this complement which restrict transitions to pairs of sub-levels with given angular momenta. These rules depend on the polarization of the incident light. Since the techniques for the production and analysis of polarized light are well developed, these selection rules have enabled the development of sensitive and powerful experimental methods. In *optical resonance* information is obtained from an analysis of the fluorescence emitted by atoms after they have been excited by quasi-resonant light. *Optical pumping* permits an assembly of atoms to be entirely pumped into a given set of Zeeman sub-levels, or to create a population inversion in a given pair of levels, both situations in which the atoms are far from thermal equilibrium. Although originally employed only in atomic ground states, optical pumping has also been used in widely separated pairs of energy levels permitting laser amplification to be obtained (see Chapter 3).

In this complement we consider only the case of a one-electron atom, for which the selection rules are easy to establish. It should be realized, however, that the results we shall derive are of quite general validity, since they can be established from symmetry arguments independent of the model that we shall use here.[1]

2B.1.2 Linearly polarized light

We consider a (hydrogenic) one-electron atom which we take to be at the origin of coordinates and suppose that it interacts with light linearly polarized along Oz for which the electric field at the origin is

$$\mathbf{E}(\mathbf{0}, t) = \mathbf{e_z}\, E_0 \cos(\omega t + \varphi), \tag{2B.6}$$

$\mathbf{e}_z$ being the unit vector in the z-direction. The interaction Hamiltonian can then be put in the form

$$\hat{H}_\text{I}(t) = \hat{W} \cos(\omega t + \varphi), \tag{2B.7}$$

with

$$\hat{W} = -qE_0\, \hat{\mathbf{r}}.\mathbf{e_z} = -qE_0\, \hat{z} = -\hat{\mathbf{D}}.\mathbf{e}_z E_0. \tag{2B.8}$$

We wish to ascertain which are the possible transitions from the atomic ground state:

$$|a\rangle = |n = 1, l = 0, m = 0\rangle\,, \tag{2B.9}$$

[1] A. Messiah, *Quantum Mechanics*, Dover (1999).

n being the principal quantum number, l the orbital angular momentum quantum number and m the magnetic quantum number.[2] For an arbitrary excited state defined by the quantum numbers n, l, m,

$$|b\rangle = |n,l,m\rangle\,, \tag{2B.10}$$

the transition matrix element is

$$W_{\mathrm{ba}} = \langle b|\hat{W}|a\rangle = -qE_0\,\langle b|\hat{z}|a\rangle. \tag{2B.11}$$

The form, in spherical polar coordinates, of the wavefunctions for a single-electron atom is[2]

$$\psi_{nlm}(r,\theta,\phi) = R_{nl}(r)Y_l^m(\theta,\phi). \tag{2B.12}$$

This expression involves the spherical harmonics $Y_l^m(\theta,\phi)$. The matrix element (2B.11) then takes the form

$$W_{\mathrm{ba}} = -qE_0 \int r^2 \sin\theta \;\mathrm{d}r\;\mathrm{d}\theta\;\mathrm{d}\phi\; R_{nl}^*(r)Y_l^{m^*}(\theta,\phi)\, z\, R_{10}(r)Y_0^0(\theta,\phi). \tag{2B.13}$$

But transformed to spherical coordinates, z becomes

$$z = r\cos\theta = r\sqrt{\frac{4\pi}{3}}\,Y_1^0(\theta,\phi). \tag{2B.14}$$

Also,

$$Y_0^0(\theta,\phi) = \frac{1}{\sqrt{4\pi}}, \tag{2B.15}$$

which permits the separation from the integral (2B.13) of the angular part:

$$I_{\mathrm{ang}} = \iint \sin\theta\;\mathrm{d}\theta\;\mathrm{d}\varphi\; Y_l^{m^*}(\theta,\phi)Y_1^0(\theta,\phi), \tag{2B.16}$$

which is simply the overlap integral of two spherical harmonics. Using the orthogonality relations of the spherical harmonics we obtain[2]

$$I_{\mathrm{ang}} = \delta_{l1}\delta_{m0} \tag{2B.17}$$

and

$$W_{\mathrm{ba}} = -q\frac{E_0}{\sqrt{3}}\delta_{l1}\delta_{m0}\int_0^\infty \mathrm{d}r\; r^3\; R_{nl}^*(r)R_{10}(r). \tag{2B.18}$$

Thus we see that only the states $|b\rangle = |n, l=1, m=0\rangle$ can be excited from the ground state $|1,0,0\rangle$ by linearly polarized light. Similarly the only states that can de-excite to the ground state under the influence of z-polarized light are of this form. The same selection rules therefore apply to absorption and stimulated emission.

If, instead of starting with an atom in the ground state, we started with an atom in an arbitrary state $|i\rangle$, we could have shown, again using the properties of spherical harmonics,

[2] See, for example, CDL1, Chapters VI and VII.

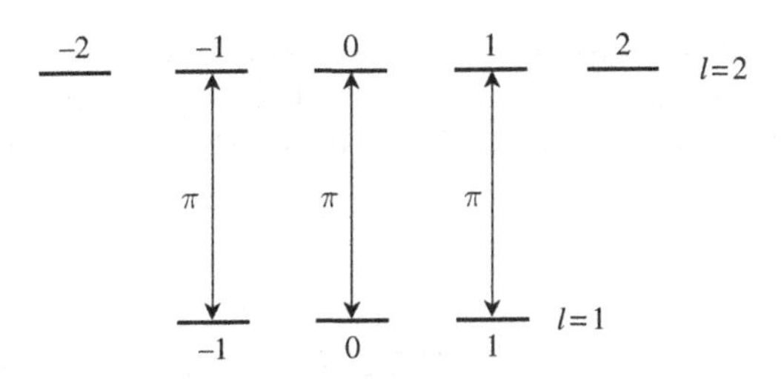

Figure 2B.1 π-transitions between a level with $l = 1$ and another with $l = 2$. For light polarized linearly parallel to the quantization axis only the indicated transitions are possible.

that the following selection rules involving the quantum numbers of the state $|i\rangle$ and the final state $|k\rangle$ are obeyed:

$$\left\{ \begin{array}{c} \text{linear polarization} \\ \text{along the quantization axis} \end{array} \right\} \Rightarrow \left\{ \begin{array}{c} l_k - l_i = \pm 1 \\ m_k = m_i. \end{array} \right. \tag{2B.19}$$

Such a transition, induced by light linearly polarized along the quantization axis, is known as a π-transition. It is shown by a vertical line on a diagram in which the Zeeman sub-levels associated with the same magnetic quantum number are shown aligned vertically (Figure 2B.1).[3]

Comments

(i) The magnetic quantum numbers m_i and m_k characterize the component, l_z, of the angular momentum of the atom about the quantization axis, Oz. The selection rule (2B.19) only applies, however, if the light polarization is parallel to this axis. If the polarization were chosen along an axis $O\xi$, the selection rule would apply to Zeeman sub-levels associated with the same axis, that is the eigenstates of the operator

$$\hat{L}_\xi = \hat{\mathbf{L}}.\mathbf{e}_\xi, \tag{2B.20}$$

where $\mathbf{e}_\xi$ is the unit vector along $O\xi$.

(ii) A wave that is linearly polarized along Oz necessarily propagates along an axis perpendicular to Oz.

(iii) For a many-electron atom, the atomic states are characterized by the quantum numbers J and m_J associated, respectively, with the operators for the total electronic angular momentum $\hat{\mathbf{J}}$ and its projection on the z-axis, $\hat{J}_z$. It is possible to show that, as a consequence of the invariance under rotation of the potential in which the electrons evolve, the selection rules for transitions coupled to light linearly polarized along the z-axis are

$$\left\{ \begin{array}{c} \text{linear polarization} \\ \text{along the quantization axis} \end{array} \right\} \Rightarrow \left\{ \begin{array}{c} J_k - J_i = \pm 1 \text{ or } 0 \\ m_k = m_i. \end{array} \right. \tag{2B.21}$$

These selection rules can be interpreted as a statement of the conservation of angular momentum in the absorption and emission of radiation, and demand that the value

$$m_z = 0 \tag{2B.22}$$

[3] We call a level an ensemble of states with the same energy. If n and l are fixed the $2l + 1$ states $|n, l, m\rangle$ with $-l \leq m \leq l$ have the same energy and are known as Zeeman sub-levels. This name is a reminder of the fact that this degeneracy is lifted in the presence of a magnetic field (Zeeman effect).

is attributed to the z-component of the angular momentum of a photon linearly polarized along Oz (such a wave propagates perpendicularly to Oz). If the atom has a nuclear spin I, the selection rules (2B.21) are generalized to this case by introducing the total angular momentum $\hat{\mathbf{F}} = \hat{\mathbf{J}} + \hat{\mathbf{I}}$ and its component $\hat{F}_z$.

2B.1.3 Circularly polarized light

Definition

Consider now an electromagnetic wave of which the electric field at $\mathbf{r} = 0$ is

$$\mathbf{E} = -\frac{E_0}{\sqrt{2}}\left[\mathbf{e_x}\cos(\omega t + \varphi) + \mathbf{e_y}\sin(\omega t + \varphi)\right], \tag{2B.23}$$

which can be written

$$\mathbf{E} = \mathrm{Re}\left\{\boldsymbol{\varepsilon}_+ \mathrm{E}_0\, \mathrm{e}^{-\mathrm{i}(\omega t + \varphi)}\right\}, \tag{2B.24}$$

with

$$\boldsymbol{\varepsilon}_+ = -\frac{\mathbf{e_x} + \mathrm{i}\mathbf{e_y}}{\sqrt{2}}. \tag{2B.25}$$

This electric field rotates at angular frequency ω about Oz in the right-hand sense. This field, circularly polarized in the (xOy) plane, is termed σ_+ polarized with respect to the Oz axis.

Light that is circularly polarized in the (xOy) plane, but which rotates in the opposite sense is known as σ_- polarized light. The electric field of σ_- polarized light is obtained by replacing, in (2B.24), the complex polarization vector $\boldsymbol{\varepsilon}_+$ by the vector

$$\boldsymbol{\varepsilon}_- = \frac{\mathbf{e_x} - \mathrm{i}\mathbf{e_y}}{\sqrt{2}}. \tag{2B.26}$$

Comments

(i) The field (2B.24), rotating in the xOy plane, belongs to a wave propagating along Oz. In the case in which the wave propagates in the positive z-direction the electric field describes a right-handed corkscrew (right-handed helicity). In the converse case, in which the wave propagates towards negative z, the helicity is left-handed. Whatever the sense of propagation along the z-axis the wave is σ_+ circularly polarized with respect to the (oriented) z-axis which has been used to define the Zeeman sub-levels of the atom. Helicity and circular polarization should not be confused.

(ii) The unit vectors $\mathbf{e}_+$ and $\mathbf{e}_-$ which characterize the circular polarizations σ_+ and σ_- are defined to within a multiplicative constant of modulus one. The choices made in (2B.25) and (2B.26) correspond to the standard definition of the components of a vector operator. This choice facilitates the use of general theorems such as the Wigner–Eckart theorem.[4]

[4] A. Messiah, *Quantum Mechanics*, Dover (1999).

Selection rules

For a σ_+ circularly polarized wave, given by (2B.24), the electric dipole interaction Hamiltonian can be written

$$\hat{H}_{\mathrm{I}} = \frac{qE_0}{\sqrt{2}}\left[\hat{x}\cos(\omega t + \varphi) + \hat{y}\sin(\omega t + \varphi)\right]. \tag{2B.27}$$

In this formula the resonant (in $e^{-i\omega t}$) and anti-resonant (in $e^{+i\omega t}$) parts can be identified on writing

$$\hat{H}_{\mathrm{I}} = \frac{1}{2}\hat{W}e^{-i(\omega t+\varphi)} + \frac{1}{2}\hat{W}^{\dagger}e^{i(\omega t+\varphi)}, \tag{2B.28}$$

where

$$\hat{W} = \frac{qE_0}{\sqrt{2}}(\hat{x} + i\hat{y}) = -qE_0\,\mathbf{r}\cdot\hat{\boldsymbol{\varepsilon}}_+ = -\hat{\mathbf{D}}\cdot\boldsymbol{\varepsilon}_+\,E_0. \tag{2B.29}$$

The similarity of these equations (2B.27–2B.28) and those of (2B.7–2B.8) is striking; the difference is entirely contained in the polarization vector $\mathbf{e}_+$ which is complex, whilst $\boldsymbol{\varepsilon}_z$ is not. Here $\hat{W}^{\dagger}$ is no longer a self-adjoint operator, and is therefore distinct from its complex conjugate $\hat{W}$. The selection rules for σ_+ polarization can be obtained by studying the transition matrix element associated with the resonant part of (2B.28):

$$W_{\mathrm{ba}} = \langle b|\hat{W}|a\rangle = \frac{qE_0}{\sqrt{2}}\langle b|\hat{x} + i\hat{y}|a\rangle. \tag{2B.30}$$

As in Section 2B.1.2 of this complement we shall look for the possible transitions to levels $|b\rangle = |n, l, m\rangle$ from the $|a\rangle = |1, 0, 0\rangle$ ground state of a hydrogenic atom. The procedure is as before: we write the wavefunctions ψ_a and ψ_b in terms of spherical harmonics, and use the relation

$$\frac{x + iy}{\sqrt{2}} = \frac{r}{\sqrt{2}}\sin\theta\, e^{i\varphi} = -r\sqrt{\frac{4\pi}{3}}Y_1^1(\theta, \phi). \tag{2B.31}$$

Finding an expression for W_{ba} then involves evaluation of the angular integral of equation

$$I_{\mathrm{ang}} = \iint \sin\theta\, d\theta\, d\varphi\; Y_l^{m^*}(\theta, \phi)Y_1^1(\theta, \phi) = \delta_{l1}\delta_{m1}. \tag{2B.32}$$

We finally obtain

$$W_{\mathrm{ba}} = -\frac{qE_0}{\sqrt{3}}\delta_{l1}\delta_{m1}\int_0^{\infty} dr\; r^3 R_{nl}^*(r)R_{10}(r), \tag{2B.33}$$

which closely resembles (2B.18). The important distinction is that only excited states $|b\rangle = |n, l = 1, m = 1\rangle$ with a magnetic quantum number $m = 1$ are coupled to the $m = 0$ ground state by σ_+ polarized light, whilst this quantum number was zero for the final states excited from $m = 0$ by π-polarized light.

By generalizing the above results, the selection rules can be deduced for transitions from a lower level which is not necessarily the hydrogenic ground state. It is then found that the

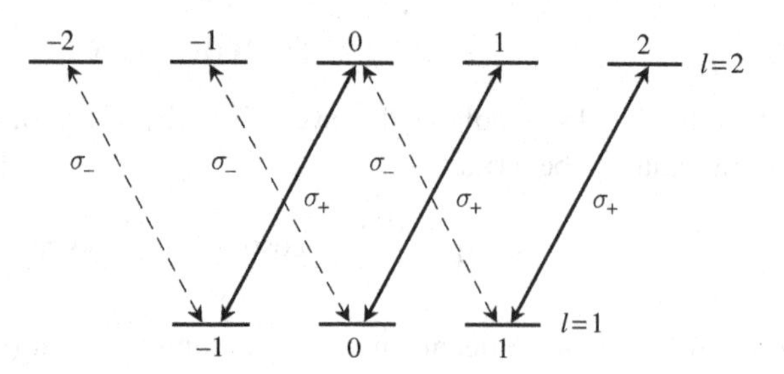

Figure 2B.2 Transitions induced by σ_+ (solid lines) and σ_- (dashed lines) circularly polarized light. The light polarization is defined relative to the axis used to define the magnetic quantum number m.

absorption of a photon, taking the atom from a level $|i\rangle$ to a higher level $|k\rangle$, is possible provided that

$$\left\{\begin{array}{c} E_k > E_i \\ \sigma_+ \text{ polarization} \end{array}\right\} \Rightarrow \left\{\begin{array}{l} l_k - l_i = \pm 1 \\ m_k = m_i + 1. \end{array}\right. \tag{2B.34}$$

If, on the other hand, we consider the possibility of a transition from an excited atomic level to one lower in energy accompanied by stimulated emission of σ_+ polarized light, we find that the magnetic quantum number m must decrease by one, with l, the angular momentum quantum number, also changing by one. With the choice of notation adopted above the selection rules in this case are identical to (2B.34).

Finally, in an energy level diagram in which the Zeeman sub-levels are aligned as described previously, transitions involving σ_+ light are indicated by diagonal lines rising to the right (Figure 2B.2).

Comments

(i) All the arguments concerning σ_+ circularly polarized light above apply equally well to light of σ_- polarization, for which the polarization vector $\boldsymbol{\varepsilon}_-$ is defined in (2B.26). The corresponding electric field is then

$$\mathbf{E} = \frac{E_0}{\sqrt{2}}\left[\mathbf{e_x}\cos(\omega t + \varphi) - \mathbf{e_y}\sin(\omega t + \varphi)\right], \tag{2B.35}$$

$$\mathbf{E} = \mathrm{Re}\left\{\boldsymbol{\varepsilon}_- E_0 \mathrm{e}^{-\mathrm{i}(\omega t + \varphi)}\right\}. \tag{2B.36}$$

We then obtain the following selection rules for transitions between a lower level $|i\rangle$ and an upper level $|k\rangle$ induced by the absorption of a σ_- photon:

$$\left\{\begin{array}{l} E_k > E_i \\ \sigma_- \text{ polarization} \end{array}\right\} \Rightarrow \left\{\begin{array}{l} l_k - l_i = \pm 1 \\ m_k = m_i - 1. \end{array}\right. \tag{2B.37}$$

In Figure 2B.2 such a transition is shown by a diagonal line rising to the left.

(ii) For a many-electron atom, for which the numbers J and m_j associated with the total electronic angular momentum $\mathbf{J}$ are good quantum numbers, the selection rules for transitions induced by light that is circularly polarized σ^+ or σ^- with respect to the Oz axis are

$$\left\{\begin{array}{c} E_k > E_i \\ \sigma_+ \text{ polarization} \end{array}\right\} \Rightarrow \left\{\begin{array}{c} J_k - J_i = \pm 1 \text{ or } 0 \\ m_k = m_i + 1 \end{array}\right. , \tag{2B.38}$$

(to be compared with Equation 2B.34), and similarly,

$$\left\{\begin{array}{c} E_k > E_i \\ \sigma_- \text{ polarization} \end{array}\right\} \Rightarrow \left\{\begin{array}{c} J_k - J_i = \pm 1 \text{ or } 0 \\ m_k = m_i - 1. \end{array}\right. \tag{2B.39}$$

For an atom of total angular momentum $\hat{\mathbf{F}} = \hat{\mathbf{I}} + \hat{\mathbf{J}}$, the selection rules are written in an analogous way with the quantum numbers F and m_F.

These selection rules can be interpreted as statements of the conservation of angular momentum since a $\sigma_\pm$ photon has an angular momentum of which the component along the quantization axis is $\pm\hbar$ (compare Complement 2A).

(iii) A linear polarization along a direction perpendicular to Oz, for example along Ox, is called a *linear* σ polarization. It can be considered as being composed of the superposition of a σ_+ and a σ_- polarized wave, as described by

$$\mathbf{e_x} = \frac{\boldsymbol{\varepsilon}_- - \boldsymbol{\varepsilon}_+}{\sqrt{2}} \tag{2B.40}$$

(see Equations (2B.25) and (2B.26)). Such a wave can excite transitions from a ground state $|a\rangle = |1, 0, 0\rangle$ to a state which mirrors the above superposition of polarizations:

$$|b\rangle = \frac{1}{\sqrt{2}} \left(|n, l = 1, m = -1\rangle - |n, l = 1, m = +1\rangle \right). \tag{2B.41}$$

It can be shown that this is identical to the state

$$|b\rangle = |n, l = 1, m_x = 0\rangle , \tag{2B.42}$$

that is to say the state associated with the eigenvalue $m_x = 0$ of the x-component, $\hat{L}_x$ of the orbital angular momentum, $\hat{\mathbf{L}}$. This result could have been obtained directly using the results of Section 2B.1.2, if we had chosen the quantization axis to lie along Ox.

It is important to realize that this decomposition of the linear polarization into two circularly polarized components is not just an academic exercise. Often it is useful to choose the quantization axis along the direction of propagation of the wave, which implies that a linear polarization is necessarily σ and not π.

2B.1.4 Spontaneous emission

Spontaneous emission can only be properly treated within the framework of a fully quantum mechanical theory of light (see Chapter 6). Here we limit ourselves to stating, without proof, the selection rules and the characteristics of the corresponding emission diagrams.

An atom in an excited state $|k\rangle$ of angular momentum (J_k, m_k) (m_k is the quantum number associated with the component $\hat{J}_z$ of $\hat{\mathbf{J}}$) can spontaneously de-excite to a state $|i\rangle$ of lower energy, characterized by angular momenta (J_i, m_i), provided that

$$J_i - J_k = \pm 1 \text{ or } 0 \tag{2B.43}$$

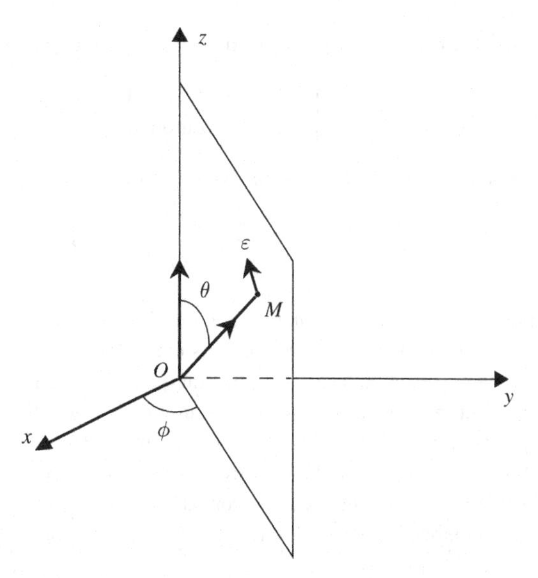

Figure 2B.3 **Spontaneous emission for a π-transition. Its characteristics are identical to those of the radiation emitted by a classical dipole oscillating along *Oz*. Radiation is emitted preferentially in directions perpendicular to *Oz*, with a linear polarization in a plane containing *Oz* and the emission direction.**

and

$$\begin{aligned} m_i &= m_k && (\pi \text{ transition}) \\ m_i &= m_k - 1 && (\sigma_+ \text{ transition}) \\ m_i &= m_k + 1 && (\sigma_- \text{ transition}). \end{aligned} \tag{2B.44}$$

These selection rules are shown diagramatically by merging Figures 2B.1 and 2B.2.

The angular distribution of the emitted radiation is given by a radiation diagram identical to that of the corresponding classical oscillating dipole (compare Complement 2A). This correspondence applies equally to the polarization of the radiation emitted in a given direction, as illustrated below.

Consider, for example, the above selection rules for the case of a π-transition (Figure 2B.3). The corresponding classical dipole oscillates along Oz. The distribution of the emitted radiation therefore varies as $\sin^2\theta$: it has a minimum of zero in the z-direction and a maximum in any direction perpendicular to this axis. The emitted light is *linearly polarized* in a direction given by the projection of the Oz-axis in a plane perpendicular to the direction of emission. This result is in agreement with the quantum treatment of spontaneous emission (Chapter 6).

In the case of a σ_+ or σ_- transition (see Figure 2B.4), the corresponding classical dipole rotates in the (xOy) plane, the sense of rotation being right-handed with respect to Oz for the σ_+ case and left-handed for the σ_- case. For emission along the z-axis, the light is circularly polarized in the sense corresponding to the rotation of the dipole. For emission perpendicular to Oz, the polarization is linear perpendicular to Oz and to the direction of emission. For any other emission direction the emitted light has an elliptical polarization,

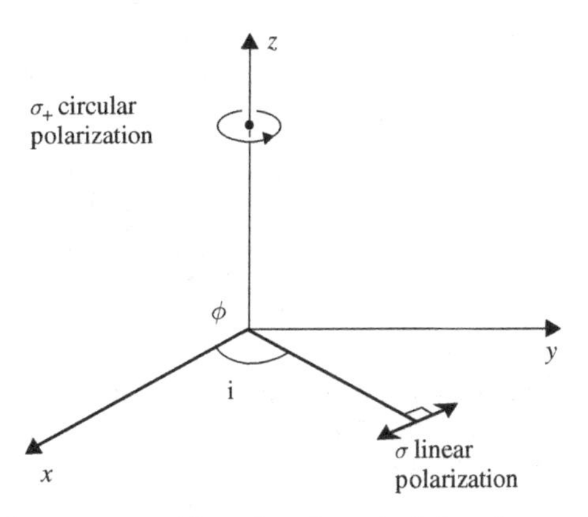

Figure 2B.4 **Spontaneous emission for a σ_+ transition. The characteristics of σ_+ polarized spontaneous emission are identical to the radiation emitted by a classical dipole rotating in the (xOy) plane. For emission in the z-direction the radiation is circularly polarized. For emission perpendicular to Oz it is linear in the plane (xOy) perpendicular to Oz (termed σ *linear* polarization). The radiation emitted in other directions is elliptically polarized.**

of which the characteristics are given by the projection of the rotating dipole on a plane perpendicular to the emission direction.

These features of spontaneous emission, in particular the manner in which the polarization of the emitted light is related to the change Δm_j of the magnetic quantum number, yield useful insights into the behaviour of atoms as will become apparent in the next sections of this complement.

2B.2 Resonance fluorescence

2B.2.1 Principle

Let us consider an atomic medium which is submitted to an electromagnetic wave that is quasi-resonant with an atomic transition linking the ground state to an excited state $|b\rangle$ of spontaneous decay rate Γ_{sp}. When the field is exactly on resonance, the light promotes the atoms to the excited state from which they de-excite by spontaneous emission. Resonant atomic fluorescence takes place (see Figure 2B.5). This is a consequence of the resonant scattering discussed in Section 2.1.4. This fluorescence is extremely bright, as $\Gamma_{sp}/2$ photons are emitted per second and per atom in the case of a saturated closed transition. It can even be observed if there is *a single atom interacting with the light beam*. Resonance fluorescence is thus the privileged means of experimentally monitoring single quantum systems having a closed transition and a significant decay rate Γ_{sp}.

Let us now assume that the ground and excited states $|a\rangle$ and $|b\rangle$ have, respectively, zero and one angular momentum ($J_a = 0, J_b = 1$), and that the incident light, which propagates in the Oy direction, is linearly polarized along Oz. We can apply the above selection rules by considering the process as comprising an excitation stage populating only the sub-state

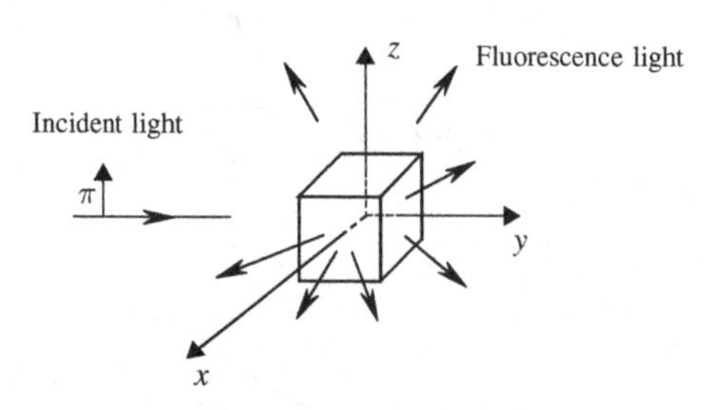

Figure 2B.5 Resonance fluorescence. A glass cell, containing an atomic vapour, is illuminated by light which is resonant with a transition from the atomic ground state. The vapour fluoresces, emitting light in all directions.

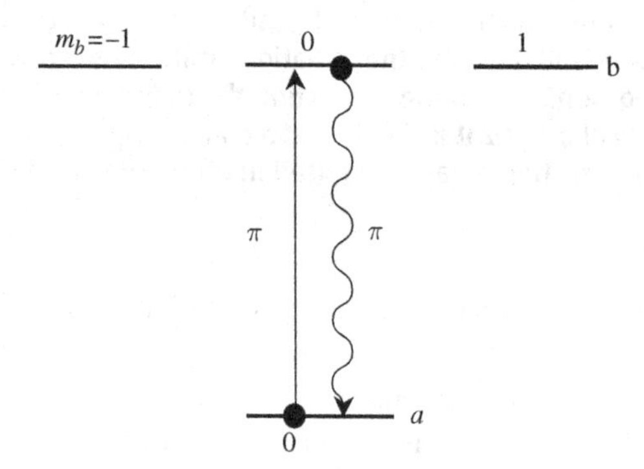

Figure 2B.6 Resonance fluorescence for an atom excited by π-polarized light. The process is resonant only for the $m_b = 0$ intermediate sub-level, so that the rescattered light is π-polarized.

$m_b = 0$ followed by a spontaneous emission stage (see Figure 2.4 and comment (i) of Section 2.1.4), which also takes place on a π-transition (Figure 2B.6). Consequently the fluorescent light observed in the Ox direction is found to be linearly polarized along Oz. This is easily verified by observing it through a polarizer; the detected light is extinguished when its transmission axis is perpendicular to the z-direction. Any alteration in the polarization of the emitted light can then be detected with great sensitivity, since it results in the appearance of light transmitted by the polarizer against zero background. In the following we shall give some examples of phenomena which do alter the polarization of the emitted light and which can therefore be studied employing this technique.

2B.2.2 Measurement of population transfers in the excited state

Suppose a mechanism were to cause the redistribution of population amongst the sub-levels of the excited state following the excitation of the atom. For example, in Figure 2B.7 the excitation of the atom to the $m_b = 0$ sub-level is followed by the transfer of population to the $m_b = \pm 1$ levels. The atom can then de-excite by the emission of σ_+ or σ_- polarized photons and the fluorescence will therefore have a component in the Oy direction that is

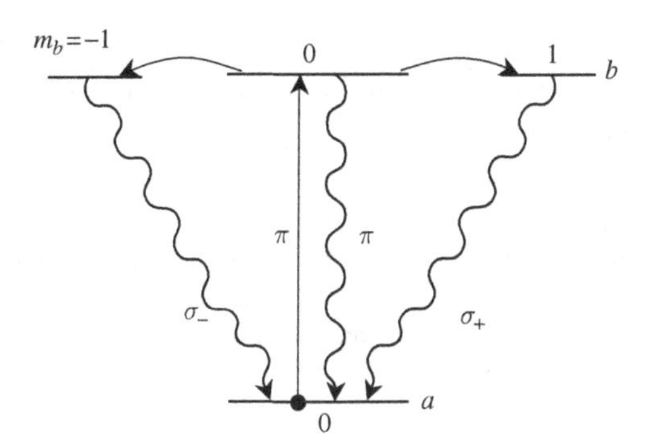

Figure 2B.7 Transfer of population within the excited state. Although the excitation, induced by a π-photon, only populates the $m_b = 0$ sub-level, the transfer of some of this population to other excited state sub-levels can be induced by collisions or the presence of a magnetic field. The de-excitation of the atom can now involve σ_+ or σ_- photons and the polarization of the fluorescence is therefore altered.

transmitted by the polarizer in the x detection path described above. Equally, the appearance of fluorescence in the Oz direction, in which none was emitted previously, could be detected.

Hanle effect

A magnetic field perpendicular to Oz can lead to such a redistribution. For example, a static magnetic field $\mathbf{B}_0$ parallel to Ox induces transitions between $m_b = 0$ and $m_b = \pm 1$. These arise from the additional Zeeman term that appears in the atomic Hamiltonian:

$$\hat{H}_x^{\text{Ze}} = -g\frac{q}{2m}B_0\hat{J}_x \tag{2B.45}$$

(g is the Landé factor, which is equal to unity for a purely orbital angular momentum, and the value of which is state-dependent when the total angular momentum has both orbital and spin contributions). This Zeeman term has non-diagonal matrix elements between magnetic sub-states expressed in the Oz basis. For example,

$$\langle J_b = 1, m_b = 1|\hat{H}_x^{\text{Ze}}|J_b = 1, m_b = 0\rangle = \frac{\hbar\omega_B}{\sqrt{2}}, \tag{2B.46}$$

which features the Larmor frequency,

$$\omega_B = -g\frac{q}{2m}B_0. \tag{2B.47}$$

Whilst the atom remains in the excited level, its population oscillates between the sub-levels with $m_b = 0$ and $m_b = \pm 1$. Light polarized along Oy can therefore be emitted in the subsequent spontaneous decay. If, however, the spontaneous decay occurs before the field-induced population transfer has occurred, the emitted light remains polarized along Oz. The important parameter in determining the degree of depolarization is therefore $\omega_B/\Gamma_{\text{sp}}$. Its value can easily be controlled by varying the value, B_0, of the applied magnetic field; increasing B_0 increases the degree of depolarization. This phenomenon, known as the Hanle effect, was observed for the first time in 1923, before the development of quantum mechanics.

Comment Apart from their technical simplicity (in allowing the determination of radiative lifetimes of the order of a few tens of nanoseconds without requiring the use of fast detectors), methods based on the Hanle effect have the decisive advantage of being insensitive to the Doppler effect induced by atomic motion, which limits the resolution of conventional spectroscopic techniques (compare Complement 3G).

Double resonance

Let us assume that an additional static magnetic field, $\mathbf{B_0}$, is applied in the Oz direction. The degeneracy of the Zeeman sub-levels in the excited state is then lifted by the Zeeman effect, and transitions between these levels can now be induced by a second external electromagnetic field. This technique, developed by Brossel and Bitter in 1949, is known as the *double resonance method.*

The experimental arrangement is shown in Figure 2B.8. The atoms are submitted to an optical excitation from the ground state with light linearly polarized along Oz, and to a radio-frequency field of frequency Ω oscillating along the Ox axis. Population transfer between the excited Zeeman levels will occur when Ω is equal to the Larmor frequency ω_B, ω_B being the energy difference between levels $m_b = 0$ and $m_b = \pm 1$. The transfer of

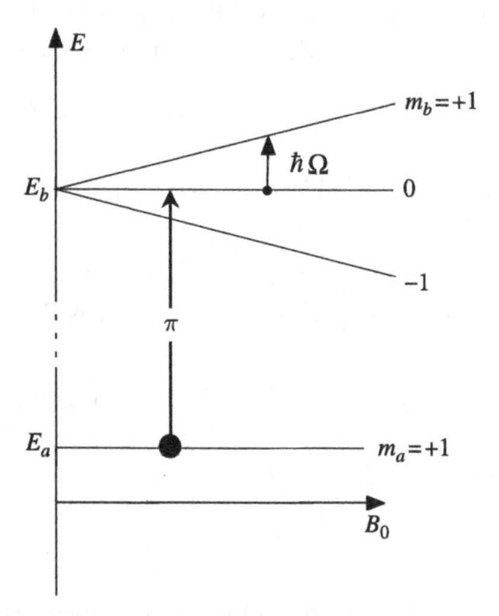

Figure 2B.8 Energy levels in a double resonance experiment. The magnetic field B_0 parallel to Oz lifts the degeneracy of the Zeeman sub-levels in the excited state. The incident light, polarized along Oz populates exclusively the sub-level $m_b = 0$. Then, by applying a radio-frequency field resonant with transitions between the Zeeman sub-levels, population is transferred to the sub-levels $m_b = \pm 1$, this being detectable through a change in the polarization of the light emitted in the subsequent spontaneous decay.

population to the $m_b = \pm 1$ sub-levels leads to the appearance in the fluorescence light of a component with polarization along Oy. If the magnetic field B_0 (or the radio-frequency Ω) is tuned, the polarization of the resonance fluorescence has a resonant behaviour which gives precise information on the structure of the excited state, including the value of its Landé factor.

Comment Like the Hanle effect, the method of double resonance has the advantage of being virtually unaffected by the Doppler motion of the atoms. In fact, for a radio-frequency wave, of which the frequency $\Omega/2\pi$ is typically 100 MHz, and for an atom moving at $v = 300\ \mathrm{ms}^{-1}$, the Doppler shift is of the order of 100 Hz, which is negligible compared to the widths of the resonances observed, of the order of a few MHz.

The same calculation carried out for visible light yields a Doppler shift of the order of 500 MHz, which is large compared to the Zeeman shifts induced by a magnetic field of a few millitesla. The selective excitation of the $m_b = 0$ sub-level (shown in Figure 2B.8) is therefore entirely due to the selection rules discussed above and cannot be due to frequency resonance effects, since the Zeeman sub-levels are not resolved when the Doppler shift exceeds their splitting.

2B.3 Optical pumping

Optical pumping stems from an extension to the ground state of the concepts discussed above. In this technique, invented by Kastler in 1949, the selection rules governing the interaction of an atom with polarized light are exploited to obtain a non-equilibrium distribution of population amongst the sub-levels of the ground state. This technique has many applications of which we shall describe a few examples.

2B.3.1 $J = 1/2 \rightarrow J = 1/2$ transition excited by circularly polarized light

Consider an atomic transition between a ground state of total angular momentum $J_a = 1/2$ and an excited state with total angular momentum $J_b = 1/2$, excited by light which has a σ_+ circular polarization around Oz (Figure 2B.9).

In the absence of incident light, the two ground-state sub-levels, which have the same energies, are equally populated: half of the atoms are initially in each of the states $m_a = -1/2, +1/2$:

$$\pi_{+1/2}(0) = \pi_{-1/2}(0) = \frac{1}{2}. \tag{2B.48}$$

Because of the selection rule $\Delta m = +1$, σ_+ light interacts only with ground-state atoms in the $m_a = -1/2$ sub-level (taking the quantization axis Oz to lie along the propagation direction of the light field) and connects this sub-level to the excited $m_b = +1/2$ sub-level. Spontaneous emission occurs from this level and brings the atom back to the ground state.

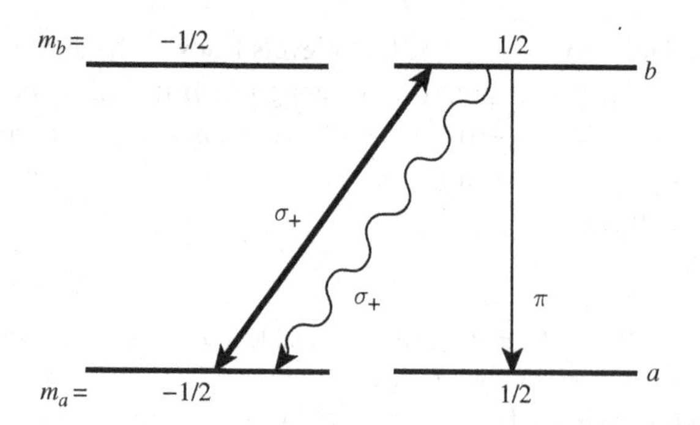

Figure 2B.9 **Optical pumping for a $J_a = 1/2 \leftrightarrow J_b = 1/2$ transition excited by circularly polarized light. The $m_a = -1/2$ ground-state sub-level is completely depopulated after a few fluorescence cycles.**

For a $1/2 \leftrightarrow 1/2$ transition this spontaneous decay transfers the atom with equal probability to each of the ground-state sub-levels. Therefore only half of the excited atoms return to the $m_a = -1/2$ sub-level after one pumping cycle. After two such cycles the fraction is 1/4 and so on. In contrast the atoms in the $m_a = +1/2$ sub-level have no way of escaping and accumulate over the cycles. The result is that the system tends to a steady state in which the atomic population is totally in the $m_a = +1/2$ 'trapping state', the population of the $m_a = -1/2$ sub-level being reduced to zero:

$$\pi_{-1/2}(\infty) = 0 \tag{2B.49}$$

$$\pi_{+1/2}(\infty) = 1. \tag{2B.50}$$

Once the steady state has been reached, the incoming light is no longer absorbed by the atoms which do not emit fluorescent light any longer. The atoms are said to be trapped in a 'dark state'. The progress towards the steady state can be monitored by measuring the transmission of the pumping light (Figure 2B.10). Initially there is a finite absorption, which is proportional to the population of the $m_a = -1/2$ sub-level. This absorption then decays with a time constant τ_p which is the reciprocal of the optical pumping rate Γ_p.

In the steady-state situation described by (2B.49) and (2B.50), the atomic vapour is no longer in thermal equilibrium. Moreover the mean value of the total atomic angular momentum $\hat{\mathbf{J}}$ is now $+\hbar/2$: the atom rotates around the Oz axis. As the absorbed σ_+ light also carried angular momentum, we see that optical pumping is an efficient process to transfer angular momentum from the light to the atoms.

However, any relaxation process will tend to equalize the populations of the ground-state sub-levels and to restore the thermal equilibrium. This in turn will lead to the reappearance of absorption of the optical pumping light. Therefore, optical pumping, like the techniques of optical resonance described above, permits the study of relaxation processes, such as collisions and the effects of magnetic fields. Here, however, we are concerned with relaxation within the ground and not the excited state. The infinite radiative lifetime of the ground state then confers an important advantage. For example, optical pumping methods can be

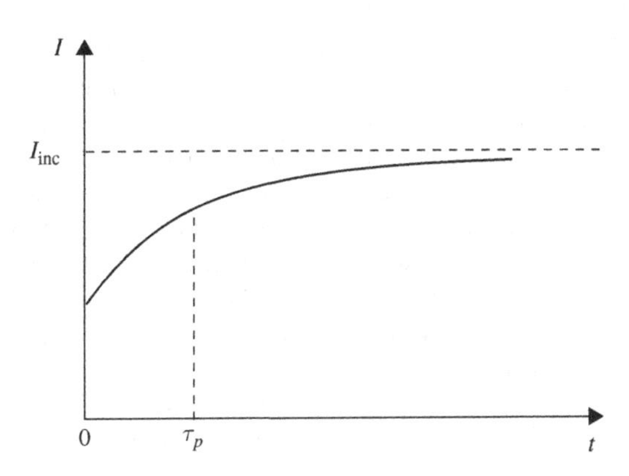

Figure 2B.10 Evolution of the transmitted intensity for the case of σ_+-polarized light incident at time $t = 0$ on an atomic vapour with a $J_a = 1/2 \leftrightarrow J_b = 1/2$ transition with which it is resonant. The vapour, which is initially absorbing, becomes transparent over a timescale of a few times the optical pumping times τ_P.

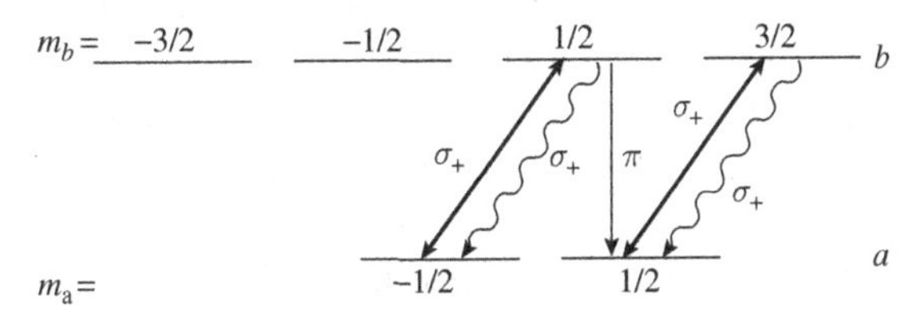

Figure 2B.11 Optical pumping of a $J_a = 1/2 \leftrightarrow J_b = 3/2$ transition excited by σ_+ circularly polarized light. The ground $m_a = -1/2$ sub-level is depopulated after a few pumping cycles and all the atoms undergo transitions between the two sub-levels $m_a = 1/2 \leftrightarrow m_b = 3/2$ only. (Double arrows correspond to absorption and stimulated emission; single arrows correspond to spontaneous emission.)

used to measure magnetic fields with a sensitivity exceeding 10^{-11} tesla, or to perform magnetic resonance imaging of the lung cavities, by using a helium gas which has been previously optically pumped and which is inhaled by the patients.[5]

Comments

(i) Optical pumping is not restricted to the interaction of atoms with circularly polarized light. In fact, it can be shown that for any polarization of the field and for any $J \to J - 1$ or $J \to J$ transition, there exist 'dark states', or 'trapping states', that do not interact with the field and into which the atoms are optically pumped.

(ii) Optical pumping selects a given sub-level, but it can also select a given two-level system. Let us consider for example a $1/2 \leftrightarrow 3/2$ transition illuminated by σ_+ light (Figure 2B.11): the two ground-state sub-levels $m_a = -1/2$ and $m_a = +1/2$ are now connected, respectively, to the excited state sub-levels $m_b = +1/2$ and $m_b = +3/2$. Fluorescence from the sub-level $m_b = +3/2$ can only generate σ_+ photons and transfers the atoms back to the ground-state sub-level $m_a = +1/2$. Consequently atoms in the two-level subspace $(m_a = +1/2, m_b = +3/2)$ cannot

[5] See, for example, T. Beardsley, Seeing the breath of life, *Scientific American*, **280** (1999).

escape from it, whereas atoms in the other sub-levels of the ground and excited states can be transferred to this subspace by excitation and spontaneous emission. At the end of the pumping process, all atoms belong therefore to this subspace, and behave as an ideal two-level system.

(iii) The steady-state population of the ground-state Zeeman sub-levels arises in practice from an equilibrium between the effects of optical pumping, which tend to impose a population balance far from thermal equilibrium, and those of relaxation processes, such as collisions, which tend to restore it. As a result, the steady-state populations and the transmission of the pumping light depend in a complicated fashion on the intensity of the latter (the asymptotic values given by (2B.49) and (2B.50) only being obtained for infinite incident intensity). This property is exploited in numerous nonlinear optical phenomena (see Complement 7B, in particular Section 7B.1.2 in which we calculate the variation of the refractive index with the incident light intensity for a $J_a = 1/2 \leftrightarrow J_b = 1/2$ transition).

2B.3.2 Rate equations for optical pumping

The steady-state populations of the ground-state sub-levels of an atomic vapour subjected to optical pumping light can be obtained from the solution of a set of rate equations which describe the evolution of these populations as a result of absorption and spontaneous emission processes.[6] In the following we illustrate this procedure on a concrete example.

We choose for the purpose of this example a vapour of atoms with a transition between a ground level $|a\rangle$ and an excited level $|b\rangle$ with $J_a = 1$ and $J_b = 2$ (Figure 2B.12). We propose to determine the steady-state distribution of population amongst the sub-levels of the ground state when the vapour is illuminated with light of a fixed polarization. If the light intensity is low the evolution of the ground sub-level populations can be described by a system of rate equations analogous to those of Section 2.6.2. This is obtained by the following procedure.

Firstly, we define for the transition $a \leftrightarrow b$ a matrix, Γ_{ij}, which characterizes the coupling strengths between all possible pairs of sub-levels composed of one sub-level from the ground state and one from the excited state. For the $J_a = 1 \leftrightarrow J_b = 2$, transition used as the example, the matrix is the following:

$$(\Gamma_{ij}) = \Gamma_{sp}\,(C_{ij}) = \Gamma_{sp}\begin{pmatrix} 1 & \frac{1}{2} & \frac{1}{6} & 0 & 0 \\ 0 & \frac{1}{2} & \frac{2}{3} & \frac{1}{2} & 0 \\ 0 & 0 & \frac{1}{6} & \frac{1}{2} & 1 \end{pmatrix}. \tag{2B.51}$$

Its columns refer to the five sub-levels of the excited state, $m_b = -2, \ldots + 2$, whilst the rows refer to those of the ground state, $m_a = -1 \ldots +1$. In this expression Γ_{sp} is the inverse of the radiative lifetime of the excited state and the coefficients C_{ij} can be evaluated using

[6] An exact treatment requires the solution of the optical Bloch equations (compare Complement 2C). In fact it can be shown that the rate equations are valid in the limit of broadband excitation and also in the limit where the relaxation rates of the relevant atomic coherences are much larger than those of the populations.

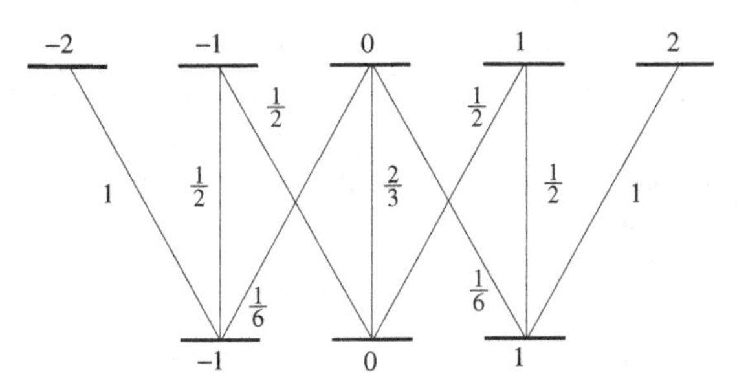

Figure 2B.12 Relative strengths C_{ij} of the transitions between Zeeman sub-levels for a $J_a = 1 \leftrightarrow J_b = 2$ transition (see 2B.51).

the Wigner–Eckart theorem, which generalizes the selection rules (the C_{ij} are in fact just the squares of the Clebsch–Gordan coefficients for the transition). These coefficients are shown in Figure 2B.12.

Each coefficient Γ_{ij} gives the spontaneous de-excitation rate from an excited sub-level m_b to a particular ground sub-level m_a. In the particular case of a transition with $J_b = J_a+1$ used here the selection rules of Section 2B.1 only permit the extreme excited state sub-levels ($m_b = \pm 2$) to decay to a single ground-state sub-level. On the other hand, the $m_b = 0$ sub-level can de-excite to the $m_a = 0$ sub-level (π-transition) with a probability of 2/3, or to the $m_a = \pm 1$ sub-levels, in each case with a probability of 1/6. Where several alternative decay routes are possible the total de-excitation rate of the sub-level j is the sum of the individual rates Γ_{ij}. As can easily be verified from (2B.51),

$$\sum_{i=1}^{3} \Gamma_{ij} = \Gamma_{\text{sp}}, \tag{2B.52}$$

so that the sub-levels of a given excited state all have the same (global) radiative lifetime.

A knowledge of the coefficients Γ_{ij} also enables one to calculate the rate of departure of atoms from each ground-state sub-level under the influence of an incident light wave. We can define a global pumping rate:

$$\Gamma_{\text{p}} = s\frac{\Gamma_{\text{sp}}}{2}, \tag{2B.53}$$

this formula being valid provided the saturation parameter,

$$s = \frac{I}{I_{\text{sat}}}\frac{1}{1+4\frac{\delta^2}{\Gamma_{\text{sp}}^2}}, \tag{2B.54}$$

introduced in Equation (2.162) is much less than one, or, equivalently, provided $\Gamma_{\text{p}} \ll \Gamma_{\text{sp}}$. This occurs when the incident wave intensity I is much smaller than the saturation intensity I_{sat}, which typically has a magnitude of a few mWcm^{-2}.

Suppose now that the polarization of the incident light causes the coupling of the sub-levels i and j. The rate of transitions from i to j is then

$$\left(\Gamma_{\text{p}}\right)_{ij} = \Gamma_{\text{p}} C_{ij}. \tag{2B.55}$$

Since the coefficients C_{ij} give the decay rates from the excited sub-levels (compare 2B.51), we are able, on combining (2B.51) and (2B.55), to obtain the rates of transfer of population between pairs of ground-state sub-levels. For example, the rate of transfer of population between $|a_i\rangle$ and $|a_{i'}\rangle$ as the result of the coupling by a light wave of the sub-levels $|a_i\rangle$ and $|b_j\rangle$ is simply

$$\left(\Gamma_{\mathrm{p}}\right)_{ii'} = \Gamma_{\mathrm{p}} C_{ij} C_{i'j}. \tag{2B.56}$$

Similarly, we can write down expressions for the rates of transfer of population between all such pairs of ground-state sub-levels. Assuming that the rate of change of population of a given sub-level is simply the sum of terms transferring population from and to other sub-levels, we can obtain a set of equations for the evolution of the populations of the entire set of ground-state sub-levels. By way of example, we give these equations for the explicit case of excitation by light linearly polarized along Oz (shown by vertical lines in Figure 2B.12). Using the coefficients of (2B.51), one obtains the following equations describing the evolution of the ground-state sub-levels:

$$\frac{1}{\Gamma_{\mathrm{p}}} \frac{\mathrm{d}}{\mathrm{d}t}(\pi_{-1}) = -\pi_{-1} C_{-1-1} C_{0-1} + \pi_0 C_{00} C_{-10} \tag{2B.57}$$

$$\frac{1}{\Gamma_{\mathrm{p}}} \frac{\mathrm{d}}{\mathrm{d}t}(\pi_0) = -\pi_0 C_{00}(C_{-10} + C_{10}) + \pi_{-1} C_{-1-1} C_{0-1} + \pi_1 C_{11} C_{01} \tag{2B.58}$$

$$\frac{1}{\Gamma_{\mathrm{p}}} \frac{\mathrm{d}}{\mathrm{d}t}(\pi_1) = -\pi_1 C_{11} C_{01} + \pi_0 C_{00} C_{10}. \tag{2B.59}$$

Substituting the numerical values for the coefficients C_{ij} from (2B.51), these become

$$\frac{1}{\Gamma_{\mathrm{p}}} \frac{\mathrm{d}}{\mathrm{d}t}(\pi_{-1}) = -\frac{1}{4}\pi_{-1} + \frac{1}{9}\pi_0 \tag{2B.60}$$

$$\frac{1}{\Gamma_{\mathrm{p}}} \frac{\mathrm{d}}{\mathrm{d}t}(\pi_0) = -\frac{2}{9}\pi_0 + \frac{1}{4}\pi_{-1} + \frac{1}{4}\pi_1 \tag{2B.61}$$

$$\frac{1}{\Gamma_{\mathrm{p}}} \frac{\mathrm{d}}{\mathrm{d}t}(\pi_1) = -\frac{1}{4}\pi_1 + \frac{1}{9}\pi_0. \tag{2B.62}$$

These equations can easily be solved for arbitrary initial conditions with the constraint that the total population be unity:

$$\pi_{-1} + \pi_0 + \pi_1 = 1. \tag{2B.63}$$

The steady state is obtained immediately by setting the left-hand sides of Equations (2B.60–2B.62) to zero. We find

$$\pi_{-1}(\infty) = \pi_1(\infty) = \frac{4}{17} \tag{2B.64}$$

$$\pi_0(\infty) = \frac{9}{17}. \tag{2B.65}$$

A solution for the transient regime is not much more difficult. The populations are given by sums of exponentially decaying terms of which the reciprocals of the time constants are found by diagonalizing the matrix associated with the right-hand sides of (2B.60–2B.62):

$$\begin{pmatrix} -\frac{1}{4} & \frac{1}{9} & 0 \\ \frac{1}{4} & -\frac{2}{9} & \frac{1}{4} \\ 0 & \frac{1}{9} & -\frac{1}{4} \end{pmatrix}, \tag{2B.66}$$

which gives

$$\begin{aligned} \Gamma' &= 0 \\ \Gamma'' &= 0.10\,\Gamma_{\mathrm{p}} \\ \Gamma''' &= 0.63\,\Gamma_{\mathrm{p}}. \end{aligned} \tag{2B.67}$$

The existence of an eigenvalue of zero simply reflects the existence of a conserved quantity, in this case the sum of the populations of the ground-state sub-levels (Equation 2B.63).

In the above example we have shown how a set of rate equations allows us to calculate the temporal evolution of the populations of the ground-state sub-levels. This procedure, suitably generalized, finds use in many situations.

Comments

(i) The rate equations are only valid if the polarization of the incident wave is such that each ground-state sub-level is coupled to at most one sub-level in the excited state. Additionally, the intensity of the pumping light must not drive the transition to saturation.

(ii) It is apparent from the above treatment that the time constants governing the evolution of the ground sub-level populations can be very long compared to the decay time of the excited state Γ_{sp}, if I is very small compared to I_{sat}. This has important consequences.

(iii) If the rate equations for the situation of Figure 2B.11 are written down, it emerges that $\pi_{-1/2}(\infty) = 0$ in agreement with our intuitive argument. Similarly, in the $J_a = 1 \rightarrow J_b = 2$ case of Figure 2B.12, a σ_+ polarized beam transfers all the population to the $m_a = +1$ sub-level and the solution $\pi_1(\infty) = 1$ is expected. It is often the case that the steady-state solution in such a situation can be inferred without the necessity of a mathematical treatment.

(iv) In many situations, the quantities governing the macroscopic behaviour of the atomic sample are not the populations of the Zeeman sub-levels but averaged quantities, such as

$$\frac{\langle \hat{J}_z \rangle}{\hbar} = \sum_{m_a} m_a \pi_{m_a}, \tag{2B.68}$$

which is known as the *orientation*, or

$$\frac{\langle \hat{J}_z^2 \rangle}{\hbar^2} - \frac{1}{3}\frac{\langle \hat{\mathbf{J}}^2 \rangle}{\hbar^2} = \sum_{m_a} m_a^2 \pi_{m_a} - \frac{J_a(J_a+1)}{3}, \tag{2B.69}$$

termed the *alignment*. In the absence of optical pumping, when all the Zeeman sub-levels are equally populated, the orientation and the alignment are both zero; the system has a spherical symmetry. Their having non-zero values breaks this spherical symmetry, and is a signature of the occurrence of optical pumping.

2C Complement 2C The density matrix and the optical Bloch equations

The arguments of Chapter 2, as well as of those of subsequent chapters, have as their foundation the formalism based on the state vector of a system of which the evolution is described by the Schrödinger equation. In fact, such an approach is badly suited to the case in which the coupling between an atom and its environment (for example through collisions with other atoms or spontaneous emission into formerly empty modes of the electromagnetic field) cannot be neglected. If the correlations induced by these interactions between the atom and its environment do not concern us and we are only interested in the evolution of the atom, the formalism of the *density matrix* must be employed. This provides a description at all times of the state of the atom, although a state vector for the atom alone cannot be defined. In this formalism the effect of the environment on the atom is accounted for by the introduction of suitable *relaxation* terms (Section 2C.1) in the equation of evolution of the density matrix. An important application of the density matrix is to the case of a two-level atomic system for which the relaxation terms lead to its de-excitation to a level of lower energy. We shall show that in this case the density matrix can be represented by a vector, known as the *Bloch vector*, which will allow us to give simple geometrical pictures of the evolution of the system.

In Section 2C.2, we show how perturbation theory, introduced in Chapter 1, can be generalized to describe the evolution of the density matrix. We analyse in this manner the interaction of an atom with a weak electromagnetic field and hence obtain expressions for the linear susceptibility of an atomic vapour. There is another case for which the equations governing the evolution of the density matrix lead to simple analytical results: this is the case in which two levels alone play a role in the interaction of the atom with incident radiation. The evolution of the density matrix is then described by a system of equations known as the *optical Bloch equations* (Section 2C.3), the steady-state solution of which one can exactly determine even if the radiation is very intense. These permit the description of a broad range of physical phenomena. In 2C.4, we introduce the Bloch vector, a geometrical representation of the evolution of a two-level system introduced first in the context of nuclear magnetic resonance, but now widely used in the context of quantum optics. We shall finally discuss in 2C.5 the relationship between the rate equations for the populations introduced in Section 2.6 and the optical Bloch equations.

2C.1 Wavefunctions and density matrices

2C.1.1 Isolated and coupled systems

The evolution of an isolated system, of which the state at $t = 0$ is $|\psi_0\rangle$, is described by the Schrödinger equation:

$$i\hbar \frac{d}{dt} |\psi(t)\rangle = \hat{H} |\psi(t)\rangle , \tag{2C.1}$$

with the initial condition

$$|\psi(0)\rangle = |\psi_0\rangle . \tag{2C.2}$$

If the system is composed of two sub-systems A and B, of which the states at $t = 0$ are $|\psi_A\rangle$ and $|\psi_B\rangle$, the solution of the Schrödinger equation with the initial condition $|\psi_A\rangle \otimes |\psi_B\rangle$ leads to an *entangled state* $|\psi(t)\rangle$ which cannot be factorized into a product of states describing separately the states of the two sub-systems:

$$|\psi(t)\rangle \neq |\psi_A(t)\rangle \otimes |\psi_B(t)\rangle .$$

This is because the interaction of the two sub-systems A and B creates *quantum correlations* between them which will remain even once the interaction between the systems is terminated. If one wants then to study the interaction between the sub-system A and a third system C it is no longer possible to write the initial state of the system composed of A and C in the form $|\psi'_A\rangle \otimes |\psi_C\rangle$. Strictly speaking it is necessary to work with the space of tensor products of states relating to the sub-systems A, B and C. When the system A interacts successively in this way with a sequence of systems $B, C, D, \ldots X$, it is apparent that the tensor product space that must be considered will grow rapidly in extent because of the resulting quantum correlations accumulated at successive interactions. For this reason, the wavefunction of a single atom A undergoing collisions with other atoms at a rate of one every few microseconds would, at the end of a second, have to take into account the states of the million or so atoms encountered in this time. Any effort to describe the state of the system by a state vector would very soon be overwhelming.

2C.1.2 The density matrix representation

In practice, for most problems, the information contained in the global wavefunction for a system and its environment exceeds that required for predicting the outcome of a measurement bearing on the system alone. Rather than describing the effect of each and every interaction it is usually sufficient to describe the average effect. Such an approach necessitates the use of a *density operator* $\hat{\sigma}$, rather than that of a state vector to describe the system. We present briefly in the following some important properties of this operator.[1]

[1] For a more detailed description of the density matrix see CDL I, Chapter III, Complement E.

In the case in which the state of system A can be represented by the state vector $|\psi_A\rangle$ (we call this case a *pure state*), the corresponding density operator is the projector,

$$\hat{\sigma} = |\psi_A\rangle\langle\psi_A|. \tag{2C.3}$$

More generally, $\hat{\sigma}$ is an *operator* acting in the space spanned by the states of A. In a given basis, it can be expressed in the form of a matrix, known equally commonly as the density matrix. In the following, we employ the terms density operator and density matrix interchangeably.

The probability of finding the system in a state $|i\rangle$ is equal to $\sigma_{ii} = \langle i\,|\,\hat{\sigma}\,|\,i\rangle$ (this is obviously true for a pure state). The probability of finding the system in one of the available basis states being unity, leads to

$$\text{Tr}\hat{\sigma} = \sum_i \sigma_{ii} = 1. \tag{2C.4}$$

The diagonal element σ_{ii} is named the *population* of the state $|i\rangle$.

For a pure state, it is straightforward to verify that the mean value of an observable represented by an operator $\hat{O}$ is

$$\langle\hat{O}\rangle = \text{Tr}\left(\hat{\sigma}\hat{O}\right). \tag{2C.5}$$

This is true also in the general case.

The evolution of the density operator of an isolated system A in a pure state can be deduced from the Schrödinger equation (2C.1). It is described by the equation

$$\frac{\mathrm{d}\hat{\sigma}}{\mathrm{d}t} = \frac{1}{\mathrm{i}\hbar}\left[\hat{H},\hat{\sigma}\right], \tag{2C.6}$$

which involves the commutator of the system Hamiltonian and the density operator. This again is easily verified for a pure state, and is valid for any system described initially by a density matrix, whether it remains isolated or interacts with other systems, provided these interactions are describable in terms of a Hamiltonian.

The density matrix formalism is especially useful in the case where the initial system is not perfectly determined and cannot be described by a single state vector. If one only knows that there is a probability p_i of preparing the system in state $|\psi_i\rangle$, one can define the following 'statistical' density matrix:

$$\hat{\sigma} = \sum_i p_i|\psi_i\rangle\langle\psi_i|. \tag{2C.7}$$

It is no longer a projector, and will be called a *statistical mixture*. Expressions (2C.5) and (2C.6) for the mean values and the evolution of the system are still valid in this case.

If, in addition, system A undergoes at random instants a succession of brief, weak collisions with other systems $B, C, \ldots X$,[2] their *average* effect can be represented by the addition to Equation (2C.6) of a relaxation operator. The relaxation terms which concern the population terms σ_{ii} are

[2] For more insight into relaxation processes see, for example, CDG II, Chapter IV. Equations of the form (2C.8–2C.9) in particular, can be obtained if the system evolves little over the course of a single collision.

$$\left\{\frac{\mathrm{d}}{\mathrm{d}t}\sigma_{ii}\right\}_{\mathrm{rel}} = -\left(\sum_{j\neq i}\Gamma_{i\to j}\right)\sigma_{ii} + \sum_{j\neq i}\Gamma_{j\to i}\sigma_{jj}. \tag{2C.8}$$

This equation expresses the fact that the population σ_{ii} of the state $|i\rangle$ decreases as the result of transitions to other states $|j\rangle$ and increases as a result of transitions from these states.

For the off-diagonal elements of the density matrix σ_{ij} (termed *coherences* since they depend on the relative phases of the $|i\rangle$ and $|j\rangle$ components of the system wavefunction), the relaxation terms are written

$$\left\{\frac{\mathrm{d}}{\mathrm{d}t}\sigma_{ij}\right\}_{\mathrm{rel}} = -\gamma_{ij}\sigma_{ij}. \tag{2C.9}$$

Note that if the coefficients $\Gamma_{i\to j}$ are real (and positive), nothing prevents the coefficients γ_{ij} from being complex (provided, however, that $\gamma_{ij} = \gamma_{ji}^*$). The coefficients $\Gamma_{i\to j}$ and γ_{ij} can be calculated if the Hamiltonian governing the interaction of system A and its environment are known. Here we do not concern ourselves with their evaluation but consider them to be introduced phenomenologically.

Comment Equation (2C.9) describing the relaxation of the coherences implicitly supposes that all the Bohr frequencies of the system are different. In the case that two or more of these are identical it is necessary to add terms coupling the coherences which have the same frequency to (2C.9). These terms, which transfer coherence between pairs of states, become very important when a symmetry of the Hamiltonian leads to the exact equality of several Bohr frequencies. A striking example is that of the harmonic oscillator for which the transfer of coherence as a result of relaxation plays an essential role in the dynamics.[3]

2C.1.3 Two-level systems

Many physical situations can be represented by an idealized two-level quantum system.[4] It is important therefore to be able to characterize the relaxation of such a system. In the following we shall describe firstly the case of a *closed* system, which corresponds to the case in which the lower in energy of the two levels is stable (for example, the ground state), and secondly the case of an *open* system, in which both levels may be unstable (excited) levels.

Closed systems

Consider a two-level atom of which the lower and upper levels are denoted, respectively, a and b (Figure 2C.1). If spontaneous emission from the upper to the lower level is the only

[3] See CDG II, Complement B_{IV}.

[4] R. P. Feynmann, R. B. Leighton, and M. Sands, *Lectures on Physics*, Vol. III, Addison-Wesley (2005), Chapter 11.

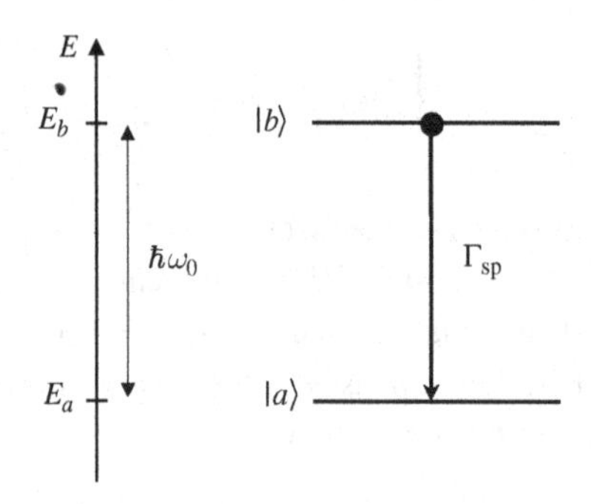

Figure 2C.1 A closed two-level system. The relaxation of the upper level b is, necessarily, to the lower level a. The sum of the populations of the two levels is constant.

source of relaxation (see Chapter 2 for a phenomenological description, or Chapter 6 for a quantitative one), Equations (2C.8–2C.9) become

$$\left\{\frac{\mathrm{d}}{\mathrm{d}t}\sigma_{bb}\right\}_{\mathrm{rel}} = -\Gamma_{\mathrm{sp}}\sigma_{bb} \tag{2C.10}$$

$$\left\{\frac{\mathrm{d}}{\mathrm{d}t}\sigma_{aa}\right\}_{\mathrm{rel}} = \Gamma_{\mathrm{sp}}\sigma_{bb}, \tag{2C.11}$$

where $\Gamma_{\mathrm{sp}}^{-1}$ is the radiative lifetime of the upper level. Equation (2C.10) describes the exponential decay of the population of the excited state at rate $\Gamma_{\mathrm{sp}}^{-1}$. Equation (2C.11), on the other hand, expresses the fact that any atom leaving state b ends up in state a. As far as relaxation effects are concerned this two-level system is closed.

Relaxation of the coherences σ_{ba} is described by

$$\left\{\frac{\mathrm{d}\sigma_{ba}}{\mathrm{d}t}\right\}_{\mathrm{rel}} = -\gamma\sigma_{ba}, \tag{2C.12}$$

with $\gamma = \Gamma_{\mathrm{sp}}/2$ when the relaxation is a result exclusively of spontaneous emission.

Comments

(i) The system described above is a prime example of one which lends itself to treatment in terms of the density matrix. A description of the global system comprising the two-level atom and the emitted photon brings into play a very large state space involving the degrees of freedom of the atom and of the electromagnetic field (See Chapter 6). In fact Equations (2C.10) and (2C.11) result from a partial trace over the variables that concern the radiation field, resulting in equations which involve a space of only two states (the number of available states of the atom).[5] The problem with this procedure is getting around the correlations that can arise between the atom and the spontaneously emitted photons.

(ii) Often spontaneous emission is not the only cause of relaxation. We have already mentioned one other source, that of interatomic collisions. Consider a system of two-level atoms interacting with a gas of perturbing atoms. Collisions with the perturbing gas can cause transitions of the

[5] See CDG II, Chapter IV, parts A and E.

two-level atoms in both directions: from b to a (*collisional quenching*) and from a to b (*collisional excitation*) but these processes are often negligible compared to spontaneous emission so that, overall, the relaxation of the populations is properly described by Equations (2C.10)–(2C.11). On the other hand, the evolution of the coherences is strongly perturbed by collisions owing to the temporal variation of the atomic Bohr frequency, ω_0, over the course of the collision as a result of the interaction potential between the atom and the perturber. These *dephasing collisions* lead to an increase in the relaxation rate of the coherences to

$$\gamma = \frac{\Gamma_{\text{sp}}}{2} + \gamma_{\text{coll}}, \tag{2C.13}$$

where the first term Γ_{sp} is the rate induced by spontaneous emission and γ_{coll} is associated with the collisional relaxation. The latter term is proportional to the number of perturber atoms per unit volume and their speed v:

$$\gamma_{\text{coll}} = \frac{N}{V}\sigma_{\text{coll}}\, v. \tag{2C.14}$$

For a given interatomic potential, the cross-section for dephasing collisions σ_{coll} (not to be confused with an element of the density matrix!) depends only on the temperature, and the rate of collisional relaxation is therefore proportional to the pressure at a given temperature.

Open systems

Suppose now that both levels a and b are excited atomic states, a being the lower in energy of the two (Figure 2C.2). The relaxation of the populations in this system is described by

$$\left\{\frac{\mathrm{d}}{\mathrm{d}t}\sigma_{bb}\right\}_{\text{rel}} = -\Gamma_b\sigma_{bb} - \Gamma_{b\to a}\sigma_{bb} \tag{2C.15}$$

$$\left\{\frac{\mathrm{d}}{\mathrm{d}t}\sigma_{aa}\right\}_{\text{rel}} = -\Gamma_a\sigma_{aa} + \Gamma_{b\to a}\sigma_{bb}. \tag{2C.16}$$

In these equations, Γ_a^{-1} and Γ_b^{-1} describe the rates of relaxation to external states and $\Gamma_{b\to a}$ is the rate of relaxation of level b to level a (which can be due to spontaneous emission). The evolution of the coherence σ_{ba} is given by

$$\left\{\frac{\mathrm{d}}{\mathrm{d}t}\sigma_{ba}\right\}_{\text{rel}} = -\gamma\sigma_{ba} \tag{2C.17}$$

where, in the case of relaxation caused exclusively by spontaneous emission,

$$\gamma = \gamma_{\text{sp}} = \left(\frac{\Gamma_b^{\text{sp}}}{2} + \frac{\Gamma_a^{\text{sp}}}{2}\right). \tag{2C.18}$$

Contrary to the situation described by Equations (2C.10–2C.11), which are consistent with the condition $Tr\hat{\sigma} = 1$ so that the total population of the two-level system is conserved, Equations (2C.15–2C.16) lead to a global population that decreases with time. Such a system can therefore only be a sub-system of a larger whole, hence the designation 'open system'. We can consider then that the excited states a and b are able to de-excite to a lower-lying state f, from which they can be re-excited by an external pumping mechanism. In the following we shall assume that this pumping, which in practice could be produced by

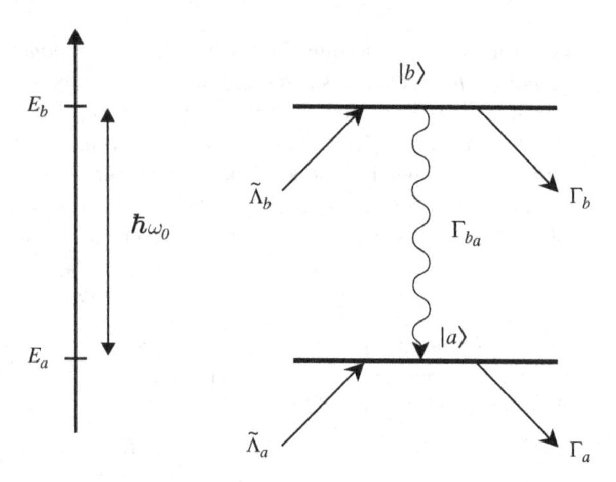

Figure 2C.2 An open two-level system. Levels a and b can de-excite to one or several lower energy levels (not shown). Equally, their populations can both be fed by an external mechanism. The sum of the two populations is no longer constant.

electron bombardment, by an external source of resonant light or by many other processes is described by the following equations:

$$\left\{\frac{\mathrm{d}}{\mathrm{d}t}\sigma_{bb}\right\}_{\text{feed}} = \tilde{\Lambda}_b\sigma_{ff} \tag{2C.19}$$

$$\left\{\frac{\mathrm{d}}{\mathrm{d}t}\sigma_{aa}\right\}_{\text{feed}} = \tilde{\Lambda}_a\sigma_{ff}. \tag{2C.20}$$

Usually the pumping rates $\tilde{\Lambda}_a$ and $\tilde{\Lambda}_b$ are small compared to Γ_a and Γ_b so that the populations in the excited state remain small compared to that of the ground state.[6] We have therefore $\sigma_{ff} \approx 1$, which means that Equations (2C.19–2C.20) are well approximated by

$$\left\{\frac{\mathrm{d}}{\mathrm{d}t}\sigma_{bb}\right\}_{\text{feed}} \approx \tilde{\Lambda}_b \tag{2C.21}$$

$$\left\{\frac{\mathrm{d}}{\mathrm{d}t}\sigma_{aa}\right\}_{\text{feed}} \approx \tilde{\Lambda}_a. \tag{2C.22}$$

The pumping of the excited-level populations does not generally give rise to non-zero coherences between different electronic states, so that there is no analogous pumping term for σ_{ba} to consider.

Comments

(i) Despite its resemblance to the model used in Sections 2.2 and 2.3, the system considered here is different in a number of respects. For example, the rates of relaxation of the two levels are no longer equal. Furthermore, we take account here of the possibility of the internal relaxation of

[6] The pumping rates $\tilde{\Lambda}_a$ and $\tilde{\Lambda}_b$ introduced here concern a single atom. They are related to those for an assembly of N atoms, as used in Chapter 2, by $\Lambda_i = N\tilde{\Lambda}_i$.

state b to state a. In this, much more general, situation the use of the optical Bloch equations is indispensable.

(ii) It is obvious that the closed two-level system is just a special case of an open one, with the relaxation and pumping rates to and from states outside the two-level system set to zero.

2C.2 Perturbative treatment

2C.2.1 Iterative solution for the evolution of the density matrix

Presentation of the problem

The goal of this section is to generalize to the density matrix formalism the time-dependent perturbation methods introduced in Chapter 1 and applied there to the evolution of the state vector. As in Chapter 1, we suppose that the Hamiltonian can be decomposed into a principal term $\hat{H}_0$, which is time-independent and has eigenstates $|n\rangle$ with energies E_n, and a perturbation term $\hat{H}_1(t)$. Like in Chapter 1, in order to render more transparent the series solution for the density matrix in terms of a hierarchy of terms of increasing order in the perturbation, we shall rewrite the perturbation part of the Hamiltonian in the form $\lambda\hat{H}'_1(t)$, the real parameter λ being small compared to unity, whilst $\hat{H}'_1(t)$ is supposed to be of the same order of magnitude as $\hat{H}_0$. The equation governing the evolution of the density matrix, which takes the place of the Schrödinger equation, here takes the form

$$\frac{\mathrm{d}\hat{\sigma}}{\mathrm{d}t} = \frac{1}{\mathrm{i}\hbar}\left[\hat{H}_0 + \hat{H}_1(t), \hat{\sigma}\right] + \left\{\frac{\mathrm{d}\hat{\sigma}}{\mathrm{d}t}\right\}. \tag{2C.23}$$

The first term is the commutator (2C.6) describing the Hamiltonian part of the evolution. The second term, $\left\{\frac{\mathrm{d}\hat{\sigma}}{\mathrm{d}t}\right\}$, is the sum of the terms relating to relaxation and pumping of the populations and coherences. To proceed we now substitute $\hat{H}_1(t) = \lambda\hat{H}'_1(t)$ in Equation (2C.23):

$$\frac{\mathrm{d}\hat{\sigma}}{\mathrm{d}t} = \frac{1}{\mathrm{i}\hbar}\left[\hat{H}_0, \hat{\sigma}\right] + \left\{\frac{\mathrm{d}\hat{\sigma}}{\mathrm{d}t}\right\} + \frac{\lambda}{\mathrm{i}\hbar}\left[\hat{H}'_1(t), \hat{\sigma}\right] \tag{2C.24}$$

and, additionally, expand the density matrix as a power series in λ:

$$\hat{\sigma} = \hat{\sigma}^{(0)} + \lambda\hat{\sigma}^{(1)} + \lambda^2\hat{\sigma}^{(2)} + \ldots \tag{2C.25}$$

which generalizes the procedure of Section 1.2. Substituting (2C.25) into (2C.24) and collecting together terms of the same order in λ, we find:

- to zeroth order,

$$\frac{\mathrm{d}\hat{\sigma}^{(0)}}{\mathrm{d}t} - \frac{1}{\mathrm{i}\hbar}\left[\hat{H}_0, \hat{\sigma}^{(0)}\right] - \left\{\frac{\mathrm{d}\hat{\sigma}^{(0)}}{\mathrm{d}t}\right\} = 0 \tag{2C.26}$$

- to first order,

$$\frac{\mathrm{d}\hat{\sigma}^{(1)}}{\mathrm{d}t} - \frac{1}{\mathrm{i}\hbar}\left[\hat{H}_0, \hat{\sigma}^{(1)}\right] - \left\{\frac{\mathrm{d}\hat{\sigma}^{(1)}}{\mathrm{d}t}\right\} = \frac{1}{\mathrm{i}\hbar}\left[\hat{H}'_1(t), \hat{\sigma}^{(0)}\right] \tag{2C.27}$$

- to order r,

$$\frac{\mathrm{d}\hat{\sigma}^{(r)}}{\mathrm{d}t} - \frac{1}{\mathrm{i}\hbar}\left[\hat{H}_0, \hat{\sigma}^{(r)}\right] - \left\{\frac{\mathrm{d}\hat{\sigma}^{(r)}}{\mathrm{d}t}\right\} = \frac{1}{\mathrm{i}\hbar}\left[\hat{H}'_1(t), \hat{\sigma}^{(r-1)}\right]. \tag{2C.28}$$

Zeroth-order solution

To zeroth order in λ, and in the basis of the eigenstates of $\hat{H}_0$, Equation (2C.26) leads to equations for the populations, $\sigma_{jj}^{(0)}$, which are identical to the relaxation equations (2C.8). The steady-state solution of these equations gives the populations of the various states in the absence of the perturbation.

The coherences $\sigma_{jk}^{(0)}$ on the other hand, are given, according to (2C.26) and (2C.9), by

$$\frac{\mathrm{d}}{\mathrm{d}t}\sigma_{jk}^{(0)} + \mathrm{i}\omega_{jk}\sigma_{jk}^{(0)} + \gamma_{jk}\sigma_{jk}^{(0)} = 0, \tag{2C.29}$$

where $\omega_{jk} = (E_j - E_k)/\hbar$ is the Bohr frequency for the transition $j \to k$. If all the coherences are initially zero ($\sigma_{jk}^{(0)}(0) = 0$) they remain so at all later times. If they differ initially from zero they decay to zero with time constants $(\mathrm{Re}\{\gamma_{jk}\})^{-1}$.

We assume in the following that $\hat{\sigma}^{(0)}$ attains its steady state before the perturbation $\hat{H}_1(t)$ is applied so that the only non-zero terms of $\hat{\sigma}^{(0)}$ are the (time-independent) population terms $\sigma_{jj}^{(0)}$. The initial state of the system is then specified by the set of values of these populations, which we presume to be known. For a system at thermal equilibrium they are distributed according to the Boltzmann distribution.

First-order solution

We apply (2C.27) first of all to the populations and obtain, on taking diagonal elements,

$$\frac{\mathrm{d}\sigma_{jj}^{(1)}}{\mathrm{d}t} - \left\{\frac{\mathrm{d}\sigma_{jj}^{(1)}}{\mathrm{d}t}\right\} = \frac{1}{\mathrm{i}\hbar}\sum_{\ell}\left[\left\langle j\left|\hat{H}'_1(t)\right|\ell\right\rangle\sigma_{\ell j}^{(0)} - \sigma_{j\ell}^{(0)}\left\langle \ell\left|\hat{H}'_1(t)\right|j\right\rangle\right]. \tag{2C.30}$$

The term in the right-hand side of (2C.30) with $\ell = j$ is zero since its two parts then cancel. The other terms also vanish since all the zeroth-order coherences $\sigma_{\ell j}^{(0)}$ are zero. The right-hand side of (2C.30) therefore vanishes. It follows that the solution of (2C.30) is simply

$$\sigma_{jj}^{(1)}(t) = 0 \tag{2C.31}$$

for all times t: *to first order the populations do not change.*

Consider now the coherences to the same order in the perturbation. Using Equations (2C.27) and (2C.9) and remembering that the zeroth-order coherences are zero, we find

$$\frac{\mathrm{d}}{\mathrm{d}t}\sigma_{jk}^{(1)} + \mathrm{i}\omega_{jk}\sigma_{jk}^{(1)} + \gamma_{jk}\sigma_{jk}^{(1)} = \frac{1}{\mathrm{i}\hbar}\left\langle j\left|\hat{H}_1'(t)\right|k\right\rangle\left(\sigma_{kk}^{(0)} - \sigma_{jj}^{(0)}\right). \tag{2C.32}$$

The solution of this equation satisfying the initial condition $\sigma_{jk}^{(1)}(t_0) = 0$ is

$$\sigma_{jk}^{(1)}(t) = \frac{\sigma_{kk}^{(0)} - \sigma_{jj}^{(0)}}{\mathrm{i}\hbar}\int_{t_0}^{t} \mathrm{e}^{-(\mathrm{i}\omega_{jk}+\gamma_{jk})(t-t')}\left\langle j\left|\hat{H}_1'(t')\right|k\right\rangle \mathrm{d}t'. \tag{2C.33}$$

Recall that the first-order term in the perturbation expansion (2C.25) is $\lambda\hat{\sigma}^{(1)}$. It follows that the first-order corrections to the density matrix $\hat{\sigma}^{(0)}$ are

$$\lambda\sigma_{jj}^{(1)}(t) = 0 \tag{2C.34}$$

and

$$\lambda\sigma_{jk}^{(1)}(t) = \frac{\sigma_{kk}^{(0)} - \sigma_{jj}^{(0)}}{\mathrm{i}\hbar}\int_{t_0}^{t} \mathrm{e}^{-(\mathrm{i}\omega_{jk}+\gamma_{jk})(t-t')}\left\langle j\left|\hat{H}_1(t')\right|k\right\rangle \mathrm{d}t'. \tag{2C.35}$$

Higher-order terms

To evaluate the higher-order terms in the expansion (2C.25) it is simply necessary to solve Equation (2C.28) for successive values of r. In this way, the second-order populations and coherences can be obtained from the first-order ones and the third-order terms can, in a like manner, be found from the second.

Comment In orders higher than the first, the coherences and populations are generally both non-zero.

2C.2.2 Atom interacting with an oscillating field: regime of linear response

Sinusoidal perturbation

We consider again in this section an atom described by a time-independent Hamiltonian $\hat{H}_0$ interacting with the oscillating electric field of an electromagnetic wave. The interaction Hamiltonian $\hat{H}_1(t)$ is of the form $\hat{W}\cos\omega t$ with $\hat{W} = -\hat{\mathbf{D}}\cdot\mathbf{E}_0$. Substituting this into (2C.35) we find in the long-time limit (that is to say for $t - t_0 \gg 1/\gamma_{jk}$):

$$\lambda\sigma_{jk}^{(1)}(t) = \frac{\sigma_{kk}^{(0)} - \sigma_{jj}^{(0)}}{2\mathrm{i}\hbar}W_{jk}\left[\frac{\mathrm{e}^{-\mathrm{i}\omega t}}{\mathrm{i}(\omega_{jk}-\omega)+\gamma_{jk}} + \frac{\mathrm{e}^{\mathrm{i}\omega t}}{\mathrm{i}(\omega_{jk}+\omega)+\gamma_{jk}}\right]. \tag{2C.36}$$

For the case of a quasi-resonant excitation (see Chapter 1, Section 1.2.4), one of the energy denominators is much smaller than the other. In the following we assume that ω_{jk} is positive (level j is higher in energy than level k) and that

$$|\omega_{jk} - \omega|, |\gamma_{jk}| \ll |\omega_{jk} + \omega|. \tag{2C.37}$$

It is then possible to neglect the anti-resonant term in (2C.36), which then takes the simplified form,

$$\lambda\sigma_{jk}^{(1)}(t) \approx \frac{\sigma_{kk}^{(1)} - \sigma_{jj}^{(0)}}{2i\hbar} W_{jk} \frac{\mathrm{e}^{-i\omega t}}{\gamma_{jk} + i(\omega_{jk} - \omega)}. \tag{2C.38}$$

These results, to order one in W_{jk} (which is proportional to the electric field) account for the atoms' *linear* response.

The mean value of the atomic electric dipole; the linear susceptibility

A knowledge of the density matrix permits evaluation of the mean value of the atomic electric dipole:

$$\langle\hat{\mathbf{D}}\rangle = \mathrm{Tr}(\hat{\sigma}\hat{\mathbf{D}}). \tag{2C.39}$$

For an isolated atom, the diagonal matrix elements $\mathbf{D}_{ii}$ are zero (because of the invariance of the Hamiltonian $\hat{H}_0$ under an inversion of spatial coordinates; see Complement 2B) so that only the coherences σ_{jk} appear in the expansion (2C.39):

$$\langle\hat{\mathbf{D}}\rangle = \sum_{j,k} \sigma_{jk}\mathbf{D}_{kj}. \tag{2C.40}$$

We deduce from the fact that the first term of the perturbation expansion for the coherences is of order one that, to lower order, the atomic electric dipole $\langle\hat{\mathbf{D}}^{(1)}\rangle$ is a linear function of the applied field $\mathbf{E}$. It is equal to

$$\langle\hat{\mathbf{D}}^{(1)}\rangle = \sum_{j,k} \sigma_{jk}^{(1)}\mathbf{D}_{kj}. \tag{2C.41}$$

We assume now, for simplicity, that the electric field is parallel to Oz and that the initial state, and therefore $\sigma^{(0)}$, is invariant under rotation. For obvious symmetry reasons the induced atomic dipole is therefore also parallel to Oz. We shall therefore drop the vector notation, bearing in mind that in the following we consider the z-components of the vector quantities. Regrouping the terms in $\sigma_{jk}^{(1)}$, we find, from (2C.36) and (2C.41) (supposing for simplicity that the relaxation rates γ_{jk} are real):

$$\begin{aligned}\langle D^{(1)}\rangle = \sum_{j>k} &\frac{\sigma_{kk}^{(0)} - \sigma_{jj}^{(0)}}{\hbar}(-D_{jk}E)D_{kj} \\ &\times \left\{ \left(\frac{\omega - \omega_{jk}}{(\omega - \omega_{jk})^2 + \gamma_{jk}^2} - \frac{\omega + \omega_{jk}}{(\omega + \omega_{jk})^2 + \gamma_{jk}^2} \right) \cos\omega t \right. \\ &\left. + \left(\frac{\gamma_{jk}}{(\omega + \omega_{jk})^2 + \gamma_{jk}^2} - \frac{\gamma_{jk}}{(\omega - \omega_{jk})^2 + \gamma_{jk}^2} \right) \sin\omega t \right\}.\end{aligned} \tag{2C.42}$$

Comments

(i) In writing Equation (2C.42) we have assumed that the rates γ_{jk} are real. If γ_{jk} has real and imaginary components, γ'_{jk} and $i\gamma''_{jk}$, then γ_{jk} must be replaced in (2C.42) by γ'_{jk} and ω_{jk} by $\omega_{jk}+\gamma''_{jk}$. The resonance is therefore displaced in frequency by the interactions giving rise to the relaxation.

(ii) The form of (2C.42) rests on the assumption that the relaxation is described by the same coefficients γ_{jk} whatever the detuning from resonance. This implies that the in quadrature part of the atomic electric dipole is a Lorentzian function of $|\omega_{jk}-\omega|$ for any value of the detuning, which is not true in general. Usually, the Lorentzian form is valid in the neighbourhood of resonance and out to detunings many times γ_{jk}, but it ceases to hold for sufficiently large values of detuning that are specific to the exact mechanism responsible for the relaxation.[7]

Using Equations (2.173–2.175) it is possible to rewrite (2C.42) in terms of the real and imaginary parts χ' and χ'' of the susceptibility. We denote by N/V the number of atoms per unit volume and $N_k^{(0)} = N\sigma_{kk}^{(0)}$ the number of atoms in state $|k\rangle$ in the absence of the electromagnetic field. In the quasi-resonant approximation (applied to the transition $j \to k$), we find

$$\chi' = \frac{N_k^{(0)} - N_j^{(0)}}{V} \frac{|D_{jk}|^2}{\varepsilon_0 \hbar} \frac{(\omega_{jk}-\omega)}{(\omega_{jk}-\omega)^2 + \gamma_{jk}^2} \tag{2C.43}$$

$$\chi'' = \frac{N_k^{(0)} - N_j^{(0)}}{V} \frac{|D_{jk}|^2}{\varepsilon_0 \hbar} \frac{\gamma_{jk}}{(\omega_{jk}-\omega)^2 + \gamma_{jk}^2}. \tag{2C.44}$$

Clearly, Equations (2C.43) and (2C.44) are generalizations of the formulae (2.177–2.178). In fact the treatment of Chapter 2 amounts to a special case of the above situation in which the relaxation rates are equal ($\Gamma_j = \Gamma_k = \gamma_{jk} = \Gamma$) and in which the rate of relaxation from state $|j\rangle$ to $|k\rangle$ is negligible ($\Gamma_{j\to k} = 0$). The formalism developed above has therefore a much wider domain of applicability than the model discussed in Chapter 2. As we shall see in the following chapter, the occurrence of a population inversion, which is fundamental to the operation of optical amplifiers, often relies on the difference of relaxation rates of the levels involved. The above treatment is therefore well adapted to the study of such a situation.

Comparison with classical theory

Close to resonance the above results (2C.43) and (2C.44) are analogous to those found from the classical model of the elastically bound electron described in the first complement of this chapter (see Section 2A.4 of Complement 2A).

Far from resonance we must return to (2C.42) which takes into account all the available atomic transitions. In the off-resonant case γ_{jk} is much smaller than $|\omega_{jk}-\omega|$ and $|\omega_{jk}+\omega|$ and can be neglected. The imaginary part of the susceptibility χ'' is then negligible and the real part is

[7] For more details see CDG II, Complement B_{VI}.

$$\chi' = \sum_{j>k} \frac{N_k^{(0)} - N_j^{(0)}}{V} \frac{|D_{jk}|^2}{\varepsilon_0 \hbar} \frac{2\omega_{jk}}{(\omega_{jk}^2 - \omega^2)}. \tag{2C.45}$$

This result can be rewritten in terms of the dimensionless quantities f_{jk} introduced in expression (2A.57) of Complement 2A (oscillator strengths):

$$f_{kj} = \frac{2m\omega_{jk}}{\hbar q^2} |D_{jk}|^2. \tag{2C.46}$$

In this way we obtain

$$\chi' = \sum_{j>k} \frac{N_k^{(0)} - N_j^{(0)}}{V} \frac{q^2}{m\varepsilon_0} \frac{f_{kj}}{(\omega_{jk}^2 - \omega^2)}. \tag{2C.47}$$

In particular, in the common situation in which the only level to be appreciably populated is the ground state, which here we denote by $|a\rangle$, we can replace all the populations $N_j^{(0)}$ by zero, except $N_a^{(0)}$ which is equal to N. The susceptibility is then given by

$$\chi' = \sum_j \frac{N}{V} \frac{q^2}{m\varepsilon_0} \frac{f_{aj}}{(\omega_{ja}^2 - \omega^2)}. \tag{2C.48}$$

The above formula is the quantum analogue of expression (2A.53) found using the Lorentz model (see Complement 2A), which is far from resonance:

$$\chi'_{cl} = \frac{N}{V} \frac{q^2}{m\varepsilon_0} \frac{1}{(\omega_0^2 - \omega^2)}. \tag{2C.49}$$

The close resemblance between formulae (2C.48) and (2C.49) is all the more striking given that the oscillator strengths obey the Reiche–Thomas–Kuhn sum rule:

$$\sum_j f_{aj} = 1. \tag{2C.50}$$

The situation described by (2C.48) then corresponds, in some sense, to that in which the classical oscillator has several resonance frequencies, each associated with a weighting factor f_{aj}. Formulae (2C.48) and (2C.50) were, in fact, introduced from empirical arguments before the advent of quantum mechanics. One of the earliest successes of the new theory was to rigorously demonstrate their validity.[8]

2C.3 Optical Bloch equations for a two-level atom

2C.3.1 Introduction

We consider an atom with two levels, a and b, interacting with a quasi-resonant electromagnetic field and aim now to treat the interaction in a non-perturbative manner, as we

[8] A. Sommerfeld, *Optics*, Academic Press, New York (1954).

did for the case of no relaxation in Section 2.3.2. We shall present here an overview of the *optical Bloch equations* which give the evolution of the density matrix in such a situation and are a basic tool of quantum optics.[9]

Consider first of all the Hamiltonian terms arising from Equation (2C.6). The total Hamiltonian $\hat{H}$ is the sum of the atomic Hamiltonian $\hat{H}_0 = \hbar\omega_0|b\rangle\langle b|$ (level a is taken to have zero energy) and the electric dipole Hamiltonian $\hat{H}_{\mathrm{I}}(\mathrm{t}) = -\hat{D}E_0 \cos\omega t$. We find:

$$\frac{\mathrm{d}\sigma_{bb}}{\mathrm{d}t} = \mathrm{i}\Omega_1 \cos\omega t\,(\sigma_{ba} - \sigma_{ab}) \tag{2C.51}$$

$$\frac{\mathrm{d}\sigma_{aa}}{\mathrm{d}t} = -\mathrm{i}\Omega_1 \cos\omega t\,(\sigma_{ba} - \sigma_{ab}) \tag{2C.52}$$

$$\frac{\mathrm{d}\sigma_{ab}}{\mathrm{d}t} = \mathrm{i}\omega_0\sigma_{ab} - \mathrm{i}\Omega_1 \cos\omega t\,(\sigma_{bb} - \sigma_{aa}) \tag{2C.53}$$

$$\frac{\mathrm{d}\sigma_{ba}}{\mathrm{d}t} = -\mathrm{i}\omega_0\sigma_{ba} + \mathrm{i}\Omega_1 \cos\omega t\,(\sigma_{bb} - \sigma_{aa}). \tag{2C.54}$$

Here, we have introduced the resonance Rabi frequency, $\Omega_1 = -D_{ab}E_0/\hbar$ (See Chapter 2, Equations (2.86) and (2.100)). Notice that these equations are not completely independent. On the one hand, $\frac{\mathrm{d}}{\mathrm{d}t}(\sigma_{aa} + \sigma_{bb}) = 0$, which is a consequence of the fact that $\mathrm{Tr}\hat{\sigma} = \sigma_{aa} + \sigma_{bb} = 1$ for a closed system. On the other, Equations (2C.53) and (2C.54) are complex conjugates, since the density matrix is Hermitian and $\sigma_{ab}^* = \sigma_{ba}$.

In the quasi-resonant approximation, only one of the complex exponential components of $\cos\omega t$ contributes appreciably to the evolution. To convince oneself of this, one should notice that as Ω_1 tends to zero, σ_{ab} varies as $\mathrm{e}^{\mathrm{i}\omega_0 t}$, as indicated by (2C.53). Therefore only the most slowly oscillating terms are retained, which yield the dominant contribution upon integration. With this simplification the above equations give

$$\frac{\mathrm{d}\sigma_{bb}}{\mathrm{d}t} = \frac{\mathrm{i}\Omega_1}{2}(\mathrm{e}^{\mathrm{i}\omega t}\sigma_{ba} - \mathrm{e}^{-\mathrm{i}\omega t}\sigma_{ab}) \tag{2C.55}$$

$$\frac{\mathrm{d}\sigma_{aa}}{\mathrm{d}t} = -\frac{\mathrm{i}\Omega_1}{2}(\mathrm{e}^{\mathrm{i}\omega t}\sigma_{ba} - \mathrm{e}^{-\mathrm{i}\omega t}\sigma_{ab}) \tag{2C.56}$$

$$\frac{\mathrm{d}\sigma_{ab}}{\mathrm{d}t} = \mathrm{i}\omega_0\sigma_{ab} - \mathrm{i}\frac{\Omega_1}{2}\mathrm{e}^{\mathrm{i}\omega t}(\sigma_{bb} - \sigma_{aa}) \tag{2C.57}$$

$$\frac{\mathrm{d}\sigma_{ba}}{\mathrm{d}t} = -\mathrm{i}\omega_0\sigma_{ba} + \mathrm{i}\frac{\Omega_1}{2}\mathrm{e}^{-\mathrm{i}\omega t}(\sigma_{bb} - \sigma_{aa}). \tag{2C.58}$$

We now specify separately the relaxation terms appropriate for closed and open systems.

2C.3.2 Closed systems

For the case of a closed system (see Section 2C.1.3) Equations (2C.10–2C.12) combined with Equations (2C.55–2C.58) lead to:

[9] For a more detailed treatment see, for example, CDG II, Chapter V, or L. Allen and J. H. Eberly, *Optical Resonance and Two-Level Atoms*, Dover (1987).

$$\frac{d\sigma_{bb}}{dt} = i\frac{\Omega_1}{2}(e^{i\omega t}\sigma_{ba} - e^{-i\omega t}\sigma_{ab}) - \Gamma_{sp}\sigma_{bb} \tag{2C.59}$$

$$\frac{d\sigma_{aa}}{dt} = -i\frac{\Omega_1}{2}(e^{i\omega t}\sigma_{ba} - e^{-i\omega t}\sigma_{ab}) + \Gamma_{sp}\sigma_{bb} \tag{2C.60}$$

$$\frac{d\sigma_{ab}}{dt} = i\omega_0\sigma_{ab} - i\frac{\Omega_1}{2}e^{i\omega t}(\sigma_{bb} - \sigma_{aa}) - \gamma\sigma_{ab} \tag{2C.61}$$

$$\sigma_{ba}^* = \sigma_{ab}. \tag{2C.62}$$

We assume that γ is real.

In order to solve this system of equations we start by eliminating the rapid time dependence by making the change of variables:

$$\sigma'_{ba} = e^{i\omega t}\sigma_{ba} \tag{2C.63}$$

$$\sigma'_{ab} = e^{-i\omega t}\sigma_{ab} \tag{2C.64}$$

$$\sigma'_{bb} = \sigma_{bb} \tag{2C.65}$$

$$\sigma'_{aa} = \sigma_{aa}. \tag{2C.66}$$

With these changes, Equations (2C.59)–(2C.62) become:

$$\frac{d\sigma'_{bb}}{dt} = i\frac{\Omega_1}{2}(\sigma'_{ba} - \sigma'_{ab}) - \Gamma_{sp}\sigma'_{bb} \tag{2C.67}$$

$$\frac{d\sigma'_{aa}}{dt} = -i\frac{\Omega_1}{2}(\sigma'_{ba} - \sigma'_{ab}) + \Gamma_{sp}\sigma'_{bb} \tag{2C.68}$$

$$\frac{d\sigma'_{ab}}{dt} = i(\omega_0 - \omega)\sigma'_{ab} - i\frac{\Omega_1}{2}(\sigma'_{bb} - \sigma'_{aa}) - \gamma\sigma'_{ab}. \tag{2C.69}$$

The steady-state solution is then found by equating to zero the time derivatives appearing in (2C.67) and (2C.69), using $\sigma'_{aa} + \sigma'_{bb} = 1$ (conservation of the total population). We obtain in this way:

$$\sigma'_{bb} = \frac{1}{2}\frac{\Omega_1^2\frac{\gamma}{\Gamma_{sp}}}{(\omega_0 - \omega)^2 + \gamma^2 + \Omega_1^2\frac{\gamma}{\Gamma_{sp}}} \tag{2C.70}$$

$$\sigma'_{aa} = 1 - \sigma'_{bb} \tag{2C.71}$$

$$\sigma'_{ab} = i\frac{\Omega_1}{2}\frac{\gamma - i(\omega - \omega_0)}{\gamma^2 + (\omega_0 - \omega)^2 + \Omega_1^2\frac{\gamma}{\Gamma_{sp}}} \tag{2C.72}$$

$$\sigma'_{ba} = \sigma'^*_{ab}. \tag{2C.73}$$

It is apparent that the population of the excited level varies from 0 to $1/2$ when the intensity (and hence Ω_1^2) is increased. It therefore remains less than that of the ground state and so a steady-state population inversion is not obtainable in a closed two-level system as the result of coupling with a light wave.

The coherence σ_{ba} to first order in the interaction (i.e. linear in Ω_1) is determined by substituting the first-order approximation of (2C.72) into (2C.63). This yields

$$\sigma_{ba}^{(1)} = -\frac{i}{2}\frac{\Omega_1}{\gamma + i(\omega_0 - \omega)}e^{-i\omega t}. \tag{2C.74}$$

This result is in agreement with (2C.38) since $\sigma_{aa}^{(0)} - \sigma_{bb}^{(0)} = 1$ and $W_{ba}/\hbar = \Omega_1$. However, the principal advantage of Equations (2C.70–2C.73) is that their validity is not restricted to small values of the amplitude of the incident field. They therefore allow a correct description of saturation phenomena, which owe their occurrence to the appearance of the term in Ω_1^2 (proportional to the incident intensity) in the denominator. By substituting the results (2C.72) and (2C.63) in (2C.40) we obtain the following expression for the mean value of the atomic electric dipole moment:

$$\langle \hat{D} \rangle = \mathrm{i}\frac{|D_{ba}|^2}{2\hbar}\frac{\gamma + \mathrm{i}(\omega - \omega_0)}{(\omega_0 - \omega)^2 + \gamma^2 + \Omega_1^2\frac{\gamma}{\Gamma_{\mathrm{sp}}}} E\,\mathrm{e}^{-\mathrm{i}\omega t} + c.c. \tag{2C.75}$$

If the medium contains N/V atoms per unit volume, the dipole moment per unit volume is $P = \frac{N}{V}\langle \hat{D} \rangle$. Using the definitions (2.173–2.175) of Chapter 2, and on writing $D_{ba} = d$, we obtain the real and imaginary parts of the susceptibility, associated with dispersion and absorption, respectively:

$$\chi' = \frac{N}{V}\frac{d^2}{\varepsilon_0\hbar}\frac{\omega_0 - \omega}{(\omega_0 - \omega)^2 + \gamma^2 + \Omega_1^2\frac{\gamma}{\Gamma_{\mathrm{sp}}}} \tag{2C.76}$$

$$\chi'' = \frac{N}{V}\frac{d^2}{\varepsilon_0\hbar}\frac{\gamma}{(\omega_0 - \omega)^2 + \gamma^2 + \Omega_1^2\frac{\gamma}{\Gamma_{\mathrm{sp}}}}. \tag{2C.77}$$

These non-perturbative expressions for the susceptibility of an assembly of atoms are useful for many problems in nonlinear optics. They permit the description of high-intensity phenomena whilst correctly taking into account the effect of relaxation in a two-level system. In the case that the relaxation is a result solely of spontaneous emission, we have $\gamma = \Gamma_{\mathrm{sp}}/2$ and we find the result (2.188) stated without proof in Chapter 2.

Note that results (2C.76–2C.77) relate to the steady-state solution. Clearly, there will also exist interesting, transient solutions of the optical Bloch equations. For example, when the electromagnetic field is applied suddenly, the atoms are initially all in the ground state and the resulting solution yields Rabi oscillations damped by the relaxation terms.[10] This phenomenon is known as a *coherent transient* (See Section 2.3.2).

2C.3.3 Open systems

We now consider the case of an open system, so as to be able to compare the results of the density matrix treatment with the simple model of Chapter 2 and to present a more realistic model of an optical amplifier. We employ the relaxation and pumping terms given by Equations (2C.15–2C.17) and (2C.21–2C.22), respectively, assuming, however, that the rate of direct transfer from state b to state a, $\Gamma_{b\to a}$, is negligible ($\Gamma_{b\to a} \ll \Gamma_b$) so that we can write $\Gamma_{b\to a} = 0$ in Equations (2C.15) and (2C.16). Using Equations (2C.55–2C.58), we obtain the following system of Bloch equations for an open system:

[10] L. Allen and J. H. Eberly, *Optical Resonance and Two-Level Atoms*, Dover (1987).

$$\frac{d\sigma_{bb}}{dt} = i\frac{\Omega_1}{2}(e^{i\omega t}\sigma_{ba} - e^{-i\omega t}\sigma_{ab}) - \Gamma_b\sigma_{bb} + \tilde{\Lambda}_b \quad (2C.78)$$

$$\frac{d\sigma_{aa}}{dt} = -i\frac{\Omega_1}{2}(e^{i\omega t}\sigma_{ba} - e^{-i\omega t}\sigma_{ab}) - \Gamma_a\sigma_{aa} + \tilde{\Lambda}_a \quad (2C.79)$$

$$\frac{d\sigma_{ab}}{dt} = i\omega_0\sigma_{ab} - i\frac{\Omega_1}{2}e^{i\omega t}(\sigma_{bb} - \sigma_{aa}) - \gamma\sigma_{ab}. \quad (2C.80)$$

As in the previous case, this system of equations may be solved with the aid of the change of variables of (2C.63–2C.66), which renders the right-hand sides of the Bloch equations independent of time.

We now introduce the steady-state probabilities p_a and p_b of finding the atom in the states a and b in the absence of the incident electromagnetic field:

$$p_b = \frac{\tilde{\Lambda}_b}{\Gamma_b} \quad \text{and} \quad p_a = \frac{\tilde{\Lambda}_a}{\Gamma_a}. \quad (2C.81)$$

The mean population relaxation rate $\overline{\Gamma}$ is defined by

$$\frac{2}{\overline{\Gamma}} = \frac{1}{\Gamma_a} + \frac{1}{\Gamma_b}. \quad (2C.82)$$

The steady-state solution of the Bloch equations then yields the following relations for the populations of the two levels:

$$\sigma_{bb} - \sigma_{aa} = (p_b - p_a)\left(1 - \frac{\Omega_1^2\frac{\gamma}{\overline{\Gamma}}}{(\omega_0 - \omega)^2 + \gamma^2 + \Omega_1^2\frac{\gamma}{\overline{\Gamma}}}\right) \quad (2C.83)$$

$$\sigma_{bb} + \sigma_{aa} = p_b + p_a + \frac{\Omega_1^2}{2}\frac{\left(\frac{\gamma}{\Gamma_a} - \frac{\gamma}{\Gamma_b}\right)(p_b - p_a)}{(\omega_0 - \omega)^2 + \gamma^2 + \Omega_1^2\frac{\gamma}{\overline{\Gamma}}}. \quad (2C.84)$$

Notice first of all that, in the special case $\gamma = \Gamma_a = \Gamma_b$, formulae (2C.83) and (2C.84) coincide exactly with those obtained with the simple model of Chapter 2. Secondly, the non-conservation of the total population $\sigma_{aa}+\sigma_{bb}$ is a consequence of the open character of the system. This is compensated for by changes in the population of the other, unspecified, levels of the system.

From the steady-state value of the coherence,

$$\sigma_{ba} = i\frac{\Omega_1}{2}\frac{(p_b - p_a)\left[\gamma + i(\omega - \omega_0)\right]}{(\omega - \omega_0)^2 + \gamma^2 + \Omega_1^2\frac{\gamma}{\overline{\Gamma}}}e^{-i\omega t}, \quad (2C.85)$$

we derive the atomic dipole moment and thus the real and imaginary parts, χ' and χ'', of the atomic susceptibility for an ensemble comprising N/V atoms per unit volume:

$$\chi' = \frac{N_b^{(0)} - N_a^{(0)}}{V}\frac{d^2}{\varepsilon_0\hbar}\frac{(\omega_0 - \omega)}{(\omega_0 - \omega)^2 + \gamma^2 + \Omega_1^2\frac{\gamma}{\overline{\Gamma}}} \quad (2C.86)$$

$$\chi'' = \frac{N_b^{(0)} - N_a^{(0)}}{V}\frac{d^2}{\varepsilon_0\hbar}\frac{\gamma}{(\omega_0 - \omega)^2 + \gamma^2 + \Omega_1^2\frac{\gamma}{\overline{\Gamma}}}, \quad (2C.87)$$

where $N_b^{(0)} = p_b N$ and $N_a^{(0)} = p_a N$ are the numbers of atoms in levels a and b in the absence of the electromagnetic field at frequency ω. Formulae (2C.86–2C.87) generalize those of Section 2.5, obtained for the simple model assuming equal relaxation rates for the two levels. They show that the medium is amplifying if the steady-state population of the upper level of the transition, $N_b^{(0)}$, is larger than that, $N_a^{(0)}$, of the lower level. From (2C.81) this population inversion can be achieved either by a sufficiently large pumping rate of the upper level or by a sufficiently rapid decay rate of the lower level.

2C.4 The Bloch vector

The interaction of a two-level atom with an oscillating electric field is formally equivalent to the interaction with an oscillating magnetic field of a spin 1/2 particle on which is imposed an additional static magnetic field in order to lift the degeneracy of the $m = \pm 1/2$ levels.[11] This situation, which occurs experimentally in nuclear magnetic resonance,[12] was studied notably by Bloch. The vector representing the angular momentum of the spin 1/2 particle can be readily extended to the general case of a two-level atom, as we shall see in this section: it is called the Bloch vector, and will allow us to give simple geometrical interpretation of the evolution of the system.

2C.4.1 Definition

The Bloch vector $\mathbf{U}$ is the mean value, evaluated in the state of the system, of a vectorial operator $\hat{\mathbf{S}}$. The three Cartesian components of $\hat{\mathbf{S}}$ are the three Pauli matrices $\hat{\sigma}_x, \hat{\sigma}_y, \hat{\sigma}_z$ (given in Section 2A.5 of Complement 2A) divided by two.[13] Its elements are therefore the quantities u, v and w, defined by:

$$u + \mathrm{i}v = \sigma_{ba} \tag{2C.88}$$

$$w = \frac{1}{2}\left(\sigma_{aa} - \sigma_{bb}\right). \tag{2C.89}$$

The third element w depends only on the populations of the two levels under consideration, while u and v constitute the real and imaginary parts of the coherence, and are therefore also linked to the mean value of the electric dipole (Equation (2C.40)).

If the system is in a pure state, it is also described by a state vector, defined to within a global phase factor, which can therefore always be written in the form

$$|\psi\rangle = \cos\frac{\theta}{2}\,|a\rangle + \mathrm{e}^{-\mathrm{i}\phi}\sin\frac{\theta}{2}\,|b\rangle\,. \tag{2C.90}$$

[11] The strong analogy between any two-level system and a spin 1/2 particle was underlined by Feynman in *Lectures in Physics*, Vol. III, Chapter 11.

[12] See, for example, A. Abragam, *The Principles of Nuclear Magnetism*, Clarendon Press, Oxford (1961).

[13] In the case where the two-level system is that of a spin 1/2, the vector is equal, to within a factor $\hbar$, to the value of the angular momentum of the system.

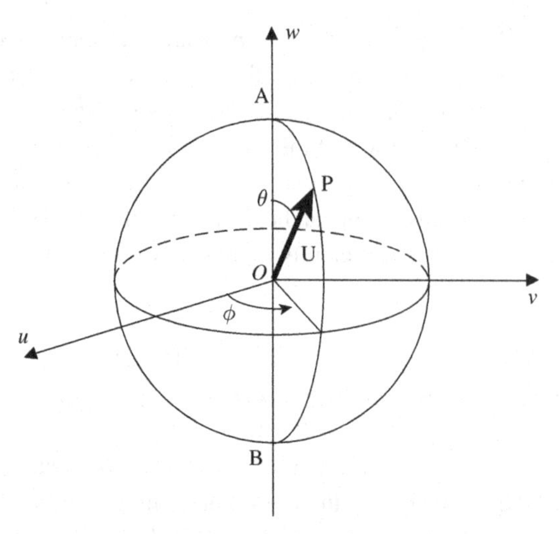

Figure 2C.3 **Bloch vector U describing a pure state of a two-level atom. Its projection on the w-axis gives the population difference. Its projection on the uOv plane ('transverse plane') gives the complex representation of the atomic coherence. Points A and B correspond to the stationary states $|a\rangle$ and $|b\rangle$.**

We can then write

$$u + \mathrm{i}v = \frac{1}{2}\sin\theta \mathrm{e}^{-\mathrm{i}\phi} \tag{2C.91}$$

$$w = \frac{1}{2}\cos\theta. \tag{2C.92}$$

For a pure state, the point P at the head of the Bloch vector is therefore on the surface of a sphere known as the Bloch sphere, of radius $\frac{1}{2}$ (Figure 2C.3). The angles ϕ and θ are the polar coordinates of the point P, or of the vector **U**. The stationary states $|a\rangle$ and $|b\rangle$ correspond, respectively, to points A and B on the axis Oz of colatitude $\theta = 0$ and $\theta = \pi$.

It is easy to see, using the inequalities relative to the elements of the density matrix, that in the general case $u^2+v^2+w^2 \leq \frac{1}{4}$: for a statistical mixture, the point P always lies inside the sphere of radius $\frac{1}{2}$. One could expect this result, since such a state is the statistical mean of a number of pure states, the corresponding representative point is the weighted centre of the points P corresponding to these pure states, all of which lie on the sphere of radius $\frac{1}{2}$. The Bloch vector for a statistical mixture therefore lies inside this sphere.

In the absence of any interaction, the populations σ_{aa} and σ_{bb} are stationary but not the coherence, which evolves at the optical frequency ω_0: $\sigma_{ba}(t) = \sigma_{ba}(0)\mathrm{e}^{-\mathrm{i}\omega_0 t}$. The Bloch vector therefore precesses about the axis Oz with angular velocity $-\omega_0$. To eliminate this evolution of the system, one makes the transformation to a *rotating reference frame*, rotating about the axis $\tilde{w}$ with the same angular velocity. In this rotating reference frame, the Bloch vector has elements $\tilde{u}$, $\tilde{v}$ and $\tilde{w}$ with

$$\tilde{u} + \mathrm{i}\tilde{v} = \sigma_{ba}(t)\mathrm{e}^{\mathrm{i}\omega_0 t} \tag{2C.93}$$

$$\tilde{w} = w = \frac{1}{2}(\sigma_{aa} - \sigma_{bb}). \tag{2C.94}$$

Comment

There exists in another domain of physics a geometrical representation analogous to that of the Bloch vector and the Bloch sphere: it is the vector representing a given polarization state on the Poincaré sphere, introduced by Poincaré to describe the polarization states of a monochromatic electromagnetic wave. This corresponds to situations which can be described in a two-dimensional Hilbert space.

2C.4.2 Effect of a monochromatic field

We now apply a resonant monochromatic field $E_0 \cos(\omega_0 t)$ to the atom, starting at time $t = 0$. In the absence of relaxation, the evolution of the density matrix of the system is governed by Equations (2C.55–2C.58). We deduce from these equations the evolution of the elements $\tilde{u}(t)$, $\tilde{v}(t)$ and $\tilde{w}(t)$ of the Bloch vector in the rotating reference frame:

$$\frac{\mathrm{d}\tilde{u}}{\mathrm{d}t} = 0, \qquad \frac{\mathrm{d}\tilde{v}}{\mathrm{d}t} = -\Omega_1 \tilde{w}, \qquad \frac{\mathrm{d}\tilde{w}}{\mathrm{d}t} = \Omega_1 \tilde{v}. \tag{2C.95}$$

These equations lead to a simple evolution for the Bloch vector $\tilde{\mathbf{U}}$ in the rotating reference frame:

$$\frac{\mathrm{d}\tilde{\mathbf{U}}}{\mathrm{d}t} = \boldsymbol{\Omega} \times \tilde{\mathbf{U}}, \tag{2C.96}$$

where the vector $\boldsymbol{\Omega}$ has elements $(\Omega_1, 0, 0)$. This is the classical equation for a precession, just like that obtained for a magnetic moment in a magnetic field, or a spinning top in a gravitational field. The Bloch vector maintains a constant modulus and sweeps out a cone around the vector $\boldsymbol{\Omega}$ at angular velocity $|\boldsymbol{\Omega}|$.

When the initial state of the system at $t = 0$ is $|a\rangle$, the Bloch vector starts at point A and remains on the sphere of radius $\frac{1}{2}$, as is normal for a pure state. It describes a circle in the (v, w) plane, thus passing periodically through points B and A: the Rabi oscillation, described in Section 2.3.2, therefore appears as a geometric precession of the Bloch vector. Both the populations and the coherence (the projection of the Bloch vector on the plane xOy) oscillate in time. If one removes the applied resonant field after a time T, the Bloch vector stops evolving in the rotating reference frame, and remains fixed at its final position. In particular, for a $\pi/2$ pulse (see Section 2.3.2), for which the interaction time is such that $\Omega_1 T = \pi/2$, the Bloch vector comes to rest along the direction v, where the coherence is maximum and the difference between the populations of the two levels is zero.

Suppose now that the applied field is of the form $E_0 \cos(\omega_0 t - \varphi)$. It is straightforward to modify Equations (2C.55–2C.58) describing the evolution of the system to take into account the phase-shift of the field. In this case, the Bloch equations in the absence of relaxation are written

$$\frac{\mathrm{d}\sigma'_{bb}}{\mathrm{d}t} = -\frac{\mathrm{d}\sigma'_{aa}}{\mathrm{d}t} = -\mathrm{i}\frac{\Omega_1}{2}\left[\mathrm{e}^{\mathrm{i}(\omega_0 t - \varphi)}\sigma'_{ba} + \mathrm{e}^{-\mathrm{i}(\omega_0 t - \varphi)}\sigma'_{ab}\right] \tag{2C.97}$$

$$\frac{\mathrm{d}\sigma'_{ab}}{\mathrm{d}t} = -\mathrm{i}\frac{\Omega_1}{2}(\sigma'_{bb} - \sigma'_{aa}). \tag{2C.98}$$

One deduces that the Bloch vector always obeys the precession equation (2C.96), but with a vector $\boldsymbol{\Omega}$ with elements $(\Omega_1 \cos(\varphi), \Omega_1 \sin(\varphi), 0)$. In the rotating reference frame, the Bloch

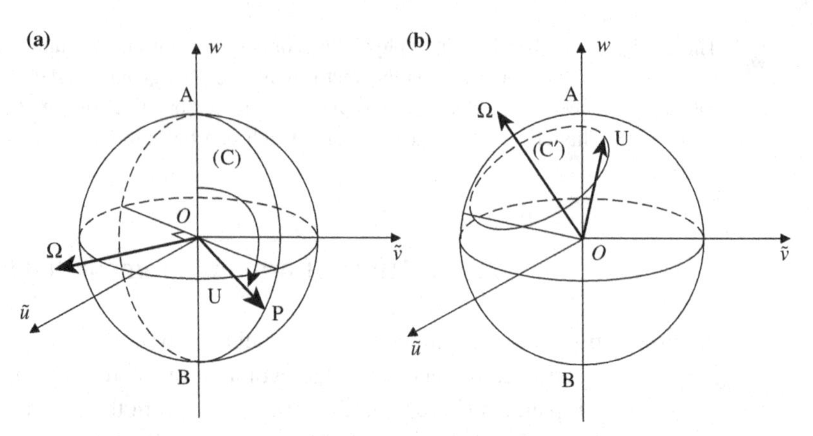

Figure 2C.4 Evolution of the Bloch vector **U** in the rotating frame under the influence of an oscillating electric field and starting from point A: a) resonant case: **U** rotates at constant velocity on the circle (C) of diameter AB and in the plane perpendicular to the vector **Ω** representing the complex field in the transverse plane; b) non-resonant case: **U** describes a cone of axis **Ω** which is no longer in the 'transverse plane' $(\tilde{u}, \tilde{v})$. Its extremity moves at constant velocity on a smaller circle (C′) which does not include point B.

vector now precesses about the direction defined by the angle φ in the (u, v) plane, that is to say, about the direction corresponding to the complex field $E_0 \exp\{-\mathrm{i}(\omega_0 t - \varphi)\}$, which is fixed in the rotating reference frame. It therefore traces a circle in the plane perpendicular to $\tilde{E}$ containing the w-axis (see Figure 2C.4).

The present discussion is only valid if the applied field is exactly resonant with the atomic transition. The phenomena become more complex in the quasi-resonant case where the frequency ω of the field is different, though remaining close, to ω_0. It can be shown that it is necessary to make the transformation to the reference frame rotating at the frequency ω of the field, in order to obtain a simple equation (of type (2C.96)) for the evolution of the system.[14] In this reference frame, the Bloch vector precesses about a vector $\mathbf{\Omega}$ which has the elements $(\Omega_1 \cos\varphi, \Omega_1 \sin\varphi, \omega - \omega_0)$, and which is no longer necessarily in the (u, v) plane (see Figure 2C.4(b)). If the system is initially in state $|a\rangle$, the Bloch vector will sweep out a cone with an opening angle smaller than $\pi/2$, which will no longer bring the vector through point B after a half-cycle, and thus the transfer from state $|a\rangle$ to state $|b\rangle$ will no longer be perfect.

2C.4.3 Effect of relaxation

In the case of an open system, the effect of relaxation is to damp the three components of the Bloch vector, generally with a different damping rate for each of the longitudinal and transverse components. The length of the Bloch vector will therefore decrease as a

[14] See CDG II, Chapter V.

function of time. The system will therefore pass from an initial pure state to a statistical mixture. If the relaxation constants Γ_a and Γ_b are equal, which corresponds to the situation envisaged in Section 2.4.1, the Bloch vector will evolve on a Bloch sphere of which the radius decreases exponentially in time. In the presence of an electromagnetic field, the Bloch vector will shorten as it precesses.

2C.4.4 Rapid adiabatic passage

Let us now consider the following situation: the system is initially in state $|a\rangle$ (the head of the Bloch vector is at point A) in the presence of a weak, far-detuned electromagnetic field, such that $\omega - \omega_0 \gg \Omega_1$. Vector $\mathbf{\Omega}$ is therefore almost aligned along the w-axis: the Rabi precession of the vector $\mathbf{U}$ therefore sweeps out a cone with a small opening angle, and the system will never stray far from its initial state.

We now vary the frequency ω of the applied field slowly with respect to the precession period $2\pi/\Omega_1$. The Bloch vector therefore precesses about a vector $\mathbf{\Omega}$ which varies in time, but always maintains a small angle with respect to $\mathbf{\Omega}$ (see Figure 2C.5): it will therefore *adiabatically follow* the evolution of $\mathbf{\Omega}$. Now consider the case where the frequency ω decreases, passes through the resonant value ω_0 and tends towards a final frequency far-detuned below resonance, such that $\omega - \omega_0 \ll -\Omega_1$. The Bloch vector will follow the evolution of vector $\mathbf{\Omega}$ (Figure 2C.5), and its head will finally arrive close to point B: there will therefore have been a *nearly complete transfer from state $|a\rangle$ to state $|b\rangle$*. It is

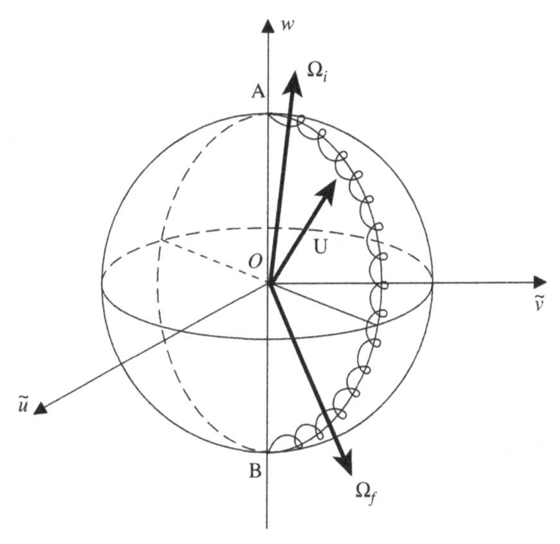

Figure 2C.5 Evolution of the Bloch vector **U** in the rotating frame under the influence of an oscillating electric field the frequency, ω, of which is swept from a value larger than the resonant frequency ω_0 to a value smaller than ω_0. **U** rotates around a vector Ω which moves from an initial position Ω_i almost aligned along OA to a final position Ω_f almost aligned along OB. The final state of the extremity of **U** is therefore very close to B.

interesting to compare this method with the application of a π pulse, because it is not necessary to control so precisely the value of the applied field and its pulse length: it is sufficient to simply sweep the frequency of the applied field across the resonance.

If one now takes into account relaxation processes, one finds that the entire process must be completed in a time that is short compared with the lifetime of the system, but sufficiently slowly that the condition of adiabaticity with respect to the precession of the Bloch vector is maintained. This explains the reason for the slightly contradictory name *rapid adiabatic passage* generally used for this technique.

2C.5 From the Bloch equations to the rate equations

If one numerically solves the Bloch equations, one finds an evolution of the coherences and populations having the form of damped oscillations, leading to the steady-state solutions given by Equations (2C.70–2C.72) (or (2C.83–2C.85), for times much longer than all of the relaxation times of the system). This would seem to conflict with the evolution of the populations derived from the Einstein equations (Equations (2.224) and (2.225)), which consist of a monotonic decay to the steady state, as expected for a set of 'rate equations'. In addition, the optical Bloch equations depend on the amplitude and phase of the electromagnetic field, whereas the Einstein equations depend only on its intensity. We shall show in this section that the rate equations introduced by Einstein can actually be derived from the Bloch equations in some limiting cases.

2C.5.1 Case of fast relaxation of coherences

Let us assume that the relaxation rate γ of the coherence is much greater than the relaxation rates Γ_a, Γ_b of the populations, and let us specifically take the example of a closed system (the same conclusions would be obtained in the case of an open system) described by Equations (2C.68) and (2C.69). Equation (2C.69) can obviously be rewritten as

$$\frac{d\sigma'_{ab}}{dt} + \gamma\sigma'_{ab} = \mathrm{i}(\omega_0 - \omega)\sigma'_{ab} - \mathrm{i}\frac{\Omega_1}{2}(\sigma'_{bb} - \sigma'_{aa}). \tag{2C.99}$$

This shows that a characteristic evolution time for σ'_{ab} is γ. Similarly, Equation (2C.68) shows that a characteristic time for the evolution of σ'_{aa} or σ'_{bb} is Γ_{sp}. If γ is very large compared to Γ_{sp}, one can consider σ'_{aa} and σ'_{bb} as constant at the scale of γ^{-1}, and look for the stationary solution of Equation (2C.99):

$$\sigma'_{ab}(t) = -\mathrm{i}\frac{\Omega_1}{2}\frac{\left[\sigma'_{bb}(t) - \sigma'_{aa}(t)\right]}{\gamma - \mathrm{i}(\omega_0 - \omega)}. \tag{2C.100}$$

This equation implies that the coherence is driven by the instantaneous value of the population difference, which is indeed a slowly varying quantity because of our assumption $\gamma \gg \Gamma_{\mathrm{sp}}$, and for which it is not possible to neglect the time derivative term with respect to the damping term in its evolution equation. We are led to the conclusion that *the slowly damped variable imposes its (slow) dynamics on the rapidly damped variable.* In addition, the 'fast' variable σ'_{ab} is now governed by an algebraic relation instead of a differential equation: it is no longer an independent dynamical variable of the problem, but a mere function of the populations. This approximation method is known as the *adiabatic elimination of fast variables* and is encountered in many branches of physics.

Inserting expression (2C.100) in Equations (2C.67) and (2C.68), one finally finds

$$\frac{\mathrm{d}\sigma'_{bb}}{\mathrm{d}t} = -\frac{\mathrm{d}\sigma'_{aa}}{\mathrm{d}t} = -\frac{\gamma}{2}\frac{\Omega_1^2}{\gamma^2 + (\omega_0 - \omega)^2}\left[\sigma'_{bb}(t) - \sigma'_{aa}(t)\right] - \Gamma_{\mathrm{sp}}\sigma'_{bb}(t). \tag{2C.101}$$

This equation links the population evolution to the value of Ω_1^2, proportional to the electromagnetic field power: the phase of the field no longer influences the evolution of the system. Recalling that the second fraction on the right-hand side of Equation (2C.101) is the saturation parameter s, and that $\gamma \equiv \Gamma_D$, one exactly retrieves the Einstein rate equation.[15] Let us note that, once we know the value of the populations by solving the rate equations, we also have Equation (2C.100), which allows us to determine the atomic coherence and therefore to calculate the atomic susceptibility.

2C.5.2 Case of an optical field of finite coherence time

We have considered so far that the electromagnetic field interacting with the atom had a well-defined phase, and was therefore perfectly monochromatic. However, for most optical sources, the phase, and sometimes also the amplitude, are random variables. The source is said to have a finite *temporal coherence*, and therefore a non-zero frequency bandwidth.

We have shown in the previous section that the Bloch vector precesses around a direction depending on the phase of the complex field. Therefore, the phase of the atomic coherence is directly related to the phase of the applied field: if it is composed of wave packets the phase of which randomly varies from one packet to the other, the Bloch vector will precess around randomly varying directions: its projection on the (u, v) plane will randomly vary and therefore average to zero over times much longer than the field coherence time. The atomic coherence is therefore rapidly destroyed in a field with small temporal coherence: one can expect that in this case the evolution will be governed only by the populations. This is what we shall see more precisely in this section.

To describe in a quantitative way the finite temporal coherence of a light source, one usually introduces the field autocorrelation function:[16]

$$G(\tau) = \overline{\mathcal{E}(t)\mathcal{E}^*(t+\tau)}, \tag{2C.102}$$

where $\mathcal{E}(t)$ is the complex field

$$\mathcal{E}(t) = \frac{E_0}{2}\,\mathrm{e}^{\mathrm{i}(\varphi-\omega t)}, \tag{2C.103}$$

[15] In contrast to Equation (2.224) there is no pumping term in Equation (2C.101) because such a process was not introduced in the present discussion.

[16] L. Mandel and E. Wolf, *Optical Coherence and Quantum Optics*, Cambridge University Press (1995).

and the overline stands for the average over the random process responsible for the field fluctuations. The autocorrelation function G depends only on $|\tau|$ because the source is assumed to be *stationary* and tends to zero when τ exceeds a characteristic time T_c, called the *correlation time* of the field fluctuations. Let us note that the *Wiener–Khintchine theorem* establishes that the spectral density of the light source is the Fourier transform of the field autocorrelation function. This means that T_c is of the order of the inverse of the source bandwidth, and is very small for most classical sources such as spectral lamps. We shall assume in the following that the light source is much broader than the resonance linewidth, equal to the dipole relaxation rate γ, which implies that $\gamma T_c \ll 1$.

Taking as an initial condition $\sigma'_{ab}(t = 0) = 0$, we can write Equation (2C.98) in the integral form:

$$\sigma'_{ab}(t) = -\frac{\mathrm{i}D_{ab}}{\hbar}\int_0^t \mathrm{d}t'\mathcal{E}(t')[\sigma'_{bb}(t') - \sigma'_{aa}(t')]\mathrm{e}^{-\gamma(t-t')}. \tag{2C.104}$$

This solution can in turn be inserted in the Bloch equation (2C.97). We then obtain the following equation for the upper state population:

$$\frac{\sigma'_{bb}}{\mathrm{d}t} = -\Gamma_{\mathrm{sp}}\sigma'_{bb} - \frac{D_{ab}^2}{\hbar^2}\left[\int_0^t \mathrm{d}t'\mathcal{E}(t)\mathcal{E}^*(t')(\sigma'_{bb}(t') - \sigma'_{aa}(t'))\mathrm{e}^{-\gamma(t-t')} + c.c.\right]. \tag{2C.105}$$

In order to get the average evolution of the population (i.e. the evolution averaged over times much longer than the coherence time), we must take the average of Equation (2C.105) over the fluctuations of the source, which amounts to replacing $\mathcal{E}(t)\mathcal{E}^*(t')$ by the correlation function $G(t-t')$, which takes non-negligible values only when $|t-t'| \le T_c$. Let us now assume that the population difference $\sigma'_{bb} - \sigma'_{aa}$ varies slowly during a field correlation time T_c, usually very short. We then make a small error by replacing $\sigma'_{bb}(t') - \sigma'_{aa}(t')$ by $\sigma'_{bb}(t) - \sigma'_{aa}(t)$ (short memory approximation, or 'Markov approximation'). If in addition we only consider times t much longer than T_c, we will add negligible integrand by replacing in the integral the boundaries $\int_0^t$ by $\int_0^\infty$. With these two approximations, we finally get the following rate equation for the populations:

$$\frac{\sigma'_{bb}}{\mathrm{d}t} = -\Gamma_{\mathrm{sp}}\sigma'_{bb} - B(\sigma'_{bb} - \sigma'_{aa}). \tag{2C.106}$$

The value of coefficient B is obtained by setting $\tau = t - t'$, using the fact that $G(\tau) = G(-\tau)$, and noting that $G(\tau)$ is narrower than $\mathrm{e}^{-\gamma|\tau|}$:

$$B = \frac{D_{ab}^2}{\hbar^2}\int_{-\infty}^{+\infty} \mathrm{d}\tau G(\tau)\mathrm{e}^{-\gamma|\tau|} \approx \frac{2D_{ab}^2}{\gamma\hbar^2}G(0). \tag{2C.107}$$

According to the definition (2C.102), $2G(0)$ is nothing else than the mean intensity of the source, so that we finally retrieve a rate equation for the population, in the form of the Einstein equations.[17]

[17] Let us mention that Einstein introduced his famous rate equations (2.224) and (2.225) in 1917 to account for the spectral distribution of thermal radiation, which is obviously a very broadband source. Its phenomenological equations were therefore perfectly correct in the case that he considered.

In conclusion, rate equations yield a very good approximation of the exact atomic evolution in at least two situations:

- when the relaxation rate of the coherences is much greater than the population relaxation rates, which is the case in dense systems, which constitute the amplifying medium of most laser systems;
- when the atom interacts with broadband light, characterized by a coherence time which is much smaller than the atomic relaxation times γ^{-1}, Γ_a^{-1}, Γ_b^{-1}.

In contrast, the interaction between a low-pressure atomic sample (for which γ, Γ_a and Γ_b have the same order of magnitude) with a monochromatic field of high temporal coherence is one of the cases where rate equations often give wrong results: some aspects of this situation, in which the atomic coherence plays a leading role, will be studied in Complement 2D.

Comment Note that in the steady state, $\mathrm{d}\sigma_{ab}/\mathrm{d}t$, is equal to zero and can obviously be neglected with respect to $\gamma\sigma_{ab}$. This implies that the steady-state solutions of the optical Bloch equations and of the rate equations coincide, whatever the values of the damping rates.

2C.6 Conclusion

The set of optical Bloch equations constitutes an essential tool in the study of atom–light interactions. In the domain of applicability of the quasi-resonant approximation, which corresponds to many frequently encountered experimental situations, they are exact, affording as they do a quantum description of the atomic evolution which properly takes into account the effects of relaxation phenomena.

They have the advantage over simpler, approximate treatments in that they describe correctly two types of phenomena that are fundamental to the interactions of atoms with laser light. These are, firstly, effects appearing at high intensity, including saturation, but more generally all kinds of nonlinear effects and, secondly, the existence of coherences between atomic levels as a result of interaction with the light field. It is in fact the coherences that determine the phase and frequency of the induced electric dipole moment (see Equation (2C.40)). Since the re-emitted field derives directly from the existence of this induced dipole moment,[18] it is the coherences that determine the precise characteristics of the radiation emitted by an atom subjected to an arbitrary incident field.[19]

The system of optical Bloch equations can readily be generalized to the case of systems comprising more than two levels, but, in practice, the resulting system of equations is often

[18] See Complement 2B.

[19] See CDG II, Chapter V.

inextricable if the number of levels under consideration is greater than three.[20] It is then necessary to return to simpler, more approximate treatments: at low intensity perturbation methods correctly give the linear susceptibility and at high intensity one can resort to rate equation methods which correctly describe phenomena such as saturation, but which do not account for phenomena related to the existence of coherences between levels. The experience obtained in manipulating the optical Bloch equations for two- and three-level systems can be of considerable use in determining the simplifying approximations that can usefully be made.

[20] Note that in this case the determination of the relaxation terms is far from obvious. There exist, for example, phenomena such as the transfer of coherence between pairs of levels as the result of spontaneous emission, which cannot occur in a two-level system. The characteristics of such processes are, however, of importance in numerous phenomena. They are described in C. Cohen-Tannoudji, Thesis, *Annales de Physique (Paris)*, **7**, 423 (1962) (and reproduced in C. Cohen-Tannoudji, *Atoms in Electromagnetic Fields*, World Scientific (1994)).

2D

Complement 2D Manipulation of atomic coherences

We have seen in Chapter 2 that an atom is not necessarily in a state of well-defined energy, such as $|a\rangle$ or $|b\rangle$. It can be in a linear superposition of these states, of type

$$|\psi(t)\rangle = \gamma_a(t)\,|a\rangle + \gamma_b(t)e^{-i\omega_0 t}\,|b\rangle\,. \tag{2D.1}$$

These states are particular cases of a more general category of states, which are not necessarily pure states and are therefore described by a density matrix $\hat{\sigma}$ (see Complement 2C) of which the non-diagonal element σ_{ab}, called the coherence, is non-zero. They are characterized by a Bloch vector which does not lie on the w-axis (see Figure 2C.3 of Complement 2C). In contrast to the stationary states of the atom, they have a non-zero electric dipole moment $\langle\hat{D}\rangle = 2\mathrm{Re}[\sigma_{\mathrm{ab}}(\mathrm{t})\mathrm{D}_{\mathrm{ba}}]$ and are therefore strongly coupled to the radiation field.

The control and manipulation of atomic states of this type are at the origin of a large number of interesting physical effects which have been the subject of active research during past decades. We will give several examples in this complement. Let us also point out that these states can be considered a resource for information processing at the quantum level. They constitute possible 'qubits', the extension of the classical two-level system, known as a 'bit', to the quantum domain. The manipulation of qubits for *quantum information processing* is the subject of a very active field of research, because it has been shown theoretically that certain operations, such as the factorization of large members, could be carried out much more efficiently using qubits than using classical bits (see Complement 5E).

Coherences can be induced and manipulated either by acting directly on the two-level system, or by using a third auxiliary level. In this complement we will study each situation in turn, giving some examples of the manipulation of these coherences: 'Ramsey fringe' spectroscopy and photon echoes for two-level systems (Section 2D.1), and coherent population trapping and electromagnetically induced transparency for the three-level system (Section 2D.2).

2D.1 Direct manipulation of a two-level system

2D.1.1 Generalities

We start with the case where the relaxation time of the coherences is sufficiently long to allow the manipulation of the coherences at leisure. For example, one can use the two hyperfine sub-levels of the ground state of an alkali atom, such as sodium or caesium, which are only separated by a few GHz. The relaxation time of the coherences is not limited by the spontaneous emission lifetime, of the order of several million years, but by parasitic effects, such as collisions or magnetic field inhomogeneities which can be suppressed. In these conditions, relaxation processes can be neglected.

To manipulate coherences it is sufficient, for example, to apply a monochromatic field (radio-frequency in this case) resonant with the transition, for a certain time (see Section 2.3). This will have the effect of making the Bloch vector of the two-level system precess through a certain angle (see Section 2C.4). The most efficient configuration is that in which an exactly resonant pulse is applied, causing the Bloch vector to turn through an angle $\pi/2$ about the vector representing the complex field in the (u, v) plane, from its initial position along the w-axis. The Bloch vector therefore comes to rest in the (u, v) plane. This creates a coherence with a phase which is directly linked to that of the applied field. By applying a sequence of pulses of defined durations to the atom, one can then make the state vector of the system evolve in a controlled manner. This technique, first developed in nuclear magnetic resonance is also currently used in the domain of atom–laser interaction. We will give two examples of this in the following sections.

2D.1.2 Ramsey fringes

We will now consider in detail the simplest field pulse configuration, invented by N. Ramsey. This configuration is frequently used and has important applications in metrology. It consists of the application of a sequence of two pulses to the atom, that is, a field of the form $E \cos(\omega_0 t)$ which is applied into two pulses: a first $\pi/2$ pulse between $t = 0$ and t_1, then a second $\pi/2$ pulse between T and $T + t_1$. The atom evolves freely during the time interval $T - t_1$ between the pulses, which is long compared to the pulse length t_1. We will study the evolution of the Bloch vector in the reference frame which rotates at the frequency of the applied field, supposing that the atom is initially in state $|a\rangle$ (see Figure 2D.1).

Let us first treat the case of exact resonance: $\omega = \omega_0$. The first $\pi/2$ pulse causes the head of the Bloch vector to pass from point A to point C on the v-axis and therefore creates a maximal atomic coherence. During the dark phase, the atom is described by point C, fixed in the rotating reference frame. The second $\pi/2$ pulse causes the head of the Bloch vector to pass from point C to point B: the two $\pi/2$ pulses give an overall rotation of an angle π. At the end of the two-pulse sequence, the system finds itself in the excited state, just as though it had been subject to a single π pulse.

We now consider the quasi-resonant case, where $\omega \neq \omega_0$, but ω remains sufficiently close to ω_0 that the two pulses are close to being $\pi/2$ pulses. After the first pulse the Bloch vector $\mathbf{U}$ is practically in the (u, v) plane and its head is close to point C. Between the first and second pulse, the vector representing the complex field $\tilde{E}$ is fixed along the u-axis, but not the Bloch vector, which turns about w (in the fixed reference frame) at the angular frequency ω_0 and not ω. In the reference frame rotating at frequency ω, at the instant T which marks the beginning of the second pulse, the Bloch vector will have turned by an angle $(\omega_0 - \omega)T$, which is non-negligible if T is sufficiently long. The following cases are particularly important:

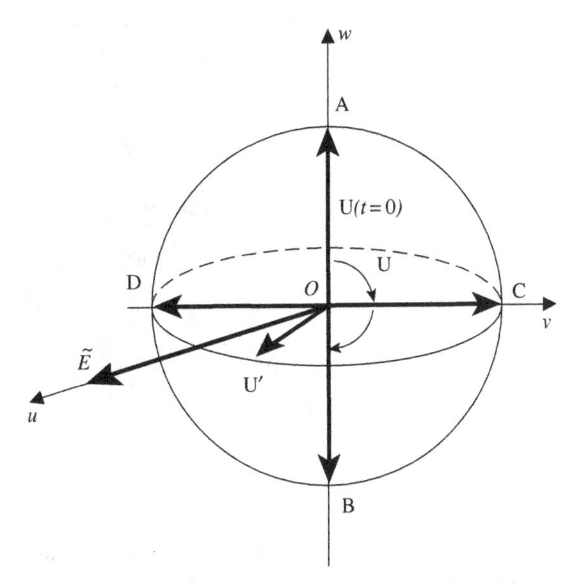

Figure 2D.1 Evolution of the Bloch vector for an atom subjected to two $\pi/2$ pulses separated by a free-evolution time T in the reference frame rotating at a frequency ω of the applied field. At the end of the first pulse, the head of the Bloch vector lies close to point C. In the resonant case the Bloch vector remains fixed between the two pulses, whereas in the non-resonant case it rotates in the plane uOv until position U′.

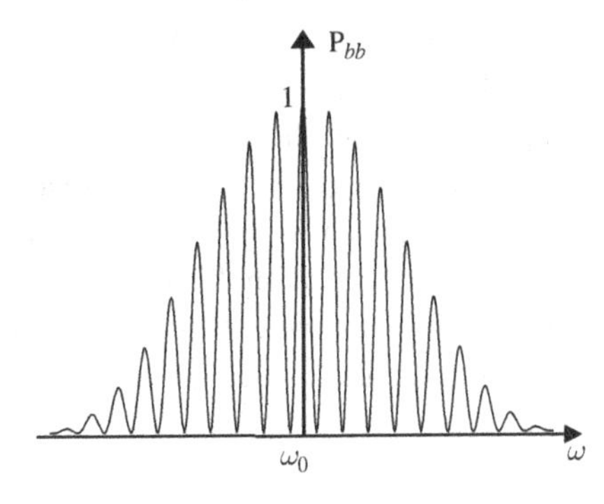

Figure 2D.2 Probability (P) of finding the system in an excited state after application of two Ramsey pulses of frequency ω.

(1) if $(\omega_0 - \omega)T \equiv \pi(2\pi)$, P will have made a half-turn between the two pulses and will be in a position diametrically opposite its original position, at point D. The second pulse therefore brings the Bloch vector from D to A: the atom returns to its original state $|a\rangle$;
(2) if $(\omega_0 - \omega)T \equiv 0(2\pi)$, P will have made a complete turn before the second pulse, and the two rotations will sum again. The atom is transferred to state $|b\rangle$.

A detailed calculation, taking into account the fact that the pulses are not exactly $\pi/2$, shows that the probability of the atom ending up in the upper state at the end of the two pulses, as a function of the frequency of the applied field, has the form given in Figure 2D.2. We observe oscillations, or 'fringes', of short period $2\pi/T$, which superpose with the broad

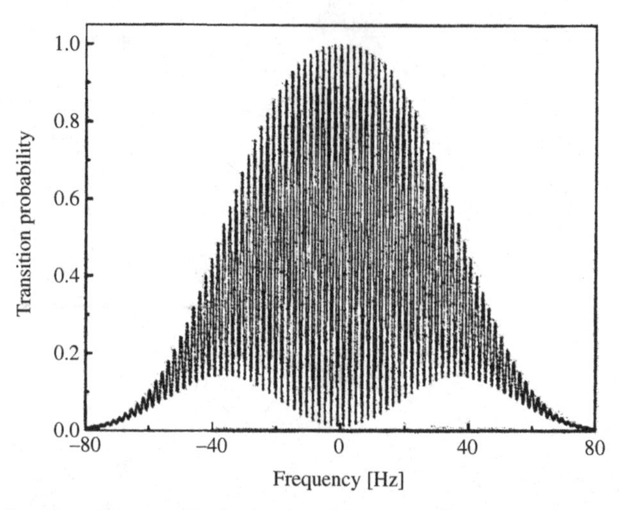

Figure 2D.3 Ramsey fringes in a fountain atomic clock. The plot represents the number of atoms in the excited state as a function of the frequency of the applied field. The width of the central fringe is 2 Hz, allowing the local oscillator to be stabilized with an exactitude better than 10^{-14}.

resonance (of width $2\pi/t_1$), centred on $\omega = \omega_0$, which we would obtain for a single π pulse of duration t_1. The width of the central peak is therefore linked, not to the atom–light interaction time, $2t_1$, but to the time between the two pulses which, respectively, create and read the atomic coherence. For example, if $T = 1$ s, the frequency width of the central peak is 1 Hz. This type of response allows us to pinpoint the resonance $\omega = \omega_0$ with a precision better than that obtained with a single pulse. We are therefore not surprised to find that this technique is used in frequency metrology. In particular, the caesium atomic clocks used to define the second have operated on this principle since the 1960s. Furthermore, the use of cold atoms has allowed the time T between the two pulses to be increased considerably. Figure 2D.3 shows a signal recorded from a caesium 'atomic fountain' clock, which uses cold atoms.[1] The pre-cooled atoms are launched vertically upwards with a small velocity and fall back down due to gravity. In this way, the atoms pass twice through a cavity containing the field at the transition frequency. The height of the atom trajectory is of the order of a metre, and the time separating the two pulses, as seen by the atoms, is of the order of a second. The 2 Hz width of the central fringe allows a clock stability of the order of $10^{-14}/\sqrt{t}$, where t is the integration time in seconds.

2D.1.3 Photon echoes

This technique was invented by E. Hahn in the context of nuclear magnetic resonance, and was later extended to the field of optics. It is applicable to an *inhomogeneous ensemble of atoms*, where each atom has a different resonance frequency. These frequencies ω_0 are

[1] A. Clairon *et al.*, Ramsey Resonance in a Zacharias Fountain, *Europhysics Letters* **16**, 165 (1991).

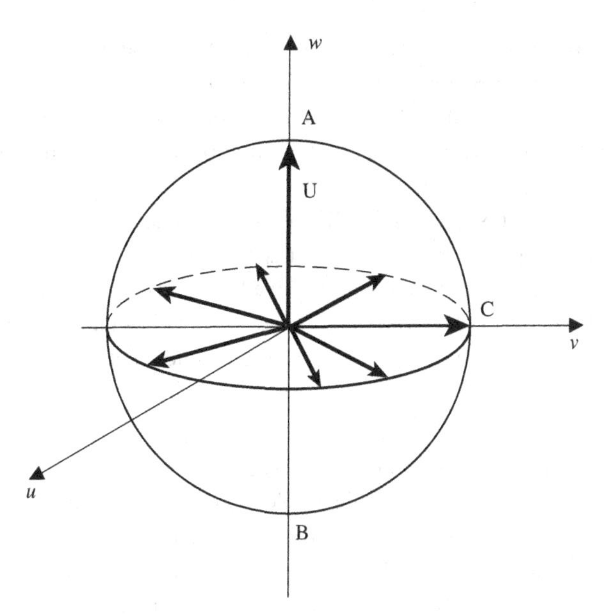

Figure 2D.4 Distribution of Bloch vectors after free precession of an inhomogeneous ensemble of atoms.

spread over a band centred on $\overline{\omega_0}$, of width Δ_0, called the inhomogeneous width. This situation is often encountered, for example in solids where atoms situated at different lattice sites of a crystal have different environments, which induce different shifts of the transition frequency. Similarly, in atomic vapours, the different velocities of the atoms result in different Doppler shifts for the atomic transition. The inhomogeneous width Δ_0 is often broader than the natural linewidth Γ of the transition and can mask single-atom phenomena. For example, the absorption spectrum of an ensemble of atoms is broadened by this effect which prevents the natural linewidth Γ from being measured. Photon echo is a way to get rid of inhomogeneous broadening in high-precision spectroscopy, even in the presence of this 'parasitic' broadening.

The sequence of pulses for photon echo comprises first a $\pi/2$ pulse at time $t = 0$, followed at a later time t_1 by a π pulse. Between the two pulses, the atoms evolve freely. We suppose that the applied field has a frequency ω equal to the centre frequency $\overline{\omega_0}$ and we consider the evolution of the Bloch vector representing an atom initially in state $|a\rangle$, for which the Bohr frequency is $\omega_0 \simeq \overline{\omega_0}$ (Figure 2D.4). We place ourselves in the reference frame rotating at the frequency $\overline{\omega_0}$ of the field: the first pulse brings the Bloch vector from the initial position A to a position close to C (exactly to this point C if $\omega_0 = \overline{\omega_0}$). During the period between the pulses, the Bloch vector evolves freely and turns through an angle $\omega_0 t_1$ about Oz in the fixed reference frame, that is through an angle

$$\phi = (\omega_0 - \overline{\omega}_0)t_1 \tag{2D.2}$$

in the rotating reference frame. These angles are different for different atoms, and in the case where $\Delta_0 t_1 \gg 1$ the different Bloch vectors are spread in a uniform way along the equator of the Bloch sphere in the (u, v) plane at time t_1: the polarization vector $\mathbf{P}$ obtained by averaging the electric dipole moments of the ensemble of atoms is therefore zero and the atomic ensemble does not emit any light.

In fact, each atom has a coherence and therefore a dipole moment, but this is masked by the effect of averaging over the inhomogeneous ensemble of dipoles with phases spread over 2π. It is the second pulse which will allow these dipole moments to rephase. The second π pulse causes the Bloch vector to make a half-turn around the u-axis. Immediately after this pulse, the Bloch vector is again in the (u, v) plane, and makes an angle $-\phi$ to the Ox-axis. It will subsequently evolve freely, just as it did during the previous period with no field, between the two pulses. When time t_2 has elapsed after the second pulse, the Bloch vector will have turned through a total angle equal to

$$\phi' = -\phi + (\omega_0 - \overline{\omega}_0)t_2. \tag{2D.3}$$

This angle is zero when $t_2 = t_1$, whatever the resonance frequency of the atom considered: at this instant, all the atomic dipole moments will point along the u direction. Their average value will therefore be non-zero and a macroscopic polarization $\mathbf{P}$ will appear transiently in the atomic ensemble. This polarization will serve as a source term in Maxwell's equations and lead to a pulse of light, termed a 'photon echo' because it appears after a time equal to the time separating the first two pulses. In fact, the processes which lead to relaxation of the atomic coherences cause the intensity of the echo to become weaker as the time t_1 becomes longer. The measurement of this decay is often used to obtain the relaxation rate of the atomic coherence.

2D.2 Use of a third level

The precise manipulation of atomic coherences by directly addressing the atomic transition is delicate, because it requires very precise control of the amplitude, homogeneity and pulse length of the applied field. However, the situation is different if one uses an auxiliary level and field at one or more frequencies to address the transitions linking the auxiliary level to the two levels between which one wishes to create a coherence. Using this method, a coherence can be produced in a *stationary state of the system.* We have already seen an example of the use of an auxiliary level in the phenomenon of optical pumping (Complement 2B), in which a population distribution far from thermodynamic equilibrium is obtained. We will see the same phenomenon used in a different configuration to produce atomic coherences.

2D.2.1 Coherent population trapping

Case of degenerate levels

Consider an atom with the energy level configuration shown in Figure 2D.5(a) where, for example, $|a_1\rangle$ and $|a_2\rangle$ are Zeeman sub-levels (and are therefore degenerate in zero

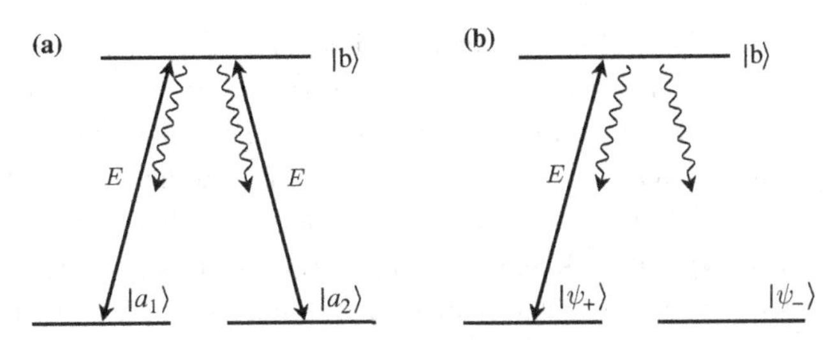

Figure 2D.5 Configuration of energy levels for coherent population trapping for the case in which a field of amplitude E couples two degenerate levels $|a_1\rangle$ and $|a_2\rangle$ to an excited state $|b\rangle$: (a) in the usual atomic basis states; (b) in the basis of coupled and non-coupled states.

magnetic field) and $|b\rangle$ is an excited state. The atom is subjected to an electromagnetic field of frequency ω and polarization $\boldsymbol{\varepsilon}$. We define:

$$d_1 = \langle a_1 | \mathbf{D}.\boldsymbol{\varepsilon} | b \rangle \tag{2D.4}$$

$$d_2 = \langle a_2 | \mathbf{D}.\boldsymbol{\varepsilon} | b \rangle , \tag{2D.5}$$

the dipole matrix elements, assumed to be real for simplicity. Let us introduce the orthogonal quantum states:

$$|\psi_+\rangle = \frac{d_1 |a_1\rangle + d_2 |a_2\rangle}{\sqrt{d_1^2 + d_2^2}} \tag{2D.6}$$

$$|\psi_-\rangle = \frac{d_2 |a_1\rangle - d_1 |a_2\rangle}{\sqrt{d_1^2 + d_2^2}}. \tag{2D.7}$$

The degenerate ground level can be described just as well using the basis of states $|\psi_+\rangle$ and $|\psi_-\rangle$ as using the basis of states $|a_1\rangle$ and $|a_2\rangle$. We have redrawn the energy level configuration in Figure 2D.5(b) in this new basis set. The corresponding dipole matrix elements are

$$\langle \psi_+ | \mathbf{D}.\boldsymbol{\varepsilon} | b \rangle = \sqrt{d_1^2 + d_2^2} \tag{2D.8}$$

$$\langle \psi_- | \mathbf{D}.\boldsymbol{\varepsilon} | b \rangle = 0. \tag{2D.9}$$

This shows that *state $|\psi_-\rangle$ is not coupled to the excited state b by the applied field.* This can be interpreted as destructive interference between the transition amplitudes connecting $|b\rangle$ to $|a_1\rangle$ and $|a_2\rangle$. In contrast, the rates of de-excitation by spontaneous emission from $|b\rangle$ to $|\psi_+\rangle$ and $|\psi_-\rangle$, Γ_+ and Γ_-, are both non-zero in principle, because the dipole matrix element does not cancel for all the possible polarizations of the spontaneously emitted photon. One is therefore in the same situation as described for optical pumping (Complement 2B): after a certain number of cycles of absorption and spontaneous emission, the system finds itself in state $|\psi_-\rangle$. In fact, the spontaneous emission from the excited state populates states $|\psi_-\rangle$ and $|\psi_+\rangle$, but no process is capable of exciting the system out of $|\psi_-\rangle$, which is therefore a 'dark' state where atoms accumulate.

One has therefore put the atom into state $|\psi_-\rangle$, that is to say, *a well-defined coherent superposition of the two ground states*, in the stationary regime, by simply illuminating the atom with a monochromatic field. Let us again point out that, once in this state, the atom no longer interacts with the applied field. The atom is no longer subject to optical pumping cycles and no longer fluoresces, hence the term 'dark state'.

This type of dark state has been exploited in an efficient method for motion-cooling ensembles of atoms and allows, in principle, arbitrarily low temperatures to be reached.[2] This method is known as velocity-selective coherent population trapping, which is the subject of Complement 8A.

If the two lower levels are not exactly degenerate, the states $|\psi_+\rangle$ and $|\psi_-\rangle$ are no longer stationary. One obtains

$$|\psi_-(t)\rangle = \frac{1}{\sqrt{|d_1|^2 + |d_2|^2}} \left(d_2^* e^{-iE_{a_1}t/\hbar} |a_1\rangle - d_1^* e^{-iE_{a_2}t/\hbar} |a_2\rangle \right). \qquad (2D.10)$$

The relative phase between the two components of the state evolve as $\exp\{i(E_{a_1} - E_{a_2})t/\hbar\} = e^{i\omega_{12}t}$. When $t = t_0 = \pi/\omega_{12}$, there is a change of sign in the superposition and the nulling of the matrix element does not occur. The state $|\psi_-(t)\rangle$ is now coupled to the field and is therefore not a dark state. One can see the fragility of the dark state created by this method, with respect to a lifting of the ground state degeneracy. In the case of Zeeman sub-levels, this could occur as a result of stray magnetic fields. This phenomenon has been used to create highly sensitive magnetometers.

Non-degenerate levels

We will now see that there is another, slightly more complicated, field configuration which allows the creation of a coherent superposition of two states of different energy.

Let us consider an atom with the configuration shown in Figure 2D.6, where the levels $|a_1\rangle$ and $|a_2\rangle$ have different energies (for example, two hyperfine sub-levels of the ground state). This atom is now subjected to two electromagnetic fields $\mathbf{E}_1(t) = E_1\boldsymbol{\varepsilon}_1 \cos(\omega_1 t)$ and $\mathbf{E}_2(t) = E_2\boldsymbol{\varepsilon}_2 \cos(\omega_2 t + \phi)$, which are quasi-resonant with the transitions $|a_1\rangle \to |b\rangle$ and $|a_2\rangle \to |b\rangle$, respectively. We introduce the two Rabi frequencies for these interactions:

$$\Omega_1 = -\langle a_1|\,\mathbf{D}.\boldsymbol{\varepsilon}_1\,|b\rangle\, E_1/\hbar \qquad (2D.11)$$

$$\Omega_2 = -\langle a_2|\,\mathbf{D}.\boldsymbol{\varepsilon}_2\,|b\rangle\, E_2/\hbar. \qquad (2D.12)$$

The interaction Hamiltonian, restricted to the levels involved, is written as a 3×3 matrix:

$$\hat{H}_1 = \begin{bmatrix} 0 & 0 & \Omega_1 \cos\omega_1 t \\ 0 & 0 & \Omega_2 \cos(\omega_2 t + \phi) \\ \Omega_1 \cos\omega_1 t & \Omega_2 \cos(\omega_2 t + \phi) & 0 \end{bmatrix}. \qquad (2D.13)$$

[2] F. Bardou, J.-P. Bouchaud, A. Aspect and C. Cohen-Tannoudji, *Lévy Statistics and Laser Cooling*, Cambridge University Press (2002).

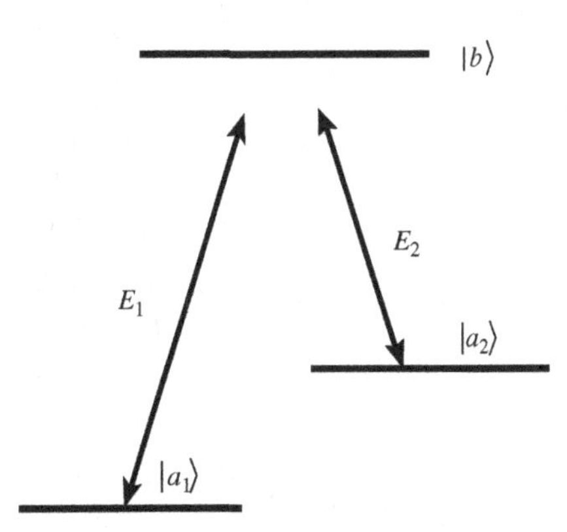

Figure 2D.6 Configuration of energy levels for coherent population trapping for the case in which two fields of amplitudes E_1 and E_2 and frequencies ω_1 and ω_2 couple two non-degenerate levels $|a_1\rangle$ and $|a_2\rangle$, respectively, to an excited state $|b\rangle$.

Now consider the state

$$|\psi_-(t)\rangle = \frac{1}{\sqrt{\Omega_1^2+\Omega_2^2}}\left(\Omega_2 e^{-iE_{a_1}t/\hbar}|a_1\rangle - \Omega_1 e^{-iE_{a_2}t/\hbar - i\phi}|a_2\rangle\right). \tag{2D.14}$$

The matrix element coupling the state $|\psi_-(t)\rangle$ to the excited state $|b(t)\rangle = e^{-iE_b t/\hbar}|b\rangle$ is given by

$$\begin{aligned}\langle\psi_-(t)|\hat{H}_1|b(t)\rangle =& \frac{\Omega_1\Omega_2}{\sqrt{\Omega_1^2+\Omega_2^2}}\left(\cos\omega_1 t\, e^{-i(E_b-E_{a_1})t/\hbar}\right.\\ &\left.-\cos(\omega_2 t+\phi)\, e^{-i(E_b-E_{a_2})t/\hbar+i\phi}\right).\end{aligned} \tag{2D.15}$$

In the quasi-resonant approximation for the two applied fields, one considers only low-frequency components of this matrix element, which are the dominant contribution to the transition probability. This results in

$$\langle\psi_-(t)|\hat{H}_1|b(t)\rangle \simeq \frac{\Omega_1\Omega_2}{2\sqrt{\Omega_1^2+\Omega_2^2}}\left(e^{-i(E_b-E_{a_1}-\hbar\omega_1)t/\hbar} - e^{-i(E_b-E_{a_2}-\hbar\omega_2)t/\hbar}\right). \tag{2D.16}$$

This matrix element is therefore zero at all times if the following condition is respected:

$$\hbar(\omega_1-\omega_2) = E_{a_2} - E_{a_1}. \tag{2D.17}$$

This condition corresponds to the same detuning from resonance for the two transitions $|a_1\rangle \to |b\rangle$ and $|a_2\rangle \to |b\rangle$, that is to say, a Raman resonance (see Section 2.1.5). When this condition is satisfied, no transition is possible, at any time, between the states $|\psi_-(t)\rangle$ and $|b(t)\rangle$. This is a situation analogous to that of the previous paragraph. The state $|\psi_-(t)\rangle$ is a non-coupled state and plays the role of the dark state: all the atoms will rapidly fall into this state if the condition of Equation (2D.17) is fulfilled. The system finds itself in

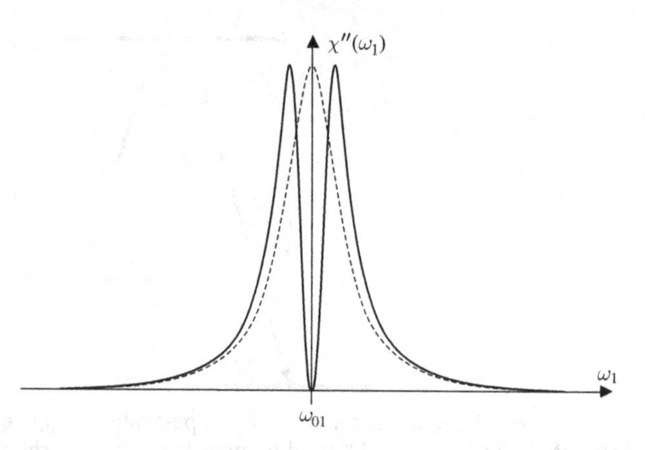

Figure 2D.7 **Absorption coefficient $\chi''(\omega_1)$ for the transition $|a_1\rangle \to |b\rangle$ in the presence (solid lines) or absence (dotted line) of a field resonant with the transition $|a_2\rangle \to |b\rangle$ as a function of the frequency ω_1 of the first field.**

a coherent superposition, which is not stationary in this case, of the two ground states of the atom. The atom ceases to interact with the fields and becomes transparent. Note that the phase of the coherence of the state created in this way is directly linked to the relative phase ϕ between the two applied fields, which should therefore have a mutual temporal coherence sufficiently long that the atomic coherence created does not average to zero during the interaction time.

2D.2.2 Electromagnetically induced transparency

Principle

We still consider the configuration of the previous paragraph, with two quasi-resonant fields applied to the three-level system. The frequency ω_1 of the first field can vary around the resonant frequency ω_{01} of the transition $|a_1\rangle \to |b\rangle$, whereas the second field is taken to be exactly on resonance with the transition $|a_2\rangle \to |b\rangle$ ($\omega_2 = \omega_{02} = (E_b - E_{a_2})/\hbar$). The variation of the absorption rate of the first field by the atomic vapour as a function of the detuning from resonance $\delta_1 = \omega_1 - \omega_{01}$ is given in Figure 2D.7, in the presence or absence of the second field.

When there is only the first field, we obtain the usual Lorentzian absorption spectrum, centred on $\delta_1 = 0$, as we have seen in Chapter 2. At exact resonance, the atomic medium is opaque because it absorbs.

A calculation using the Bloch equations for the three-level system allows the absorption rate in the presence of the second field to be determined for all values of δ_1, as shown in the figure. Note that there is a very narrow dip at $\delta_1 = 0$, which corresponds to a suppression of the absorption. *There is therefore no absorption when the two fields are*

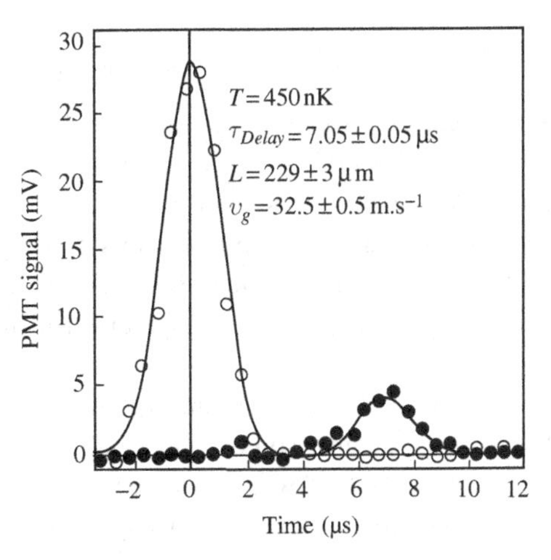

Figure 2D.8 Experimental recording of a light pulse which has propagated through an atomic ensemble with two lasers resonant on a three-level system, as in Figure 2D.6. Open circles: in the absence of the second field. Filled circles: in the presence of the second field and under the conditions of induced electromagnetic transparency: the group velocity of light is reduced by seven orders of magnitude [after L.V. Hau, S. Harris, Z. Dutton, C. Behroozi, Nature **397**, 594 (1999)].

exactly at resonance with each of the two transitions. In fact, it is this zero detuning ($\delta_1 = 0$) that satisfies the condition of Equation (2D.17) when the second field is resonant, and leads to coherent population trapping, giving transparency. This phenomenon, paradoxical at first sight since we usually associate exact resonance with an absorption maximum rather than a null, is called *electromagnetically induced transparency (EIT).*

Application: slow light

In any system that responds linearly to an excitation, the causality constraint (that an effect follows its cause) introduces a mathematical relation between the real and imaginary parts of the linear response, that is to say, between the absorption rate, which depends on χ'', and the refractive index, which depends on χ'. It is the Kramers–Krönig relation:

$$\chi'(\omega) = \frac{2}{\pi} \int_0^{\infty} \omega' \mathrm{d}\omega' \frac{\chi''(\omega')}{\omega'^2 - \omega^2}. \tag{2D.18}$$

This relation implies, for example, that if the absorption has a Lorentzian form (Figure 2.15(a)), the refractive index will have a dispersion profile (Figure 2.15(b)). Note that the refractive index varies rapidly in the neighbourhood of the resonance $\delta_1 = 0$. Applied to the case of electromagnetically induced transparency, Equation (2D.18) implies that the refractive index for the field $\mathbf{E}_1$ also varies rapidly close to $\delta_1 = 0$, that is to say, in the region where induced electromagnetic transparency occurs. Recalling that the group velocity of a light pulse v_g is given by the expression

$$v_g = \frac{c}{n + \omega \mathrm{d}n/\mathrm{d}\omega}, \tag{2D.19}$$

we see that v_g can be smaller in the regions where $\mathrm{d}n/\mathrm{d}\omega$ is very large, that is to say, in the immediate neighbourhood of the resonance. When only the field $\mathbf{E}_1$ is applied, this region close to resonance is also that for which the absorption is maximum, and the pulse is therefore absorbed at the same time as its group velocity is modified. However, in the presence of the second field $\mathbf{E}_2$, the resonance lies in a region where the medium is transparent: the light is not absorbed and the effect of the slowing down of the light can be measured.

Figure 2D.8 shows the results of an experiment carried out on a dense cloud of sodium atoms. It allows the comparison of a light pulse which has propagated through a medium perturbed by an intense field with the same light pulse recorded under the same conditions, but in the absence of the intense perturbing field. A delay of 7 μs was observed between the two pulses, from which the group velocity of the pulse in the atomic medium was determined to be just 32 ms^{-1}, or around 100 kmh^{-1}! This demonstrates that the slowing of a light pulse propagating through the medium illuminated by the second field can be extremely spectacular, since in this particular case the group velocity was reduced by 7 orders of magnitude. In addition, if the field $\mathbf{E}_2$ is switched off while the light pulse is propagating through the atomic medium, the atoms that are in the dark state will remain in this state, because this state is stationary. *The coherence is therefore stored in the medium*, as it is between the two field pulses in the Ramsey experiment, at least for as long as the coherence relaxation effects remain small. If one switches the field $\mathbf{E}_2$ back on again, the atomic medium will again emit, by the stimulated Raman effect, the original field pulse at the first transition frequency, which will therefore have been transmitted by the medium with a controllable delay. Thereby, one has a very promising method for writing, storing and reading atomic coherences for applications to quantum information processing.

Complement 2E The photoelectric effect

The photoelectric effect concerns the ejection of electrons from matter illuminated by incident electromagnetic radiation. This phenomenon was first observed by Hertz for metals (the electrodes of his spark-gap discharged when illuminated by ultra-violet radiation), but also occurs with atoms or molecules, which can be ionized by incident ultra-violet light or X-rays.

The existence of a *threshold frequency* for the incident radiation, ω_S, below which the emission of electrons does not occur is a characteristic of the photoelectric effect that classical physics, or, more precisely, classical electrodynamics, is unable to account for. It is this feature that led Einstein, in 1905, to postulate that monochromatic light of frequency ω is composed of particles, later to be known as *photons*,[1] each having an energy

$$E_{\text{photon}} = \hbar\omega. \tag{2E.1}$$

Einstein then interpreted the photoelectric effect in terms of a collision between a bound electron and a photon leading to the disappearance of the latter coupled with the transfer of its energy to the electron (Figure 2E.1). If we denote by E_I the ionization energy for an atom (the energy required for the electron to be ejected from an atom), or, for the case of a metal, the work-function, Einstein's expression for the kinetic energy of the ejected electron, E_e, is

$$E_e = \hbar\omega - E_I. \tag{2E.2}$$

Since this kinetic energy must be positive, the radiation frequency must be greater than the threshold frequency:

$$\omega_S = \frac{E_I}{\hbar}. \tag{2E.3}$$

After the unambiguous experimental confirmation, in 1915, of the correctness of this model by Millikan,[2] Einstein received, in 1921, the Nobel Prize for 'services rendered to theoretical physics and particularly for his explanation of the photoelectric effect'.

In this complement, however, we shall study the photoelectric effect utilizing a semi-classical model in which the light field is *not* quantized. Nowadays, it is realized that the phenomenon can actually be understood quite well without the necessity of introducing

[1] A. Einstein, *Annalen de Physik* **17**, 132 (1905). Note that Einstein's hypothesis is much more radical than that proposed by Planck in 1900 in order to account for the form of the spectrum of black-body radiation. Planck supposed only that the energy *exchanged* between matter and radiation was quantized.

[2] Several quotations show that Millikan fully expected to disprove Einstein's hypothesis. Like the majority of physicists of his time, he considered this incompatible with all the known wavelike properties of light (interference, diffraction). See, for example, A. Pais, *Subtle is the Lord, The Science and Life of Albert Einstein*, Oxford University Press (1984).

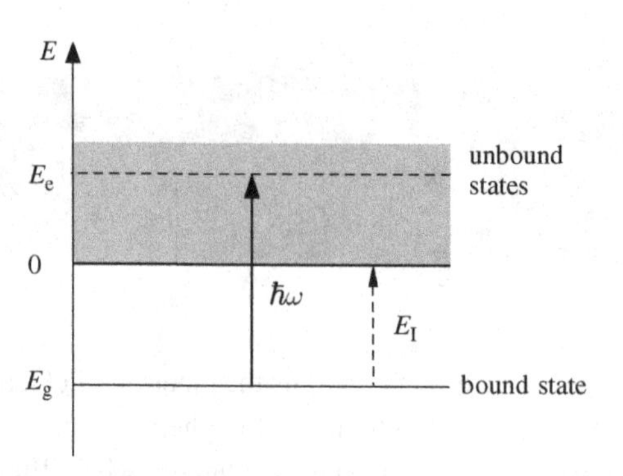

Figure 2E.1 Einstein's interpretation of the photoelectric effect. The bound electron, initially with a negative energy $E_g = -E_I$, can be transferred to an unbound state by the absorption of a photon of energy $\hbar\omega$ greater than E_I.

the hypothesis of the existence of photons, provided that the *matter* is treated as a quantum system. (This does not diminish the achievement of Einstein in recognizing the need for quantization – in 1905 the quantization of matter was no more understood than that of light, since the Bohr model of the atom was not introduced until 1913.) It is interesting to realize that the results that we shall establish in this complement are exact, even if the photoelectric effect can be treated more elegantly in the framework of a completely quantum mechanical treatment of the matter–light interaction, which will be described in Chapter 6.

The model of this complement provides a concrete example of the coupling between a discrete level and quasi-continuum by a perturbation depending sinusoidally on time (compare Section 1.3.4 of Chapter 1). In this respect it can be considered as a worked example, illustrating the procedure for obtaining quantitative results with the Fermi golden rule.

2E.1 Description of the model

2E.1.1 The bound atomic state

Consider an atomic electron in a bound state $|g\rangle$. It could be, for example, that of a hydrogen atom in its ground state, or in a bound excited state, or the valence electron of an alkali-metal atom. The following arguments could apply equally well to an electron in a deeply bound shell of a many-electron atom. If we take as the origin of energy, the potential energy of an electron very far from the nucleus, the energy of the bound state is negative,

$$E_g = -E_I, \tag{2E.4}$$

E_I being the ionization energy, a positive number. We start off by giving a few orders of magnitude. Since the ground state of hydrogen has an ionization energy of 13.6 eV, the associated photoelectric threshold frequency is

$$\frac{\omega_s}{2\pi} = \frac{E_I}{h} = 3.3 \times 10^{15}\,\text{Hz}, \tag{2E.5}$$

which corresponds to a wavelength of $\lambda_s = 91$ nm, in the far ultra-violet.

Whilst the excited states of the hydrogen atom have smaller ionization energies, by contrast, the internal states of heavy (many-electron) atoms have much higher ionization energies, exceeding often a keV. These correspond to threshold wavelengths of less than a nanometre (X-rays).

When in the following we have need to use an explicit expression for the wavefunction of a bound electron (or its Fourier transform), we shall employ those for a hydrogenic atom, which are well known.[3]

2E.1.2 Unbound states: the density of states

The final states of the photoionization processes are of positive energy and correspond to unbound states of the electron. These are the states of the ionized atom. In the case of hydrogen, explicit expressions for these states can be provided.[3] However, in order to provide a simplified discussion of matters concerning the quasi-continuum, we shall only consider high-energy unbound states, which are very little affected by the attractive Coulombic potential of the nucleus. These states can be represented to a good approximation by plane de Broglie waves:

$$\psi_e(\mathbf{r}) = A \exp i\mathbf{k}_e.\mathbf{r}. \tag{2E.6}$$

They are eigenstates of the momentum operator:

$$\hat{\mathbf{P}}\,|\psi_e\rangle = \hbar\mathbf{k}_e\,|\psi_e\rangle\,. \tag{2E.7}$$

The energy corresponding to these states is

$$E_e = \frac{\hbar^2 k_e^2}{2m} \tag{2E.8}$$

(m being the mass of the electron). Notice that the approximation (2E.6) to (2E.8) is better the larger the energy E_e, since the electron's Coulomb energy is then small compared to its kinetic energy.[4]

In order to obtain quantitative results, it is necessary to normalize the wavefunctions. The means of achieving this are not self-evident for the plane waves of (2E.6). Here we adopt a procedure analogous to that of Section 1.3.2a of Chapter 1, according to which we suppose that the electron is confined in a cubical volume of side L, large compared to atomic dimensions and of which the plane faces are normal to the three Cartesian axes

[3] See, for example, H. A. Bethe and E. E. Salpeter, *Quantum Mechanics of One- and Two-Electron Atoms*, Plenum New York (1977).

[4] For the hydrogen atom, this amounts to $\alpha c \ll v$, where $v = \hbar k_e/m$ and $\alpha \approx 1/137$ is the fine-structure constant (Equation 2E.47).

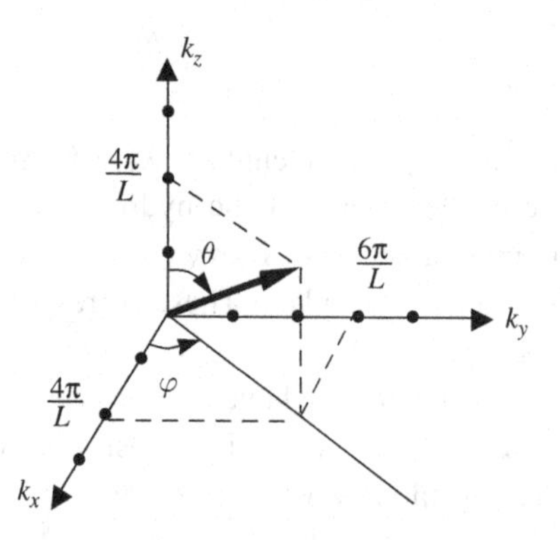

Figure 2E.2 **Representation of the wavevectors of free electrons in reciprocal space.** Owing to the periodic boundary conditions (the period being L) the allowed values for the head of the **k** vector constitute the points of a cubic lattice of side $2\pi/L$. By way of example, we illustrate in the figure the wavevector $\mathbf{k}_e$ associated with ($n_x = 2, n_y = 3, n_z = 2$). This wavevector could, equivalently, be described by its modulus k_e and its direction (θ, ϕ).

$\mathbf{e_x}$, $\mathbf{e_y}$ and $\mathbf{e_z}$. The wavefunction $\psi_e(\mathbf{r})$ is therefore zero outside this cube, from which the normalization condition,

$$|A| = \frac{1}{L^{3/2}}, \tag{2E.9}$$

can easily be derived. We assume for simplicity that A is real and positive.

We know that the wavefunctions must satisfy boundary conditions on the cube faces, which leads to the discrete spectra of the allowed energies and momenta. Instead of the 'true' situation of a single, isolated potential well (with $\psi_e(\mathbf{r})$ zero on the cube faces), we shall adopt in the following, periodic boundary conditions, which lead to more convenient expressions for future manipulations.[5] We therefore write

$$\psi_e(\mathbf{r} + L\mathbf{e}_x) = \psi_e(\mathbf{r}) \tag{2E.10}$$

and similar expressions imposing periodicity on the wavefunction along the other two Cartesian axes. The allowed values of $\mathbf{k}_e$ are, therefore

$$\mathbf{k}_e = \frac{2\pi}{L}(n_x\mathbf{e}_x + n_y\mathbf{e}_y + n_z\mathbf{e}_z), \tag{2E.11}$$

with (n_x, n_y, n_z) relative integers. In wavevector (reciprocal) space these values constitute the points of a *cubic lattice* of side $2\pi/L$ (See Figure 2E.2).

The discretization procedure described above leads, then, to the transformation of the continuum of free-electron states to a quasi-continuum, the energy separation of the quasi-continuum states tending to zero with increasing L. Note that this dimension is in fact arbitrary and should cancel from expressions for physical quantities derived using this treatment.

[5] See also Chapter 6, where an analogous procedure is employed to calculate the mode-density of the electromagnetic field.

In order to apply the Fermi golden rule, we must express the density of states in the form $\rho(\beta, E)$ introduced in Section 1.3.3c of Chapter 1. Notice first that the density of states in reciprocal space is uniform and is equal to

$$\rho(\mathbf{k}_e) = \left(\frac{L}{2\pi}\right)^3, \tag{2E.12}$$

since each state is associated with a cubical cell of volume $2\pi/L^3$ (see Figure 2E.2). To rewrite the density of states in terms of the energy, we first represent an electronic wavevector $\mathbf{k}_e$ by its modulus, the square of which is proportional to the energy (Equation 2E.8), and its direction (θ, ϕ). An elementary volume in reciprocal space is then the intersection of a solid angle $d\Omega$ with a spherical shell with inner and outer radii k_e and $k_e + dk_e$, respectively. This volume is

$$d^3k_e = k_e^2\, dk_e\, d\Omega, \tag{2E.13}$$

and the number of allowed states contained in it is

$$d^3N = \left(\frac{L}{2\pi}\right)^3 k_e^2\, dk_e\, d\Omega = \left(\frac{L}{2\pi\hbar}\right)^3 \sqrt{2E_e m^3} dE_e\, d\Omega. \tag{2E.14}$$

Thus the density of states associated with (E_e, Ω) is

$$\rho(E_e, \theta, \varphi) = \frac{d^3N}{dE_e d\Omega} = \left(\frac{L}{2\pi\hbar}\right)^3 \sqrt{2E_e m^3} = \left(\frac{L}{2\pi\hbar}\right)^3 m\hbar k_e. \tag{2E.15}$$

We shall now use this expression to evaluate the probability that a photoelectron is ejected with a given energy within an element of solid angle $d\Omega$ about a given direction.

Comment If instead of characterizing the element of solid angle by its magnitude $d\Omega$ one used instead the coordinates (θ, ϕ), the corresponding density of states $\frac{d^3N}{dE_e d\theta d\phi}$ would be given by (2E.15) multiplied by a factor of $\sin\theta$.

2E.1.3 The interaction Hamiltonian

The atom is irradiated by an electromagnetic wave described, in the Coulomb gauge, by the potentials

$$\mathbf{A}(\mathbf{r}, t) = \boldsymbol{\varepsilon} A_0 \cos(\omega t - \mathbf{k}.\mathbf{r}), \tag{2E.16}$$

$$U(\mathbf{r}, t) = 0. \tag{2E.17}$$

The interaction of the atom with the field will be described to lowest order by the Hamiltonian (compare Equation (2.54) of Chapter 2):

$$\hat{H}_{I1} = -\frac{q}{m}\hat{\mathbf{p}}.\mathbf{A}(\hat{\mathbf{r}}, t). \tag{2E.18}$$

Notice that we do not make the long-wavelength approximation here, so that our treatment will be equally valid for X-ray wavelengths, which are not large compared to atomic dimensions. It will remain valid also as a description of photoionization from highly excited atomic states, of which the spatial extent is much larger than low-lying ones.

We evaluate now the matrix element $\langle e\,|\hat{H}_{11}|\,g\rangle$ responsible for transitions between the bound state $|\mathrm{g}\rangle$ and an ionized state $|\mathrm{e}\rangle$:

$$\begin{aligned}\left\langle \mathrm{e}\middle|\hat{H}_{\mathrm{I1}}\middle|\mathrm{g}\right\rangle &= \left\langle \mathrm{e}\middle|-\frac{q}{m}\hat{\mathbf{p}}.\mathbf{A}(\hat{\mathbf{r}},t)\middle|\mathrm{g}\right\rangle \\ &= -\frac{q}{2m}A_0\,\mathrm{e}^{-\mathrm{i}\omega t}\left\langle \mathrm{e}\middle|\hat{\mathbf{p}}.\boldsymbol{\varepsilon}\,\mathrm{e}^{\mathrm{i}\mathbf{k}.\hat{\mathbf{r}}}\middle|\mathrm{g}\right\rangle - \frac{q}{2m}A_0\mathrm{e}^{\mathrm{i}\omega t}\left\langle \mathrm{e}\middle|\hat{\mathbf{p}}.\boldsymbol{\varepsilon}\,\mathrm{e}^{-\mathrm{i}\mathbf{k}.\hat{\mathbf{r}}}\middle|\mathrm{g}\right\rangle.\end{aligned} \tag{2E.19}$$

Notice that, because the energy of the final state is larger than that of the initial state, only the term in $\exp(-\mathrm{i}\omega t)$ of (2E.19) is capable of inducing a resonant transition and so giving rise to a non-negligible amplitude for the final state of energy E_{e} where

$$E_{\mathrm{e}} = \hbar\omega + E_{\mathrm{g}} = \hbar\omega - E_{\mathrm{I}} \tag{2E.20}$$

(compare Section 1.2.4 of Chapter 1). We adopt the quasi-resonant approximation and keep only the first term of the right-hand side of (2E.19). This approximation is valid provided the interaction time is much larger than $1/\omega_{\mathrm{s}}$ (which equals 10^{-16} s for photoionization from the ground state of hydrogen). We therefore write

$$\left\langle \mathrm{e}\middle|\hat{H}_{I1}\middle|\mathrm{g}\right\rangle \approx \frac{1}{2}\mathrm{e}^{-\mathrm{i}\omega t}W_{\mathrm{eg}}, \tag{2E.21}$$

with

$$W_{\mathrm{eg}} = -\frac{q}{m}A_0\left\langle \mathrm{e}\middle|(\hat{\mathbf{p}}_{\mathrm{e}}.\boldsymbol{\varepsilon})\mathrm{e}^{\mathrm{i}\mathbf{k}.\hat{\mathbf{r}}}\middle|\mathrm{g}\right\rangle. \tag{2E.22}$$

The plane wave approximation to the unbound electronic states that we adopted in (2E.7) allows (2E.22) to be re-expressed in the form

$$\begin{aligned}W_{\mathrm{eg}} &\approx -\frac{q}{m}A_0\hbar(\mathbf{k}_{\mathrm{e}}.\boldsymbol{\varepsilon})L^{-\frac{3}{2}}\int \mathrm{d}^3r\; r\mathrm{e}^{\mathrm{i}(\mathbf{k}-\mathbf{k}_{\mathrm{e}}).\mathbf{r}}\psi_{\mathrm{g}}(\mathbf{r}) \\ &= -\left(\frac{2\pi}{L}\right)^{\frac{3}{2}}\frac{qA_0}{m}\hbar k_{\mathrm{e}}(\mathbf{u}_{\mathrm{e}}.\boldsymbol{\varepsilon})\tilde{\psi}_{\mathrm{g}}(\mathbf{k}_{\mathrm{e}}-\mathbf{k}).\end{aligned} \tag{2E.23}$$

In this expression we have introduced the Fourier transform $\tilde{\psi}_{\mathrm{g}}(\mathbf{q})$ of the wavefunction $\psi_{\mathrm{g}}(\mathbf{r})$:

$$\tilde{\psi}_{\mathrm{g}}(\mathbf{q}) = (2\pi)^{-\frac{3}{2}}\int \mathrm{d}^3r\,\mathrm{e}^{-\mathrm{i}\mathbf{q}.\mathbf{r}}\psi_{\mathrm{g}}(\mathbf{r}). \tag{2E.24}$$

(Care must be taken here not to confuse the reciprocal space dummy vector $\mathbf{q}$, appearing in the Fourier transform, with the electronic charge.) To within a factor of $\hbar^{-3/2}$, $\tilde{\psi}_{\mathrm{g}}(\mathbf{q})$ is the electronic wavefunction in the momentum representation. Additionally,

$$\mathbf{u}_{\mathrm{e}} = \frac{\mathbf{k}_{\mathrm{e}}}{k_{\mathrm{e}}} \tag{2E.25}$$

is the unit vector in the direction of emission of the photoelectron.

In the case of a hydrogen atom we can use exact expressions for the Fourier transforms $\tilde{\psi}_g(\mathbf{q})$ of the bound states.[6] For the ground state we have

$$\tilde{\psi}_{n=1,l=0}(\mathbf{q}) = \frac{2\sqrt{2}}{\pi} \frac{1}{(\mathbf{q}^2 a_0^2 + 1)^2} a_0^{\frac{3}{2}}, \tag{2E.26}$$

a_0 being the Bohr radius:

$$a_0 = \frac{4\pi\varepsilon_0}{q^2} \frac{\hbar^2}{m} = 0.53 \times 10^{-10}\,\mathrm{m}. \tag{2E.27}$$

The Fourier transforms of the excited states can be written in terms of functions which, for states of spherical symmetry ($l = 0$), have the following asymptotic values:

$$\tilde{\psi}_{n,l=0}(\mathbf{q} = 0) = (-1)^{n-1} \frac{2\sqrt{2}}{\pi} a_0^{\frac{3}{2}} n^{\frac{5}{2}} \tag{2E.28}$$

and, for $n|\mathbf{q}|a_0 \gg 1$,

$$\tilde{\psi}_{n,l=0}(\mathbf{q}) \approx \frac{2\sqrt{2}}{\pi} \left(\frac{a_0}{n}\right)^{\frac{3}{2}} \left(\frac{1}{|\mathbf{q}|\, a_0}\right)^4. \tag{2E.29}$$

These expressions will enable us to evaluate explicitly the matrix element of (2E.22), provided that the amplitude A_0 of the electromagnetic wave is known. For an incident travelling wave with an irradiance (power per unit area) Π, we know that

$$\Pi = \varepsilon_0 c \frac{\omega^2 A_0^2}{2}, \tag{2E.30}$$

since the amplitude of the electric field is ωA_0. As suggested by Einstein's interpretation of the photoelectric effect, it is useful also to introduce the flux of photons (per unit area per unit time):

$$\frac{\Pi}{\hbar\omega} = \varepsilon_0 c \frac{\omega A_0^2}{2\hbar}. \tag{2E.31}$$

We now have at our disposal all the numerical values necessary for an evaluation of the photoionization probability.

2E.2 The photoionization rate and cross-section

2E.2.1 Ionization rate

As we saw in Section 1.3, a perturbative treatment of the coupling between a discrete level and a quasi-continuum leads, after integration over the accessible final states, to a transition probability that increases linearly in time. This enables the definition of a *transition rate* (a probability of transition per unit time) that is proportional to the square modulus of

[6] See Footnote 3, this complement.

the coupling matrix element taken between the initial state and the final quasi-continuum states that are resonantly coupled to it. In the present case these are the states which satisfy (2E.20).

To calculate the photoionization rate it is sufficient now to apply Fermi's golden rule. More precisely, we must use the form relevant to the case of a degenerate continuum introduced in Equation (1.97), extended to the case of a sinusoidal perturbation. This leads to the introduction of an additional factor of 1/4 (compare Equation (1.101)). The photoionization rate per unit solid angle is thus,

$$\frac{d\Gamma}{d\Omega} = \frac{\pi}{2\hbar} \left| W_{eg} \right|^2 \rho(E_e, \theta, \varphi), \tag{2E.32}$$

in which the matrix element W_{eg} (2E.23), and the density of final states (2E.15) are evaluated for a final energy (or for the corresponding value of k_e) given by Equation (2E.20):

$$E_e = \frac{\hbar^2 k_e^2}{2m} = \hbar\omega + E_s = \hbar\omega - E_I = \hbar(\omega - \omega_s). \tag{2E.33}$$

As expected, L cancels from the final result, and we obtain for the rate of photo-ejection of electrons in a direction $\mathbf{u}_e$,

$$\frac{d\Gamma}{d\Omega}(\mathbf{u}_e) = \frac{\pi q^2 A_0^2 k_e^3}{2m\hbar} (\mathbf{u}_e.\boldsymbol{\varepsilon})^2 \left| \tilde{\psi}_g(\mathbf{k}_e - \mathbf{k}) \right|^2. \tag{2E.34}$$

On integrating over all possible emission directions we obtain the total photoionization rate from the state $|g\rangle$:

$$\Gamma = \frac{\pi q^2 A_0^2}{2m\hbar} k_e^3 \int d\Omega (\mathbf{u}_e.\boldsymbol{\varepsilon})^2 \left| \tilde{\psi}_g(\mathbf{k}_e - \mathbf{k}) \right|. \tag{2E.35}$$

The above results account completely for all the characteristics of photoionization as observed experimentally. Firstly, Equation (2E.33), which gives the kinetic energy of the emitted electron is none other than Einstein's equation, which is in perfect agreement with the observations of Millikan. Notice, however, that this equation does not arise, as in Einstein's reasoning, from the a-priori assumption of the conservation of energy in an elementary collision between a photon and the bound electron. It appears here as a consequence of the resonant variation of the transition amplitude when the energies of the initial and final states are related by (2E.20).

Secondly, we find, as expected, a transition rate that is proportional to the square of the amplitude of the electromagnetic wave and therefore to the irradiance (2E.30). We find also that the rate of photoionization in a direction $\mathbf{u}_e$ varies as $(\mathbf{u}_e.\boldsymbol{\varepsilon})^2$, that is as the square of the cosine of the angle between the electric field vector and the emission direction. This, again, is in agreement with the results of experiments, which demonstrate the existence of a preferred emission direction parallel to that of the electric field.

In this section we have been able to account, without introducing the notion of a photon, for all the essential characteristics of the photoelectric effect. Many other useful results can be extracted from the above treatment, of which we give a few examples in the following.

Comment It is sometimes incorrectly argued that the rapid appearance of photoelectrons on switching on the light field, even in the case of this being of very low intensity, cannot be accounted for by the above model. The reasoning here is that the energy flux, as given by the Poynting vector, across a surface of atomic dimensions is then insufficient to provide the energy required for ejection of an electron on so short a timescale. In fact, it should be borne in mind that the transition rates calculated here are quantum probabilities. If a series of experiments is carried out and the distribution of arrival times of the first emitted electron is plotted, this distribution will be found to be flat from time zero, in agreement with Equations (2E.34) and (2E.35). The possibility of photoemission occurring at short delay times can be interpreted as a mere consequence of the quantum nature of atoms.

2E.2.2 The photoionization cross-section

Even if it is not necessary to the validity of our present treatment, the description of photoionization in terms of the collision of a photon with an atom provides a useful physical picture. It is then helpful to consider the photoionization rate per unit photon flux (Equation (2E.31)). This quantity has the dimensions of area and is known as the *photoionization cross-section* σ_g. The photoionization rate then takes the form

$$\Gamma = \sigma_g \frac{\Pi}{\hbar\omega}. \tag{2E.36}$$

Using (2E.35) we obtain

$$\sigma_g = \frac{\pi q^2}{m\varepsilon_0 c\omega} k_e^3 \int d\Omega (\mathbf{u}_e.\boldsymbol{\varepsilon})^2 \left|\tilde{\psi}_g(\mathbf{k}_e - \mathbf{k})\right|^2, \tag{2E.37}$$

which can be written

$$\sigma_g = 2\pi r_0 \lambda k_e^3 \int d\Omega (\mathbf{u}_e.\boldsymbol{\varepsilon})^2 \left|\tilde{\psi}_g(\mathbf{k}_e - \mathbf{k})\right|^2. \tag{2E.38}$$

In the above expression we have introduced the wavelength λ of the incident radiation and the classical electron radius:

$$r_0 = \frac{q^2}{4\pi\varepsilon_0 mc^2} = \frac{1}{a_0}\frac{\hbar^2}{m^2c^2} \simeq 2.8 \times 10^{-15}\,\text{m} \tag{2E.39}$$

and we have used expression (2E.27) for the Bohr radius, a_0.

Returning to the photoionization rate in a given direction (2E.34) we can define the *differential cross-section*:

$$\frac{d\sigma_g}{d\Omega}(\mathbf{u}_e) = 2\pi r_0 \lambda k_e^3 (\mathbf{u}_e.\boldsymbol{\varepsilon})^2 \left|\tilde{\psi}_g(\mathbf{k}_e - \mathbf{k})\right|. \tag{2E.40}$$

2E.2.3 Long-time behaviour

Finally, we shall take time to verify the validity of the approximations used in this section. In particular, we must be sure that it is possible to find a time-scale long enough that the quasi-resonant approximation should be justified (compare Section 2E.1.3), but short enough that the ionization probability per atom, Γt, remains small compared to unity. This consideration imposes an upper limit on the intensity of the incident wave. In practice, this limit is sufficiently great that it is not encountered, except with the most intense laser sources.

Even if the above condition is satisfied, one might ask what occurs when the interaction time is sufficiently long that the perturbation treatment no longer applies. In this case it is possible to show, by a treatment analogous to that of Section 1.3.2 of Chapter 1 (the Wigner–Weisskopf method), that the probability of finding an atom in the initial state $|g\rangle$ decreases exponentially according to

$$P_g(t) = e^{-\Gamma t}. \tag{2E.41}$$

2E.3 Application to the photoionization of hydrogen

Using the known expressions for the hydrogenic wavefunctions, it is possible to evaluate exactly the corresponding photoionization cross-sections. These values have been compared successfully with experiments. Let us show how the calculations can be completed. Since we have made a plane-wave approximation for the wavefunction of the ejected electron (Equation 2E.6), we limit our discussion to the case in which its kinetic energy is large, that is to frequencies of the incident radiation much larger than the threshold frequency. For the case of hydrogen this corresponds to illumination of the atom by X-rays. We then have

$$k_e \approx \sqrt{\frac{2m\omega}{\hbar}} \gg \sqrt{\frac{2m\omega_S}{\hbar}}. \tag{2E.42}$$

For the ground state of hydrogen the threshold frequency is

$$\omega_S = \frac{E_I}{\hbar} = \frac{q^2}{4\pi\varepsilon_0}\frac{1}{2a_0\hbar} = \frac{\hbar}{2ma_0^2}. \tag{2E.43}$$

(We have used here expression (2E.27) for the Bohr radius, a_0.) Expression (2E.42) tells us that

$$k_e a_0 \gg 1, \tag{2E.44}$$

that is, we remain in the regime in which the creation of electron–positron pairs is energetically forbidden ($\hbar\omega \ll 0.5\,\text{MeV}$). From (2E.42) and (2E.43) we have

$$\frac{k_e}{k} \approx \sqrt{\frac{2mc^2}{\hbar\omega}} \gg 1. \tag{2E.45}$$

Under these conditions, replacing $\mathbf{k}_e - \mathbf{k}$ by $\mathbf{k}_e$ and using the asymptotic form (2E.28), expression (2E.37) leads to

$$\sigma_{n=1,l=0} = 64\alpha a_0^2\left(\frac{\omega_S}{\omega}\right)^{7/2}\int(\mathbf{u}_e.\boldsymbol{\varepsilon})^2 d\Omega = \frac{256\pi}{3}\alpha a_0^2\left(\frac{\omega_S}{\omega}\right)^{7/2}, \tag{2E.46}$$

α being the fine-structure constant:

$$\alpha = \frac{q^2}{4\pi\varepsilon_0 \hbar c} = \frac{1}{a_0}\frac{\hbar}{mc} = \frac{1}{137.04}. \tag{2E.47}$$

Notice that the photoionization cross-section is much smaller than the area πa_0^2 of a disc of radius equal to the Bohr radius.

Comments

(i) Having obtained numerical values for the photoionization cross-sections we can return to the question of the validity of the treatment of (2E.2). We know that the resonant approximation is valid if $t \gg \omega^{-1}$. The requirement that the transition probability should remain small compared to unity can be written

$$\frac{\Pi}{\hbar\omega}\sigma t \ll 1. \tag{2E.48}$$

Using (2E.46) σ has an upper limit equal to $\pi a_0^2 \approx 10^{-20}\,\text{m}^2$, so condition (2E.48) gives:

$$t \ll 10^{20}\frac{\hbar\omega}{\Pi}. \tag{2E.49}$$

This is compatible with the condition $t \gg \omega^{-1}$ provided that $\Pi \ll 10^{20}\hbar\omega^2$, or, taking $\omega \approx 10\,\omega_S$, $\Pi \ll 10^{20}\,\text{W.m}^{-2}$. For the majority of available X-ray sources, this condition is satisfied. Synchrotron sources are capable of supplying intensities of the order of $10^{13}\,\text{W.m}^{-2}$ at wavelengths of order 10 nm. Plasmas obtained by the intense irradiation of matter can emit up to $10^{15}\,\text{W.m}^{-2}$. However, high harmonic generation (of order greater than 100) driven by an infra-red laser is capable of producing coherent beams of soft X-rays (of wavelength of order 10 nm) which can be focused. In this way still higher intensities can be obtained, as well, or by other methods such as ultra-short light pulses production.

Once these sources become more practical, it is to be expected that the coherent nature of the emission will enable much higher intensities to be obtained than are implied by the limit of $10^{20}\,\text{Wm}^{-2}$. A new regime of matter–light interactions will then be attained, in which the photoionization time is shorter than the period of the incident radiation. A new theoretical approach will then be required.

(ii) The process of radiative capture is the inverse of the photoelectric effect. Here the initial state consists of a free electron and a nucleus and the final state a bound atom accompanied by an emitted photon. This process can only be modelled by a full quantum theory of the matter–light interaction.

3 Principles of lasers

In this chapter we shall describe the principle of the operation of lasers, their common features and the properties of the light they emit. Our aim is not to provide an exhaustive catalogue of the types of laser available at the time of writing. Such an account would, in any case, soon be obsolete. Rather, we shall use concrete examples of existing systems to illustrate important features or general principles. We do not want either to give an extensive theoretical description of a laser's properties and of its dynamics. We restrict ourselves here to a rather simplified approach to its main features and refer the reader to more specialized handbooks for further information (see the further reading section at the end of the main chapter).

The physical principles accounting for laser operation can appear quite straightforward. This impression stems from the fact that the essential concepts are now well understood, whilst the detail and some incorrect notions are passed over in silence.[1] It is interesting to note, however, how painstaking our progress in understanding lasers has been. It is usually considered that the prehistory of the laser commenced in 1917 when Einstein introduced the notion of *stimulated emission*. In fact, Einstein was led to the conclusion that such a phenomenon must occur from considerations of the thermodynamic equilibrium of the radiation field and a sample of atoms at a finite temperature T. He found that the *Planck formula* for the *black-body* energy distribution could only be derived if, in addition to *absorption* and *spontaneous emission*,[2] stimulated emission was assumed to occur. Much work, both experimental and theoretical, was prompted by Einstein's prediction of stimulated emission. This came to fruition in 1928 with the demonstration of the phenomenon in a neon discharge by Ladenburg and Kopferman.[3] Curiously, interest in stimulated emission declined thereafter, as few applications were foreseen in view of the belief that was widely held at the time that no system could be prepared in a state sufficiently far from *thermal equilibrium* that significant optical gain could be obtained. Furthermore, few physicists were sufficiently familiar with electronics to be aware of the possibility of using *feedback* to maintain oscillation in a system exhibiting gain. This meant

[1] For more information on the development of our understanding of lasers see M. Bertolotti, *Masers and Lasers, An Historical Approach*, Hilger (1983), C. H. Townes in *Centennial Papers*, IEEE Journal of Quantum Electronics, **20**, 545-615 (1984) or J. Hecht, *Laser Pioneers*, Academic Press (1992).

[2] The existence of these processes was already demonstrated experimentally, although they were not accounted for by any existing theoretical framework.

[3] These authors did not, in fact, observe the amplification of an incident wave, but the modification of the refractive index of the gas in the vicinity of the frequency of a resonance as a function of the difference in population of the two levels involved in the transition, which constitutes an indirect proof of the existence of stimulated emission (see Section 2.5).

that the only conceivable applications applied to a gain much larger than one, which was believed to be unobtainable.

It wasn't until 1954, when Townes, Gordon and Zeiger demonstrated the ammonium maser, that the situation changed. Townes' contribution was to realize that if the amplifying medium were placed in a *resonant cavity*, oscillation could occur, even if the single-pass gain was small, provided only that the gain was sufficient to compensate for the (small) losses of the cavity.[4]

The generalization of the *maser* ('microwave amplification by stimulated emission of radiation') of the microwave domain to the *laser* ('light amplification by stimulated emission of radiation') in the optical domain, was not immediate and gave rise to a number of controversies. It wasn't until 1958 that Schawlow and Townes published their suggestion that, in the optical domain, *two aligned mirrors* could perform the function of the resonant cavity. Their article prompted furious activity amongst experimentalists which resulted in the realization of the first laser by Maiman in 1960. This device used ruby as the active medium, and settled a dispute with other physicists who believed they had demonstrated that laser action was not possible in such a material. As we shall show later, the gain mechanism in ruby is in some respects unique and is not self-evident. The advent of the ruby laser was closely followed by demonstrations of laser action in a helium–neon gas mixture, by Javan, and in carbon dioxide by Patel. Amongst the striking advances that have followed have been the demonstration of high-power lasers such as those employing gain media of neodymium-doped glass or of doped fibres, and tunable systems employing dyes or solid-state materials such as titanium-doped sapphire. The semiconductor revolution has provided lasers remarkable for their miniaturization, efficiency and price which has enabled laser technology to become very widely spread, even in mass-produced consumer products (such as CDs, DVDs, bar-code readers, printers ...).

It was shown in the last chapter that it is possible for a medium in which a population inversion has been created to amplify an incident light wave. In Section 3.1 of this chapter we show how a light amplifier can be turned into a light *generator* when the amplifying medium is placed in a cavity which reinjects the amplified light into the gain medium and we study the influence of the cavity on the emitted light. In the second section of the chapter we present some general techniques for obtaining population inversion in a material in order to achieve optical gain. We consider general schemes for three- and four-level systems and give, as concrete examples of each, descriptions of laser systems having played an important role in the field.

In Section 3.3, we analyse the spectral properties of the light emitted by a laser. We show that, depending on the characteristics of the cavity and the spectral width of the gain curve, the laser can have either a well-defined frequency (single-mode operation) or can emit several frequencies simultaneously (multi-mode operation). In the latter case, we show how it is possible to force single-frequency output and give an estimate of the degree of spectral purity thus obtained. In Section 3.4 we describe some of the characteristics of *pulsed* laser systems. These emit light in intense pulses of duration from a

[4] The maser was discovered independently by the Russian physicists Basov and Prokhorov.

nanosecond (10^{-9} s) to as little as a femtosecond (10^{-15} s). The power achieved at the peak of the pulse can be as great as a terawatt (10^{12} W) or even a petawatt (10^{15} W). In the Conclusion we review the properties of laser light and show how they differ from those of light emitted by a classical source. This will allow for a deeper understanding of the applications of lasers, in particular those which are described in the complements to this chapter.

Complement 3A provides an overview of the properties of Fabry–Perot cavities, which were an important tool in spectroscopy before they were employed to provide feedback in laser systems. Whilst in this chapter and in Complement 3A we rely upon a plane-wave description of the light field, in Complement 3B we consider the non-uniform transverse intensity distribution in a real laser beam. We consider in particular the case when this corresponds to the fundamental transverse mode of the laser cavity, which has a Gaussian intensity distribution. We discuss also the *spatial coherence* of the emitted laser light. Complement 3C gives elements of *energetic photometry* of light sources which are essential for a deep understanding of the difference between incoherent light and laser light. Complement 3D treats the question of the spectral width of a laser and shows that there is a fundamental lower limit related to the random diffusion of the phase of the emitted field because of the occurrence of spontaneous emission. This is known as the Schawlow–Townes limit.

Finally, Complements 3E to 3G are devoted to descriptions of some important applications of lasers in various domains. Complements 3E and 3F show, respectively, how the energy and the coherence of laser light can be used, whereas Complement 3G describes the spectacular advances that the laser has permitted in spectroscopy.

3.1 Conditions for oscillation

3.1.1 Lasing threshold

Lasers, like electronic oscillators, rely on the application of *feedback* to an amplifier for their operation. For a laser the amplifier is the inverted atomic or molecular medium, in which optical gain is obtained (see Section 2.5 of Chapter 2). In a single pass through the amplifying medium, the optical intensity is increased by a factor $G = \exp gL_A$, known as the single-pass gain, where L_A is the length of the amplifying zone and g the gain per unit length, assumed here uniform inside the amplifying medium and given, according to the simplified model of Chapter 2, by Equation (2.214). The necessary feedback is realized by the mirrors which define the *laser cavity* (Figure 3.1). The laser cavity of Figure 3.1 is formed by three mirrors which cause a light ray to circulate in a closed triangular path. We assume that two of these mirrors are totally reflecting and that the third, the output mirror, has transmission and reflection coefficients R and T, respectively, with $R + T = 1$.

Consider a weak light ray circulating in the cavity from a point A close to the output mirror (see Figure 3.2). If the intensity of the wave at this point is I_A, well below the saturation intensity I_{sat} of the amplifier (see Section 2.4.2) then its intensity just before

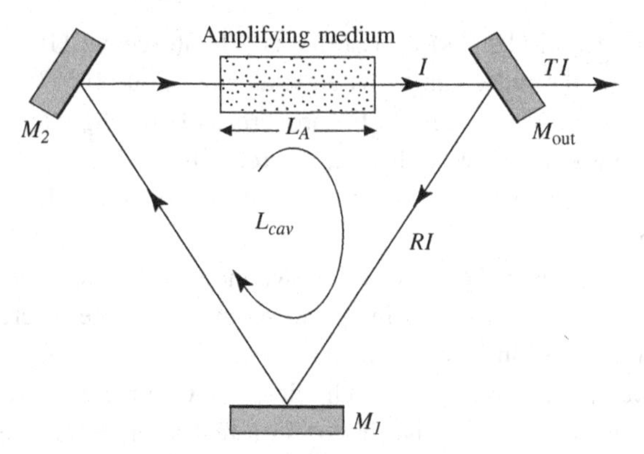

Figure 3.1 Ring-cavity laser. The partially transmitting output mirror M_{out} reflects a part of the incident radiation back into the amplifier via the totally reflecting mirrors M_1 and M_2. If the gain G of the amplifying medium is sufficient to overcome the losses, a circulating light wave is established in the cavity.

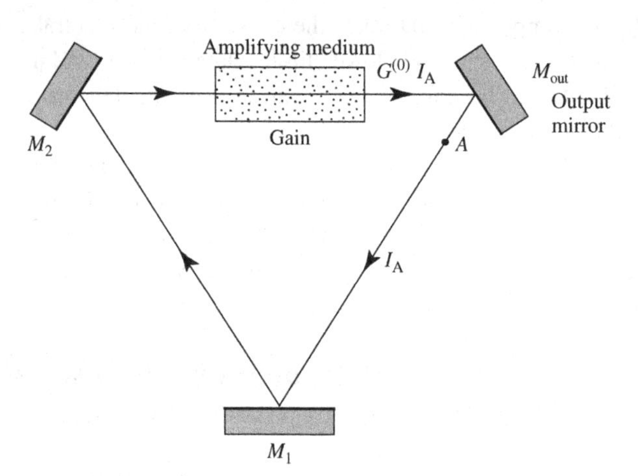

Figure 3.2 Diagram of the path followed by a light beam in a ring cavity showing the evolution of its intensity.

it strikes M_{out} on its return from one pass around the cavity is $G^{(0)}I_A$, where $G^{(0)}$ is the unsaturated gain per pass in the amplifier:

$$G^{(0)} = \exp\{g^{(0)}L_A\}, \tag{3.1}$$

$g^{(0)}$ being the unsaturated gain per unit length (Section 2.5.4). After reflection on M_{out} the intensity is reduced to $RG^{(0)}I_A$.

In order for oscillation to occur, the intensity of the beam after one cavity round trip must exceed that with which it started out. For a weak input wave the *threshold condition* is given by

$$G^{(0)}R = G^{(0)}(1 - T) > 1. \tag{3.2}$$

In practice, it is necessary to consider, as well as the loss by transmission at the output mirror, other losses arising, for example, by absorption or light scattering, either within the cavity or on the mirror surfaces. We shall denote the sum of all these losses by an absorption coefficient A, so that condition (3.2) becomes

$$G^{(0)}(1-T)(1-A) > 1, \tag{3.3}$$

or

$$G^{(0)} > \frac{1}{(1-T)(1-A)}. \tag{3.4}$$

In the small loss and small gain limit $A, T, g^{(0)}L_A \ll 1$, often encountered in practice, this threshold condition becomes

$$G^{(0)} - 1 \simeq g^{(0)}L_A > T + A. \tag{3.5}$$

This condition simply expresses the fact that the unsaturated gain of the amplifier must be *greater than the total losses* in order for oscillation to commence. It is important to realize that *no incident beam* is necessary for this to occur; oscillation is initiated by noise, generally by photons spontaneously emitted in the gain region.

3.1.2 The steady state: intensity and frequency of the laser output

The steady-state field amplitude

Above threshold the power circulating in the laser cavity cannot continue to increase with each round trip. In fact, the gain is usually a decreasing function of the intensity of the circulating wave because of saturation effects (see Section 2.5.4 of Chapter 2). It follows that G is less than $G^{(0)}$:

$$G(I) < G^{(0)}. \tag{3.6}$$

In the steady state the net result after a complete cavity round trip is that the optical intensity is unchanged, so the gain exactly compensates for the losses. If we write for the electric field of the circulating wave at point A:

$$E_A(t) = E_0 \cos(\omega t + \varphi), \tag{3.7}$$

then, after the light has travelled once around the cavity, the field at the same point becomes

$$E'_A(t) = \sqrt{R(1-A)G(I)}\, E_0 \cos(\omega t + \varphi - \psi), \tag{3.8}$$

with $I = E_0^2/2$. The factor preceding E_0 in (3.8) expresses the effect of the absorption, gain and the reflections on the wave in the course of its circuit. The phase change of the wave in making the round trip is denoted by

$$\psi = 2\pi \frac{L_{\text{cav}}}{\lambda}, \tag{3.9}$$

for a light wave of which the vacuum wavelength is $\lambda = 2\pi c/\omega$ and for a cavity of *optical* length L_{cav}. The condition that in the steady state the field is unchanged after one round trip then leads to

$$E'_A(t) = E_A(t) \tag{3.10}$$

at any time. This condition implies the equality of both the amplitudes and the phases of the fields $E_A(t)$ and $E'_A(t)$. We shall examine the implications of these two equalities in the next paragraph.

Comments

(i) The optical length L_{cav} of the cavity that appears in Equation (3.9) differs from its geometrical length for a number of reasons. Firstly, in a medium of refractive index n and of length l, the optical length is nl. Secondly, the optical length includes the effects of phase changes accumulated, for example, on reflection at the mirrors and on passing through the waists of the light beam propagating in the cavity.

(ii) We have implicitly assumed that the fields circulating in the cavity are plane waves so that their amplitude and phase are independent of the transverse coordinate. In practice, the fields have always a finite transverse extent, which needs to be taken into account. The consequences of this are considered in Complement 3B. Notice, however, that these can be described by a similar approach to that used in the derivation of Equation (3.10). Suppose that the field at point A is $E_A(x, y, t)$, where x and y are coordinates in a plane perpendicular to the propagation direction. The condition for the steady state is then simply

$$E'_A(x, y, t) = E_A(x, y, t)$$

for all x and y. Since a spatially limited wave tends to spread because of the occurrence of diffraction, it is impossible to satisfy the above condition unless focusing elements, such as lenses or spherical mirrors, are placed in the cavity to counter its effects.

Diffraction is also at the root of the *spatial coherence* of a laser beam. According to the Fresnel–Huygens principle,[5] the field at any point (x', y') is a result of the superposition of amplitudes originating at all points (x, y) on the previous cavity round trip. This determines the *relative phase* of the field at two points in a plane transverse to the propagation direction and hence results in the coherence of the fields at these points. This property of coherence distinguishes spatially extended laser sources from conventional light sources, for which the phase relationship between two points in a transverse plane is indeterminate. It is also worth noting that the same mechanism is responsible for the transverse amplitude distribution (see Complement 3B on the case of Gaussian beams).

Steady-state intensity

According to condition (3.10), for the steady state to be established, the circulating intensity must satisfy

$$R(1 - A)G(I) = 1, \tag{3.11}$$

[5] See Kirchhoff's theory of diffraction in M. Born and E. Wolf, *Principles of Optics*, Cambridge University Press (1999), § 8.2 and § 8.3; J.-P. Jackson, *Classical Electrodynamics*, § 9.8, Wiley (1998).

or, equivalently,

$$G(I) = \frac{1}{(1-T)(1-A)}, \tag{3.12}$$

i.e. the intensity of the circulating field settles at a value which causes an exact *equilibrium between the gain and losses* in the cavity. Note that Equation (3.12) determines the steady-state intensity I in the laser cavity, since $G(I)$ is a monotonic function of I.

Comments

(i) The gain in the amplifying medium is due to the rate differences that occur of stimulated emission and absorption from the lower level of the laser transition. In expressions (3.5) and (3.12) (as in all the results of Sections 2.4 and 2.5 of Chapter 2), the loss caused by absorption *in the amplifying medium* is included in $G(I)$, the net value of the gain.

(ii) In the ring cavity of Figure 3.1 we have supposed that light circulates in the cavity in a given sense (here in the direction $M_{out}M_1M_2$). In the general case, propagation in the other sense is equally possible, so that oscillation can take place in both senses. However, it is usual to insert elements in the cavity which induce different losses for the two propagation directions. This ensures a lower lasing threshold for one of the modes, which alone oscillates.

(iii) It is easy to envisage ring laser cavities having more than three mirrors. In fact, it is usual to employ four.

(iv) Many lasers employ linear as opposed to ring cavities (see Figure 3.3). If one considers such a cavity composed of a totally reflecting mirror, M_1, and a partially transmitting mirror, M_{out}, two passes of the amplifying medium will occur for each reflection on M_{out} (one in the direction M_sM_1 and one on the return trip) so that condition (3.12) has to be replaced by

$$G^2(I) = \frac{1}{(1-T)(1-A)^2}. \tag{3.13}$$

Actually, the situation is more complicated, because the two waves travelling in opposite directions give rise to a stationary wave inside the laser medium. The resulting spatial variation of the gain (due to gain saturation) is responsible for subtle effects in such linear cavity lasers.

As for the round-trip optical length L_{cav}, it is now close to twice the distance between the two mirrors (the exact value depending on the index of the medium and on the mirror phase-shifts).

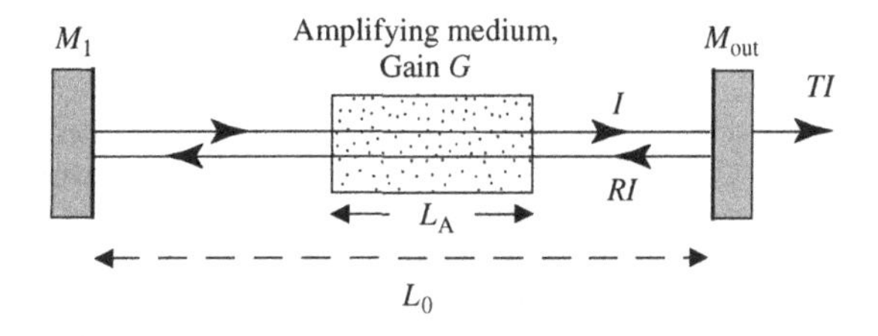

Figure 3.3 Linear cavity laser.

Condition on the phase

The requirement on the phase of the circulating wave resulting from steady-state condition (3.10) is

$$\psi = 2p\pi, \tag{3.14}$$

where p is an integer. Expression (3.9) implies that this can be rewritten in the form

$$L_{\text{cav}} = p\lambda_p, \tag{3.15}$$

or, since $\lambda = 2\pi c/\omega$,

$$\frac{\omega_p}{2\pi} = p\frac{c}{L_{\text{cav}}}. \tag{3.16}$$

Condition (3.15) states that the optical length of the cavity should be an *integer multiple of the wavelength.* This multiple is, in general, very large (of the order of 10^6) since laser cavity lengths are typically of order 1 m, whereas wavelengths are of the order of a micron.[6]

Let us note that condition (3.10) constrains the round-trip phase-shift, but not in any way the absolute phase ϕ of the laser output, which can take any value. We show in Complement 3D that this property is responsible for a phenomenon of *phase-diffusion* of the output field, which in turn leads to the existence of a finite spectral width of the laser output field.[7]

Oscillation frequency

The unsaturated gain coefficient of the amplifying medium, $G^{(0)} = \exp g^{(0)}L_{\text{A}}$, is a function of frequency,[8] which usually has a bell-shaped form, centred on some frequency ω_{M}, as shown in Figure 3.4. (For the case of the simple model of Chapter 2, the explicit form of this curve can be deduced from Equation (2.215).) If we assume that the maximum gain, which occurs when $\omega = \omega_{\text{M}}$, is larger than $1 + T + A$ then there will be a limited band of frequencies $[\omega', \omega'']$ for which the threshold condition (3.4) is satisfied.

In fact, the laser does not emit light over the whole range of frequencies $[\omega', \omega'']$. This is because the frequencies that oscillate must satisfy the additional condition (3.16) (see Figure 3.5). These discrete frequencies ω_p are associated with different *longitudinal modes* and satisfy (3.16) for different values of the integer p. The frequency separation of these modes is equal to c/L_{cav} (which typically has a value of 5×10^8 Hz, for a cavity of length 60 cm, and which is therefore much smaller than the optical frequency, which is of the order of 10^{14} Hz).

[6] Semiconductor lasers (see Section 3.2.3), which have much shorter cavities, are notable exceptions, especially for Vertical Cavity Surface Emitting Lasers (VCSELs) for which p is of the order of 1.

[7] It is important to realize the distinction between the absolute phase of the field, which can have any value, and its time derivative, the frequency, which is subject to the constraint (3.16).

[8] Here, we follow a convention, which is not without danger, of referring to an *angular frequency* ω as a *frequency.* Let us recall that, strictly speaking, the frequency is equal to $\omega/2\pi$ and has units of Hz whilst the angular frequency has units of rad·s^{-1}.

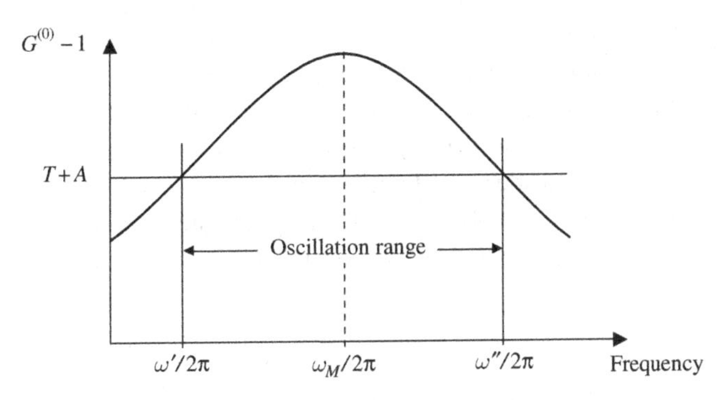

Figure 3.4 The gain–frequency curve for a typical laser transition. The laser can only operate over the frequency range $[\omega', \omega'']$ for which the gain exceeds the losses.

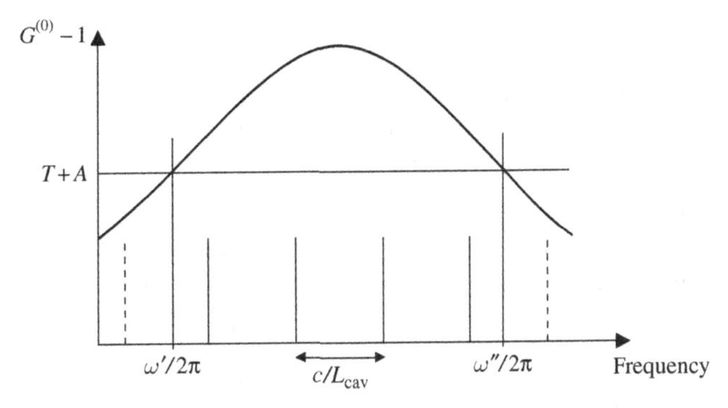

Figure 3.5 Laser oscillation frequencies. The laser can oscillate at the frequencies of the longitudinal cavity modes falling within the frequency band $[\omega', \omega'']$ for which the gain exceeds the losses.

The number of longitudinal modes for which (3.4) is satisfied can vary between 1 and 10^5, depending on the nature of the gain medium. We shall return to this point in Section 3.3, where we shall discuss also the possibility of the coexistence of, or competition between, simultaneously oscillating modes.

3.2 Description of the amplifying media of some lasers

3.2.1 The need for population inversion

In Sections 2.4 and 2.5 of Chapter 2 we described a first example of a system capable of amplification of an incident wave. It is a dilute medium composed of atoms with two excited states a and b, of energies E_a and E_b, respectively, in which the upper level, b, is more strongly populated than the lower level, a, so that the competition between stimulated emission and absorption leads to net gain. Here we generalize the results of that discussion to more complicated, but more realistic systems.

For the systems that we shall consider in this section most of the results of Chapter 2 remain valid, notably the symmetric relationship between absorption and stimulated emission. If we consider a medium of thickness dz with a density of N_a/V atoms (or molecules) in the lower state and of N_b/V in the upper state,[9] a wave of frequency ω propagating in the direction of positive z has its intensity changed according to (compare Equation (2.205)),

$$\frac{1}{I(z)}\frac{\mathrm{d}I(z)}{\mathrm{d}z} = g, \tag{3.17}$$

the gain per unit length g being proportional to the population difference (compare Equation (2.245)):

$$g = \frac{N_b - N_a}{V}\sigma_{\mathrm{L}}(\omega). \tag{3.18}$$

This equation shows the competition between absorption and stimulated emission processes, the absorption being proportional to N_a and the stimulated emission to N_b. The laser cross-section, $\sigma_{\mathrm{L}}(\omega)$ (see Section 2.6), is a positive number (having the dimensions of area) which, generally, manifests a resonant behaviour, that is it has a maximum at a frequency that we shall denote by ω_{M}. For a homogeneous, non-saturated medium, the non-saturated gain $g^{(0)}$ does not depend on the intensity, which therefore varies according to

$$I(z) = I(0)\exp\{g^{(0)}z\}. \tag{3.19}$$

If stimulated emission dominates over absorption, the gain is positive and the wave is amplified. This is obviously the case that is of relevance to the operation of lasers. It occurs in the case of population inversion:

$$N_b > N_a. \tag{3.20}$$

Such a situation differs from that when the system is in thermodynamic equilibrium, when the Boltzmann equation,

$$\left(\frac{N_b}{N_a}\right)_{\text{th.equ.}} = \exp\left(-\frac{E_b - E_a}{k_{\mathrm{B}}T}\right), \tag{3.21}$$

is satisfied. Realization of the required non-equilibrium state can only be achieved by exploiting the *kinetics*. Ideally, one would pump the population of the excited level b only and the lower level of the laser transition a would de-excite very quickly. However, in reality one encounters *relaxation processes* which transfer population from the excited state to the lower state. Progress in the development of new laser systems has essentially relied on the discovery of means of circumventing this problem in order to obtain population inversion. In the following we give some examples which illustrate how this has been achieved.

[9] Here V is the volume common to the amplifying medium and the laser cavity.

Comments

(i) Spontaneous emission from level b to level a is inevitable and has a deleterious effect on efforts to achieve population inversion. Spontaneous emission increases in importance as the frequency is increased (see Chapter 6, comment (ii) of Section 6.4.4), which explains why it is relatively easy to obtain laser oscillation in the infra-red, but considerably more difficult in the ultra-violet. X-ray lasers exist, but require extremely intense pumping processes.

(ii) The *optical parametric oscillator*, studied in Complement 7A, is another kind of optical oscillator. The amplification process relies on the phenomenon of parametric mixing, which is an exchange of energy between a pump light wave and the emitted wave in a nonlinear medium.

3.2.2 Four-level systems

Description of the four-level scheme

The simplified model of the laser atomic medium of Chapter 2 (Sections 2.4–2.6) supposed the existence of at least three atomic energy levels: the ground state and two excited levels a and b. We considered the optical gain on the transition $b \rightarrow a$ when a population inversion existed between this pair of levels. In practice, however, most lasers involve at least four accessible levels. We shall outline briefly the mode of operation of a number of examples including the neodymium laser, the helium–neon laser, tunable dye lasers, the titanium-doped sapphire laser and molecular lasers.

The energy level scheme of a four-level laser is depicted in Figure 3.6. A *pumping* mechanism (optical pumping driven by the light emitted by a lamp or by an auxiliary laser, excitation by an electric discharge etc.) transfers atoms from their ground state g to an excited state e. The *rapid relaxation* of this level then transfers its population to the upper laser level b. The spontaneous radiative decay between b and a is assumed to be slow compared to the relaxation of level e to level b. Finally, the atoms in the lower level of the laser transition a decay to the ground state by a rapid relaxation process. The characteristic

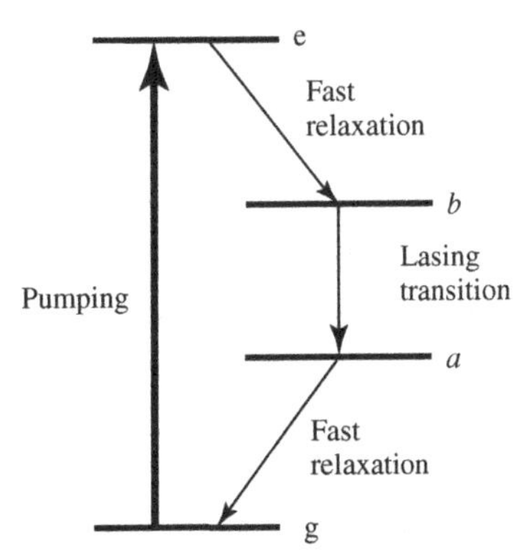

Figure 3.6 Level scheme for a four-level system. The laser operates on the transition $b \rightarrow a$. Levels e and a decay rapidly to levels b and g, whilst a pumping mechanism populates e from the ground state g.

de-excitation times of the excited levels are denoted by τ_e, τ_b and τ_a, on which the assumptions detailed earlier impose the relations $\tau_e, \tau_a \ll \tau_b$. The level e is pumped at a constant rate w. Using these rates, we can write equations describing the kinetics of the system (also known as rate equations; see Section 2.6.2) and from these deduce the population difference $N_b - N_a$ on which the amplification depends.

When laser intensity is sufficiently weak that the effect of absorption and stimulated emission on the transition $b \to a$ can be neglected (i.e. in the unsaturated regime), the rate equations can be written:

$$\frac{\mathrm{d}}{\mathrm{d}t}N_e = w(N_g - N_e) - \frac{N_e}{\tau_e} \tag{3.22}$$

$$\frac{\mathrm{d}}{\mathrm{d}t}N_b = \frac{N_e}{\tau_e} - \frac{N_b}{\tau_b} \tag{3.23}$$

$$\frac{\mathrm{d}}{\mathrm{d}t}N_a = \frac{N_b}{\tau_b} - \frac{N_a}{\tau_a} \tag{3.24}$$

$$\frac{\mathrm{d}}{\mathrm{d}t}N_g = \frac{N_a}{\tau_a} - w(N_g - N_e). \tag{3.25}$$

These equations automatically assure the conservation of the total population:

$$\frac{\mathrm{d}N}{\mathrm{d}t} = 0 \tag{3.26}$$

$$N_a + N_b + N_e + N_g = N. \tag{3.27}$$

Let us write the steady-state equations ($\mathrm{d}N_i/\mathrm{d}t = 0$ with $i = g, a, b, e$), neglecting small terms associated with $\tau_a, \tau_e \ll \tau_b$, and assuming a weak pumping regime ($w\tau_e \ll 1$). We obtain

$$N_e \simeq w\tau_e N_g, \tag{3.28}$$

$$N_b \simeq w\tau_b N_g, \tag{3.29}$$

$$N_a \simeq w\tau_a N_g, \tag{3.30}$$

hence the relative population inversion,

$$\frac{N_b - N_a}{N} \simeq \frac{w\tau_b}{1 + w\tau_b}. \tag{3.31}$$

The population inversion can be appreciable if level b has a long lifetime. Such a four-level system is therefore ideal for laser operation, as we shall see in the following.

Comments

(i) If the pumping mechanism arises from the absorption of light, Equations (3.22–3.25) are the rate equations discussed at the end of Chapter 2. When the pumping occurs by electronic collisions, as in an electric discharge, it is an incoherent process, leading also to rate equations.

(ii) In practice, laser media are almost never perfect four-level systems like the one described in the simple model above. In particular, extra decay mechanisms hinder the creation of the population inversion.

(iii) When the intensity circulating in the laser amplifier is large enough, absorption and spontaneous emission are taken into account by adding to (3.23) and subtracting from (3.24) $\sigma_L(N_a - N_b)\,\Pi/\hbar\omega$, when σ_L is the laser cross-section, and Π is the intensity expressed in units of $W.\mathrm{m}^{-2}$ (see Equations (2.232–2.233)). With the same approximations as above, Equation (3.31) then becomes

$$\frac{N_b - N_a}{N} \simeq \frac{w\,\tau_b}{1 + w\,\tau_b}\,\frac{1}{1 + \frac{\Pi}{\Pi_{\mathrm{sat}}}}, \tag{3.32}$$

where the saturation intensity is

$$\Pi_{\mathrm{sat}} = \hbar\omega_L\,\frac{1 + w\,\tau_b}{\sigma_L\,\tau_b}. \tag{3.33}$$

Using (3.18), we then find a gain of the form, similar to (2.214),

$$g = \frac{g^{(0)}}{1 + s}, \tag{3.34}$$

with $s = \Pi/\Pi_{\mathrm{sat}}$ the saturation parameter.

The neodymium laser

Figure 3.7 shows the energy levels of a Nd^{3+} (neodymium) ion when embedded at low concentration in glass (neodymium-doped glass) or in crystalline YAG (neodymium-doped yttrium aluminium garnet). Pumping is achieved using an exterior light source, usually either an arc lamp or a diode laser, which excites ions into excited bands from which they decay rapidly into the upper laser level, thanks to a *non-radiative* relaxation process due to coupling with the vibrations of the material.

Laser emission occurs principally on the transition at 1.06 μm in the infra-red. Other transitions are capable of laser action, but with much reduced efficiency. They are consequently little used. In the pulsed operating mode, the energy output can reach several joules

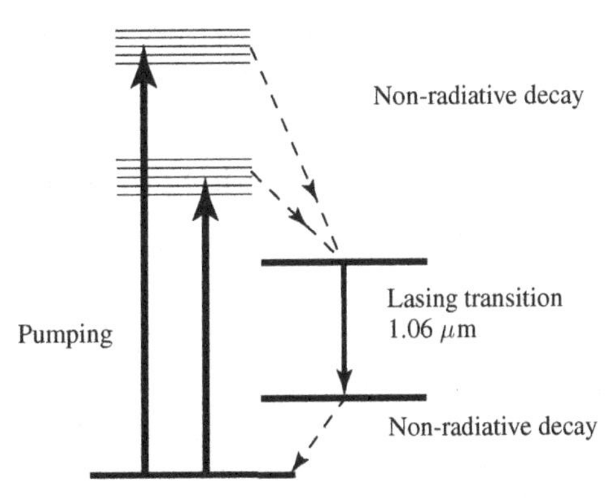

Figure 3.7 Energy-level scheme for a neodymium ion embedded in a solid such as glass or crystalline YAG, showing discrete levels and bands. The transitions used for pumping have wavelengths in the regions of 0.5 μm and 0.8 μm. Laser emission occurs at 1.06 μm.

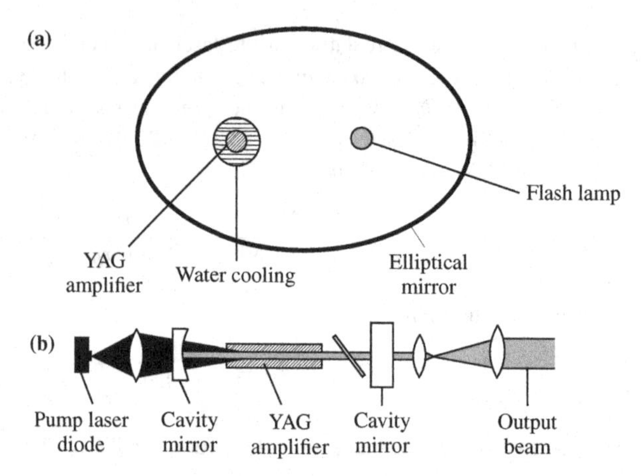

Figure 3.8 **(a) Cross-section of the amplifying zone of a flash-lamp pumped neodymium YAG laser. The lamp and gain rod are situated at the foci of an ellipse. Such a laser can deliver 1 to 100 W in the infra-red. It consumes an electrical power in the range of 0.1 to 10 kW. (b) Laser diode pumped Nd:YAG laser. Such a laser can have an efficiency exceeding 50%.**

in a pulse of duration 1 ps to 1 ns, whilst in continuous-wave (cw) operation powers exceeding 100 W can be obtained. For many applications a nonlinear crystal (see Chapter 7) is placed at the laser exit in order to produce a coherent beam of the second harmonic of the laser output wavelength, at 532 nm (in the green), or of the third harmonic at 355 nm in the ultra-violet.

Neodymium lasers are highly efficient (the efficiency ranging from 1% for a flash-lamp pumped system, to 50% for some laser diode pumped systems) and have a vast range of applications. Because of their high power in pulsed operation (when pumped by a flash lamp, see Figure 3.8(a)) they are used, for example, in experiments aimed at initiating thermonuclear fusion (see Complement 3E). These lasers are also used as cw sources (Figure 3.8(b)). In this case pumping by a semiconductor laser at 0.8 μm is possible (see Section 3.2.3 below). Frequency-doubled in a non-linear crystal, they are frequently used as pump sources for tunable lasers (see below). These devices are widely used because of their small physical size and good efficiency.

The helium–neon laser

The helium–neon laser employs a pumping mechanism of some subtlety. A continuous electric discharge in a mixture of helium and neon gases excites the helium to a number of *metastable* levels, which have a very long lifetime because they have no electric dipole coupling to the atomic ground state. In subsequent collisions of an excited helium atom with a neon atom in its ground state the stored internal energy can be transferred, putting the neon atom into an excited state of similar energy. As shown in Figure 3.9, laser action can then be obtained on transitions of which the excited levels are those pumped by the collisional energy transfer. Lasing can be obtained on lines at 3.39 μm and 1.15 μm as well as on the well-known, red transition at 633 nm.

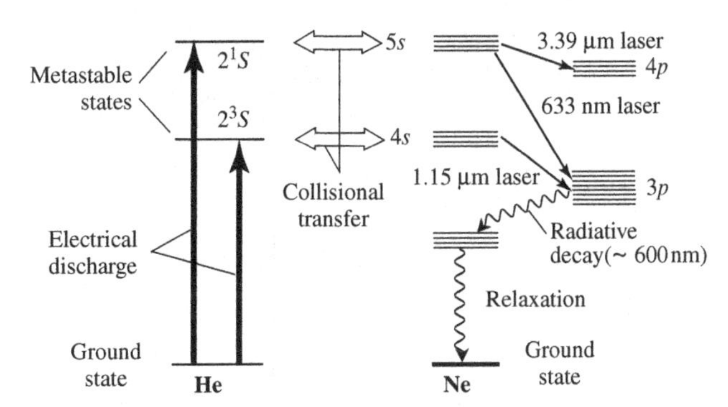

Figure 3.9 **The energy levels of helium and neon involved in the operation of the helium–neon laser. The energy stored by the electronic excitation of metastable levels of helium is transferred to the neon atoms in collisions.**

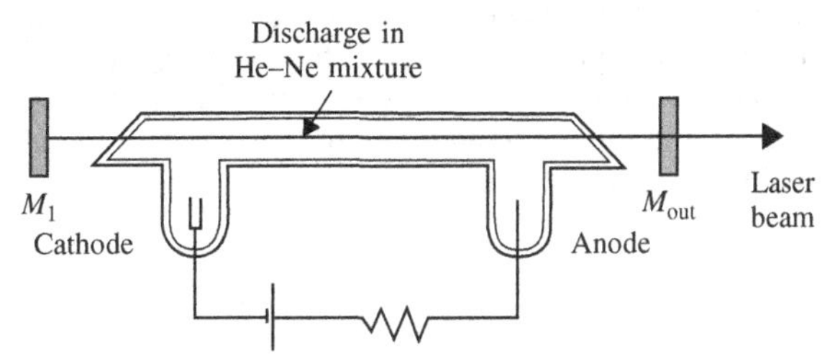

Figure 3.10 **Construction of a helium–neon laser producing a linearly polarized output beam.**

Figure 3.10 shows a possible configuration for the cavity of a helium–neon laser. The gas is enclosed in a tube between a pair of end windows inclined at *Brewster's angle* which cause no reflection loss for light passing through them with its linear polarization vector in the plane of the paper. The orthogonal light polarization does experience a reflection loss and consequently has a higher threshold gain. The laser therefore oscillates only on the low-loss polarization mode and the output is *linearly polarized.* The laser transition that oscillates is selected by the mirrors which are made reflecting only over the required narrow band of wavelengths. The output power of such a laser is typically of the order of a few mW. Some industrial versions differ from the configuration shown in that the mirrors are internal to the vacuum envelope so that the output is unpolarized.

Tunable lasers

The lasers that we have described in the previous paragraphs, and many others, have a very narrow gain curve, which extends over a range of some gigahertz. It is therefore not possible to tune the output wavelength very far (visible light extends over a bandwidth of some 3×10^{14} Hz). It was to meet the need for light sources with a broader spectral range that the tunable lasers were developed. These are lasers the output frequency of which can be tuned over a relatively wide range because of the extended width of their gain curves. This can arise, for example, if the lower level of the laser transition belongs to a *continuum*

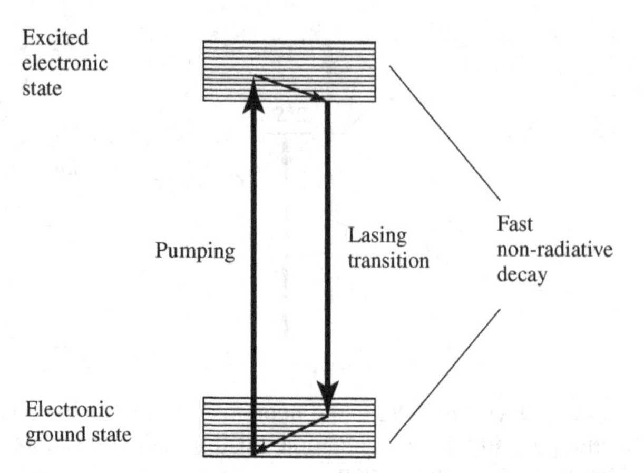

Figure 3.11 Energy-level scheme showing the mechanism of operation of a tunable laser. The energy levels of the active medium (a dye molecule or an ion strongly perturbed by its environment) comprise bands separated by electronic transitions. Non-radiative relaxation within a given band towards the lowest level takes place very quickly (on a timescale of order 1 ps). The tuning range of the output is determined by the energy width of the lower band of the laser transition.

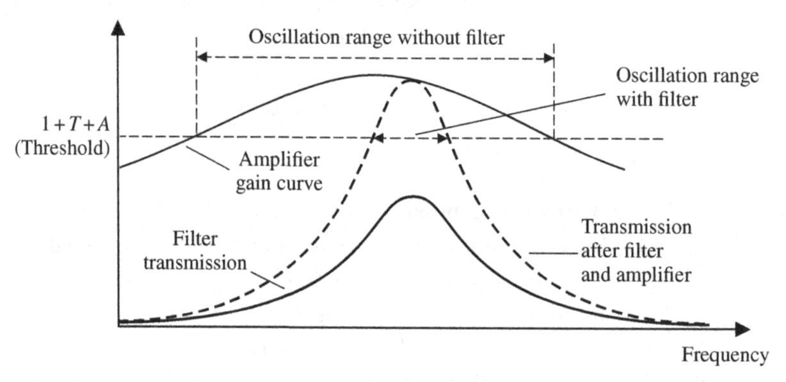

Figure 3.12 Wavelength selection by an optical filter. The transmission curve of the filter is narrower than the gain curve of the amplifying medium and approaches unity at its peak. The round-trip gain of the laser at a given frequency is the product of the gain in the absence of the filter and the transmission of the filter at that frequency. Oscillation can occur when this product exceeds one.

(or dense quasi-continuum) of levels. This configuration is encountered in dye molecules in a liquid solvent or for some metallic ions in crystals, such as the titanium ion embedded in a sapphire crystal (see Figure 3.11).

Tuning of these lasers is achieved by the positioning in the laser cavity of *frequency-selective* elements, which induce losses for wavelengths outside the desired operating range (see Figure 3.12). A titanium:sapphire laser, for example, pumped by a frequency-doubled YAG laser at 0.53 μm, is tunable over the range [700 nm, 1100 nm] (that is, over a range of 200 THz which is considerably greater than the precision of about 1 MHz to which the linewidth can be set). For a pump intensity of 10 W the useful output power is more than 1 W in single-mode operation.

Molecular lasers

Population inversion of a carbon dioxide laser (CO_2 laser) is achieved in an electrical discharge. These systems can be very powerful (10 kW or more) and they have numerous industrial and military uses. The laser transition, which is at a wavelength of 10.6 μm in the far infra-red, involves two distinct molecular vibrational states (i.e. states involving different modes of excitation for the relative motion of the atoms constituting the molecule).

There are many other molecular laser systems that emit in the infra-red. In some of these systems the active molecules of the gain medium are created directly in the excited state by a chemical reaction, which automatically ensures the existence of a population inversion. One such system is the hydrogen fluoride (HF) laser in which excited molecules are created in the reactions

$$H + F_2 \rightarrow HF^* + F$$
$$H_2 + F \rightarrow HF^* + H.$$

Such a laser works in a pulsed mode and can emit immense power: typically 4 kJ in a pulse of duration 20 ns, giving a peak power of 200 GW. Note that such a laser can operate without a supply of electricity, since energy is converted from a chemical form; it is necessary only to have a supply of the reactant gases.

If the laser transition is between distinct *electronic* levels, the wavelength of the emission is often in the ultra-violet. The most commonly used molecular lasers are the *excimer* lasers, employing ArF or KrF molecules, which emit in the ultra-violet, at 195 nm and 248 nm, respectively. In these systems the lower level of the laser transition is unstable (see Figure 3.13) because there is no stable bound state of the two atoms (only the excited

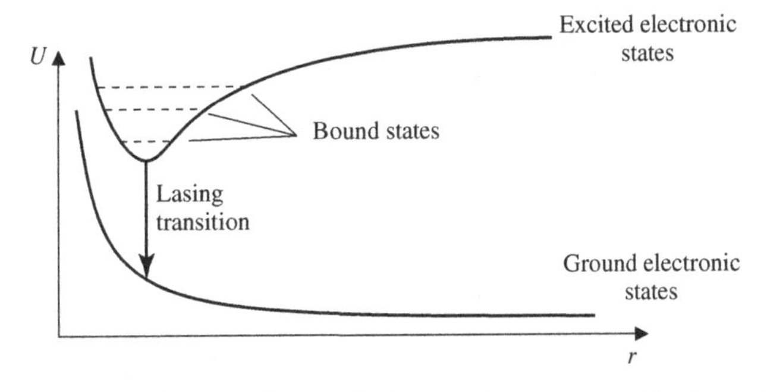

Figure 3.13 Energy-level scheme for an excimer laser. The interaction energy of the two atoms is shown as a function of their separation. In the excited state, the interaction potential has a well which supports bound vibrational–rotational states. A discharge in a mixture containing a rare gas and a halogen (for example Ar and F_2) leads to the formation of stable ArF molecules in the excited state. The upper laser level is one such bound state of ArF. The lower level of the laser transition is a dissociative state, which leads to the fission of the molecule into atomic Ar and F.

states are stable). The question of the depopulation of the lower laser level a of the laser transition (Figure 3.6) is therefore elegantly answered since the ground-state molecule very quickly fragments, liberating its two constituent atoms.

Semiconductor lasers

Presently, semiconductor lasers (also known as diode lasers) are the cheapest and most widespread lasers in use. They operate essentially in the red and infra-red spectral regions. Every CD or DVD player contains several (emitting at a wavelength, respectively, around 780 nm or 650 nm) and fibre-optics based telecommunications rely heavily on devices operating at 1.3 μm and 1.5 μm (the spectral region in which optical fibres exhibit minimum dispersion and maximum transmission, see Section 3F.4 of Complement 3F).

The emission of light in a diode laser occurs in the junction region of a forward-biased semiconductor diode, formed of heavily doped material (see Figure 3.14). The *electron–hole recombination* occurring at the junction liberates energy in the form of a photon of energy $\hbar\omega \approx E_g$, where E_g is the energy separating the top of the valence band from the lower edge of the conduction band (this band-gap being of the order of 1.4 eV for gallium arsenide, AsGa, equivalent to a wavelength between 800 nm and 900 nm). In the regime in which only spontaneous emission occurs one has a light-emitting diode (LED), which is the basis of many instrument displays. If, however, the injected current is increased, a regime can be reached in which stimulated emission is predominant: the system then exhibits a very large optical gain.

The first diode lasers were operational as early as 1962, but for a long time their use was restricted to specialized laboratories: it was necessary to maintain their temperature at that of liquid nitrogen in order to limit the non-radiative relaxation rates. Additionally, a very high current density was required for threshold to be reached: 5×10^3 A·cm^{-2} at 77 K. In the 1970s, developments in semiconductor components fabricated from gallium arsenide and similar materials allowed considerable progress to be made. It became possible using

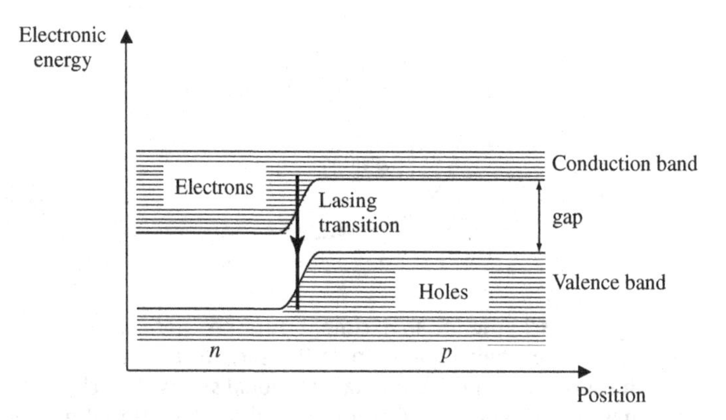

Figure 3.14 Energy-level scheme for a semiconductor laser. Light emission occurs in association with electron–hole recombination, when a positive potential difference is applied between zones p and n.

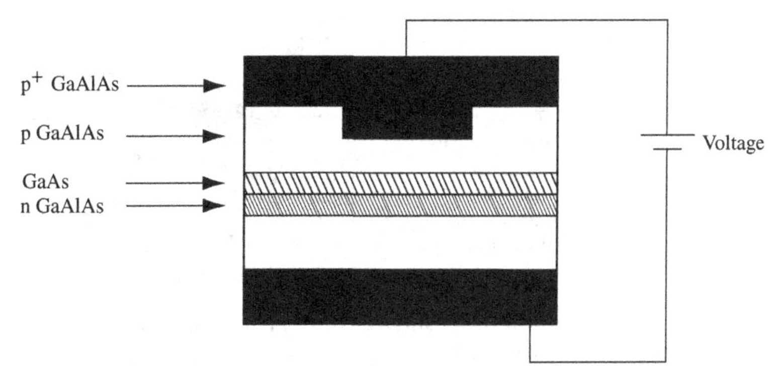

Figure 3.15 Schematic diagram of the various layers in the active region of a diode laser. Several layers each with different gaps are superimposed.

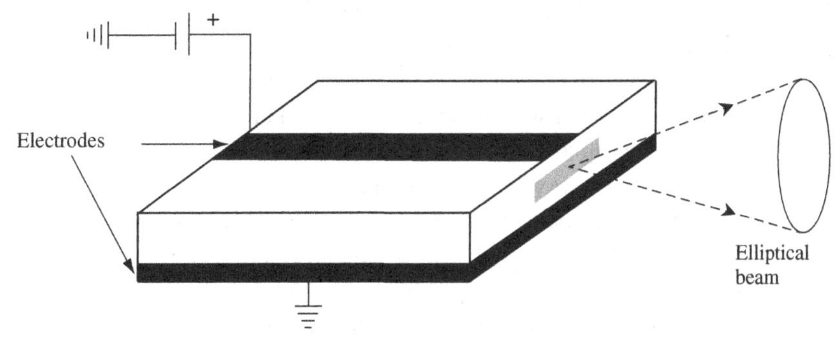

Figure 3.16 Light emission from a diode laser occurs from the end facets (of typical dimensions 1 μm × 10 μm). Because of diffraction, the emitted beam has a far-field elliptical cross-section, with a vertical divergence of the order of 30° and a lateral divergence of a few degrees.

multi-layer structures (termed *heterostructures*, see Figure 3.15) to confine the recombination to a layer of thickness 0.1 μm (gallium arsenide/gallium aluminium arsenide lasers: $GaAs/Al_xGa_{1-x}As$). The threshold current density was in this way greatly reduced.

Stimulated emission occurs in the thin active layer, within which the light is guided, as in an optical fibre (the refractive index here being higher than in the surrounding material). Two parallel facets, cleaved orthogonal to the active channel, serve as the mirrors of a monolithic linear cavity of length typically 400 μm (see Figure 3.16). Finally, a transverse confinement (of typical width 10 μm) allows the current required for the production of a population inversion to be reduced still further (see Figures 3.15 and 3.16). Nowadays, commercially available diode lasers operate at room temperature and provide coherent light at powers of several hundreds of milliwatts in the c.w. regime for an input current of the order of 1 A. By using semiconductors of different gaps, one is able now to cover the whole near infra-red and visible spectrum with a set of diode lasers. Let us quote in particular the blue diode lasers, using InGaN/AlGaN junctions, which operate around 400 nm and are used in more advanced generations of optical storage discs and RGB video projectors.

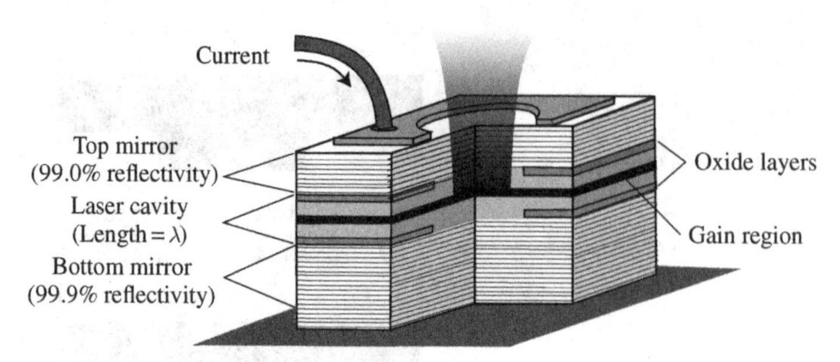

Figure 3.17 **Scheme of a vertical cavity surface emitting laser. The total length of the laser cavity is of the order of a micrometre.**

Because of diffraction the emitted beam is elliptical and highly divergent (see Figure 3.16); collimation must be achieved with the aid of external optical components. These lasers are nevertheless widely used because of their ease of operation, their overall efficiency which exceeds 50%, and their compactness. Assembled by the hundreds in tight packages, they constitute laser arrays of high brightness and efficiency in the c.w. or quasi-c.w. regime (up to 100 W c.w. with a conversion efficiency from electrical power of about 75%) that are now widely used in industry.

VCSELs (Vertical Cavity Surface Emitting Lasers) are another kind of semiconductor lasers (see Figure 3.17). The laser cavity is made of different layers of semiconductors, so that the output beam is perpendicular to the surface of the device. The mirrors of the laser cavity are made by many layers of alternated indices, which constitute an optical grating that reflects the light at a chosen wavelength. The gain medium is a layer of semiconductor which is so thin that it constitues a quantum well, having discrete levels between which the inversion takes place when the system is electrically or optically pumped. The VCSEL generally oscillates in the fundamental mode of the cavity, for which the intracavity standing wave has a single antinode. These lasers are interesting for their good optical quality and the possibility of integrating many of them on the same chip.

3.2.3 Laser transition ending on the ground state: the three-level scheme

For most lasers the laser transition has as its lower level different from the ground state of the active atom or molecule. It is, in fact, difficult to obtain oscillation if this is not the case, since it is necessary then to ensure that an excited state has a larger population than the ground state, which is difficult to achieve. Curiously, however, the first operational laser (the ruby laser) was precisely of this type. Decades later, a new three-level system, that of the erbium ion, Er^+, embedded in silica, has found use as an amplifying medium in fibre-optic based telecommunications. We propose, therefore, to study the distribution of population in two- and three-level systems, in order to ascertain the conditions in which laser amplification ending on the ground state can be obtained.

Two-level scheme

Firstly we show that for a closed two-level system (of which the lower and upper levels are denoted a and b, respectively) it is not possible to create a steady population inversion by way of a pumping mechanism describable in terms of rate equations analogous to those of Section 3.2.2:

$$\frac{\mathrm{d}}{\mathrm{d}t}N_b = w(N_a - N_b) - \frac{N_b}{\tau_b}, \tag{3.35}$$

$$\frac{\mathrm{d}}{\mathrm{d}t}(N_a + N_b) = 0. \tag{3.36}$$

In the steady state we find, from (3.35),

$$\frac{N_b}{N_a} = \frac{w\tau_b}{1 + w\tau_b}, \tag{3.37}$$

which shows indeed that the population of the excited state is always less than that of the ground state.

Comments

(i) We saw in the preceding chapter (Section 2.3.2) that when a coherent, monochromatic wave interacts with a two-level system which is initially in the ground state, it is possible, after a period of time, to find a probability of one that the system is in the excited state (Rabi precession). A population inversion is therefore realized transiently. However, this inversion necessitates a coherent incident wave identical to that which would be created by laser action.

(ii) In the ammonium (NH_3) and hydrogen masers,[10] the transition on which laser action occurs is related to a two-level system. In these devices a molecular beam is prepared in the excited state by a Stern–Gerlach type method, allowing selection of the excited molecules only. This beam of totally inverted molecules then traverses a resonant cavity which performs a role analogous to that of the optical cavity of a laser. A significant fraction of the molecules is then transferred to the ground state by stimulated emission. The molecules then leave the cavity. The ground state molecules are in this way physically removed from the cavity, thus ensuring that the population inversion is maintained. Such a system is, in fact, well described by the equations of Section 2.4 of Chapter 2, provided Γ_{D} is interpreted as the inverse of the mean time that the molecules spend in the cavity.

Three-level scheme

Figure 3.18 illustrates the energy-level scheme for a three-level system, which, as we shall show, can give rise to laser emission on the transition linking the intermediate level b and the ground state a. The pumping operates between level a and an excited state e which rapidly decays (with time-constant τ_{e}) to level b. The spontaneous radiative de-excitation of level b to level a is, in the absence of stimulated emission, much slower: $\tau_{\mathrm{e}} \ll \tau_b$.

[10] See C. Townes and A. Schawlow, *Microwave Spectroscopy*, § 15.10 and 17.7, Dover (1975).

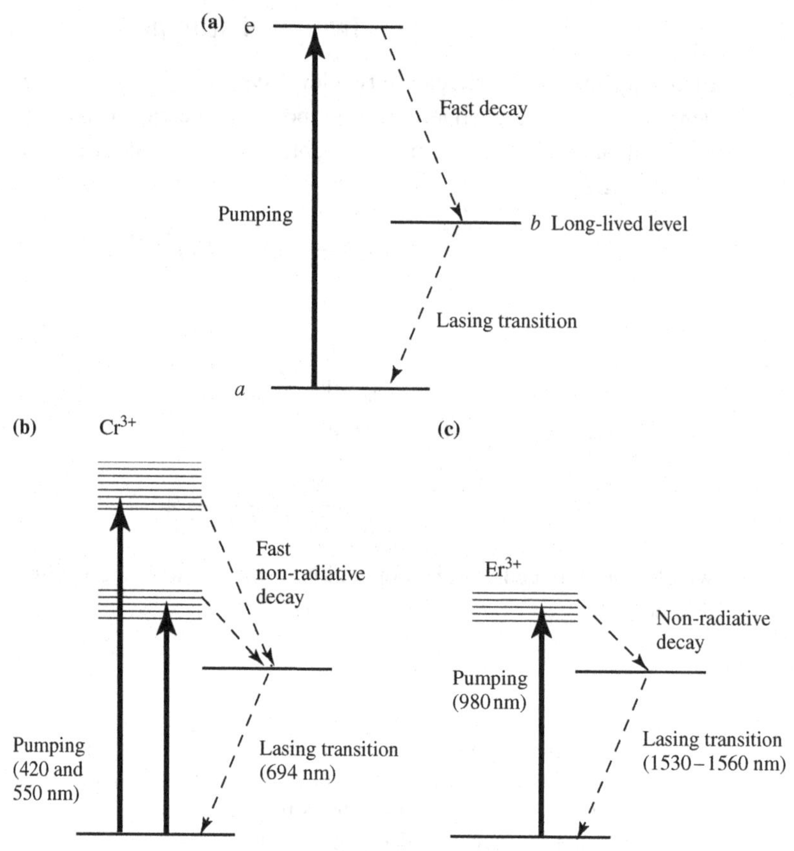

Figure 3.18 (a) Model energy-level scheme for a three-level system, (b) its practical realization in the case of the Cr^{3+} ion in a ruby crystal and (c) in the case of the Er^{3+} ion.

The rate equations describing the pumping and relaxation processes in a three-level laser are the following:

$$\frac{\mathrm{d}}{\mathrm{d}t}N_\mathrm{e} = w(N_a - N_\mathrm{e}) - \frac{N_\mathrm{e}}{\tau_\mathrm{e}} \tag{3.38}$$

$$\frac{\mathrm{d}}{\mathrm{d}t}N_b = \frac{N_\mathrm{e}}{\tau_\mathrm{e}} - \frac{N_b}{\tau_b} \tag{3.39}$$

$$N_a + N_b + N_\mathrm{e} = N, \tag{3.40}$$

where w is the pumping rate from level a to level e. In the limit (commonly realized) in which $w\tau_\mathrm{e} \ll 1$, the steady-state ratios of the populations (that is for which the time-derivatives in Equations (3.38) and (3.39) are zero) are

$$\frac{N_b}{N_\mathrm{e}} = \frac{\tau_b}{\tau_\mathrm{e}} \gg 1, \tag{3.41}$$

$$\frac{N_e}{N_a} = w\tau_e \ll 1. \tag{3.42}$$

The population inversion is given by

$$\frac{N_b - N_a}{N} = \frac{w\tau_b - 1}{w\tau_b + 1}. \tag{3.43}$$

This result can be compared with that obtained for the four-level laser configuration (Equation (3.31)). Whilst a population inversion is easy to obtain for a four-level system, for a three-level system this can only be achieved *if the pumping is sufficiently strong* that

$$w \gg \frac{1}{\tau_b}. \tag{3.44}$$

The necessity of a large pumping rate is one disadvantage of three-level systems. A second is associated with the difficulty of obtaining continuous operation of some such systems. In fact, once laser action has commenced it is also necessary to account in the rate equations for the rate of stimulated emission transferring atoms from level b to level a. This leads to an effective reduction in the lifetime of level b and the population inversion given by (3.43) will then decrease to a value such that laser action terminates. The pumping process will then be able to cause the accumulation of more atoms in the excited state with the result that the gain rises until laser action is once again initiated. As a result the laser emits a train of pulses, originating from transient relaxation oscillations (see Figure 3.19). The detailed dynamical behaviour depends on various characteristic timescales in the system such as the atomic relaxation times and the cavity lifetime, and unless it is controlled (see for instance Section 3.4.2) this phenomenon is a serious limitation to the use of this type of laser source.

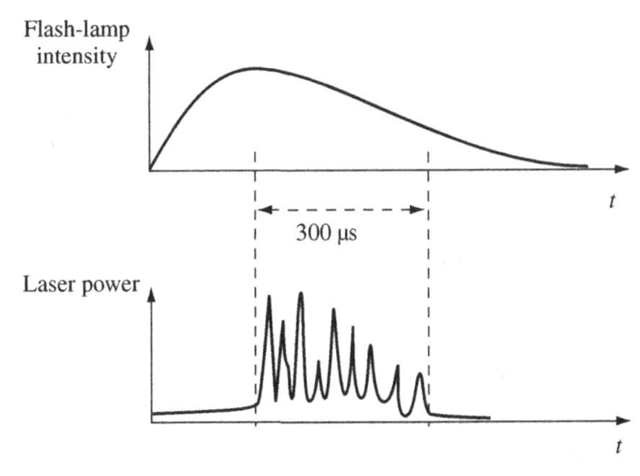

Figure 3.19 Schematic representation of the temporal evolution of (a) the flash-lamp pumping of a ruby laser and (b) the power output from the laser.

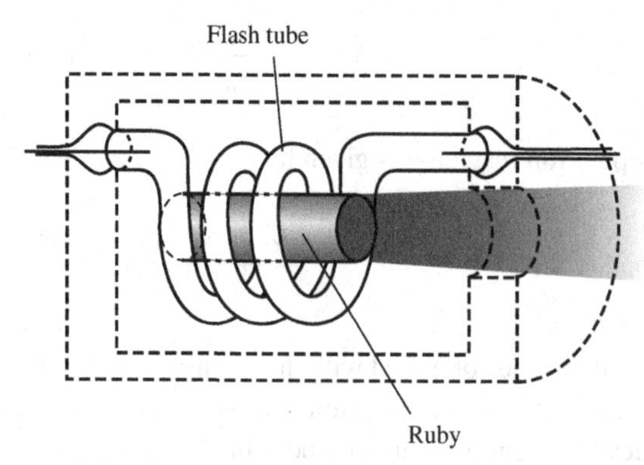

Figure 3.20 **Schematic diagram of a ruby laser of the type built by Maiman.**

The ruby laser

In the case of the ruby laser, the active element is the Cr^{3+} ion. Figure 3.18 shows the energy levels of the Cr^{3+} ion in the crystal. Notice the existence of two absorption bands which can be excited by the light emitted by a flash-lamp.

Figure 3.20 shows a sketch of the laser built by Maiman, the first to produce a coherent laser beam, in May 1960. The amplifying medium is a rod of ruby pumped by a flash-lamp in which it is enclosed. The end facets of the ruby rod are polished and coated with a semi-reflecting silver coating to form a linear laser cavity. The laser emits pulses in the red at a wavelength of 694 nm.

The erbium laser

The most common laser system that corresponds to the energy-level scheme of Figure 3.18 is the erbium laser. This laser, which amplifies infra-red light in the wavelength range 1.52–1.56 μm has given a new lease of life to three-level lasers. Two particularly noteworthy characteristics of these lasers are that the emission wavelength corresponds to the minimum of the absorption in optical fibres, and that it has a rather broad gain bandwidth. It is therefore ideally suited for applications in *telecommunications*. Erbium ions can, in fact, be implanted into silica which can be used to fabricate *erbium-doped fibres*. When excited by pump light, for example from a semiconductor laser at 1.48 μm or 0.98 μm, these fibres function as extremely efficient optical amplifiers. When such lengths of amplifying fibre are inserted into long fibre-optic cables they can regenerate the light pulses attenuated by the absorption losses in a rather broad frequency range, enabling many communication channels of slightly different wavelengths to be carried and amplified simultaneously (WDM: Wavelength Difference Multiplexing).

Laser oscillation is obtained if the erbium-doped fibre is placed in a suitable cavity. For example, the mirrors can be integrated onto the end facets of the fibre itself. One obtains in such a way very convenient fibre lasers, which are totally reliable against mechanical vibrations, and turn out to be amazingly powerful, considering the small size of the fibre core (more than a kW c.w. in an almost TEM_{00} mode) and efficient (80% conversion efficiency from pump light to laser light). They also have a wide range of industrial applications. Let us mention that other rare-earth ions can be used for the same purpose and provide other fibre lasers in other parts of the near infra-red spectrum.

3.3 Spectral properties of lasers

3.3.1 Longitudinal modes

We showed in Section 3.1.2 that the oscillation condition (that the unsaturated gain should surpass the losses) could be satisfied for several longitudinal modes of the laser cavity (Figure 3.5). The number of modes on which the laser can oscillate is then of the order of the ratio of the width of the gain curve and the frequency interval c/L_{cav} between successive modes (L_{cav} is the optical round-trip cavity length). For example, for the red line of the helium–neon laser the width of the gain curve is 1.2 GHz, whilst for a typical round-trip cavity length of 0.6 m, the longitudinal mode spacing is 0.5 GHz. The laser therefore oscillates on two or three modes. It should be realized, however, that whilst the position of the gain curve is fixed in frequency, the comb of longitudinal cavity modes is displaced when the cavity length changes (for example, as a result of thermal expansion of the mechanical structure holding the mirrors). Thus the laser, which might oscillate at a given instant on two modes might oscillate on three moments later.

This ability of the helium–neon laser to oscillate simultaneously on all the longitudinal modes lying within the range of its gain curve is by no means universal, however. For other lasers, such as the Nd:YAG laser, fewer modes oscillate than one might expect. The difference in behaviour is directly related to the nature of the mechanism responsible for the broadening of the gain curve. In particular, whilst the gain curve of the helium–neon laser is *inhomogeneously broadened*, that of the Nd:YAG system is essentially *homogeneously broadened*. A spectral line is homogeneously broadened if all the active centres (atoms, ions, molecules) have the same spectral response. In contrast, a spectral line is said to be inhomogeneously broadened if the appearance of different frequencies in the response of the global gain medium is associated with active centres that have different spectral responses. For instance, neon atoms with different velocities have different resonance frequencies, in the laboratory frame, because of the Doppler effect. Thus an inhomogeneously broadened curve is simply a juxtaposition of a number of homogeneous lineshapes centred on different frequencies which reflect the different environments of the various active centres.

The distinction between homogeneous and inhomogeneous broadening is vital in determining the number of modes that oscillate. Recall that the steady-state operation of a laser arises because gain saturation reduces the gain until it is precisely equal to the losses from the cavity (see Section 3.1.2). In the case of inhomogeneous broadening (Figure 3.21) the *gain saturation* (that is, the decrease in the gain as the intensity increases) only affects those atoms (or molecules) of which the resonant frequency is that of the light field; the

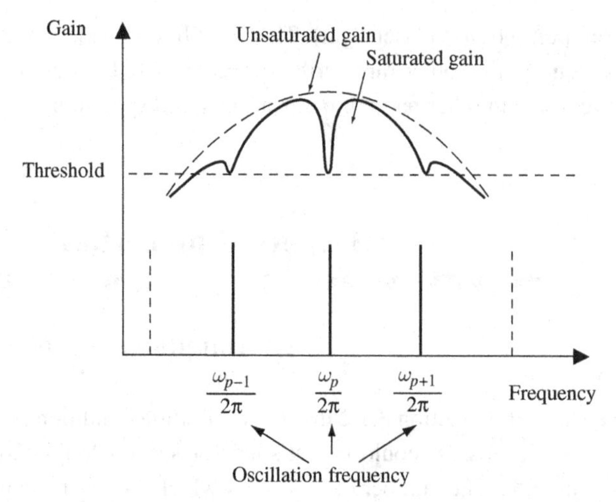

Figure 3.21 Effect of saturation on an inhomogeneously broadened gain curve: holes are burnt in the gain curve only at the laser oscillation frequencies. This allows a number of modes to oscillate simultaneously without competing against each other.

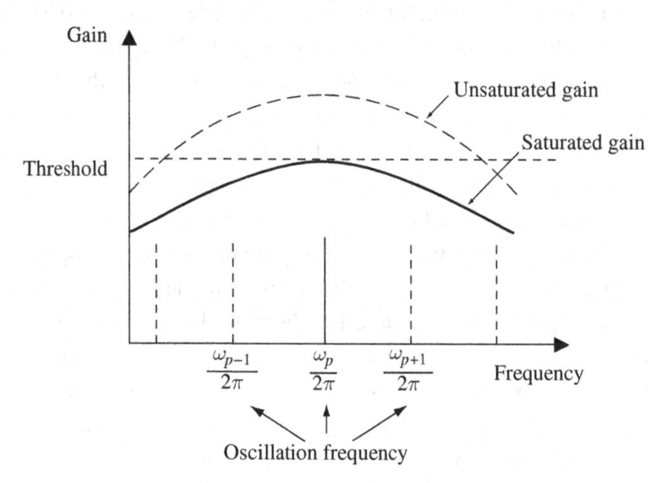

Figure 3.22 Effect of gain saturation in the case of a homogeneously broadened gain curve. The entire gain curve is depressed by the phenomenon of gain saturation, so that a single mode, that nearest in frequency to the maximum of the gain curve, can oscillate.

laser oscillation has no effect on the gain appearing at other frequencies. In this case oscillation can occur on all the cavity modes at the frequencies of which the gain is above the threshold value. This is the case with the helium–neon laser, where inhomogeneous broadening is dominant because of the Doppler motion of the atoms.

Consider now the case of a laser for which the gain curve reflects a homogeneous broadening of the laser transition (Figure 3.22). In this case the response of each atom is affected in the same manner by the gain saturation. This leads therefore to the entire gain curve being displaced towards lower gain as the intensity in the laser cavity increases so that,

Table 3.1 Characteristics of some common continuous-wave lasers including the approximate number of modes that can possibly oscillate.

	Gain linewidth	Typical frequency interval between modes	Number of possible longitudinal modes	Dominating broadening mechanism
He–Ne	1.2 GHz	500 MHz	3	Inhomogeneous (Doppler)
CO_2 (high pressure)	0.5 GHz	100 MHz	5	Homogeneous
Neodymium:YAG	120 GHz	300 MHz	400	Homogeneous
Dye (Rhodamine 6G)	25 THz	250 MHz	10^5	Homogeneous
Titanium-doped sapphire	100 THz	250 MHz	4×10^5	Homogeneous
Semiconductor	1 THz	100 GHz	10	Homogeneous

finally, oscillation occurs on a single longitudinal mode: that lying closest in frequency to the maximum of the gain curve. Oscillation on this mode effectively prevents laser emission on the other modes. This phenomenon is called 'mode competition'.

In practice, it is unusual for the broadening of a laser gain curve to be purely homogeneous or inhomogeneous in nature. For the continuous-wave lasers, mentioned in Table 3.1, one of these types of broadening is dominant as indicated. However, very often the homogeneous and inhomogeneous widths can have comparable values, and the situation is not as clear-cut as suggested by the simple reasoning sketched on Figures 3.21 and 3.22.

Comment Some systems can have homogeneous or inhomogeneous broadening, depending on some parameter such as the pressure in a gas laser. For example, the linewidth of the transition at 10.6 μm in CO_2 gas at moderate pressure is extremely small (less than 1 MHz), but the thermal motion of the molecules of the gas leads to a Doppler width which is considerably larger (50 MHz at room temperature). This Doppler broadening is inhomogeneous, because their different velocities endow molecules with distinct resonance frequencies in the laboratory frame. In high pressure CO_2 laser systems, however, collisional broadening dominates the spectral width of the laser emission. This broadening is homogeneous since the average effect of collisions is the same for all molecules and their transition frequencies are affected in an identical manner.[11]

3.3.2 Single longitudinal mode operation

For many applications (such as spectroscopy,[12] holography, metrology) it is desirable that the laser output should be as near monochromatic as possible. It is therefore often necessary

[11] Homogeneous and inhomogeneous line broadening mechanisms are contrasted also in Complement 3G, devoted to laser spectroscopy.

[12] See Complement 3G.

to ensure that a laser capable of oscillating on many modes simultaneously is constrained to oscillate on a single longitudinal mode (single-mode operation). As we have shown, such a situation occurs naturally when the dominant line broadening mechanism is homogeneous in nature. This, however, is rarely the case in practice and it is usually necessary to take steps to ensure single-mode operation. A particularly straightforward method consists of employing a laser cavity sufficiently short that the longitudinal mode separation c/L_{cav} is greater than the width of the gain curve. Such a technique works for short helium–neon lasers ($c/L_{cav} = 1$ GHz for $L_{cav} = 0.3$ m) and for some semiconductor lasers ($c/L_{cav} = 1000$ GHz). However, this method is not always applicable.

The surest manner of ensuring that a laser operates on a single longitudinal mode is to introduce into the cavity a frequency-selective filter, of which the transmission profile is narrower than the gain curve, and which ensures that the net gain coefficient (the product of the gain coefficient in the absence of the filter and the transmission of the filter) is above threshold for a single cavity mode only (Figure 3.12). Of course, it is also necessary to ensure that the peak transmission of the filter is as close to one as practically possible, in order not to introduce extra losses at the operating wavelength, and therefore to ensure, usually by the means of a servo mechanism, that the maximum of the transmission of the filter is tuned accurately to the position of the laser cavity mode that oscillates (preferably a mode close to the maximum of the gain curve).

Comments

(i) Often it is necessary, in order to achieve the required degree of frequency selectivity, to employ a filter composed of a number of separate elements with successively narrower transmission peaks (all of which must be accurately set to the same wavelength). For example (Figure 3.12), a single-mode titanium–sapphire laser includes a triple birefringent filter, which selects a frequency band 300 GHz wide, followed by a 'thin' Fabry–Perot étalon (1 mm wide), which has a transmission curve 3 GHz wide and a 'thick' étalon (5 mm wide), which transmits a single cavity mode. Such a system necessitates the use of nested servo-locking mechanisms which can be difficult to set up.

(ii) It is generally more difficult to make a linear cavity laser work on a single longitudinal mode than one with a ring cavity. In the case of a linear cavity the intracavity light field is a standing wave, which comprises an alternating sequence of nodes and antinodes. At the antinodes gain saturation occurs and diminishes the gain. However, a given mode does not reduce the gain in the region of its nodes, which can lead to the oscillation of a second mode for which the antinodes are *spatially separated* from those of the first. It is easy to show, for the case of the amplifying medium occupying the central region of a linear cavity, that two adjacent modes ω_p and ω_{p+1} satisfy precisely this requirement, which explains the tendency of such systems to oscillate on several modes in the absence of frequency selective elements.

The spatial gain modulation described above does not occur for a ring-cavity laser in which the light circulates in a single sense; the intracavity light field is then a travelling wave, which has a spatially uniform intensity. This explains why a dye laser or a titanium:sapphire laser (which are essentially homogeneously broadened) should operate spontaneously on a single mode in a ring cavity whilst its output is multi-mode when they have a linear cavity.

Once single-mode operation has been achieved, it is often desirable to be able to control the output frequency to a precision far exceeding the inter-mode separation c/L_{cav}. The frequency of the pth mode is given by (3.16). It is a function of the cavity length L_{cav} and a change in frequency equal to the inter-mode spacing is obtained for a length change equal to the optical wavelength:

$$\delta L_{\text{cav}} = \frac{L_{\text{cav}}}{p} = \lambda_p. \tag{3.45}$$

Therefore, in order to fix the laser frequency to a precision better than the cavity free-spectral range it is necessary to control the cavity length to better than one optical wavelength. This can be achieved by mounting one of the cavity mirrors on a piezoelectric transducer which allows the control of the cavity length to within 0.1 nm.

Comments

(i) With semiconductor lasers the optical length of the cavity can be controlled by varying the temperature of the device or the electric current (on which the refractive index depends).

(ii) In some semiconductor lasers single-mode operation is ensured by the use of distributed feed-back. The walls of the optical waveguide are modulated in the longitudinal direction with a period of a multiple of the half-wavelength of desired operation: in this way a single mode is selected.

3.3.3 Spectral width of the laser output

We now address the question of the precision to which the frequency of the output of a single-mode laser is determined: that is, what is the linewidth of its spectral power density? This width depends on several factors, which we now discuss.

Technical broadening

As we saw above, any alteration in the optical length of the laser cavity causes a corresponding change in the frequency ω_p of the selected cavity mode p. Thus, for a typical laser cavity length of 1 m, the displacement of a mirror through $\lambda/600$ (1 nm for a laser operating in the yellow region of the visible spectrum) leads to a frequency change of 1 MHz. Such displacements can arise for numerous reasons, for example, *thermal expansion* of the cavity structure (a temperature change of 2 mK being sufficient to give such a length change for the structure of quartz, which has a particularly small thermal expansion coefficient) or *changes of pressure* (either atmospheric or caused by acoustic waves), which give rise to a change in the refractive index of the air in the laser cavity.

These phenomena give rise to a more or less random modulation of the laser frequency, which is known as *jitter*. Because of this, the short-term frequency stability of a laser, in the absence of active stabilization techniques, is not usually better than a few MHz. In the longer term (more than one minute) laser frequency variations arise mainly as a result of slow temperature changes so that the laser frequency is rarely defined to better than c/L_{cav}, the longitudinal mode separation. For many applications this situation will not suffice and active frequency stabilization techniques must be employed.

These function in the following manner: the laser frequency is compared to a fixed reference frequency and the cavity length L_{cav} is then controlled in order to minimize the recorded frequency difference. The stability achieved is then limited by that of the reference frequency itself and by the noise present in the servo-locking system. Often the reference frequency is a selected mode of a Fabry–Perot cavity (see Complement 3A) and much care is taken to ensure its stability, for example, by regulating its temperature, enclosing it in a vacuum chamber and isolating it from vibrations. The short-term stability of the laser can then approach 1 MHz, but longer-term variations are more difficult to eliminate, in particular those caused by the ageing of materials employed in the laser cavity.

The long-term stabilization of a laser usually relies on the comparison of its frequency with an absolute frequency reference, such as an atomic or molecular resonant transition. Of course, it is then necessary to find a resonance that is sufficiently narrow and to eliminate additional sources of broadening such as the Doppler motion (see Complement 3G and Chapter 8) and stray electric and magnetic fields. However, such schemes allow the frequency locking of lasers with extraordinary precision. A long-term stability to within 0.1 Hz for a laser operating on an optical transition, equivalent to a precision of one part in 10^{16} has been obtained in the laboratory (or, the ratio of 1 minute to the lifetime of the universe). Such a laser could serve as a time standard, which would be 10^3 times more stable than the caesium atomic clocks currently in use. The precision of the caesium atomic clock is already demanded by some stringent applications in telecommunications and satellite based navigation systems and it is to be expected that an improved precision will find application.

The fundamental linewidth limit: the Schawlow–Townes limit

Suppose that all the causes of frequency jitter for a single-mode laser are eliminated and that the cavity length is precisely fixed. What then determines the limit to the precision to which the laser frequency can be defined? This question was answered by Schawlow and Townes and the fundamental limit to the frequency width of a laser's output is consequently known as the Schawlow–Townes limit. Its cause is spontaneous emission, which is inseparable from stimulated emission in matter–light interactions and which therefore occurs too in the active medium in a laser cavity. Whenever a spontaneous emission event occurs in the cavity it adds a small contribution to the electromagnetic field present, a contribution which has a random amplitude and phase. The resultant amplitude fluctuation is rapidly corrected by the phenomenon of gain saturation, but the small change to the phase persists since condition (3.10) does not constrain the phase φ of the output wave in any way. The result is that the phase of the laser output diffuses in time as a random-walk process. It leads to the laser output having a finite frequency width, $\Delta\omega_{ST}$, which is of the order of the inverse of the phase correlation time (the time after which the memory of the initial phase is lost). A simplified calculation of this process is presented in Complement 3D. Very approximately, for a laser operating far above threshold, the width $\Delta\omega_{ST}$ is given by the cavity linewidth $\Delta\omega_{cav}$ (in the absence of gain) divided by the *number of photons* in the laser cavity under operating conditions.[13]

For a helium–neon laser one has typically $\Delta\omega_{ST}/2\pi \approx 10^{-3}$ Hz, whilst for a semiconductor laser emitting a power of 1 mW, $\Delta\omega_{ST}/2\pi \approx 1$ MHz. Because of the small number of photons in the small semiconductor laser cavity the corresponding Schawlow–Townes limit is easily accessible to experiments whilst the technical broadening is dominant for the case of the helium–neon laser.

3.4 Pulsed lasers

Apart from continuous-wave operation that we have mostly considered so far, lasers provide the possibility of functioning in a pulsed mode. Whilst these devices generally produce an optical output of similar average power, the maximum instantaneous power, and hence the maximum electric field, obtained at the peak of a pulse can be very large. This has permitted the study of new phenomena in matter–light interactions, for example in *nonlinear optics* (see Chapter 7), *multi-photon ionization* and in *laser-induced plasmas* amongst others. Whilst some lasers, such as Maiman's ruby system, function spontaneously in pulsed mode, because of the relaxation oscillations inherent in the three-level laser configuration (see Section 3.2.3 and Figure 3.19) others have to be forced to work in a pulsed mode. We show in Section 3.4.2 how it is possible to harness the relaxation oscillations occurring in some lasers to produce very intense pulses. Firstly, however, we consider the *mode-locked* operation of continuously pumped laser systems. With some refinements this technique has allowed the generation of intense pulses with durations as little as a few femtoseconds, equivalent to just a few optical periods in the visible spectral region and has opened up new possibilities for the study of *ultra-fast phenomena.* Furthermore, mode-locked lasers have led to the development of frequency combs, which constitute an absolute reference for optical frequencies.[14]

3.4.1 Mode-locked lasers

Mode locking is an elegant technique which allows the generation of extremely brief laser pulses. In order to understand the principle consider a continuous-wave laser oscillating on many longitudinal modes of the laser cavity. The emitted light then contains several (nearly) monochromatic components, corresponding to the various modes that oscillate, of which the phases are a priori *uncorrelated.* The electric field at the output mirror of such a laser can be written

[13] Recall that $\Delta\omega_{cav}$ and the lifetime τ_{cav} of radiation in the cavity are related by an expression of the form $\Delta\omega_{cav}\tau_{cav} \approx 1$ (see Complement 3A).

[14] J. L. Hall, Nobel Lecture, Defining and Measuring Optical Frequencies, *Reviews of Modern Physics* **78**, 1279 (2006). T. W. Hänsch, Nobel Lecture, Passion for Precision, *Reviews of Modern Physics* **78**, 1297 (2006). Available on-line at: http://nobelprize.org/nobel-prizes/physics/laureates/2005.

$$E(t) = \sum_{k=0}^{N-1} E_0 \cos(\omega_k t + \phi_k), \tag{3.46}$$

where N is the number of oscillating modes and the ϕ_k are their phases, supposed initially to be random and uncorrelated. We have assumed for simplicity that the amplitudes of all the modes are equal and given by E_0. The frequency of the kth mode is given by

$$\omega_k = \omega_0 + k\Delta, \tag{3.47}$$

where $\Delta/2\pi = c/L_{cav}$ (the inter-mode separation, see Section 3.1.2) and ω_0 is the frequency of the mode $k = 0$.

We characterize the laser output by its time-dependent intensity, i.e. the quantity measured by a fast photodetector, with a sufficiently short response time, θ (Complement 2E). Generalizing definition (2.90) to the case of an electric field with a spectrum of bandwith less than $1/\theta$, as for instance (3.46) with $N\Delta < \theta^{-1}$,

$$\begin{aligned} E^2(t) &= \left[\sum_k \frac{E_0}{2} \mathrm{e}^{-\mathrm{i}(\omega_k t + \phi_k)} + \text{c.c.}\right]\left[\sum_j \frac{E_0}{2} \mathrm{e}^{-\mathrm{i}(\omega_j t + \phi_j)} + \text{c.c.}\right] \\ &= \frac{E_0^2}{4} \sum_k \sum_j \left[\mathrm{e}^{-\mathrm{i}[(\omega_k + \omega_j)t + \phi_k + \phi_j]} + \text{c.c.}\right] \\ &+ \frac{E_0^2}{2} \sum_k \sum_j \left[\mathrm{e}^{-\mathrm{i}[(\omega_k - \omega_j)t + \phi_k - \phi_j]} + \text{c.c.}\right]. \end{aligned} \tag{3.48}$$

When averaged over time interval θ, the first term gives a null contribution. In contrast, the second term varies slowly at the scale of θ, and the intensity is thus

$$\begin{aligned} I(t) &= 2\left[\sum_k \frac{E_0}{2} \mathrm{e}^{-\mathrm{i}(\omega_k t + \phi_k)}\right]\left[\sum_j \frac{E_0}{2} \mathrm{e}^{\mathrm{i}(\omega_j t + \phi_j)}\right] \\ &= 2\, E^{(+)}(t)\, E^{(-)}(t) = 2|E^{(+)}(t)|^2, \end{aligned} \tag{3.49}$$

where we have introduced the analytic signal (Section 4.3.4):

$$E^{(+)}(t) = \sum_k \frac{E_0}{2} \mathrm{e}^{-\mathrm{i}\phi_k}\, \mathrm{e}^{-\mathrm{i}\omega_k t}. \tag{3.50}$$

Expression (3.49) can be transformed as

$$I(t) = \frac{NE_0^2}{2} + E_0^2 \sum_{j>k} \cos\left[(\omega_j - \omega_k)t + \phi_j - \phi_k\right], \tag{3.51}$$

which is the sum of the average intensity,

$$\bar{I} = \frac{NE_0^2}{2} \tag{3.52}$$

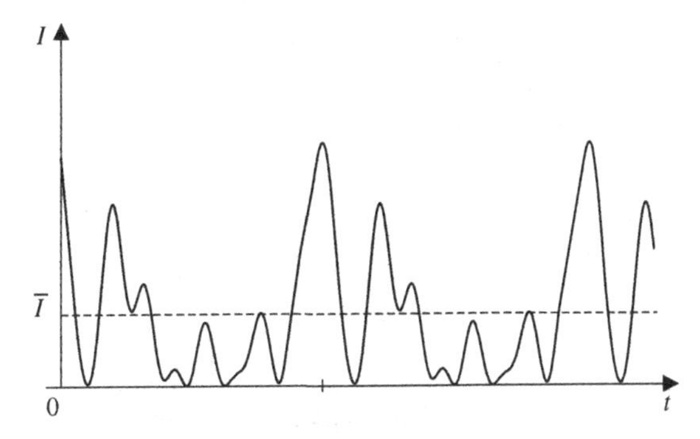

Figure 3.23 **Temporal variation of the output intensity of a multi-mode laser. The phases of the oscillating modes are independent random variables. The intensity fluctuates around its mean value.**

and of a fluctuating term of which the root-mean-square amplitude, $\Delta I = \sqrt{\overline{(I(t) - \bar{I})^2}}$, for random, uncorrelated ϕ_k and in the limit $N \gg 1$ is

$$\Delta I = \bar{I}. \tag{3.53}$$

The output intensity of a multi-mode laser therefore has significant temporal fluctuations of which the magnitude is of the same order as the mean intensity. A plot showing the temporal evolution of the intensity in a typical example of such a system is shown in Figure 3.23.

The origin of the intensity fluctuations seen above is the *interference of the fields corresponding to the various modes* oscillating at different frequencies; interference which is at some instants constructive and at others destructive. Clearly, if all the modes could be made to interfere constructively at some instant the briefest, highest intensity pulses would be obtained. Since the intensity varies as the square of the net electric field and the average intensity is just $\bar{I}$ (the second term of Equation (3.51) averaging to zero whatever assumptions are made about the variation of ϕ_k), it follows that the high-intensity pulses must be extremely brief.

Consider, therefore, the case in which all the phases are strongly correlated. For the sake of simplicity, we assume that the phases ϕ_k all have the same value, $\phi_k = 0$. Using (3.49), we now have

$$I(t) = \frac{1}{2} \left| \sum_{k=0}^{N-1} E_0 e^{i(\omega_k t + \phi_k)} \right|^2, \tag{3.54}$$

which gives

$$I(t) = \frac{E_0^2}{2} \left| \sum_{k=0}^{N-1} e^{i\omega_k t} \right|^2. \tag{3.55}$$

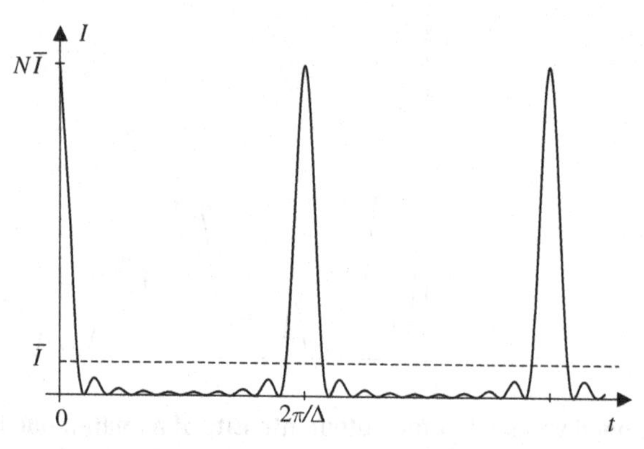

Figure 3.24 **Temporal variation of the output intensity of a mode-locked laser. This variation results from the constructive interference of a comb of *N* cavity modes, equally spaced in frequency (separation Δ) and having the same phase.**

Equation (3.47) then implies

$$I(t) = \frac{E_0^2}{2}\left|\sum_{k=0}^{N-1} e^{ik\Delta t}\right|^2. \tag{3.56}$$

Performing the summation, we obtain for the intensity,

$$I(t) = \frac{E_0^2}{2}\left|\frac{\sin\left(N\Delta\frac{t}{2}\right)}{\sin\left(\Delta\frac{t}{2}\right)}\right|^2. \tag{3.57}$$

Figure 3.24 shows the temporal variation of $I(t)$. The laser emits *pulses* at a repetition frequency of

$$T = \frac{2\pi}{\Delta} = \frac{L_{\text{cav}}}{c}, \tag{3.58}$$

which is just the round-trip time in the cavity. This implies that a single light pulse circulates in the cavity so that a pulse of light is emitted each time this optical wave packet bounces off the output mirror. The intensity at the peak of each pulse is given by

$$I_{\text{max}} = \frac{N^2 E_0^2}{2}, \tag{3.59}$$

or, using (3.52),

$$I_{\text{max}} = N\bar{I}. \tag{3.60}$$

The temporal width of each pulse is

$$\frac{2\pi}{N\Delta} = \frac{T}{N}. \tag{3.61}$$

Thus, the greater the number N of modes that oscillate, the greater is the peak intensity and the shorter is the duration of each pulse. As we saw in Section 3.1.2, this number of modes is proportional to the spectral width of the gain curve. In order to generate very short pulses this width must be large.

Employing this technique, it is possible to obtain from a commercial titanium:sapphire laser pulses of duration as short as 100 fs and of peak power in the 100 kW range. The duration of these pulses, once suitably amplified, can then be further reduced to less than one femtosecond, and the peak power brought to the petawatt (10^{15} W) range in some devices.

Mode locking can be achieved in practice by modulating the cavity losses, using an acousto-optic or an electro-optic modulator, at frequency Δ. Consider, for example, a particular mode at frequency ω_k of which the amplitude is modulated at this frequency,

$$E(t) = E_0[1 + m\cos(\Delta t)]\cos(\omega_k t + \phi_k). \tag{3.62}$$

It can be transformed to

$$E(t) = E_0\left\{\cos(\omega_k t + \phi_k) + \frac{m}{2}\cos\left[(\omega_k + \Delta)t + \phi_k\right] + \frac{m}{2}\cos\left[(\omega_k - \Delta)t + \phi_k\right]\right\}, \tag{3.63}$$

in which the frequencies $(\omega_k \pm \Delta)$ of the two neighbouring modes $k-1$ and $k+1$ appear. Under certain conditions these components, which have the same phase as mode k, lock the phases of the $(k-1)$th and $(k+1)$th mode to that of the kth mode. This occurs over the whole comb of oscillating modes with the result that all the phases are locked to the same value. The phenomenon can also be understood if we look at the situation in the time-domain. Consider a light pulse circulating in the cavity. This will experience the smallest attenuation if it passes through the modulator at the moment, once in each cycle, when this imposes the smallest loss. Thus the period of the modulation must precisely equal the cavity round-trip time for sustained oscillation to occur, which is precisely the requirement of (3.58).

Alternatively, it is possible to mode lock the laser *passively* by placing into the cavity a *saturable absorber*. This is a material for which the transmission increases as does the incident light intensity: it is bleached by the light and is practically transparent at high intensity. For example, an ensemble of two-level atoms possesses this property (see Chapter 2, Section 2.4.3). The loss of the material at low intensity is chosen to ensure that the laser can *only* oscillate when the cavity modes cooperate (oscillate in phase) to ensure the maximum pulse intensity. Thus mode locking occurs spontaneously.

The laser medium itself can also be used to achieve passive mode locking. In the Kerr-lens method (also known as *magic mode locking*), which works remarkably well in the titanium:sapphire laser, a pinhole introduced into the cavity between the amplifier and the mirror prevents the laser from oscillating in the continuous regime, because of the losses originating from the pinhole. However, under appropriate conditions, a laser pulse of high intensity undergoes self-focusing (see Section 7B.5.1) in the laser medium. As a result, the optical beam is narrower at the position of the pinhole and the losses introduced by it are reduced. The laser can thus operate in a pulsed mode.

Mode-locked lasers provide trains of ultra-short pulses which are useful to monitor and control very fast processes. Because of their high peak power, they induce very spectacular nonlinear effects. Furthermore, when the cavity length is suitably servo-controlled with reference to the caesium clock, they provide a set of equally spaced modes spanning the whole visible spectrum, having frequencies known with a relative accuracy of 10^{-14}. Such a 'frequency comb'[15] is a kind of ruler which is now widely used in spectroscopy (see Complement 3G).

3.4.2 Q-switched lasers

Figure 3.19 shows the temporal evolution of the output intensity of a ruby laser when the crystal is pumped by a flash-lamp. This consists of a train of pulses of temporal width about 0.1 μs separated by a few microseconds. For a total emitted energy over the pulse train of 1 J the average power is of the order of a kilowatt, whilst the peak power is an order of magnitude larger.

The form of the temporal evolution of the ruby laser output is determined by relaxation oscillations: when the population inversion is sufficient, laser oscillation starts. However, the burst of stimulated emission which results decreases the population inversion and hence the gain, until oscillation can no longer be maintained. The population inversion can then recover until it is sufficient for oscillation to again be initiated.

The emission by a laser of a train of pulses with more or less random temporal separations limits the usefulness of such devices for many applications. Although it is possible to include in the cavity an electro-optic device for selecting out a single pulse of such a train (known as *cavity dumping*), the problem of timing remains. A much more elegant solution involves increasing the cavity loss to prevent laser oscillation during the beginning of the pump pulse, when the population inversion is building up, and then suddenly reducing it so that the accumulated energy is released in a giant burst of stimulated emission (Figure 3.25). A device performing this function is known as a *Q-switch* since it modulates

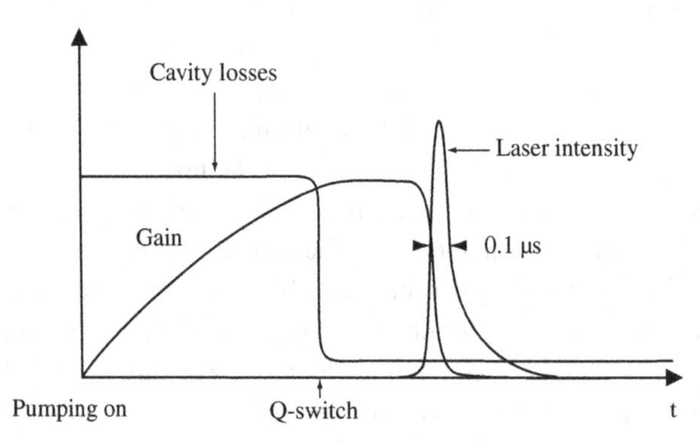

Figure 3.25 **Operation of a Q-switched laser. The cavity loss is maintained at a high value during the pumping phase in which the population inversion is built up. It is then suddenly switched to a small value, resulting in the emission of the stored energy in a single giant pulse of which the peak power can exceed 10^7 W.**

[15] See Footnote 14, this chapter.

the cavity quality factor. The peak power obtained using such a Q-switched laser, which in practice can exceed a megawatt, is up to two orders of magnitude larger than that possible in the absence of Q-switching.

Practical implementations of Q-switches depend on a range of devices. The first of these simply employed a rotating mirror in the laser cavity, which only briefly completed the light circuit once per revolution. However, this arrangement is now obsolete, having been abandoned because of the problems involved in synchronizing the rotation to the flash-lamp pulse, and with vibrations caused by the rotation mechanism. Nowadays, acousto-optic modulators or electro-optic modulators are employed. They have switching times as short as 10 ns.

Another method that deserves attention concerns passive Q-switching arising from insertion into the laser cavity of a saturable absorbing material, also used for mode locking. Whilst the population inversion is still small, in the early stages of the pump pulse, the light circulating in the cavity is too weak to bleach the absorber, which prevents oscillation. On the other hand, when the population inversion reaches its peak the initiation of laser oscillation provokes a rapid decrease in the loss caused by the saturable absorber, and a giant pulse is emitted. In this case the opening of the Q-switch occurs spontaneously; there is no longer a problem of synchronization.

Q-switched solid-state lasers (ruby or neodymium) are frequently employed for two distinct types of applications: firstly, when high peak powers are required at a moderate average power (for example, for some nonlinear optical effects) and secondly, applications requiring the emission of a single pulse on demand, such as laser range finders (LIDAR). These are capable of determining the earth–moon separation to a precision of 3 mm, or of studying the drifts of the continents by measuring the round-trip time of a light pulse relayed by a satellite (See Complement 3F).

3.5 Conclusion: lasers versus classical sources

At the beginning of the year 2000, a frequency-doubled YAG laser capable of emitting 15 watts costs around 50,000 euros, a 150 W incandescent lamp, on the other hand, costs 1 euro. That lasers are employed at all indicates, therefore, that the light they emit must have some rather remarkable properties. The principal one of these is that the admittedly smaller energy of the laser can be concentrated in an extremely narrow region of space and time, or of the spectrum. This leads to the generation of energy densities that are enormously superior to those obtainable with ordinary sources.

3.5.1 Classical light sources: a few orders of magnitude

According to Planck's law, a black body at a temperature of 3000 K emits approximately 500 $W.cm^{-2}$ into the surrounding space, in a spectral region extending from the infra-red (1.5 μm) to the green (0.5 μm), that is, a range of over 4×10^{14} Hz. These figures are typical for the filament of an incandescent lamp; a high pressure arc lamp has a radiance (the power emitted per unit surface area per unit solid angle, see Complement 3C) ten times higher.

Is it possible to concentrate all this power onto a small area? The answer provided by the laws of optics (Complement 3C) is unequivocal; whatever the optical system employed the luminance cannot be increased beyond its initial value: a system capable of doing this would violate the Second Law of Thermodynamics. Therefore, under optimum conditions (using lossless optics of large numerical aperture) the highest irradiance (power per unit surface) that could be obtained would be 500 $W.cm^{-2}$.

Suppose a requirement to excite an atomic transition with a linewidth of 1 MHz with such a source. Only a small fraction of the light emitted by a classical source will be in the relevant spectral region. In the example above, the useful light flux is around 1 $\mu W.cm^{-2}$. This is insufficient to observe nonlinear optical effects, for which a flux on resonance of 1 $mW.cm^{-2}$ is typically required. Pulsed flash-lamps are capable of reaching peak radiances two or three orders of magnitude larger than incandescent sources but they remain by a large margin incapable of attaining the strong-field domain.

3.5.2 Laser light

Consider now the laser beam in the above example, delivering a continuous optical power of 15 watts. Because of its spatial coherence, it can be focused to a diffraction-limited spot of which the diameter is of the order of the optical wavelength, that is an area of 1 μm^2. Thus the irradiance at the focus is 10^9 $W\,cm^{-2}$ as opposed to the 500 $W\,cm^{-2}$ that we had before. Moreover, this power is concentrated into a spectral range which could be less than 1 MHz wide. Thus, when the laser is tuned to the frequency of an atomic transition, the atom responds to the total power; the irradiance provided by such a source is therefore well into the strong-field regime.

In the temporal domain, the mode-locking techniques described in Section 3.4.1 allow the concentration of the laser's energy output into extremely brief pulses of high intensity and at a high repetition rate. The peak power in such a system is orders of magnitude larger than the average power. Mode locking of dye or titanium:sapphire lasers allows peak powers up to 10^{15} $W\,cm^{-2}$ in even shorter pulses. Under such conditions the electric field associated with the laser pulse is larger than the nuclear Coulomb field acting on the electron of a hydrogen atom in its ground state. This opens the way to studies of matter–light interactions at energies not hitherto accessible.

In conclusion, thanks to its coherence properties, laser light has the capability of being concentrated to a degree impossible with light from any other source. Its *temporal coherence* allows its energy to be concentrated either in time or in the frequency domain, whilst its *spatial coherence* allows it to be focused onto a small area or to be transmitted in a highly parallel beam. It is these possibilities that have enabled the laser to find such wide-ranging applications from those in fundamental research to those in technology designed for the mass market.

Further reading

There exists a vast literature on lasers. The following can be consulted for more details on issues passed over in this chapter, but the list cannot be considered exhaustive.

G. P. Agrawal and N. K. Dutta, *Semiconductor Lasers*, Van Nostrand (1993).

M. Bertolotti, *Masers and Lasers, a Historical Approach*, Institute of Physics Publishing (1983).

L. A. Coldren and S. W. Corzine, *Diode Lasers and Photonic Integrated Circuits*, Wiley (1995).

H. Haken, *Laser Theory*, Springer (1983).

W. Koechner, *Solid-state Laser Engineering*, Springer (2006).

L. Mandel and E. Wolf, *Optical Coherence and Quantum Optics*, Cambridge University Press (1995).

K. Petermann, *Laser Diode Modulation and Noise*, Kluwer (1991).

E. Rosencher and B. Vinter, *Optoelectronics*, Cambridge (2002).

M. Sargent, M. O. Scully, W. E. Lamb, *Laser Physics*, Addison-Wesley (1974).

A. E. Siegman, *Lasers*, University Science Books (1986).

W. Silfvast, *Laser Fundamentals*, Cambridge University Press (2004).

O. Svelto, *Principles of Lasers*, Plenum Press (1998).

J. T. Verdeyen, *Laser Electronics*, Prentice Hall (1995).

A. Yariv, *Quantum Electronics*, Wiley (1989).

3A Complement 3A The resonant Fabry–Perot cavity

Fabry–Perot interferometers have had a long and rich history. Originally invented to enable high resolution spectroscopic measurements, they have turned out to play a fundamental role in the development of lasers. Often a laser is nothing more than an amplifying medium enclosed in such a cavity. This cavity performs the dual functions of providing the optical feedback required to maintain oscillation and of selecting the wavelength within the range of the gain curve at which oscillation occurs. The role of the Fabry–Perot cavity does not stop there, however. Additional Fabry–Perot devices can be inserted into the laser cavity to narrow still further the frequency width of the laser's output. In tunable laser systems several filters of this type are often found in series, with successively narrower spectral widths. Finally, a Fabry–Perot étalon external to the laser cavity often provides a frequency reference to which the laser is locked.

3A.1 The linear Fabry–Perot cavity

Consider two plane, partially transmitting and non-absorbing mirrors M_1 and M_2 (of which we denote the amplitude transmission and reflection coefficients by t_1, t_2 and r_1, r_2, respectively) which are made accurately parallel. As we shall see, such a system constitutes a resonant cavity for light. Consider an electromagnetic wave $\mathbf{E}_\mathrm{i}(\mathbf{r}, t)$ of polarization $\boldsymbol{\varepsilon}$ propagating along the Oz direction perpendicular to the plane of the mirrors:

$$\mathbf{E}_\mathrm{i}(\mathbf{r}, t) = E_\mathrm{i}\boldsymbol{\varepsilon}\cos(\omega t - kz + \varphi_\mathrm{i}) = \left[\boldsymbol{\varepsilon}\mathcal{E}_\mathrm{i}\mathrm{e}^{\mathrm{i}(kz-\omega t)} + \mathrm{c.c.}\right]. \tag{3A.1}$$

E_i and $\mathcal{E}_\mathrm{i}$ are, respectively, the real and complex amplitudes of the incident electric field and the surface of M_1 is at $z = 0$. We denote in the same way, by $E_\mathrm{r}, E_\mathrm{t}, E_\mathrm{c}$ and E'_c the amplitudes of the reflected and transmitted fields propagating between the two mirrors. These waves are shown in Figure 3A.1.

Often we shall represent the field by its power per unit area (energy flux), Π (which is related to the mean value of the Poynting vector), rather than its amplitude. The energy flux due to the incident field is, for example,

$$\Pi_\mathrm{i} = \varepsilon_0 c \frac{E_\mathrm{i}^2}{2} = 2\varepsilon_0 c|\mathcal{E}_\mathrm{i}|^2. \tag{3A.2}$$

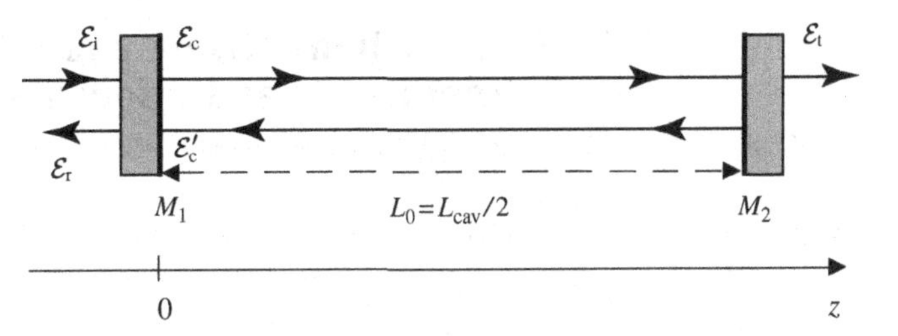

Figure 3A.1 Electric fields incident on, transmitted by and circulating in the cavity. The values $\mathcal{E}_i$, $\mathcal{E}_r$, $\mathcal{E}_c$, $\mathcal{E}'_c$ and $\mathcal{E}_t$ apply at the surface of M_1 ($z = 0$).

In the following we shall calculate the intensity transmission and reflection coefficients of the Fabry–Perot cavity:

$$T = \frac{\Pi_t}{\Pi_i} = \left|\frac{\mathcal{E}_t}{\mathcal{E}_i}\right|^2 \tag{3A.3}$$

$$R = \frac{\Pi_r}{\Pi_i} = \left|\frac{\mathcal{E}_r}{\mathcal{E}_i}\right|^2. \tag{3A.4}$$

A suitable choice of the origins of the phases allows the relations linking the amplitudes of the various fields meeting at mirror M_1 to be cast in the form

$$\mathcal{E}_c = t_1\mathcal{E}_i + r_1\mathcal{E}'_c \tag{3A.5}$$

$$\mathcal{E}_r = -r_1\mathcal{E}_i + t_1\mathcal{E}'_c, \tag{3A.6}$$

with r_1 and t_1 real and obeying the equation

$$r_1^2 + t_1^2 = R_1 + T_1 = 1. \tag{3A.7}$$

Note that the minus sign in (3A.6) guarantees that the field transformation matrix is unitary (compare Section 5.1.1), which in turn assures the conservation of energy flux: $|\mathcal{E}_c|^2 + |\mathcal{E}_r|^2 = |\mathcal{E}_i|^2 + |\mathcal{E}'_c|^2$ or $\Pi_c + \Pi_r = \Pi_i + \Pi'_c$.

Similarly, at mirror M_2 we find the expressions

$$\mathcal{E}_t = t_2\mathcal{E}_c \exp(ikL_0) \tag{3A.8}$$

$$\mathcal{E}'_c = r_2\mathcal{E}_c \exp(2ikL_0), \tag{3A.9}$$

where L_0 is the distance between the two mirrors. The coefficients r_2 and t_2 are real and satisfy

$$r_2^2 + t_2^2 = R_2 + T_2 = 1. \tag{3A.10}$$

Here again, the conservation of energy flux is ensured: $|\mathcal{E}_t|^2 + |\mathcal{E}'_c|^2 = |\mathcal{E}_c|^2$, or $\Pi_t + \Pi'_c = \Pi_c$.

Comment The sign of the reflection term in (3A.9) or (3A.5), the opposite of that written in Equation (3A.6), is related to the fact that the *reflecting surface* of the mirrors is directed towards the interior of the cavity in the former case, and towards the exterior in (3A.6).

3A.2 Cavity transmission and reflection coefficients and resonances

Equations (3A.5) and (3A.8) give, immediately on introducing the round-trip cavity length $L_{cav} = 2L_0$,

$$\frac{\mathcal{E}_t}{\mathcal{E}_i} = \frac{t_1 t_2 \exp(ikL_{cav}/2)}{1 - r_1 r_2 \exp(ikL_{cav})}, \tag{3A.11}$$

from which we deduce the transmission coefficient of the device,

$$T = \left|\frac{\mathcal{E}_t}{\mathcal{E}_i}\right|^2 = \frac{T_1 T_2}{1 + R_1 R_2 - 2\sqrt{R_1 R}\cos(kL_{cav})}. \tag{3A.12}$$

Expression (3A.12) shows that T reaches a maximum when $kL_{cav} = 2p\pi$ (see Figure 3A.2), that is, when

$$\omega = \omega_p = 2\pi p \frac{c}{L_{cav}} \quad \text{with integer } p. \tag{3A.13}$$

The transmission coefficient is then

$$T_{max} = \frac{T_1 T_2}{\left(1 - \sqrt{R_1 R_2}\right)^2}. \tag{3A.14}$$

If $R_1 = R_2$ we find $T_{max} = 1$, even if $T_1 = T_2$ is arbitrarily close to zero. In this case, the cavity, made of two highly reflecting mirrors, transmits the incident field *totally*! Of course, in practice imperfections in the mirrors, such as losses or imperfect flatness, place an upper limit on the value of T_{max}.

The frequencies ω_p defined by (3A.13) are the resonant frequencies of the cavity. They correspond to the situation where the cavity length L_0 is a multiple of the half wavelength,

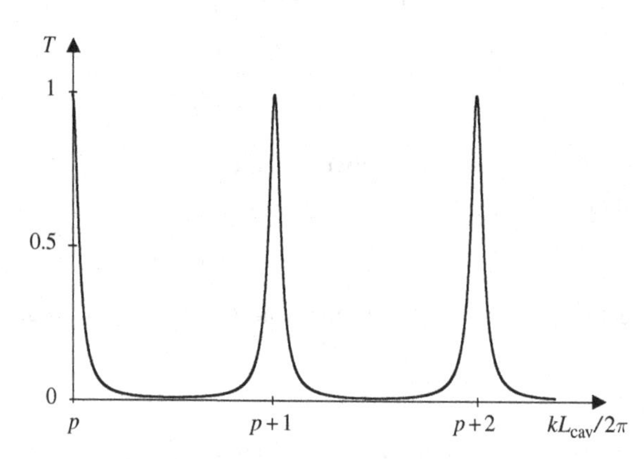

Figure 3A.2 Transmission of the cavity. Marked resonances appear when the phase accumulated in a cavity round trip, kL_{cav}, is a multiple of 2π; the curve here is calculated for $R_1 = R_2 = 0.8$.

like for the resonances of a vibrating string. In fact, since $|\mathcal{E}_t|^2 = T_2|\mathcal{E}_c|^2$, we find for the intracavity field, using (3A.12):

$$\left|\frac{\mathcal{E}_c}{\mathcal{E}_i}\right|^2 = \frac{T_1}{1 + R_1R_2 - 2\sqrt{R_1R_2}\cos(kL_{cav})}, \tag{3A.15}$$

which shows that for a given incident field E_i, the power circulating in the cavity is maximal for $\omega = \omega_p$ (or, equivalently, $kL_{cav} = 2p\pi$). For example, for a symmetric cavity, the power circulating in the cavity is larger than the incident intensity by a factor of $1/T_2$. This *enhancement* factor can be very large if T_2 is small. We now understand why the overall transmission of this cavity can be close to 1: the mirror M_2 indeed transmits only a small part of the intracavity power, but this one is much larger than the incident power when the cavity is resonant, and the two effects exactly balance each other.

Note finally that the cavity reflection coefficient can be derived from expression (3A.12):

$$R = \left|\frac{\mathcal{E}_r}{\mathcal{E}_i}\right|^2 = 1 - T. \tag{3A.16}$$

It passes through a minimum at resonance. In particular, in the case of a symmetric cavity, ($R_1 = R_2$), and for perfect mirrors we find $R = 0$: the reflection coefficient passes through zero at the resonant frequencies.

Comments

(i) As an alternative method for deriving Equation (3A.12) we could, instead of writing expressions (3A.5) and (3A.8) for the global amplitudes, have considered the Fabry–Perot cavity as a multiple-beam interferometer and calculated the sum of all the transmitted and reflected waves. The treatment presented here is simpler and just as rigorous.

(ii) Very often experiments in spectroscopy or nonlinear optics which require large intensities (such as two-photon transitions; see Complement 3G, or frequency-doubling; see Chapter 7) are actually enclosed in a resonant Fabry–Perot cavity, which allows the intensity experienced by the active medium to be significantly larger than that available from the incident laser beam. Depending on the application, this could be a ring or linear cavity. In the former case, the intensity is constant, while it has nodes and antinodes in the case of a linear cavity.

(iii) When the cavity is filled with a medium of refractive index n, L_{cav} must be taken as the round-trip optical length, i.e. $2nL_0$. This is easily generalized to the case of a component of thickness e and refractive index n, inserted in the cavity. One has then, $L_{cav} = 2L_0 + 2(n-1)e$.

(iv) The complex amplitude of the reflected field is given by

$$\frac{\mathcal{E}_r}{\mathcal{E}_i} = \frac{-r_1 + r_2 e^{ikL_{cav}}}{1 - r_1 r_2 e^{ikL_{cav}}}.$$

In the case of a symmetric cavity ($r_1 = r_2$), the phase of the reflected field exhibits a change of π at cavity resonance.

3A.3 Ring Fabry–Perot cavity with a single coupling mirror

Most of the results derived above for the case of a linear Fabry–Perot cavity can be extended to the case of a ring-type cavity. Consider that shown in Figure 3A.3. It consists of a single partially reflecting mirror M (with coefficients r_1, t_1) and two totally reflecting mirrors.

The relations linking the incident amplitude $\mathcal{E}_i$, the reflected amplitude $\mathcal{E}_r$ and the field circulating in the cavity $\mathcal{E}_c$ are analogous to (3A.5–3A.6):

$$\mathcal{E}_c = t_1\mathcal{E}_i + r_1\mathcal{E}'_c \tag{3A.17}$$

$$\mathcal{E}_r = -r_1\mathcal{E}_i + t_1\mathcal{E}'_c \tag{3A.18}$$

$$\mathcal{E}'_c = \mathcal{E}_c \exp(ikL_{cav}), \tag{3A.19}$$

where L is the round-trip cavity length. Combining the first and last relations, we find:

$$\mathcal{E}_c = \frac{t_1\mathcal{E}_i}{1 - r_1 \exp(ikL_{cav})} \tag{3A.20}$$

and

$$\left|\frac{\mathcal{E}_c}{\mathcal{E}_i}\right|^2 = \frac{T_1}{1 + R_1 - 2\sqrt{R_1}\cos(kL_{cav})}. \tag{3A.21}$$

This expression coincides with (3A.15) when $R_2 = 1$. For a ring cavity with a single output mirror, the value of the intracavity intensity is therefore identical to that obtained for the linear case in which one of the mirrors has unit reflectivity. This applies to all the results obtained above for the linear cavity. This includes in particular the appearance of sharp resonances which occur for the same values of the angular frequency, given by (3A.13).

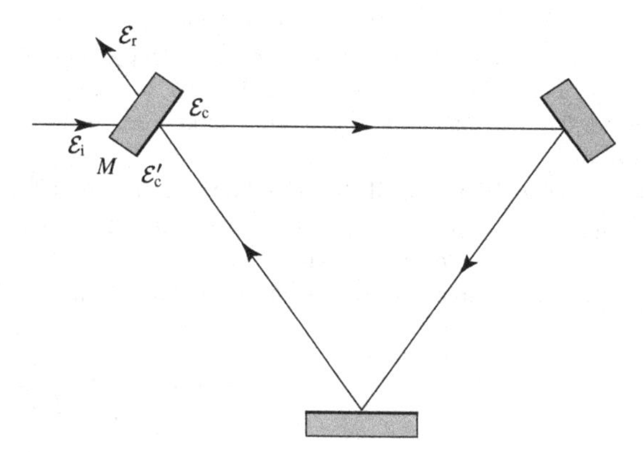

Figure 3A.3 Ring Fabry–Perot cavity. The amplitudes $\mathcal{E}_i$, $\mathcal{E}_r$ and $\mathcal{E}'_c$ are evaluated at the surface of the partially transmitting mirror M. The length of the ring is L_{cav}.

As far as the reflected field is concerned, we find, on combining Equations (3A.18), (3A.19) and (3A.20),

$$\frac{\mathcal{E}_r}{\mathcal{E}_i} = \frac{1 - r_1 \exp(-ikL_{cav})}{1 - r_1 \exp(ikL_{cav})} \exp(ikL_{cav}), \tag{3A.22}$$

which implies that the *reflection coefficient* $|\mathcal{E}_r/\mathcal{E}_i|^2$ *is equal to 1, whatever the cavity length.* Thus for the ring cavity of Figure 3A.3, all the incident intensity is reflected (in the absence of cavity losses). This does not mean that the cavity resonances are not manifest in the reflected field, simply that *they appear only on its phase.*

3A.4 The cavity finesse

It is instructive to rewrite the resonant denominator D of expressions (3A.12) and (3A.15) in the form

$$\begin{aligned} D &= 1 + R_1R_2 - 2\sqrt{R_1R_2} + 4\sqrt{R_1R_2}\sin^2\left(k\frac{L_{cav}}{2}\right) \\ &= (1 - \sqrt{R_1R_2})^2\left[1 + m\sin^2\left(k\frac{L_{cav}}{2}\right)\right], \end{aligned} \tag{3A.23}$$

in which the parameter m is given by

$$m = \frac{4\sqrt{R_1R_2}}{(1 - \sqrt{R_1R_2})^2}. \tag{3A.24}$$

The maximum value of $1/D$ increases as R_1R_2 approaches 1. Additionally, $1/D$ is a function of the round-trip phase kL_{cav}, which attains its maximum value on resonance (Equation 3A.13). The width of these resonances can be characterized by finding the values of kL_{cav} for which $1/D$ falls to one half of its maximum value. In this way we find, for $m \gg 1$,

$$k\frac{L_{cav}}{2} \approx p\pi \pm \frac{1}{\sqrt{m}}.$$

The cavity *finesse* is defined as the ratio of the interval of π between two adjacent resonance peaks to the width $2/\sqrt{m}$ of one of the peaks:

$$\mathcal{F} = \pi\frac{\sqrt{m}}{2} = \pi\frac{(R_1R_2)^{1/4}}{1 - \sqrt{R_1R_2}}. \tag{3A.25}$$

If the frequency of the incident wave is scanned, the resonance peaks have a width

$$\frac{\Delta\omega_{cav}}{2\pi} = \frac{1}{\mathcal{F}}\frac{c}{L_{cav}}. \tag{3A.26}$$

Clearly, the resonances become increasingly narrow as $\mathcal{F}$ increases.

A common way of characterizing a resonant system (usually either in electronics or mechanics) is to study the rate of decay of oscillation once the excitation is ceased. The energy stored in a linear cavity of cross-sectional area S is

$$W = \frac{L_0}{c}(\Pi_c + \Pi'_c)S, \tag{3A.27}$$

where Π_c and Π'_c are the circulating energy flux in the cavity in the positive and negative z-directions, respectively (see Figure 3A.1). The power emitted at the output mirrors is

$$-\frac{d}{dt}W = (T_2\Pi_c + T_1\Pi'_c)S. \tag{3A.28}$$

Using the fact that $\Pi'_c = R_2\,\Pi_c$, we obtain

$$-\frac{d}{dt}W = \frac{T_2 + T_1R_2}{1 + R_2}\frac{c}{L_0}W. \tag{3A.29}$$

For a resonant system with a natural oscillation frequency ω, one defines the *quality factor* Q by

$$\frac{dW}{dt} = -\frac{\omega}{Q}W. \tag{3A.30}$$

Comparing Equations (3A.29) and (3A.30) and using $\omega = 2\pi c/\lambda$, we find

$$Q = \pi\frac{1 + R_2}{T_2 + T_1R_2}\frac{L_{cav}}{\lambda}. \tag{3A.31}$$

Because of the appearance of the factor L_{cav}/λ, which commonly exceeds 10^5 for optical wavelengths and when R_1 and R_2 are both close to 1, Q can be very large: the cavity field undergoes many oscillations before it decays away.

3A.5 Cavity with a large finesse

It is interesting to consider the case in which the mirrors have large reflection coefficients. Then we can use the lowest order terms in an expansion in T_1 and T_2 of (3A.25) and (3A.31). We find, in this way,

$$\mathcal{F} \approx \frac{2\pi}{T_1 + T_2} \tag{3A.32}$$

$$Q \approx \mathcal{F}(L_{cav}/\lambda). \tag{3A.33}$$

Mirrors can routinely be made with negligible absorption and a transmission coefficient of the order of 1%, which leads to a finesse exceeding 100. With high-quality dielectric coatings and very flat mirror surfaces it is possible to attain finesse values larger than 10^5.

In this limit (R_1 and R_2 very close to 1), Equation (3A.15) for the intracavity intensity gives, using (3A.23), (3A.24) and (3A.25),

$$\left|\frac{\mathcal{E}_c}{\mathcal{E}_i}\right|^2 = \frac{4T_1}{(T_1+T_2)^2\left(1+\frac{4\mathcal{F}^2}{\pi^2}\sin^2\left(\frac{kL_{cav}}{2}\right)\right)}. \tag{3A.34}$$

To conclude, we give a few quantitative results for a linear cavity of length $L_0 = 5$ cm and of finesse $\mathcal{F} \approx 300$ (obtained for $T_1 = T_2 = 0.01$): its free spectral range (i.e. the inter-mode spacing), c/L_{cav}, is equal to 3 GHz; the frequency width of a cavity resonance, $\Delta\omega_{cav}/2\pi$, is equal to 10 MHz; the quality factor at $\lambda = 500$ nm, in the visible, is $Q = 6 \times 10^7$; the intracavity circulating power is a hundred times larger than the input one. Such a cavity can be used for very high resolution spectroscopy: if its length L_0 is changed the transmitted frequencies are also altered. A $\lambda/2$ change in L_0 scans the transmitted frequency through one free-spectral range, or 3 GHz in this case. Thanks to the small value of $\Delta\omega_{cav}$, one could, for example, study the structure of the spectral line emitted by a helium–neon laser (3 modes separated by 500 MHz).

Comments

(i) Transmission losses are not the only ones that affect a Fabry–Perot cavity. Losses at the mirror surfaces must be taken into account as must possible absorption or scattering losses occurring if a material medium is placed in the cavity. Consider the linear cavity of Figure 3A.1 and suppose that the fractional intensity loss for propagation from one mirror to the other is A_1, so that Equations (3A.8–3A.9) become:

$$\mathcal{E}_t = t_2\mathcal{E}_c\sqrt{1-A_1} \tag{3A.35}$$

$$\mathcal{E}'_c = r_2\mathcal{E}_c(1-A_1)\exp(2ikL_0), \tag{3A.36}$$

whilst Equations (3A.5–3A.6) are unchanged. The preceding analysis can then easily be adapted to include this extra loss term. We find, for example, in the limit $T_1, T_2, A_1 \ll 1$, resonant transmission and reflection coefficients and a finesse given by

$$T_{max} = \frac{4T_1T_2}{(Ab+T_1+T_2)^2} \tag{3A.37}$$

$$R_{res} = \frac{(Ab-T_1+T_2)^2}{(Ab+T_1+T_2)^2} \tag{3A.38}$$

$$\mathcal{F} = \frac{2\pi}{Ab+T_1+T_2}, \tag{3A.39}$$

where $Ab = 2Ab_1$ is the additional round-trip loss.

(ii) The case of an amplifying medium (such as that found in a laser) corresponds to $A < 0$. It is then possible to obtain $T_{max} = \infty$ which shows that an output wave can appear even in the absence of an incident wave.

3A.6 Linear laser cavity

The preceding analysis can be applied to a linear laser cavity by putting $T_2 = 0$ ($R_2 = 1$), which corresponds in Figure 3A.1 to a laser beam emitted to the left. In this case, the transmission of the cavity is, of course, zero, but the formulae from (3A.12) onwards remain valid. In particular, the resonant behaviour of the intracavity intensity still applies. The finesse is now (for the case in which $T_1 \ll 1$),

$$\mathcal{F} = \frac{2\pi}{T_1}. \tag{3A.40}$$

The energy fluxes circulating in the cavities Π'_c and Π_c are now equal:

$$\Pi_c = \Pi'_c, \tag{3A.41}$$

and they are related to the incident flux Π_i by

$$\frac{\Pi_c}{\Pi_i} = \frac{2\mathcal{F}}{\pi\left(1 + \frac{4\mathcal{F}^2}{\pi^2}\sin^2\left(\frac{kL_{cav}}{2}\right)\right)}. \tag{3A.42}$$

The energy stored in the cavity is equal, according to (3A.27) and (3A.41), to

$$W = \frac{L_{cav}}{c}\Pi_c S, \tag{3A.43}$$

which becomes, on using (3A.26) and (3A.42),

$$W = \frac{4\Pi_i S}{\Delta\omega_{cav}\left(1 + \frac{4\mathcal{F}^2}{\pi^2}\sin^2\left(\frac{kL_{cav}}{2}\right)\right)}. \tag{3A.44}$$

This shows that, at resonance and for a given input intensity, the energy in the cavity is larger the narrower the width of a resonance.

Comments

(i) In the case of a laser, the situation differs in the respect that there is no input light beam, $\Pi_i = 0$. However, energy circulates in the cavity ($\Pi_c \neq 0$) because of the presence of the amplifying medium. In this case Equation (3A.43) remains valid and the light flux Π_o emitted by the laser is equal to

$$\Pi_o = T_1 \Pi_c, \tag{3A.45}$$

which, on using (3A.26), (3A.27) and (3A.43), becomes

$$W = \frac{\Pi_o S}{\Delta\omega_{cav}}, \tag{3A.46}$$

which relates the output power to the power circulating in the cavity.

(ii) The results derived in Section 3A.6 may be straightforwardly applied to the case of a ring cavity (see Section 3A.3) because of the similarity between the ring-cavity case and that of a linear cavity with one mirror, M_2, having a reflection coefficient of unity.

3B Complement 3B The transverse modes of a laser: Gaussian beams

The simplified description of the operation of lasers, presented in Section 3.1 of Chapter 3, rests on the assumption of the infinite transverse extent of the laser cavity, so that the circulating light fields can be represented by plane waves. This is obviously a somewhat unrealistic assumption; the various components of a laser cavity are of limited spatial extent, more usually transverse dimensions are of the order of a centimetre. If the light waves were really plane waves, the diffraction at one of these aperture-limiting components would make impossible reproduction of the form of the wavefront after a complete cavity round trip and would, furthermore, introduce severe losses. In practice diffraction losses are compensated for by the use of focusing elements such as concave mirrors, but a theoretical treatment of the light field based on plane waves is then inappropriate.

A more useful description of the intracavity light field is one in terms of a wave with a non-uniform transverse spatial distribution which also takes into account the question of the stability of the wave propagating in the cavity. This description must also account for diffraction effects and the reflections on the cavity mirrors. Such a stable light field is known as a *transverse mode* of the cavity.

In general, finding an expression for the transverse modes of an arbitrary cavity is a complicated problem. Fortunately for the cavity geometries most commonly employed in continuous-wave lasers (and most particularly for a linear cavity composed of two concave mirrors) classes of simple solutions exist: the *transverse Gaussian modes*.

3B.1 Fundamental Gaussian beam

It is possible, as we shall see in Section 3B.3, to find an approximate solution of the equations governing the propagation in a vacuum of an electromagnetic radiation which has a finite extent around the propagation axis Oz, and which can thus describe a light beam of finite transverse section. Its electric field is given by

$$\mathbf{E}(\mathbf{r},t) = \boldsymbol{\varepsilon} E(\mathbf{r},t), \tag{3B.1}$$

with

$$E(\mathbf{r},t) = 2\mathrm{Re}\left\{E_0 \exp\left(-\frac{x^2+y^2}{w^2}\right)\exp\left(\mathrm{i}k\frac{x^2+y^2}{2R}\right)\exp\left[\mathrm{i}(kz-\omega t-\phi)\right]\right\}. \tag{3B.2}$$

It describes a wave of angular frequency ω propagating in the Oz direction and with cylindrical symmetry about this axis. The unit vector $\boldsymbol{\varepsilon}$, perpendicular to Oz is the polarization of the wave. In Equation (3B.2) E_0, w, R and ϕ are all real functions of z which have simple interpretations; w is an order of magnitude of the transverse extent of the wave. It is easy to show that the second exponential in expression (3B.2) describes a transverse variation of the phase that is characteristic, for small values of x and y, of a spherical wave of radius of curvature R. For propagation in the direction of positive z, the wave is *divergent* if R is positive, as will be realized from a consideration of the form of the surfaces on which the phase is constant. Conversely, the wave is *convergent* for $R < 0$.

Taking as the origin of z the point at which the radius of the beam is minimum (the radius of the beam at this point is known as the beam *waist* as quite commonly is the location itself) we find (as will be shown in Section 3B.3),

$$w(z) = w_0\sqrt{1 + \frac{z^2}{z_R^2}}, \tag{3B.3}$$

$$E_0(z) = E_0(0)\frac{w_0}{w(z)}, \tag{3B.4}$$

$$R(z) = z + \frac{z_R^2}{z}, \tag{3B.5}$$

$$\phi = \tan^{-1}\left(\frac{z}{z_R}\right). \tag{3B.6}$$

The radius at the beam waist, w_0, characterizes the minimum transverse extent of the beam, in the region of $z = 0$. The *Rayleigh length* z_R, equal to

$$z_R = \pi\frac{w_0^2}{\lambda}, \tag{3B.7}$$

characterizes the distance along the beam axis over which this extent varies appreciably. In fact, the transverse extent increases by a factor of $\sqrt{2}$ between $z = 0$ and $z = z_R$ (from formula (3B.3)). Thus, for a distance z from the origin that is small compared to z_R, the light beam has a transverse extent that is practically constant. Since in this region the radius of curvature is very large, the beam approximates closely to a plane wave propagating in a cylindrical volume parallel to the Oz-axis.

According to the laws of diffraction, the region in which the beam has a cylindrical cross-section is shorter the smaller is the beam waist. This is expressed in (3B.7) by the fact that z_R is proportional to w_0^2. Thus for $\lambda = 633$ nm the Rayleigh length, $z_R = 5$ m for $w_0 = 1$ mm being reduced to $z_R = 0.5$ mm for $w_0 = 0.01$ mm.

At a distance from the waist that is large compared to z_R, we find

$$w(z) \approx \frac{\lambda}{\pi w_0}z, \tag{3B.8}$$

$$R(z) \approx z. \tag{3B.9}$$

In this domain the beam appears to diverge uniformly from the point $x, y, z = 0$ and is bounded by a cone of half-angle at this point $\lambda/\pi\, w_0$ (which have the values 2×10^{-4} and

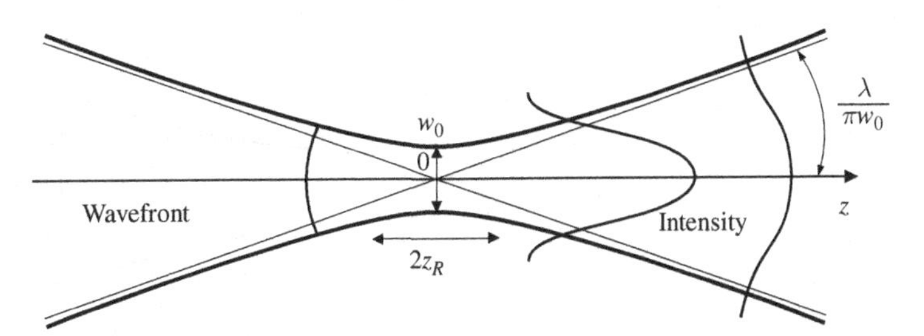

Figure 3B.1 Diagram of a Gaussian beam showing schematically the form of the transverse intensity profile and of the wavefronts as a function of propagation.

2×10^{-2}, respectively in the numerical examples above). This *divergence*, which occurs simply as a consequence of diffraction, increases as the waist size is reduced. This is why, in some applications, it is necessary to use a large area beam (see Complement 3F, Section 3F.1). Figure 3B.1 shows a schematic representation of a Gaussian beam.

Equation (3B.6) shows that the field accumulates, in addition to the usual propagation phase-shift kz, a supplementary phase-shift as one passes from $-\infty$ to $+\infty$. This phase-shift, which is well known in optics, is known as the *Gouy phase*. Such a phase-shift appears in crossing the waist of any focused light beam.

It is possible to relate E_0 to the power Φ carried by the wave and to the cross-section of the beam w, since this energy flux must be the same for any plane perpendicular to the direction of propagation. From the relation

$$\frac{\varepsilon_0 c}{2} \iint \mathrm{d}x\, \mathrm{d}y\, |E(x, y, z, t)|^2 = \Phi, \tag{3B.10}$$

we deduce

$$E_0(z) = \frac{1}{w(z)}\sqrt{\frac{\Phi}{\varepsilon_0 \pi c}}. \tag{3B.11}$$

This relation is consistent with that of (3B.4).

3B.2 The fundamental transverse mode of a stable cavity

Consider a resonant linear cavity composed of a plane mirror M_1 and of a concave mirror M_2 of radius of curvature R_2, separated by a distance L_0 (Figure 3B.2).

It is intuitively obvious that, if there exists a Gaussian beam solution for which the wavefronts coincide exactly with the surfaces of the two mirrors, then this solution will be a stable mode of the cavity. Such a solution will therefore have its waist at the plane mirror and will also satisfy

$$R(L_0) = R_2, \tag{3B.12}$$

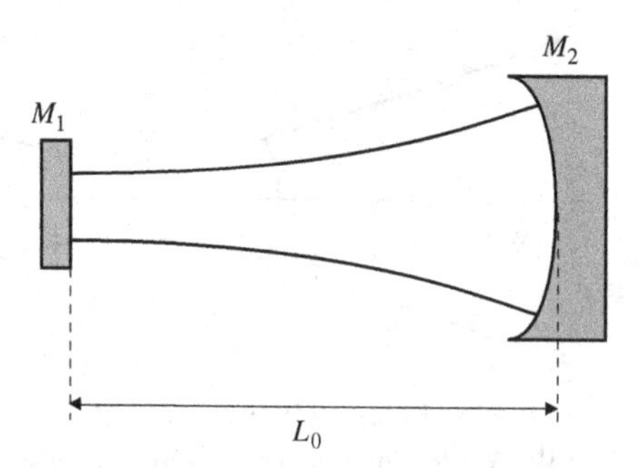

Figure 3B.2 Diagram of a stable optical cavity formed by a concave mirror of radius of curvature R_2 (with $R_2 \geq L_0$) and by a plane mirror showing its fundamental (Gaussian) transverse mode.

from which we deduce, with the help of (3B.5),

$$z_R = \sqrt{(R_2 - L_0)L_0}. \tag{3B.13}$$

This expression only has a solution if

$$R_2 \geq L_0. \tag{3B.14}$$

This is a statement of the *stability condition* for the cavity sketched in Figure 3B.2. If it is satisfied, a stable Gaussian mode exists of which the wavefronts exactly match the mirror surfaces. Notice also that expressions (3B.13) and (3B.7) taken together give the waist size w_0 as a function of R_2 and L_0.

The above reasoning can be generalized to the case of any linear cavity enclosed between two mirrors with radii of curvature R_1 and R_2. The stability condition then becomes

$$0 \leq \left(1 - \frac{L_0}{R_1}\right)\left(1 - \frac{L_0}{R_2}\right) \leq 1. \tag{3B.15}$$

The parameter z_R (or w_0) characterizing the mode is then uniquely determined by demanding that the wavefronts exactly match the mirror surfaces.

3B.3 Higher-order Gaussian beams

In this section we shall find the general expression of an electromagnetic wave having a finite extent around the Oz-axis. We start from the equation governing the propagation of an electric field in a vacuum, deduced from Maxwell's equations:

$$\Delta \mathbf{E} - \frac{1}{c^2}\frac{\partial^2 \mathbf{E}}{\partial t^2} = 0. \tag{3B.16}$$

We proceed in the paraxial approximation, that is we assume that the electromagnetic wave propagates approximately parallel to Oz (the normals to the wavefronts only make a small angle to Oz) and that the intensity is only non-negligible close to the same axis (at distances small compared to the radius of curvature but large compared to the wavelength). The transversality condition on the electric field then permits us to neglect the component of **E** along Oz, and we can write

$$\mathbf{E}(\mathbf{r},t)=\boldsymbol{\varepsilon}E(\mathbf{r},t)=2\boldsymbol{\varepsilon}\mathrm{Re}\left\{\mathcal{E}(x,y,z)\mathrm{e}^{\mathrm{i}(kz-\omega t)}\right\}, \tag{3B.17}$$

where $\boldsymbol{\varepsilon}$ is the polarization in the xOy plane and $\mathcal{E}$ is supposed to vary on length scales which are much larger than the wavelength, the fast oscillation of the wave, described by the factor $\exp(ikz)$ having been put aside. Inserting into Equation (3B.16), and neglecting $\frac{\partial^2}{\partial z^2}\mathcal{E}$ in front of $k\frac{\partial}{\partial z}\mathcal{E}$, we obtain for the envelope function $\mathcal{E}$ the following propagation equation at the paraxial, and 'slowly varying envelope approximation':

$$2\mathrm{i}k\frac{\partial\mathcal{E}}{\partial z}=\frac{\partial^2\mathcal{E}}{\partial x^2}+\frac{\partial^2\mathcal{E}}{\partial y^2}. \tag{3B.18}$$

This equation will be very useful in various parts of the book, for example in Chapter 7 on nonlinear optics.

The mathematical derivation of the general solution of the linear equation (3B.18) can be found in various texts.[1] One finds that it is a linear combination of functions $\mathcal{E}_{mn}(x,y,z)$ (m and n being positive integers or zero) forming a discrete basis of solutions, denoted TEM_{mn} and given by

$$\begin{aligned}\mathcal{E}_{mn}(x,y,z)=&\frac{C}{w(z)}\exp\left\{-\frac{x^2+y^2}{w^2(z)}\right\}\exp\left\{\mathrm{i}k\frac{x^2+y^2}{2R(z)}\right\}\\&\times H_m\left[\frac{x\sqrt{2}}{w(z)}\right]H_n\left[\frac{y\sqrt{2}}{w(z)}\right]\mathrm{e}^{-\mathrm{i}\phi_{mn}(z)}.\end{aligned} \tag{3B.19}$$

In this expression, C is an arbitrary constant, $w(z)$ and $R(z)$ are as given in Equations (3B.3) and (3B.5). H_m and H_n are *Hermite polynomials* of degree m and n, that are also encountered in the stationary wavefunctions of the harmonic oscillator in quantum mechanics. For example, the first Hermite polynomials are $H_0(x)=1$, $H_1(x)=2x$, $H_2=4x^2-1$. The longitudinal Gouy phase-shift is now given by

$$\phi_{mn}(z)=(m+n+1)\tan^{-1}\left(\frac{z}{z_R}\right). \tag{3B.20}$$

These solutions are called *Hermite–Gauss modes*. The fundamental Gaussian mode, that we have given in Equation (3B.2) and studied in the first section of this complement, is just the TEM_{00} mode. Note that all TEM_{mn} modes have the same transverse dependence of the phase. The higher modes are differentiated from the fundamental mode by a different Gouy phase and by the transverse variation of their amplitude which is the product of a Gaussian and of a pair of Hermite polynomials. The physical consequence of this is that the energy flux through a surface normal to the beam propagation direction, instead of decreasing uniformly away from the Oz-axis, possesses nodal lines and decreases on average more

[1] A. E. Siegman, *Lasers*, University Science Books (1986); for more details see, for example, H. Kogelnik and T. Li, Laser Beams and Resonators, *Applied Optics* **5**, 1550 (1966).

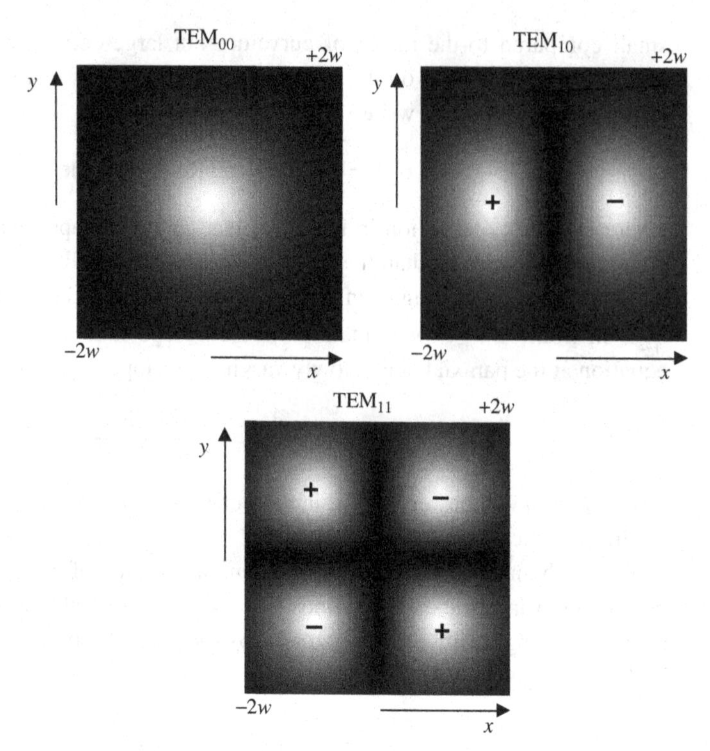

Figure 3B.3 Density plots of the intensity distribution for a few Hermite–Gauss modes for $-2w < x < 2w$ and $-2w < y < 2w$. The + and − signs indicate the sign of the electric field in the different lobes of the TEM_{10} and TEM_{11} modes. One sees that the fundamental TEM_{00} mode decreases away from the Oz-axis more rapidly than the higher-order modes.

slowly. Figure (3B.3) gives as an example the appearance of a section of the beam in a few low-order Gaussian modes.

Comment It is also possible to solve the paraxial equation (3B.18) in cylindrical coordinates (ρ, θ, z). One obtains a set of linearly independent solutions of the form

$$\mathcal{E}_{lm}(\rho,\theta,z) = \frac{C'}{w(z)}\left[\frac{\rho\sqrt{2}}{w(z)}\right]^{|m|} \exp\left\{-\frac{\rho^2}{w^2(z)}\right\} \exp\left\{\mathrm{i}\frac{\rho^2}{2R(z)}\right\} L_l^{|m|}\left[\frac{2\rho^2}{w^2(z)}\right] \mathrm{e}^{\mathrm{i}m\theta}\,\mathrm{e}^{-\mathrm{i}\phi_{lm}(z)}, \tag{3B.21}$$

where l and m are integers, with $l > 0$, and $L_l^{|m|}$ is a generalized Laguerre polynomial. The evolution of the Gouy phase ϕ_{lm} is given by the following equation:

$$\phi_{lm}(z) = (2l + |m| + 1)\tan^{-1}\left(\frac{z}{z_R}\right). \tag{3B.22}$$

These modes are called the *Laguerre–Gauss modes* TEM^*_{lm}. When $m \neq 0$, because of their azimuthal dependence $\mathrm{e}^{\mathrm{i}m\theta}$, they have a helicoidal wavefront around the z-axis. We will see in Complement 4B that this implies that they carry an *orbital angular momentum* proportional to the integer m.

3B.4 Longitudinal and transverse modes of a laser

Consider a stable linear cavity, for example the plane-concave cavity of Figure 3B.2. As all the Gaussian modes have the same transverse dependence of the phase, and therefore the same wavefronts, they can all be solutions of the Maxwell equations (at the paraxial approximation) with the boundary conditions imposed on the wavefronts by the presence of the two mirrors, provided that $z_R = \pi w_0^2/\lambda$ is given by expression (3B.13) which was derived for the TEM_{00} mode.

We saw in Chapter 3 (Section 3.1.2) that the laser will oscillate on any cavity mode situated in the frequency region in which the gain curve exceeds the threshold value of the gain. We now extend the discussion in the chapter which considered uniquely the longitudinal modes to also include transverse cavity modes. In general, an oscillating mode is now determined by three integers: m and n determining the transverse mode in the Hermite–Gauss basis, and p the longitudinal one. More precisely, since the phase must accumulate a change of an integral multiple of 2π radians in a cavity round trip, condition (3.16) of Chapter 3, becomes, when suitably generalized,

$$\omega_{mnp}\frac{L_{\text{cav}}}{c} - 2\left[\phi_{mn}(L_0) - \phi_{mn}(0)\right] = 2p\pi, \tag{3B.23}$$

with $L_{\text{cav}} = 2L_0$. To the term $\omega_{mnp}L_{\text{cav}}/c$ already encountered in Chapter 3, is also added the change over the cavity length of the Gouy phase, this phase change being multiplied by two, because of the two passes of the cavity in one round trip.

Using Equation (3B.21), adapted to the case of the plane-concave cavity (the cross-section of the beam is smallest at the surface of the plane mirror); we can rewrite Equation (3B.23) in the form

$$\omega_{mnp} = \frac{c}{L_{\text{cav}}}\left[2p\pi + 2(n+m+1)\tan^{-1}\left(\frac{L_0}{z_R}\right)\right], \tag{3B.24}$$

which becomes, on using the value of z_R given by Equation (3B.13),

$$\frac{\omega_{mnp}}{2\pi} = \frac{c}{2L_0}\left[p + \frac{1}{\pi}(n+m+1)\cos^{-1}\left(1 - \frac{2L_0}{R_2}\right)\right], \tag{3B.25}$$

which only depends on the length of the cavity $L_0 = L_{\text{cav}}/2$ and the radius of curvature of the concave mirror, R_2. The latter result shows that the transverse modes, characterized by the values of n and m, generally *oscillate at different frequencies.*

The discussion of Chapter 3 must therefore be generalized to take into account the additional degrees of freedom associated with the transverse modes. It would appear a priori that several transverse modes could oscillate simultaneously. This is problematic, since it implies that the laser will no longer operate on a single mode and be quasi-monochromatic. Additionally, the oscillation of a number of transverse modes implies that the transverse intensity distribution of the output beam will not be uniform, but will have nodes and maxima whereas, for many applications, an intensity profile that is as uniform as possible is desirable. In practice, however, the fundamental TEM_{00} mode is privileged (as long as the laser cavity is well aligned), as is clear from Figure 3B.2, since this mode is the most compact spatially. The laser cavity always contains limited apertures, whether the transverse extent of the field is limited deliberately, or whether the limit is imposed by the size of the cavity mirrors or other intracavity components. These apertures increase losses for the higher-order modes, which decay more slowly away from the Oz-axis (see Figure 3B.3). Thus if the round-trip gain is just sufficient to ensure oscillation, usually the fundamental transverse mode alone will contribute. As the gain is increased, however, higher transverse modes will start to appear in the output beam unless suitable precautions are taken. These could include the insertion of an aperture in the cavity to increase the cavity losses in a way that favours the fundamental mode. The result is single-mode operation, but at the cost of reduced output power. Whether beam quality is preferred over optical output power often depends on the envisaged application.

Comments

(i) The above Gaussian mode solutions are those most commonly encountered. These are the modes associated with an empty cavity enclosed by spherical mirrors. The presence of an amplifying medium which, by virtue of their nonlinear properties, can couple different transverse modes, or the inclusion in the cavity of masks, can cause the oscillation of modes with very different spatial structures.

(ii) When the cavity length is half the radius of curvature of the curved mirror ($L_0 = R_2/2$), a configuration called 'semi-confocal cavity', the quantity $\cos^{-1}(1 - 2L_0/R_2)$ in (3B.25) is 1. This implies that the only possible frequencies generated by the laser are multiples of $c/4L_0$ (instead of $c/2L_0$ for a cavity with planar mirrors), and that many different transverse modes, or any combination of these modes, may oscillate simultaneously. The 'confocal cavity', made of two identical curved mirrors of radii of curvature equal to the cavity length ($L_0 = R$) is an 'unfolded' semi-confocal cavity. It has similar properties: the only possible frequencies are multiples of $c/4L_0$, and the emitted field may have any transverse shape, provided it is even or odd with respect to the change ($x \to -x, y \to -y$). It is then the transverse properties of the gain medium which impose the transverse shape on the laser beam.

Complement 3C Laser light and incoherent light: energy density and number of photons per mode

The difference between laser light and the light emitted by an incoherent source can only be fully appreciated with reference to certain notions of energetic photometry, which are spelt out in the first part of this complement. It will be shown in Section 3C.2 how the laws of photometry drastically reduce the energy density that can be obtained from a conventional incoherent source (such as a heated filament, or a discharge lamp) in comparison with a laser source (Section 3C.3). Far from being merely circumstantial, these laws for classical sources are of a fundamental kind that can be deduced from the basic principles of thermodynamics. Another way to relate these properties of light to the fundamental principles of physics is to examine them in the context provided by the statistical physics of photons, as will be discussed in Sections 3C.4 and 3C.5.

3C.1 Conservation of radiance for an incoherent source

3C.1.1 Étendue and radiance

An incoherent source comprises a large number of independent, elementary emitters, emitting electromagnetic waves with a random distribution of uncorrelated phases. It emits light in every direction. A light beam produced by this source can be decomposed into elementary pencils of light. Since the light is incoherent, the total power carried by the beam is the sum of the powers carried by the elementary pencils.

An elementary pencil is defined by the element $\mathrm{d}S$ of the source from which it originates, and a second surface element $\mathrm{d}S'$, as shown in Figure 3C.1. The value of the pencil étendue (sometimes referred to as 'geometric extent') is given by

$$\mathrm{d}U = \frac{\mathrm{d}S\cos\theta\,\mathrm{d}S'\cos\theta'}{MM'^2}, \tag{3C.1}$$

where θ and θ' are the angles between the average direction MM' of the pencil and the normals $\mathbf{n}$ and $\mathbf{n}'$ to the two surface elements. Introducing solid angles,

$$\mathrm{d}\Omega = \frac{\mathrm{d}S'\cos\theta'}{MM'^2} \tag{3C.2}$$

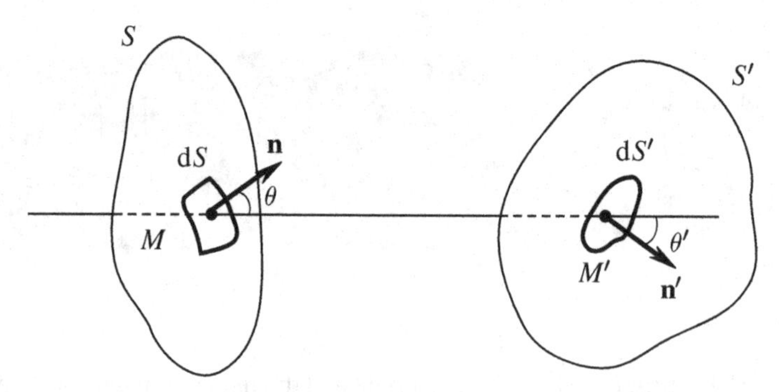

Figure 3C.1 Light beams and light pencils. The two surface elements dS and dS' determine a light pencil, i.e. the set of rays passing from dS to dS'. The whole set of pencils passing from S to S' makes up the beam.

and

$$\mathrm{d}\Omega' = \frac{\mathrm{d}S\cos\theta}{MM'^2}, \tag{3C.3}$$

the étendue may be written in the form

$$\mathrm{d}U = \mathrm{d}S\cos\theta\,\mathrm{d}\Omega = \mathrm{d}S'\cos\theta'\mathrm{d}\Omega'. \tag{3C.4}$$

The radiant flux dΦ carried by the radiation in the pencil, i.e. the transported power, is given by

$$\mathrm{d}\Phi = L(M,\theta)\,\mathrm{d}U, \tag{3C.5}$$

where $L(M,\theta)$ is called the radiance at the point M. For many types of sources (and in particular, a black body with uniform temperature), the radiance depends on neither the point M nor the direction θ, and this is the case considered hereafter in order to simplify the notation.

A light beam is defined by two apertures (one of which can be identified with the source), and the power Φ it transports is clearly the sum of the powers $d\Phi$ carried by the elementary pencils of which it comprises. If the radiance L is uniform and independent of the direction, it follows that

$$\Phi = LU, \tag{3C.6}$$

where the étendue U is a purely geometrical quantity, obtained by double integration of (3C.4).

The definitions and properties discussed here have been given implicitly for a monochromatic beam. In the case of a polychromatic beam, a differential spectral irradiance $\mathcal{L}(\omega)$ is defined. It is then understood that the above properties apply to each spectral element dω characterized by the radiance $\mathcal{L}(\omega)$dω. Consequently, for a spectral interval dω and a pencil with étendue dU,

$$\mathrm{d}\Phi = \mathcal{L}(\omega)\mathrm{d}\omega\mathrm{d}U. \tag{3C.7}$$

Since the different spectral elements of an incoherent source are independent, their contributions to the total power are additive and expression (3C.7) can be integrated over the variable ω.

In the following, we shall need to know the relationship between the étendue $\mathrm{d}U$ of a pencil and the differential elements $\mathrm{d}x$, $\mathrm{d}y$ and $\mathrm{d}k_x$, $\mathrm{d}k_y$ in the direct and reciprocal spaces (the latter being the space of wavevectors $\mathbf{k}$). In Cartesian coordinates and for a propagation direction close to the Oz-axis, we find

$$\mathrm{d}\Omega = \frac{1}{k^2}\mathrm{d}k_x\,\mathrm{d}k_y = \frac{c^2}{\omega^2}\mathrm{d}k_x\,\mathrm{d}k_y. \tag{3C.8}$$

For a surface element $\mathrm{d}S = \mathrm{d}x\,\mathrm{d}y$, perpendicular to Oz, we thus have

$$\mathrm{d}U = \frac{c^2}{\omega^2}\mathrm{d}x\,\mathrm{d}k_x\,\mathrm{d}y\,\mathrm{d}k_y. \tag{3C.9}$$

3C.1.2 Conservation of radiance

If a light beam propagates through a non-absorbent medium, and if the dioptric components it encounters have been given an antireflective treatment, the power is conserved. It can also be shown that, for a perfect (aberration-free) optical system, the étendue is conserved during propagation. More exactly $n\mathrm{d}U$ is conserved, where n is the refractive index, but here we consider only beams in a vacuum. Equation (3C.4) then shows that the radiance is conserved during propagation. The beam remains characterized by an unchanging radiance, which makes it easy to calculate the photometric quantities required to understand an optical instrument, as we shall see below.

Note that if the beam encounters absorbent media, or partially reflecting surfaces, the radiance will decrease even if the optical system is aberration free. Moreover, if the system is not aberration free, the étendue can only grow, and this generally leads to a decrease in radiance.

To sum up, during the propagation of a beam produced by an incoherent source, the radiance is conserved if the optical systems it passes through are perfect. Otherwise it can only decrease.

Comments

(i) This property is closely related to the second law of thermodynamics which would be violated if it were possible to increase the specific intensity during propagation. There is also a connection with statistical mechanics, which can be made clear if we observe that the conservation of throughput is nothing other than the analogue of Liouville's theorem, i.e. the conservation of volume in phase space during a Hamiltonian evolution.[1]

(ii) The properties discussed above are no longer valid if the media involved have nonlinear behaviour. Note, for example, that in this case frequency changes may occur, and the spectral elements may no longer be independent. Also, nonlinear processes such as parametric amplification (Complement 7A) may increase irradiance.

[1] See, for example, C. Kittel, *Elementary Statistical Physics*, Dover (2004).

3C.2 Maximal irradiance by an incoherent source

The irradiance is the radiant power received per unit area of a surface, i.e.

$$E = \frac{d\Phi}{dS'}. \tag{3C.10}$$

Let us consider the maximal irradiance that can be obtained from an incoherent source of specific intensity L.

Consider first the case shown in Figure 3C.1, where the irradiated surface area dS' is directly opposite the source. Using (3C.4) and (3C.6), it can be seen that the irradiance into the solid angle $d\Omega'$ due to the source element dS is given by

$$dE = L \cos\theta' \, d\Omega'. \tag{3C.11}$$

The total irradiance is found by integrating over the whole source. If the source is very large compared with the distance MM', the total solid angle will be roughly equal to 2π, but in the integration the term $\cos\theta'$ will give a factor of 1/2. In the end, one finds that

$$E \leq \pi L, \tag{3C.12}$$

with equality only if the source viewed from M' subtends a solid angle of 2π.

Is it possible to go beyond this limiting value of the irradiance by using some kind of optical instrument? The answer is negative, since conservation of radiance means that (3C.11) remains true even if the pencil of light is passed through a perfect instrument. When this formula is integrated, the integration limits bring in the angle subtended by the output aperture of the instrument. The irradiance therefore depends only on the aperture of the instrument and increases as the aperture increases, without ever being able to exceed the limiting value given by (3C.12).

So whatever instrument is used, the radiance determines the order of magnitude of the maximal irradiance that can be obtained from a given source, i.e. in the final count, the maximal electric field, which is the important quantity for the interaction between matter and radiation.

Comments

(i) Using systems of elliptical or parabolic mirrors, point M' can be irradiated by a solid angle greater than 2π up to a limit of 4π. In this case, the bound given by (3C.12) must be doubled.

(ii) The above considerations show how it is possible to burn dry tinder by focusing the sun's rays using a magnifying glass: one is increasing the sun's apparent diameter, which goes from 2α (less than 10^{-2} radians) to $2u$ (which can easily reach 60° or more) (see Figure 3C.2). The irradiance is then multiplied by $(1 - \cos u)/\alpha^2$ (close to 10^4 in our example). The same principle is at work in solar ovens placed at the top of a tower and in sight of a large number of mirrors which reflect the sun's rays onto the target. The temperature obtained can reach several thousand kelvins, but never go above the temperature of the sun's surface (5000 K).

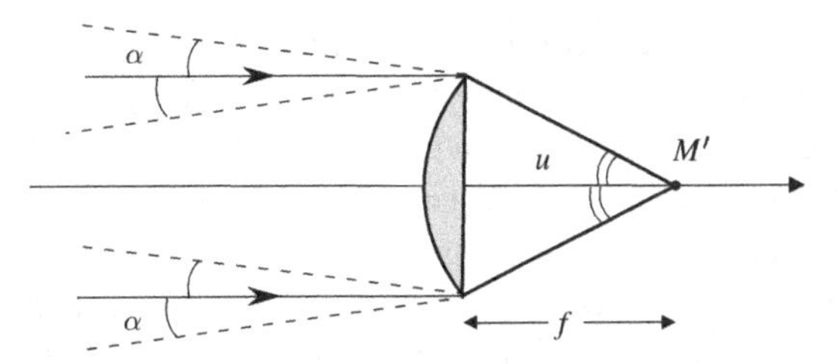

Figure 3C.2 Obtaining high temperatures by focusing the sun's rays. The irradiance given by a stigmatic and perfectly transparent instrument of aperture $2u$ is greater than the direct irradiance due to the sun by a factor of $(u/\alpha)^2$, where 2α is the angular diameter of the sun ($2\alpha = 30' \approx 8 \times 10^{-3}$ rad). Dry tinder can be set alight by means of a magnifying glass.

Orders of magnitude

According to Planck's law, a black body at 3000 K emits about 500 W.cm^{-2} into a half-space, over a spectral band extending from the infra-red (1.5 μm) to the green (0.5 μm), a range of some 4×10^{14} Hz. This is roughly the output of an incandescent lamp. The maximal irradiance that can be obtained is thus of the order of 500 W.cm^{-2}. A high-pressure arc lamp or the sun will give at best ten times more, i.e. 5 kW.cm^{-2}.

The maximal irradiance cited here is the total power density received per unit area. Another important quantity is the power per unit area and per unit frequency. It is equal to about 10^{-12} $\text{W.cm}^{-2}.\text{Hz}^{-1}$ in our example. To understand the relevance of this quantity, suppose for example that one wishes to excite an atomic transition. Only one frequency band, of the order of the natural width of the atomic transition will be used, i.e. a few MHz, and the useful irradiance is of the order of 10^{-5} W.cm^{-2}. Such a value is much too low to reach the strong-field regime, and it is even too low to saturate an atomic transition. Indeed, saturation intensities, leading to a saturation parameter s at resonance (see Equation (2.160) or (2.189)) of the order of unity, are closer to the value 10^{-3} W.cm^{-2}.

3C.3 Maximal irradiance by laser light

Let us now consider a 10 W laser beam. Since it is spatially coherent, it can be focused on an elementary diffraction spot, with size of the order of the wavelength, i.e. an area less than 1 μm^2. We now have available some 10^9 W.cm^{-2}, rather than at most a few times 10^3 W.cm^{-2} that can be obtained from an arc lamp. Moreover, this power is supplied in a spectral band that can be of the order of, or even less than, 1 MHz. If one aims to excite a narrow atomic transition, all the power can be used and the strong-field regime can be reached (saturation parameter 10^{12} for our example).

In the time domain, the technique of mode locking described in Section 3.4.1 can be used to concentrate the laser power in time in the form of pulses as short as 10 fs (10^{-14} s), repeated every 10 ns. The maximal power is 10^6 times greater than the average power and the irradiance reaches 10^{15} W.cm^{-2}. Under these conditions, the electric field of the electromagnetic wave is *stronger than the Coulomb field* exerted by the proton on the electron of a hydrogen atom in its ground state. A new energy scale is thus attained in the interaction between light and matter.[2]

What characterizes laser light is thus the possibility of concentrating it both in space and in time (or in the spectrum), whereupon enormous energy densities can be obtained, as attested by a great many laser applications. This possibility is related to the fact that a laser mode contains a very large number of indistinguishable photons, while in the case of thermal radiation, the number of photons per elementary mode of the radiation is less than unity. We shall discuss this point first for cavity modes, then for freely propagating beams.

3C.4 Photon number per mode

3C.4.1 Thermal radiation in a cavity

The energy of the thermal radiation contained in a cavity, e.g. a cube of side L, at temperature T, is proportional to the volume L^3 of the cavity. Planck's law then gives the energy in the frequency band of width $\mathrm{d}\omega$ around ω as

$$\mathrm{d}E = L^3 \frac{\hbar\omega^3}{\pi^2 c^2} \frac{1}{\exp(\hbar\omega/k_\mathrm{B}T) - 1} \mathrm{d}\omega. \tag{3C.13}$$

In such a cavity, discrete modes can be defined, characterized by a wave vector $\mathbf{k}$ (or by a frequency $\omega = ck$ and a direction of propagation $\mathbf{k}/k$), and a polarization. The density of modes in $\mathbf{k}$ space is also proportional to the volume L^3, and the number of modes with frequencies lying in the band of width $\mathrm{d}\omega$ about ω is equal to (Section 6.4.2)

$$\mathrm{d}N = L^3 \frac{\omega^2}{\pi^2 c^2} \mathrm{d}\omega. \tag{3C.14}$$

Dividing $\mathrm{d}E$ by $\hbar\omega\mathrm{d}N$ shows that the number of photons in a mode of frequency ω is given by the Bose–Einstein distribution[3]

$$\mathcal{N}(\omega) = \frac{1}{\exp(\hbar\omega/k_\mathrm{B}T) - 1}. \tag{3C.15}$$

This number of photons per mode, independent of the volume and shape of the cavity, is called the *photon degeneracy parameter*, where the photons are treated as indistinguishable bosons, since photons in the same mode cannot be distinguished by either their frequency, their propagation direction or their polarization. For a heat source at 3000 K, the number of photons per mode $\mathcal{N}(\omega)$ is less than unity, of the order of 3×10^{-4} in the middle of the

[2] Suitable amplification and compression of such laser pulses can lead to another increase of several orders of magnitude. See for instance http://extreme-light-infrastructure.com/.

[3] This well-known result is interpreted by considering the heat radiation in each mode as an ensemble of bosons at thermodynamic equilibrium (see, for example, C. Kittel, *Elementary Statistical Physics*, Dover (2004)).

visible spectrum (5×10^{14} Hz, or $\lambda = 0.6\,\mu$m). For solar radiation (temperature 5800 K), $\mathcal{N}(\omega)$ is of the order of 10^{-2} in the middle of the visible spectrum. In both cases, *there are far fewer than one photon per mode.*

3C.4.2 Laser cavity

Consider a single-mode laser emitting a power Φ. If the output mirror has transmission coefficient T and the cavity a total length L_{cav}, as shown in Figure 3.1, the number of photons contained in the laser cavity is equal to

$$\mathcal{N}_{\text{cav}} = \frac{\Phi}{\hbar\omega}\frac{1}{T}\frac{L_{\text{cav}}}{c}. \tag{3C.16}$$

As an example, consider a helium–neon laser delivering 1 mW, with a cavity of length 1 m and coefficient $T = 10^{-2}$. Equation (3C.16) shows that there are about 2×10^9 photons in the mode of the cavity laser. *This very high degeneracy is an essential feature of laser light.* It is intimately related to the possibility of obtaining considerably higher power densities than from thermal radiation, where the degeneracy parameter is less than unity for practically accessible temperatures.

3C.5 Number of photons per mode for a free beam

3C.5.1 Free propagative mode

In the last section, we discussed the idea of a radiation mode trapped in a cavity. The definition of an elementary radiative mode is a little more subtle in the case of freely propagating beams. Consider a pencil of light emitted into free space, e.g. from a small hole in the wall of a closed box (black-body radiation, see Figure 3C.3). How can the

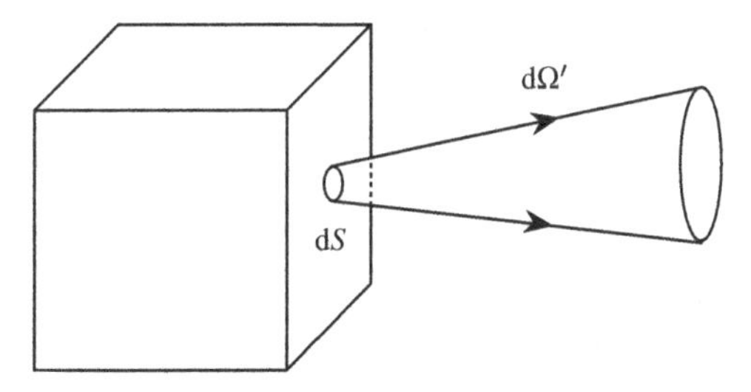

Figure 3C.3 **Pencil of radiation** obtained by making a small hole of cross-section $\mathrm{d}S$ in the wall of a cavity containing radiation at temperature T and considering only a solid angle $\mathrm{d}\Omega'$. The specific intensity is that of a black body at temperature T. A free mode is associated with a minimal pencil, such that $\mathrm{d}U = \mathrm{d}S\mathrm{d}\Omega' = \lambda^2$, where $\lambda = 2\pi c/\omega$, and a minimal wave packet of duration $\Delta t = 2\pi/\Delta\omega$.

notion of a mode be generalized to this case where there is no longer a boundary condition to discretize the problem?

We define a *free mode* by first considering a pencil of *minimal étendue compatible with the wave nature of the radiation*. Such an elementary pencil, with average wavelength $\lambda = 2\pi c/\omega$ and cross-sectional area dS, will have an angular divergence due to diffraction equal to $\lambda/\sqrt{S}$. Its étendue is given by

$$\mathrm{d}U_{\mathrm{min}} = S\left(\frac{\lambda}{\sqrt{S}}\right)^2 = \lambda^2 = 4\pi^2\frac{c^2}{\omega^2}. \tag{3C.17}$$

Using (3C.9), this equation can be considered as a consequence of the relations describing the effects of diffraction:

$$\Delta x\, \Delta k_x \geq 2\pi, \tag{3C.18}$$

$$\Delta y\, \Delta k_y \geq 2\pi. \tag{3C.19}$$

Regarding the direction of propagation z (longitudinal), we consider the conjugate variables time and frequency, and in particular, a wave packet of duration Δt. Its spectral width then satisfies the relation

$$\Delta t\, \Delta\omega \geq 2\pi, \tag{3C.20}$$

and, for a minimal wave packet, we have

$$(\Delta t\, \Delta\omega)_{\mathrm{min}} = 2\pi. \tag{3C.21}$$

A dispersion,

$$\Delta k_z = \frac{\Delta\omega}{c}, \tag{3C.22}$$

in the component k_z of the wavevector is associated with the dispersion $\Delta\omega$, while the longitudinal extent,

$$\Delta z = c\Delta t, \tag{3C.23}$$

is associated with the duration Δt. The analogous relation to (3C.18) and (3C.19) is then deduced, i.e.

$$\Delta z\, \Delta k_z \geq 2\pi, \tag{3C.24}$$

with equality holding for a minimal wave packet.

From Equations (3C.18), (3C.19) and (3C.24), we conclude that a free-space mode is characterized by a minimal extension in phase space (direct product of the real space and the wavevector space). This minimal unit is called an elementary cell in the phase space. It is not possible to specify an electromagnetic wave packet more precisely with respect to position and wave vector. The corresponding phase space volume is

$$(\Delta x\, \Delta k_x)_{\mathrm{min}}(\Delta y\, \Delta k_y)_{\mathrm{min}}(\Delta z\, \Delta k_z)_{\mathrm{min}} = (2\pi)^3. \tag{3C.25}$$

Comment

In the case of a material particle, the phase space is the direct product of the position and momentum spaces, and the elementary cell has volume h^3, the cube of Planck's constant. Equation (3C.25) can be recovered by using the relation $\mathbf{p} = h\mathbf{k}/2\pi$ for the photon.

3C.5.2 Pencil of heat radiation

The spectral radiance $\mathcal{L}(\omega)$ of a black body, or of a hole in the wall of a closed box containing heat radiation, is

$$\mathcal{L}(\omega) = \frac{c}{4\pi}\frac{1}{L^3}\frac{\mathrm{d}E}{\mathrm{d}\omega}, \tag{3C.26}$$

where the energy density $(1/L^3)\mathrm{d}E/\mathrm{d}\omega$ is given by Planck's law (3C.13).

The energy contained in a mode, characterized by (3C.17) and (3C.21), or (3C.25) is given by

$$\begin{aligned} E_{\text{mode}} &= \frac{1}{2}\mathcal{L}(\omega)\,\mathrm{d}U_{\text{min}}(\Delta\omega\Delta t)_{\text{min}} \\ &= 4\pi^3\frac{c^2}{\omega^2}\mathcal{L}(\omega). \end{aligned} \tag{3C.27}$$

The factor of 1/2 accounts for the existence of two orthogonal polarizations, and hence two distinguishable modes for a minimal wave packet. Using (3C.26) and (3C.13), this implies that

$$E_{\text{mode}} = \frac{\hbar\omega}{\exp(\hbar\omega/k_{\text{B}}T) - 1}, \tag{3C.28}$$

whereupon the *number of photons per free propagative mode of the black-body radiation* $E_{\text{mode}}/\hbar\omega$ is still given by the Bose–Einstein distribution (3C.14).

3C.5.3 Beam emitted by a laser

The divergence of a transverse single-mode laser beam is entirely due to diffraction. It is thus a minimal pencil for transverse coordinates. Regarding longitudinal coordinates, we consider as above a wave packet with minimal duration compatible with the spectral width $\Delta\omega$ of the line, according to (3C.21), i.e.

$$\Delta t = \frac{2\pi}{\Delta\omega}. \tag{3C.29}$$

The number of photons per free mode is thus,

$$\mathcal{N}_{\text{laser}} = \frac{\Phi}{\hbar\omega}\frac{2\pi}{\Delta\omega}. \tag{3C.30}$$

The linewidth $\Delta\omega$ is often much less than $2\pi c/L_{\text{cav}}$, and comparing (3C.30) with (3C.16), it can be seen that the number of photons per mode in a laser beam is very much greater than unity. If the Schawlow–Townes limit is taken for the linewidth (see Section 3.3.3b and Complement 3D), which is typically of the order of 10^{-3} Hz, the number of photons per mode can exceed 10^{15} for a laser of a few mW. Even if more realistic line widths are considered, of the order of MHz, values greater than 10^9 photons per mode are found for cavities of length around 1 m.

For freely propagating beams, we find once again that *a laser beam has a much higher photon number per mode than unity, in contrast with a beam produced by a heat source.*

3C.6 Conclusion

It has been shown that laser light offers the possibility of focusing in space as well as in time, or in a complementary way, in angle (highly collimated beam) and in spectrum (highly monochromatic beam). These possibilities are fundamentally related to the fact that all the photons are in the same mode: since they are indistinguishable bosons, they are spatially and temporally coherent, and they can be concentrated on the minimal dimensions allowed by the size × divergence relation (diffraction), and by the time × frequency relation.

Hence, we may say that what characterizes laser light most fundamentally, as compared with ordinary light, is a photon number per mode much higher than unity.

3D Complement 3D The spectral width of a laser: the Schawlow–Townes limit

The spectral width of the output of most single-mode lasers is determined by technical limitations associated with the stability of the optical length of the laser cavity (see Section 3.3.3). However, in the absence of these, there is a more fundamental limit to the degree of monochromaticity that can be achieved. This limit, known as the Schawlow–Townes limit is, in fact, rather narrower than the passive bandwidth of the laser cavity or the width of the gain curve of the active medium it contains. We calculate in a heuristic fashion in this complement the Schawlow–Townes limit for a laser operating far above threshold.[1]

The fundamental mechanism for the spectral broadening of a laser output beam is the *spontaneous emission* by the gain medium of photons into the laser mode. Spontaneous emission adds to the complex field of the laser mode $\mathcal{E}_L$ a fluctuating field, $\mathcal{E}_{sp}$ corresponding to the addition of a single photon with a random phase. The total field therefore undergoes amplitude and phase fluctuations. The fluctuations of the amplitude are damped by the gain saturation of the amplifying medium and only the phase fluctuations persist, because the mechanism responsible for laser oscillation does not impose any phase to the generated field. Thus, in the course of successive spontaneous emission events, the *phase* of the laser field undergoes a *random walk*. After a time τ_c (the field correlation time) the phase of the laser field can no longer be predicted; it has lost all memory of its initial value. As a consequence, the laser frequency, which is just the time derivative of the phase, cannot be determined to better than $1/\tau_c$. Thus the laser spectral linewidth is of the order of

$$\Delta\omega_{ST} \approx \frac{1}{\tau_c}. \tag{3D.1}$$

In order to evaluate the correlation time τ_c we consider the evolution of the complex amplitude of the laser field $\mathcal{E}_L$ in the complex plane. Each spontaneous emission process changes $\mathcal{E}_L$ by an amount $\mathcal{E}_{sp}$, a vector ('phasor') in the Fresnel plane (see Figure 3D.1) of amplitude E_{sp} and of random direction. If the field amplitude increases, the gain decreases because of gain saturation and becomes smaller than the losses, which subsequently restore the amplitude of the laser field to its initial value. A similar restoring process occurs when the field amplitude happens to decrease. As a result of successive spontaneous emissions the extremity of the vector $\mathcal{E}_L$ undergoes a random walk, in so doing tracing out a

[1] The original derivation of this limit can be found in A. L. Schawlow and C. H. Townes, Infrared and Optical Masers, *Physical Review* **112**, 1940 (1958). See also M. Sargent III, M. O. Scully, and W. E. Lamb, Jr, *Laser Physics*, Addison-Wesley (1974) and L. Mandel and E. Wolf, *Optical Coherence and Quantum Optics*, Cambridge (1995), for more rigorous approaches.

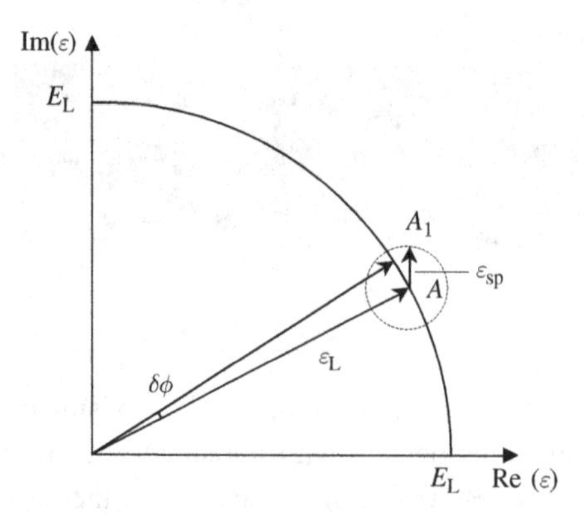

Figure 3D.1 Diagram showing the evolution of the laser mode complex amplitude $\mathcal{E}_L$ in the complex plane. Each spontaneous emission process adds a vector of amplitude E_{sp} and of random phase to the vector representing the laser field (evolution from A to A_1). The amplitude of the laser field is then corrected by the phenomenon of gain saturation, leaving only the effect of the phase-shift $\delta\phi$ which is of the order of E_{sp}/E_L. This is the fundamental process behind phase diffusion of a laser field.

trajectory close to a circle of radius E_L. After N_{sp} spontaneous emission processes, the tip of the vector $\mathcal{E}_L$ travels, on average, through an angle $\Delta\phi$ given by

$$(\Delta\phi)^2 \approx N_{sp}\frac{E_{sp}^2}{2E_L^2}. \tag{3D.2}$$

The factor of two arises here because spontaneous emission contributes to the diffusion of the phase only through the component of $\mathcal{E}_{sp}$ that is tangential to the circle of radius E_L.

The correlation time τ_c is that over which $\Delta\phi$ changes by an amount of order 1 radian. This time corresponds to a mean number of spontaneous emission processes:

$$\overline{N_{sp}} \approx 2\left(\frac{E_L}{E_{sp}}\right)^2. \tag{3D.3}$$

The optical energy in the laser cavity, which has volume V, is $\varepsilon_0 E_L^2 V/2 = \mathcal{N}\hbar\omega$ ($\mathcal{N}$ being the mean number of photons in the laser cavity), whilst the energy of a spontaneous photon is $\hbar\omega = \varepsilon_0 E_{sp}^2 V/2$. It follows that Equation (3D.3) can be rewritten,

$$\overline{N_{sp}} \approx 2\mathcal{N}. \tag{3D.4}$$

Denoting by Γ_{mod} the probability per unit time that an atom in the excited state b spontaneously emits a photon into the laser mode, and by N_b the number of atoms in the excited state, we find that the number of spontaneous photons emitted into the laser mode in time τ_c is

$$\overline{N_{sp}} = \Gamma_{mod} N_b \tau_c. \tag{3D.5}$$

Combining Equations (3D.1), (3D.4) and (3D.5) we obtain a first expression for the spectral width of the laser output:

$$\Delta\omega_{\mathrm{ST}} \approx \frac{1}{\tau_{\mathrm{c}}} \approx \frac{\Gamma_{\mathrm{mod}} N_b}{2\mathcal{N}}. \tag{3D.6}$$

It is equally possible to relate the rate of spontaneous emission into the laser mode to the output power Φ. In fact, it can be shown (see Chapter 6, Section 6.3) that the rate of stimulated emission per atom in state b, which is just the same as the rate of absorption per atom in state a, is equal to $\Gamma_{\mathrm{mod}}\mathcal{N}$ (for a mean number of photons in the laser cavity, $\mathcal{N} \gg 1$). When the laser operates far above threshold, the loss of energy caused by spontaneous emission is negligible and the output power is simply that resulting from the competition between stimulated emission and absorption in the laser gain medium. Applying the conservation of energy we can, therefore, show that

$$\Phi = \Gamma_{\mathrm{mod}}\mathcal{N}(N_b - N_a)\hbar\omega. \tag{3D.7}$$

Eliminating Γ_{mod} between Equations (3D.6) and (3D.7) we obtain

$$\Delta\omega_{\mathrm{ST}} \approx \frac{N_b}{2(N_b - N_a)} \frac{\Phi}{\hbar\omega} \frac{1}{\mathcal{N}^2}. \tag{3D.8}$$

This expression can be transformed by noticing that the output power Φ is related to the number of photons $\mathcal{N}$ in the laser mode in the cavity. For the case of a linear laser cavity, with a partially reflecting output mirror, we find, using Equation (3A.46), with $W = \mathcal{N}\hbar\omega$ and $\Phi = \Pi_{\mathrm{o}} S$,

$$\mathcal{N}\hbar\omega = \frac{\Phi}{\Delta\omega_{\mathrm{cav}}}, \tag{3D.9}$$

where $\Delta\omega_{\mathrm{cav}}$ is the width at half-maximum of a mode of the laser cavity (see Complement 3A, Equation (3A.26)). We obtain finally for the spectral width of the laser output,

$$\begin{aligned} \Delta\omega_{\mathrm{ST}} &\approx \frac{N_b}{2(N_b - N_a)} \frac{\hbar\omega}{\Phi} (\Delta\omega_{\mathrm{cav}})^2 \\ &\approx \frac{N_b}{2(N_b - N_a)} \frac{\Delta\omega_{\mathrm{cav}}}{\mathcal{N}}. \end{aligned} \tag{3D.10}$$

This result agrees, to within a numerical factor, with the more rigorous result derived by Schawlow and Townes:

$$\Delta\omega_{\mathrm{ST}} = \pi \frac{N_b}{N_b - N_a} \frac{\hbar\omega}{\Phi} (\Delta\omega_{\mathrm{cav}})^2. \tag{3D.11}$$

Notice that expression (3D.11) combined with (3D.9) yields

$$\frac{\Delta\omega_{\mathrm{ST}}}{\Delta\omega_{\mathrm{cav}}} = \frac{\pi}{\mathcal{N}} \frac{N_b}{N_b - N_a}. \tag{3D.12}$$

Since the photon number $\mathcal{N}$ is much larger than one and the inversion factor is of the order of one, this expression shows that the Schawlow–Townes limit for the *laser linewidth* is *much smaller* than the (already small) *width of a cavity mode*. It also expresses the fact that the more photons there are in the laser mode the more stable is its phase with respect to the fluctuations induced by spontaneous emission.

Comment

For numerical applications of expression (3D.11) it is necessary to have an estimate for $\Delta\omega_{\text{cav}}$. This can be achieved using the results (3A.26) and (3A.40) of Complement 3A, which lead to

$$\Delta\omega_{\text{cav}} = T\frac{c}{L_{\text{cav}}}, \tag{3D.13}$$

where $L_{\text{cav}}/2$ is the separation of the mirrors of the linear cavity. For example, for a helium–neon laser emitting a power 1 mW, of cavity length of 1 m (so that $L = 2$ m) and with an output mirror of transmission T equal to 2%, we find

$$\frac{\Delta\omega_{\text{cav}}}{2\pi} \approx 5.10^5 \text{ Hz}; \quad \frac{\Delta\omega_{\text{ST}}}{2\pi} \approx 10^{-3} \text{ Hz}.$$

The quality factor of an oscillator with such a narrow linewidth would be enormous. In fact technical sources of line broadening, arising from fluctuations in the laser cavity length, usually give rise to a practical linewidth that is many times the fundamental limit. In this case, reaching a frequency stability as good as the Schawlow–Townes linewidth requires considerable efforts in stabilizing the cavity length.

A second interesting example is that of a diode laser. We consider a device also emitting a power of 1 mW; the difference here is that the laser cavity is very short and of low finesse. In this case we find

$$\frac{\Delta\omega_{\text{cav}}}{2\pi} \approx 2.10^{10} \text{ Hz}; \quad \frac{\Delta\omega_{\text{ST}}}{2\pi} \approx 10^{6} \text{ Hz}.$$

Again, in practice, a linewidth larger than the latter value is measured. The origin of this extra broadening is in the fluctuations of the refractive index of the gain medium arising from intensity fluctuations in the optical field. More generally, many other physical mechanisms, which are beyond the scope of this book, lead to modifications of the laser linewidth with respect to the simple Schawlow–Townes equation (3D.11).

In the case of the diode laser, the short length of the cavity is responsible for the fact that the laser linewidth is now in the MHz range and strongly limits the use of such a laser for applications requiring a good coherence, such as interferometry, high-resolution spectroscopy, etc. However, several strategies can be used to reduce the linewidth of such a laser. They are all based on the use of an external cavity to increase the effective cavity length, and thus reduce $\Delta\omega_{\text{cav}}$ and $\Delta\omega_{\text{ST}}$. For example, one can couple the laser to an external cavity, or add an extra feedback mirror or grating which re-injects part of the laser output light into the laser, thus creating some kind of 'composite' cavity with a smaller $\Delta\omega_{\text{cav}}$ as well as increasing the number $\mathcal{N}$ of photons in the cavity.

Complement 3E The laser as energy source

We have seen in the present chapter that the light emitted by a laser has properties that are radically different from those of the light emitted by classical sources. These properties have been the basis for the myriad applications found for lasers since their advent in the 1960s; they have escaped the confines of the research laboratory to become ubiquitous in industrial production and modern consumer society. Lasers now have innumerable applications in such disparate areas as medicine, metallurgy and telecommunications and are at the heart of new developments in commercial and consumer electronics (CD and DVD players, bar-code readers and printers, to name but a few examples).

The total market in the mid 2000s was estimated to be almost 6 billion dollars. It was dominated by the domains of optical storage (30% of the total amount) and communication (20%), which are mass production markets. In contrast, material processing (25%) and medical applications (8%) involve a smaller number of very expensive lasers. Research and instrumentation amount to 6% of the total sales. The significant fraction of laser sales related to research and development is a testament to the relative youth of the technology. New applications are still coming to light, some of which may have profound economic consequences for the future.

It will not be possible to provide an exhaustive account of these applications here. We shall, therefore, concentrate on a few significant examples selected from the broad categories introduced above. We are interested here only in applications which make use of the ability of the laser beam to *deliver energy* in a precise and controllable fashion. Complement 3F will deal with applications in which other properties, such as directivity and monochromaticity, play the dominant role. Finally, in Complement 3G we discuss spectroscopic applications of lasers.

3E.1 Laser irradiation of matter

Consider a pulsed laser beam (wavelength λ, constant power P_0 during the pulses) focused by an optical system onto a target in order to obtain localized irradiation on a small area S. The irradiance P_0/S can be very large so that the electric field at the irradiated surface is sufficient to profoundly alter the structure of the material. This can cause heating, melting or even ionization, leading to the formation of a plasma. The precise effect depends, of course, on the nature of the material of the target and the characteristics of the laser pulse.[1] However, these processes have some features that are quite general to matter–light interactions that we shall discuss in the following.

3E.1.1 The light–matter coupling

The optical energy incident on the target can be broadly divided into three parts: the fraction R that is reflected or, more generally, back-scattered, the fraction T that is transmitted or forward scattered into the material and the remaining fraction $A = 1 - R - T$ that is absorbed in the medium. The first two parts are obviously of little use and so must be minimized.

The coefficients R and T vary considerably as a function of the material, of its surface quality and of the wavelength of the incident radiation. Metals are almost perfect conductors at long wavelengths. Consequently, they reflect most of the incident light in the infra-red and microwave domains. The coefficient R exceeds 90% for $\lambda > 1\,\mu$m for most metals, whilst it is less than 50% in the visible and ultra-violet spectral regions. Conversely, non-metallic materials, organic and inorganic, strongly absorb laser radiation at most wavelengths. The transmission coefficient T is negligible for opaque materials such as metals, which absorb the incident radiation within a layer less than a micron thick. The transmission plays an important role, however, in the interaction of laser light with *biological material*, for which the penetration depth is typically of the order of a centimetre. These considerations are only of relevance to the early moments of the light–matter interaction. This is soon dominated by the state reached by the material under the action of the laser light. If the material is brought close to its melting point, it behaves to a good approximation like a black body, having an absorption coefficient close to unity. The energy transfer then becomes much more efficient and is practically independent of the material. If the surface material is ionized by the laser light it is the resulting plasma, which expands above the focal region, that continues to absorb energy from the light field. The liberated charged particles are then forced to oscillate by the applied optical field and radiate in turn. This process leads to the total reflection of the incident light if the electron density of the plasma N exceeds a threshold value corresponding to the equality of the frequency of plasma oscillations $\omega_p = (Nq^2/m\varepsilon_0)^{1/2}$ with that of the optical field.[2] This threshold density is $N_T = 4\pi^2\varepsilon_0 mc^2/q^2\lambda^2$. The formation of this *plasma barrier* must therefore be avoided at all costs if efficient energy transfer is to be achieved. The expression for N_T above implies that this will be achieved more easily *the shorter the wavelength* of the laser employed, thus increasing the threshold density.

[1] For a review of the subject see, for example, H. Bass, Laser-material interactions in *Encyclopedia of Lasers and Optical Technology*, Academic Press, San Diego (1991).

[2] See, for example, R. P. Feynman, R. B. Leighton and M. Sands, *The Feynman Lectures on Physics*, Addison-Wesley (2005).

3E.1.2 Energy transfer

The power AP_0 absorbed by the material at the focal point of the laser beam causes an increase in the local temperature of the material, that we shall now study. When the absorption depth of the laser light is very small, the heat diffusion equation within the irradiated material may be expressed:

$$\rho C\frac{\partial T}{\partial t} = K\Delta T + \frac{AP_0}{\pi r^2}\delta(z), \tag{3E.1}$$

where r is the radius of the laser spot, ρ the material density, C its specific heat capacity and K its thermal conductivity. The quantity ΔT is the three-dimensional Laplacian of the temperature ($\Delta T = \partial^2 T/\partial x^2 + \partial^2 T/\partial y^2 + \partial^2 T/\partial z^2$). The one-dimensional Dirac function $\delta(z)$ indicates an absorption in a very thin layer. The solution of Equation (3E.1) then depends on the material geometry, the spatial and temporal characteristics of the laser pulse and on a parameter $\kappa = K/\rho C$, known as the *thermal diffusivity*, which characterizes the thermal behaviour of the material (for example, κ has the value 0.05 $cm^2.s^{-1}$ for steel and 1 $cm^2.s^{-1}$ for copper).

Let us assume first that the laser pulse is short enough so that the heat introduced by it into the material has diffused a distance that is small compared to the extent of the irradiated surface. For this case, a dimensional analysis of Equation (3E.1) shows that the heat diffuses into the depth of the material, as well as laterally over a distance equal to $\sqrt{\kappa t}$ at a time t after the end of the laser pulse.

During the pulse itself, after a time $t \gg \frac{r^2}{\kappa}$, a steady state is reached at the centre of the laser spot in which the power deposited at the surface by the laser balances the power dissipated by heat conduction. From Equation (3E.1), the steady-state temperature is $T_S \sim \frac{AP_0}{Kr}$.

As an example, in the case of steel ($\rho = 7800$ kg.m^{-3}, $C = 500$ J.kg^{-1}.K^{-1}) the fusion temperature (1700 K) is reached for an incident beam from a CO_2 laser ($\lambda = 10.6\,\mu$m) of power approximately 100 W focused on 30 μm. At this wavelength, $A = 0.01$. A YAG laser ($\lambda = 1.06\,\mu$m, $A = 0.1$) requires a power of 1 W when focused on 3 μm. The times required to reach the steady state for these two examples are approximately 200 μs and 2 μs.

In real systems, the absorption coefficient depends on the surface state of the material and the power needed to melt the surface is significantly larger, though still well within reach of existing lasers.

In contrast, at short times ($t \ll r^2/\kappa$), Equation (3E.1) shows that the surface temperature increases rapidly, proportionally to t, and to the instantaneous power absorbed at the surface, $AP_0/\pi r^2$. With the focusing discussed earlier, a 1 kW CO_2 laser is capable of heating steel to its fusion point in about 500 ns. One can then employ very brief laser pulses of quite low average power to melt a metal surface locally. Very short pulses (in the picosecond range) have the additional advantage of reducing the size of the heated zone to just the irradiated zone, as they allow no time for significant thermal conduction to occur. For pulse durations at the femtosecond scale, the energy is transferred only to the electronic degrees of freedom of the atoms and molecules composing the target, as there is no time for coupling to the vibrational degrees of freedom of the lattice to take place: ionization and dissociation processes predominate over heating in this time range.

3E.1.3 Mechanical effects

Once the boiling point of the material is passed, the laser irradiation induces a new mechanical effect. Fragments of solid or vaporized material or ions, in the case that a plasma is formed, are ejected at high velocity from the heated region. The material can be ejected at high speeds. This results in the cleaning of impurities from the heated zone.

Finally, since momentum is conserved overall, the substrate can experience a considerable compression normal to its irradiated surface, this force greatly exceeding that related to the radiation pressure alone. We shall show in Section 3E.4 of this complement how this *inertial effect* can be profitably employed.

Also to be mentioned in this part is the removal of a polluting layer on materials like stone: the intense laser pulse ionizes the first microns of the layer, generating a surface shock wave that ejects the remainder. This technique has found many interesting applications in the preservation of cultural heritage by providing a cost effective and non-abrasive way of cleaning statues and ancient buildings.

3E.1.4 Photo-chemical effects and photo-ablation

The laser pulse can also induce chemical effects related to *photo-dissociation* or *photo-fragmentation* of the molecules of which the medium is composed. Lasers of short wavelength, which emit photons of higher energy are the most efficient at producing such effects at low levels of the energy flux. Relatively low-intensity laser sources can also initiate polymerization reactions of the surface molecules of some materials.

Another effect, known as *photo-ablation* occurs when ultra-violet light, such as that produced by an excimer laser, is focused onto materials. When the deposited energy exceeds a threshold value (of the order of 100 $mJ.cm^{-2}$ for organic materials), material is removed from the interaction region. Figure 3E.1 shows, by way of example, a hole dug in a stainless steel foil 200 μm thick for different laser pulse durations. One notices its extremely clean edges and bottom, due to the threshold effect, and that there is almost no visible damage to the surrounding material.

The phenomenon of photo-ablation is very complex and not yet completely understood. It roughly occurs in the following manner: the ultra-violet photons break chemical bonds and lead to the formation of large numbers of fragments. The explosive expansion of the medium then results in its being ejected from the site of irradiation. Since the function of the incident photons is principally to break chemical bonds rather than cause heating, the surrounding material is virtually unaffected. This explains the lack of burning around the ablated zones in Figure 3E.1(c).

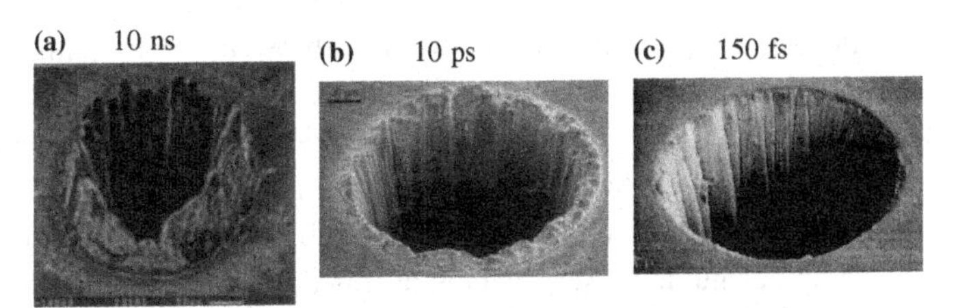

Figure 3E.1 Holes made in a stainless steel foil 200 μm thick by a series of laser pulses of different durations.

3E.2 Machining and materials processing using lasers

3E.2.1 Thermal effects

We explained in the last section how, by suitable choice of laser wavelength, pulse energy and degree of focusing of the laser beam, it is possible, with good precision, to raise a region of material of well-defined extent to a desired high temperature. The effect is therefore comparable to that of a blow-torch but with the advantage of not causing undesirable heating of surrounding material. This limits the amount of energy needlessly consumed and obviates the need for costly re-machining of the material after treatment. The laser beam has the additional advantage that it may be delivered by a simple optical system the mirrors of which, for example, can be operated under computer control. For industrial uses it is very often the cost, efficiency and ease of use of laser systems that are of overriding importance. CO_2 lasers, emitting in the far infra-red (10.6 μm) where the absorption is weak, are nevertheless commonly employed in the treatment of metals, because of their very high efficiency. Semiconductor lasers and fibre lasers, which can reach the kW range in the c.w. or quasi c.w. regime with good efficiency, are more and more often used in such applications.

Depending on the temperature reached by the substrate material in the irradiated zone a number of different effects can occur. In order of ascending temperature these are:

- $T < T_{\text{fusion}}$: *surface treatment*. This might be of a chemical nature (such as the formation of an alloy with a metal deposited on the surface), or structural (for example, the *vitrification* of the surface). The resulting material is often tougher and more resistant to impact or chemical or thermal attack. This kind of treatment is much used in the automobile industry, for example in the manufacture of pistons, valves and crank-shafts.
- $T_{\text{fusion}} < T < T_{\text{boiling}}$: *welding*. This is achieved without the addition of solder to the joint and without the necessity of any physical contact with the surfaces to be joined. It can be applied to work-pieces to which access with traditional blow-torches is difficult (such as the interior of piping). A 10 kW CO_2 laser, for example, is capable of welding 1 cm thick steel plates at a rate of 1 m per minute. This technique is also more and more used in the fabrication of car bodies for which it replaces the traditional technique of

spot-welding, because it can weld in the same process high-strength steel, aluminium and plastics and in a faster and highly reproducible way.

- $T > T_{\text{boiling}}$: *engraving, drilling, cutting*. Here the effect depends on the depth and the lateral extent of the vaporized region. It is possible to machine materials with very high melting points, such as ceramics or ruby. There are countless examples of the use of computer-guided laser systems in this domain, including the engraving of trademarks, the modification and repair of microprocessor chips, the cutting of car dashboards and the waste-free cutting of cloth for clothes manufacture, to name but a few.

So far we have only considered effects which are linear as a function of laser intensity. These are basically surface effects, because of the small penetration depth of most materials. If the material is more transparent, the interaction and heating process takes place in the total region illuminated by the focused Gaussian laser beam in the material, starting from its surface. But at high intensities, nonlinear effects take place, in which the temperature increase, for example, is proportional to the Nth power of the laser intensity (typically $2 < N < 10$).[3] These effects favour the regions where the light intensity is the highest: they will take place essentially *in the focal region of the laser beam*. Among many applications, one widely spread in the public domain is the marking of 3D images inside glass blocks: a tightly focused, pulsed laser beam creates by nonlinear interaction in this transparent material micro-cracks exactly at the focal spot, which locally changes its refractive index. One then moves the focal point by a computer-aided beam-steering system to reconstruct in the glass bulk any 3D object the shape of which is stored in the computer memory.

3E.2.2 Transfer of material

It is possible to put to good use the ejection of matter that occurs when a surface is subjected to an intense laser beam. For example, this makes possible the controlled vaporization of materials, potentially with very high melting points, and their subsequent deposition on another surface placed in the vicinity. This technique is employed for the deposition of thin layers of superconducting YBaCuO-type material, amongst others. These thin layers are composed of a regular crystalline structure of excellent quality, as witnessed by their large critical currents.

3E.3 Medical applications

If the phenomenology of the medical applications of lasers bears some similarity to that of the industrial applications described above, medical applications are distinguished by the fact that the material worked on is living tissue.[4] This has the following consequences:

[3] See Chapter 7 on non-linear optics.

[4] See, for example, H. -P. Berlien and G. J. Müller, *Applied Laser Medicine*, Springer-Verlag, Berlin, Heidelberg, New York (2003).

- The optical energy required is significantly less and so low-power lasers can often be used.
- Heat conduction to healthy tissue surrounding that to be treated must be limited to the minimum possible.
- The penetration of the laser beam is significantly higher than in metals. It depends strongly on the wavelength employed, being typically 50 μm for a CO_2 laser, 800 μm for a Nd:YAG laser at 1.06 μm and 200 μm for a green laser at 0.53 μm. However, these distances are still not very great and so treatment is limited to relatively thin tissues which can be easily reached by the laser beam.
- Since the treated tissue is living it can react to the trauma induced by the laser in a manner which could potentially counteract the treatment. Thus trauma must be minimized.

The effect of laser irradiation here also depends on the temperature attained during the laser pulse. If this is very high, the tissue can be locally vaporized, which allows the making of incisions and the destruction of unhealthy cells. If, however, the maximum temperature is between those at which coagulation and carbonization occur, localized photocoagulation is induced, which can be very useful in surgery. Finally, processes like photo-ablation by picosecond or femtosecond pulses makes possible very precise elimination of tissue without damage to surrounding organs, and has been used, for example, to remove brain tumours. These properties explain the great interest in the *laser scalpel* for procedures in surgery and odontology. However, in spite of the obvious advantages of the laser scalpel in terms of hygiene and their ability to simultaneously cut and cauterize tissue, practical devices still suffer from difficulties in precisely adapting the laser power to the highly variable local conditions of the tissue.

One of the earliest applications of lasers in medicine was in *dermatology*. Many skin disorders such as wine-stain blemishes and angiomas can be treated. Lasers are used also in *ophthalmology*, where the wavelength employed is of crucial importance; a detached retina is 'spot-welded' using a frequency-doubled YAG laser emitting visible radiation which is only absorbed at the retina itself. Short-sightedness on the other hand is treated with lasers operating in the ultra-violet ('Laser-Assisted in SItu Keratomileusis' or LASIK). This radiation is absorbed at the surface of the cornea and does not penetrate inside the eye globe. It creates fine incisions which modify its radius of curvature and change the eye focal length. Femtosecond lasers are more and more used in such applications, because of their much better cutting precision and reduction of collateral effects.

The natural partner of the laser in medicine is the catheter, a flexible tube which can be introduced into the body (via the digestive tract or blood vessels) and which contains an optical fibre for the transport of the laser beam and another for optical inspection (the endoscope). In addition, a further channel allows the injection of drugs or the removal of debris from the laser treatment. This device allows very precise and delicate surgery to be performed, whilst causing the minimum possible trauma to the patient in contrast to the effect of a classical operation. One example concerns the treatment of coronary obstructions (angioplasty): the catheter is introduced into the coronary artery and passed up to the atheroma responsible for the obstruction of the vessel. This is then destroyed by photo-ablation. This kind of treatment, as many others, is not, however without its side-effects. Heating effects can in fact lead to serious complications from the burning of nearby tissues and might result in the artery again becoming obstructed. This treatment is presently used as a complement to other techniques such as balloon angioplasty.

Laser treatments relying on heating or photo-ablation have the medical disadvantage of not being selective with regard to healthy and unhealthy tissue. However, in some instances it is possible to make use of the different spectral characteristics of the absorption of the two species of cells to attack only the unhealthy ones. To this end research is currently in progress on the use of several molecular species, in particular derivatives of hemato-porphyrine. These molecules have the property of preferentially becoming attached to cancerous or atheromatous cells. If the tissue treated with this chemical is irradiated at a wavelength near the peak of its absorption (635 nm), the molecule decomposes into a toxic substance which kills only the cell to which it is attached. This technique, which might sound very promising, in fact poses some serious problems: ingestion of the compound renders the entire organism sensitive to the effects of light exposure. Furthermore, a laser beam at the required wavelength does not penetrate far into tissue, which limits the depth to which cells can be destroyed.

To sum up, the development of medical treatments based on lasers is still in its infancy. Following initial enthusiasm for the new kinds of treatment in the early 1990s, the number of treatments for which they are used has stabilized or even fallen slightly. If in some applications laser techniques are firmly established (as in ophthalmology, for example), in others their usefulness is debatable (for example, in the treatment of inflammations by low-power continuous-wave lasers). This hiatus in new developments should allow a reorientation of the research effort to finding laser sources to meet medical requirements and not the converse as has often been the case. In the long term there is little doubt that the increasing miniaturization of lasers, improvements in their reliability, the ability to spatially and temporally shape their output and the parallel developments in fibre-optics will ensure that lasers are more and more widely used as medical tools.

Whilst in this section we have discussed the use of lasers in the treatment of illness, their applications also extend to diagnostics, for example, in the measurement of blood flow by Doppler velocimetry, of the blood protein content or measurements by diffractometry of the size of living cells.

3E.4 Inertial fusion

Research on laser-induced thermonuclear fusion is being actively pursued world-wide. Although this research is still very far from producing practical devices, we describe it here because of its immense potential importance. The basis for laser-induced fusion is the reaction between a nucleus of deuterium and one of tritium in which a neutron is emitted:

$$\mathrm{D} + \mathrm{T} \quad \rightarrow \quad {}^{4}\mathrm{He}\,(3.5\ \mathrm{MeV}) + n\,(14.1\ \mathrm{MeV}),$$

(between parenthesis is indicated the kinetic energy of the products, i.e. the released heat).

It produces an energy of 3×10^{11} J per gram of matter, to be compared to the energy of 3×10^4 J produced by the combustion of 1g of carbon. In order to release this enormous quantity of energy it is necessary firstly to obtain the fuel. Deuterium is abundantly available from sea water (1g per 30 litres), whilst tritium is synthesized by bombarding lithium, also obtained from sea water, with thermal neutrons. Secondly, it is necessary to provide sufficient thermal energy to the nuclei that they can overcome their mutual electrostatic repulsion and collide. For this to be achieved a temperature of the order of 10^8 K is required. Finally, the rate of such collisions must be sufficiently great that a sustained reaction occurs. This requires that the product of the particle density N and the confinement time τ should be greater than 10^{14} particles $\text{cm}^{-3}.\text{s}^{-1}$ (the *Lawson criterion*). Two approaches to achieving these conditions are currently under investigation: magnetic confinement, for which τ is of the order of a second, and laser or heavy-ion beam inertial confinement, for which τ is of the order of a nanosecond. With the latter method it is therefore necessary for much higher densities to be achieved for the reaction to be triggered.

In the conceptually simplest configuration envisaged for laser inertial confinement, called 'direct attack', a target containing 1 mg of deuterium and tritium, of diameter 0.1 mm, is placed at the focus of several very high-power laser beams. The exterior layer of the target is ionized leading to the formation of a plasma corona which efficiently absorbs the incident radiation. The energy absorbed heats the target and vaporizes the outer layers which expand explosively. As a result of the action–reaction law, a strong radial force is directed on the remainder of the target which is compressed to a density a thousand times larger than that of a solid. A 'hot point' is created at the centre, where the fusion reaction starts, producing energetic α-particles. These α-particles ignite fusion reactions in the remainder of the sphere and result in its thermonuclear combustion. This process is termed 'inertial' because it is the mechanical inertia that maintains the density of the compressed target, before it explodes, for a confinement time $\tau \simeq 100$ ps, fulfilling the Lawson condition. During the reaction, energetic neutrons escape from the target and are absorbed by the walls of the reaction chamber, thus providing the useful energy output of the process in the form of heat. If 30% of the target burns, the energy release is roughly 100 MJ.

A 1 GW power station operating on this principle would require a laser producing pulses of roughly 2 MJ energy, of duration 10 ns and at a rate of the order of ten per second. In order for the scheme described above to be realized outside the domain of science fiction it will be necessary to resolve a number of technical difficulties. Firstly, the coupling of the radiation to the target needs to be made as efficient as possible. It turns out that most of the incident energy is absorbed by the surrounding plasma layer when it has a density slightly smaller than the critical density N_S (given in Section 3E.1.1). Thus the coupling efficiency is optimal for lasers of short wavelength, as we saw earlier in the chapter. This efficiency can reach 90% for wavelengths in the region of 300 nm. Next, the energy absorbed by the target should be employed to produce compression, rather than exciting oscillatory modes of the plasma or prematurely heating the combustible material at the centre of the target. This would increase the local pressure at the core and so oppose the inertial compression. These factors imply that the target bead must have a quite sophisticated layer structure and that the temporal form of the laser pulse must be optimized in order for efficient compression to be achieved. The laser illumination must also be highly uniform which necessitates the use of very many uniformly intense laser beams and requires that the effects of nonlinear processes, such as self-defocusing, are minimized. Another approach relaxing this constraint, called 'indirect attach' is under investigation.

Three large projects are currently under development: The National Ignition Facility in the USA, the Laser MégaJoule project in France and the FIREX project in Japan. In those facilities, the starting point is a 'small' neodymium laser, emitting well-controlled weak pulses, of nJ power and in the infra-red range (1.05 μm). The output beam is then amplified by a large factor in a series of neodymium:glass amplifiers of increasing diameter which are pumped by powerful flash-lamps to a level of several tens of kJ. Large area nonlinear crystals produce efficient frequency doubling, then frequency mixing between the resulting second harmonic pulse and the infra-red initial pulse, thus generating intense UV pulses at 330 nm.[5] Several hundreds of analogous devices produce perfectly synchronized pulses of total energy of several MJ in the UV, which are focused on a 0.6 mm diameter focal spot in a vacuum chamber of large dimensions enclosing the target and designed to collect the released fusion energy. The total system is enclosed in a 300-m-long building. These devices are intended to reach target ignition, and hopefully to deliver a fusion energy larger than the light energy that has been deposited in the target ('scientific break even'). However, their very low repetition rate (a few shots per day) is a major drawback. Their main present application is indeed military: they recreate the conditions pertaining within the core of an exploding thermonuclear bomb and can therefore provide important information regarding their characteristics without the drawbacks inherent in full-scale nuclear tests. New lasers, more efficient and likely to work at much higher repetition rates are needed in the perspective of energy production. The question of whether one day lasers will provide the answer to the world's increasing energy demands will not therefore be answered until well into the twenty-first century.

[5] See Chapter 7 on nonlinear optics.

Complement 3F The laser as source of coherent light

In the previous complement we considered only applications for which the energy delivered by a laser beam is of principal importance. Many other applications rely on the unique *coherence properties* of the light lasers emit. We discuss some of these applications in this complement.

3F.1 The advantages of laser light sources

3F.1.1 Geometrical properties

As we saw in Complement 3B, the *spatial coherence* of a laser beam ensures its directivity. In this respect it is the closest approximation that we possess to the light-ray of geometrical optics: a parallel pencil of light that can be focused to a point. We shall describe in the following paragraph how this property can be used profitably. Of course, the laws of diffraction make impossible realization of the ideal limit of geometrical optics; in practice the laser beam has a finite divergence and can only be focused to a spot of finite size. More precisely, if the laser beam has a Gaussian transverse profile (see Complement 3B), characterized by the radius of its waist w_0, its divergence is given by

$$\alpha = \lambda / \pi w_0, \tag{3F.1}$$

where λ is the laser wavelength. To give an idea of orders of magnitude, this implies that a helium–neon laser, for which $\lambda = 632$ nm and, typically, $w_0 = 2$ mm has a divergence of 4×10^{-4} rad. This value might seem small, but it leads all the same to a spot of diameter 20 cm at a distance of 300 m, which is insufficiently small for some applications. It is apparent from (3F.1) that in order to reduce the divergence a shorter wavelength laser could be employed, or a larger beam diameter, by inserting appropriate optics in the path. For example, by using a telescope with a magnification of 10 to 25, a beam that is well collimated on the scale of a few kilometres can be obtained. Beyond such distances other effects come into play related to the variation of the refractive index of the air, for example the curvature of the beam in the presence of a refractive index gradient, or scattering caused by atmospheric turbulence.

Conversely, when a laser beam, assumed parallel, is focused by a lens of focal length f, the radius of the spot (the beam waist) so produced is given by[1]

[1] This expression is valid only under appropriate circumstances. For an exact treatment of the focusing of a laser see A. E. Siegman, *Lasers*, p. 675, University Science Books (1986).

$$w_0 = \lambda f \,/\, \pi w. \tag{3F.2}$$

By using large numerical aperture lenses for which the induced aberrations have been minimized (for example, microscope objectives) and illuminating their entire aperture by a well-collimated laser beam, it is possible to obtain *a focal spot of size of the order of an optical wavelength*. This is a lower limit that can only be achieved, however, for lasers operating on the fundamental TEM_{00} transverse mode. This can be difficult to realize, particularly with pulsed laser systems.

3F.1.2 Spectral and temporal properties

In Chapter 3, we showed that the *temporal coherence* of a single-mode laser ensures its monochromaticity. A laser beam is therefore also the closest possible approximation to a classical monochromatic electromagnetic wave, of which the electric field is described by

$$\mathbf{E}(\mathbf{r}, t) = \mathbf{E}_0 \cos\left[\omega t + \phi(\mathbf{r}) + \varphi(t)\right] \tag{3F.3}$$

and in which the phase $\phi(\mathbf{r})$ characterizes the spatial dependence of the wavefront ($\phi(\mathbf{r}) = -\mathbf{k} \cdot \mathbf{r}$ for a plane wave). For an ordinary light source, the phase $\varphi(t)$ varies randomly, and can be considered to be constant only within a time τ, which characterizes the temporal coherence of the source. This time τ is of the order of the inverse of the spectral width of the light emitted. It is very short for classical sources but can be longer than a millisecond for the most monochromatic lasers. The wave-train emitted can be coherent over lengths as great as several kilometres. As a result, the characteristic phenomena of wave optics, such as interference and diffraction, are easily observed with lasers, even over large areas and for long path differences. Laser illumination is easily recognized from the *speckled pattern* produced when it is incident on any surface, which occurs as a result of the interference in the eye of waves scattered by the irregular surface.[2]

The unique coherence properties of laser light eventually led to the practical development of *holography*, which was initially conceived by Gabor in 1948.[3] A hologram permits a record to be made simultaneously of the intensity and the phase of an optical wave. This is achieved by recording on the hologram the interference figure resulting from the superposition of a wave scattered by an object and a reference wave identical to the initial object wave.

The spectral purity of laser sources has applications in spectroscopy, in the LIDAR techniques (see Section 3F.3), or in making measurements that rely on the detection of small changes in frequency (such as laser gyroscopes that will be described in the next sections).

[2] See, for example, J. Goodman, *Statistical Optics*, Wiley InterScience (1985).

[3] See, for example, R. J. Collier, C. B. Burckhardt and L. H. Lin, *Optical Holography*, New York, Academic Press (1971).

3F.1.3 The manipulation of laser beams

Because of their small diameters and good degree of collimation, laser beams are exceptionally easy to manipulate. This advantage, whilst not being of such a fundamental nature as those described above, is, however, at the root of a number of applications, for example in laser shows. These are made possible by the fact that very small, rapidly responding beam-steering devices can be employed. Of these adjustable or rotating mirrors are the simplest and slowest (these are usable for frequencies up to a few kHz). For this purpose, micro-mirrors fabricated using MEMS technology (Micro Electro Mechanical Systems) are more and more used for beam scanning and addressing. When it is desired to modulate the light at frequencies up to a few MHz, use can be made of the *acousto-optical* effect. This relies on the scattering of the light on an acoustic wave set up in a transparent crystal. At still higher frequencies (up to 60 GHz) electro-optic devices can be used. These are crystals the refractive index of which is modulated by an applied oscillating electric field. They are capable of controlling not only the direction and intensity of a laser beam, but also its phase or polarization. They are, in fact, an indispensable complement to lasers in many of the systems that we shall describe in the following.

Finally, we should point out that in the case of semiconductor lasers, it is possible to modulate the intensity of the output at extremely high frequencies (higher than 10 GHz) simply by modulating the drive current. The great convenience of this method is an additional advantage of diode lasers that has been put to good use, for example in telecommunications (see Section 3F.4).

3F.2 Laser measurement of distances

The techniques employed here are numerous and precisely which is appropriate depends on the length scale and on the degree of precision required.

Interferometric techniques offer the possibility of the precise measurement of small displacements. In Figure 3F.1, the object whose displacement is required to be measured carries a mirror which closes one of the two arms of an interferometer, for example, of the

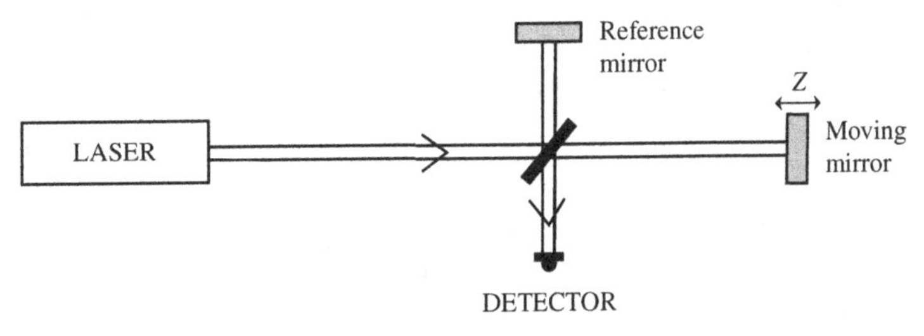

Figure 3F.1 Interferometric measurement of a distance: example of a Michelson interferometer. The detector counts the interference fringes induced by the displacement Z of the mirror.

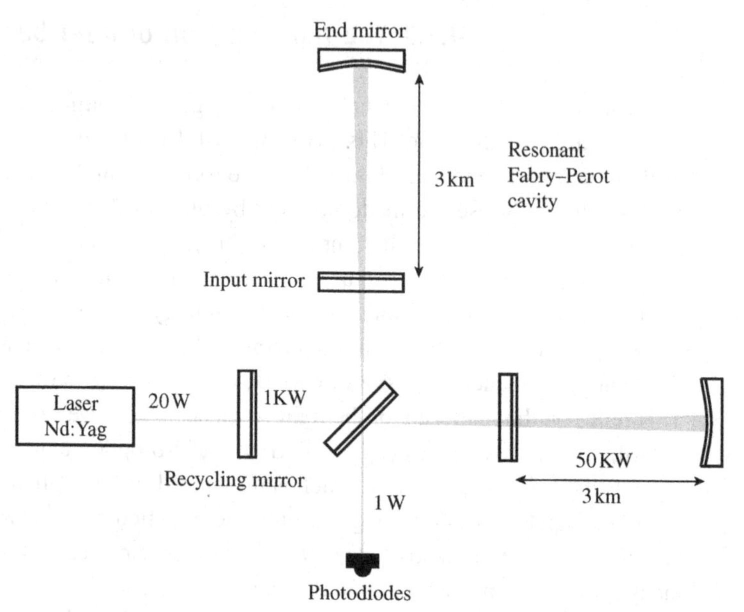

Figure 3F.2 Gravitational wave detector. The setup consists of a Michelson interferometer in which each arm is replaced by a Fabry–Perot cavity. The interferometer itself is made resonant by the use of a recycling mirror.

Michelson type. As the object moves, the order of interference at the detector changes and the detector counts fringes. Such devices, used in conjunction with machine tools enable the position of these to be set to a precision of 0.3 μm over a range of some tens of centimetres.

The technique can be improved even further by inserting two Fabry–Perot cavities in the arms of the Michelson interferometer, as shown in Figure 3F.2. Moreover, by adding a so-called recycling mirror at the input of the interferometer, the whole setup is made resonant.

Several interferometers of this type are currently under development across the world in order to detect the *gravitational waves* generated by remote supernovae which induce on their passage very weak vibration of all masses. In particular they are expected to induce a vibration of path difference in a Michelson interferometer of the order of $10^{-23}L$, where L is the length of one arm of the interferometer, at a frequency of about 1 kHz. The facilities constructed in the VIRGO (Figure 3F.3) or the LIGO projects have arm lengths of several kilometres, and there is a good hope that they will indeed be able to detect gravitational waves. The longer term LISA project of gravitational waves detection is based on a space interferometer between space craft separated by millions of kilometres.

Of course, only the remarkable temporal and spatial coherence properties of a laser make such a resonant interferometer usable. Moreover, in order to maintain the coherence properties of light over the many round trips it experiences in the interferometer, the two arms are actually made of two 3-km-long, 1.2-m-diameter tubes pumped at ultra-low

Figure 3F.3 Gravitational wave interferometer VIRGO, situated near Pisa (Italy). The arm length is 3 km. The building in the front contains the beamsplitter and the complete laser system. (http://virgo.web.lal.in2p3.fr).

pressure. Moreover, the mirrors have been specially designed to exhibit reflectivities larger than 99.999% with a nanometric surface control. Finally, all the optical components are isolated from seismic vibrations by 10-m-high systems of compound pendulums.

3F.3 Remote sensing using lasers: the LIDAR

The acronym LIDAR stands for 'Laser Detection and Ranging', in analogy with RADAR. Similarly to radar, it involves sending a light beam at remote distances (from a few tens of metres to tens of kilometres or even more) and in analysing the back-scattered light. Depending on the nature of the back-scattering object (target vehicle, aerosol cloud, water droplets, molecules from the atmosphere, etc.), the type of light pulse emitted by the laser (short pulse, long coherent pulse, two-wavelength light, etc.), and the type of detection (coherent, incoherent etc.), the general LIDAR concept can be dedicated to a large variety of measurements. In the following, we take two different examples.

3F.3.1 Atmospheric LIDAR

In this type of LIDAR, a short (typically a few ns long) light pulse (typical energy: 10 mJ) is emitted into the atmosphere. There, it encounters atmospheric particles and molecules. At every distance along the line of sight, a small fraction of the incident radiation is back-scattered. A small part of this back-scattered radiation is collected by the detection system.

By observing the detected signal versus time, one can deduce information about the structure of the atmosphere (aerosols, clouds, atmosphere layers, absorption, pollution, etc.) along the line of sight. Then by scanning the line of sight, a 3D image of the atmosphere can be constructed. Thanks to the high spatial coherence of the laser light, a space transverse resolution of the order of 1 m can be obtained, with a vertical resolution of the order of a few metres for a range typically between 10 m and 10 km. All these figures, together with the kind of measurement achieved, depend strongly on the atmospheric conditions and can be adjusted by choosing the wavelength of the laser. Extra measurements can be achieved by analysing the spectrum of the back-scattered light as, for example, in the case of Raman back-scattering of water droplets (Raman LIDAR), or by sending two close wavelengths simultaneously into the atmosphere, only one of which is absorbed by a given species (DIAL: Differential Absorption LIDAR).

3F.3.2 Coherent LIDAR

The preceding example illustrated the spatial coherence of laser light used to image atmospheric properties. In coherent LIDAR, one also uses the temporal coherence of the emitted light. Indeed, this system is based on a continuous light source with a coherence length longer than the LIDAR range. Part of the light emitted by the laser is temporally modulated (for example, cut into pulses by a modulator), amplified and emitted towards the target. The back-scattered light is then mixed with the light emitted by the laser before detection. The detector is then sensitive to the beatnote between the so-called 'local oscillator' and the back-scattered light. If the light has been back-scattered by a target moving at velocity along the line of sight (for example, a plane or a cloud), the back-scattered light is then frequency shifted by the Doppler shift $2v/\lambda$, where λ is the wavelength of the emitted light. The measurement of the beatnote frequency permits measurement of the target velocity. For example, for a plane flying at $v = 100\,\mathrm{ms}^{-1}$, the Doppler frequency shift is equal to 130 MHz for near-infra-red light ($\lambda = 1.5\,\mu\mathrm{m}$). If one wants to measure this velocity with a precision of $1\,\mathrm{ms}^{-1}$ in a single pulse, this requires a precision of the beatnote frequency measurement of about 1 MHz. This imposes a pulse duration longer than $1\,\mu\mathrm{s}$.

Such a coherent LIDAR using heterodyne detection can lead to much more complex measurements than just range and velocity measurements. For example, using quasi-random intensity and phase modulation of the input pulse, the shape, velocity and distance of the target can be determined. This is similar to the coherent treatment of radar pulses, but using the spatial directivity of spatially coherent laser light.

3F.3.3 Measurement of angular velocities

In this section we shall describe a scheme which permits the determination of angular velocity. This concerns the laser gyroscope which is able to perform measurement of rotation with respect to an inertial frame.

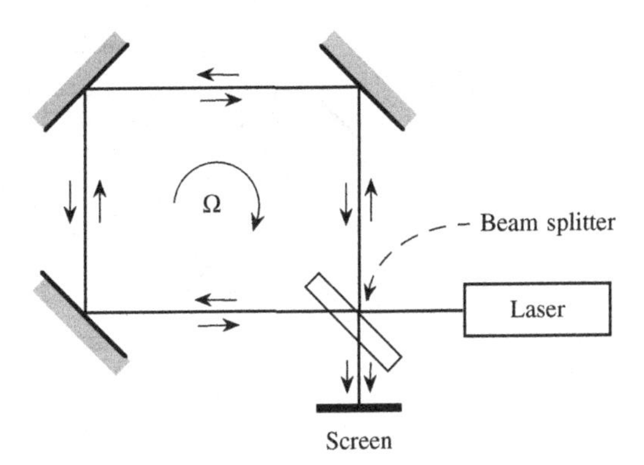

Figure 3F.4 Sagnac interferometer.

The Sagnac effect, discovered by Sagnac in 1914, enables the rotation to be measured of a frame with respect to any inertial frame by interferometry. Consider the ring interferometer of Figure 3F.4. When this apparatus rotates with respect to an inertial frame a path difference δx appears between two laser beams propagating in opposite senses in the interferometer, given by

$$\delta x = c\delta t = \frac{4A\Omega}{c}, \tag{3F.4}$$

where δt is the difference in the round-trip time for light travelling in the two possible directions and A the area enclosed by the optical path in the interferometer. The principle of this Sagnac interferometer is used in the fibre-optic gyroscope,[4] in which the effective area of the interferometer is considerably increased by winding a large number of turns of the optical fibre. Otherwise, an active system, the *gyrolaser*, can be used.

A laser gyro is a ring-cavity laser in which oscillation can occur simultaneously on the two modes, (1) and (2), propagating in opposite directions around the cavity (see Figure 3F.5). The two output beams are then made to interfere on a photodetector.[5] Given that the wavelength of the laser oscillation is an integer time c/L and that the effective length L differs for the two modes by the quantity $c\delta t$ derived earlier, their frequency difference is given by

$$\delta\omega/\omega = c\,\delta t\,/\,L, \tag{3F.5}$$

from which,

$$\frac{\delta\omega}{2\pi} = \frac{4A}{\lambda L}\Omega. \tag{3F.6}$$

The beatnote frequency $\delta\omega$ is proportional to the angular speed of rotation, and the proportionality factor depends only on the geometrical parameters of the laser: the area A and

[4] Hervé C. Lefèvre, *The Fiber-Optic Gyroscope*, Artech House (1993).

[5] For a more detailed account see, for example, W. W. Chow *et al.*, The Ring Laser Gyro, *Reviews of Modern Physics* **57**, 61 (1985).

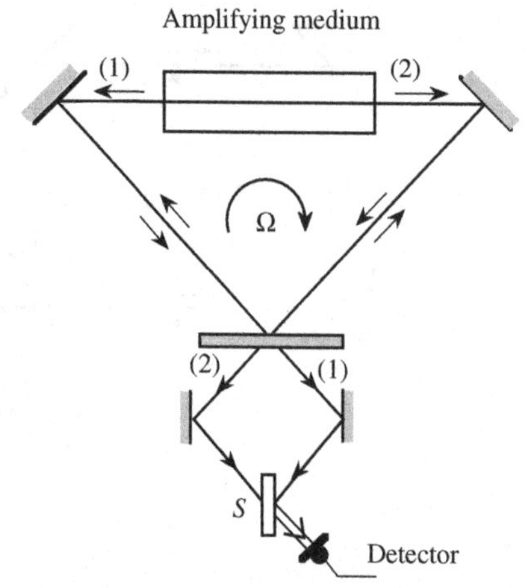

Figure 3F.5 Schematic diagram of a ring-laser gyro.

the optical path length L. The factor $4A/L\lambda$ is known as the 'scale factor'. Its order of magnitude is about L/λ. Taking a cavity length of 1 m we find that this gives a frequency difference $\delta\omega$, larger than Ω by a factor of the order of 10^6. It is the magnitude of this factor that enables very small angular velocities to be measured. Thus for a square cavity laser gyro of side 1 m and operating at a wavelength of 0.6 μm, the Earth's rotation induces a frequency difference of about 100 Hz. The measurement is achieved by a photodetector which produces a signal proportional to the time-averaged total light intensity obtained by the superposition on the beamsplitter of the two waves emitted by the laser (see Figure 3F.5). As in coherent LIDAR (Section 3F.3.2) this produces a term at the beat frequency $\delta\omega$, which can be measured electronically to good precision.

In this way an all-optical gyroscope is achieved. Nowadays such systems are manufactured on a large scale and equip aeroplanes, rockets and missiles. Compared to classical inertial systems, which employ mechanical gyroscopes, they have the advantages of containing only a few moving parts and therefore being insensitive to the effects of vibration and acceleration, of a large dynamic range and good precision. In fact this is only limited by the intrinsic fluctuations affecting the magnitude of $\delta\omega$, any other effect on the laser cavity affects the two modes in an identical manner and thus changes identically their emission frequencies ω_1 and ω_2. Ultimately, the precision of the determination of Ω is limited only by the intrinsic linewidth of the laser (Schawlow–Townes limit, see Complement 3D). Thus the resolution of such systems can be as good as 1 mHz, allowing the determination of Ω to better than 10^{-3} °/ hr. Notice that this corresponds to a relative variation of laser frequency of the order of 10^{-17}.

The operation of the laser gyro can, however, be disturbed by one phenomenon which prevents the measurement of very small angular velocities. This concerns the mutual frequency locking of the two counter-propagating modes. Because of the existence of weak back-scattering of radiation at the cavity mirrors, there is a mutual coupling between the two modes which is most effective when they are very close in frequency. This effectively

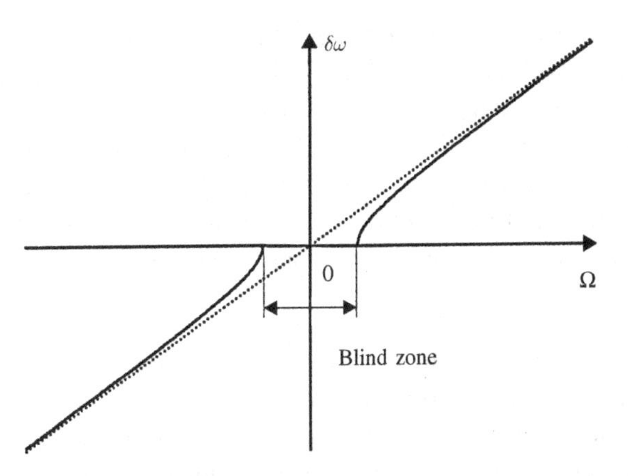

Figure 3F.6 Variation of the beat frequency $\delta\omega$ in a laser gyroscope as a function of the angular velocity of rotation for small angular velocities. Dashed line: ideal response. Full line: actual response in the presence of lock-in, showing a blind zone.

forces the two modes to oscillate *at the same frequency* (see Figure 3F.6). Thus there is a 'lock-in region', in which $\delta\omega$ is zero even for finite Ω. One solution to this problem consists of modulating Ω (by vibrating the support of the laser, for example) so that the laser passes the minimum period of time in the blind zone. This makes it possible to measure rotations of the order of 10^{-2} °/ hr, making the laser gyro suitable for application to inertial navigation.

3F.4 Optical telecommunications

Light is now the most important conveyor of information between people all over the world. This is due to the simultaneous availability of the following components:

- Semiconductor lasers, described in Section 3.2.3, which emit monochromatic light radiation of controllable wavelength and the intensity of which can be easily and rapidly modulated, for example by controlling the electric current, up to frequencies of several tens of GHz.
- Single mode fused silica optical fibres, which have ultra-low absorption coefficients (less than 0.3 dB or 5% per km around a wavelength of 1.55 μm). Moreover, the usefulness of these fibres in carrying information at a very high rate (several tens of Gbits.s^{-1}) is reinforced by the fact that the wavelength of 1.5 μm is close to the minimum dispersion of the fibre. Typical fibres have a dispersion of 15 ps.nm^{-1}.km^{-1} at 1.5 μm, allowing fast signals to propagate along long distances without experiencing strong deformations. Light from semiconductor lasers can be efficiently injected into such single-mode fibres thanks to the good spatial coherence of the laser.

- Semiconductor photodiodes which convert the incident optical signal into an electrical one, with excellent efficiency, short response time and low dark current, and which can be operated in conjunction with tunable interferometric frequency filters, able to separate reproducibly light with a wavelength difference of less than a nanometre.
- Broadband light amplifiers, and chiefly the Erbium Doped Fibre Amplifier, which can amplify, directly in the fibre and with a high gain, light pulses of different wavelengths in a bandwidth of 30 THz around 1.55 μm.

Because of the perfection and low cost of these components, optically based telecommunications systems have surpassed classical technologies based on electric signals in wires or cables or Hertzian waves. Another reason is that optical systems have by far the greatest carrier frequency, which means that more channels can be accommodated with an increased bandwidth per channel, allowing modulation formats as high as 40 Gbits.s^{-1}. They are also relatively insensitive to external perturbations and interference.

Currently existing optical communications systems rely on the coding of data into bits, which are pulses of light of different wavelengths (more than one hundred) which are simultaneously launched into the fibre (wavelength multiplexing). The maximum rate at which information can be transmitted, which is determined by the smallest acceptable time between bits, is limited by the *chromatic dispersion* of the fibre. This arises because the light pulse is not monochromatic because of the laser linewidth and of its short duration; its different spectral components propagate at different velocities, which results in a temporal broadening of the transmitted pulses. By way of example the submarine links installed in the early 2000s employ 80 diode-lasers of different wavelengths operating around 1550 nm at a rate of 10 Gbits.s^{-1} and separated by less than a nanometre. The power launched into the fibre by each diode-laser is a few mW, which is sufficient, given the low absorption experienced at that wavelength, to transmit data over a distance of 50 km between amplifiers (for classical coaxial cables the equivalent distance was only 1.5 km !). Transoceanic cables have been installed by the hundreds across the globe, with capacities reaching 1Tbits.s^{-1} per fibre, each fibre being able to simultaneously carry roughly 100 million compressed voice telephone circuits. Together with the terrestrial fibre links, they constitute a dense network which is the physical basis of the world-wide web. It permits the exchange of an amazing flow of information between almost all places on earth.

3F.5 Laser light and other information technologies

This section concerns a different kind of application of lasers in information technologies. As an example, we see how a laser beam is used to read the information stored in an optical storage device.

The simplest instance of the reading of information by lasers involves information recorded in the form of a *bar-code*. A diode-laser beam is scanned across the printed code at a lateral speed of a few ms^{-1}. The back-scattered light is then measured by a

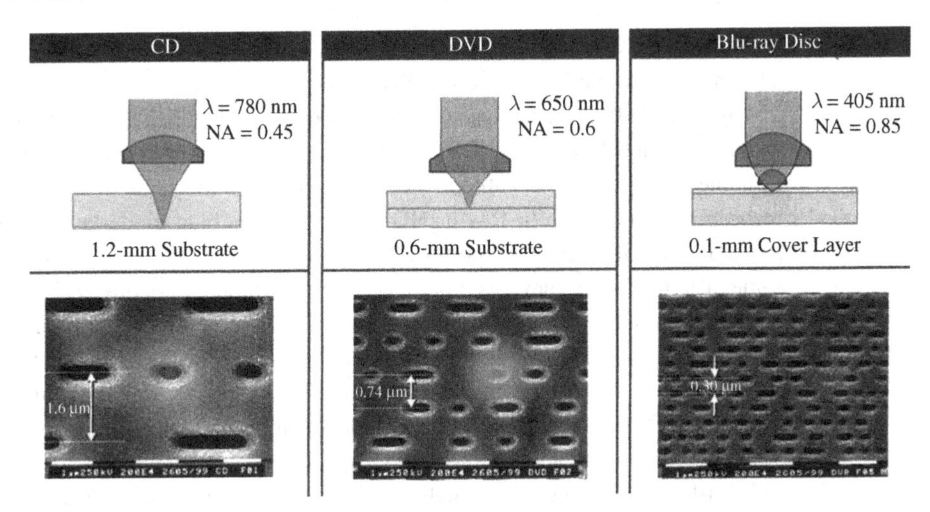

Figure 3F.7 The three steps of optical data storage (Courtesy of Philips Research).

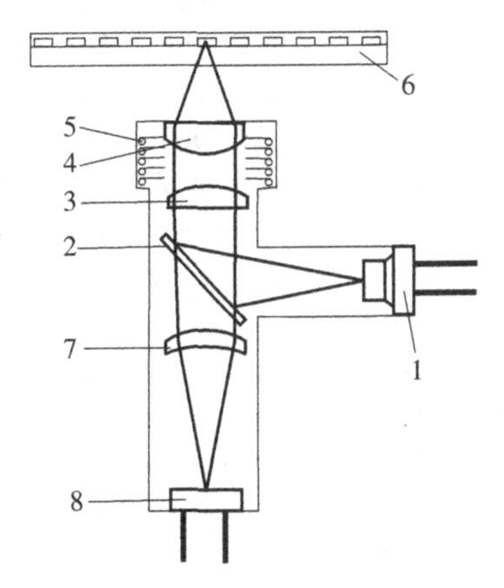

Figure 3F.8 Cross-sectional view of the optical system of a compact disc player: 1) diode-laser, 2) beamsplitter, 3) and 4) focusing lenses, 5) system for adjusting the position of the focal spot, 6) optical disc, 7) astigmatic optical system, 8) quadrant photodiode detector.

photodetector, this receiving a succession of pulses of high and low intensity that reflect the light and dark bands of the code.

Lasers are also widely employed in the reading of *optical discs*, which are capable of storing immense quantities of data in a small volume, and used for audio, video and computer data storage. As explained in Section 3F.1.1, a laser beam can be focused by a lens of high numerical aperture to a spot with a diameter of the order of the optical wavelength i.e. of the order of 1 μm. This, coupled with the fact that optical reading systems are capable of following recorded data tracks very precisely, means that data can be stored at a density of the order of one bit per wavelength squared. A further advantage is that the small inertia of the reading head ensures that optically stored data can be accessed very quickly. At the end of the first decade of the twenty-first century, several generations of optical memories were available (see Figure 3F.7). The oldest one is the Compact Disc (CD) which uses infra-red diode-lasers at a wavelength of $\lambda = 0.78\,\mu\text{m}$ and a focusing lens of numerical aperture 0.45, able to store 650 Mbytes on a single-sided 12-cm-diameter disc. Then came the Digital Video Disc (DVD) which uses red diode-lasers at $\lambda = 0.65\,\mu\text{m}$ and a lens of numerical aperture 0.6, able to store 4.7 Gbytes on the same disc consisting of several recording layers. The Blu-Ray or HD-DVD disc uses blue diode-lasers at $\lambda = 0.405\,\mu\text{m}$ and a doublet of numerical aperture 0.85, able to store between 20 and 27 GBytes on the same disk. In all these devices digital data consist of 'pits' of depth $\lambda/4$ and variable length, and the intensity of the back-reflected light is then monitored by a photodiode. The light amplitude reflected by the pits is shifted by a factor π with respect to the light reflected by the surface, so that the reflected light intensity changes suddenly when the laser spot arrives at the edge of a pit. It is this sudden change in intensity that constitutes a '1' bit, whereas a uniform scattered intensity signifies a '0'.

In order to read the data reliably, the position of the optical system with respect to the surface of the disc must remain fixed to a sub-wavelength precision. An active electromechanical actuator is used which moves the lens and locks all three coordinates of the focal point of the laser. As far as the distance to the surface of the disc is concerned, this is achieved by using a cylindrical lens which makes the optical system astigmatic. This results in the shape of the light spot on the disc depending on its distance. The scattered light is then detected by a quadrant photodiode detector, to give a signal which is sensitive to this shape. This provides the error signal which enables the distance to be locked. The radial locking to the data tracks is often achieved using two supplementary light beams. The difference in the intensity of the light back-scattered from these two beams provides a signal which enables the laser to be centred on the data track to the required degree of accuracy. As for the tangential positioning, this is achieved by adjusting the rate of rotation of the disc using synchronizing marks recorded alongside the data. It is therefore the tangential velocity of the reading head that is locked (at a speed of $1.25\ \text{m.s}^{-1}$), rather than the rotation rate of the disc.

The system described above is only capable of reading data written beforehand on the disc. In the writable systems, the laser power can be increased to a value which induces the ablation of a thin layer of material from the disc surface. The information is then stored irreversibly (CD-R discs). Other discs may be written and erased many times (CD-RW discs). These work by exploiting the *phase transition* between the crystallized and amorphous phases of the disc surface, which have different reflection coefficients. Data are written by using a high-power laser pulse which heats the surface until it locally melts. The material then cools rapidly below the crystallization temperature, preventing the formation of the crystalline phase. Information is erased by heating the amorphous regions above the crystallization temperature but below the melting point long enough to regain the crystalline state.

It seems that the 'race to short wavelengths' which has taken place over more than 30 years to improve data storage capacity is close to its end with the advent of the blue and violet diode-lasers. New progress in the development of optical memories will probably come from two techniques which are the object of intense research: 3D holographic storage in bulk materials, and the storage of more than two levels of information (0, 1) in a region of dimension λ^2.

3G Complement 3G Nonlinear spectroscopy

The advent of lasers has revolutionized the methods and possibilities of spectroscopy. Their monochromaticity permits the study of extremely narrow spectral features which would not have been resolved using classical techniques. Moreover, their extreme spectral power density, in giving rise to the field of nonlinear optics, which will be studied in Chapter 7, has resulted in the development of new and powerful spectroscopic techniques.[1] These techniques, of which we describe briefly some examples in this complement, have as their basis the fact that an atom does not respond linearly when subjected to a very intense light field.

First of all we shall describe the broadening mechanisms that explain the fact that the experimentally measured widths of atomic or molecular resonances are larger than those that would be observed if the atoms were isolated and at rest. We shall introduce in Section 3G.1 the concept of *homogeneous* width, which can be considered as the intrinsic width of a transition, and *inhomogeneous width* which concerns the variation of the frequency of a given transition that occurs because of the state of motion of the atom, or because of factors in its environment. We shall explain why it is possible, using nonlinear optical techniques to eliminate the effects of inhomogeneous broadening on recorded spectra, resulting in an improvement over the resolution of conventional techniques. We shall then concentrate more specifically on *Doppler broadening*, which is the most commonly encountered source of inhomogeneous broadening, describing two nonlinear optical techniques, *saturated absorption spectroscopy* (in Section 3G.2) and *two-photon spectroscopy* (in Section 3G.3), which enable it to be overcome. To conclude, we shall discuss in Section 3G.4 the progress that has been made, thanks to nonlinear spectroscopy, in the study of the spectrum of atomic hydrogen, the most fundamental atomic system.

3G.1 Homogeneous and inhomogeneous broadening

A resonance transition of a stationary, isolated atom has a Lorentzian frequency profile of which the width, γ, is determined by the lifetime of the atomic dipole involved (see Equation 2C.44). However, the spectral lines measured in the laboratory very often have much larger frequency widths. This can arise if the atoms under study are in the gaseous phase, so that their velocities are distributed over a wide range: if an atom has velocity $\mathbf{v}$,

[1] Laser sources producing ultra-short pulses, which therefore have a large frequency width, can be employed to study the dynamics of atomic or molecular systems in the time domain: the system is excited by one pulse, its evolution being probed subsequently with a second. For the sake of brevity, we do not describe these time-resolved spectroscopic techniques here; we only discuss techniques for spectroscopy in the frequency domain.

the frequency of the light emitted in the direction **u** is displaced from that of light emitted by a stationary atom by an amount $\omega_0 \frac{\mathbf{v.u}}{c}$ where c is the speed of light (this is the Doppler effect). The spectral lineshape of the emitted light as recorded by a stationary observer will then be a Gaussian of width (the Doppler width) much larger than γ, reflecting the Maxwellian distribution of the velocities of the particles of a gas in thermal equilibrium. In the visible spectral region the Doppler width has a magnitude of the order of $\omega_0\overline{v}/c$, where $\overline{v}$ is the root-mean-square speed of the atoms of the gas, $\overline{v} = \sqrt{k_B T/M}$. At room temperature this Doppler width is typically a hundred times the natural linewidth. It follows that the thermal motion of the atoms of a gas limits the precision of the spectroscopic measurements that can be made in studying, for example, the spectral distribution of the light emitted or absorbed by the gas; any features which occur on a frequency scale that is small compared to the Doppler width will be swamped by it and will not be resolved.

A similar kind of broadening is associated with the spectra of ions implanted in a glassy or crystalline medium. There is no Doppler width as such, since the ions are held immobile.[2] However, because of their random spatial distribution in the host matrix, the energy levels of each ion experience different Stark shifts arising from the electric field of the surrounding ions. The transition frequencies for the ions are therefore distributed over a range of values determined by the spread of possible environments and this is reflected in the shape of the resonance line registered for the sample as a whole.

The *homogeneous* width of a transition is the width that would be measured when any atom of a sample is isolated and probed independently. The most obvious source of homogeneous broadening is the natural width related to the finiteness of the lifetimes of the atomic levels involved in the transition; it is this that defines the fundamental limit of spectroscopic resolution and it is only realized in practice if all extraneous perturbations are eliminated. The natural linewidth of a transition can be calculated on the basis of a fully quantum mechanical theory of the matter–light interaction (see Chapter 6). Other sources of homogeneous broadening do exist, for example that arising from the collisions of the atoms of a gas. Like homogeneous broadening caused by spontaneous decay, collisional broadening in a gas gives rise to a Lorentzian spectral profile.

3G.2 Saturated absorption spectroscopy

Very soon after the advent of the laser the first sub-Doppler spectroscopic technique (allowing the resolution of features on a frequency scale that is small compared to the Doppler width) was demonstrated. This method, known as *saturated absorption spectroscopy*

[2] In fact, the atoms are not at rest in the solid, but vibrate, because of their thermal motion. When the amplitude of this motion is small compared to the wavelength, it does not induce Doppler broadening (Lamb–Dicke effect).

revolutionized the field of spectroscopy and is still widely employed.[3] We shall describe it in detail in the following paragraph.

3G.2.1 Holes in a population distribution

Consider the example of an ensemble of ions trapped in a crystalline host material. Defects in the crystalline structure lead to spatial variations of the crystal field so that the Stark shifts of energy levels and hence transition frequencies differ from ion to ion. The difference in the transition frequency for an ion at a site labelled i and the transition frequency for an isolated ion can be written in the form

$$\hbar\omega_i = \hbar\omega_0 + \hbar\delta\omega_i, \tag{3G.1}$$

where $\hbar\omega_0$ is the transition energy for an isolated ion and $\hbar\delta\omega_i$ is the energy difference for the ion at site i. Suppose that the ions are subjected to an incident laser field of frequency ω' and amplitude E' and that the absorption of this beam is measured. Obviously, this absorption will be due solely to those ions that occupy lattice sites such that

$$\hbar\omega_i = \hbar\omega'. \tag{3G.2}$$

By tuning the laser frequency, the range of transition frequencies ω_i can be mapped. The form of the absorption lineshape will therefore provide a measure of the distribution of the atoms over the sites i (this constituting a source of inhomogeneous broadening; see Figure 3G.1(a)).

However, a second laser beam of angular frequency ω and a large amplitude E can be used to *saturate* the transition. This laser has the effect of equalizing the populations (see Chapter 2, expression (2.159)) in the ground and excited states of the class of ions at sites j for which

$$\hbar\omega_j = \hbar\omega. \tag{3G.3}$$

When the frequency ω' of the probe laser beam is tuned, the absorption it experiences is identical to that observed in the absence of the strong field E, except for frequencies in the vicinity of ω. When $\omega' \approx \omega$ the two laser beams interact with the same class of atoms and the probe beam then experiences a smaller absorption because the number of ions of class j in the ground state has been reduced by the strong laser field. The form of the spectral dependence of the probe absorption is then as shown in Figure 3G.1(b). The width of the narrow component of the absorption spectrum is of the order of $2\sqrt{\gamma^2 + \Omega_1^2\gamma/\Gamma_{sp}}$, where 2γ is the (homogeneous) natural width of the transition, Γ_{sp} the reciprocal of the natural lifetime of the excited state and $\Omega_1 = -dE/\hbar$ the Rabi frequency for the strong wave E. In the case of a closed two-level system, that width is $\sqrt{\Gamma_{sp}^2 + 2\Omega_1^2}$ (see Equation (2.188)).

[3] The phenomenon of saturated absorption is at the root of the observation in the gain curve of a gas laser of a narrow dip in the gain (the 'Lamb dip') situated at the resonance frequency. The technique of saturated absorption spectroscopy was perfected by C. Bordé and T. Hänsch and has been considerably refined. For more details see the article by T. Hänsch, A. Schawlow and G. Series in *Scientific American* **240**, p. 72 (March 1979) or *Nonlinear Laser Spectroscopy* by V. Letokhov and V. Chebotayev, p. 72, Springer (1977).

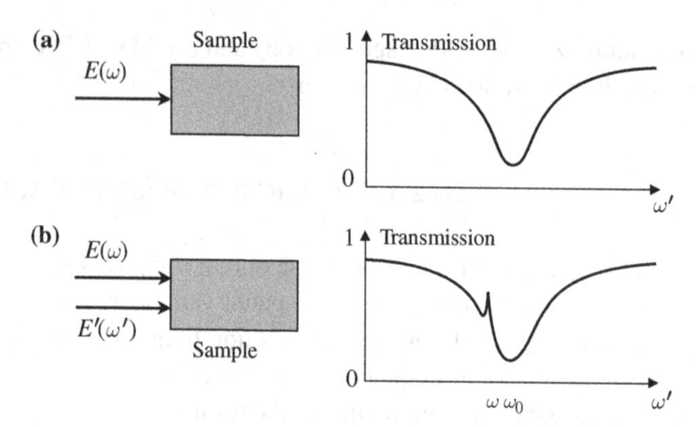

Figure 3G.1 **Absorption lineshape for a probe beam as a function of its frequency ω'. If the probe beam alone is incident on the medium (a), the absorption profile yields information on the inhomogeneous linewidth of the transition. The same profile recorded in the presence of a strong wave of frequency ω exhibits an increase in the probe transmission when $\omega' = \omega$, because of the saturation of the transition by the strong wave (b). The width of the narrow peak is related to the homogeneous width of the transition.**

The increase in the probe transmission when its frequency equals that of the saturating wave is clearly visible in Figure 3G.1(b). As the intensity of the saturating wave is decreased the width of this peak becomes smaller, tending, in the low-intensity limit, to the natural linewidth of the transition. Thus this technique presents a significantly enhanced resolution compared to a simple absorption measurement. It does, however, suffer from the demerit that the position of the narrow peak is determined by the frequency of the saturating light, and is not related to a fundamental property of the species being probed. We shall see how this disadvantage can be overcome in the following section, in which we consider how the inhomogeneous broadening that arises from the Doppler effect can be eliminated.

3G.2.2 Saturated absorption in a gas

The Maxwellian velocity distribution

Consider an ensemble of two-level atoms (with ground and excited levels a and b and with $E_b - E_a = \hbar\omega_0$). When the atoms are in thermal equilibrium, by far the majority are in the ground internal state. The atoms have a distribution of velocities, the probability that the x-component of the velocity of a given atom lies in the range v_x to $v_x + dv_x$ being given by the Maxwell–Boltzmann distribution:

$$f(v_x) = \frac{1}{\bar{v}\sqrt{2\pi}} \exp\left(-\frac{v_x^2}{2\bar{v}^2}\right), \tag{3G.4}$$

where $\bar{v} = \sqrt{\frac{2k_B T}{M}}$ (the one-dimensional root-mean-square speed).

Excitation of a single velocity class

Suppose that these atoms interact with a monochromatic incident light wave at frequency ω close to that of the atomic resonance:

$$\mathbf{E}(\mathbf{r}, t) = \mathbf{E}\cos(\omega t - \mathbf{k}.\mathbf{r}). \tag{3G.5}$$

An atom of velocity $\mathbf{v}$ traces out a trajectory

$$\mathbf{r} = \mathbf{r}_0 + \mathbf{v}t \tag{3G.6}$$

and experiences, in its own rest frame, an electric field,

$$\mathbf{E}\cos\left[(\omega - \mathbf{k}.\mathbf{v})t - \mathbf{k}.\mathbf{r}_0\right], \tag{3G.7}$$

of which the frequency is displaced from that of the incident light by the *Doppler effect*. For a fixed incident frequency ω, only atoms with velocities $\mathbf{v}$ such that

$$\hbar(\omega - \mathbf{k}.\mathbf{v}) = \hbar\omega_0 \tag{3G.8}$$

are resonant with the incident light and can be excited. If the light propagates in the x-direction, the excited atoms will all have velocities with an x-component v_x satisfying

$$\frac{v_x}{c} = \frac{\omega - \omega_0}{\omega} \tag{3G.9}$$

(since for a dilute medium $k \approx \omega/c$). In the presence of the wave $E(\mathbf{r}, t)$, a hole appears in the ground-state atoms' velocity distribution at the position of the velocity class for which v_x satisfies (3G.9). The atoms transferred to the excited state therefore have a well-defined value of v_x (see Figure 3G.2). For two-level atoms, the velocity distribution of the excited-state atoms is just the complement of that of the ground-state atoms.

Comment The complementarity of the two velocity distributions in Figure 3G.2 is no longer assured when the atoms undergo collisions. These change the velocities of the interacting particles and so tend to fill in the hole in the ground-state velocity distribution and broaden and flatten the peak in the excited-state distribution. The inequality of the collision cross-sections of atoms in the ground and in the excited

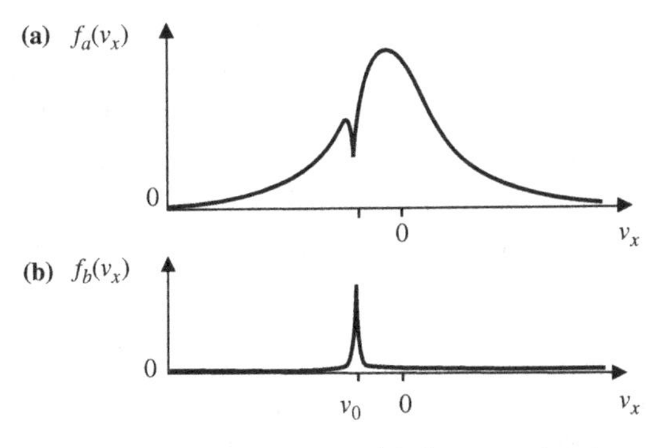

Figure 3G.2 The distribution of atomic velocities for atoms in (a) the ground state and (b) the excited internal state, when the atoms are subjected to a laser field at angular frequency ω. The velocity v_0 is as given by (3G.8).

states then results in collisionally induced modifications of the two distributions which are different. Suppose, for example, that the collision cross-section is much larger in the excited state than in the ground state. The hole in the ground-state velocity distribution $f_a(v_x)$ will then persist much longer than the peak in the excited-state distribution $f_b(v_x)$, which will rapidly evolve into a Gaussian peak centred on $v_x = 0$. As a consequence of the non-complementarity of the two velocity distributions, the body of the gas *acquires a non-zero net velocity along the x-direction.* In the example discussed above this *light-induced drift* velocity is in the opposite direction to v_0. This effect has been observed in several laboratories.[4] It should be noted that it is not a mechanical effect arising from the transfer of momentum to an atom on the absorption of a photon (radiation pressure). In fact, the light-induced drift is in the opposite direction to the propagation of the light causing it if the laser is tuned to the high-frequency side of the resonance.

Saturated absorption spectroscopy

In saturated absorption spectroscopy a laser beam of frequency ω is divided into two parts at a beamsplitter, which are then passed through the gaseous medium in directions that are very nearly opposite. The finite angular separation of the two beam directions is, in practice, just sufficient to allow the transmitted intensity of the weaker of the two beams to be recorded by a detector.

The strong beam B interacts with the atoms of velocity classes that satisfy (3G.9). Beam B', which nearly counterpropagates with B, of which the phase therefore varies as $(\omega t + \mathbf{k}.\mathbf{r})$, excites atoms of which the velocities satisfy the relation

$$\hbar(\omega + \mathbf{k}.\mathbf{v}') = \hbar\omega_0, \tag{3G.10}$$

which gives

$$\frac{v'_x}{c} = -\frac{\omega - \omega_0}{\omega}c. \tag{3G.11}$$

A comparison of expressions (3G.11) and (3G.9) shows that if $\omega \neq \omega_0$ the waves B and B' interact with *different velocity classes* and therefore experience an absorption independent of the presence of the other beam. If $\omega = \omega_0$, however, both waves interact with the atoms having $v_x = 0$ (see Figure 3G.4). It follows that if the strong wave B saturates the atomic transition, the transmission of wave B' will be increased for $\omega \approx \omega_0$.

The absorption lineshape of the probe wave, of which the long-range form is determined by the Doppler profile of the transition therefore exhibits a narrow inverted peak around $\omega = \omega_0$. The width of this peak is of the order of $\sqrt{\Gamma_{\text{sp}}^2 + 2\Omega_1^2}$ where Ω_1 is the Rabi frequency associated with the saturating wave B. As the intensity of the waves is reduced this width tends to its limiting value of the natural linewidth.[5]

[4] See, for example, G. Nienhuis, Impressed by Light: Mechanical Action of Radiation on Atomic Motion, *Physics Reports* **138**, 151 (1986).

[5] When the saturating wave is very intense the populations of levels a and b are equalized. One might expect therefore that the absorption of the probe wave would be practically zero. In fact this is not the case. A more careful analysis, taking into account the shifts in the energy levels induced by the high-intensity field, shows that the absorption, though greatly decreased, does not tend to zero. For more details see CDG II, exercise 20.

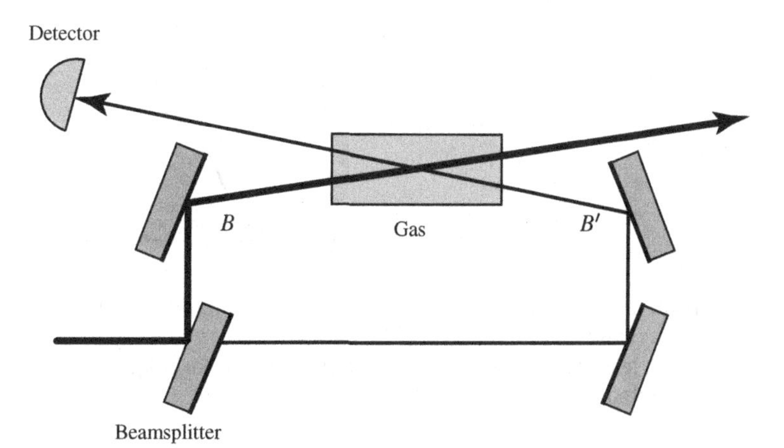

Figure 3G.3 Schematic representation of a saturated absorption experiment. The incident beam is divided into two at a beamsplitter. The transmitted intensity of the weaker of the two resulting beams B' is then recorded as a function of its frequency. The angle between the two beams is, in practice, much smaller than is shown in the figure.

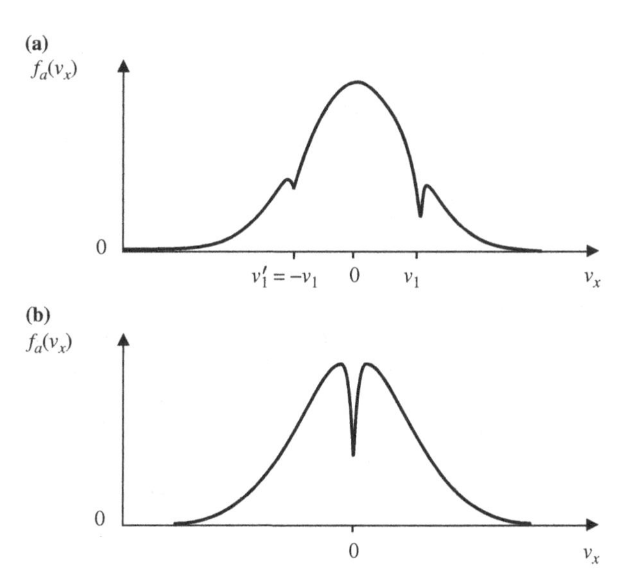

Figure 3G.4 Velocity distribution of the ground-state atoms. (a) When $\omega \neq \omega_0$ the beams B and B' excite distinct groups of atoms at velocities symmetrically disposed about $v_x = 0$. (b) When $\omega = \omega_0$ both beams interact with the same velocity class: $v_x = 0$.

The facts that the narrow feature in the absorption spectrum has a width approaching the limit of the natural linewidth and is centred at the atomic resonance frequency endow this technique with considerable advantages as far as spectroscopy is concerned. Notably, it is often used to provide an absolute frequency reference to which a laser frequency can be locked.

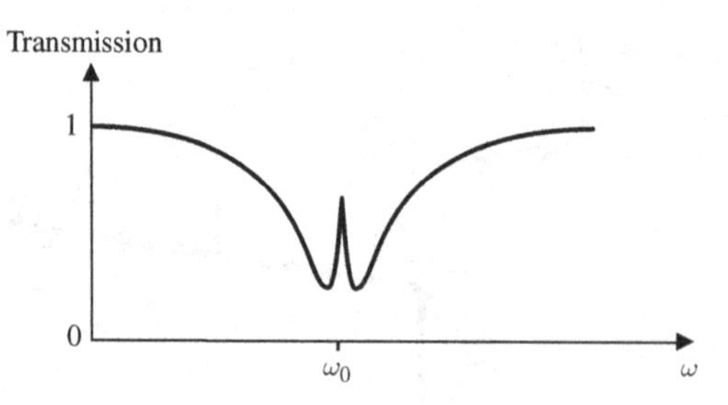

Figure 3G.5 **Form of the saturated absorption signal for an atomic vapour: the transmission of the weak probe wave B' as a function of its frequency ω. (In real experiments, the narrow feature is much less pronounced so that noise suppression techniques are often used to enhance the saturated absorption signal.)**

Comment When the atom has several sub-levels, either in a or in b, other narrow, resonant signals, *cross-over resonances* appear. Consider, for example, the case of an atom with two sub-levels a and a', with an energy spacing less than the Doppler width, in the ground state, and a single excited state b. If the saturating beam is resonant on the transition $a \rightarrow b$, it will strongly modify the population of level b, which will alter the absorption experienced by the weak beam when it is tuned to either of the transitions $a \rightarrow b$ and $a' \rightarrow b$. In the configuration of Figure 3G.3 where the strong and the weak beam have the same frequency but opposite directions, the absorption will undergo a resonant change when the same velocity class interacts with both counter-propagating beams on either of the two transitions. It is easy to show that such resonances will occur when the frequency of the optical field ω is given by either of the three frequencies,

$$\omega = \omega_{a\rightarrow b}, \ \omega = (\omega_{a\rightarrow b} + \omega_{a'\rightarrow b})/2, \ \omega = \omega_{a'\rightarrow b}.$$

The resonance at $\frac{1}{2}(\omega_{a\rightarrow b} + \omega_{a'\rightarrow b})$ is a cross-over resonance.

In most atoms and molecules the energy levels have a hyperfine structure composed of several sub-levels. The saturated absorption signal obtained in the configuration of Figure 3G.3 is then complicated. However this complexity turns out to be very useful, as it provides a grid of absolute frequencies, which can be determined to a precision limited only by the natural linewidth, to which a laser frequency can be locked. Notice finally that it is possible for an interaction between the two optical beams to arise from a mechanism other than saturation. For example, under suitable conditions, this can result from optical pumping between the ground-state sub-levels a and a' (See Complement 2B).

3G.3 Doppler-free two-photon spectroscopy

3G.3.1 Two-photon transitions

We saw in Chapter 2 Section 2.3.3 that an atom interacting with two electromagnetic waves of angular frequencies ω_1 and ω_2 can be transferred to an excited state by the absorption of a photon from each beam (see Figure 3G.6). The resonance condition for this process is

$$E_b - E_a = \hbar(\omega_1 + \omega_2). \tag{3G.12}$$

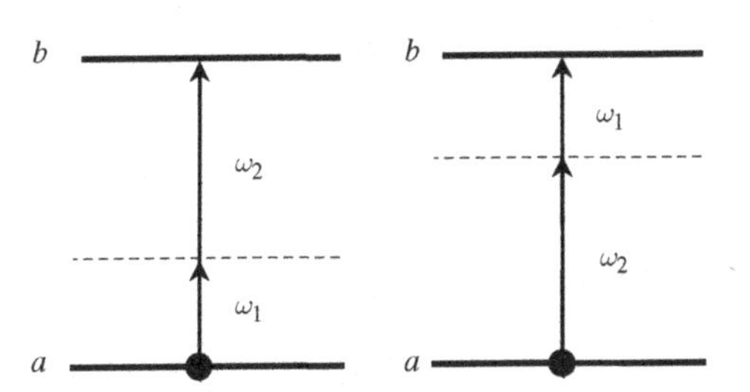

Figure 3G.6 Two-photon transition between levels a and b. The transition amplitude is the sum of two terms, one corresponding to the absorption of a photon at frequency ω_1, followed by the absorption of one at ω_2 and the second corresponding to the inverse temporal order. The resonance condition (3G.12) is the same for the two terms.

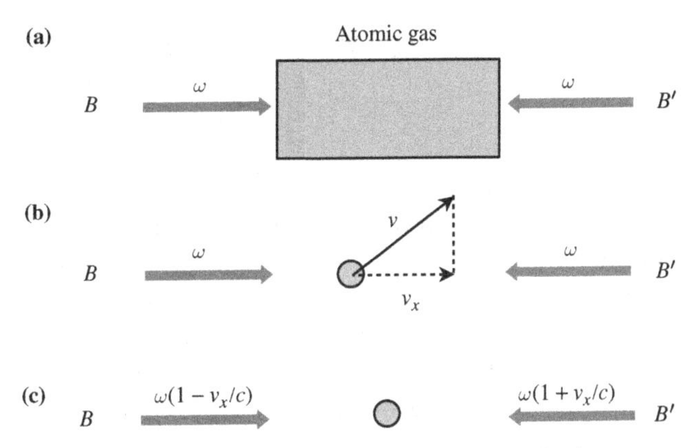

Figure 3G.7 Doppler-free two-photon spectroscopy scheme. (a) Experimental scheme: an atomic vapour interacts with two counterpropagating beams of the same frequency. (b) In the laboratory frame, an atom of speed v interacts with two counterpropagating waves of the same frequency. (c) In the atom's rest frame it interacts with two waves of which the frequencies are symmetrically shifted by the Doppler effect.

3G.3.2 Elimination of Doppler broadening

Consider atoms in a gas interacting with two counterpropagating travelling waves B and B' of the same angular frequency ω (see Figure 3G.7(a)). If in the laboratory frame (Figure 3G.7(b)) an atom of speed $\mathbf{v}$ interacts with two waves of frequency ω, in its own rest frame (Figure 3G.7(c)) it experiences two waves of frequencies ω_1 and ω_2 symmetrically shifted from ω by the Doppler effect. Thus for $|\mathbf{v}| \ll c$,

$$\omega_1 = \omega\left(1 - \frac{v_x}{c}\right), \tag{3G.13}$$

$$\omega_2 = \omega\left(1 + \frac{v_x}{c}\right). \tag{3G.14}$$

Consider the case in which the atom absorbs one photon from each of the counterpropagating beams. Using expressions (3G.13) and (3G.14) the resonance condition can be cast in the form

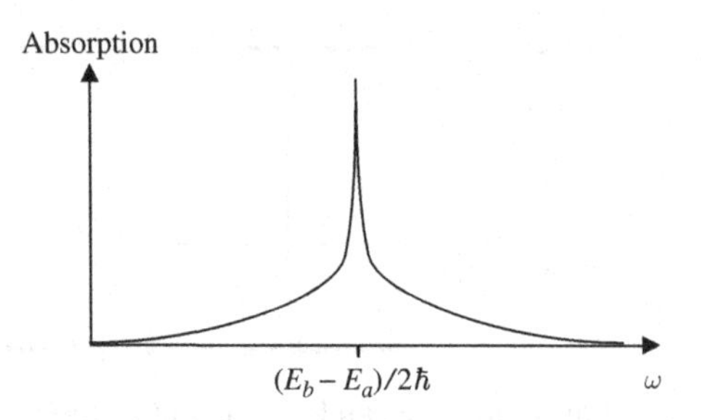

Figure 3G.8 **Two-photon absorption on the 5s to 5d transition of Rb, measured in a vapour cell (courtesy Vincent Jacques, ENS Cachan).**

$$E_b - E_a = \hbar\omega\left(1 - \frac{v_x}{c}\right) + \hbar\omega\left(1 + \frac{v_x}{c}\right), \tag{3G.15}$$

or

$$E_b - E_a = 2\hbar\omega. \tag{3G.16}$$

Notice that *the velocity-dependent terms in the resonance condition cancel.* Therefore atoms of *any* velocity $\mathbf{v}$ will be transferred to the excited state when condition (3G.16) is satisfied. It should be realized that this elimination of the Doppler broadening of the transition is only achieved when the excitation is by two counterpropagating waves of the same frequency. If the two-photon transition is driven by a double excitation by a single travelling wave of frequency ω, the resonance condition is

$$E_b - E_a = 2\hbar\omega_1 = 2\hbar\omega\left(1 - \frac{v_x}{c}\right), \tag{3G.17}$$

which therefore depends on the atomic velocity. In this case, for a given value of ω a single velocity class, that for which (3G.17) is satisfied, is excited. Thus the lineshape of the resonance is Doppler broadened.

Since in general the atoms undergo with equal probabilities two-photon transitions arising from the absorption of two photons from a single incident beam and from the absorption of one photon from each of the counterpropagating beams, the absorption spectrum as a function of ω is the superposition of two resonances of roughly equal areas (if the intensities of B and B' are approximately equal). Thus the narrow Doppler-free resonance due to the absorption of a photon from each of the counterpropagating beams is superimposed on a Doppler broadened background resonance arising from transitions in which both photons come from one or other of the beams.

Comment The elimination of the Doppler broadening in two-photon transitions can be understood in terms of the simultaneous conservation of energy and momentum. Consider an atom of mass M and velocity $\mathbf{v}$ in the laboratory frame. The energies and momenta of the photons in the two incident beams in this frame are $\hbar\omega$ and $\hbar\mathbf{k}$ for one beam and $\hbar\omega$ and $-\hbar\mathbf{k}$ for the other. The conservation of momentum in the two-photon absorption process implies that

$$M\mathbf{v}' = M\mathbf{v} + \hbar\mathbf{k} - \hbar\mathbf{k}, \tag{3G.18}$$

where $\mathbf{v}'$ is the velocity of the atom at the end of the process. Thus the velocity is unchanged as a result of the two-photon absorption. The conservation of energy can be expressed by

$$E_b + \frac{1}{2}Mv'^2 = E_a + \frac{1}{2}Mv^2 + 2\hbar\omega. \tag{3G.19}$$

Since $v' = v$ we recover the resonance condition (3G.16). Notice that the above expressions imply that the atom does not recoil when it absorbs two photons of the same frequency from each of two beams propagating in opposite directions. This treatment can straightforwardly be generalized to higher-order multi-photon transitions. Consider, by way of example, the absorption of three photons from beams of wavevectors $\mathbf{k}_1, \mathbf{k}_2$ and $\mathbf{k}_3$ of frequencies ω_1, ω_2 and ω_3 (in the laboratory frame). If the beams are directed such that the sum of the momenta of the absorbed photons is zero:

$$\sum_{i=1}^{3} \hbar\mathbf{k}_i = 0, \tag{3G.20}$$

the momentum of the atom, and hence its kinetic energy, will be unchanged as a result of the absorption process. It follows that the resonance condition will be identical for all atoms irrespective of their velocities:

$$E_b - E_a = \sum_{i=1}^{3} \hbar\omega_i. \tag{3G.21}$$

The absorption profile obtained by tuning the frequency of one of the beams will then exhibit a resonance free of Doppler broadening.

3G.3.3 Properties of Doppler-free two-photon spectroscopy

Two-photon spectroscopy permits the study of features of atomic and molecular spectra with a precision limited only by the natural linewidth of the transitions concerned. It is a powerful tool, which in some respects is complementary to that provided by saturated absorption spectroscopy, since it permits transitions between levels of the same parity to be studied whilst the latter can only be used to study transitions linking the ground state with levels of opposite parity (see Complement 2B). It might be thought that the excitation of a two-photon transition might require the use of intense laser beams which could lead to appreciable light-shifts (see Chapter 2, Section 2.3.3c) of the atomic levels involved, which would be a significant disadvantage for the method as a tool in high-resolution spectroscopy. In practice, however, this is not usually the case. The fact that all the atoms contribute to the resonance, as opposed to the single velocity class $v_x = 0$ that does so in saturated absorption spectroscopy, at least partly compensates for the smaller transition amplitude for the two-photon process compared to one involving a single photon.[6]

[6] For more details on multi-photon transitions and on the elimination of Doppler broadening in these processes see the chapter 'Multi-Photon Resonant Processes in Atoms' by G. Grynberg, B. Cagnac and F. Biraben, in *Coherent Nonlinear Optics: Recent Advances*, edited by M. S. Feld and V. S. Letokhov, Springer-Verlag (1980).

3G.4 The spectroscopy of the hydrogen atom

3G.4.1 A short history of hydrogen atom spectroscopy

The hydrogen atom is the simplest atomic system, being composed of a single electron and a single proton. This simplicity makes the hydrogen atom an ideal testing ground for theory and much effort has therefore been directed at its experimental study. Thus, it was at the close of the nineteenth century, that the matching of the observed spectrum of hydrogen with the formula of Balmer was a catalyst to the emergence of quantum theory. The earliest and most striking successes, firstly of Bohr's theory, and then of Schrödinger's equation were in putting Balmer's formula, which was inexplicable in terms of classical mechanics, on a sound theoretical basis. The improvements in spectroscopic techniques that were achieved in the early years of the twentieth century quickly led to the realization that the hydrogenic spectral lines possessed a *fine structure*, the interpretation of which eventually led to the introduction of the concept of the intrinsic spin of an electron. This was first included phenomenologically in the description of the hydrogen atom by Uhlenbeck and Goudsmit and later shown by Dirac to arise naturally in a *relativistic* quantum mechanical description. Later, at the close of the 1940s, Lamb and Retherford showed, using radio-frequency spectroscopy that the $2S_{1/2}$ and $2P_{1/2}$ energy levels were not, as predicted by Dirac's theory, degenerate in energy, and therefore that this theory did not provide a complete description of the hydrogen atom. The interpretation of this small difference in energy (less than 10^{-6} of the energy of the states) occupied some of the greatest physicists of the mid twentieth century (Bethe, Dyson, Feynmann, Schwinger, Tomonaga, Weisskopf etc.) and finally led to the development of one of the most sophisticated and certainly the most precise physical theory: *quantum electrodynamics*.

This brief summary is intended to show how studies on the hydrogen atom have played a leading role in the development of physics in the twentieth century, and how successive improvements in the precision of the spectroscopic tools employed have proven the inadequacy of existing theories and have permitted important theoretical advances to be made as a consequence. In this respect the measurements made by nonlinear spectroscopy since 1970 have been of singular importance; they have enabled the precision with which the optical spectrum of the hydrogen atom is known to be increased by five orders of magnitude.[7] More precise values for the transition frequencies between hydrogen levels, obtained by Doppler-free two-photon spectroscopy, were, in 2005:

$$\nu_{2S_{1/2}-12D_{5/2}} = 799\ 191\ 727\ 402.8\ (\pm 6.7)\ \text{kHz}$$

$$\nu_{1S_{1/2}-2S_{1/2}} = 2\ 466\ 061\ 413\ 187.103(\pm 0.046)\ \text{kHz}$$

[7] A very complete account of the current state of knowledge of the hydrogen atom, both theoretical and experimental, is given in *The Spectrum of Atomic Hydrogen: Advances*, edited by G. W. Series, World Scientific, Singapore (1988), and in B. Cagnac *et al.*, *Reports on Progress in Physics* **57**, 853 (1994). See also the Nobel lecture by T. W. Hänsch, *Reviews of Modern Physics* **78**, 1297 (2006).

The latter measurement, with a relative uncertainty of 2×10^{-14}, is the most precise ever performed in physics! It uses a revolutionary technique to measure optical frequencies, invented by T. Haensch and J. Hall:[8] the 'frequency comb' generated by a mode-locked laser having a free spectral range stabilized on the caesium atomic clock frequency is used as the ticks on a ruler to measure with unprecedented accuracy the optical frequencies of the lasers used for the Doppler-free two-photon spectroscopy.

3G.4.2 The hydrogen atom spectrum

From Balmer to Dirac

Non-relativistic quantum theory shows that the energy levels of an electron in a Coulomb potential can be described by three quantum numbers n, l and m and that the energies of the allowed levels depend on n only:[9]

$$E_{n\ell m} = -\frac{Ry}{n^2}, \tag{3G.22}$$

where Ry is the Rydberg constant which is a function of the mass m and charge q of the electron:[10]

$$Ry = \frac{mq^4}{32\pi^2\varepsilon_0^2\hbar^2}. \tag{3G.23}$$

The principal result of Dirac's relativistic quantum theory is the relativistic wave equation.[11] This equation leads naturally to the introduction of the electron spin s, and to energy levels characterized by the total angular momentum quantum number j ($\mathbf{j} = \mathbf{l} + \mathbf{s}$):

$$E_{n\ell jm} = -\frac{R_y}{n^2}\left[1 + \frac{\alpha^2}{n}\left(\frac{1}{j+1/2} - \frac{3}{4n}\right)\right]. \tag{3G.24}$$

One of the remarkable predictions of the Dirac equation is that, within a multiplet of states with fixed n, sub-levels with the same value of j but differing values of l will have the same energy. Thus, of the states with $n = 2$, the level $2S_{1/2}(l = 0, j = 1/2)$ and the level $2P_{1/2}(l = 1, j = 1/2)$ will be degenerate (see Figure 3G.9(a)).

[8] J. L. Hall, Nobel Lecture, Defining and Measuring Optical Frequencies, *Reviews of Modern Physics* **78**, 1279 (2006). T. W. Hänsch, Nobel Lecture, Passion for Precision; *Reviews of Modern Physics* **78**, 1297 (2006). Available on-line at: http://nobelprize.org/nobel-prizes/physics/laureates/2005.

[9] Note that the degeneracy in l is unique to the Coulomb potential.

[10] The Rydberg constant Ry corresponds strictly speaking to the case of an infinitely heavy nucleus. In the case of the hydrogen atom, it must be corrected by replacing the electron mass m by the reduced electron mass $m(1 + m/M)^{-1}$, where M is the mass of the proton.

[11] See, for example, H. A. Bethe and E. E. Salpeter, *Quantum Mechanics of One- and Two-Electron Atoms*, Plenum (1977).

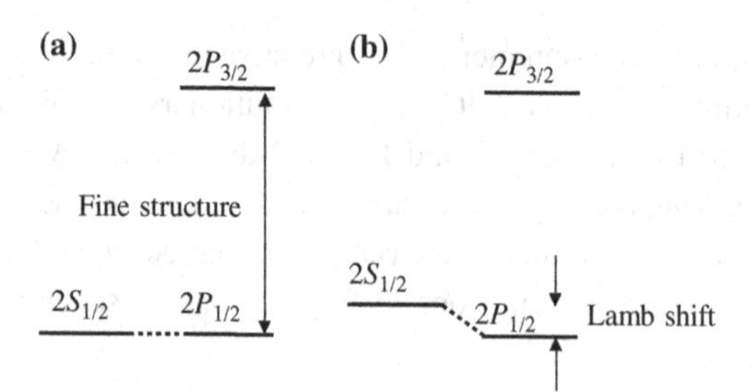

Figure 3G.9 Structure of the $n = 2$ multiplet of the hydrogen atom (a) according to the Dirac equation and (b) in reality. The energy separation of the $2S_{1/2}$ and $2P_{1/2}$ levels is known as the Lamb shift. The numerical value of this shift (in units of frequency) is about 1.057 GHz compared to the fine-structure separation between the $2P_{1/2}$ and $2P_{3/2}$ levels, which is 9.912 GHz.

The Lamb shift

Experiments carried out by Lamb and Retherford at the close of the 1940s demonstrated that the levels $2S_{1/2}$ and $2P_{1/2}$ do not have exactly the same energy (Figure 3G.9(b)): the Dirac equation does not provide a complete description of the hydrogen atom. The origin of the deviation of reality from the predictions of the Dirac equation lies in the detailed nature of the interaction of the electron with the quantized electromagnetic field (see Chapter 6), and more specifically with 'vacuum fluctuations'. It is this coupling, even in the absence of an externally applied field, that is responsible for the de-excitation of atomic levels by spontaneous decay. It is responsible also for the fact that the electron in a hydrogen atom is capable of emitting and absorbing virtual photons. Such processes lead to small shifts in the energy levels, which are different for the $2S_{1/2}$ and $2P_{1/2}$ states. The energy difference between these states arising from these differing radiative shifts is known as the *Lamb shift*. It was a success of the Quantum ElectroDynamics (QED), developed by Feynmann, Schwinger and Tomonaga, to calculate a value of the Lamb shift equal to the one measured in the experiments by Lamb and Retherford.

The increasing precision of the measurements of Lamb shifts has allowed very stringent tests to be made of the theory of quantum electrodynamics and has prompted a parallel effort to refine the methods used in the calculations of which the complexity increases rapidly with the degree of precision required. At the beginning of the twenty-first century, good agreement had been obtained between theoretical and experimental values of these shifts to a relative precision of 10^{-4}.

3G.4.3 Determination of the Rydberg constant

High-resolution spectroscopy of atomic hydrogen also allows spectroscopists to give an accurate determination of the Rydberg constant (which is a product involving e, m_e and $\hbar$). Other measurements yield values for other combinations of these quantities and therefore, together with the results of measurements of the Rydberg constant, allow precise values

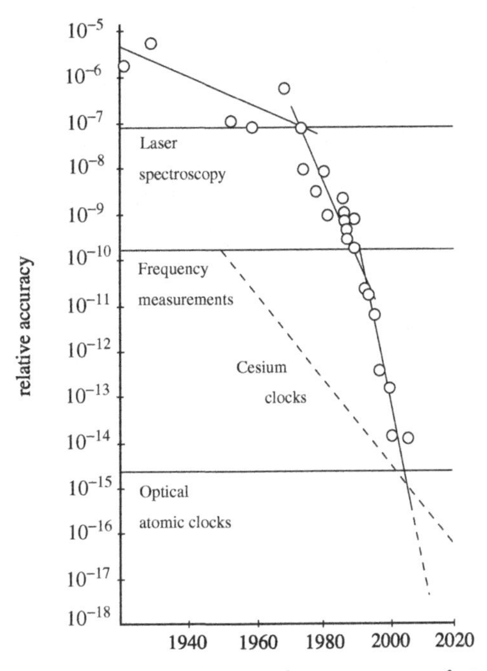

Figure 3G.10 **Evolution in time of the relative accuracy in optical spectroscopy of atomic hydrogen [taken from T. W. Hänsch, *Rev. Mod. Phys.* 78, 1297 (2006)].**

for the fundamental constants e and m_e to be inferred.[12] The procedure currently used to determine the Rydberg constant consists of correcting calculated relativistic and quantum electrodynamical effects for a $nlj \to n'l'j'$ transition of atomic hydrogen for other effects linked to the finite nuclear mass, and equating the result to experimentally derived energy differences. The consistency of the theory is then verified by ensuring that the value for R_∞ so obtained is independent of the transition considered.

We show in Figure 3G.10 the variation with time over the last seventy years of precision measurements based on optical spectroscopy of atomic hydrogen. The abrupt change of slope in 1970 occurred as a result of introduction of the techniques of nonlinear spectroscopy, described previously in this complement. These have allowed the precision to be increased since this date from one part in 10^7 to less than one part in 10^{11}. The current best value (in 2005) is

$$R_\infty/hc = 10\,973\,731.568\,525(84)\ \text{m}^{-1}$$

(a relative precision of 8×10^{-12}). The accuracy reached now is such that one must take into account the details of the proton charge distribution in the comparison between theory and experiment.

[12] For more details, see, for example, B. Cagnac, M. D. Plimmer, L. Julien and F. Biraben, The Hydrogen Atom, a Tool for Metrology, *Reports on Progress in Physics* **57**, 853 (1994).

PART II

QUANTUM DESCRIPTION OF LIGHT AND ITS INTERACTION WITH MATTER

4 Quantization of free radiation

Many processes, including absorption and stimulated emission occurring in lasers, can be handled using a semi-classical model for the atom–radiation interaction, in which the matter is given a quantum description, but the radiation is represented as a classical electromagnetic field (see Chapter 2). There are other phenomena that cannot be adequately described without quantizing the radiation. For example, it has been known since the 1930s that spontaneous emission can only be treated correctly using a fully quantum framework for the interaction, in which both the matter **and** the radiation are quantized, as we shall see in Chapter 6.

However, it was not until the 1970s that situations were found in which a *free electromagnetic field*, far from sources, exhibited properties and behaviour that could not be described by a classical field, but which could be perfectly well interpreted in terms of a quantized field. This chapter is devoted to the quantization of the free electromagnetic field, far from the charges and currents sourcing it. This free electromagnetic field will be called *radiation*, and in Chapter 6 we shall specify exactly what is meant by radiation when sources are present.

The *canonical quantization* procedure used here starts from a description of the classical dynamics of the field in the framework of the Hamiltonian formalism, the basic features of which are discussed in Section 4.1. Then Sections 4.2 and 4.3 describe how to expand the free classical electromagnetic field in decoupled normal modes, with the dynamics of each mode being described by two dynamical variables, which are the real and imaginary parts of a normal variable, up to a multiplicative constant. Section 4.4 uses this decomposition to show that the radiation energy can be written in the Hamiltonian form as the sum of the energies associated with each normal mode. This allows us to identify pairs of conjugate canonical variables. In Section 4.5, we shall then be able to quantize the radiation, finding a Hamiltonian that is formally identical to the one describing a set of independent material harmonic oscillators. The notion of the photon then arises naturally as the elementary excitations of the quantized electromagnetic field, whose ground state is the vacuum (Section 4.6). A more detailed study of quantized free radiation will be left to the next chapter, together with a discussion of some typical phenomena in quantum optics.

The chapter is followed by complements containing material closely related to what is discussed here, but which would unnecessarily dilute the main argument leading to quantization of radiation if included along the way. Complement 4A gives a non-trivial example of the classical Hamiltonian formalism, treating the case of a charged particle moving in an electromagnetic field. This is an interesting example for obtaining a better appreciation of the subtleties of the classical Hamiltonian formalism, and it will also be used in Chapter 6. Complement 4B introduces certain fundamental physical quantities associated with electromagnetic radiation, namely its momentum and angular momentum. These are expressed in a form suitable for determining the associated quantum observables. It will be noted that the question of the angular momentum of the radiation raises questions that have still not been completely elucidated. These are discussed briefly in this complement. Finally, Complement 4C returns to the decomposition of the radiation, showing that modes other than monochromatic plane travelling waves can be used. It is also shown in this complement how to define *wave packets containing a single photon.*

4.1 Classical Hamiltonian formalism and canonical quantization

4.1.1 Quantizing a system of material particles

It is well known how to quantize a system of material particles. The classical problem is first formulated in Hamilton's canonical form, in which the energy is expressed as a function of the particle positions **r** and the conjugate canonical momenta **p**. It is thus written $H(x_1, \ldots, x_i, \ldots ; p_1, \ldots, p_i, \ldots ; t)$, where x_1, x_2, x_3 are the position coordinates of the first particle, and p_1, p_2, p_3 are the corresponding conjugate canonical momenta. The quantum Hamiltonian for the system is then $\hat{H}(\hat{x}_1, \ldots, \hat{x}_i, \ldots ; \hat{p}_1, \ldots, \hat{p}_i, \ldots ; t)$, where the classical variables have been replaced by operators obeying the canonical commutation relations:

$$[\hat{x}_i, \hat{p}_j] = \mathrm{i}\hbar\, \delta_{ij}, \tag{4.1}$$

where $\delta_{ij} = 1$ if $i = j$, and $\delta_{ij} = 0$ if $i \neq j$. Hence, for a particle of mass m located in a potential $V(x, y, z)$, the Hamiltonian is given in Cartesian coordinates by

$$\hat{H}(\hat{x}, \hat{y}, \hat{z}\,; \hat{p}_x, \hat{p}_y, \hat{p}_z) = \frac{\hat{p}_x^2 + \hat{p}_y^2 + \hat{p}_z^2}{2m} + V(\hat{x}, \hat{y}, \hat{z}). \tag{4.2}$$

In the representation in which the state of the quantum system is described by a wavefunction, we have

$$\hat{H} = -\frac{\hbar^2}{2m}\left(\frac{\partial^2}{\partial x^2} + \frac{\partial^2}{\partial y^2} + \frac{\partial^2}{\partial z^2}\right) + V(x, y, z), \tag{4.3}$$

since the observables $\hat{x}$ and $\hat{p}_x$ then take the form x and $\dfrac{\hbar}{\mathrm{i}}\dfrac{\partial}{\partial x}$, respectively, operators satisfying (4.1).

As soon as we can write down the quantum Hamiltonian of a system, we can also formulate the Schrödinger equation that determines its dynamics:

$$\mathrm{i}\hbar\frac{\mathrm{d}}{\mathrm{d}t}|\psi\rangle = \hat{H}|\psi\rangle. \tag{4.4}$$

The general results obtained from the theory of quantum mechanics can then be applied.

Taken as we have just described it, the whole procedure looks rather straightforward. However, we have not specified how to identify pairs of conjugate canonical variables. Now this step is of course crucial, and it is not at all obvious when we turn to less elementary situations than a point particle moving in a potential, described using Cartesian coordinates. For example, we may wish to consider a charged particle moving in a magnetic field. We may also wish to use cylindrical or spherical coordinates, or we may be interested in combinations of coordinates, such as the position of a centre of mass. In all these cases, one must be able to determine the conjugate canonical momenta associated with the coordinates used to describe the system.

This problem has a general solution presented in any treatise on analytical mechanics.[1] Starting with a Lagrangian depending on generalized coordinates and their time derivatives, it is shown how to deduce the dynamical equations by means of a variational principle. It is also shown how to move to the *classical Hamiltonian formulation* of the problem, which uses coordinates and their conjugate canonical momenta to give the Hamiltonian expression for the energy of the system. In order to provide a complete and rigorous quantization procedure, one must therefore start with a Lagrangian.[2] In this book, we shall simply describe the basics of the Hamiltonian formulation of a problem, so that we can identify pairs of conjugate canonical variables, without reference to the Lagrangian.

4.1.2 Classical Hamiltonian formulation: Hamilton's equations

Let us consider an arbitrary system comprising material particles subject to constraints, e.g. a vibrating string or coupled pendulums, an electromagnetic field or a charged particle in a magnetic field (see Complement 4A). In the classical Hamiltonian formulation, the energy H of the system is written in terms of generalized coordinates q_i and their conjugate momenta p_i in the form, $H(q_1, \ldots, q_i, \ldots; p_1, \ldots, p_i, \ldots; t)$, and the dynamics of the system is fully described by the first-order differential equations,

$$\frac{\mathrm{d}q_i}{\mathrm{d}t} = \frac{\partial H}{\partial p_i}, \tag{4.5}$$

$$\frac{\mathrm{d}p_i}{\mathrm{d}t} = -\frac{\partial H}{\partial q_i}. \tag{4.6}$$

These are known as Hamilton's equations. For example, in the case of a particle of mass m in a potential $V(x, y, z)$, the Hamiltonian can be written

$$H = V(x, y, z) + \frac{1}{2m}(p_x^2 + p_y^2 + p_z^2). \tag{4.7}$$

[1] See, for example, L. Landau and E. Lifshitz, *Mechanics*, Pergamon Press (1982).

[2] See, for example, CDG I.

Hamilton's equations for the pair (x, p_x) are, therefore,

$$\frac{\mathrm{d}x}{\mathrm{d}t} = \frac{p_x}{m}, \tag{4.8}$$

$$\frac{\mathrm{d}p_x}{\mathrm{d}t} = -\frac{\partial V}{\partial x}. \tag{4.9}$$

These are indeed the usual dynamical equations. *We shall take this as a justification* of the fact that x and p_x constitute *a pair of conjugate canonical variables.*

In this book, we shall generalize the above approach. Suppose we have a set of real variables $\{q_1, \ldots, q_N; p_1, \ldots, p_N\}$ characterizing a given physical system at each instant of time and from which we may obtain its energy:

$$E = H(q_1, \ldots, q_N; p_1, \ldots, p_N). \tag{4.10}$$

If the equations describing the system dynamics can be written in the form (4.5–4.6), we shall say that *the system has been expressed in canonical form* and that the (p_i, q_i), $i = 1, \ldots, N$ are *pairs of conjugate canonical variables.*

Comment If the energy (4.7) is written in the form

$$E = V(x, y, z) + \frac{1}{2}m(v_x^2 + v_y^2 + v_z^2), \tag{4.11}$$

then the variables x_i, v_i are not pairs of conjugate canonical variables, since Hamilton's equations for the pair (x, v_x) are not the correct dynamical equations.

4.1.3 Canonical quantization

This is the generalization of the procedure described in Section 4.1.1 to pairs of conjugate variables which are not necessarily the position and momentum of material particles. It consists in associating observables $\hat{q}_1, \ldots, \hat{q}_N; \hat{p}_1, \ldots, \hat{p}_N$ with the conjugate canonical variables identified by the procedure in Section 4.1.2, and imposing the commutation relations

$$[\hat{q}_i, \hat{p}_j] = \mathrm{i}\hbar\, \delta_{ij}. \tag{4.12}$$

The quantum Hamiltonian is then $\hat{H} = H(\hat{q}_1, \ldots, \hat{q}_N; \hat{p}_1, \ldots, \hat{p}_N; t)$.

4.1.4 Hamiltonian formalism for radiation: stating the problem

To quantize the electromagnetic field, we wish to follow a procedure analogous to the one discussed above for material particles. We would thus like to write the dynamical equations in the same way as Hamilton's equations. This means that, for the electromagnetic field, we must identify pairs of conjugate canonical variables for which the dynamical equations are expressed as coupled first-order differential equations of the form (4.5–4.6). However, a priori, the equations describing the dynamics of electromagnetic radiation are partial differential equations, namely Maxwell's equations, which thus constitute a continuously infinite system of *coupled* differential equations. In the following sections, we shall reformulate the equations of electromagnetism in such a way as to identify, for the electromagnetic field in vacuum, pairs of conjugate canonical variables with *decoupled* dynamics described by Hamilton's equations.

4.2 Free electromagnetic field and transversality

4.2.1 Maxwell's equations in vacuum

In this chapter, we are concerned with the free field and therefore write Maxwell's equations *in vacuum*, in the *absence of charges and currents*:

$$\nabla \cdot \mathbf{E}(\mathbf{r}, t) = 0 \tag{4.13}$$

$$\nabla \cdot \mathbf{B}(\mathbf{r}, t) = 0 \tag{4.14}$$

$$\nabla \times \mathbf{E}(\mathbf{r}, t) = -\frac{\partial}{\partial t}\mathbf{B}(\mathbf{r}, t) \tag{4.15}$$

$$\nabla \times \mathbf{B}(\mathbf{r}, t) = \frac{1}{c^2}\frac{\partial}{\partial t}\mathbf{E}(\mathbf{r}, t). \tag{4.16}$$

Since this is a system of first-order linear differential equations, it is enough to give the value of the fields at a given time t_0 and everywhere in space in order to uniquely determine their whole future evolution. The value of the fields at t_0 can also be used to calculate various quantities such as the energy or angular momentum at this instant of time. The whole set of six components of the electromagnetic field at all points of space at a given instant thus constitutes *a complete set of dynamical variables*. However, this set contains a continuous infinity of coupled variables, since Maxwell's equations are partial differential equations, and it is difficult to quantize directly. We shall now show how to reduce the problem to a countable set of variables that are decoupled. This simplifies many problems in electromagnetism, but what we shall be particularly interested in here is the fact that this procedure allows us to identify *decoupled pairs of conjugate canonical variables*.

4.2.2 Spatial Fourier expansion

The system we are interested in is assumed finite. It is enclosed in a finite volume V that is much greater than the volume occupied by the system. For simplicity, it is assumed that the volume is a cube of side L. We then define the *spatial Fourier components* $\tilde{\mathbf{E}}_{\mathbf{n}}(t)$ of the field $\mathbf{E}(\mathbf{r}, t)$ by the relation

$$\tilde{\mathbf{E}}_{\mathbf{n}}(t) = \frac{1}{L^3}\int_V \mathrm{d}^3r\, \mathbf{E}(\mathbf{r}, t)\, \mathrm{e}^{-\mathrm{i}\mathbf{k}_{\mathbf{n}}\cdot\mathbf{r}}. \tag{4.17}$$

These Fourier components can in turn be used to calculate the complex field at any point within the volume V, since[3]

$$\mathbf{E}(\mathbf{r},t) = \sum_{\mathbf{n}} \tilde{\mathbf{E}}_{\mathbf{n}}(t)\, \mathrm{e}^{\mathrm{i}\mathbf{k}_{\mathbf{n}}\cdot\mathbf{r}}. \tag{4.18}$$

In the above equations, the integral is over the volume V and $\sum_{\mathbf{n}}$ is an abbreviated notation for $\sum_{n_x,n_y,n_z}$, where the (positive or negative) integer numbers n_x, n_y, n_z define the three components of the vector $\mathbf{k}_{\mathbf{n}}$ according to

$$(\mathbf{k}_{\mathbf{n}})_x = n_x\frac{2\pi}{L} \quad ; \quad (\mathbf{k}_{\mathbf{n}})_y = n_y\frac{2\pi}{L} \quad ; \quad (\mathbf{k}_{\mathbf{n}})_z = n_z\frac{2\pi}{L}. \tag{4.19}$$

The ends of the vectors $\mathbf{k}_{\mathbf{n}}$ form a cubic lattice with a repeat distance $2\pi/L$ which becomes smaller as the volume V increases. Equation (4.18) shows that any complex field in the volume $V = L^3$ can be expanded in terms of the basis of functions $\exp\{\mathrm{i}\mathbf{k}_{\mathbf{n}}\cdot\mathbf{r}\}$, where $\mathbf{k}_{\mathbf{n}}$ is given by (4.19). Equation (4.17) gives the amplitude $\tilde{\mathbf{E}}_{\mathbf{n}}(t)$ of each component. The space containing the vectors $\mathbf{k}$ is called the *reciprocal space* or **k**-*space*, and the functions $\tilde{\mathbf{E}}_{\mathbf{n}}(t)$ are the components of the field in the reciprocal space. Since the functions $\tilde{\mathbf{E}}_{\mathbf{n}}(t)$ depend only on the time, the notation will sometimes be abbreviated to $\tilde{\mathbf{E}}_{\mathbf{n}}$ in the following, leaving the time dependence implicit.

Note a very important relation, expressing the reality of the field $\mathbf{E}(\mathbf{r},t)$:

$$\tilde{\mathbf{E}}^*_{\mathbf{n}}(t) = \tilde{\mathbf{E}}_{-\mathbf{n}}(t). \tag{4.20}$$

It is readily obtained by taking the complex conjugate of Equation (4.17), and noting that $\mathbf{k}_{-\mathbf{n}} = -\mathbf{k}_{\mathbf{n}}$.

A major advantage in going to the reciprocal space is that differential operators are replaced by algebraic relations. Table 4.1 sums up the correspondence.

Table 4.1 Various mathematical quantities in direct space and the corresponding quantity in reciprocal space, for a real scalar field $F(\mathbf{r})$ and a real vector field $\mathbf{V}(\mathbf{r})$. Time dependences have been omitted to simplify the notation, but it should be remembered that all quantities are evaluated at a given instant of time.

Direct space	Reciprocal space
$F(\mathbf{r})$, $\mathbf{V}(\mathbf{r})$	$\tilde{F}_{\mathbf{n}}$, $\tilde{\mathbf{V}}_{\mathbf{n}}$
$F(\mathbf{r}) = F^*(\mathbf{r})$	$\tilde{F}^*_{\mathbf{n}} = \tilde{F}_{-\mathbf{n}}$
$\nabla F(\mathbf{r})$	$\mathrm{i}\mathbf{k}_{\mathbf{n}}\,\tilde{F}_{\mathbf{n}}$
$\nabla\cdot\mathbf{V}(\mathbf{r})$	$\mathrm{i}\mathbf{k}_{\mathbf{n}}\cdot\tilde{\mathbf{V}}_{\mathbf{n}}$
$\nabla\times\mathbf{V}(\mathbf{r})$	$\mathrm{i}\mathbf{k}_{\mathbf{n}}\times\tilde{\mathbf{V}}_{\mathbf{n}}$

[3] Note that there is no general agreement in the literature about the normalization of Equations (4.17) and (4.18). With the choice made here, $\tilde{\mathbf{E}}_{\mathbf{n}}$ has the physical dimensions of an electric field.

Comments

(i) Equations (4.19) express 'periodic boundary conditions' on the walls of the fictitious cube of side L for the plane waves used in the expansion (4.18). This type of condition has already been encountered in Complement 2E.

(ii) If formula (4.18) is extended outside volume V, copies of the relevant system are obtained, each one given by a translation by a multiple of the elementary cell dimension L along the three axes. This produces a periodic function in space, decomposing naturally into a Fourier series.

(iii) Volume V, introduced for mathematical reasons, remains entirely fictitious here. However, there are physical situations in which the particles are really contained in a cavity with reflecting walls (see Complement 6B). To investigate this problem, it is often convenient to take the actual volume of the cavity for volume V appearing in (4.17) and (4.18), and to consider standing wave modes (Complement 4C).

4.2.3 Transversality of the free electromagnetic field and polarized Fourier components

Taking into account the expressions for $\boldsymbol{\nabla} \cdot \mathbf{E}$ and $\boldsymbol{\nabla} \cdot \mathbf{B}$ in the reciprocal space, Maxwell's equations (4.13) and (4.14) imply the following relations for the Fourier components:

$$\mathbf{k_n} \cdot \tilde{\mathbf{E}}_\mathbf{n} = 0, \tag{4.21}$$

$$\mathbf{k_n} \cdot \tilde{\mathbf{B}}_\mathbf{n} = 0. \tag{4.22}$$

They are thus perpendicular to the wavevector $\mathbf{k_n}$ at all points $\boldsymbol{n}$ of the reciprocal space. Any field with zero divergence, whose Fourier components obey this type of relation, is called a *transverse field.*

Equation (4.21) shows that each component $\tilde{\mathbf{E}}_\mathbf{n}$ belongs to a two-dimensional space, orthogonal to $\mathbf{k_n}$ (see Figure 4.1). Two mutually orthogonal unit vectors $\boldsymbol{\varepsilon}_{\mathbf{n},1}$ and $\boldsymbol{\varepsilon}_{\mathbf{n},2}$ can be chosen (in infinitely many different ways) in the plane orthogonal to $\mathbf{k_n}$, whence we may write

$$\tilde{\mathbf{E}}_\mathbf{n} = \tilde{E}_{\mathbf{n},1}\boldsymbol{\varepsilon}_{\mathbf{n},1} + \tilde{E}_{\mathbf{n},2}\boldsymbol{\varepsilon}_{\mathbf{n},2}. \tag{4.23}$$

Equations (4.18) and (4.23) show that, since it is transverse, the field $\mathbf{E}(\mathbf{r},t)$ can be expanded in terms of a basis of polarized Fourier components with wavevector $\mathbf{k_n}$ and polarizations $\boldsymbol{\varepsilon}_{\mathbf{n},s}$ orthogonal to $\mathbf{k_n}$. Each component is then labelled by a set of four indices $\{n_x, n_y, n_z; s\}$. The first three are integers defining $\mathbf{k_n}$ (see Equations (4.19)), while the fourth can take two values, $s = 1$ or 2, which characterize the basis of transverse polarizations associated with $\mathbf{k_n}$. We use the index ℓ to denote this set of four numbers:

$$\ell = (n_x, n_y, n_z; s) = (\mathbf{n}; s) \tag{4.24}$$

and we shall write

$$\mathbf{E}(\mathbf{r},t) = \sum_\ell \boldsymbol{\varepsilon}_\ell \tilde{E}_\ell(t)\,\mathrm{e}^{\mathrm{i}\mathbf{k}_\ell\cdot\mathbf{r}}, \tag{4.25}$$

with

$$\tilde{E}_\ell(t) = \frac{1}{L^3}\int_V \mathrm{d}^3r\, \boldsymbol{\varepsilon}_\ell \cdot \mathbf{E}(\mathbf{r},t)\,\mathrm{e}^{-\mathrm{i}\mathbf{k}_\ell\cdot\mathbf{r}}. \tag{4.26}$$

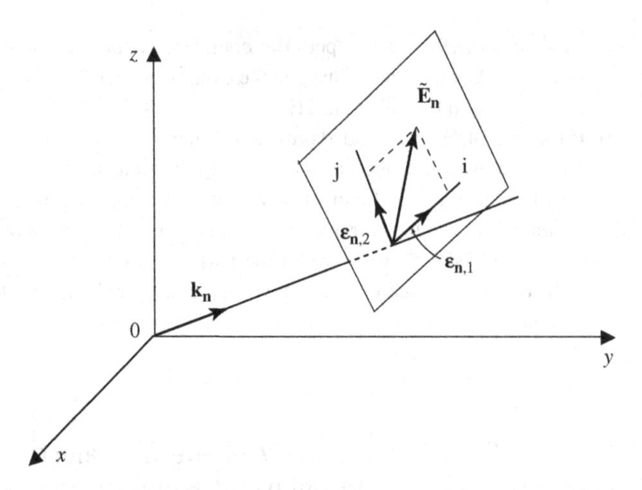

Figure 4.1 Two unit vectors in the plane perpendicular to $\mathbf{k_n}$ chosen as an orthonormal basis for the space of transverse fields in the reciprocal space. (The choice of orthogonal vectors $\boldsymbol{\varepsilon}_{\mathbf{n},1}$ and $\boldsymbol{\varepsilon}_{\mathbf{n},2}$ in this plane is a priori arbitrary.) The Fourier component $\tilde{\mathbf{E}}_\mathbf{n}(t)$ lies in this plane and can thus be expressed in terms of the basis $\{\boldsymbol{\varepsilon}_{\mathbf{n},1}, \boldsymbol{\varepsilon}_{\mathbf{n},2}\}$.

The vector $\mathbf{k}_\ell$ is defined by

$$k_{\ell_x} = n_x \frac{2\pi}{L} \quad ; \quad k_{\ell_y} = n_y \frac{2\pi}{L} \quad ; \quad k_{\ell_z} = n_z \frac{2\pi}{L}, \tag{4.27}$$

where the numbers n_x, n_y and n_z are positive or negative integers. Moreover, we have

$$\boldsymbol{\varepsilon}_\ell \cdot \mathbf{k}_\ell = 0 \tag{4.28}$$

and, denoting the two values of ℓ associated with the same $\mathbf{n}$ by ℓ_1 and ℓ_2,

$$\boldsymbol{\varepsilon}_{\ell_1} \cdot \boldsymbol{\varepsilon}_{\ell_2} = 0 \tag{4.29}$$

where $\ell_1 = (n_x, n_y, n_z; s = 1)$ and $\ell_2 = (n_x, n_y, n_z; s = 2)$. Finally, we define $-\ell$ as

$$-\ell = (-n_x, -n_y, -n_z; s), \tag{4.30}$$

so that

$$\boldsymbol{\varepsilon}_{-\ell} = \boldsymbol{\varepsilon}_\ell \tag{4.31}$$

and

$$\mathbf{k}_{-\ell} = -\mathbf{k}_\ell. \tag{4.32}$$

So to sum up, the field $\mathbf{E}(\mathbf{r}, t)$, being transverse, can be expressed as a sum of ℓ-components in the form (4.25). It is also convenient here to introduce the unit vector $\boldsymbol{\varepsilon}'_\ell$ which forms a right-handed triad with $\mathbf{k}_\ell$ and $\boldsymbol{\varepsilon}_\ell$:

$$\boldsymbol{\varepsilon}'_\ell = \frac{\mathbf{k}_\ell}{k_\ell} \times \boldsymbol{\varepsilon}_\ell. \tag{4.33}$$

As shown by (4.15), to each component $\boldsymbol{\varepsilon}_\ell \tilde{E}_\ell$ we can associate a component of **B** along $\boldsymbol{\varepsilon}'_\ell$, and we decompose $\mathbf{B}(\mathbf{r},t)$ using $\boldsymbol{\varepsilon}'_\ell$, whence

$$\mathbf{B}(\mathbf{r},t) = \sum_\ell \boldsymbol{\varepsilon}'_\ell \, \tilde{B}_\ell(t)\, \mathrm{e}^{\mathrm{i}\mathbf{k}_\ell\cdot\mathbf{r}}. \tag{4.34}$$

The Maxwell equations (4.15) and (4.16) will then become simple equations coupling $\tilde{E}_\ell(t)$ and $\tilde{B}_\ell(t)$, as we will see in Section 4.3.1.

Comments

(i) These decompositions of a transverse field into two orthogonal polarizations can be generalized to the case of left- and right-circular polarizations by introducing complex unit vectors whose orthogonality is expressed by the vanishing of the Hermitian product (see, for example, Complement 2B).

(ii) Let us consider the two polarized Fourier components ℓ and $-\ell$. One should notice that

$$\boldsymbol{\varepsilon}'_{-\ell} = -\boldsymbol{\varepsilon}'_\ell, \tag{4.35}$$

as follows from (4.31), (4.32) and (4.33). As a consequence, one has

$$\tilde{B}^*_\ell = -\tilde{B}_{-\ell}. \tag{4.36}$$

4.2.4 Vector potential in the Coulomb gauge

We know from Chapter 2 that an electromagnetic field can be described in terms of vector and scalar potentials $\mathbf{A}(\mathbf{r},t)$ and $U(\mathbf{r},t)$, respectively. The complex magnetic and electric fields are then given by

$$\mathbf{B}(\mathbf{r},t) = \nabla \times \mathbf{A}(\mathbf{r},t), \tag{4.37}$$

$$\mathbf{E}(\mathbf{r},t) = -\frac{\partial \mathbf{A}(\mathbf{r},t)}{\partial t} - \nabla\, U(\mathbf{r},t). \tag{4.38}$$

When given in this form, in terms of the potentials, the fields $\mathbf{E}(\mathbf{r},t)$ and $\mathbf{B}(\mathbf{r},t)$ automatically satisfy Maxwell's equations (4.13) and (4.14), which require these fields to be transverse. However, the potentials themselves are not uniquely defined. There is in principle an infinity of pairs $\{\mathbf{A}, U\}$ that correspond to the same electromagnetic field. This degree of freedom can be put to use to lay down a further condition called a *gauge condition*. In low-energy physics of the kind that interests us here, it is convenient to use the *Coulomb gauge* defined by the condition

$$\nabla \cdot \mathbf{A}(\mathbf{r},t) = 0. \tag{4.39}$$

Because of Equation (4.13), U is then constant, and we take

$$U(\mathbf{r},t) = 0. \tag{4.40}$$

Condition (4.39) states the transversality of $\mathbf{A}(\mathbf{r},t)$, which can thus be expanded in an analogous way to expansion (4.25):

$$\mathbf{A}(\mathbf{r},t) = \sum_{\ell} \boldsymbol{\varepsilon}_{\ell}\, \tilde{A}_{\ell}(t)\, \mathrm{e}^{\mathrm{i}\mathbf{k}_{\ell}\cdot\mathbf{r}}, \tag{4.41}$$

with

$$\tilde{E}_{\ell}(t) = -\frac{\mathrm{d}}{\mathrm{d}t}\tilde{A}_{\ell}(t). \tag{4.42}$$

Expansion (4.34) of the magnetic field now follows from (4.41) by application of (4.37), and we have the simple algebraic relation,

$$\tilde{B}_{\ell}(t) = \mathrm{i}k_{\ell}\,\tilde{A}_{\ell}(t). \tag{4.43}$$

Once the Coulomb gauge condition is imposed, one can use $\mathbf{A}(\mathbf{r},t)$ rather than $\mathbf{B}(\mathbf{r},t)$ in the description of the electromagnetic field.

4.3 Expansion of the free electromagnetic field in normal modes

4.3.1 Dynamical equations of the polarized Fourier components

The field dynamics is given by the Maxwell equations (4.15) and (4.16). Using expansions (4.25) and (4.34), we obtain

$$\frac{\mathrm{d}}{\mathrm{d}t}\tilde{B}_{\ell}(t) = -\mathrm{i}k_{\ell}\,\tilde{E}_{\ell}(t) \tag{4.44}$$

$$\frac{\mathrm{d}}{\mathrm{d}t}\tilde{E}_{\ell}(t) = -\mathrm{i}c^2k_{\ell}\,\tilde{B}_{\ell}(t). \tag{4.45}$$

Solutions of this set of coupled differential equations obviously involve oscillatory components at frequencies $\pm\,\omega_{\ell}$, with

$$\omega_{\ell} = c\,k_{\ell}. \tag{4.46}$$

Using this definition, and the substitution (4.43), we have an equivalent system,

$$\frac{\mathrm{d}}{\mathrm{d}t}\tilde{A}_{\ell}(t) = -\tilde{E}_{\ell}(t) \tag{4.47}$$

$$\frac{\mathrm{d}}{\mathrm{d}t}\tilde{E}_{\ell}(t) = \omega_{\ell}^2\,\tilde{A}_{\ell}(t). \tag{4.48}$$

It is quite remarkable that these two equations (or equivalently Equations 4.44 and 4.45) should form a closed system, and that they do not couple components with different values of ℓ. It may then appear that the electromagnetic field has been decomposed into components with independent dynamics. Actually, this is not completely true since Fourier components ℓ and $-\ell$ are linked by the relations (see Equation 4.20)

$$\tilde{E}_{-\ell} = \tilde{E}_{\ell}^{*} \tag{4.49}$$

$$\tilde{A}_{-\ell} = \tilde{A}_{\ell}^{*}. \tag{4.50}$$

We will now see how to obtain a complete decoupling, by use of normal variables.

4.3.2 Normal variables

The two equations (4.47) and (4.48), determining the dynamics of the ℓ-component of the electromagnetic field, are coupled first-order differential equations, equivalent to a second-order differential equation. The solution involves two complex integration constants, i.e. four independent real variables, to characterize a given situation. In fact, as we shall now show, they can be separated into two sets of independent pairs of real dynamical variables. To do this, we introduce the quantities $(\omega_\ell \tilde{A}_\ell \mp \mathrm{i}\tilde{E}_\ell)$, which decouple Equations (4.47–4.48). More precisely, we define

$$\alpha_\ell = \frac{1}{2\mathcal{E}_\ell^{(1)}}(\omega_\ell \tilde{A}_\ell - \mathrm{i}\tilde{E}_\ell) \tag{4.51}$$

$$\beta_\ell = \frac{1}{2\mathcal{E}_\ell^{(1)}}(\omega_\ell \tilde{A}_\ell + \mathrm{i}\tilde{E}_\ell), \tag{4.52}$$

where $\mathcal{E}_\ell^{(1)}$ is a constant, the value of which will be fixed later. The system (4.47–4.48) is then equivalent to

$$\frac{\mathrm{d}\alpha_\ell}{\mathrm{d}t} + \mathrm{i}\omega_\ell \alpha_\ell = 0 \tag{4.53}$$

and

$$\frac{\mathrm{d}\beta_\ell}{\mathrm{d}t} - \mathrm{i}\omega_\ell \beta_\ell = 0. \tag{4.54}$$

The solutions are

$$\alpha_\ell(t) = \alpha_\ell(0)\,\mathrm{e}^{-\mathrm{i}\omega_\ell t} \tag{4.55}$$

and

$$\beta_\ell(t) = \beta_\ell(0)\,\mathrm{e}^{\mathrm{i}\omega_\ell t}. \tag{4.56}$$

Inverting (4.51–4.52) we then obtain the most general evolution of the polarized Fourier component ℓ,

$$\tilde{A}_\ell(t) = \frac{\mathcal{E}_\ell^{(1)}}{\omega_\ell}[\alpha_\ell(t) + \beta_\ell(t)] \tag{4.57}$$

$$\tilde{E}_\ell(t) = \mathcal{E}_\ell^{(1)}[\mathrm{i}\alpha_\ell(t) - \mathrm{i}\beta_\ell(t)], \tag{4.58}$$

where $\alpha_\ell(t)$ and $\beta_\ell(t)$ are given by (4.55–4.56). Evolution of the ℓ-component thus depends on four real dynamical variables, the real and imaginary parts of $\alpha_\ell(t)$ and $\beta_\ell(t)$, but these variables can be associated in pairs $(\text{Re}\{\alpha_\ell\}, \text{Im}\{\alpha_\ell\})$ and $(\text{Re}\{\beta_\ell\}, \text{Im}\{\beta_\ell\})$ with independent dynamics.

Finally, thanks to the use of complex *normal variables* $\alpha_\ell(t)$ and $\beta_\ell(t)$, the free electromagnetic field can be expressed as a function of decoupled pairs of real dynamical variables. We shall see that each member of this pair is itself a pair of conjugate canonical variables. But first, let us show how normal variables can be used to expand the free field in normal modes.

4.3.3 Expansion of the free field in normal modes

Using the solutions (4.57–4.58) of Maxwell's equations, we can express the electromagnetic field as (see Equations (4.25) and (4.41))

$$\mathbf{A}(\mathbf{r},t) = \sum_\ell \boldsymbol{\varepsilon}_\ell \frac{\mathcal{E}_\ell^{(1)}}{\omega_\ell}[\alpha_\ell(t) + \beta_\ell(t)]\mathrm{e}^{\mathrm{i}\mathbf{k}_\ell\cdot\mathbf{r}} \tag{4.59}$$

$$\mathbf{E}(\mathbf{r},t) = \sum_\ell \boldsymbol{\varepsilon}_\ell \mathcal{E}_\ell^{(1)}[\mathrm{i}\alpha_\ell(t) - \mathrm{i}\beta_\ell(t)]\mathrm{e}^{\mathrm{i}\mathbf{k}_\ell\cdot\mathbf{r}}. \tag{4.60}$$

We now use the reality conditions (4.49–4.50). Inserting them in (4.51) and comparing with (4.52), we obtain

$$\beta_\ell^*(t) = \alpha_{-\ell}(t). \tag{4.61}$$

As stated earlier, the four real dynamical variables associated with the Fourier components ℓ and $-\ell$ are linked, and we can use just one of the two series of normal variables α_ℓ or β_ℓ. More precisely (4.59) can be expressed as

$$\mathbf{A}(\mathbf{r},t) = \sum_\ell \boldsymbol{\varepsilon}_\ell \frac{\mathcal{E}_\ell^{(1)}}{\omega_\ell}[\alpha_\ell(t) + \alpha_{-\ell}^*(t)]\,\mathrm{e}^{\mathrm{i}\mathbf{k}_\ell\cdot\mathbf{r}}. \tag{4.62}$$

In the summation, ℓ is a dummy index, and can be changed into $-\ell$. Doing this in the second sum of (4.62), we obtain, using (4.32),

$$\mathbf{A}(\mathbf{r},t) = \sum_\ell \boldsymbol{\varepsilon}_\ell \frac{\mathcal{E}_\ell^{(1)}}{\omega_\ell}\left[\alpha_\ell(t)\mathrm{e}^{\mathrm{i}\mathbf{k}_\ell\cdot\mathbf{r}} + \alpha_\ell^*(t)\mathrm{e}^{-\mathrm{i}\mathbf{k}_\ell\cdot\mathbf{r}}\right], \tag{4.63}$$

a form showing explicitly the reality of $\mathbf{A}(\mathbf{r},t)$. Similarly,

$$\mathbf{E}(\mathbf{r},t) = \sum_\ell \boldsymbol{\varepsilon}_\ell \mathcal{E}_\ell^{(1)}\left[\mathrm{i}\,\alpha_\ell(t)\mathrm{e}^{\mathrm{i}\mathbf{k}_\ell\cdot\mathbf{r}} - \mathrm{i}\,\alpha_\ell^*(t)\mathrm{e}^{-\mathrm{i}\mathbf{k}_\ell\cdot\mathbf{r}}\right]. \tag{4.64}$$

Finally, from (4.37), (4.33) and (4.63), we obtain

$$\mathbf{B}(\mathbf{r},t) = \sum_\ell \boldsymbol{\varepsilon}_\ell' \frac{\mathcal{E}_\ell^{(1)}}{c}\left[\mathrm{i}\,\alpha_\ell(t)\mathrm{e}^{\mathrm{i}\,\mathbf{k}_\ell\cdot\mathbf{r}} - \mathrm{i}\,\alpha_\ell^*(t)\mathrm{e}^{-\mathrm{i}\mathbf{k}_\ell\cdot\mathbf{r}}\right]. \tag{4.65}$$

The electromagnetic field is now expanded as a sum of real components $\mathbf{A}_\ell, \mathbf{E}_\ell$ and $\mathbf{B}_\ell$, labelled by ℓ, each of which is nothing other than a polarized monochromatic travelling wave. Indeed, if we write

$$\alpha_\ell = |\alpha_\ell| \, e^{i\varphi_\ell} \, e^{-i\omega_\ell t}, \tag{4.66}$$

we obtain

$$\mathbf{A}_\ell(\mathbf{r}, t) = \boldsymbol{\varepsilon}_\ell \frac{\mathcal{E}_\ell^{(1)}}{\omega_\ell} 2|\alpha_\ell| \cos\left(\mathbf{k}_\ell \cdot \mathbf{r} - \omega_\ell t + \varphi_\ell\right) \tag{4.67}$$

$$\mathbf{E}_\ell(\mathbf{r}, t) = -\boldsymbol{\varepsilon}_\ell \mathcal{E}_\ell^{(1)} 2|\alpha_\ell| \sin\left(\mathbf{k}_\ell \cdot \mathbf{r} - \omega_\ell t + \varphi_\ell\right) \tag{4.68}$$

$$\mathbf{B}_\ell(\mathbf{r}, t) = -\boldsymbol{\varepsilon}'_\ell \frac{\mathcal{E}_\ell^{(1)}}{c} 2|\alpha_\ell| \sin\left(\mathbf{k}_\ell \cdot \mathbf{r} - \omega_\ell t + \varphi_\ell\right). \tag{4.69}$$

Each complex normal variable $\alpha_\ell(t)$ therefore corresponds to a component of the field expressed relative to a basis of travelling polarized monochromatic plane waves, with associated wavevector $\mathbf{k}_\ell$ and polarization $\boldsymbol{\varepsilon}_\ell$. The decomposition of the field into dynamically independent components is called a *normal mode decomposition*, and the polarized monochromatic travelling plane waves form a set of normal modes.

Comments

(i) One should not confuse the normal mode component $(\mathcal{E}_\ell^{(1)}/\omega_\ell)\alpha_\ell$ of $\mathbf{A}(\mathbf{r}, t)$ (Equation 4.63) with the Fourier component $\tilde{A}_\ell$, which has the value (see Equations 4.62 and 4.41)

$$\tilde{A}_\ell = \frac{\mathcal{E}_\ell^{(1)}}{\omega_\ell}\left(\alpha_\ell + \alpha^*_{-\ell}\right). \tag{4.70}$$

Similarly,

$$\tilde{E}_\ell = \mathcal{E}_\ell^{(1)}\left(i\alpha_\ell - i\alpha^*_{-\ell}\right) \tag{4.71}$$

and

$$\tilde{B}_\ell = \frac{\mathcal{E}_\ell^{(1)}}{c}\left(i\alpha_\ell + i\alpha^*_{-\ell}\right) \tag{4.72}$$

(the change in sign stems from $\boldsymbol{\varepsilon}'_{-\ell} = -\boldsymbol{\varepsilon}'_\ell$). Each of the Fourier components involves the two independent normal variables α_ℓ and $\alpha_{-\ell}$. As stated earlier, the normal mode components α_ℓ and $\alpha_{-\ell}$ are independent (one could have for instance $\alpha_{-\ell} = 0$ with $\alpha_\ell \neq 0$), while the polarized Fourier components $\tilde{A}_\ell$ and $\tilde{A}_{-\ell}$ are not independent.

To illustrate this point, let us take a field of the form

$$\mathbf{E}(\mathbf{r}, t) = \boldsymbol{\varepsilon}_\ell E_1 \cos(\omega t - \mathbf{k}_\ell \cdot \mathbf{r}) + \boldsymbol{\varepsilon}_\ell E_2 \cos(\omega t + \mathbf{k}_\ell \cdot \mathbf{r}), \tag{4.73}$$

which is the sum of two independent travelling waves propagating along $\mathbf{k}_\ell$ and $-\mathbf{k}_\ell$, respectively. One can check that $\tilde{E}_\ell$ involves both E_1 and E_2, while α_ℓ only depends on E_1 and $\alpha_{-\ell}$ only depends on E_2.

(ii) The idea of an expansion in normal modes may be traced back to mechanical systems. For example, two decoupled normal modes can be found to describe the dynamics of two coupled harmonic oscillators. More generally, the motion of a vibrating string can also be described in terms of decoupled modes in a quite analogous way to the expansion for the electromagnetic field.

4.3.4 Analytic signal

The expression (4.64) of $\mathbf{E}(\mathbf{r},t)$ can be written as

$$\mathbf{E}(\mathbf{r},t) = \mathbf{E}^{(+)}(\mathbf{r},t) + \mathbf{E}^{(-)}(\mathbf{r},t), \tag{4.74}$$

with

$$\mathbf{E}^{(+)}(\mathbf{r},t) = \mathrm{i}\sum_{\ell} \boldsymbol{\varepsilon}_\ell \mathcal{E}_\ell^{(1)}\, \alpha_\ell(t)\mathrm{e}^{\mathrm{i}\mathbf{k}_\ell\cdot\mathbf{r}} \tag{4.75}$$

and

$$\mathbf{E}^{(-)}(\mathbf{r},t) = [\mathbf{E}^{(+)}(\mathbf{r},t)]^*. \tag{4.76}$$

Since $\alpha_\ell(t)$ evolve as $\exp(-\mathrm{i}\omega_\ell t)$, the quantity $\mathbf{E}^{(+)}(\mathbf{r},t)$ appears as the *generalization of the complex amplitude* associated with a quantity varying sinusoidally in time. It is called the analytic signal, or the complex field, or also 'the positive frequency part' of $\mathbf{E}(\mathbf{r},t)$. It can indeed also be obtained by taking the (time–frequency) Fourier transform of $\mathbf{E}(\mathbf{r},t)$,

$$\tilde{\mathbf{E}}(\mathbf{r},\omega) = \frac{1}{\sqrt{2\pi}}\int_{-\infty}^{+\infty} \mathrm{d}t\, \mathrm{e}^{\mathrm{i}\omega t}\, \mathbf{E}(\mathbf{r},t), \tag{4.77}$$

and keeping only the positive frequency part of the reverse Fourier transform,

$$\mathbf{E}^{(+)}(\mathbf{r},t) = \frac{1}{\sqrt{2\pi}}\int_0^{\infty} \mathrm{d}\omega\, \tilde{\mathbf{E}}(\mathbf{r},\omega)\, \mathrm{e}^{-\mathrm{i}\omega t}. \tag{4.78}$$

Similar quantities can obviously be defined for $\mathbf{A}(\mathbf{r},t)$ and $\mathbf{B}(\mathbf{r},t)$.

4.3.5 Other normal modes

The expansion in polarized, travelling plane-wave modes is not the only one possible. Within volume V, the field $\mathbf{E}(\mathbf{r},t)$ can be expanded in terms of any orthonormal basis of complex vector functions $\mathbf{u}_\ell(\mathbf{r})$, satisfying

$$\int_V \mathrm{d}^3r\, \mathbf{u}_\ell(\mathbf{r})^* \cdot \mathbf{u}_{\ell'}(\mathbf{r}) = V\,\delta_{\ell\ell'}. \tag{4.79}$$

The electric field then takes the form

$$\mathbf{E}(\mathbf{r},t) = \sum_\ell \mathcal{E}_\ell(t)\mathbf{u}_\ell(\mathbf{r}), \tag{4.80}$$

where

$$\mathcal{E}_\ell(t) = \frac{1}{V} \int d^3r \, \mathbf{u}_\ell^*(\mathbf{r}) \cdot \mathbf{E}(\mathbf{r}, t). \tag{4.81}$$

Of course, the need to satisfy Maxwell's equations lays down constraints on functions $\mathbf{u}_\ell(\mathbf{r})$, but there still remains a degree of freedom that can be exploited by imposing conditions at the boundaries of volume V. Hence, by fixing periodic boundary conditions on the faces of the cube with side L, the functions $\mathbf{u}_\ell(\mathbf{r})$ recover the form $\boldsymbol{\varepsilon}_\ell \exp(i\mathbf{k}_\ell \cdot \mathbf{r})$, and this leads to the expansion in terms of monochromatic travelling plane waves. This is a judicious choice when the space is in fact unbounded and volume V is a simple computational device whose dimensions are allowed to tend to infinity at the end of the calculation. But when there are real boundary conditions, as would be imposed for example by mirrors, other conditions can be laid down, leading to a basis of standing waves, for example (see Complement 4C).

We have also seen that a laser beam can conveniently be expanded in Gaussian modes in the paraxial approximation. There are many other expansions that are designed to suit specific problems, e.g. vectorial spherical harmonics, when one is concerned with the angular momentum of the radiation, and multipolar waves when the radiation is emitted by a set of charges. In all cases, after expanding the electric field in the form (4.80), the full structure and dynamics of each mode are obtained by requiring the fields to satisfy Maxwell's equations.

4.4 Hamiltonian for free radiation

4.4.1 Radiation energy

The energy H_R of the radiation (the free electromagnetic field) is the integral of the energy density over volume V, given by

$$H_R = \frac{\varepsilon_0}{2} \int_V d^3r \left[\mathbf{E}^2(\mathbf{r}, t) + c^2 \mathbf{B}^2(\mathbf{r}, t) \right]. \tag{4.82}$$

Using expansion (4.64) of $\mathbf{E}(\mathbf{r}, t)$, we have

$$\int_V d^3r \, \mathbf{E}^2(\mathbf{r}, t) = - \sum_\ell \sum_{\ell'} \mathcal{E}_\ell^{(1)} \mathcal{E}_{\ell'}^{(1)} \boldsymbol{\varepsilon}_\ell \cdot \boldsymbol{\varepsilon}_{\ell'} \int d^3r \left(\alpha_\ell e^{i\mathbf{k}_\ell \cdot \mathbf{r}} - \alpha_\ell^* e^{-i\mathbf{k}_\ell \cdot \mathbf{r}} \right)$$
$$\left(\alpha_{\ell'} e^{i\mathbf{k}_{\ell'} \cdot \mathbf{r}} - \alpha_{\ell'}^* e^{-i\mathbf{k}_{\ell'} \cdot \mathbf{r}} \right). \tag{4.83}$$

Since each integral is carried out over an interval of length L, with each component of $\mathbf{k}_\ell$ or $\mathbf{k}_{\ell'}$ an integer multiple of $2\pi/L$ (Equation 4.27), and $\boldsymbol{\varepsilon}_{\ell_1}$ and $\boldsymbol{\varepsilon}_{\ell_2}$ associated with the same $\mathbf{n}$ are mutually orthogonal (Equation 4.29), the integrals are zero unless $\ell' = \ell$ or $\ell' = -\ell$. We thus obtain

$$\int_V d^3r \, \mathbf{E}^2(\mathbf{r}, t) = L^3 \sum_\ell \left[\mathcal{E}_\ell^{(1)} \right]^2 \left(\alpha_\ell \alpha_\ell^* + \alpha_\ell^* \alpha_\ell - \alpha_\ell \alpha_{-\ell} - \alpha_\ell^* \alpha_{-\ell}^* \right). \tag{4.84}$$

An analogous argument for the term in $c^2\mathbf{B}^2(\mathbf{r},t)$ leads to the analogous result,

$$\int_V \mathrm{d}^3r\, c^2\mathbf{B}^2(\mathbf{r},t) = L^3 \sum_\ell \left[\mathcal{E}_\ell^{(1)}\right]^2 \left(2\alpha_\ell\alpha_\ell^* + \alpha_\ell\alpha_{-\ell} + \alpha_\ell^*\alpha_{-\ell}^*\right). \tag{4.85}$$

In (4.85) plus-signs in $\alpha_\ell\alpha_{-\ell}$ and $\alpha_\ell^*\alpha_{-\ell}^*$ arise because the unit vectors $\boldsymbol{\varepsilon}'_\ell$ and $\boldsymbol{\varepsilon}'_{-\ell}$ given by (4.33) have opposite signs (Equation 4.35). When substituted into (4.82), the last two terms of (4.84) and (4.85) cancel out and we are left with

$$H_\mathrm{R} = 2\varepsilon_0 L^3 \sum_\ell \left[\mathcal{E}_\ell^{(1)}\right]^2 |\alpha_\ell|^2. \tag{4.86}$$

The radiation energy is thus given by a simple sum of the energies associated with each normal mode, without cross-terms.

4.4.2 Conjugate canonical variables for a radiation mode

Consider a single radiation mode labelled by ℓ. According to (4.86), its energy is given by

$$H_\ell = 2\varepsilon_0\, L^3 |\mathcal{E}_\ell^{(1)}|^2\, |\alpha_\ell|^2. \tag{4.87}$$

The evolution equation of α_ℓ, (4.53), derived from the Maxwell equations can also be written as

$$\frac{\mathrm{d}}{\mathrm{d}t}\mathrm{Re}\{\alpha_\ell\} = \omega_\ell\, \mathrm{Im}\{\alpha_\ell\} \tag{4.88}$$

$$\frac{\mathrm{d}}{\mathrm{d}t}\mathrm{Im}\{\alpha_\ell\} = -\omega_\ell\, \mathrm{Re}\{\alpha_\ell\}. \tag{4.89}$$

These equations can be identified with the Hamilton equations:

$$\frac{\mathrm{d}Q_\ell}{\mathrm{d}t} = \frac{\partial H_\ell}{\partial P_\ell} \tag{4.90}$$

$$\frac{\mathrm{d}P_\ell}{\mathrm{d}t} = -\frac{\partial H_\ell}{\partial Q_\ell} \tag{4.91}$$

for the Hamiltonian (4.87) and the conjugate canonical variables defined by

$$Q_\ell = \sqrt{\frac{4\varepsilon_0 L^3}{\omega_\ell}}\, \mathcal{E}_\ell^{(1)}\, \mathrm{Re}\{\alpha_\ell\} \tag{4.92}$$

$$P_\ell = \sqrt{\frac{4\varepsilon_0 L^3}{\omega_\ell}}\, \mathcal{E}_\ell^{(1)}\, \mathrm{Im}\{\alpha_\ell\}. \tag{4.93}$$

We have therefore identified the conjugate canonical variables of the problem as being, up to a scaling factor, the real and imaginary parts of the normal variables of the normal modes.

4.5 Quantization of radiation

4.5.1 Canonical commutation relations

The method of canonical quantization of the free electromagnetic field proceeds in three steps:

- The conjugate canonical variables $Q_\ell(t)$ and $P_\ell(t)$ identified in Section 4.4.2 are associated with time-independent Hermitian operators $\hat{Q}_\ell$ and $\hat{P}_\ell$.
- Canonical commutation relations are imposed on these operators.
- Zero commutators are associated with operators corresponding to different modes, since these modes are decoupled.

Hence, by construction,

$$\left[\hat{Q}_\ell, \hat{P}_{\ell'}\right] = \mathrm{i}\hbar\,\delta_{\ell\ell'}, \tag{4.94}$$

$$\left[\hat{Q}_\ell, \hat{Q}_{\ell'}\right] = \left[\hat{P}_\ell, \hat{P}_{\ell'}\right] = 0. \tag{4.95}$$

The normal-mode complex amplitude $\alpha_\ell(t)$ is therefore associated with an operator that we will call $\hat{a}_\ell$. Using (4.92) and (4.93), we can write

$$\hat{Q}_\ell + \mathrm{i}\,\hat{P}_\ell = \sqrt{\frac{4\varepsilon_0 L^3}{\omega_\ell}}\,\mathcal{E}_\ell^{(1)}\,\hat{a}_\ell. \tag{4.96}$$

The operator $\hat{a}_\ell$ is not Hermitian and obeys the following commutation relations, derived from (4.94) and (4.95):

$$[\hat{a}_\ell, \hat{a}_\ell^\dagger] = \frac{\hbar\omega_\ell}{2\varepsilon_0 L^3}\frac{1}{\left[\mathcal{E}_\ell^{(1)}\right]^2}\delta_{\ell\ell'} \tag{4.97}$$

$$[\hat{a}_\ell, \hat{a}_{\ell'}] = 0. \tag{4.98}$$

We can now use the freedom of choice of constant $\mathcal{E}_\ell^{(1)}$ to simplify these commutation relations. Taking

$$\mathcal{E}_\ell^{(1)} = \sqrt{\frac{\hbar\omega_\ell}{2\varepsilon_0 L^3}}, \tag{4.99}$$

we have the following *commutation relations, which constitute the starting point for the whole of quantum optics*:

$$[\hat{a}_\ell, \hat{a}_{\ell'}^\dagger] = \delta_{\ell\ell'} \tag{4.100}$$

$$[\hat{a}_\ell, \hat{a}_{\ell'}] = 0. \tag{4.101}$$

Using expression (4.99) of $\mathcal{E}_\ell^{(1)}$, (4.96) yields

$$\hat{a}_\ell = \frac{1}{\sqrt{2\hbar}}(\hat{Q}_\ell + \mathrm{i}\hat{P}_\ell) \tag{4.102}$$

$$\hat{a}_\ell^\dagger = \frac{1}{\sqrt{2\hbar}}(\hat{Q}_\ell - \mathrm{i}\hat{P}_\ell), \tag{4.103}$$

which implies that

$$\hat{a}_\ell\hat{a}_\ell^\dagger + \hat{a}_\ell^\dagger\hat{a}_\ell = \frac{1}{\hbar}(\hat{Q}_\ell^2 + \hat{P}_\ell^2). \tag{4.104}$$

Comments

(i) We can now give an interpretation of $\mathcal{E}_\ell^{(1)}$. Let us take $\alpha_\ell = 1$ in expression (4.68) for a classical field in mode ℓ. As shown by (4.75), it describes a monochromatic wave with a complex amplitude of the electric field of modulus $\mathcal{E}_\ell^{(1)}$. Integrated over volume V, the electromagnetic energy of that wave has a value $\hbar\omega_\ell$, i.e. the energy of a photon of frequency ω_ℓ (see Section 4.6). This justifies the notation which reminds us that $\mathcal{E}_\ell^{(1)}$ is the amplitude of the classical field with energy equal to that of a single photon in mode ℓ (Section 4.6).

(ii) The commutation relations (4.100–4.101) are identical to those of annihilation and creation operators of a set of independent material harmonic oscillators. We will therefore make use of all of the mathematical results concerning these operators derived for the quantum harmonic oscillator.

4.5.2 Hamiltonian of the quantized radiation

Let us write the classical Hamiltonian H_R (Equation 4.86) in terms of the conjugate variables introduced in (4.92):

$$H_\mathrm{R} = \sum_\ell \hbar\omega_\ell|\alpha_\ell|^2 = \sum_\ell \frac{\omega_\ell}{2}(Q_\ell^2 + P_\ell^2). \tag{4.105}$$

The quantum operator associated with this classical quantity is, therefore,

$$\hat{H}_\mathrm{R} = \sum_\ell \frac{\omega_\ell}{2}(\hat{Q}_\ell^2 + \hat{P}_\ell^2), \tag{4.106}$$

which can also be written using (4.104) and (4.100) in the form

$$\hat{H}_\mathrm{R} = \sum_\ell \frac{\hbar\omega_\ell}{2}(\hat{a}_\ell\hat{a}_\ell^\dagger + \hat{a}_\ell^\dagger\hat{a}_\ell) = \sum_\ell \hbar\omega_\ell\left(\hat{a}_\ell^\dagger\hat{a}_\ell + \frac{1}{2}\right). \tag{4.107}$$

This is formally identical to the Hamiltonian of an assembly of decoupled quantum harmonic oscillators. This Hamiltonian, together with the commutation relation (4.100), completely defines the structure of the radiation state, as we shall see in Section 4.6.

4.5.3 Field operators

In order to obtain the expression for the quantum observables associated with the classical fields, one only has to replace the classical normal variables α_ℓ and their conjugates α_ℓ^* in the classical expressions, by the corresponding quantum operators $\hat{a}_\ell$ and $\hat{a}_\ell^\dagger$. One thus associates the following Hermitian operators with the fields (4.63–4.65):

$$\hat{\mathbf{A}}(\mathbf{r}) = \sum_\ell \boldsymbol{\varepsilon}_\ell \frac{\mathscr{E}_\ell^{(1)}}{\omega_\ell} \left(\mathrm{e}^{\mathrm{i}\mathbf{k}_\ell \cdot \mathbf{r}} \hat{a}_\ell + \mathrm{e}^{-\mathrm{i}\mathbf{k}_\ell \cdot \mathbf{r}} \hat{a}_\ell^\dagger \right) = \hat{\mathbf{A}}^{(+)}(\mathbf{r}) + \hat{\mathbf{A}}^{(-)}(\mathbf{r}) \tag{4.108}$$

$$\hat{\mathbf{E}}(\mathbf{r}) = \sum_\ell \mathrm{i}\, \boldsymbol{\varepsilon}_\ell \mathscr{E}_\ell^{(1)} \left(\mathrm{e}^{\mathrm{i}\mathbf{k}_\ell \cdot \mathbf{r}} \hat{a}_\ell - \mathrm{e}^{-\mathrm{i}\mathbf{k}_\ell \cdot \mathbf{r}} \hat{a}_\ell^\dagger \right) = \hat{\mathbf{E}}^{(+)}(\mathbf{r}) + \hat{\mathbf{E}}^{(-)}(\mathbf{r}) \tag{4.109}$$

$$\hat{\mathbf{B}}(\mathbf{r}) = \sum_\ell \mathrm{i}\, \frac{\mathbf{k}_\ell \times \boldsymbol{\varepsilon}_\ell}{\omega_\ell} \mathscr{E}_\ell^{(1)} \left(\mathrm{e}^{\mathrm{i}\mathbf{k}_\ell \cdot \mathbf{r}} \hat{a}_\ell - \mathrm{e}^{-\mathrm{i}\mathbf{k}_\ell \cdot \mathbf{r}} \hat{a}_\ell^\dagger \right) = \hat{\mathbf{B}}^{(+)}(\mathbf{r}) + \hat{\mathbf{B}}^{(-)}(\mathbf{r}). \tag{4.110}$$

Expressions (4.108–4.110) also show a decomposition of $\hat{\mathbf{A}}(\mathbf{r})$ into the two non-Hermitian operators $\hat{\mathbf{A}}^{(+)}(\mathbf{r})$ and $\hat{\mathbf{A}}^{(-)}(\mathbf{r})$, which are Hermitian conjugates, and similarly for $\hat{\mathbf{E}}(\mathbf{r})$ and $\hat{\mathbf{B}}(\mathbf{r})$. The operators $\hat{\mathbf{A}}^{(+)}(\mathbf{r}), \hat{\mathbf{E}}^{(+)}, \hat{\mathbf{B}}^{(+)}$, are called the positive frequency parts of the field operators. They are the quantum quantities associated with the classical analytic signal introduced in Section 4.3.4.

Comments

(i) It should come as no surprise that time-dependent classical dynamical variables like $\mathbf{E}(\mathbf{r}, t)$ have been replaced by quantum observables that do not depend on the time. Indeed, we are using the Schrödinger formalism of quantum mechanics, in which observables are not generally time dependent, since the temporal evolution of the system is represented through the state vectors.

There is another formulation of quantum mechanics, the Heisenberg representation, in which the observables are time dependent. We shall not be using this in the present book, unless explicitly stated.

(ii) If a basis of circular or elliptical polarizations is used, the vectors $\boldsymbol{\varepsilon}_\ell$ are complex. The complex conjugates $\boldsymbol{\varepsilon}_\ell^*$ of the $\boldsymbol{\varepsilon}_\ell$ then arise in the expressions for the operators $\hat{\mathbf{A}}^{(-)}$, $\hat{\mathbf{E}}^{(-)}$, and $\hat{\mathbf{B}}^{(-)}$.

4.6 Quantized radiation states and photons

The quantum properties of physical systems are described using two mathematical tools: Hermitian *operators* associated with measurable quantities called observables, satisfying well-defined commutation relations, and *state vectors*, belonging to a Hilbert space, which describe the specific state of the system. The canonical quantization procedure has allowed us to construct the operators relating to the quantized free electromagnetic field. It remains to introduce the quantum states of this field. A basis to expand these states will be found by seeking the eigenstates of the radiation Hamiltonian.

4.6.1 Eigenstates and eigenvalues of the radiation Hamiltonian

As noted above, we have cast the dynamics of the radiation, in the Hamiltonian formalism, in a form analogous to the one for an ensemble of independent material harmonic oscillators. To obtain the eigenvalues and eigenstates of the quantized Hamiltonian, we will then use results available in standard textbooks on quantum mechanics about the quantum mechanical harmonic oscillators.[4] More specifically, we use the formalism introduced by Dirac, based on the properties of non-Hermitian operators $\hat{a}_\ell$ and $\hat{a}_\ell^\dagger$, completely determined by the commutation relations (4.100–4.101).

In Dirac's method, the Hamiltonian $\hat{H}_\text{R}$ for the radiation is expressed as a linear combination of operators $\hat{N}_\ell$, defined by

$$\hat{N}_\ell = \hat{a}_\ell^\dagger \hat{a}_\ell. \tag{4.111}$$

It takes the form

$$\hat{H}_\text{R} = \sum_\ell \hbar\omega_\ell \left(\hat{N}_\ell + \frac{1}{2}\right). \tag{4.112}$$

The commutation relation,

$$[\hat{a}_\ell, \hat{a}_\ell^\dagger] = 1, \tag{4.113}$$

which is a special case of (4.100), is sufficient to establish that the set of eigenvalues of each operator $\hat{N}_\ell$ is precisely the set of *non-negative integers*.[4] There are therefore eigenvectors $|n_\ell\rangle$, such that

$$\hat{N}_\ell |n_\ell\rangle = n_\ell |n_\ell\rangle, \quad \text{with} \quad n_\ell = 0, 1, 2, \ldots \tag{4.114}$$

These vectors form a basis for the Hilbert space $\mathcal{F}_\ell$ of the radiation states in mode ℓ. Let us summarize the main properties of these so-called *number states*, under the effect of $\hat{a}_\ell$ and $\hat{a}_\ell^\dagger$:

$$\hat{a}_\ell |n_\ell\rangle = \sqrt{n_\ell}\, |n_\ell - 1\rangle \quad \text{if} \quad n_\ell > 0 \tag{4.115}$$

$$\hat{a}_\ell |0_\ell\rangle = 0 \tag{4.116}$$

$$\hat{a}_\ell^\dagger |n_\ell\rangle = \sqrt{n_\ell + 1}\, |n_\ell + 1\rangle. \tag{4.117}$$

The lowest-energy state $|0_\ell\rangle$ plays a particular role, as will be emphasized later. Note in particular that all the eigenstates $|n_\ell\rangle$ derive from $|0_\ell\rangle$, by a repeated application of (4.117), which yields

$$|n_\ell\rangle = \frac{(a_\ell^+)^n}{\sqrt{n_\ell!}} |0_\ell\rangle. \tag{4.118}$$

[4] See, for example, BD, Chapter 7, or CDL, Chapter V.

Let us now address the question of the eigenstates and eigenvalues of $\hat{H}_{\mathrm{R}}$. Since the $\hat{N}_\ell$ are mutually commuting, the eigenstates of $\hat{H}_{\mathrm{R}}$ are the tensor products of the states $|n_\ell\rangle$ over all possible modes ℓ, i.e. states of the form $|n_1\rangle \otimes |n_2\rangle \otimes \ldots \otimes |n_\ell\rangle \otimes \ldots$, which will be written in the abbreviated form $|n_1, n_2, \ldots, n_\ell, \ldots\rangle$. We then have

$$\hat{H}_{\mathrm{R}}|n_1, n_2, \ldots, n_\ell, \ldots\rangle = \sum_\ell \left(n_\ell + \frac{1}{2}\right) \hbar\omega_\ell |n_1, n_2, \ldots, n_\ell, \ldots\rangle. \tag{4.119}$$

The *ground state* of the radiation, called the *radiation vacuum*, corresponds to the state where all integers n_ℓ are zero. It will be denoted by $|0\rangle$ to simplify:

$$|0\rangle = |n_1 = 0, n_2 = 0, \ldots, n_\ell = 0, \ldots\rangle. \tag{4.120}$$

Its energy E_{V}, which is the minimal energy of the quantized electromagnetic field, is given by

$$E_{\mathrm{V}} = \sum_\ell \frac{1}{2} \hbar\omega_\ell. \tag{4.121}$$

Note that it is actually infinite. We shall return to this point.

Any eigenstate of $\hat{H}_{\mathrm{R}}$ can be obtained from the ground state $|0\rangle$ by applying some product of the creation operators, $\hat{a}_\ell^\dagger$. Indeed, extending (4.118), we have

$$|n_1, n_2, \ldots, n_\ell \ldots\rangle = \frac{(\hat{a}_1^\dagger)^{n_1} (\hat{a}_2^\dagger)^{n_2} \ldots (\hat{a}_\ell^\dagger)^{n_\ell} \ldots}{\sqrt{n_1! \, n_2! \ldots n_\ell! \ldots}} |0\rangle. \tag{4.122}$$

4.6.2 The notion of a photon

Equation (4.119) shows that, compared with the energy (4.121) of the radiation ground state $|0\rangle$, the stationary state $|n_1, n_2, \ldots, n_\ell, \ldots\rangle$ has an extra energy,

$$E_{n_1, n_2, \ldots, n_\ell} - E_{\mathrm{V}} = \sum_\ell n_\ell \, \hbar\omega_\ell. \tag{4.123}$$

In fact, it looks exactly as though state $|n_1, n_2, \ldots, n_\ell, \ldots\rangle$ contains n_1 particles of energy $\hbar\omega_1$, and n_2 particles of energy $\hbar\omega_2$, and in general, n_ℓ particles of energy $\hbar\omega_\ell$, for each value of the label ℓ, while the ground state contains no particles and is thus referred to as the *vacuum*. The particles we are talking about here are *photons*. They are the elementary excitations of the quantized electromagnetic field. Relations (4.115) and (4.117) show that the operator $\hat{a}_\ell$ reduces the number of photons in mode ℓ by one, while the operator $\hat{a}_\ell^\dagger$ increases the number of photons in mode ℓ by one. This explains why these operators are called *annihilation* and *creation operators*, respectively.

The operator,

$$\hat{N}_\ell = \hat{a}_\ell^\dagger \hat{a}_\ell, \tag{4.124}$$

is the quantum observable characterizing the *number of photons in mode* ℓ in volume V. The operator,

$$\hat{N} = \sum_{\ell} \hat{N}_{\ell}, \tag{4.125}$$

then gives the *total number of photons* in volume V. This observable can, in principle, be measured using photon counters. In the case where volume V coincides with a real cavity, for example, the cavity could be opened up suddenly to allow the radiation to propagate into photon counters. The observable $\hat{N}_{\ell}$ could be measured using a photon counter equipped with a filter to select the relevant frequency ω_{ℓ}, a polarizer to select polarization $\boldsymbol{\varepsilon}_{\ell}$, and a collimator to select the direction of propagation $\mathbf{k}_{\ell}$.

Photons can have other well-defined properties related to the type of normal modes in terms of which the field has been expanded. For example, for the travelling plane waves introduced in Section 4.3.3, the momentum of the radiation can be written in the form (see Complement 4B):

$$\hat{\mathbf{P}}_{\mathrm{R}} = \sum_{\ell} \hbar\mathbf{k}_{\ell}\, \hat{a}_{\ell}^{\dagger}\hat{a}_{\ell}. \tag{4.126}$$

Note that the eigenstates $|n_1, \ldots n_{\ell}, \ldots\rangle$ of the Hamiltonian are also eigenstates of the operator $\hat{\mathbf{P}}_{\mathrm{R}}$, since

$$\hat{\mathbf{P}}_{\mathrm{R}}|n_1, \ldots n_{\ell}, \ldots\rangle = \Big(\sum_{\ell} n_{\ell}\hbar\mathbf{k}_{\ell}\Big)|n_1, \ldots n_{\ell}, \ldots\rangle. \tag{4.127}$$

It is exactly as though one had n_1 particles of momentum $\hbar\mathbf{k}_1$, n_2 particles of momentum $\hbar\mathbf{k}_2$ and n_{ℓ} particles of momentum $\hbar\mathbf{k}_{\ell}$ in the general mode ℓ. Hence, by quantizing the radiation relative to a basis of travelling plane waves – which is a natural basis for free radiation – it turns out that the eigenstates $|n_1, n_2, \ldots, n_{\ell}, \ldots\rangle$ of the Hamiltonian correspond to n_1 particles of energy $\hbar\omega_1$ *and* momentum $\hbar\mathbf{k}_1$, and so on. A photon of such a mode ℓ has energy $\hbar\omega_{\ell}$ and momentum $\hbar\mathbf{k}_{\ell}$. Note that, for each particle, we have

$$(\hbar\omega_{\ell})^2 - (\hbar^2 k_{\ell}^2)\, c^2 = 0. \tag{4.128}$$

The particle is therefore relativistic with zero mass, moving at the speed of light.

Photons can have a well-defined angular momentum, if the modes used to quantize the field correspond to eigenstates of the angular momentum observables. Hence for circularly polarized travelling plane waves, the orbital angular momentum is zero, and the intrinsic angular momentum of the radiation can be written in the form (see Complement 4B):

$$\hat{\mathbf{S}}_{\mathrm{R}} = \sum_{\ell} \hat{a}_{\ell}^{\dagger}\hat{a}_{\ell}\varepsilon_{\ell}\hbar\frac{\mathbf{k}_{\ell}}{k_{\ell}}, \tag{4.129}$$

provided that the modes ℓ are circularly polarized, and the number ε_{ℓ} is equal to ± 1, depending on whether the polarization is right circular or left circular with respect to $\mathbf{k}_{\ell}$, respectively. The states $|n_1, n_2, \ldots, n_{\ell}, \ldots\rangle$ are then eigenstates of the component $\hat{S}_{\mathbf{k}_{\ell}}$ of $\hat{\mathbf{S}}_{\mathrm{R}}$ in the direction of $\mathbf{k}_{\ell}$:

$$\hat{S}_{\mathbf{k}_{\ell}}|n_1, \ldots n_{\ell} \ldots\rangle = \sum_{\ell} \varepsilon_{\ell}\hbar|n_1, \ldots n_{\ell} \ldots\rangle. \tag{4.130}$$

It is as though a photon in mode ℓ had a well-defined angular momentum component along $\mathbf{k}_\ell$, with value $\pm\hbar$, depending on the sign of ε_ℓ.

We thus find that *the properties of photons are inextricably tied to the modes into which the classical radiation is decomposed*, before quantization is carried out. Insofar as a mode is associated with a well-defined frequency ω_ℓ, the photons in this mode have a well-defined energy $\hbar\omega_\ell$. They can have other properties, but they depend on the type of modes in terms of which the electromagnetic field has been quantized. Some caution is therefore advisable before associating the classical notions of momentum and angular momentum too closely with this notion of the photon. For example, a photon in a rectilinearly polarized plane-wave mode does not have a well-defined angular momentum. But as can be seen from the way they are defined here, such notions are nevertheless useful and often lead to simple physical interpretations. For example, the selection rules relating to absorption or stimulated emission of circularly polarized light (see Complement 2B) can be interpreted as the conservation of the total angular momentum of the system (atom + photon). The same goes for the change in the wavefunction of an atom that absorbs or emits a travelling plane-wave photon, interpreted in terms of the conservation of momentum.

Of course, it is tempting to pursue the analogy with a material particle and ask whether there might be a *photon wavefunction*. Strictly speaking, the answer is negative, although there are situations in which the complex amplitude of a classical electric field plays an analogous role to the wavefunction of a material particle. We shall return to this delicate question in Chapter 5, after discussing the probability of detecting a photon, a notion that plays a somewhat analogous role to the squared modulus of the wavefunction of a material particle, i.e. the probability density of finding the particle at a given point.

Comments

(i) In quantum mechanics, there are many other examples of physical systems behaving classically as an ensemble of independent oscillators, and in the quantum context as an ensemble of independent particles. One example is provided by phonons, which are the vibrations in a crystal lattice.

(ii) Since the number of photons in a mode ℓ can be any non-negative whole number, photons are *bosons*. Actually the formalism presented here, based on creation and annihilation operators with commutation relations (4.100–4.101), automatically satisfies the requirement that the state of a set of identical bosons remains unchanged under permutation of two particles.[5] It is sometimes called the formalism of 'second quantization'.

4.6.3 General radiation state

The eigenstates $|n_1, n_2, \ldots, n_\ell, \ldots\rangle$ of the Hamiltonian form a basis for the space $\mathcal{F}$ of radiation states, which is the tensor product of the spaces $\mathcal{F}_\ell$ for each of the normal modes. The most general quantum state $|\psi\rangle$ of radiation can thus be expanded in the form

[5] See for instance CDL, Chapter XIV, or BD, Chapter 16.

$$|\psi\rangle = \sum_{n_1=0}^{+\infty} \sum_{n_2=0}^{+\infty} \cdots \sum_{n_\ell=0}^{+\infty} \cdots C_{n_1 n_2 \ldots n_\ell \ldots} |n_1, n_2, \ldots, n_\ell, \ldots\rangle, \tag{4.131}$$

where the $C_{n_1 n_2 \ldots n_\ell \ldots}$ are arbitrary complex numbers, the only restriction here being the usual normalization condition $\langle\psi|\psi\rangle = 1$. Each of the indices n_ℓ can assume an infinite number of values, and there are infinitely many such indices. *The variety of possible states for the quantized electromagnetic field is thus enormous. It is quite incommensurate with the variety of possible states available to a classical electromagnetic field,* for which an arbitrary state is completely determined by the sequence of complex numbers $\mathcal{A}_\ell(0)$ describing the radiation in each mode ℓ, as we have seen in Section 4.2.4.

To make this clear, suppose we consider only a finite number M of modes. Classical radiation is then totally described by M complex numbers. Consider now quantized radiation, and restrict to a maximum of N photons per mode. The space of states (4.131) has dimension $(N+1)^M$, varying exponentially with M. In comparison, the classical state space has dimension M.

Exploration of this truly enormous uncharted territory is far from completed. It began with investigation of the number states that we have just introduced, and Glauber's coherent, or quasi-classical states, to be discussed in the next chapter. More and more exotic states like single photon states or squeezed states, with no classical counterpart, are now under investigation. Entangled states, which exhibit non-local correlation properties between different measurements, are at the root of quantum information developments. All these will be exemplified in the following chapters. No doubt many other states with weird and wonderful quantum properties remain to be discovered.

4.7 Conclusion

In this chapter, we have seen how the classical free electromagnetic field, far from charges, i.e. radiation, can be described by a set of pairs of canonical conjugate variables, by expanding in terms of normal modes. The radiation can then be quantized, by use of the canonical quantization procedure. Thanks to the Dirac method, and using the formal analogy with a set of quantum harmonic oscillators, we have obtained the eigenvalues and eigenstates of the Hamiltonian of radiation, and introduced the notion of a photon.

We shall see in the next chapter how to implement the quantized radiation formalism in order to describe a certain number of important phenomena in quantum optics. We shall also show how to link these results with classical electromagnetism, and we shall explain why most optical phenomena known up until the 1970s can be described without difficulty in the framework of classical optics and the semi-classical model of the light–matter interaction. It will also allow us to clearly identify phenomena which *cannot* be described in that framework, and which demand the use of a full quantum description.

4A

Complement 4A Example of the classical Hamiltonian formalism: charged particle in an electromagnetic field

To illustrate the Hamiltonian formalism presented in Section 4.1 with a non-trivial example, we consider the case of a particle with charge q and mass m moving in an electromagnetic field described by scalar and vector potentials $U(\mathbf{r}, t)$ and $\mathbf{A}(\mathbf{r}, t)$, respectively (see Chapter 2). It is assumed that the Hamiltonian takes the form

$$H = \frac{1}{2m}[\mathbf{p} - q\mathbf{A}(\mathbf{r}, t)]^2 + q\,U(\mathbf{r}, t), \tag{4A.1}$$

as given in any advanced textbook on classical electrodynamics.

The first Hamilton equation (4.5) for the Hamiltonian in (4A.1) is then

$$\frac{\mathrm{d}x}{\mathrm{d}t} = \frac{1}{m}[p_x - qA_x(\mathbf{r}, t)], \tag{4A.2}$$

and likewise for the two other components. We deduce that, in the presence of a magnetic field deriving from a vector potential $\mathbf{A}$, the canonical momentum conjugate to the position $\mathbf{r}$ is

$$\mathbf{p} = m\frac{\mathrm{d}\mathbf{r}}{\mathrm{d}t} + q\mathbf{A}(\mathbf{r}, t). \tag{4A.3}$$

The components v_i of the velocity ($i = x, y, z$) are, according to (4A.3),

$$v_i = \frac{1}{m}(p_i - qA_i). \tag{4A.4}$$

It thus transpires that the first term in the Hamiltonian (4A.1) is the kinetic energy.

We now turn to the second Hamilton equation (4.6). In the Ox direction,

$$\frac{\mathrm{d}p_x}{\mathrm{d}t} = -\frac{\partial H}{\partial x} = \frac{q}{m}\sum_{i=x,y,z}(p_i - qA_i)\frac{\partial A_i}{\partial x} - q\frac{\partial U}{\partial x}. \tag{4A.5}$$

Using (4A.4) for the velocity components, Equation (4A.5) can be written

$$\frac{\mathrm{d}p_x}{\mathrm{d}t} = q\sum_{i=x,y,z} v_i\frac{\partial A_i}{\partial x} - q\frac{\partial U}{\partial x}. \tag{4A.6}$$

Treating p_x as a function of v_x and $A_x(\mathbf{r}, t)$, as given by (4A.4), we now find that

$$\begin{aligned}\frac{\mathrm{d}p_x}{\mathrm{d}t} &= m\frac{\mathrm{d}v_x}{\mathrm{d}t} + q\frac{\partial A_x}{\partial x}\frac{\mathrm{d}x}{\mathrm{d}t} + q\frac{\partial A_x}{\partial y}\frac{\mathrm{d}y}{\mathrm{d}t} + q\frac{\partial A_x}{\partial z}\frac{\mathrm{d}z}{\mathrm{d}t} + q\frac{\partial A_x}{\partial t}\\ &= m\frac{\mathrm{d}v_x}{\mathrm{d}t} + q\frac{\partial A_x}{\partial x}v_x + q\frac{\partial A_x}{\partial y}v_y + q\frac{\partial A_x}{\partial z}v_z + q\frac{\partial A_x}{\partial t}.\end{aligned} \tag{4A.7}$$

Substituting into (4A.6) and simplifying, this yields

$$m\frac{\mathrm{d}v_x}{\mathrm{d}t} = q\left(-\frac{\partial U}{\partial x} - \frac{\partial A_x}{\partial t}\right) + qv_y\left(-\frac{\partial A_y}{\partial x} - \frac{\partial A_x}{\partial y}\right) + qv_z\left(-\frac{\partial A_z}{\partial x} - \frac{\partial A_x}{\partial z}\right). \quad (4A.8)$$

Now the electric and magnetic fields associated with the potentials $U(\mathbf{r}, t)$ and $\mathbf{A}(\mathbf{r}, t)$ are

$$\mathbf{E} = -\nabla U - \frac{\partial \mathbf{A}}{\partial t}, \quad (4A.9)$$

and

$$\mathbf{B} = \nabla \times \mathbf{A}. \quad (4A.10)$$

We now observe that (4A.8) is the x component of the Newton–Lorentz force law:

$$m\frac{\mathrm{d}\mathbf{v}}{\mathrm{d}t} = q\mathbf{E} + \mathbf{v} \times \mathbf{B}. \quad (4A.11)$$

The other two components of this equation are obviously obtained in the same way. Hence, starting with the Hamiltonian (4A.1) and writing down Hamilton's equations, we do indeed arrive at the usual dynamical equation (4A.11). We have also identified the canonical momentum p_x conjugate to position x, given in (4A.3) along with the other two components p_y and p_z.

As discussed in Chapter 2 (see Section 2.2.2), we may now quantize the motion of the particle by replacing x and p_x in (4A.1) by observables $\hat{x}$ and $\hat{p}_x$, respectively, with commutator

$$[\hat{x}, \hat{p}_x] = \mathrm{i}\hbar, \quad (4A.12)$$

and likewise for (y, p_y) and (z, p_z).

4B Complement 4B Momentum and angular momentum of radiation

4B.1 Momentum

4B.1.1 Classical expression

The momentum $\mathbf{P}_\mathrm{R}$ of electromagnetic radiation is proportional to the integral of the Poynting vector over the volume V:[1]

$$\mathbf{P}_\mathrm{R} = \varepsilon_0 \int_V \mathrm{d}^3r\, \mathbf{E}(\mathbf{r},t) \times \mathbf{B}(\mathbf{r},t), \tag{4B.1}$$

where $\mathbf{E}(\mathbf{r},t)$ and $\mathbf{B}(\mathbf{r},t)$ are the real fields. This quantity is conserved in time for a free field.

The real fields can now be expanded in terms of polarized plane-wave modes, and we may proceed as with energy H_R in Section 4.4.1. Using the expansions (4.64) and (4.65) of $\mathbf{E}(\mathbf{r},t)$ and $\mathbf{B}(\mathbf{r},t)$,

$$\begin{aligned}\mathbf{P}_\mathrm{R} = &-\varepsilon_0 \sum_\ell \sum_m \frac{\mathcal{E}_\ell^{(1)}\mathcal{E}_m^{(1)}}{c} \boldsymbol{\varepsilon}_\ell \times \boldsymbol{\varepsilon}'_m \int_V \mathrm{d}^3r \left(\alpha_\ell \mathrm{e}^{\mathrm{i}\mathbf{k}_\ell\cdot\mathbf{r}} - \alpha_\ell^* \mathrm{e}^{-\mathrm{i}\mathbf{k}_\ell\cdot\mathbf{r}}\right) \\ &\left(\alpha_m \mathrm{e}^{\mathrm{i}\mathbf{k}_m\cdot\mathbf{r}} - \alpha_m^* \mathrm{e}^{-\mathrm{i}\mathbf{k}_m\cdot\mathbf{r}}\right).\end{aligned} \tag{4B.2}$$

When carried out over L^3, the integrals are zero unless $\ell = m$ or $\ell = -m$. We thus obtain

$$\begin{aligned}\mathbf{P}_\mathrm{R} &= \varepsilon_0 L^3 \sum_\ell \frac{\left[\mathcal{E}_\ell^{(1)}\right]^2}{c}\left[\boldsymbol{\varepsilon}_\ell \times \boldsymbol{\varepsilon}'_\ell\, 2|\alpha_\ell|^2 - \boldsymbol{\varepsilon}_\ell \times \boldsymbol{\varepsilon}'_{-\ell}(\alpha_\ell\alpha_{-\ell} + \alpha_\ell^*\alpha_{-\ell}^*)\right] \\ &= \varepsilon_0 L^3 \sum_\ell \frac{\left[\mathcal{E}_\ell^{(1)}\right]^2}{c}\left[2|\alpha_\ell|^2 + (\alpha_\ell\alpha_{-\ell} + \alpha_\ell^*\alpha_{-\ell}^*)\right]\frac{\mathbf{k}_\ell}{k_\ell},\end{aligned} \tag{4B.3}$$

where we have used

$$\boldsymbol{\varepsilon}_\ell \times \boldsymbol{\varepsilon}'_\ell = \frac{\mathbf{k}_\ell}{k_\ell} \tag{4B.4}$$

and

$$\boldsymbol{\varepsilon}_\ell \times \boldsymbol{\varepsilon}'_{-\ell} = -\frac{\mathbf{k}_\ell}{k_\ell}. \tag{4B.5}$$

[1] J. D. Jackson, *Classical Electrodynamics*, Wiley (1998).

In the sum over ℓ, the term in round brackets in the last expression of (4B.3) is the same for ℓ and $-\ell$, but $\mathbf{k}_{-\ell} = -\mathbf{k}_\ell$, and the terms cancel out. Replacing $\mathcal{E}_\ell^{(1)}$ by its value, we obtain

$$\mathbf{P}_\mathrm{R} = \sum_\ell |\alpha_\ell|^2 \hbar \mathbf{k}_\ell. \tag{4B.6}$$

The total momentum of the radiation is thus expressed as a sum of the momenta associated with the different modes labelled by ℓ. The momentum of a given mode lies along its wavevector $\mathbf{k}_\ell$ and thus describes a radiation pressure effect. Like the energy, it is proportional to $|\alpha_\ell|^2$, which thus characterizes the degree of excitation of the mode, as it did for the energy.

4B.1.2 Momentum operator

By an analogous quantization procedure to the one given for the energy in Section 4.5.2, we obtain the following expression for the momentum operator of the radiation field:

$$\hat{\mathbf{P}}_\mathrm{R} = \sum_\ell \frac{\hbar \mathbf{k}_\ell}{2}\left(\hat{a}_\ell \hat{a}_\ell^\dagger + \hat{a}_\ell^\dagger \hat{a}_\ell\right) = \sum_\ell \hbar \mathbf{k}_\ell \hat{a}_\ell^\dagger \hat{a}_\ell. \tag{4B.7}$$

Note that there is no term 1/2 in the summand of the last expression in (4B.7). The reason is that the terms carrying the factor of 1/2, which arise when the commutation relations for $\hat{a}$ and $\hat{a}^\dagger$ are used, cancel out in pairs between mode ℓ and mode $-\ell$ because $\mathbf{k}_{-\ell} = -\mathbf{k}_\ell$.

Comments

(i) For each mode ℓ, note that $|\mathbf{P}_\ell| = H_\ell/c$, since $\omega_\ell = ck_\ell$. This relation reflects the idea of a relativistic particle, in fact with speed c, which has zero rest mass and energy H_ℓ.

(ii) Decomposition of the total momentum of the field into a sum of momenta for each mode taken separately corresponds specifically to expansion of the field in terms of a basis of polarized *travelling* plane waves. The decoupling between modes does not necessarily occur when other bases are used (see Complement 4C). For example, the expression for the momentum decomposed relative to standing or Gaussian wave modes involves a double sum over ℓ and ℓ' which contains cross-terms.

4B.2 Angular momentum

4B.2.1 Classical expression

Total angular momentum

The angular momentum $\mathbf{J}_\mathrm{R}$ of the radiation is another constant of the motion for a free field. It is proportional to the integral over volume V of the moment of the Poynting vector about the coordinate origin:

$$\mathbf{J}_\mathrm{R} = \varepsilon_0 \int_V \mathbf{r} \times [\mathbf{E}(\mathbf{r},t) \times \mathbf{B}(\mathbf{r},t)]\, \mathrm{d}^3 r. \qquad (4B.8)$$

Using the usual formula for a double vector product and integrating by parts, bearing in mind the assumption that the fields are zero at the surface of volume V introduced for the mode expansion, one finds that $\mathbf{J}_\mathrm{R}$ can be written as a sum of two terms:

$$\mathbf{J}_\mathrm{R} = \mathbf{L}_\mathrm{R} + \mathbf{S}_\mathrm{R}, \qquad (4B.9)$$

given by

$$\mathbf{L}_\mathrm{R} = \varepsilon_0 \sum_{j=(x,y,z)} \int_V \mathrm{d}^3 r\, E_j(\mathbf{r},t)(\mathbf{r} \times \nabla) A_j(\mathbf{r},t), \qquad (4B.10)$$

$$\mathbf{S}_\mathrm{R} = \varepsilon_0 \int_V \mathrm{d}^3 r\, \mathbf{E}(\mathbf{r},t) \times \mathbf{A}(\mathbf{r},t). \qquad (4B.11)$$

The term $\mathbf{L}_\mathrm{R}$, which depends on the choice of origin, is called the orbital angular momentum, while $\mathbf{S}_\mathrm{R}$, which does not depend on this choice, is called the intrinsic angular momentum or spin angular momentum.

Comment It is easy to show that this separation is gauge invariant. The qualifiers 'orbital', 'intrinsic', or 'spin' are used by analogy with the case for a material particle. This analogy should not be taken too literally, however. For example, in the paraxial approximation, $\mathbf{L}_\mathrm{R}$ remains unchanged when the coordinate origin is moved. Hence it too is intrinsic to some extent.

Intrinsic angular momentum

Consider an expansion of the fields as in (4.63) and (4.64), where we explicitly write the two polarization components associated with the same $\mathbf{k}_n$:

$$\mathbf{A}(\mathbf{r},t) = \sum_\mathbf{n} \frac{\mathcal{E}_n^{(1)}}{\omega_n} \left(\boldsymbol{\varepsilon}_{\mathbf{n},1} \alpha_{\mathbf{n},1}\, \mathrm{e}^{\mathrm{i}\mathbf{k}_\mathbf{n}\cdot\mathbf{r}} + \boldsymbol{\varepsilon}_{\mathbf{n},2} \alpha_{\mathbf{n},2}\, \mathrm{e}^{\mathrm{i}\mathbf{k}_\mathbf{n}\cdot\mathbf{r}} \right) + \text{c.c.} \qquad (4B.12)$$

and

$$\mathbf{E}(\mathbf{r},t) = \mathrm{i} \sum_\mathbf{m} \mathcal{E}_m^{(1)} \left(\boldsymbol{\varepsilon}_{\mathbf{m},1} \alpha_{\mathbf{m},1}\, \mathrm{e}^{\mathrm{i}\mathbf{k}_\mathbf{m}\cdot\mathbf{r}} + \boldsymbol{\varepsilon}_{\mathbf{m},2} \alpha_{\mathbf{m},2}\, \mathrm{e}^{\mathrm{i}\mathbf{k}_\mathbf{m}\cdot\mathbf{r}} \right) + \text{c.c.}, \qquad (4B.13)$$

where c.c. stands for 'complex conjugate'. Recall that

$$\boldsymbol{\varepsilon}_{\mathbf{n},1} \cdot \boldsymbol{\varepsilon}_{\mathbf{n},2} = 0 \qquad (4B.14)$$

$$\boldsymbol{\varepsilon}_{\mathbf{n},1} \times \boldsymbol{\varepsilon}_{\mathbf{n},2} = \frac{\mathbf{k}_\mathbf{n}}{k_n}. \qquad (4B.15)$$

As above, when the expansions (4B.12) and (4B.13) are inserted in (4B.11), only terms with $\mathbf{m} = \mathbf{n}$ or $\mathbf{m} = -\mathbf{n}$ yield non-null integrals over V. Using (4B.15) we then obtain

$$\mathbf{S}_\mathrm{R} = \mathrm{i}\, 2\varepsilon_0 L^3 \sum_\mathbf{n} \frac{\left[\mathcal{E}_n^{(1)}\right]^2}{\omega_n} \frac{\mathbf{k}_\mathbf{n}}{k_n} \left(\alpha_{\mathbf{n},1} \alpha_{\mathbf{n},2}^* - \alpha_{\mathbf{n},2} \alpha_{\mathbf{n},1}^* \right). \qquad (4B.16)$$

If we have only one linearly polarized wave, we take $\boldsymbol{\varepsilon}_{\mathbf{n},1}$ along that polarization, and $\alpha_{\mathbf{n},2} = 0$. Hence $\mathbf{S}_{\mathrm{R}}$ is zero. *A linearly polarized plane wave has zero intrinsic angular momentum.*

Consider now the basis of circular polarizations, $\boldsymbol{\varepsilon}_{\mathbf{n},+}$ and $\boldsymbol{\varepsilon}_{\mathbf{n},-}$ (see, for example, Section 2B.1.3):

$$\boldsymbol{\varepsilon}_{\mathbf{n},\pm} = \mp \frac{\boldsymbol{\varepsilon}_{\mathbf{n},1} \pm \mathrm{i}\boldsymbol{\varepsilon}_{\mathbf{n},2}}{\sqrt{2}}. \tag{4B.17}$$

For this basis,

$$\boldsymbol{\varepsilon}_{\mathbf{n},+} \times \boldsymbol{\varepsilon}_{\mathbf{n},+} = \boldsymbol{\varepsilon}_{\mathbf{n},-} \times \boldsymbol{\varepsilon}_{\mathbf{n},-} = 0, \tag{4B.18}$$

$$\boldsymbol{\varepsilon}^*_{\mathbf{n},+} \times \boldsymbol{\varepsilon}_{\mathbf{n},-} = 0, \tag{4B.19}$$

$$\boldsymbol{\varepsilon}^*_{\mathbf{n},+} \times \boldsymbol{\varepsilon}_{\mathbf{n},+} = \mathrm{i}\frac{\mathbf{k}_{\mathbf{n}}}{k_{\mathbf{n}}}, \tag{4B.20}$$

$$\boldsymbol{\varepsilon}^*_{\mathbf{n},-} \times \boldsymbol{\varepsilon}_{\mathbf{n},-} = -\mathrm{i}\frac{\mathbf{k}_{\mathbf{n}}}{k_{\mathbf{n}}}. \tag{4B.21}$$

We can expand **A** and **E** as in (4B.12) and (4B.13), with $\boldsymbol{\varepsilon}_{\mathbf{n},+}$ and $\boldsymbol{\varepsilon}_{\mathbf{n},-}$ instead of $\boldsymbol{\varepsilon}_{\mathbf{n},1}$ and $\boldsymbol{\varepsilon}_{\mathbf{n},2}$. Note then that $\boldsymbol{\varepsilon}^*_{\mathbf{n},\pm}$ appear in the complex conjugate terms. A calculation similar to the previous one then yields

$$\mathbf{S}_{\mathrm{R}} = 2\varepsilon_0 L^3 \sum_{\mathbf{n}} \frac{\left[\mathcal{E}_n^{(1)}\right]^2}{\omega_n} \frac{\mathbf{k}_{\mathbf{n}}}{k_{\mathbf{n}}} \left(|\alpha_{\mathbf{n},+}|^2 - |\alpha_{\mathbf{n},-}|^2\right). \tag{4B.22}$$

Replacing $\mathcal{E}_n^{(1)}$ by its value (4.99), we have finally,

$$\mathbf{S}_{\mathrm{R}} = \hbar \sum_{\mathbf{n}} \frac{\mathbf{k}_{\mathbf{n}}}{k_{\mathbf{n}}} \left(|\alpha_{\mathbf{n},+}|^2 - |\alpha_{\mathbf{n},-}|^2\right). \tag{4B.23}$$

In the circular polarization basis, the contributions decouple. A wave with polarization σ_+ along $\mathbf{k}_n$ has positive intrinsic angular momentum in the $\mathbf{k}_n$ direction, while a wave with polarization σ_- along $\mathbf{k}_n$ has negative intrinsic angular momentum in the $\mathbf{k}_n$ direction.

Comparing with the expression for the Hamiltonian, expanded in terms of the circular polarization basis, one observes that a circularly polarized wave of energy H_{R} carries angular momentum $\pm H_{\mathrm{R}}/\omega$. This effect can be demonstrated by shining circularly polarized light onto a half-wave plate which reverses the direction of polarization. The plate is then subjected to a couple $2P/\omega$, where P is the incident power.[2]

[2] R. A. Beth, Direct Detection of the Angular Momentum of Light, *Physical Review* **48**, 471 (1935).

Orbital angular momentum

A simple calculation shows that $\mathbf{L}_\mathrm{R}$ is zero for any travelling plane wave with constant amplitude in a transverse plane. The quantity $\mathbf{L}_\mathrm{R}$ only differs from zero if the wave is non-uniform. In fact, this is a rather delicate matter which still gives rises to theoretical and experimental research, and some degree of controversy.

The quantity $\mathbf{L}_\mathrm{R}$ can be unambiguously interpreted in the context of the *paraxial approximation*, in which one only considers waves propagating near an optical axis, e.g. the Oz-axis, with small angles relative to this axis. This is the case when the field varies in the plane xOy over characteristic distances that are much longer than the wavelength. Diffraction then plays only a minor role, the wave is 'quasi-planar', and the vector potential can be written in the form

$$\mathbf{A}(\mathbf{r},t) = \boldsymbol{\varepsilon}\, U(x,y,z)\,\mathrm{e}^{\mathrm{i}(kz-\omega t)} + \text{c.c.}, \tag{4B.24}$$

where $\boldsymbol{\varepsilon}$ is the unit polarization vector and $U(x,y,z)$ is a 'slowly varying envelope', the variation being 'slow' on the scale of the wavelength. The Gaussian modes introduced in Complement 3B form a basis for the space of functions of this kind. Suppose now that $U(x,y,z)$ has the form

$$U(x,y,z) = u(r)\,\mathrm{e}^{\mathrm{i}m\phi}, \tag{4B.25}$$

where (r,ϕ) are polar coordinates in the transverse plane xOy. The number m is necessarily an integer so that $U(x,y,z)$ takes the same value for ϕ and $\phi + 2\pi$. The expression for the orbital angular momentum of such a field is obtained from (4B.10), which yields

$$\mathbf{L}_R = \frac{H_R}{\omega} m\mathbf{e}_z, \tag{4B.26}$$

where $\mathbf{e}_z$ is the unit vector along the Oz-axis.

We find that $\mathbf{L}_\mathrm{R}$ only differs from zero if m is non-zero. The wave-front of the field must be 'helical' if the wave is to carry orbital angular momentum. Indeed, in the wave described by (4B.25), the phase is constant for helices with axis Oz and distance λ/m between consecutive turns, which thus effect m turns per wavelength. Note that, when $m \neq 0$, the field must vanish on the Oz-axis in order to avoid discontinuities as one crosses the Oz-axis.

The *Laguerre–Gauss modes* $\mathrm{TEM}^*_{\mathrm{lm}}$ (see Complement 3B) have the kind of ϕ dependence given in (4B.25). They thus carry the orbital angular momentum given by (4B.26) (see Figure 4B.1). Moreover, they form a basis for the space of fields in the paraxial approximation. If the complex vector potential is expanded in terms of such a basis, it can be shown that the orbital angular momentum is given by

$$\mathbf{L}_\mathrm{R} = \hbar\,\mathbf{e}_z \sum_\ell m|\alpha_\ell|^2, \tag{4B.27}$$

where the collective index ℓ is in this case the list of indices required to define the basis, i.e. the indices ℓ and m to specify the Laguerre–Gauss modes, an integer n_z expressing the discretization of the basis along the Oz-axis and the index $s = 1,2$ for the polarization, which can be taken to be linear.

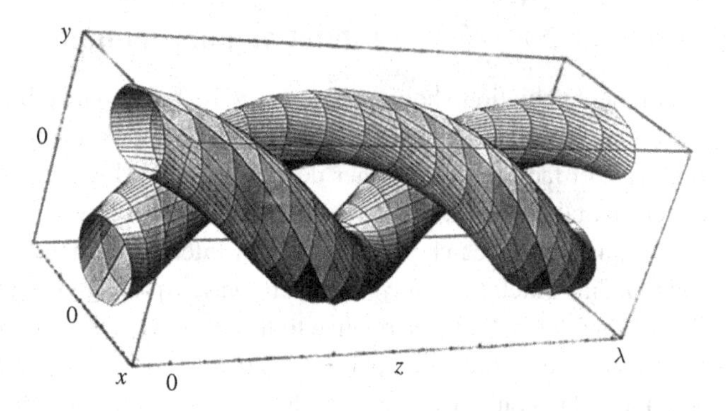

Figure 4B.1 **Helical structure** of a Laguerre–Gauss mode TEM_{01}^{*} $[(r/w)\exp\{-r^2/w^2\}e^{i\phi}e^{ikz}]$. The figure represents the surface on which the intensity is equal to half its maximal value. Note that the polarization is linear. Such a beam has non-zero orbital angular momentum. Figure taken from H. He *et al.*[3]

A certain number of experiments have been carried out to measure this orbital angular momentum.[3] However, the interpretation of these experiments remains somewhat delicate insofar as the paraxial approximation is non-rigorous.

4B.2.2 Angular momentum operators

The classical expressions are easily quantized by following the procedure outlined in Section 4.5.2 for the energy and using

$$[\hat{a}_{\mathbf{n},i}, \hat{a}_{\mathbf{n},i}^{\dagger}] = 1. \tag{4B.28}$$

For the intrinsic (or spin) angular momentum, expressed relative to a circular polarization basis (Equation (4B.23)), this leads to

$$\hat{\mathbf{S}}_{\mathrm{R}} = \sum_{\mathbf{n}} \left(\hat{a}_{\mathbf{n},+}^{\dagger}\hat{a}_{\mathbf{n},+} - \hat{a}_{\mathbf{n},-}^{\dagger}\hat{a}_{\mathbf{n},-}\right)\hbar\frac{\mathbf{k}_n}{k_n}. \tag{4B.29}$$

Factors of 1/2 due to symmetrization and use of the commutation relations cancel out.

In a like manner, the orbital angular momentum operator can be written in terms of the basis of Laguerre–Gauss modes, whence

$$\hat{\mathbf{L}}_{\mathrm{R}} = \sum_{\ell} m\hat{a}_{\ell}^{\dagger}\hat{a}_{\ell}\,\hbar\mathbf{e}_z. \tag{4B.30}$$

[3] See, for example, N. B. Simpson, K. Dholakia, L. Allen and M. J. Padgett, Mechanical Equivalence of Spin and Orbital Angular Momentum of Light: An Optical Spanner, *Optics Letters* **22**, 52 (1997); H. He, M. E. J. Friese, N. R. Heckenberg and H. Rubinsztein-Dunlop, Direct Observation of Transfer of Angular Momentum to Absorptive Particles from a Laser Beam with a Phase Singularity, *Physical Review Letters* **75**, 826 (1995).

We know that each operator of the form $\hat{a}_j^\dagger \hat{a}_j$ (with $[\hat{a}_j, \hat{a}_j^\dagger] = 1$) has for eigenvalue spectrum the whole set of natural numbers (see Section 4.6.1). We thus observe that the elementary quantum of the intrinsic angular momentum, or the orbital angular momentum, is equal to the reduced Planck constant $\hbar$, this angular momentum being taken along the direction of propagation of the relevant wave.

4C Complement 4C Photons in modes other than travelling plane waves

The quantization procedure discussed in this chapter is based on decomposition of the free classical electromagnetic field into polarized travelling plane waves. These constitute *normal modes*, with the same dynamics as a set of decoupled harmonic oscillators. The quantization of each of these harmonic oscillators led us to introduce the idea of *a photon with well-defined energy and momentum for each of these modes.*

In fact, there are other possible decompositions for free radiation, into normal modes other than travelling plane waves, and one can thereby define photons with other types of property, e.g. stationary-wave photons, which do not have a well-defined momentum. In this complement, we shall show how to define photons for other normal modes, using a unitary transformation equivalent to a change of normal mode basis, and we shall identify those properties that are invariant under such a change of basis. We shall thus address the question as to whether some bases might be better suited to the solution of certain specific problems.

Finally, we shall show that there are transformations that are not basis changes, but which allow one to define other types of photon of some importance, e.g. photons associated with a wave packet.

4C.1 Changing the normal mode basis

4C.1.1 Unitary transformation of creation and annihilation operators

In Chapter 4, quantization of the free electromagnetic field was based on a normal mode expansion onto polarized travelling waves (Equations 4.63–4.65). Actually, as shown by Equations (4.74–4.75), this expansion of the field $\mathbf{E}(\mathbf{r}, t)$ is equivalent to an expansion of the analytic signal $\mathbf{E}^{(+)}(\mathbf{r}, t)$ relative to the set of functions

$$\mathbf{u}_\ell(\mathbf{r}) = \boldsymbol{\varepsilon}_\ell \, \mathrm{e}^{\mathrm{i}\mathbf{k}_\ell \cdot \mathbf{r}}, \tag{4C.1}$$

where $\mathbf{k}_\ell$ and $\boldsymbol{\varepsilon}_\ell$ satisfy (4.27–4.29). In the quantum formalism, the complex components α_ℓ in this expansion are replaced by the non-Hermitian operators $\hat{a}_\ell$, obeying the commutation relations (4.100–4.101).

Let us now carry out the following transformation on the operators $\hat{a}_\ell^\dagger$, which leads to a new set of operators $\hat{b}_m^\dagger$:

$$\hat{b}_m^\dagger = \sum_\ell U_m^\ell \, \hat{a}_\ell^\dagger, \tag{4C.2}$$

where U_m^ℓ is an element of a unitary matrix $\mathbf{U}$, i.e. with $\mathbf{U}^\dagger = \mathbf{U}^{-1}$. The Hermitian conjugate of (4C.2) can be written

$$\hat{b}_m = \sum_\ell (U_m^\ell)^* \, \hat{a}_\ell = \sum_\ell (U^{-1})_\ell^m \, \hat{a}_\ell \tag{4C.3}$$

and the commutators are now found to be

$$\left[\hat{b}_m, \hat{b}_{m'}^\dagger\right] = \sum_\ell \sum_{\ell'} (U^{-1})_\ell^m \, U_{m'}^{\ell'} \left[\hat{a}_\ell, \hat{a}_{\ell'}^\dagger\right] = \sum_\ell (U^{-1})_\ell^m \, U_{m'}^\ell = \delta_{m,m'}. \tag{4C.4}$$

This yields the relation analogous to (4.100). Likewise, the analogous relation to (4.101) is found:

$$\left[\hat{b}_m, \hat{b}_{m'}\right] = \left[\hat{b}_m^\dagger, \hat{b}_{m'}^\dagger\right] = 0. \tag{4C.5}$$

4C.1.2 New normal modes

Suppose further that transformation (4C.2) only affects modes ℓ associated with the same frequency $\omega_\ell = \omega_m$. Using the unitarity of $\mathbf{U}$, the radiation Hamiltonian becomes

$$\hat{H}_\mathrm{R} = \sum_\ell \hbar\omega_\ell \left(\hat{a}_\ell^\dagger \hat{a}_\ell + \frac{1}{2}\right) = \sum_m \hbar\omega_m \left(\hat{b}_m^\dagger \hat{b}_m + \frac{1}{2}\right). \tag{4C.6}$$

This is a new decomposition of the radiation into a new set of normal modes. Their spatial structure can be specified by applying the transformation to the electric field $\hat{\mathbf{E}}^{(+)}(\mathbf{r})$ (Equation 4.109). Inverting (4C.3), we obtain

$$\hat{\mathbf{E}}^{(+)}(\mathbf{r}) = \mathrm{i} \sum_\ell \mathcal{E}_\ell^{(1)} \mathbf{u}_\ell(\mathbf{r}) \hat{a}_\ell = \mathrm{i} \sum_m \mathcal{E}_m^{(1)} \mathbf{v}_m(\mathbf{r}) \hat{b}_m, \tag{4C.7}$$

with

$$\mathbf{v}_m(\mathbf{r}) = \sum_\ell U_m^\ell \mathbf{u}_\ell(\mathbf{r}) \tag{4C.8}$$

and

$$\mathcal{E}_m^{(1)} = \sqrt{\frac{\hbar\omega_m}{2\varepsilon_0 L^3}}. \tag{4C.9}$$

Here we have used the fact that the transformation only mixes modes of the same frequency.

The functions $\mathbf{v}_m(\mathbf{r})$ form a new orthogonal basis with respect to which the analytic signal of the classical radiation can be decomposed. The canonical quantization procedure could have been carried out using this basis. Here, we shall continue to use the unitary transformation to discuss the properties of photons in mode m.

4C.1.3 Invariance of the vacuum and photons in mode m

The quantization procedure based on the modes ℓ leads to a vacuum $|0\rangle$ defined by the property

$$\hat{a}_\ell|0\rangle = 0, \quad \text{for all } \ell. \tag{4C.10}$$

Using (4C.3), it follows that

$$\hat{b}_m|0\rangle = 0, \quad \text{for all } m, \tag{4C.11}$$

which means that $|0\rangle$ is also the vacuum for the modes m. We may then deduce a formula for the eigenstates of the m mode number operator $\hat{N}_m = \hat{b}_m^\dagger \hat{b}_m$, which are denoted by $|m : n_m\rangle$ to avoid ambiguity:

$$|m : n_m\rangle = \frac{1}{\sqrt{n_m!}}\left(\hat{b}_m^\dagger\right)^{n_m}|0\rangle = \frac{1}{\sqrt{n_m!}}\left(\sum_\ell U_m^{\ell}\hat{a}_\ell\right)^{n_m}|0\rangle. \tag{4C.12}$$

This expression takes on a particularly simple form for one-photon states:

$$|m : 1\rangle = \sum_\ell U_m^\ell|\ell : 1\rangle. \tag{4C.13}$$

4C.1.4 Invariance of the total photon number

In the basis of modes ℓ, the observable $\hat{N}_{\text{tot}}$ associated with the total photon number is given by

$$\hat{N}_{\text{tot}} = \sum_\ell \hat{a}_\ell^\dagger \hat{a}_\ell.$$

The unitary transformation (4C.2–4C.3) can be used to express $\hat{N}_{\text{tot}}$ relative to the basis of modes m, whence,

$$\hat{N}_{\text{tot}} = \sum_m \hat{b}_m^\dagger \hat{b}_m = \sum_\ell \hat{a}_\ell^\dagger \hat{a}_\ell, \tag{4C.14}$$

and it follows that the total number of photons in a given radiation state is invariant under change of basis.

This property is particularly interesting when we consider an eigenstate of the total photon number. We have already encountered this in Section 4C.1.3, where we showed the invariance of the vacuum, a non-degenerate zero-photon state. Likewise, (4C.14) can be used to show that a one-photon state remains a one-photon state in all such bases.

4C.1.5 Properties of photons in different bases

The photons introduced in Section 4.6.2 of this chapter are associated with the travelling plane-wave modes used to quantize the field. The eigenstates $|\ell : n_\ell\rangle$ of the number operators $\hat{N}_\ell = \hat{a}_\ell^\dagger \hat{a}_\ell$ associated with each mode ℓ are eigenstates of the Hamiltonian $\hat{H}_\mathrm{R}$, but also of the momentum $\hat{\mathbf{P}}_\mathrm{R}$, and this led us to attribute an energy $\hbar\omega_\ell$ and momentum $\hbar\mathbf{k}_\ell$ to each mode ℓ photon. Moreover, if we choose a basis of circular polarizations, the mode ℓ photons will have a well-defined intrinsic angular momentum $\hat{S}_{\mathbf{k}_\ell}$ along $\mathbf{k}_\ell$. What happens when we carry out a change of basis of the type considered here?

As we saw above, if we consider a transformation that only combines modes associated with the same frequency, the Hamiltonian takes the form (4C.6), and it commutes with each number operator $\hat{N}_m = \hat{b}_m^\dagger \hat{b}_m$. The mode m photons thus have a well-defined energy $\hbar\omega_m$. However, there is in general no reason why momentum $\hat{\mathbf{P}}_\mathrm{R}$ should be expressible in terms of operators $\hat{N}_m$, so there is no reason why the states $|m : n_m\rangle$ should be eigenstates of $\hat{\mathbf{P}}_\mathrm{R}$. This comment applies equally well to the intrinsic angular momentum $\hat{\mathbf{S}}_\mathrm{R}$ and the orbital angular momentum $\hat{\mathbf{L}}_\mathrm{R}$ introduced in Complement 4B. As indicated in this chapter, *a photon does not necessarily have a well-defined momentum or intrinsic angular momentum.*

4C.1.6 Example: 1D standing wave modes

To simplify the example, we restrict ourselves here to waves depending on only one coordinate z and linearly polarized along Ox. This is therefore a scalar field, whose analytic signal can be expanded in terms of a basis,

$$u_\ell(z) = \mathrm{e}^{\mathrm{i}k_\ell z}, \tag{4C.15}$$

where

$$k_\ell = \ell\frac{2\pi}{L}, \quad \ell \text{ a positive or negative integer.} \tag{4C.16}$$

A state $|\ell : n_\ell\rangle$ describes n_ℓ photons of energy $\hbar\omega_\ell$ and momentum $\hbar k_\ell$.

For the positive integer number m, we construct

$$\hat{b}_{m_+}^\dagger = \frac{1}{\sqrt{2}}\left(\hat{a}_{\ell=m}^\dagger + \hat{a}_{\ell=-m}^\dagger\right) \tag{4C.17}$$

$$\hat{b}_{m_-}^\dagger = \frac{1}{\sqrt{2}}\left(\hat{a}_{\ell=m}^\dagger - \hat{a}_{\ell=-m}^\dagger\right). \tag{4C.18}$$

These relations define a unitary transformation of the form discussed in Section 4C.1.1. The operators $\hat{b}_{m_+}^\dagger$ and $\hat{b}_{m_-}^\dagger$ are creation operators for photons in the modes defined by

$$v_{m_+} = \frac{1}{\sqrt{2}}\left(e^{ik_m z} + e^{-ik_m z}\right) = \sqrt{2}\cos k_m z \tag{4C.19}$$

$$v_{m_-} = \frac{1}{\sqrt{2}}\left(e^{ik_m z} - e^{-ik_m z}\right) = i\sqrt{2}\sin k_m z. \tag{4C.20}$$

These functions are standing-wave modes, with respect to which any function on the interval $[0, L]$ can be expanded, taking

$$k_m = m\frac{2\pi}{L}, \quad m \text{ a positive integer.} \tag{4C.21}$$

What are the properties of the photons for this new basis? Consider, for example, a one-photon state of the form

$$|m_+ : 1\rangle = \hat{b}^\dagger_{m_+}|0\rangle = \frac{1}{\sqrt{2}}\left(\hat{a}^\dagger_{\ell=m}|0\rangle + \hat{a}^\dagger_{\ell=-m}|0\rangle\right). \tag{4C.22}$$

This photon has energy $\hbar\omega_m$. But if we measure its momentum, we find either $+\hbar k_m$ or $-\hbar k_m$, with equal probability 1/2.

Consider now the squared magnitude of the electric field, since this is the quantity detected by a *photoelectric detector* or photographic plate. The electric field operator has the form

$$\hat{E}^{(+)}(z) = i\sum_{m_+} \mathcal{E}^{(1)}_{m_+} v_{m_+}(z)\,\hat{b}_{m_+} + i\sum_{m_-} \mathcal{E}^{(1)}_{m_-} v_{m_-}(z)\,\hat{b}_{m_-}. \tag{4C.23}$$

The signal to which the detector is sensitive at point z has the form (see Section 5.1 of Chapter 5)

$$w^{(1)}(z) = \langle\psi|\hat{E}^{(-)}(z)\,\hat{E}^{(+)}(z)|\psi\rangle = \|\hat{E}^{(+)}(z)|\psi\rangle\|^2. \tag{4C.24}$$

Applied to the state in (4C.22), this yields

$$w^{(1)}(z) = \left[\mathcal{E}^{(1)}_{m_+}\right]^2 2\cos^2\frac{2\pi m}{L}z. \tag{4C.25}$$

A photoelectric detector will reveal a spatial modulation, and in particular the nodes and antinodes of the standing wave. This is very different from a travelling plane wave, for which the photoelectric signal is uniform throughout space.

4C.1.7 Choosing the best mode basis to suit a physical situation

We have just calculated the photodetection signal for a photon in state $|m_+ : 1\rangle$. In the stationary-wave basis, this calculation turned out to be particularly simple, but it is obvious that it could easily have been done in the plane-wave basis. Indeed, the field is given by

$$\hat{E}^{(+)}(z) = i\sum_{\ell} \mathcal{E}^{(1)}_{\ell}\, e^{ik_\ell z}\hat{a}_\ell, \tag{4C.26}$$

whence,

$$\hat{E}^{(+)}(z)\left(\frac{1}{\sqrt{2}}\hat{a}^{\dagger}_{\ell=m}|0\rangle+\frac{1}{\sqrt{2}}\hat{a}^{\dagger}_{\ell=-m}|0\rangle\right)=\mathrm{i}\frac{\mathcal{E}^{(1)}_{m}}{\sqrt{2}}\left(\mathrm{e}^{\mathrm{i}k_m z}+\mathrm{e}^{-\mathrm{i}k_m z}\right)|0\rangle, \tag{4C.27}$$

which yields

$$w^{(1)}(z)=2\left[\mathcal{E}^{(1)}_{m}\right]^{2}\cos^{2}k_m z, \tag{4C.28}$$

that is, the same result as in (4C.25).

In principle, any basis can be used to carry out a calculation, but in practice, it is worth trying to find a basis suited to the task at hand. Hence, if we have a photon corresponding to a stationary wave, use of the basis (4C.19–4C.20) gives the $\cos^2 kz$ modulation of the photodetection signal directly, while this modulation only appears in (4C.28) after adding amplitudes that give an interference term. In more complex situations, a judicious choice of basis is advisable. Hence, in the case of a source emitting photons in a particular mode, determined for example by a resonant cavity, as happens for a single-mode laser, it is most convenient to quantize the field with respect to this mode, and it would be cumbersome to try to do otherwise. Generally speaking, choice of basis is guided by symmetry considerations. Hence, in principle, in free space, invariant under translation, a travelling plane-wave basis constitutes a judicious choice. But if a perfect mirror imposes a boundary condition at $z=0$, it is better to use stationary-wave modes. If we are interested in emission from a source located at the coordinate origin, a multipole expansion generally proves to be an excellent choice. These considerations are particularly useful when dealing with spontaneous emission, which can in principle be produced in all accessible modes; it often happens that an appropriate choice of basis can considerably limit the number of modes that need to be taken into account.

4C.2 Photons in a wave packet

Consider the transformation (4C.2), but assume now that we no longer restrict to a situation in which we only combine modes with the same frequency ω_ℓ. The expression for the radiation Hamiltonian $\hat{H}_{\mathrm{R}}$ in terms of $\hat{b}_m$ and $\hat{b}^{\dagger}_m$ will contain cross-terms of the form $\hat{b}^{\dagger}_m\hat{b}_{m'}$ with m' different from m, and the radiation can no longer be considered as a sum of decoupled oscillators. We are no longer dealing with normal modes here. This does not mean that such transformations have no use. If the transformation is unitary, the commutation relations still hold, i.e.

$$\left[\hat{b}_m,\hat{b}^{\dagger}_{m'}\right]=\delta_{m,m'}, \tag{4C.29}$$

and we may consider the eigenstates of $\hat{N}_m=\hat{b}^{\dagger}_m\hat{b}_m$, generated in the usual way from the vacuum:

$$|m:n_m\rangle=\frac{1}{\sqrt{n_m!}}(\hat{b}^{\dagger}_m)^{n_m}\,|0\rangle. \tag{4C.30}$$

Such a state can be considered to contain 'n_m photons' of type m. A particularly interesting state is the one-photon state, which is given by (4C.13) at time $t = 0$. At some later time t, it becomes

$$|m:1\rangle(t) = \sum_\ell U_m^\ell \, \mathrm{e}^{-\mathrm{i}\omega_\ell t} |\ell:1\rangle. \tag{4C.31}$$

Although it is not an eigenstate of the total Hamiltonian $\hat{H}_\mathrm{R}$, such a state remains at any time an eigenstate of the total photon number operator $\hat{N} = \sum_\ell \hat{a}_\ell^\dagger \hat{a}_\ell$ (Equation 4.125), with eigenvalue 1. It is therefore a one-photon state.

As in Section 4C.1.2, a profile of the classical electromagnetic field can be associated with each particular value of m, but this profile will be time dependent. To find it, let us calculate the action of $\hat{\mathbf{E}}^{(+)}(\mathbf{r}, t)$ on $|m:1\rangle(t)$:

$$\hat{\mathbf{E}}^{(+)}(\mathbf{r})|m:1\rangle(t) = \left[\sum_\ell \mathcal{E}_\ell^{(1)} \, \boldsymbol{\varepsilon}_\ell \, \mathrm{e}^{\mathrm{i}(\mathbf{k}_\ell\cdot\mathbf{r}-\omega_\ell t)} \, U_m^\ell \right] |0\rangle = \hat{\mathbf{E}}^{(+)}(\mathbf{r}, t)|0\rangle. \tag{4C.32}$$

The analytic signal $\hat{\mathbf{E}}^{(+)}(\mathbf{r}, t)$ describes the generalized mode of which $|m:1\rangle$ is a one-photon state. Since there is a distribution of frequencies ω_ℓ, this profile evolves in time. In Chapter 5 and Complement 5B, we discuss the case where $\hat{\mathbf{E}}^{(+)}(\mathbf{r}, t)$ describes a propagating wave packet, and $|m:1\rangle$ then describes a one-photon wave packet.

5 Free quantum radiation

In this chapter we shall be using the formalism of Chapter 4 to describe some properties of the quantized free electromagnetic field, emphasizing the analogies and differences with classical electromagnetism. The analogies explain why classical optics was so successful, to the extent that, even though light was not quantized, it was able to describe almost all known optical phenomena up until the 1970s. The differences highlight typical quantum behaviour, with no classical equivalent, which lies at the heart of modern quantum optics.

The chapter begins by presenting formal ways of describing the main components of quantum optics experiments, namely, photodetectors and semi-reflecting mirrors. One can then understand homodyne detection, used to measure field quadrature components, even at optical frequencies for which no detector is fast enough to follow the oscillations of the field itself.

Section 5.2 deals with the quantum vacuum where, in contrast to the classical vacuum, radiation has properties, in particular, fluctuations, with which one can associate physical effects.

We then discuss in Section 5.3 some quantum radiation states in the simple case where a single mode ℓ of the field is excited (non-vacuum). We introduce the idea of a quasi-classical state, or coherent state, used to make the link with classical optics. We shall also study other states, those with a definite number of photons, or squeezed states, which exhibit typically quantum properties, quite inconceivable in classical electromagnetism, and we shall show how these properties can be understood by reference to the Heisenberg relations, with an extremely useful graphical representation.

We then discuss multimode states (Section 5.4). It is still possible to define quasi-classical states to make a link with classical optics in the general case. We also describe a state containing a single multimode photon, which has no classical equivalent. There is a very large class of other non-classical multimode states called entangled states. We shall only outline their definition, leaving explicit examples to the complements.

Returning to the one-photon and quasi-classical multimode states, Section 5.5 contains a detailed examination of the behaviour of these states in an interferometer. This case study will help us to understand why classical optics has been so successful, while at the same time illustrating the need for quantum optics in order to account for phenomena observed with non-classical states. After tackling the question of a possible wavefunction for the photon (Section 5.6), we end this chapter by stressing the strength of the quantum optics formalism, which provides a unified description of both particle and wave aspects of light, and which lies at the heart of applications of optics to the field of quantum information.

This chapter is rather long, but the issues discussed here are key to the study of quantum optics, and the reader who wishes to get to grips with this field is encouraged to neglect none of them. However, for a first reading, Section 5.4 on multimode states could be skipped. On the other hand, it will be absolutely essential to attend to this section before moving on to the complements and subsequent chapters.

The ideas introduced in this chapter, and in particular the typically quantum properties of various non-classical states, are exemplified in several complements. Complement 5A is devoted to squeezed states. These allow one to carry out measurements with better accuracy than would be possible according to the 'standard quantum limits'. Squeezed states are likely to considerably increase the range of large gravitational wave interferometers. Complement 5B develops the quantum formalism for both quasi-classical and one-photon wave packets. Some of the results in this complement are used in Sections 5.4 and 5.5 of this chapter. There is also a description of a remarkable two-photon quantum effect: the coalescence of two one-photon wave packets on a semi-reflecting mirror.

Complement 5C gives a simple but striking example of an entangled state, namely, two photons with entangled polarizations. This type of state has been used experimentally to resolve the problem known as the 'Einstein–Podolsky–Rosen paradox'. By observing a violation of Bell's inequalities for pairs of photons prepared in such a state, the long-standing disagreement between Einstein and Bohr, running since 1935, was finally decided. Complement 5D describes another striking example of entangled states, namely the two-mode states, which are formally rather simple, but have intriguing properties. Finally, Complement 5E turns to a very active field of research, quantum information, where entanglement is exploited in new ways of processing and transmitting data.

5.1 Photodetectors and semi-reflecting mirrors. Homodyne detection of the quadrature components

The formalism set up in the last chapter can be used to describe quantized radiation in a homogeneous space and to determine the result of measurements of the various fields **A** (in Coulomb gauge), **E** or **B** at the point **r**. But this formalism is inadequate to describe a standard optical experiment, where the space is not uniform but contains elements like mirrors or semi-reflecting mirrors able to reflect all or part of the incident waves. Furthermore, in contrast to Hertzian waves where the field detected by an antenna can be visualized on an oscilloscope, there is still no detector fast enough to measure the fields of visible light, which oscillate at frequencies greater than 10^{14} Hz, as a function of time. However, there are photodetectors sensitive to the average of the square of the electric field. In order to handle quantum optics experiments, we explain in this section how to describe both photodetectors and semi-reflecting mirrors in the framework of quantum optics. We then apply these new elements of the formalism to a technique known as homodyne detection, used to determine the quadrature components of the electric field, even when the detectors are too slow to monitor the oscillations of the field itself. This section appeals to some results established later in the chapter, and while the reader might return to this after reading the whole chapter, it is not recommended to skip it on a first reading.

5.1.1 Photodetection

Photocurrent and count rate. Classical radiation

Photodetectors available at the present time, e.g. the eye, photographic plates, photodiodes, photomultipliers and CCD cameras, all measure a 'light intensity', i.e. the time average of the square of the electric field. A classical model of photodetection, in which the detector is quantized but the light is described by a classical electromagnetic field (Chapter 2), shows that the signal appears in the form of an electric current $i(\mathbf{r}, t)$, the 'photocurrent', with average value equal to

$$\overline{i(\mathbf{r}, t)} = s_d |\mathbf{E}^{(+)}(\mathbf{r}, t)|^2 = s_d\, \mathbf{E}^{(-)}(\mathbf{r}, t)\, \mathbf{E}^{(+)}(\mathbf{r}, t). \tag{5.1}$$

The quantity $\mathbf{E}^{(+)}(\mathbf{r}, t)$ is the analytic signal (or complex amplitude, see Section 4.3.4) of the classical electric field and, for a monochromatic wave, $\mathbf{E}^{(-)}\mathbf{E}^{(+)}$ is the time average of the square of the field, up to a factor of 2. The detector, assumed very small, is placed at point $\mathbf{r}$. It is characterized by the sensitivity s_d, which depends on the kind of detector and its size. The average indicated by the overbar must be understood in a statistical sense, imagining the experiment repeated many times, fluctuations being inherent in the photoelectric detection process.

At low intensities, photodetectors like photomultipliers or avalanche photodiodes with very low intrinsic noise can operate in another way. *Very short surges of current* with a random distribution are observed, like the ones appearing on a particle counter. These are isolated events, or 'clicks', that can be counted, and this is known as the *photon counting regime*. Despite the name, this regime can still be described semi-classically by classical radiation interacting with a quantum detector, e.g. an ionizable atom placed at point $\mathbf{r}$ (see Complement 2E). The electron released by this process triggers a cascade that can build up to a detectable macroscopic pulse. Using the Fermi golden rule, the ionization probability of the atom per unit time is found to be

$$w(\mathbf{r}, t) = s\, \mathbf{E}^{(-)}(\mathbf{r}, t)\cdot \mathbf{E}^{(+)}(\mathbf{r}, t). \tag{5.2}$$

The quantity $w(\mathbf{r}, t)\mathrm{d}t$ is the probability of detecting a click near point $\mathbf{r}$, between t and $t + \mathrm{d}t$. It is equal to the average number of pulses counted between t and $t + \mathrm{d}t$, with the average being taken over a large number of identical experiments. The sensitivity s in the counting regime is equal to $s_\mathrm{d}/q_\mathrm{e}$, where q_e is the charge on the electron, and s_d is the sensitivity of the detector when there is no cascade process (see 5.1).

Comments

(i) In the semi-classical model, the fluctuations between successive outcomes are due to the quantum nature of the detector. For a given electromagnetic field, the probability per unit time (5.2) of producing a charge q_e at any point of the detector can be calculated. This is used to find the average current over the whole surface of the detector, i.e.

$$\bar{i} = q_\mathrm{e} \int_\mathrm{det} \mathrm{d}^2 r\, w(\mathbf{r}, t). \tag{5.3}$$

Since the precise moments when the charges are produced have a random distribution, the current fluctuates, and these fluctuations are modelled by taking into account the fact that the elementary detectors are quantum and independent.

(ii) If the field has spectral components distributed over a band $\Delta\omega$, the photocurrent $i(\mathbf{r}, t)$ and the photodetection rate $w(\mathbf{r}, t)$ vary with a characteristic time that cannot be shorter than $1/\Delta\omega$. These variations appear in the product $\mathbf{E}^{(-)}(\mathbf{r}, t)\mathbf{E}^{(+)}(\mathbf{r}, t)$. We assume that the detector is fast enough to monitor these variations. If not, a time average must be taken over a window equal to the response time of the detector.

Quantized radiation

The above model (a detector atom with a tendency to ionize under the effect of the radiation, together with a cascade process allowing detection of individual photoelectrons) can be adapted to the case of quantized radiation. The interaction with the detector is described with the formalism to be discussed in Chapter 6, and it is found that the average count rate at time t for a detector at $\mathbf{r}$ is given by[1]

$$w(\mathbf{r}, t) = s\langle\psi(t)|\hat{\mathbf{E}}^{(-)}(\mathbf{r}) \cdot \hat{\mathbf{E}}^{(+)}(\mathbf{r})|\psi(t)\rangle, \tag{5.4}$$

where $|\psi(t)\rangle$ is the radiation state at time t, the operator $\hat{\mathbf{E}}^{(+)}(\mathbf{r})$ is the positive frequency part of the electric field observable, and $\hat{\mathbf{E}}^{(-)}(\mathbf{r})$ is the Hermitian conjugate. This formula is clearly the quantum analogue of the semi-classical result (5.2). There are analogous expressions for more sophisticated quantities. One very important quantity is *the probability of two simultaneous detections* at points $\mathbf{r}_1$ and $\mathbf{r}_2$ at time t (the double or *coincidence click*), given by

$$w^{(2)}(\mathbf{r}_1, \mathbf{r}_2, t) = s^2 \sum_{i,j=x,y,z} \langle\psi(t)|\hat{E}_i^{(-)}(\mathbf{r}_1)\cdot\hat{E}_j^{(-)}(\mathbf{r}_2)\cdot\hat{E}_j^{(+)}(\mathbf{r}_2)\cdot\hat{E}_i^{(+)}(\mathbf{r}_1)|\psi(t)\rangle. \tag{5.5}$$

More precisely, $w^{(2)}(\mathbf{r}_1, \mathbf{r}_2, t)\mathrm{d}t_1\mathrm{d}t_2$ gives the probability of recording a click at $\mathbf{r}_1$ between times t and $t + \mathrm{d}t_1$ and also a click at $\mathbf{r}_2$ between times t and $t + \mathrm{d}t_2$. This probability can be measured with two photodetectors and an electronic circuit working as a coincidence counter.

Comments

(i) If the radiation is intense enough, the pulses will overlap and one detects a photocurrent $i(\mathbf{r}, t)$ with average value given by (5.4) up to a multiplicative factor. For its part, expression (5.5) is replaced by the correlation function $\overline{i(\mathbf{r}_1, t)\, i(\mathbf{r}_2, t)}$, where the overbar indicates a statistical average over a large number of runs.

(ii) For quantized radiation and an ideal detector, the randomness is related to the quantum nature of the radiation. There are in fact detectors with quantum efficiency close to 100%, which transform each photon into an elementary charge with absolute certainty. In this case, the fluctuations in the current correspond to situations where the number of photons interacting with the detector is fluctuating (see, for example, Section 5.3.4).

[1] See for example CDG II, Complement A_{II}.

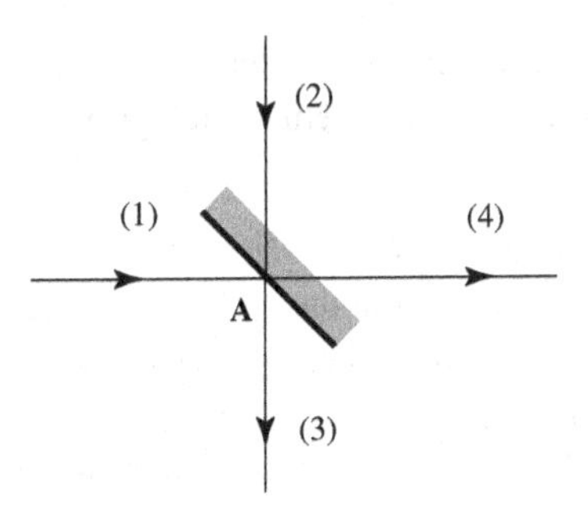

Figure 5.1 The semi-reflecting mirror couples at point A the two fields entering via ports (1) and (2) with the fields leaving by ports (3) and (4). (Suitable transparent films have been deposited on the first face of the mirror so that it has a reflection coefficient of 50%, while an anti-reflecting coating has been deposited on the second face, which plays no role.)

5.1.2 Semi-reflecting mirror

Consider a light beam (1) incident on a semi-reflecting mirror (see Figure 5.1). After the encounter, the beam divides into a transmitted beam (4) and a reflected beam (3). There is a second input port (2), symmetric to (1) with respect to the plane of the mirror, on which a second incident beam would give the same output beams (4) and (3) by reflection and transmission, respectively. To simplify, we assume that all the beams are polarized perpendicularly to the plane of the figure, and we thus treat the electric fields as scalar quantities. We also assume that each beam corresponds to a travelling plane-wave mode, all modes having the same frequency.

In classical optics, continuity relations are imposed on the classical electromagnetic field at the interface of the various layers deposited on the surface of the mirror. This leads to the following relations between the complex amplitudes of the incoming and outgoing plane waves, evaluated at the same point of the mirror, e.g. at A:

$$E_3^{(+)} = \frac{1}{\sqrt{2}}\left(E_1^{(+)} + E_2^{(+)}\right), \tag{5.6}$$

$$E_4^{(+)} = \frac{1}{\sqrt{2}}\left(E_1^{(+)} - E_2^{(+)}\right). \tag{5.7}$$

These relations are valid for a lossless mirror, where the reflected and transmitted beams have the same intensity. The negative sign in the second equation ensures equality of the light power incident on the mirror, proportional to $|E_1^{(+)}|^2 + |E_2^{(+)}|^2$, and the outgoing light power, proportional to $|E_3^{(+)}|^2 + |E_4^{(+)}|^2$.

On the quantum level, the state of the field after the mirror must be expressed as a function of the state of the field before the mirror. The transformation is described by a linear operator acting on the state vector characterizing the incoming radiation. Measurement results are then obtained by taking expectation values of the outgoing field operators in the transformed state. This approach turns out to be rather long-winded and often leads to difficult calculations.

There is a second method for handling this problem, and other similar problems of transformation by an optical system, which proves to be much more convenient. The procedure is as follows:

(1) The field operators $\hat{E}_3$ and $\hat{E}_4$ in the outgoing modes are expressed in terms of the field operators $\hat{E}_1$ and $\hat{E}_2$ in the incoming modes.
(2) The radiation is described by its state in the incoming space.
(3) Measurement results are calculated by taking quantum expectation values of the outgoing field operators expressed in terms of the incoming field operators, as described in (1), while the radiation is still described by the incoming state vector, as stated in (2).

The point about this method is that *the transformation of the field operators $\hat{E}_i^{(+)}$ referred to in (1) is often the same as the transformation of the corresponding classical fields $E_i^{(+)}$.* One can then go straight to the relations known from classical optics when treating the corresponding problem in quantum optics. This is indeed the case for the semi-reflecting mirror. Equations (5.6–5.7) for the complex classical fields at the beamsplitter do indeed transpose to the complex field operators in the form

$$\hat{E}_3^{(+)} = \frac{1}{\sqrt{2}}\left(\hat{E}_1^{(+)} + \hat{E}_2^{(+)}\right), \tag{5.8}$$

$$\hat{E}_4^{(+)} = \frac{1}{\sqrt{2}}\left(\hat{E}_1^{(+)} - \hat{E}_2^{(+)}\right). \tag{5.9}$$

Even though we have not proven these relations, they are plausible enough, because they imply that the expectation values of the quantum fields transform in exactly the same way as the classical fields. They also guarantee that the commutators of the creation and annihilation operators of the outgoing modes, which are proportional to $\hat{E}^{(-)}$ and $\hat{E}^{(+)}$, respectively (see formula (4.109)), do indeed take the values required for the quantized radiation, i.e. $[\hat{a}_3, \hat{a}_3^\dagger] = 1$ and $[\hat{a}_4, \hat{a}_4^\dagger] = 1$.

Comments

(i) The above method is justified by observing that the expressions for the signals involve the operators $\hat{O}$ via expectation values $\langle\psi|\hat{O}|\psi\rangle$, which are invariant under the effect of a unitary transformation carried out on both $|\psi\rangle$ and $\hat{O}$. This is analogous to the argument justifying the equivalence of the Heisenberg and Schrödinger pictures in quantum mechanics.
(ii) Relations (5.6–5.7) imply a particular choice for the phases of the fields. In general, phase factors can be included in the coefficients, provided that the transformation remains unitary. The same applies to (5.8–5.9). The reader should not therefore be surprised to find different forms of these relations.

5.1.3 Homodyne detection

With today's technology, photodetectors are unable to follow the extremely fast oscillation of the electric field in a light wave, and hence to obtain the phase of this field directly. The device we are about to describe beats the observed field with a reference field, called the local oscillator, thereby providing indirect access to this information.

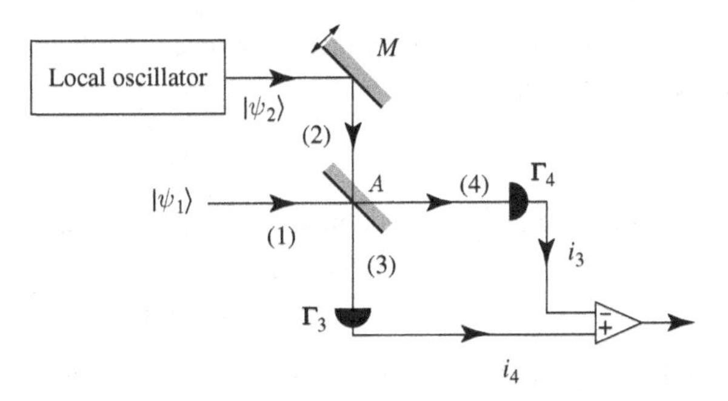

Figure 5.2 Balanced homodyne detection device. A semi-reflecting mirror at *A* produces interference between a local oscillator and radiation in mode (1). The radiation entering ports (1) and (2) is single-mode and the same frequency in each. The difference between the photocurrents provides a way of measuring the quadrature components of the radiation in the incoming state $|\psi_1\rangle$. The phase of the wave produced by the local oscillator is controlled by translating the mirror *M*, so that one can choose the quadrature to be measured.

The device is illustrated schematically in Figure 5.2. The field to be measured, in an arbitrary state $|\psi_1\rangle$ of mode 1, is mixed on a semi-reflecting mirror with the field produced by an intense single-mode laser (well above threshold), which produces a field in a quasi-classical state $|\psi_2\rangle = |\alpha_2\rangle$, an eigenstate of the operator $\hat{a}_2$ with eigenvalue α_2. We shall see in Section 5.3.4 that this kind of radiation state has properties very close to a classical monochromatic field of complex amplitude:

$$E_2^{(+)}(\mathbf{r}, t) = \mathrm{i}\,\mathcal{E}_2^{(1)}\alpha_2\,\mathrm{e}^{-\mathrm{i}\omega_2 t}\,\mathrm{e}^{\mathrm{i}\mathbf{k}_2\cdot\mathbf{r}}, \tag{5.10}$$

where ω_2 and $\mathbf{k}_2$ are the frequency and wavevector of mode 2, $\mathcal{E}_2^{(1)}$ is the 'one-photon amplitude' for mode 2 (see Section 4.5.1), and $\alpha_2 = |\alpha_2|\,\mathrm{e}^{\mathrm{i}\varphi_2}$ is a dimensionless complex number. The frequency ω_2 is the same as that of mode 1, which is being analysed. The phase φ_2 of the incoming field on the semi-reflecting mirror can be adjusted by moving the mirror M, e.g. by means of a piezoelectric stage. Two photodetectors placed at $\mathbf{r}_3$ and $\mathbf{r}_4$ measure the photocurrents i_3 and i_4, while a differential amplifier determines the averaged difference $\overline{i_3 - i_4}$, proportional to (see Section 5.1.1)

$$\overline{d} = \langle\psi|\hat{E}^{(-)}(\mathbf{r}_3)\,\hat{E}^{(+)}(\mathbf{r}_3)|\psi\rangle - \langle\psi|\hat{E}^{(-)}(\mathbf{r}_4)\,\hat{E}^{(+)}(\mathbf{r}_4)|\psi\rangle. \tag{5.11}$$

This expression of the difference signal is calculated using the procedure discussed in Section 5.1.2. The transformation (5.8–5.9) of the fields on the semi-reflecting mirror yields, using expression (4.109) to write the input fields,

$$\hat{E}^{(+)}(\mathbf{r}_3) = \frac{\mathrm{i}}{\sqrt{2}}\,\mathcal{E}^{(1)}\,\mathrm{e}^{\mathrm{i}\mathbf{k}_3\cdot\mathbf{r}_3}(\hat{a}_1 + \hat{a}_2) \tag{5.12}$$

$$\hat{E}^{(+)}(\mathbf{r}_4) = \frac{\mathrm{i}}{\sqrt{2}}\,\mathcal{E}^{(1)}\,\mathrm{e}^{\mathrm{i}\mathbf{k}_4\cdot\mathbf{r}_4}(\hat{a}_1 - \hat{a}_2). \tag{5.13}$$

Since all modes have the same frequency ω_2, we use the same constant $\mathcal{E}^{(1)}$ for all the fields. These expressions can be substituted into (5.11), provided that the state of radiation

$|\psi\rangle$ has been expressed in the incoming space. As the two beams are independent, $|\psi\rangle$ is the tensor product:

$$|\psi\rangle = |\psi_1\rangle \otimes |\alpha_2\rangle. \tag{5.14}$$

Substituting (5.12) and (5.13) into (5.11), we obtain the average difference signal as

$$\overline{d} = -\left[\mathcal{E}^{(1)}\right]^2 \langle\psi(t)|\hat{a}_1^\dagger\hat{a}_2 + \hat{a}_1\hat{a}_2^\dagger|\psi(t)\rangle. \tag{5.15}$$

With the help of (5.75), to be discussed in Section 5.3.4, the term relating to the quasi-classical state of mode 2 is found to be

$$\langle\alpha_2(t)|\hat{a}_2|\alpha_2(t)\rangle = \alpha_2\,\mathrm{e}^{-\mathrm{i}\omega_2 t} = |\alpha_2|\,\mathrm{e}^{\mathrm{i}\varphi_2}\,\mathrm{e}^{-\mathrm{i}\omega_2 t}. \tag{5.16}$$

Likewise, for a state in a single mode of the field, with frequency ω_2, the expression (5.52) for $|\psi_1(t)\rangle$ gives

$$\langle\psi_1(t)|\hat{a}_1|\psi_1(t)\rangle = \mathrm{e}^{-\mathrm{i}\omega_2 t}\langle\psi_1(0)|\hat{a}_1|\psi_1(0)\rangle. \tag{5.17}$$

It follows that $\overline{d}$ *does not depend on the time* and can be written as a function of state $|\psi_1(0)\rangle$ of the mode 1 radiation at time $t = 0$:

$$\overline{d} = -\left[\mathcal{E}^{(1)}\right]^2 |\alpha_2|\langle\psi_1(0)|(\mathrm{e}^{-\mathrm{i}\varphi_2}\hat{a}_1 + \mathrm{e}^{\mathrm{i}\varphi_2}\hat{a}_1^\dagger)|\psi_1(0)\rangle. \tag{5.18}$$

We now introduce the mode 1 *quadrature operators* $\hat{E}_{Q1}$ and $\hat{E}_{P1}$:

$$\hat{E}_{Q1} = \mathcal{E}^{(1)}(\hat{a}_1 + \hat{a}_1^\dagger) \tag{5.19}$$

$$\hat{E}_{P1} = -\mathrm{i}\,\mathcal{E}^{(1)}(\hat{a}_1 - \hat{a}_1^\dagger). \tag{5.20}$$

The difference signal (5.18) is then

$$\overline{d} = -|\alpha_2|\mathcal{E}^{(1)}\Big(\cos\varphi_2\langle\psi_1(0)|\hat{E}_{Q1}|\psi_1(0)\rangle + \sin\varphi_2\langle\psi_1(0)|\hat{E}_{P1}|\psi_1(0)\rangle\Big). \tag{5.21}$$

By taking phase φ_2 equal to either 0 or $\pi/2$ (mod π), we can measure the expectation value of either $\hat{E}_{Q1}$ or $\hat{E}_{P1}$, respectively. Later we shall examine the significance of these observables, which are proportional to the conjugate canonical observables $\hat{Q}_\ell$ and $\hat{P}_\ell$ of the radiation introduced in Section 4.5.1. More precisely, using (4.102) and (4.103), it can be checked that (5.19–5.20) become

$$\hat{E}_{Q\ell} = \mathcal{E}_\ell^{(1)}(\hat{a}_\ell + \hat{a}_\ell^\dagger) = \mathcal{E}_\ell^{(1)}\sqrt{\frac{2}{\hbar}}\hat{Q}_\ell \tag{5.22}$$

$$\hat{E}_{P\ell} = -\mathrm{i}\,\mathcal{E}_\ell^{(1)}(\hat{a}_\ell - \hat{a}_\ell^\dagger) = \mathcal{E}_\ell^{(1)}\sqrt{\frac{2}{\hbar}}\hat{P}_\ell. \tag{5.23}$$

An analogous calculation shows that the average of the square of the difference between the photocurrents, i.e.

$$\overline{(i_3 - i_4)^2} = \overline{i_3^2} + \overline{i_4^2} - 2\,\overline{i_3 i_4}, \tag{5.24}$$

becomes

$$\overline{d^2} = \left[\mathcal{E}^{(1)}\right]^4 \left(|\alpha_2|^2 \langle \psi_1(0)|(\mathrm{e}^{-\mathrm{i}\varphi_2}\hat{a}_1 + \mathrm{e}^{\mathrm{i}\varphi_2}\hat{a}_1^\dagger)^2|\psi_1(0)\rangle + \langle \psi_1(0)|\hat{a}_1^\dagger \hat{a}_1|\psi_1(0)\rangle \right). \quad (5.25)$$

This expression shows that the quantum fluctuations of the quadratures $\hat{E}_{Q1}$ and $\hat{E}_{P1}$ can be measured if the second term is negligible, i.e. if $|\alpha_2|$ is big enough. For example, if we take $\varphi_2 = 0$, the measurement gives $\overline{d^2} = \left[\mathcal{E}^{(1)}|\alpha_2|\right]^2 \langle \psi_1(0)|\hat{E}_{Q1}^2|\psi_1(0)\rangle$.

The homodyne method is extremely useful. Indeed, it provides a way of measuring the field quadrature observables, and even their fluctuations, bypassing the problem of the slow response time of the detector. Moreover, if the local oscillator is intense enough, $|\alpha_2|$ is much bigger than 1 and there is a corresponding amplification of the signal associated with the mode 1 field, compared with the signal that would be obtained by placing the photodetector in mode 1. It is thus possible to have a much stronger signal than the technical noise of the detectors, and thus gain access to the quantum fluctuations of the radiation in mode 1.

Comments

(i) In the above calculations, time-independent results are obtained because the local oscillator and the analysed mode have the same frequency. If the local oscillator had a frequency ω_2 different from the frequency ω_1 of the analysed mode, an oscillation at frequency $\omega_1 - \omega_2$ would appear in the signal $\overline{d}$, for example, and its amplitude could be measured with a suitable electronic filter. This is known as heterodyne detection.

(ii) By repeating the measurements on field 1 many times for various values of phase φ_2 of the local oscillator, assuming these measurements to be reproducible, the mode 1 radiation state can be completely characterized and the state vector (or its density matrix when the state is not pure) can be reconstructed. This is known as quantum tomography.[2]

(iii) Homodyne detection can be analysed using semi-classical theory starting from (5.1). If a weak field, described classically by

$$E_1^{(+)} = \mathcal{E}_1\,\mathrm{e}^{-\mathrm{i}\omega t} = |\mathcal{E}_1|\,\mathrm{e}^{\mathrm{i}\varphi_1}\,\mathrm{e}^{-\mathrm{i}\omega t} \quad (5.26)$$

interferes with a strong local oscillator field,

$$E_2^{(+)} = \mathrm{i}|\mathcal{E}_2|\,\mathrm{e}^{\mathrm{i}\varphi_2}\,\mathrm{e}^{-\mathrm{i}\omega t}, \quad (5.27)$$

one obtains, in a balanced detection setup,

$$\overline{i_3 - i_4} = s_d|E_3^{(+)}|^2 - s_d|E_4^{(+)}|^2. \quad (5.28)$$

Using (5.6–5.7), this signal is

$$\overline{i_3 - i_4} = s_d|E_1^{(+)}\,E_2^{(-)} + \text{c.c.}| = s_d|\mathcal{E}_2|\left(-\mathrm{i}\mathcal{E}_1\mathrm{e}^{-\mathrm{i}\varphi_2} + \mathrm{i}\mathcal{E}_1^*\mathrm{e}^{\mathrm{i}\varphi_2}\right). \quad (5.29)$$

If the phase of the local oscillator is fixed at $\varphi_2 = 0$, one measures $E_{Q1} = -\mathrm{i}(\mathcal{E}_1 - \mathcal{E}_1^*)$, i.e. the amplitude of the field at times $t = \pi/2\omega, 5\pi/2\omega$, etc. Likewise, taking $\varphi_2 = \frac{\pi}{2}$, one obtains $E_{P1} = -\mathrm{i}(\mathcal{E}_1 + \mathcal{E}_1^*)$, i.e. the amplitude of the field $E_1(t)$ at times $t = 0, 2\pi/\omega$, etc. Equation (5.29) can still be written in the form

$$\overline{i_3 - i_4} = s_d|\mathcal{E}_2|\left(\cos\varphi_2 E_{Q1} + \sin\varphi_2 E_{P1}\right), \quad (5.30)$$

analogous to (5.21).

[2] See, for example, H. Bachor and T. Ralph, *A Guide to Experiments in Quantum Optics*, Wiley (2004).

This classical calculation allows one to interpret E_{Q1} or E_{P1} as a kind of stroboscopic view of the field at times $0, T, 2T$, etc., or $T/4, 5T/4$, etc., where T is the period $2\pi/\omega$. It also shows the amplification effect proportional to the amplitude $|\mathcal{E}_2|$ of the local oscillator.

5.2 The vacuum: ground state of quantum radiation

In this section, we discuss the properties of the ground state $|n_1 = 0, \ldots, n_\ell = 0, \ldots\rangle$, which contains zero photons in each mode. This is denoted more simply by $|0\rangle$. It is the radiation state when there is no light source. It is traditionally called the vacuum, but it might be better referred to as darkness. Indeed, we shall see that the vacuum is not in fact nothing. It has specific properties, like the ground state of any other quantum system.

The vacuum was defined in Section 4.6.1 by the property (4.116) applied to the full set of modes: in the vacuum, one cannot destroy a photon. However, the properties that we shall discuss here result from the fundamental commutation relations of the field observables (see Section 4.5.1), which are in fact closely related to the non-commutativity of $\hat{a}_\ell$ and $\hat{a}_\ell^\dagger$ specified by (4.100).

5.2.1 Non-commutativity of the field operators and Heisenberg relations for radiation

For a material harmonic oscillator, we know that the Heisenberg relation, $\Delta x \, \Delta p \geq \hbar/2$, a direct consequence of the non-commutativity of the operators $\hat{x}$ and $\hat{p}$, means that the system cannot reach a state of total rest in which both x and p are simultaneously zero with absolute certainty. The lowest-energy state results from a compromise between the kinetic energy, proportional to p^2, and the potential energy, proportional to x^2, which must remain compatible with the Heisenberg relation. This is why this state has non-zero energy $\hbar\omega/2$ ('zero point energy'), and finite extension in terms of both position and momentum around a zero average value, which can be attributed to quantum fluctuations in x and p.

The same is true for the electromagnetic field. Indeed, the canonical commutation relation (4.94) implies the following Heisenberg relation, *valid for absolutely any state of quantized radiation*:

$$\Delta Q_\ell \, \Delta P_\ell \geqslant \frac{\hbar}{2}. \tag{5.31}$$

This relation implies in particular that the real and imaginary components of the vector potential cannot be simultaneously zero with certainty. This is why, even in the vacuum $|0\rangle$, the energy, but also the variances of the fields, cannot be zero. This is a *purely quantum effect*. Similar relations can be written for the real and imaginary parts of the electric field or field quadrature components (5.19) and (5.20), for which, as a consequence of (5.22–5.23) and (5.31),

$$\Delta E_{Q\ell}\,\Delta E_{P\ell} \geqslant \left[\mathcal{E}_\ell^{(1)}\right]^2. \tag{5.32}$$

5.2.2 Vacuum fluctuations and their physical consequences

For any mode ℓ, the following relations hold (see Equation 4.116):

$$\hat{a}_\ell|0\rangle = 0 \quad ; \quad \langle 0|\hat{a}_\ell^\dagger = 0. \tag{5.33}$$

Referring to (4.108–4.110) and (4.115–4.117), it follows that

$$\langle 0|\hat{\mathbf{E}}(\mathbf{r})|0\rangle = \langle 0|\hat{\mathbf{B}}(\mathbf{r})|0\rangle = \langle 0|\hat{\mathbf{A}}(\mathbf{r})|0\rangle = 0. \tag{5.34}$$

In the vacuum, as might be expected, the average values of the fields are zero.

Let us now calculate the variances of the fields in the vacuum. Applying the same relations and also

$$\langle 0|\hat{a}_j\hat{a}_k^\dagger|0\rangle = \delta_{jk}, \tag{5.35}$$

which follows from (4.116) and (4.117), one finds

$$(\Delta\mathbf{A})^2 = \langle 0|\left(\hat{\mathbf{A}}(\mathbf{r})\right)^2|0\rangle = \sum_\ell \frac{1}{\omega_\ell^2}\left[\mathcal{E}_\ell^{(1)}\right]^2 \tag{5.36}$$

$$(\Delta\mathbf{E})^2 = \langle 0|\left(\hat{\mathbf{E}}(\mathbf{r})\right)^2|0\rangle = \sum_\ell \left[\mathcal{E}_\ell^{(1)}\right]^2 = \sum_\ell \frac{\hbar\omega_\ell}{2\varepsilon_0 L^3} \tag{5.37}$$

$$(\Delta\mathbf{B})^2 = \langle 0|\left(\hat{\mathbf{B}}(\mathbf{r})\right)^2|0\rangle = \sum_\ell \frac{1}{c^2}\left[\mathcal{E}_\ell^{(1)}\right]^2 = \frac{1}{c^2}(\Delta\mathbf{E})^2. \tag{5.38}$$

The quantity $\mathcal{E}_\ell^{(1)}$ (which one might call the field for one photon in a volume L^3) was introduced in Section 4.5.1. The above expressions are time independent, since the vacuum is an eigenstate of the Hamiltonian, i.e. a stationary state.

Quantum theory thus tells us that, even in its ground state, the radiation field is the seat of fluctuations, called '*vacuum fluctuations*'. This is a quite new fundamental property of the quantum field which cannot be reduced to the existence of light quanta, and it underlies several physical phenomena.

To begin with, vacuum fluctuations explain the fact that the properties of an atom are modified by its coupling with quantized radiation, even if no photon is present in the field. These fluctuations can induce transitions between atomic levels in the phenomenon known as *spontaneous emission*, to be examined in more detail in Chapter 6. They can also shift the atomic energy levels. In particular, the Lamb shift, i.e. the tiny energy difference between levels $2S_{1/2}$ and $2P_{1/2}$ of the hydrogen atom (see Complement 3G), cannot be explained by a theory in which the atom is isolated, even using Dirac's relativistic theory. To explain the Lamb shift, one must take into account the inevitable interaction between the hydrogen atom and the radiation vacuum. We saw in Chapter 2 that, when an atom interacts with an oscillating classical field, its energy levels are shifted by an amount proportional to the square of the field. The Lamb shift arises in an analogous way as the differential shift of the hydrogen levels $2S_{1/2}$ and $2P_{1/2}$ under the effect of vacuum fluctuations.[3]

There is an analogous physical effect for an isolated electron. Very accurate experiments have shown that the magnetic moment of a free electron is not exactly $2\mu_B$ (where μ_B is the Bohr magneton), as predicted by Dirac's theory, but differs by about one thousandth. This small discrepancy can also be explained by including the coupling between the electron and radiation in its ground state. It can be interpreted as resulting from a differential effect due to fluctuations in the magnetic field of the vacuum, affecting both the cyclotron motion and the spin precession of the electron placed in a static magnetic field.[4]

Another physical effect arising from vacuum fluctuations is the Casimir effect, which refers to a force between two electrically uncharged physical objects that are nevertheless capable of influencing the distribution of the electromagnetic field. An example is two perfectly flat metal mirrors placed parallel to one another, in which case there is an attractive force F_{Casimir}, given by[5]

$$F_{\text{Casimir}} = -\frac{\pi^2 \hbar c}{240} \frac{S}{d^4}, \tag{5.39}$$

where S is the area of the mirrors and d their separation. This force dominates all others for conducting objects separated by nanometric distances. To interpret the Casimir effect, one can carry out a calculation analogous to the one at the beginning of this section, and it is found that the variance $(\Delta \mathbf{P}_R)^2$ of the radiation momentum (Complement 4B) is not zero in the vacuum. It follows that the vacuum exerts a radiation pressure on each face of a mirror. By symmetry, the total effect is zero on a single, isolated mirror, because this pressure applies equally to each side of it. However, this symmetry is broken when two mirrors are placed one opposite the other, because their presence modifies the structure of the field modes. Indeed, between the two mirrors, there are resonant and non-resonant Fabry–Perot modes due to the two mirrors (see Complement 3A). Outside, there are free space modes of arbitrary frequency. The vacuum fluctuations depend on the structure of the modes, and the average pressure effect on each of the mirrors is not zero, creating the apparent attractive force calculated by Casimir.

Beyond these qualitative considerations, it is often a delicate matter to carry out calculations involving vacuum fluctuations. Indeed, when the volume L^3 of the box tends to infinity, the sum over ℓ in (5.36–5.38) can be replaced by an integral. It is then found that the variances of the different fields are given by *divergent integrals*.[6] Likewise, if

[3] An approximate calculation of the Lamb shift can be found in CDG II, Exercise 7. The full calculation is given in detail in H. Bethe and E. Salpeter, *Quantum Mechanics of One- and Two-Electron Atoms*, Academic Press, NY (1957).

[4] J. Dupont-Roc, C. Fabre and C. Cohen-Tannoudji, Physical Interpretations for Radiative Correction in the Non Relativistic Limit, *Journal of Physics B: atomic, molecular, and optical physics* **11**, 563 (1978).

[5] See, for example, L. Mandel and E. Wolf, *Optical Coherence and Quantum Optics*, Cambridge University Press, p. 508 (1995).

[6] However, it can be shown that, if these variances are not calculated at a point **r**, but over a region of finite extent Δr about **r**, the variances are then finite.

the Lamb shift is calculated without further precaution on the basis of the expressions (5.37–5.38) for the variances of the fields in the vacuum, the sum over all field modes introduces divergent integrals, and an infinite value is obtained. In 1947, Schwinger, Feynman and Tomonoga showed that it is possible, by the procedure known as 'renormalization', to get around the problem of infinities arising in the quantum theory of radiation. They were able to calculate accurate finite values for all observable effects. The Lamb shift, which had just been measured by Lamb and Retherford, could be calculated using their method, as could the deviation of the electron magnetic moment from $2\mu_B$, just measured by Kusch. The excellent agreement between the theoretical and experimental values constitutes a spectacular vindication of the quantum theory of radiation. At the beginning of the twenty-first century, the Lamb shift and the anomalous electron magnetic moment feature among those quantities that can be measured with the greatest accuracy (relative values of 10^{-14}), and comparison between theory and experiment remains a privileged way of testing the validity of quantum electrodynamics.

In this and the following chapters, we shall restrict our discussion to problems in which sums over an infinite number of modes, i.e. those that gave rise to the divergences, are not needed. This will not hinder us in treating one of the most spectacular consequences of vacuum fluctuations, namely spontaneous emission. The reader interested in finding out more about quantum electrodynamics is referred to more specialized literature.[7]

5.3 Single-mode radiation

There are many physical situations that can be described by a single field mode. To a first approximation, a collimated light beam from a single-mode laser can be treated as a travelling plane wave of polarization $\boldsymbol{\varepsilon}_\ell$, angular frequency $\omega_\ell = c|k_\ell|$ and direction of propagation along $\mathbf{k}_\ell/|k_\ell|$. The fact that the wave has infinite transverse dimensions may lead to calculational difficulties. In this case, the plane wave can be considered as bounded within a finite transverse region $S_\perp$, large enough in comparison with the wavelength to be able to neglect diffraction effects.

More generally, and depending on the problem under consideration, it may be preferable to use normal modes other than travelling waves. For example, the Gaussian modes TEM_{pq} are well suited to a more precise description of a laser beam, including diffraction effects due to the finite transverse dimensions of the beam. In other cases, the basis of standing wave modes may turn out to be more convenient for a given physical problem (see Complement 4C). The single mode ℓ considered in a given physical situation can therefore be of many different kinds. For the purposes of simplicity, the following discussion will be restricted to the travelling plane wave described at the beginning of this section, but most of the conclusions will remain valid for any other type of mode, when a single mode is excited.

[7] See, for example, J. M. Jauch and F. Rohrlich, *The Quantum Theory of Photons and Electrons*, Addison-Wesley (1955). J. D. Bjorken and S. D. Drell, *Relativistic Quantum Mechanics*, McGraw-Hill (New York) (1964). P. Roman, *Advanced Quantum Theory*, Addison-Wesley, Reading, MA (1965). C. Itzykson and J. B. Zuber, *Quantum Field Theory*, McGraw-Hill, NY (1980).

5.3.1 Classical description: phase, amplitude and quadratures

The classical electric field in mode ℓ can be written at any point $\mathbf{r}$:

$$\mathbf{E}_\ell(\mathbf{r},t) = \boldsymbol{\varepsilon}_\ell \mathcal{E}_\ell(0)\,\mathrm{e}^{\mathrm{i}(\mathbf{k}_\ell\cdot\mathbf{r}-\omega_\ell t)} + \text{c.c.} = \mathbf{E}_\ell^{(+)}(\mathbf{r},t) + \mathbf{E}_\ell^{(-)}(\mathbf{r},t). \tag{5.40}$$

Since there is a well-defined polarization, one can simplify the expressions by use of scalar quantities $E_\ell(\mathbf{r},t)$ and $E_\ell^{(+)}(\mathbf{r},t)$ for the field and for the analytic signal.

They are defined by

$$\mathbf{E}_\ell(\mathbf{r},t) = \boldsymbol{\varepsilon}_\ell\, E_\ell(\mathbf{r},t) = \boldsymbol{\varepsilon}_\ell \left[E_\ell^{(+)}(\mathbf{r},t) + E_\ell^{(-)}(\mathbf{r},t)\right]. \tag{5.41}$$

Furthermore, we define the complex amplitude of the field in the mode ℓ by $\mathcal{E}_\ell(0)$,

$$\mathcal{E}_\ell(0) = E_\ell^{(+)}(\mathbf{r}=0, t=0). \tag{5.42}$$

The field

$$E_\ell(\mathbf{r},t) = \mathcal{E}_\ell(0)\,\mathrm{e}^{\mathrm{i}(\mathbf{k}_\ell\cdot\mathbf{r}-\omega_\ell t)} + \text{c.c.}, \tag{5.43}$$

can also be rewritten in the form

$$E_\ell(\mathbf{r},t) = -E_{P\ell}\cos(\mathbf{k}_\ell\cdot\mathbf{r} - \omega_\ell t) - E_{Q\ell}\sin(\mathbf{k}_\ell\cdot\mathbf{r} - \omega_\ell t). \tag{5.44}$$

The quantities $E_{Q\ell}$ and $E_{P\ell}$ are called the *quadrature components* of the single-mode classical field. They have already been introduced in Section 5.1.3, and can be measured by homodyne detection. They are related to the complex amplitude $\mathcal{E}_\ell(0)$ by

$$-E_{P\ell} = \mathcal{E}_\ell(0) + \mathcal{E}_\ell(0)^*, \tag{5.45}$$

$$E_{Q\ell} = -\mathrm{i}\,[\mathcal{E}_\ell(0) - \mathcal{E}_\ell(0)^*], \tag{5.46}$$

$$\mathcal{E}_\ell(0) = -\frac{E_{P\ell}}{2} + \mathrm{i}\frac{E_{Q\ell}}{2}. \tag{5.47}$$

The classical field in a mode is thus determined at any point and time, if one knows the value of the complex amplitude $\mathcal{E}_\ell(0)$, or the two real numbers $E_{Q\ell}$ and $E_{P\ell}$, or the amplitude $|\mathcal{E}_\ell(0)|$ and the phase φ_ℓ of the wave, defined by

$$\mathcal{E}_\ell(0) = |\mathcal{E}_\ell(0)|\,\mathrm{e}^{\mathrm{i}\varphi_\ell}. \tag{5.48}$$

As a function of φ_ℓ and $|\mathcal{E}_\ell(0)|$, the field (5.43) can be written

$$E_\ell(\mathbf{r},t) = 2\,|\mathcal{E}_\ell(0)|\,\cos(\mathbf{k}_\ell\cdot\mathbf{r} - \omega_\ell t + \varphi_\ell). \tag{5.49}$$

To describe this field, it is convenient to introduce a vector in the two-dimensional complex plane representing the complex amplitude $\mathcal{E}_\ell(0)$ (see Figure 5.3(a)). Its Cartesian coordinates are $-\frac{E_{P\ell}}{2}$ and $\frac{E_{Q\ell}}{2}$, i.e. the quadrature components up to a factor, see (5.47). This vector, widely used in classical electromagnetism, is known as the *phasor*.

If we could measure the field directly (as we can for radio waves), we would be interested in its value $E_\ell(\mathbf{r},t)$, which is twice the projection on the real axis of the rotating phasor

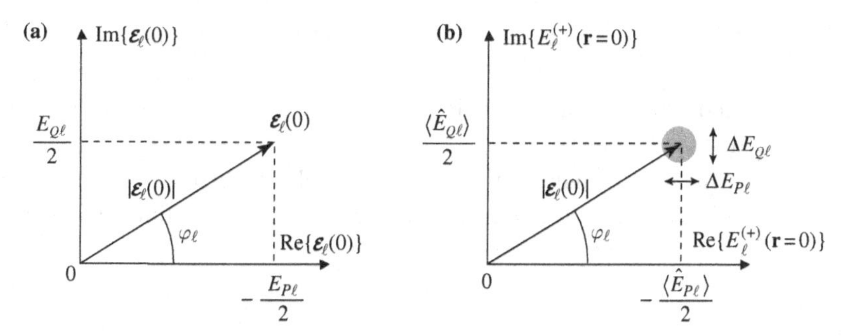

Figure 5.3 Phasor representation of the electric field in mode ℓ. **(a)** Classical field with complex amplitude $\mathcal{E}_\ell(0) = |\mathcal{E}_\ell(0)|\,e^{i\varphi_\ell}$. The quadrature components are, up to a factor of ± 2, the real and imaginary parts of $\mathcal{E}_\ell(0)$. Homodyne measurements allow one to measure the projection of $\mathcal{E}_\ell(0)$ on any axis. **(b)** Quantum field in a state where $\langle \hat{E}_\ell^{(+)}(\mathbf{r}=0)\rangle_{t=0} = |\mathcal{E}_\ell(0)|\,e^{i\varphi_\ell}$ and where the quadrature components have equal dispersions $\Delta E_{Q\ell}$ and $\Delta E_{P\ell}$ that are small compared with $\mathcal{E}_\ell$. Homodyne measurements allow one to obtain the projection of $\hat{E}_\ell^{(+)}(\mathbf{r})$ at $t = 0$ on any axis, and the results have a spread (for instance $\Delta E_{Q\ell}$ on the vertical axis). The whole set of projection measurements lead one to represent the state of the field by a random phasor, $E_\ell^{(+)}(\mathbf{r}=0)$, the end of which may lie anywhere in the grey disc. Figure 5.6 graphs the time dependence of $\langle \hat{E}_\ell(\mathbf{r}=0)\rangle_t$, and the dispersion of the measurements of its value for the case shown here.

$\mathcal{E}_\ell(0)\,e^{-i\omega_\ell t}$, i.e. a vector rotating clockwise with angular speed, ω_ℓ. However, at optical frequencies, there is no detector fast enough to follow such rapid time evolution, and we must resort to homodyne measurements (see Section 5.1.3). By choosing phase φ_2 of the local oscillator, we can then measure either quadrature component of the field in the mode, i.e. the projection of the fixed phasor $\mathcal{E}_\ell(0)$ on any axis, in particular the real or imaginary axis.

5.3.2 Single-mode quantum radiation: quadrature observables and phasor representation

Single-mode radiation states

A single-mode quantum state is by definition a state of the general form (4.131) with all photon numbers $n_1, n_2, \ldots$, equal to zero except for n_ℓ, where ℓ is the mode under consideration. The radiation state thus assumes the simple form,

$$|\psi\rangle = \sum_{n_\ell=0}^{\infty} C_{n_\ell} |n_1 = 0, \ldots, n_{\ell-1} = 0, n_\ell, n_{\ell+1} = 0, \ldots\rangle. \tag{5.50}$$

Let $\mathcal{F}_\ell$ be the Hilbert space of these mode ℓ states and write the state $|\psi\rangle$ in the simplified notation,

$$|\psi\rangle = \sum_{n_\ell=0}^{\infty} C_{n_\ell} |n_\ell\rangle. \tag{5.51}$$

The most general mode ℓ state is thus characterized by the countably infinite set of complex parameters C_{n_ℓ} (with the normalization condition $\langle\psi|\psi\rangle = 1$), while the classical single-mode state is defined by a single complex number $\mathcal{E}_\ell(0)$. A much greater variety of quantum fields is thus expected.

In order to compare with a classical field, we first calculate the expectation value of the field as a function of $\mathbf{r}$ and t. The state $|\psi\rangle$ is given at time t by

$$|\psi(t)\rangle = \mathrm{e}^{-\mathrm{i}\omega_\ell t/2} \sum_{n_\ell=0}^{\infty} C_{n_\ell} \mathrm{e}^{-\mathrm{i}n_\ell\omega_\ell t} |n_\ell\rangle. \tag{5.52}$$

This can be used to calculate the expectation value of the electric field operator (4.109). Equations (4.115–4.117) imply

$$\left\langle\psi(t)|\hat{\mathbf{E}}(\mathbf{r})|\psi(t)\right\rangle = \mathrm{i}\mathcal{E}_\ell^{(1)}\,\boldsymbol{\varepsilon}_\ell \left(\sum_{n_\ell=0}^{\infty} C^*_{n_\ell-1} C_{n_\ell}\sqrt{n_\ell}\right) \mathrm{e}^{\mathrm{i}(\mathbf{k}_\ell\cdot\mathbf{r}-\omega_\ell t)} + \text{c.c.} \tag{5.53}$$

The expectation of the electric field is thus found to vary spatio-temporally in the same way for a single-mode quantum field as for a single-mode classical field, see (5.40). The wider variety of quantum states of the field does not therefore lead to a greater range of possible time evolutions for the expectation value of the field. It is in fact in the higher-order moments of the probability distribution for the measured values of the field that this variety will manifest itself, and in particular it is in the *fluctuations* of the field measured at a given point, and in the *correlations* between measurements carried out at different points, that the most remarkable quantum effects will be observed.

Comment Note that, if a single coefficient $C_{n\ell}$ differs from 0, the expectation value of the quantum field in this state is always zero. We shall return to this point in Section 5.3.3.

Quadrature operators and the phasor representation

The interesting electric field operator for a single-mode field is the restriction $\hat{\mathbf{E}}_\ell(\mathbf{r})$ of the operator $\hat{\mathbf{E}}(\mathbf{r})$ defined in (4.109) to the subspace $\mathcal{F}_\ell$:

$$\hat{\mathbf{E}}_\ell(\mathbf{r}) = \mathrm{i}\boldsymbol{\varepsilon}_\ell\, \mathcal{E}_\ell^{(1)}\, \mathrm{e}^{\mathrm{i}\mathbf{k}_\ell\cdot\mathbf{r}}\, \hat{a}_\ell + \text{h.c.} \tag{5.54}$$

where h.c. stands for 'hermitan conjugate'. In $\mathcal{F}_\ell$, the canonical commutation relations (4.100–4.101) reduce to

$$\left[\hat{a}_\ell, \hat{a}_\ell^\dagger\right] = 1. \tag{5.55}$$

Again, since there is only one polarization, the notation can be further simplified by using the scalar operators $\hat{E}_\ell^{(+)}(\mathbf{r})$ and $\hat{E}_\ell(\mathbf{r})$:

$$\hat{E}_\ell^{(+)}(\mathbf{r}) = \mathrm{i}\,\mathcal{E}_\ell^{(1)}\hat{a}_\ell \mathrm{e}^{\mathrm{i}\mathbf{k}_\ell\cdot\mathbf{r}}, \tag{5.56}$$

$$\hat{E}_\ell(\mathbf{r}) = \mathrm{i}\,\mathcal{E}_\ell^{(1)}\left(\hat{a}_\ell \mathrm{e}^{\mathrm{i}\mathbf{k}_\ell\cdot\mathbf{r}} - \hat{a}_\ell^\dagger\, \mathrm{e}^{-\mathrm{i}\mathbf{k}_\ell\cdot\mathbf{r}}\right). \tag{5.57}$$

The Hermitian operators (observables) $\hat{E}_{Q\ell}$ and $\hat{E}_{P\ell}$ given by

$$\hat{E}_{Q\ell} = \mathcal{E}_\ell^{(1)} \left(\hat{a}_\ell + \hat{a}_\ell^\dagger\right) \tag{5.58}$$

$$\hat{E}_{P\ell} = -\mathrm{i}\,\mathcal{E}_\ell^{(1)} \left(\hat{a}_\ell - \hat{a}_\ell^\dagger\right), \tag{5.59}$$

were introduced in (5.22–5.23). They are associated with the classical single-mode field quadratures introduced in (5.45–5.46). The operators $\hat{E}_\ell^{(+)}(\mathbf{r})$ and $\hat{E}_\ell(\mathbf{r})$ can be written

$$\hat{E}_\ell^{(+)}(\mathbf{r}) = \left(-\frac{\hat{E}_{P\ell}}{2} + \mathrm{i}\frac{\hat{E}_{Q\ell}}{2}\right) \mathrm{e}^{\mathrm{i}\mathbf{k}_\ell\cdot\mathbf{r}} = \mathrm{i}\mathcal{E}_\ell^{(1)}\,\hat{a}_\ell\,\mathrm{e}^{\mathrm{i}\mathbf{k}_\ell\cdot\mathbf{r}} \tag{5.60}$$

$$\hat{E}_\ell(\mathbf{r}) = -\hat{E}_{Q\ell}\sin\mathbf{k}_\ell\cdot\mathbf{r} - \hat{E}_{P\ell}\cos\mathbf{k}_\ell\cdot\mathbf{r}. \tag{5.61}$$

As already noted in (5.22–5.23), $\hat{E}_{Q\ell}$ and $\hat{E}_{P\ell}$ are, up to a multiplicative constant, the conjugate canonical observables $\hat{Q}_\ell$ and $\hat{P}_\ell$ introduced in the quantization procedure. Their commutator is obtained from (5.55) as

$$[\hat{E}_{Q\ell}, \hat{E}_{P\ell}] = 2\mathrm{i}\left[\mathcal{E}_\ell^{(1)}\right]^2. \tag{5.62}$$

It was explained in Section 5.1.3 that the average and dispersion of field quadratures $\hat{E}_{P\ell}$ and $\hat{E}_{Q\ell}$ could be measured at time $t = 0$ (Equations (5.21) and (5.25)). The results of such measurements can then be used in a phasor diagram of the kind shown in Figure 5.3(a). Since $\hat{E}_{P\ell}$ and $\hat{E}_{Q\ell}$ do not commute, the values obtained are inevitably affected by fluctuations, or 'quantum noise', and the possible measurement results are spread over a dispersion region (see Figure 5.3(b)). A Heisenberg relation for the dispersion in these measurements is associated with the commutator (5.62):

$$\Delta E_{P\ell}\cdot\Delta E_{Q\ell} \geqslant \left[\mathcal{E}_\ell^{(1)}\right]^2. \tag{5.63}$$

The dispersion region thus has a minimal area given by (5.63). It can be considered to represent the possible values of the complex field $\hat{E}_\ell^{(+)}(\mathbf{r} = 0)$ obtained as the result of measurements, at $\mathbf{r} = 0$ and time $t = 0$. The phasor representation in Figure 5.3(b) can be used to picture what is expected when we change the phase φ_2 of the local oscillator in a homodyne measurement. This amounts to projecting $\hat{E}_\ell^{(+)}(\mathbf{r} = 0)$ onto a different axis, and the result lies in the projection of the dispersion region onto the corresponding axis. When the dispersion region is a disk, as in Figure 5.3(b), the dispersions are the same for all quadrature components. However, there are field states (see Section 5.3.5) for which the dispersion region is an ellipse, so that the dispersions are different for different quadrature components.

This representation amounts to simulating the quantum field by a random classical field, for which the various possible outcomes generate the scatter of values shown on the phasor diagram. This description is often convenient, but it should be remembered that it is only a picture of what is actually happening, and that the quantum fields lead to behaviours that would be quite incomprehensible in terms of classical fields. Several examples will be given in this and the next chapter, as well as in the complements.

Comment If we had a detector fast enough to measure the value of $\hat{E}(\mathbf{r})$ at time t, the result would be spread over a dispersion interval obtained as twice the projection on the real axis, of the phasor of Figure 5.3(b), rotating clockwise with its dispersion region attached to it, at angular frequency ω_ℓ. It is therefore clear that a homodyne measurement amounts to freezing the rapid oscillation, just like in a stroboscopic observation of a vibrating phenomenon.

5.3.3 Single-mode number state

Among all the possible states (5.51), let us begin by examining the properties of the single-mode state $|n_\ell\rangle$ for which only one coefficient C_{n_ℓ} differs from zero. This is an eigenstate of the Hamiltonian, hence stationary, and it is called a *Fock state* or *number state*. The state $|n_\ell\rangle$ is indeed an eigenstate of the photon number operator $\hat{N}_\ell$ (see Section 4.6.2), and a measurement of the number of photons by means of an ideal photodetector would give the result n_ℓ with certainty.

Now consider a measurement of the field. Since the expansion (5.52) contains only one term, the expectation value (5.53) is zero, and the same is true of the fields $\hat{\mathbf{A}}$ and $\hat{\mathbf{B}}$. For a number state, one thus has

$$\langle n_\ell|\hat{E}_\ell(\mathbf{r})|n_\ell\rangle = \langle n_\ell|\hat{B}_\ell(\mathbf{r})|n_\ell\rangle = \langle n_\ell|\hat{A}_\ell(\mathbf{r})|n_\ell\rangle = 0, \tag{5.64}$$

and this property remains true at all times. It should not be concluded that the fields are zero. Indeed, if we calculate the variance of $\hat{E}_\ell(\mathbf{r})$ given in (5.57) in the state $|n_\ell\rangle$, using (5.55), and the fact that $|n_\ell\rangle$ is an eigenstate of $\hat{a}_\ell^\dagger\hat{a}_\ell$, we obtain

$$\Delta^2 E_\ell = \langle n_\ell|\hat{E}_\ell^2(\mathbf{r})|n_\ell\rangle = (2n_\ell + 1)\left[\mathcal{E}_\ell^{(1)}\right]^2 = \frac{1}{\varepsilon_0 L^3}\left(n_\ell + \frac{1}{2}\right)\hbar\omega_\ell. \tag{5.65}$$

The variance thus increases with n_ℓ, like the energy of the state.

How can such a field be represented in a phasor diagram like the one in Figure 5.3(b)? It is tempting to plot a disc centred on the origin and of diameter ΔE_ℓ, but this would not be a good choice because the fluctuations are not small compared with the expectation values. Up to now, we have only been considering measurements of the field quadrature components, associated with a projection of the phasor on an arbitrary axis. But we know that a direct photoelectric measurement (see Section 5.1.1) corresponds to the observable $\hat{\mathbf{E}}^{(-)}(\mathbf{r})\cdot\hat{\mathbf{E}}^{(+)}(\mathbf{r})$, whose restriction to mode ℓ is

$$\hat{E}_\ell^{(-)}(\mathbf{r})\,\hat{E}_\ell^{(+)}(\mathbf{r}) = \left[\mathcal{E}_\ell^{(1)}\right]^2 \hat{a}_\ell^\dagger\hat{a}_\ell = \left[\mathcal{E}_\ell^{(1)}\right]^2 \hat{N}_\ell. \tag{5.66}$$

For a number state, this measurement gives a certain result, since $|n_\ell\rangle$ is an eigenstate of $\hat{N}_\ell$. It is tempting to consider (5.66) as the quantum observable associated with the squared modulus of the classical phasor $|\mathcal{E}_\ell(0)|^2$. But recalling that operator expressions must be symmetrized when the operators do not commute, it is more correct to consider the correspondence,

$$|\mathcal{E}_\ell(0)|^2 \quad \longleftrightarrow \quad \left[\mathcal{E}_\ell^{(1)}\right]^2 \frac{\hat{a}_\ell^\dagger\hat{a}_\ell + \hat{a}_\ell\hat{a}_\ell^\dagger}{2} = \left[\mathcal{E}_\ell^{(1)}\right]^2 \left(\hat{N}_\ell + \frac{1}{2}\right). \tag{5.67}$$

In a number state, the squared modulus of the electric field thus has a precise value, namely $\left(n_\ell + \frac{1}{2}\right)\left[\mathcal{E}_\ell^{(1)}\right]^2$.

Note that we arrive at the same conclusion by considering the sum of the squares of the quadrature components. By (5.58–5.59), this is given by

$$\hat{E}_{Q\ell}^2 + \hat{E}_{P\ell}^2 = 4\left[\mathcal{E}_\ell^{(1)}\right]^2\left(\hat{N}_\ell + \frac{1}{2}\right), \tag{5.68}$$

an expression equivalent to (5.67).

In the number state $|n_\ell\rangle$, the radiation can thus be represented by a phasor with perfectly well-defined modulus $\sqrt{n_\ell + 1/2}\,\mathcal{E}_\ell^{(1)}$, but random phase, evenly distributed between 0 and 2π (see Figure 5.4(a)). This random representation should be understood in the sense of a statistical ensemble. We imagine a large number of identically prepared systems, described by the same state $|\psi\rangle$. Each measurement is associated with another random outcome, with different phase each time. The associated time diagram is that of a statistical ensemble of sinusoidal curves, each with the same amplitude, but random phase at the origin (Figure 5.4(b)).

The number states $|n_\ell\rangle$, eigenstates of the Hamiltonian, which form a basis for the space of mode ℓ radiation states, thus have certain properties which in no way correspond to the intuitive idea of a single-mode electromagnetic wave, since the expectation values do not oscillate at frequency ω_ℓ. On the other hand, (5.65) relating the energy and the variance of the field is the same as for a classical field.

The number states, which are so convenient from a theoretical point of view, are difficult to produce experimentally. Indeed, it is no simple matter to produce a state of the field in which the number of photons is equal with absolute certainty to n, rather than $n-1$ or $n+1$, especially when n is large. Experimental setups able to produce one-photon states $|n = 1\rangle$ exploit a *single quantum emitter*, such as an isolated atom, ion or molecule in an excited state. Such emitters in fact produce a multimode one-photon state (see Section 5.4.3 and Complement 5B). Number states with small n larger than 1 have also been produced by more sophisticated experimental arrangements.

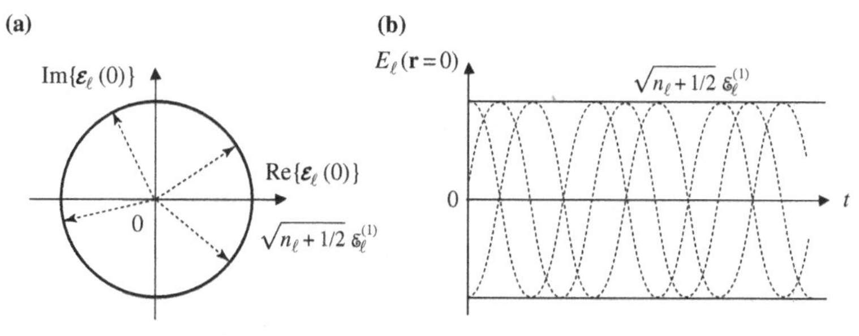

Figure 5.4 **(a)** Representation of a number state $|n_\ell\rangle$ by a classical phasor $\mathcal{E}_\ell(0)$ with perfectly well-specified modulus $\sqrt{n_\ell + 1/2}\,\mathcal{E}_\ell^{(1)}$, but random phase evenly distributed between 0 and 2π. **(b)** Corresponding time diagram, showing a statistical ensemble of sinusoidal curves, each with the same amplitude, but random phases evenly distributed between 0 and 2π.

Comments

(i) It is quite remarkable that there is no dispersion in photon number measurements when the system is in a number state. The semi-classical description of the photoelectric effect, in which the detector is quantized but the radiation is not, cannot account for this absence of fluctuations. Indeed, for a field of strictly constant magnitude, the number of photoelectrons emitted in a given interval of time is a random variable with a Poisson distribution, and the standard deviation is equal to the square root of the average value, i.e. $\Delta N = \sqrt{\bar{N}}$.

(ii) We shall see below that, for a quasi-classical state, the photons can be represented as randomly distributed in the volume of the mode, and this leads to a Poisson distribution for detection. However, in a number state, the photons can be considered to be regularly distributed.

(iii) The time diagram of Figure 5.4(b) represents the time evolutions of the classical fields associated with the phasor diagram of Figure 5.4(a). If we could measure $\hat{E}(\mathbf{r}=0,t)$, and repeat the measurement many times on identically prepared systems, we would obtain values scattered in the interval $\pm\sqrt{n_\ell + 1/2}\,\mathcal{E}_\ell^{(1)}$, for all values of t.

5.3.4 Quasi-classical states $|\alpha_\ell\rangle$

We shall now discuss a class of quantum states called *quasi-classical states* (or Glauber coherent states). In contrast to the number states, their properties are similar to those of the classical field.

Definition and properties

A quasi-classical state $|\alpha_\ell\rangle$ of mode ℓ is the eigenstate of the annihilation operator $\hat{a}_\ell$ associated with eigenvalue α_ℓ, i.e.

$$\hat{a}_\ell|\alpha_\ell\rangle = \alpha_\ell|\alpha_\ell\rangle. \tag{5.69}$$

Since the operator $\hat{a}_\ell$ is non-Hermitian, one must prove that such eigenstates do actually exist. We expand the state $|\alpha_\ell\rangle$ in terms of the basis of number states $|n_\ell\rangle$ of the given mode:

$$|\alpha_\ell\rangle = \sum_{n_\ell=0}^{\infty} c_{n_\ell}|n_\ell\rangle. \tag{5.70}$$

Using the relation $\langle n_\ell|\hat{a}_\ell^+ = \sqrt{n_\ell - 1}\,\langle n_\ell - 1|$, we obtain a recurrence relation between the coefficients of this expansion:

$$\sqrt{n_\ell}\,c_{n_\ell} = \alpha_\ell c_{n_\ell - 1}, \tag{5.71}$$

whence it follows that, up to an overall factor and after normalizing the state vector,

$$|\alpha_\ell\rangle = \mathrm{e}^{-|\alpha_\ell|^2/2}\sum_{n_\ell=0}^{\infty}\frac{\alpha_\ell^{n_\ell}}{\sqrt{n_\ell!}}|n_\ell\rangle. \tag{5.72}$$

The expression (5.72) for $|\alpha_\ell\rangle$ shows that such a state does exist for any complex α_ℓ. The reader can check using (4.115–4.116) that it is indeed an eigenstate of $\hat{a}_\ell$ with eigenvalue α_ℓ.

This state, a superposition of eigenstates of the Hamiltonian associated with different values of the energy, is not an eigenstate of the Hamiltonian. It will thus change in time. Assume that a system is in the state $|\alpha_\ell\rangle$ given by (5.72) at time $t = 0$. At a later time t, it will be in the state

$$|\psi(t)\rangle = e^{-|\alpha_\ell|^2/2} \sum_{n_\ell=0}^{\infty} \frac{\alpha_\ell^{n_\ell}}{\sqrt{n_\ell!}} e^{-i\left(n_\ell+\frac{1}{2}\right)\omega_\ell t} |n_\ell\rangle, \tag{5.73}$$

since $(n_\ell + 1/2)\hbar\omega_\ell$ is the energy of the state $|n_\ell\rangle$. The state $|\psi(t)\rangle$ can be written in the form

$$|\psi(t)|\rangle = e^{-i\omega_\ell t/2} e^{-|\alpha_\ell|^2/2} \sum_{n_\ell=0}^{\infty} \frac{\left(\alpha_\ell e^{-i\omega_\ell t}\right)^{n_\ell}}{\sqrt{n_\ell!}} |n_\ell\rangle. \tag{5.74}$$

Comparing (5.74) with (5.72) reveals that $|\psi(t)\rangle$ is an eigenvector of $\hat{a}_\ell$ with eigenvalue $\alpha_\ell\, e^{-i\omega_\ell t}$. This property could have been obtained directly by calculating $\hat{a}_\ell|\psi(t)\rangle$:

$$\hat{a}_\ell|\psi(t)\rangle = \alpha_\ell\, e^{-i\omega_\ell t}|\psi(t)\rangle. \tag{5.75}$$

As it evolves, a quasi-classical state maintains its quasi-classical character, but with an eigenvalue $\alpha_\ell\, e^{-i\omega_\ell t}$ which changes in time. We may thus write

$$|\psi(t)\rangle = |\alpha_\ell\, e^{-i\omega_\ell t}\rangle. \tag{5.76}$$

Comment Starting with (5.72), calculation shows that

$$|\langle\alpha_\ell|\alpha_{\ell'}\rangle|^2 = e^{-|\alpha_\ell - \alpha_{\ell'}|^2}. \tag{5.77}$$

The quasi-classical states are not therefore strictly orthogonal to one another, even though their scalar product is very small whenever $|\alpha_\ell - \alpha_{\ell'}| \gg 1$. They do not therefore form an orthonormal basis for the space $\mathcal{F}_\ell$ of single-mode states.[8]

Number of photons in a quasi-classical state $|\alpha_\ell\rangle$

The state $|\alpha_\ell\, e^{-i\omega_\ell t}\rangle$ is a linear superposition of states $|n_\ell\rangle$. The photon number does not therefore have a certain value. Equation (5.72) shows that the probability $P(n_\ell)$ of finding the value n_ℓ for the number of photons is

[8] A more detailed study of these states can be found in CDG I, or in L. Mandel and E. Wolf, *Optical Coherence and Quantum Optics*, Cambridge University Press (1995).

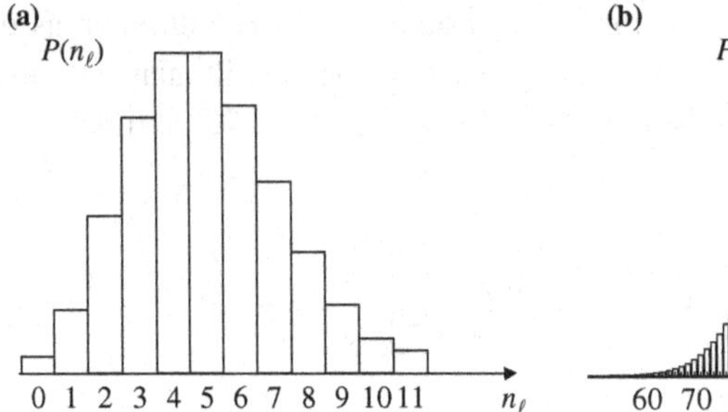

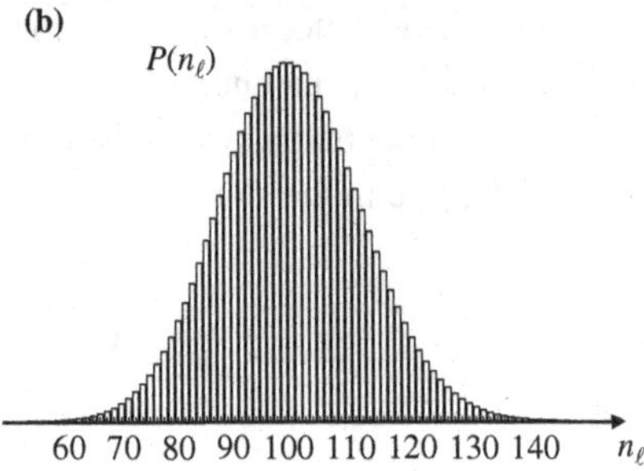

Figure 5.5 Histogram for the value of the photon number when the field is in the quasi-classical state $|\alpha_\ell\rangle$, with (a) $|\alpha_\ell|^2 = 5$ and (b) $|\alpha_\ell|^2 = 100$.

$$P(n_\ell) = |c_{n_\ell}|^2 = \mathrm{e}^{-|\alpha_\ell|^2} \frac{(|\alpha_\ell|^2)^{n_\ell}}{n_\ell!}. \tag{5.78}$$

This is a *Poisson distribution* with average $|\alpha_\ell|^2$, shown schematically in Figure 5.5. Equation (5.74) shows that this result is time independent.

The expectation and variance of the photon number in such a state can be calculated directly, without using the distribution (5.78). From the adjoint of (5.69), i.e. $\langle\alpha_\ell|\hat{a}_\ell^\dagger = \alpha_\ell^*\langle\alpha_\ell|$, the expectation of the photon number $\hat{N}_\ell = \hat{a}_\ell^\dagger\hat{a}_\ell$ is found to be

$$\langle\hat{N}_\ell\rangle = \langle\alpha_\ell|\hat{a}_\ell^\dagger\hat{a}_\ell|\alpha_\ell\rangle = |\alpha_\ell|^2. \tag{5.79}$$

Likewise, the variance $(\Delta N_\ell)^2$ of the photon number is given by

$$\begin{aligned}(\Delta N_\ell)^2 &= \langle\hat{N}_\ell^2\rangle - \langle\hat{N}_\ell\rangle^2 = \langle\alpha_\ell|\hat{a}_\ell^\dagger\hat{a}_\ell\hat{a}_\ell^\dagger\hat{a}_\ell|\alpha_\ell\rangle - |\alpha_\ell|^4 \\ &= \langle\alpha_\ell|\hat{a}_\ell^\dagger\left(\hat{a}_\ell^\dagger\hat{a}_\ell + 1\right)\hat{a}_\ell|\alpha_\ell\rangle - |\alpha_\ell|^4 = |\alpha_\ell|^2.\end{aligned} \tag{5.80}$$

The root-mean-squared deviation ΔN_ℓ in a quasi-classical state is thus,

$$\Delta N_\ell = \sqrt{\langle\hat{N}_\ell\rangle}. \tag{5.81}$$

This dependence of the fluctuations on $\sqrt{\langle\hat{N}_\ell\rangle}$ is a standard property of the Poisson distribution. For large values of the average photon number $\langle\hat{N}_\ell\rangle$ (and hence large values of $|\alpha_\ell|$), the distribution $P(N_\ell)$ is characterized by a large absolute value of the variance, since $(\Delta N_\ell \to \infty)$, but a small relative value, since $(\Delta N_\ell/\langle\hat{N}_\ell\rangle \to 0)$.

The photon number distribution (5.78) in the quasi-classical state suggests an ensemble of independent particles distributed randomly with uniform average density $|\alpha_\ell|^2/L^3$. The number of photons contained in volume L^3 is then a Poisson random variable with expected value $|\alpha_\ell|^2$. This purely particle-like picture of light in a quasi-classical state is clearly not complete, because it cannot account for wave aspects of the field. However, it is extremely useful in any problem where the phase of the field is not relevant. In particular, it can be used to make statistical predictions of the results of many photon counting experiments.

Value of the electric field in a single-mode quasi-classical state

The adjoint of (5.75) is

$$\langle\psi(t)|\hat{a}_\ell^\dagger = \alpha_\ell^* \, e^{i\omega_\ell t}\langle\psi(t)|. \tag{5.82}$$

Using (5.75), (5.82) and expression (5.57) for the field operator $\hat{E}_\ell(\mathbf{r})$, the following expression is found for the expectation value of the electric field:

$$\langle\psi(t)|\hat{E}_\ell(\mathbf{r})|\psi(t)\rangle = i\,\mathcal{E}_\ell^{(1)}\left(\alpha_\ell e^{i(\mathbf{k}_\ell.\mathbf{r}-\omega_\ell t)} - \alpha_\ell^* e^{-i(\mathbf{k}_\ell.\mathbf{r}-\omega_\ell t)}\right). \tag{5.83}$$

The expectation of the field oscillates in time at mode frequency ω_ℓ. Although this frequency is too high to be able to observe the field directly, its quadrature components can be measured by homodyne detection (see Equations 5.22 and 5.23). For a quasi-classical state, we obtain

$$\langle\psi(0)|\hat{E}_{Q\ell}|\psi(0)\rangle = \mathcal{E}^{(1)}(\alpha_\ell + \alpha_\ell^*) = 2\mathcal{E}_\ell^{(1)}\mathrm{Re}\,\{\alpha_\ell\} \tag{5.84}$$

$$\langle\psi(0)|\hat{E}_{P\ell}|\psi(0)\rangle = -i\,\mathcal{E}^{(1)}(\alpha_\ell - \alpha_\ell^*) = 2\mathcal{E}_\ell^{(1)}\mathrm{Im}\,\{\alpha_\ell\}. \tag{5.85}$$

Repeated measurements of $E_{Q\ell}$ and $E_{P\ell}$ thus give the complex number α_ℓ, which completely characterizes the quasi-classical state.

Let us now calculate the variance of the electric field at a given point $\mathbf{r}$ and time t. Using the commutation relation $[\hat{a}_\ell, \hat{a}_\ell^\dagger] = 1$, we find

$$\langle\psi(t)|\hat{E}_\ell^2(\mathbf{r})|\psi(t)\rangle = \langle\psi(t)|\hat{E}_\ell(\mathbf{r})|\psi(t)\rangle^2 + \left[\mathcal{E}_\ell^{(1)}\right]^2. \tag{5.86}$$

The variance of the electric field is thus independent of $\mathbf{r}$ and t, with the value

$$(\Delta E_\ell)^2 = \langle\hat{E}_\ell^2\rangle - \langle\hat{E}_\ell\rangle^2 = \left[\mathcal{E}_\ell^{(1)}\right]^2. \tag{5.87}$$

In the same way, for the quadrature operators $\hat{E}_{Q\ell}$ and $\hat{E}_{P\ell}$, it can be shown that

$$(\Delta E_{P\ell})^2 = (\Delta E_{Q\ell})^2 = \left[\mathcal{E}_\ell^{(1)}\right]^2. \tag{5.88}$$

The variance of the field in the quasi-classical state is thus independent of the value of α_ℓ. It is minimal with regard to the Heisenberg condition (5.63) and coincides with the amplitude of the vacuum fluctuations.[9]

The results just established are summarized in Figure 5.6. The dotted curve in Figure 5.6(b) gives the expectation value (5.83) of the field at a fixed point $\mathbf{r}$, e.g. $\mathbf{r} = 0$. This oscillates as a function of time with amplitude $2|\alpha_\ell|\mathcal{E}_\ell^{(1)}$. The dispersion around this average value, independent of time and equal to $\mathcal{E}_\ell^{(1)}$, is shown by the band containing the dotted curve. This band indicates the region in which there is a high probability of finding the electric field value at time t. The phasor representation in Figure 5.6(a), where the

[9] The vacuum $|n_\ell = 0\rangle$ is both a number state associated with $n_\ell = 0$ and a quasi-classical state associated with eigenvalue $\alpha_\ell = 0$.

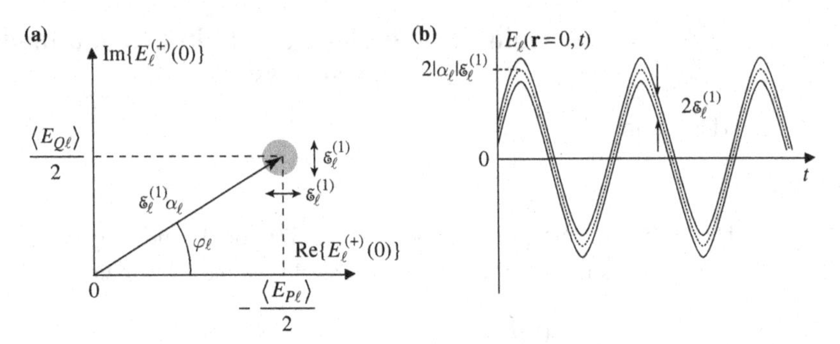

Figure 5.6 Representations of the electric field in a quasi-classical state. **(a)** Phasor diagram. **(b)** Representation of the time-dependence at point r = 0. Measurements will give values of the field in the band containing the dotted curve with a high probability. The dotted curve gives the (ensemble) average value of such measurements. The dispersion band has constant height $2\mathcal{E}_\ell^{(1)}$ normally to the time axis. The figure is plotted for an average photon number greater than unity. In fact, $|\alpha_\ell| = 5$ or $\langle N_\ell \rangle = 25$.

fluctuations ΔE_P and ΔE_Q are equal (shaded disc), is the phasor representation for a quasi-classical state. The plot in Figure 5.6(b) can be obtained by projecting the phasor rotating at $e^{-i\omega_\ell t}$ onto the real axis, with the shaded disc in tow.

A quasi-classical state is characterized by a field amplitude of the order of $|\alpha_\ell|\mathcal{E}_\ell^{(1)}$, with fluctuations of the order of $\mathcal{E}_\ell^{(1)}$. For a 'macroscopic' field, $|\alpha_\ell|^2$ is much bigger than unity and the error made in replacing the quantum field by a classical field having the same average becomes negligible. This explains why these states $|\alpha_\ell\rangle$ are referred to as quasi-classical. However, it should not be thought that these states have exactly the same properties as a classical field. The quantum fluctuations, even though minimal in such a state, can be measured experimentally. In addition, the fundamental nature of these fluctuations can be put to use. For example, there are quantum cryptography protocols (see Complement 5E) which use the quasi-classical states with small values of α_ℓ, where the information to be transmitted is 'hidden' by the quantum fluctuations.

The quasi-classical states are very easy to produce experimentally. It can be shown that a single-mode laser well above threshold, or a Hertzian wave generator, produce quasi-classical states with high values of $|\alpha_\ell|$. More generally, when a monochromatic light beam with arbitrary initial quantum state is sufficiently attenuated, a quasi-classical state is obtained. It may then happen that the value of α_ℓ is small, of order unity or even much less. Such states can be written

$$|\alpha_\ell\rangle \simeq |0\rangle + \alpha_\ell|1\rangle + \frac{\alpha_\ell^2}{\sqrt{2}}|2\rangle + \ldots \tag{5.89}$$

Even if they contain on average a very small number of photons, they have different properties to the number state $|1\rangle$, because they are mainly 'composed' of vacuum, and contain, in addition to the one-photon component, some photon pairs, triplets, etc.

5.3.5 Other quantum states of single-mode radiation: squeezed states and Schrödinger cats

Quasi-classical states are not the only minimal states. There are states known as squeezed states for which $\Delta E_{Q\ell}$, for example, is smaller than $\mathcal{E}_\ell^{(1)}$. The variance $\Delta E_{P\ell}$ is then necessarily greater than $\mathcal{E}_\ell^{(1)}$, because the product of the variances must remain equal to $\left[\mathcal{E}_\ell^{(1)}\right]^2$. Figure 5.7 shows the appearance of the field in the special case of a squeezed state. As for a quasi-classical state, the expectation value oscillates sinusoidally, but the dispersion about the average also varies in time. The band representing the fluctuations of the possible values is narrowest at the intersections A_i with the time axis, and broadens for determinations at points B_i, a quarter-phase later. The phase of this field is thus specified with greater accuracy than that of a quasi-classical field. On the other hand, the amplitude of the field is noisier than that of a quasi-classical state. This complementarity between phase and amplitude will be further discussed in the next section. Complement 5A gives an overview of the properties of these squeezed states and methods for producing them as well as applications. Let us emphasize again that, although we cannot directly see the time evolution displayed in Figure 5.7(b), a series of homodyne measurements allows us to reconstruct that diagram or, equivalently, to reconstruct the phasor diagram in Figure 5.7(a).

There are many other single-mode quantum states with specifically quantum properties, impossible to describe in the semi-classical model of the matter–radiation interaction. One could mention the 'Schrödinger cat' states, i.e.

$$|\psi_1\rangle = |0\rangle + |\alpha_\ell\rangle, \tag{5.90}$$

or

$$|\psi_2\rangle = |\alpha_\ell\rangle + |-\alpha_\ell\rangle \tag{5.91}$$

up to a normalization constant. These are linear superpositions of quasi-classical states, which for $|\alpha_\ell| \gg 1$ are different macroscopic states. State $|\psi_1\rangle$, superposition of the radiation emitted by an off source and radiation emitted by an on source, closely resembles the

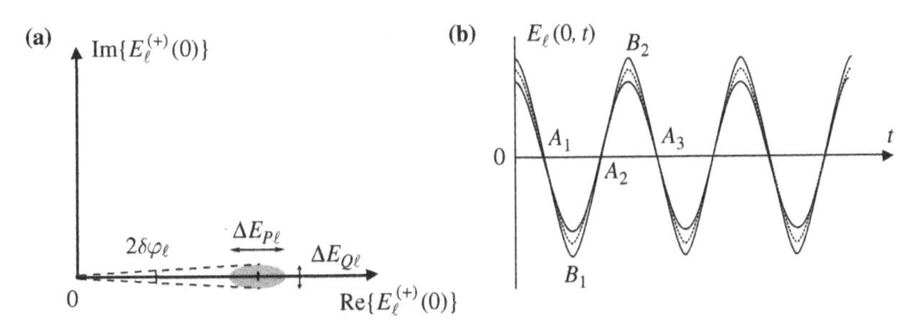

Figure 5.7 Electric field in a phase squeezed state. **(a)** Phasor diagram. **(b)** Ensemble of measured values at $r = 0$, versus time. The dispersion at $t = 0, \pi/\omega, \ldots$ is greater than in a quasi-classical state, but it is less at times $t = \pi/2\omega, 3\pi/2\omega \ldots$ Such a state is well-suited to accurate measurement of the field phase.

state imagined by Schrödinger, the superposition of a dead cat and a living cat. State $|\psi_2\rangle$ of (5.91) is a superposition of two classical fields with opposite phases. By studying these, and in particular their behaviour with regard to dissipation phenomena, one can investigate the frontier between the quantum and classical domains.[10] States like $|\psi_2\rangle$, with low values of $|\alpha_\ell|$, have been produced experimentally. To grasp the highly non-intuitive nature of these states, consider the analogous state of a pendulum: the pendulum oscillates 'at the same time' with two motions that have opposite phase. There are therefore moments when the pendulum is 'simultaneously' at two points distant from one another.

5.3.6 The limit of small quantum fluctuations and the photon number–phase Heisenberg relation

We have seen that, for a macroscopic quasi-classical state with a large value of $|\alpha_\ell|$, the quantum fluctuations of the field, of order $\mathcal{E}_\ell^{(1)}$, are small compared with the expectation values, of order $\alpha_\ell\mathcal{E}_\ell^{(1)}$. This property does in fact hold for a great many macroscopic states of the field, such as the squeezed states. It is interesting because it means that expansions can be made to first order in the fluctuations.

To be more precise, we introduce the quadrature and photon number fluctuation operators:

$$\delta\hat{E}_{Q\ell} = \hat{E}_{Q\ell} - \langle\hat{E}_{Q\ell}\rangle \tag{5.92}$$

$$\delta\hat{E}_{P\ell} = \hat{E}_{P\ell} - \langle\hat{E}_{P\ell}\rangle \tag{5.93}$$

$$\delta\hat{N}_\ell = \hat{N}_\ell - \langle\hat{N}_\ell\rangle. \tag{5.94}$$

Relation (5.68) between the quadrature and number operators can be expanded to first order in the fluctuation operators to give

$$\langle\hat{E}_{Q\ell}\rangle\delta\hat{E}_{Q\ell} + \langle\hat{E}_{P\ell}\rangle\delta\hat{E}_{P\ell} \approx 2\left[\mathcal{E}_\ell^{(1)}\right]^2\delta\hat{N}_\ell. \tag{5.95}$$

To simplify, consider a quantum state of the single-mode field which gives a real value for the average complex amplitude at $\mathbf{r} = 0$ and $t = 0$ (see Figure 5.7). Unless otherwise stated, all expectation values will be taken at $\mathbf{r} = 0$ and $t = 0$, and we shall not specify $\mathbf{r} = 0$ and $t = 0$, in order to simplify the notation. According to (5.60), such a field satisfies

$$\langle\hat{E}_{Q\ell}\rangle = 0 \tag{5.96}$$

$$\langle\hat{E}_{P\ell}\rangle = 2\langle\hat{E}_\ell^{(+)}\rangle. \tag{5.97}$$

Equation (5.95) can then be used to express the energy fluctuations in the form

$$\delta\hat{H} = \hbar\omega_\ell\delta\hat{N}_\ell = \hbar\omega_\ell\frac{\langle\hat{E}_\ell^{(+)}\rangle}{\left[\mathcal{E}_\ell^{(1)}\right]^2}\delta\hat{E}_{P\ell}. \tag{5.98}$$

[10] S. Haroche and J. M. Raimond, *Exploring the Quantum*, Oxford University Press (2006).

The energy fluctuations are thus proportional to the fluctuations in the quadrature component $\hat{E}_{P\ell}$, which corresponds to the oscillation amplitude of the average field. We shall now show that the phase fluctuations are proportional to the fluctuations in the quadrature component $\hat{E}_{Q\ell}$, which is $\pi/2$ out of phase with the average field.

The rigorous definition of a phase operator in quantum optics is a delicate matter.[11] The problem is greatly simplified by considering only macroscopic states, the subject of this section. We shall follow a simple phenomenological approach here, based on the analogy between quantum fluctuations of the classical field and statistical fluctuations of a random classical field, discussed in Section 5.3.2. We thus represent the above quantum field by a classical field of the form (5.48), affected by amplitude fluctuations $\delta|\mathcal{E}_\ell(0)|$ and phase fluctuations $\delta\varphi_\ell$, around average values $\overline{\mathcal{E}_\ell(0)}$ and $\overline{\varphi}_\ell$, respectively (see Figure 5.7(a)). For small fluctuations, the fluctuations of the complex field are

$$\delta\mathcal{E}_\ell(0) = \delta|\mathcal{E}_\ell(0)|\,\mathrm{e}^{\mathrm{i}\overline{\varphi}_\ell} + \mathrm{i}\delta\varphi_\ell\,|\overline{\mathcal{E}_\ell(0)}|\,\mathrm{e}^{\mathrm{i}\overline{\varphi}_\ell}. \tag{5.99}$$

Moreover, we assume that $\overline{\mathcal{E}_\ell(0)}$ is real, i.e. $\overline{\varphi_\ell} = 0$. Writing (5.99) as a sum of real and imaginary parts and using (5.45–5.47), the fluctuations in the quadrature components are found to be

$$\delta E_{P\ell} = 2\,\delta|\mathcal{E}_\ell(0)| \tag{5.100}$$

$$\delta E_{Q\ell} = 2\,|\mathcal{E}_\ell(0)|\,\delta\varphi_\ell. \tag{5.101}$$

This situation is shown in Figure 5.7(a).

Since the classical energy of the field is $H_\ell = 2\varepsilon_0 L^3|\mathcal{E}_\ell(0)|^2$ (see Equation 4.86), its fluctuations are given by

$$\delta H_\ell = 4\varepsilon_0 L^3 \overline{E}_\ell\,\delta E_\ell = 2\varepsilon_0 L^3 \overline{E}_\ell\,\delta E_{Q\ell}. \tag{5.102}$$

Given the expression (4.99) for $\mathcal{E}_\ell^{(1)}$, Equation (5.102) is the exact analogue of the quantum relation (5.98), provided that we identify the classical average amplitude $\overline{E}_\ell$ with the quantum expectation $\langle\hat{E}_\ell^{(+)}(0)\rangle$. If we then use the same correspondence in (5.101), the quantum fluctuations in the phase can be defined by

$$\delta\hat{\varphi}_\ell = \frac{1}{2\langle\hat{E}_\ell^{(+)}(0)\rangle}\,\delta\hat{E}_{Q\ell}. \tag{5.103}$$

They are indeed proportional to the fluctuations in the quadrature $\hat{E}_{Q\ell}$. From (5.98) and (5.103), we deduce the root-mean-squared deviations of the phase and photon number to be

$$\Delta\varphi_\ell = \frac{1}{2|\langle\hat{E}_\ell^{(+)}(0)\rangle|}\,\Delta E_{Q\ell} \tag{5.104}$$

$$\Delta N_\ell = \frac{|\langle\hat{E}_\ell^{(+)}(0)\rangle|}{(\mathcal{E}_\ell^{(1)})^2}\,\Delta E_{P\ell}. \tag{5.105}$$

[11] For more details, see S. Barnett and D. Pegg, On the Hermitian Optical Phase Operator, *Journal of Modern Optics* **36**, 7 (1989).

The phase fluctuations can thus be determined by homodyne detection (see Section 5.1.3), where the phase of the oscillator is chosen to give $\hat{E}_{Q\ell}$, while ΔN_ℓ is measured by choosing the phase of the local oscillator giving $\hat{E}_{P\ell}$.

The Heisenberg relation for the quadratures (5.63) thus implies that

$$\Delta N_\ell \cdot \Delta\varphi_\ell \geqslant \frac{1}{2}. \tag{5.106}$$

In the limit of macroscopic fields, for small quantum fluctuations, *phase and photon number look like complementary variables in the usual sense of quantum mechanics*. There is therefore no state of the field in which these two quantities are simultaneously determined in an infinitely precise way. The better defined the energy (or photon number) of a field state, the less well defined its phase will be. Note that the 'strong' quasi-classical field, with $|\alpha_\ell| \gg 1$, is a minimal state with regard to the inequalities (5.63) and (5.106). From (5.81), the deviation of the phase is given as

$$\Delta\varphi_\ell = \frac{1}{2\sqrt{\langle N_\ell\rangle}}. \tag{5.107}$$

The values (5.81) and (5.107) of the fluctuations associated with a quasi-classical state are called *standard quantum noise*. It was long thought that they constituted an impenetrable barrier for optical measurements, because no other minimal radiation states other than the quasi-classical states were known. But since the discovery of squeezed states, it has been realized that the quantum fluctuations in one of the two observables, photon number or phase, can be reduced, provided of course that one accepts an increase in the fluctuations of the complementary observable.

Comment A number state (see Section 5.3.3) is a state in which ΔN_ℓ is zero and the phase is indeterminate, a result compatible with the relation (5.106). This was not obvious a priori, insofar as the assumption of small fluctuations in comparison with the average is not satisfied for number states. As already remarked, a number state is also a non-classical state, for which the standard quantum limit (5.81) is violated.

5.3.7 Light beam propagating in free space

The field states we have been considering in this section are associated with plane waves occupying the volume $V = L^3$ used to discretize the reciprocal (momentum) space. When this volume is real (resonant optical cavity), states of this kind have an undeniable physical reality, and one can indeed imagine a state comprising n_ℓ photons in a normal mode of this cavity, or a quasi-classical state associated with a photon number expectation value $\langle\hat{N}_\ell\rangle = |\alpha_\ell|^2$ in mode ℓ. On the other hand, these states would not appear well-suited to describe a *light beam* produced by a source, e.g. a laser, *propagating freely in space* until it reaches some detector or an absorbing or scattering obstacle in an a priori undetermined position. This kind of radiation state is not characterized by a given energy, or as would be equivalent, by a well-defined photon number. We shall see that it is nevertheless possible to use the quantum states studied above to describe such propagating beams. This will be done here in the context of single-mode states.[12]

The intrinsic energy quantities characterizing a uniform cylindrical beam of cross-sectional area S are the transported power Φ, expressed in watts, or its power per unit area $\Pi = \Phi/S$, also called the irradiance, expressed in W.m^{-2} and equal to the magnitude of the Poynting vector. Another relevant quantity is the energy per unit volume $D = \Pi/c$ in J.m^{-3}. In the context of quantum optics, photon quantities are defined by dividing these energy quantities by the energy $\hbar\omega_\ell$ of one photon. This yields the photon current Φ^{phot}, i.e. the number of photons crossing area S of the beam per unit time, the photon current density per unit area Π^{phot} and the photon density per unit volume D^{phot}:

$$\Pi^{\text{phot}} = \frac{\Pi}{\hbar\omega_\ell}, \tag{5.108}$$

$$\Phi^{\text{phot}} = \frac{\Phi}{\hbar\omega_\ell} = S\frac{\Pi}{\hbar\omega_\ell}, \tag{5.109}$$

$$D^{\text{phot}} = \frac{1}{c}\frac{\Pi}{\hbar\omega_\ell}. \tag{5.110}$$

These are intrinsic physical quantities characterizing the propagating beam. To relate them to the photon number of a quantum state, we observe that the photon number operator $\hat{N}_\ell$ is in fact a quantity which depends on the quantization volume L^3, and which is only useful for intermediate calculations, since in general, L^3 disappears from the final result. We shall use this arbitrariness by replacing volume L^3 by a volume ScT in the expressions of quantum optics. This corresponds to the volume of the beam passing through the area S in a time T. The observable $\hat{\Phi}^{\text{phot}}$ characterizing the number of photons measured per unit time with a perfect detector intercepting the whole beam is then given by

$$\hat{\Phi}^{\text{phot}} = \frac{\hat{N}_\ell}{T}. \tag{5.111}$$

Consider, for example, a quasi-classical state $|\alpha_\ell\rangle$ emitted by a laser well above threshold, and identify $|\alpha_\ell|^2$ with $\langle\hat{N}_\ell\rangle = T\langle\hat{\Phi}^{\text{phot}}\rangle$. The dispersion in the values of the photon number in the quasi-classical state (given by the distribution 5.78) causes temporal fluctuations, or *quantum noise*, in measurements of the instantaneous photon current. We identify $\Delta\hat{N}_\ell$ in the volume ScT with the fluctuation in the number of photons counted over a time T. Since the root-mean-squared deviation in a state $|\alpha_\ell\rangle$ is $\Delta N_\ell = |\alpha_\ell| = \sqrt{\langle\hat{N}_\ell\rangle}$ (Equation 5.81), it follows that the root-mean-squared deviation of the photon current fluctuations $\Delta\Phi^{\text{phot}}$, measured by a photodetector over a time T, is

$$\Delta\Phi^{\text{phot}} = \frac{\Delta N_\ell}{T} = \sqrt{\frac{\langle\hat{\Phi}^{\text{phot}}\rangle}{T}} = \sqrt{\frac{\overline{\Phi}}{T\hbar\omega_\ell}}, \tag{5.112}$$

where $\overline{\Phi}$ is the average power of the beam.

[12] For more details, see H. J. Kimble, Les Houches session 56, p. 603, J. Dalibard and J. M. Raimond eds. North Holland, Amsterdam (1992), or C. Fabre, Les Houches 1995, *Quantum Fluctuations*, S. Reynaud and E. Giacobino eds. North Holland, Amsterdam (1996).

The photodetector, assumed to be perfect, converts every photon into an electron of charge q_e. The average photocurrent detected $\bar{i}$ and its fluctuations Δi are thus given by

$$\bar{i} = q_e \langle \hat{\Phi}^{\text{phot}} \rangle = \frac{q_e \overline{\Phi}}{\hbar \omega_\ell}, \tag{5.113}$$

$$\Delta i = q_e \Delta \Phi^{\text{phot}} = \sqrt{q_e \frac{\bar{i}}{T}} = \sqrt{2 q_e \bar{i} B}. \tag{5.114}$$

The bandwidth of the measurement is $B = 1/2\text{T}$, corresponding to a time average over a time T.

For a light beam in a quasi-classical state, we thus find the usual form (5.114) for *shot noise*, proportional to the square root of the average intensity and the bandwidth. As an example, consider a light beam of wavelength $\lambda = 1\,\mu\text{m}$ and average power $\overline{\Phi} = 1\,\text{mW}$ (low power continuous laser), described by a quasi-classical state. The corresponding average photon current Φ^{phot} is about 5×10^{15} photons per second. Consider a typical measurement time $T = 1\,\mu\text{s}$. The average number of photons counted is thus 5×10^9. According to (5.112) and (5.109), there is therefore a noise $\Delta\Phi = 14\,\text{nW}$ in the detected power. This value can be measured by a high-quality photodiode. This example shows that fluctuations of quantum origin can indeed be detected in a free beam.[13]

The above argument extends to other observables, and there are Heisenberg relations for all quantities relating to a propagating beam. Consider for example the Heisenberg inequality (5.106) between the phase and the photon number. Replacing the photon number operator $\hat{N}_\ell$ by $T\hat{\Phi}_{\text{phot}}$, we obtain the inequality

$$\Delta \Phi^{\text{phot}} \Delta \varphi_\ell \geq \frac{1}{2T} = B. \tag{5.115}$$

This inequality is valid for any state, whether it be quasi-classical or squeezed, provided that the fluctuations remain small compared with the expectation values (see Section 5.3.6). For the quasi-classical state examined above (power 1 mW, detector with a response time of $1\,\mu\text{s}$), the value of $|\alpha_\ell|$ is the square root of the photon number, 7×10^4, and the quadrature fluctuations, of the order of $\mathcal{E}_\ell^{(1)}$, are therefore very small compared with the expectation value of the field, of the order of $|\alpha_\ell|\,\mathcal{E}_\ell^{(1)}$, which justifies the procedure of expanding to first order in Section 5.3.6. The standard quantum noise for this situation is $\Delta\Phi^{\text{phot}} = 7 \times 10^{10}$ photons.s^{-1}, and $\Delta\varphi_\ell = 7 \times 10^{-6}$ rad.

For a squeezed state of the same average power, the inequality (5.115) remains valid. For a phase-squeezed state with fluctuations smaller than the above value, the intensity fluctuations will then be greater than the standard limit, but this presents no problem for measurements involving the phase, e.g. in an interferometer.

[13] It is much more difficult to reach this regime for position and momentum measurements on a particle. Optics is thus a privileged area for testing measurement limits in quantum theory.

Comments

(i) There is shot noise for independent classical material point particles, described by Poisson statistics, arriving at a detector, e.g. sand grains hitting a wall. In a quasi-classical state, we thus have a picture of a beam made up of photons that can be treated as independently propagating particles.

(ii) The shot noise arising in photodetection has been obtained here using the specifically quantum features of the light beam, with the photodetector merely converting them into electrical properties. Some accounts present shot noise in photodetection as a consequence of quantum properties of the photodetection process itself (the number of photoelectrons released for a given radiation energy is a Poisson random variable). It then arises as measurement noise, independent of the state of the field arriving at the detector. Could this description, coming in fact from the semi-classical model of photodetection where light is not quantized (Complement 2E), be correct? In fact, everything depends on the quantum efficiency η of the photodetector, i.e. the percentage of photons actually converted into electrons. The fully quantum approach described here is necessary when η is close to unity, while the semi-classical approach (fluctuations associated with the photodetection process itself) correctly describes the situation when $\eta \ll 1$, since it agrees with the quantum description in that case. It should be mentioned that, while the photomultipliers and avalanche photodiodes used for photon counting usually have efficiencies below 10%, there are commercially available photodiodes for the visible and near-infra-red frequency ranges with quantum efficiencies greater than 90%. In this case, the approach described in this chapter is essential for a correct description of observations concerning states other than quasi-classical states.

5.4 Multimode quantum radiation

5.4.1 Non-factorizable states and entanglement

As indicated in the conclusion of Chapter 4, the state space for quantum radiation is enormous, and we are far from having achieved a systematic exploration. The first question to ask when faced with a radiation state is whether or not it is factorizable. A factorizable state can be written in the form

$$|\psi\rangle = |u_1\rangle \otimes |u_2\rangle \otimes \ldots |u_\ell\rangle \otimes \ldots, \tag{5.116}$$

where $|u_\ell\rangle$ denotes a mode ℓ radiation state, which can itself be written in the general form (5.50–5.51). In the rest of this chapter, we shall see several important examples of factorizable radiation states. However, it should be remembered that there is a much larger class of states called *entangled states*, which cannot be factorized. This is too big a subject to be covered in detail in the present chapter. We shall simply give some examples, beginning with the case of two entangled photons, in a state which has given rise to sophisticated experiments to test the foundations of quantum physics (see Complement 5C). In these experiments, a source emits a photon at frequency ω_1 and a photon at frequency ω_2, and with the help of filters and diaphragms, the following state is isolated:

$$|\psi_{\text{EPR}}\rangle = \frac{1}{\sqrt{2}}\Big(|1_{\ell_1}, 1_{\ell_2}\rangle + |1_{m_1}, 1_{m_2}\rangle\Big), \tag{5.117}$$

where the states $|1_{\ell_1}\rangle$, $|1_{\ell_2}\rangle$, $|1_{m_1}\rangle$, and $|1_{m_2}\rangle$ are single-mode number states (see Section 5.3.3) with $n_\ell = 1$. The modes ℓ_1 and m_1 might, for example, correspond to two different directions, or to two orthogonal polarizations of photons of frequency ω_1 (see Complement 5C), and likewise for ℓ_2 and m_2. We shall show that the state $|\psi_{\text{EPR}}\rangle$ is not factorizable, i.e. it cannot be written in the form $|\psi_1\rangle \otimes |\psi_2\rangle$, where $|\psi_1\rangle$ is a state of the photon $\hbar\omega_1$, while $|\psi_2\rangle$ is a state of the photon $\hbar\omega_2$. The most general factorized state for the two photons considered here can be written

$$|\psi\rangle = \left(\lambda_1|1_{\ell_1}\rangle + \mu_1|1_{m_1}\rangle\right) \otimes \left(\lambda_2|1_{\ell_2}\rangle + \mu_2|1_{m_2}\rangle\right), \tag{5.118}$$

where λ_1, λ_2, μ_1 and μ_2, are complex numbers. To write (5.117) in the form (5.118), one would have to have

$$\lambda_1\lambda_2 = \mu_1\mu_2 = \frac{1}{\sqrt{2}} \tag{5.119}$$

and

$$\lambda_1\mu_2 = \lambda_2\mu_1 = 0. \tag{5.120}$$

The two systems of equations here are clearly incompatible, as can be seen by forming the product $\lambda_1 \cdot \lambda_2 \cdot \mu_1 \cdot \mu_2$ either on the basis of (5.119) or on the basis of (5.120).

Complement 5D discusses another example of entangled radiation states involving only two modes, ℓ and ℓ', but with the possibility of having various number states $|n_\ell\rangle$ and $|n_{\ell'}\rangle$ in each mode. For example, we discuss the twin photon state given by

$$|\psi\rangle = \sum_n C_n|n, n\rangle, \tag{5.121}$$

where $|n, n\rangle$ is $|n_\ell, n_{\ell'}\rangle$ with $n_\ell = n_{\ell'}$. If at least two coefficients C_n differ from zero, then the state (5.121) cannot be factorized.

The entangled states lead to very strong correlations. For example, in the two-mode state (5.121), if we find n photons in mode ℓ, we will then find n photons in mode ℓ' with absolute certainty. For the two photons of state (5.117), if we find the photon $\hbar\omega_1$ in mode ℓ_1, we are certain to find the photon $\hbar\omega_2$ in mode ℓ_2 (and likewise for m_1 and m_2). These strong correlations are directly related to the non-factorizability, which endures even if the two entangled systems are separated. For example, Complement 5C discusses situations where the two photons can move an arbitrary distance apart while remaining strongly correlated. The impossibility of factorizing the state of a system comprising two parts means that it is effectively a single physical entity, whose properties cannot be expressed merely as the sum of the properties of its parts, even if these parts are moved apart. This is sometimes referred to as quantum nonlocality.

In entangled systems, one finds stronger correlations than any allowed in classical physics, as attested by the violation of Bell's inequalities (Complement 5C). It is easy to understand therefore that entangled systems offer possibilities that would be quite inconceivable in the classical context. These possibilities are put to use in the field of quantum information, where secure data transfer and data processing are based on quantum physics (see Complement 5E).

After this brief glance at entangled states, we return in the rest of the chapter to factorized states.

5.4.2 Multimode quasi-classical state

A multimode quasi-classical state is a factorized state of the form (5.116), in which each state $|u_\ell\rangle$ is a quasi-classical state $|\alpha_\ell\rangle$ (see Section 5.3.4). Its form at time t is obtained from (5.76):

$$|\psi_{\mathrm{qc}}(t)\rangle = |\alpha_1 \mathrm{e}^{-\mathrm{i}\omega_1 t}\rangle \otimes \ldots |\alpha_\ell \mathrm{e}^{-\mathrm{i}\omega_\ell t}\rangle \otimes \ldots \tag{5.122}$$

It is an eigenstate of

$$\hat{\mathbf{E}}^{(+)}(\mathbf{r}) = \mathrm{i}\sum_\ell \mathcal{E}_\ell^{(1)} \hat{a}_\ell \boldsymbol{\varepsilon}_\ell\, \mathrm{e}^{\mathrm{i}\mathbf{k}_\ell\cdot\mathbf{r}}, \tag{5.123}$$

$$\hat{\mathbf{E}}^{(+)}(\mathbf{r})|\psi_{\mathrm{qc}}(t)\rangle = \left(\mathrm{i}\sum_\ell \mathcal{E}_\ell^{(1)} \alpha_\ell\, \boldsymbol{\varepsilon}_\ell \mathrm{e}^{\mathrm{i}(\mathbf{k}_\ell.\mathbf{r}-\omega_\ell t)}\right)|\psi_{\mathrm{qc}}(t)\rangle, \tag{5.124}$$

and the eigenvalue is denoted by

$$\mathbf{E}_{\mathrm{cl}}^{(+)}(\mathbf{r},t) = \mathrm{i}\sum_\ell \mathcal{E}_\ell^{(1)} \alpha_\ell \boldsymbol{\varepsilon}_\ell\, \mathrm{e}^{\mathrm{i}(\mathbf{k}_\ell\cdot\mathbf{r}-\omega_\ell t)}. \tag{5.125}$$

The subscript ‘cl’ reminds us that it has the form of the analytic signal of a classical field. However, it should not be concluded from (5.124) that $|\psi_{\mathrm{qc}}\rangle$ is an eigenstate of $\hat{\mathbf{E}} = \hat{\mathbf{E}}^{(+)} + \hat{\mathbf{E}}^{(-)}$, because the Hermitian conjugate version of (5.124) is

$$\langle\psi_{\mathrm{qc}}(t)|\hat{\mathbf{E}}^{(-)}(\mathbf{r}) = \mathbf{E}_{\mathrm{cl}}^{(-)}(\mathbf{r},t)\langle\psi_{\mathrm{qc}}(t)|, \tag{5.126}$$

where $\mathbf{E}_{\mathrm{cl}}^{(-)}(\mathbf{r},t)$ is the complex conjugate of $\mathbf{E}_{\mathrm{cl}}^{(+)}(\mathbf{r},t)$.

On the other hand, the expectation value of $\hat{\mathbf{E}}(\mathbf{r})$ in the state $|\psi_{\mathrm{qc}}(t)\rangle$ is

$$\langle\psi_{\mathrm{qc}}(t)|\hat{\mathbf{E}}(\mathbf{r})|\psi_{\mathrm{qc}}(t)\rangle = \mathbf{E}_{\mathrm{cl}}^{(+)}(\mathbf{r},t) + \mathbf{E}_{\mathrm{cl}}^{(-)}(\mathbf{r},t). \tag{5.127}$$

In a multimode quasi-classical state, the expectation value of $\hat{\mathbf{E}}(\mathbf{r})$ assumes the form of a multimode classical field.

A multimode quasi-classical state has many properties that can be obtained directly from the properties discussed in Section 5.3.4 for single-mode states. For example, the probability of observing N photons with a detector that is not frequency selective is given by composing the Poisson distributions describing the statistics of the photons in each mode.

The quantum expression for the photodetection signal (5.4) takes the value

$$\begin{aligned} w(\mathbf{r},t) &= s\langle\psi_{\mathrm{qc}}(t)|\hat{\mathbf{E}}^{(-)}(\mathbf{r})\hat{\mathbf{E}}^{(+)}(\mathbf{r})|\psi_{\mathrm{qc}}(t)\rangle \\ &= s\mathbf{E}_{\mathrm{cl}}^{(-)}(\mathbf{r},t)\mathbf{E}_{\mathrm{cl}}^{(+)}(\mathbf{r},t) = s|\mathbf{E}_{\mathrm{cl}}^{(+)}(\mathbf{r},t)|^2, \end{aligned} \tag{5.128}$$

which is identical to the expression associated with the classical field $\mathbf{E}_{\mathrm{cl}}(\mathbf{r},t)$ in the semi-classical model of photodetection (Equation 5.2). The same goes for the coincidence counting rate (5.5), for which one finds

$$w^{(2)}(\mathbf{r}_1,\mathbf{r}_2,t) = s^2|\mathbf{E}_{\mathrm{cl}}^{(+)}(\mathbf{r}_1,t)|^2|\mathbf{E}_{\mathrm{cl}}^{(+)}(\mathbf{r}_2,t)|^2, \tag{5.129}$$

that is, the product of the individual count rates (5.128) at points $\mathbf{r}_1$ and $\mathbf{r}_2$. This is what would have been obtained with a semi-classical calculation, based on classical fields, where the random character is a result of photodetection phenomena which are independent processes at points $\mathbf{r}_1$ and $\mathbf{r}_2$, once the field $\mathbf{E}_{\mathrm{cl}}(\mathbf{r},t)$ has been specified.

The results just established show us why the semi-classical model, in which the fields are classical and the photodetectors are described by the semi-classical photodetection model, is so successful. Indeed, the radiation emitted by all known light sources up until the 1970s can be described by quasi-classical states, or incoherent statistical mixtures of quasi-classical states, whence the quantum calculation yields the same result as the semi-classical calculation for single or coincidence photodetection probabilities. So an ensemble of distinct laser beams, or the various modes of a multimode laser, are perfectly described by a state of the form (5.122) or a statistical mixture of such states. Consider, for example, two laser beams in the state

$$|\psi\rangle = |\alpha_1\rangle\,|\alpha_2\rangle, \tag{5.130}$$

with modes $1(\omega_1,\mathbf{k}_1,\boldsymbol{\varepsilon}_1)$ and $2\,(\omega_2,\mathbf{k}_2,\boldsymbol{\varepsilon}_2)$, at time $t=0$. To simplify, we take $\boldsymbol{\varepsilon}_1=\boldsymbol{\varepsilon}_2$, normal to the plane $(\mathbf{k}_1,\mathbf{k}_2)$ and use scalar fields. Let us calculate the probability of photodetection at point $\mathbf{r}$ and time t:

$$\begin{aligned}
w(\mathbf{r},t) &= s\|\hat{E}^{(+)}(\mathbf{r})|\psi(t)\rangle\|^2 \\
&= s\left[\mathcal{E}_{\omega_1}^{(1)}\mathcal{E}_{\omega_2}^{(1)}\right]^2 |\alpha_1 \mathrm{e}^{-\mathrm{i}\omega_1 t}\mathrm{e}^{\mathrm{i}\mathbf{k}_1\cdot\mathbf{r}} + \alpha_2\mathrm{e}^{-\mathrm{i}\omega_2 t}\mathrm{e}^{\mathrm{i}\mathbf{k}_2\cdot\mathbf{r}}|^2 \\
&= s\left[\mathcal{E}_{\omega_1}^{(1)}\mathcal{E}_{\omega_2}^{(1)}\right]^2\left[|\alpha_1|^2+|\alpha_2|^2+\alpha_1\alpha_2^*\,\mathrm{e}^{-\mathrm{i}(\omega_1-\omega_2)t}\,\mathrm{e}^{\mathrm{i}(\mathbf{k}_1-\mathbf{k}_2)\cdot\mathbf{r}}+\text{c.c.}\right].
\end{aligned} \tag{5.131}$$

We recognize the beat between two classical waves. In the case $\omega_1=\omega_2$, the result (5.131) describes an interference pattern for two classical waves. The calculation generalizes without difficulty to an arbitrary number of beams in quasi-classical states.

It often happens that the two waves in the above calculation have phases that fluctuate randomly and independently for the various modes. For lasers, phase fluctuations are due either to technical noise or to spontaneous emission in the laser mode. In this case, we have $\alpha_1=|\alpha_1|\,\mathrm{e}^{\mathrm{i}\varphi_1}$ and $\alpha_2=|\alpha_2|\,\mathrm{e}^{\mathrm{i}\varphi_2}$, where φ_1 and φ_2 are independent stochastic processes with coherence times τ_{c_1} and τ_{c_2}, respectively, i.e. φ_1 and φ_2 will have assumed all possible values between 0 and 2π after times τ_{c_1} and τ_{c_2}. If observation of the photocurrent involves an average over a lapse of time longer than $\min\{\tau_{\mathrm{c}_1},\tau_{\mathrm{c}_2}\}$, the averaging will erase the beat (or interference) between the two beams. This is what usually happens with two distinct lasers, unless they are particularly well stabilized. In the latter case, the beat can be observed, provided that the detector is fast enough and the difference $|\omega_1-\omega_2|$ small enough.

Incoherent classical sources such as incandescence or discharge lamps emit radiation that can be described by a multimode quasi-classical state, where each mode has an extremely short coherence time τ_c. Once again, it is easy to see why the quantum theory gives exactly the same results as classical statistical optics, which describes the radiation from incoherent classical sources as an ensemble of incoherent single-mode classical waves.

5.4.3 One-photon multimode state

It was not until the 1970s that radiation states were produced that could not be described by coherent or incoherent, single- or multimode quasi-classical states. Such states have properties that the semi-classical theory cannot account for. We briefly mentioned the single-mode squeezed states (see Section 5.3.5 and Complement 5A) for which amplitude or phase fluctuations can be less than the standard quantum limit, which is the minimal value predicted by the semi-classical theory. But the simplest non-classical state of light is undoubtedly the multimode one-photon state.

A multimode one-photon state is an eigenstate of the *total* photon number operator $\hat{N} = \sum_\ell \hat{a}_\ell^\dagger \hat{a}_\ell$ with eigenvalue 1. We denote such a state by $|1\rangle$. The single-mode number states $|0, \dots, 0, n_\ell = 1, 0, \dots\rangle$ are clearly one-photon states, but they are not the only ones. In particular, states of the form

$$|1\rangle = \sum_\ell c_\ell |0, \dots, 0, n_\ell = 1, 0, \dots\rangle, \tag{5.132}$$

with $\sum_\ell |c_\ell|^2 = 1$, are eigenstates of $\hat{N}$, since $\hat{N}|1\rangle = |1\rangle$, but not of each $\hat{N}_\ell$, since $\hat{N}_\ell|1\rangle = c_\ell|0, \dots n_\ell = 1, \dots 0\rangle$. Note that, unlike the single-mode number states, they are not eigenstates of the Hamiltonian, i.e. the state (5.132) is not stationary. At time t, it is given by

$$|1(t)\rangle = \sum_\ell c_\ell \mathrm{e}^{-\mathrm{i}\omega_\ell t} |0, \dots, 0, n_\ell = 1, 0, \dots\rangle, \tag{5.133}$$

measuring energies with respect to the vacuum energy E_v. Complement 5B investigates the one-photon states in more detail. They might be referred to as quasi-particle states, because they are the quantum states whose properties most closely resemble those of an isolated particle propagating at the speed of light, just as the classical state is the quantum state closest to a classical electromagnetic wave. Let us just say that this kind of state can be produced experimentally. It is in fact the final state of the field in a spontaneous emission process for a single atom in an excited state.

When the field is in such a state, the single detection count rate on a single photodetector located at $\mathbf{r}_1$ is given by

$$w(\mathbf{r}_1, t) = s \left\| \hat{\mathbf{E}}^{(+)}(\mathbf{r}_1)|1(t)\rangle \right\|^2 = s \left| \sum_\ell c_\ell \mathcal{E}_\ell^{(1)} \boldsymbol{\varepsilon}_\ell \mathrm{e}^{\mathrm{i}(\mathbf{k}_\ell \cdot \mathbf{r}_1 - \omega_\ell t)} \right|^2. \tag{5.134}$$

The precise value of $w(\mathbf{r}_1, t)$ thus depends on the coefficients c_ℓ and the position of the detector. In principle, nothing can distinguish this single detection count rate from the result obtained using a classical field with analytic signal $\mathbf{E}^{(+)}(\mathbf{r}, t)$ given by

$$\mathbf{E}^{(+)}(\mathbf{r}_1, t) = \sum_\ell c_\ell \mathscr{E}_\ell^{(1)} \boldsymbol{\varepsilon}_\ell \mathrm{e}^{\mathrm{i}(\mathbf{k}_\ell \cdot \mathbf{r}_1 - \omega_\ell t)}. \tag{5.135}$$

Such a conclusion does not hold if we consider the coincidence count rate on two photodetectors, given by (5.5), and expressed here as

$$w^{(2)}(\mathbf{r}_1, \mathbf{r}_2, t) = s^2 \left\| \sum_{i,j=x,y,z} \hat{E}_j^{(+)}(\mathbf{r}_2) \hat{E}_i^{(+)}(\mathbf{r}_1) |1(t)\rangle \right\|^2. \tag{5.136}$$

The mode expansion of the above expression gives a sum of terms proportional to $\hat{a}_{\ell'} \hat{a}_{\ell''} |1_\ell\rangle$. Now each of these terms is zero. Indeed, if ℓ' or ℓ'' differs from ℓ, this is obvious because $|1_\ell\rangle$ is a simplified notation for $|1_\ell\rangle \otimes |0_{\ell'}\rangle \otimes |0_{\ell''}\rangle \ldots$ Regarding the term $\hat{a}_\ell \hat{a}_\ell |1_\ell\rangle$, we have

$$\hat{a}_\ell \hat{a}_\ell |1_\ell\rangle = \hat{a}_\ell |0_\ell\rangle = 0. \tag{5.137}$$

So in the end, for a one-photon state,

$$w^{(2)}(\mathbf{r}_1, \mathbf{r}_2, t) = s^2 \left\| \sum_{i,j=x,y,z} \hat{E}_j^{(+)}(\mathbf{r}_2) \hat{E}_i^{(+)}(\mathbf{r}_1) |1(t)\rangle \right\|^2 = 0. \tag{5.138}$$

The probability of double detection is strictly zero because *we are applying an annihilation operator twice* to a state containing at most a single photon. This quantum result does indeed correspond to the intuitive idea of a one-photon state, namely, it is impossible to detect simultaneously a single photon at two photodetectors placed at different points. *Such a property would be inconceivable in a classical description of electromagnetic fields*, where $w^{(2)}(\mathbf{r}_1, \mathbf{r}_2, t)$ is non-zero when the field is non-zero at each of the detectors. It is a specifically quantum property, characteristic of the one-photon states (see Complement 5B). Such a state is radically different from a highly attenuated quasi-classical state, but the difference appears only if we consider more sophisticated signals than single detection rates.

We may ask whether the quantum nature of one-photon states is related to an entanglement property, i.e. the impossibility of factorizing the state (5.132). We shall see that this is not the case, and that state (5.132) can in fact be considered as the state $|1_m\rangle$, i.e. a single-mode one-photon state in a generalized mode m. As in Complement 4C, we can define a creation operator:

$$\hat{b}_m^\dagger = \sum_\ell c_\ell \hat{a}_\ell^\dagger. \tag{5.139}$$

With its Hermitian conjugate $\hat{b}_m = \sum_\ell c_\ell^* \hat{a}_\ell$, it satisfies the canonical commuation relation,

$$\left[\hat{b}_m, \hat{b}_m^\dagger\right] = \sum_\ell |c_\ell|^2 = 1. \tag{5.140}$$

Clearly state (5.132) can now be written

$$|1\rangle = \hat{b}_m^\dagger |0\rangle, \tag{5.141}$$

that is, $|1_m\rangle$, the mode m being associated with $\hat{b}_m^\dagger$.

The generalized mode m is associated with a classical electromagnetic field that is not monochromatic, but which can be written, generalizing (4C.8),

$$\mathbf{E}_m^{(+)}(\mathbf{r},t) = \sum_\ell c_\ell \mathcal{E}_\ell^{(1)} \boldsymbol{\varepsilon}_\ell \, \mathrm{e}^{\mathrm{i}(\mathbf{k}_\ell \cdot \mathbf{r} - \omega_\ell t)}. \tag{5.142}$$

This is a classical wave packet (see Complement 5B). To conclude, *the non-classical nature* of the multimode one-photon state *is not the result of entanglement*, and must be considered as a simple generalization of the non-classical nature of a single-mode number state.

Comment There is a necessary and sufficient condition for a radiation state $|\psi\rangle$ to be expressed in the form of a single-mode state of a generalized mode m. The state $|\psi\rangle$ can be written in the form $|1_m\rangle$ if and only if all the vectors $\hat{a}_\ell|\psi\rangle$ are proportional to one another (or zero), with the modes ℓ forming a complete basis. For (5.132), we do indeed have $\hat{a}_\ell|1\rangle = c_\ell|0\rangle$, i.e. a vector proportional to $|0\rangle$ if $c_\ell \neq 0$.

Note further that the multimode quasi-classical state (5.122) can itself be considered as a single-mode state, for a generalized mode constructed from the associated classical field. More precisely, it is a quasi-classical state in the mode whose spatial distribution is proportional to $\alpha_1 \, \mathrm{e}^{\mathrm{i}\mathbf{k}_1 \cdot \mathbf{r}} + \alpha_2 \, \mathrm{e}^{\mathrm{i}\mathbf{k}_2 \cdot \mathbf{r}}$.

Such states, which reduce to the form $|1_m\rangle$, are called intrinsic single-mode states. The radiation they describe is intrinsically coherent.[14]

5.5 One-photon interference and wave-particle duality. An application of the formalism

5.5.1 Mach–Zehnder interferometer in quantum optics

The Mach–Zehnder interferometer (Figure 5.8) is the prototype for all amplitude division two-wave interferometers. The incoming mode at port (1) is divided by the beamsplitter S_{in}. Mode 2 is combined with mode 1 on S_{in}. The mirrors $M_{3'}$ and $M_{4'}$ can then recombine the two waves $3'$ and $4'$ on the beamsplitter S_{out}. Two detectors D_3 and D_4 are placed at the output ports. The mirrors $M_{3'}$ and $M_{4'}$ can be displaced to control the path length difference:

$$\delta L' = L_{3'} - L_{4'} = [S_{\text{in}} M_{3'} S_{\text{out}}] - [S_{\text{in}} M_{4'} S_{\text{out}}]. \tag{5.143}$$

[14] C. Fabre, *Quantum Optics, from One Mode to Many Modes*, in Cours des Houches (2007), available online at http://hal-sfo.ccsd.cnrs.fr/docs/00/27/05/37/PDF/CoursLesHouchesFabre2.pdf and N. Treps *et al.*, *Physical Review* A**71**, 013820 (2005).

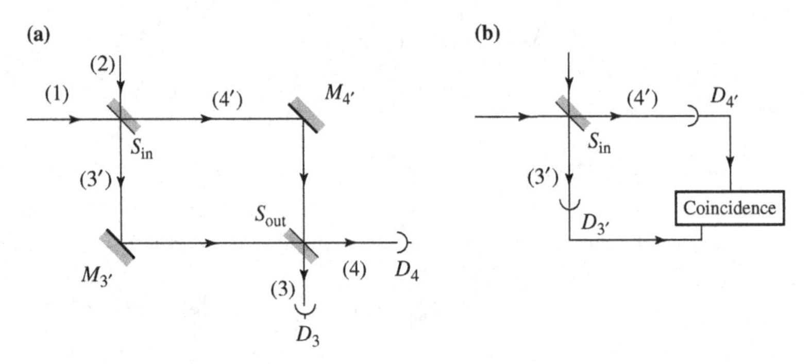

Figure 5.8 **Mach–Zehnder two-wave interferometer. (a) The photodetection signals measured by photodetectors D_3 and D_4 depend on the optical path difference $[S_{in}M_{3'}S_{out}] - [S_{in}M_{4'}S_{out}]$. (b) Two detectors $D_{3'}$ and $D_{4'}$ can be inserted into the arms of the interferometer, monitored by a coincidence circuit to check for the single-photon characteristic.**

Square brackets denote the optical paths, taking into account the refractive index of the paths and any phase-shift at the mirrors. To simplify, we consider a monochromatic incoming mode (frequency ω), polarized normally to the figure. All resulting waves $(3', 4', 3, 4)$ have the same frequency and polarization. We can then use scalar fields.

To calculate the photodetection signals at D_3 and D_4, we follow the prescription given in Section 5.1.2, i.e. we express the outgoing fields $\hat{E}_3^{(+)}$ and $\hat{E}_4^{(+)}$ as a function of the incoming fields $\hat{E}_1^{(+)}$ and $\hat{E}_2^{(+)}$, using the same relations as those relating the outgoing classical fields $E_3^{(+)}$ and $E_4^{(+)}$ to the incoming classical fields $E_1^{(+)}$ and $E_2^{(+)}$. The origin of the phases of the outgoing fields is taken on the beamsplitter S_{out}, while that of the incoming fields is taken at S_{in}. We then apply (5.8–5.9) at S_{in}, to relate $\hat{E}_{3'}^{(+)}$ and $\hat{E}_{4'}^{(+)}$ to $\hat{E}_1^{(+)}$ and $\hat{E}_2^{(+)}$. At S_{out}, we take into account the fact that the beamsplitter is reversed by putting a minus sign in the coefficient relating $\hat{E}_3^{(+)}$ and $\hat{E}_{3'}^{(+)}$ (reflection inside the glass). Multiplying by the propagation factors between S_{in} and S_{out} along paths $3'$ and $4'$, we thus have

$$\hat{E}_3^{(+)} = \frac{1}{\sqrt{2}}\left(-\hat{E}_{3'}^{(+)}\,\mathrm{e}^{\mathrm{i}kL_{3'}} + \hat{E}_{4'}^{(+)}\,\mathrm{e}^{\mathrm{i}kL_{4'}}\right) \tag{5.144}$$

$$\hat{E}_4^{(+)} = \frac{1}{\sqrt{2}}\left(\hat{E}_{3'}^{(+)}\,\mathrm{e}^{\mathrm{i}kL_{3'}} + \hat{E}_{4'}^{(+)}\,\mathrm{e}^{\mathrm{i}kL_{4'}}\right), \tag{5.145}$$

where $k = \omega/c$. After a straightforward calculation, we obtain

$$\hat{E}_3^{(+)} = \mathrm{e}^{\mathrm{i}k\overline{L'}}\left(-\mathrm{i}\sin k\frac{\delta L'}{2}\hat{E}_1^{(+)} - \cos k\frac{\delta L'}{2}\hat{E}_2^{(+)}\right)\mathrm{e}^{\mathrm{i}kr_3} \tag{5.146}$$

$$\hat{E}_4^{(+)} = \mathrm{e}^{\mathrm{i}k\overline{L'}}\left(\cos k\frac{\delta L'}{2}\hat{E}_1^{(+)} + \mathrm{i}\sin k\frac{\delta L'}{2}\hat{E}_2^{(+)}\right)\mathrm{e}^{\mathrm{i}kr_4}, \tag{5.147}$$

where $\overline{L'} = (L_{3'} + L_{4'})/2$, and the factors $\exp\{\mathrm{i}kr_3\}$ and $\exp\{\mathrm{i}kr_4\}$ stand for propagation from S_{out} to the detectors.

These relations can be used to calculate the photodetection signals w_3 and w_4 for various incoming radiation states in mode 1, with incoming mode 2 empty, so that the incoming radiation has the form

$$|\psi_{\text{in}}\rangle = |\varphi_1\rangle \otimes |0_2\rangle. \tag{5.148}$$

The photodetection signal $w_3(r_3, t)$ is obtained by writing

$$w_3(r_3, t) = s\|\hat{E}_3^{(+)}(r_3)|\psi_{\text{in}}\rangle\|^2 \tag{5.149}$$

and expressing $\hat{E}_3^{(+)}$ and $|\psi_{\text{in}}\rangle$ by (5.146) and (5.148), respectively. As the incoming state at 2 is empty, the term $\hat{E}_2^{(+)}|\psi_{\text{in}}\rangle$ is zero. What remains is

$$w_3(t) = s\sin^2\left(k\frac{\delta L'}{2}\right)\|\hat{E}_1^{(+)}|\varphi_1(t)\rangle\|^2 \tag{5.150}$$

and likewise,

$$w_4(t) = s\cos^2\left(k\frac{\delta L'}{2}\right)\|\hat{E}_1^{(+)}|\varphi_1(t)\rangle\|^2. \tag{5.151}$$

The output signal from the interferometer thus depends on neither the position of the detectors, nor the average length $\overline{L'}$ of the interferometer arms. As in the classical calculation, it depends only on the path difference $\delta L'$. However, it may in principle also depend on the type of field state entering by port (1). We shall now examine various cases for the incoming field.

5.5.2 Quasi-classical incoming radiation

Consider first the case in which the incoming field at port (1) is in a quasi-classical state:

$$|\varphi_1(t)\rangle = |\alpha_1\, \mathrm{e}^{-i\omega t}\rangle. \tag{5.152}$$

Recalling that $|\varphi_1(t)\rangle$ is an eigenstate of $\hat{E}_1^{(+)}$, we obtain

$$w_3 = w_1 \sin^2\left(k\frac{\delta L'}{2}\right), \tag{5.153}$$

where

$$w_1 = s\left[\mathcal{E}_1^{(1)}\right]^2 |\alpha_1|^2 \tag{5.154}$$

is the photodetection probability that would be obtained by placing the detector at the incoming port (1). Likewise, for port (4),

$$w_4 = w_1 \cos^2\left(k\frac{\delta L'}{2}\right). \tag{5.155}$$

The result is strictly identical to what would be obtained from a calculation in classical optics, agreeing with the general claim made in Section 5.4.2 that quantum optics gives the same results as classical optics for quasi-classical states.

The result can be generalized without difficulty (see Complement 5B) to the case of a multimode quasi-classical state of the form (5.122). If the coefficients α_ℓ only differ from zero over a frequency interval $\Delta\omega_\ell$ about ω_0, and if $\Delta\omega \cdot \delta L'/c$ is small compared with 2π, the results (5.153) and (5.155) remain valid with k replaced by k_0. If $\delta L'$ is greater than $1/\Delta\omega$, a calculation exactly like the classical interference calculation for polychromatic light shows that the fringe contrast is less than unity.

5.5.3 Particle-like incoming state

Consider now the case in which the incoming state at port (1) is a one-photon state of mode (1), i.e.

$$|\varphi_1\rangle = |1_1\rangle, \tag{5.156}$$

with port (2) remaining empty. The general expressions (5.150) and (5.151) show that the results (5.153) and (5.155) do in fact remain valid, provided that we take the photodetection probability at the incoming port to be the one-photon photodetection probability

$$w_1 = s\left[\mathcal{E}_1^{(1)}\right]^2. \tag{5.157}$$

Note that the result has exactly the same form as for a monochromatic classical wave, i.e. the photodetection probability depends sinusoidally on the path difference, and interference is observed.

If we consider a multimode one-photon state of the form (5.132) at port (1) and nothing at port (2), we obtain

$$\hat{E}_3^{(+)}(r_3)|\psi_{\text{in}}(t)\rangle = \sum_\ell \mathcal{E}_\ell^{(1)} c_\ell \mathrm{e}^{\mathrm{i}[k_\ell(\overline{L'}+r_3)-\omega_\ell t]} \sin k_\ell \frac{\delta L'}{2}|0\rangle. \tag{5.158}$$

The photodetection signal will thus be

$$w_3(r_3,t) = s\left|\sum_\ell \mathcal{E}_\ell^{(1)} c_\ell \mathrm{e}^{\mathrm{i}[k_\ell(\overline{L'}+r_3)-\omega_\ell t]} \sin k_\ell \frac{\delta L'}{2}\right|^2. \tag{5.159}$$

This expression can be simplified if the distribution of the frequencies ω_ℓ has finite width $\Delta\omega$ and the path difference L' is small compared with $c/\Delta\omega$. Denoting the central frequency of the wave packet by ω_0 and writing $k_0 = \omega_0/c$, $\cos\left(k_\ell \frac{\delta L'}{2}\right)$ can be replaced by $\cos\left(k_0 \frac{\delta L'}{2}\right)$ in (5.159). The photodetection probability then assumes the form

$$w_3(r_3,t) = F(t,r_3)\sin^2\left(k_0\frac{\delta L'}{2}\right), \tag{5.160}$$

where the function $F(t,r_3)$ is a slowly varying envelope describing the wave packet (see Complement 5B). The term $\cos^2\left(k_0\frac{\delta L'}{2}\right)$ is the interference term corresponding to frequency ω_0. Integrating over time t for the whole wave packet, we then obtain the total probability of photodetection at 3, keeping the path difference $\delta L'$ fixed:

$$\mathcal{P}_3 = \mathcal{P}_1 \sin^2\left(k_0 \frac{\delta L'}{2}\right). \tag{5.161}$$

Here $\mathcal{P}_1$ is the total photodetection probability for a one-photon wave packet at the interferometer input. The interference signal is clear. An analogous calculation for port (4) would yield

$$\mathcal{P}_4 = \mathcal{P}_1 \cos^2\left(k_0 \frac{\delta L'}{2}\right). \tag{5.162}$$

As a consequence, even for a one-photon incoming state, which we called a quasi-particle-like state, interference fringes will be observed, as is found with a quasi-classical state or classical optical waves. This prediction of quantum optics has been checked experimentally (see Figure 5.9). It illustrates the wavelike behaviour of a single photon, or the claim that a 'photon can interfere with itself'.

5.5.4 Wave-particle duality for a particle-like state

Consider once again a one-photon state of the form (5.132) arriving at the semi-reflecting mirror S_{in} (see Figure 5.8), but this time introduce photon-counting photodetectors at

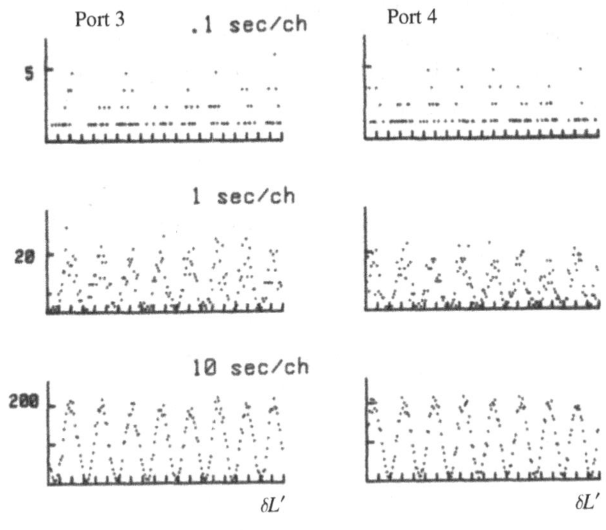

Figure 5.9 Experimental results giving the counts at ports (3) and (4) of a Mach–Zehnder interferometer as a function of the path length difference $\delta L'$ in the case where the incoming field is a one-photon state. The successive plots correspond to increasing observation times for each value of the path difference (fixed mirrors). These observation times vary from 0.1 to 10 seconds, corresponding to an average number of detections of 1 to 100 at each mirror position. Note the complementarity of the two interference patterns at each port. (Figure, P. Grangier and A. Aspect.)

ports (3′) and (4′) (see Figure 5.8(b)). A coincidence circuit is used to measure the probability of joint detections at (3′) and (4′). We write each field $\hat{E}^{(+)}_{3'}$ and $\hat{E}^{(+)}_{4'}$ as a function of the incoming fields, using (5.8–5.9). As in (5.136), the probability of joint detections is zero, because it only contains terms of the form $(\hat{a}_\ell)^2|1_\ell\rangle$ (for a detailed calculation, see Complement 5B, Section 5B.2). There is thus a distinctly particle-like behaviour here, in the sense that the photon is detected on one side or the other of S_{in}, but not both sides at once. It is tempting to conclude that the photon goes on either one side or the other. But as we saw in Section 5.5.3, the same one-photon wave packet in the interferometer of Figure 5.8(a) gives rise to interference, and this is naturally interpreted by imagining an incoming wave that divides on S_{in} and propagates in both 3′ and 4′. What we are faced with here is an example of wave–particle duality: the one-photon wave packet behaves sometimes like a wave and sometimes like a particle, depending on whether it is analysed with the setup of Figure 5.8(a) or that of Figure 5.8(b).

The Bohr complementarity principle is often cited to make the existence of these apparently contradictory types of behaviour less disconcerting. According to this principle, contradictory classical behaviour actually arises in incompatible experimental setups. It is quite true that the setups of Figures 5.8(a) and 5.8(b) cannot be implemented at the same time. When we analyse the situation, we find that the complementary quantities result in fact on the one hand from observation of interference – associated with simultaneous passage through the two ports of the interferometer – and on the other from a precise determination of the path followed – unambiguously revealed by the setup of Figure 5.8(b), since only one of the two detectors $D_{3'}$ or $D_{4'}$ gives a signal.

More refined versions of the experiment can be devised,[15] in which some limited information I can be obtained concerning the path followed, without preventing the appearance of interference with limited contrast C. For these complementary quantities, an analysis of such situations then leads to an inequality,

$$C^2 + I^2 \leq 1. \tag{5.163}$$

The two situations discussed above correspond to the limiting cases ($C = 0, I = 1$) or ($C = 1, I = 0$), respectively.

Comment Determination of the path requires the use of a particle-like state, like a one-particle wave packet. Indeed, note that, if a quasi-classical radiation state is used, the experimental arrangement of Figure 5.8(b) gives a non-zero probability for the coincidence count, and the path followed cannot be unambiguously determined.

5.5.5 Wheeler's delayed-choice experiment

In conformity with the idea of complementarity, contradictory behaviour is associated with incompatible experimental arrangements, and one does indeed find that the observed

[15] V. Jacques *et al.*, Delayed-Choice Test of Quantum Complementarity with Interfering Single Photons, *Physical Review Letters* **100**, 220402 (2008) and references therein.

behaviour depends on the chosen arrangement. One might thus be led to the following interpretation of complementarity: the quantum system 'adopts a different behaviour' depending on the apparatus it interacts with.

Such a claim should not be interpreted too naively, e.g. by thinking that the incoming photon at the beamsplitter S_{in} will adopt wavelike or particle-like behaviour, depending on whether the detectors $D_{3'}$ and $D_{4'}$ are absent or present. The point is that one can imagine, as suggested by J. A. Wheeler, that the decision to introduce these detectors might be made only after the arrival of the photon at the incoming beamsplitter S_{in}. But the setup proposed by Wheeler is even more radical. His idea was, rather than introducing further detectors $D_{3'}$ and $D_{4'}$, to use a removable outgoing beamsplitter S_{out} in the setup of Figure 5.8(a). When it is in place, we have an interferometer, but when it is withdrawn, a photodetector at D_3 or D_4 indicates unambiguously which path has been followed. If the light travel time $\overline{L}/c$ in the interferometer is longer than the duration of the one-photon wave packet, there is in principle no difficulty in deciding to withdraw or introduce S_{out} after the passage of the photon at S_{in}. More precisely, it is enough for the two events to be separated by a spacelike interval, in the relativistic sense, to be sure that the decision to withdraw or introduce the beamsplitter cannot influence the behaviour of the photon entering the interferometer, unless one accepts the possibility of some influence travelling faster than light.

The experiment has been carried out, and the result is unambiguous: if the beamsplitter S_{out} is in place when the one-photon wave packet arrives there, interference fringes are observed, whatever the situation when it entered the interferometer.[16] And conversely, if S_{out} is absent at the time of passage of the wave packet, no fringe is observed and one can say without ambiguity whether the photon has followed path $3'$ or path $4'$. If one insisted on using a classical picture of the wave – associated with the idea of travelling in both arms at the same time – or the classical picture of a particle – choosing one path or the other, but not both – then one must conclude, to paraphrase Wheeler's own words, that the behaviour adopted in the interferometer depends on the choice made at the moment of leaving it. This spectacular manifestation of wave–particle duality is what Feynman called the great mystery of quantum physics. But it should be stressed that the formalism of quantum optics provides a consistent account of these different kinds of behaviour, whatever difficulty we may find in accepting them in the light of our experience of classical waves and particles.

5.6 A wave function for the photon?

For a one-photon state of the form (5.132) in free space, the probability of photodetection at $\mathbf{r}$ (5.134) takes the form

$$w(\mathbf{r}, t) = s|\mathbf{U}(\mathbf{r}, t)|^2, \tag{5.164}$$

[16] V. Jacques *et al.*, Experimental Realization of Wheeler's Delayed-Choice Gedanken Experiment, *Science* **315**, 966 (2007).

where

$$\mathbf{U}(\mathbf{r},t) = \mathrm{i}\sum_{\ell} \boldsymbol{\varepsilon}_\ell \mathcal{E}_\ell^{(1)} c_\ell \, \mathrm{e}^{\mathrm{i}(\mathbf{k}_\ell\cdot\mathbf{r}-\omega_\ell t)} = \langle 0|\hat{E}^{(+)}(\mathbf{r})|\psi(t)\rangle. \tag{5.165}$$

In an analogous way, the probabilities of photodetection (5.158) and (5.159) at the output of the Mach–Zehnder interferometer are given as the squared modulus of a complex-valued function on space and time. It is thus reasonable to ask whether this function can be considered as the wave function of the photon, whose squared modulus, suitably normalized, would give the probability density for the presence of the photon, measured by a photodetector. We shall see that, for a one-photon state, the answer is affirmative, with some reservations, but that the formalism of quantum optics is much richer than that of single-particle wave functions.

Note that the vectorial nature of $\mathbf{U}(\mathbf{r},t)$ should not come as a surprise, because, for a particle of non-zero spin s, the wave function is a $(2s+1)$-component spinor. For example, an electron, with spin 1/2, is described by a 2-component spinor $\boldsymbol{\psi}(\mathbf{r},t)$. The probability of detecting the electron at $\mathbf{r}$ in spin state i is then given by $|\psi_i(\mathbf{r},t)|^2$, and the total probability of detection is proportional to $\sum_i |\psi_i(\mathbf{r},t)|^2$. Likewise, if a polarizing filter selecting a polarization component i is placed in front of the photodetector at $\mathbf{r}$, the photodetection probability is indeed $|U_i(\mathbf{r},t)|^2$, up to a normalization constant. This means that $\mathbf{U}(\mathbf{r},t)$ can indeed be considered as a spinorial wave function, and that the probability of photodetection at $\mathbf{r}$ and t can be associated with a probability density for the presence of the photon calculated using this wave function.

However, it should not be thought that there is a position operator $\hat{\mathbf{r}}$ for the photon. In contrast to a massive particle, there is no perfectly localized state in 3D space. The reason is that a free electromagnetic field is transverse (with zero divergence), and this imposes constraints. The reader will find that it is not possible to construct a wave packet localized in three dimensions by generalizing the procedure outlined in Complement 5B, even if one accepts a very broad frequency spectrum.[17] So $|\mathbf{U}(\mathbf{r},t)|^2$ should not be considered as the probability of finding a photon exactly at point $\mathbf{r}$, but rather as an average probability over some small volume which cannot be smaller than λ_{M}^3, where λ_{M} is the wavelength associated with the maximum frequency to which the detector is sensitive.

A major limitation with the idea of defining a wave function for one-photon states, as compared with the formalism of quantum optics, is that one cannot ask for the probability of double detection, nor calculate fluctuations in photodetection signals in the more common case of many-photon states. But these signals are important quantities in quantum optics, revealing the specific quantum features of light.

There is no general definition of a wave function for states other than one-photon states. This should come as no surprise, since there is no elementary wave function for N electrons, described in general by an entangled quantum state in a truly gigantic state space. In fact, the formalism of quantum optics presented here is nothing other than a special case of the general formalism known as second quantization, which describes N indistinguishable quantum particles. This formalism replaces one-particle wave functions by operators

[17] A localized vector-valued function $\boldsymbol{\phi}_0\delta(\mathbf{r}-\mathbf{r}_0)$ cannot have zero divergence.

that are combinations of $\hat{a}$ and $\hat{a}^\dagger$ obeying commutation relations which, for bosons, are precisely the canonical relations (4.100–4.101). One might therefore say that quantum optics describes an ideal gas of non-interacting bosons, namely photons, and that the formalism results from second quantization of the electromagnetic field, considered as the wave function of the one-photon states.

Comment There is a special case of a many-photon state for which a classical function can be defined in such a way that its squared modulus gives the photodetection probability. This is the quasi-classical state, which is by definition an eigenstate of $\hat{\mathbf{E}}^{(+)}(\mathbf{r})$, the eigenvalue being a complex function that plays the role of wave function (see Sections 5.3.4 and 5.4.2). However, as for the one-photon states, double photodetection signals and fluctuations cannot be calculated without recourse to the full quantum formalism. On the other hand, we have shown that classical optics gives correct results in this case. There is thus little interest in using a wave function formalism, and it is preferable to move directly to a classical description of the electromagnetic field in the case of quasi-classical states.

5.7 Conclusion

Throughout history, there have been two contradictory conceptions of light: the wave theory put forward by Huygens, then developed by Young and Fresnel, with the electromagnetic nature of these waves being understood by Maxwell; and the particle conception defended by Descartes, Newton, Laplace, and then later, in a different context, by Einstein. With progress in experimental observations and theoretical models came a series of more or less short-lived victories for one side over the other.

By the end of the nineteenth century, classical electromagnetism, reviewed in the first part of Chapter 4, provided a wave description of almost all known optical phenomena (adding the postulate that the quantity measured in optics, called the light intensity, is proportional to the average of the squared electric field of the Maxwellian wave). At the beginning of the twentieth century, it turned out that this impressive edifice was not without its shortcomings, and that certain phenomena, such as the black-body radiation spectrum, the photoelectric effect or the Compton effect, could not be given a satisfactory explanation. By inventing the photon, Einstein revived the particle model, which provided a simple explanation for all these effects. But the price was high, since this model was powerless to account for interference and diffraction effects. There was no option but to accept that light behaved sometimes as a particle, sometimes as a wave.[18] After the advent of quantum mechanics, it was found that most of the phenomena of quantum optics, and in particular the photoelectric effect, could be accounted for by a model based on the interaction between quantized matter and a classical electromagnetic field. It is this type of approach that was presented in Chapter 2. It was true that the photon, and more generally

[18] Einstein himself stressed this problem before the development of quantum mechanics. See, for example, the text of the conference given in Salzburg in 1909: A. Einstein, Über die Entwicklung unserer Anschauungen über das Wesen und die Konstitution der Strahlung, *Physikalische Zeitschrift* **10**, 817 (1909).

the quantization of light, led to simple pictures for interpreting the photoelectric effect, Compton scattering or the selection rules for atomic spectra, but the quantization of radiation did not appear to be absolutely essential, except for a very small number of phenomena like spontaneous emission or the Lamb shift in the hydrogen spectrum. But again, these phenomena were all concerned with radiation interacting with a microscopic source, and the quantization of free radiation seemed hardly necessary.

It was not until the last few decades of the twentieth century that clarifications, due in particular to the work of R. Glauber, led to the discovery of situations that can only be interpreted using the quantum description of the free field, e.g. double detection of a single photon, measurements of quantum fluctuations in squeezed-radiation states and experiments involving the violation of Bell's inequalities with pairs of entangled photons. To describe this kind of situation, there was no option but to use the quantum theory of radiation, developed step by step between 1920 and 1960, in particular by Dirac, Heisenberg, Jordan, Pauli, Kramers, Bethe, Schwinger, Dyson, Tomonaga and Feynman, initially to provide a consistent description of interacting matter and radiation, but then to account for subtle effects such as the Lamb shift or spontaneous emission.

As far as free radiation is concerned, quantum optics provides a unified framework to describe the wave and particle aspects of light. It also provides a way of understanding why classical optics has been so successful, because we now know that it is the optics of radiation described by quasi-classical states.

But above all, by guiding us in the investigation of phenomena that are quite unprecedented in classical optics – associated with non-classical radiation states – it has opened up a whole new field of research, leading to new applications of optics. One could mention the example of measurements that go beyond the standard quantum limit, which should for example, with the use of squeezed states, considerably improve the range of the large interferometers used as gravitational wave detectors. Quantum optics has also seen the emergence of a vast new field of research, quantum information, in which secure data transmission and data processing are based on radically different concepts to those used in classical information and communication technologies. Only the future will tell if the quantum optics revolution in fundamental research will be followed by a technological revolution, in which the applications we have just mentioned are only the first examples.

Complement 5A Squeezed states of light: the reduction of quantum fluctuations

We showed in Section 5.2 of the present chapter that according to quantum theory the value of the electric field of an electromagnetic wave could not be predicted to arbitrarily high precision, this being a consequence of the uncertainty relation satisfied by the two quadrature components of the field (Equation 5.32), which imposed a finite limit on the product of their variances. This limit is more than an abstract theoretical limit; it is often the overriding factor determining the resolution of high-precision optical measurements. In such situations measurements by photodetectors on the field exhibit uncontrollable fluctuations of quantum origin known as *quantum noise*. Fortunately, as we shall show in this complement, it is possible to overcome this by using non-classical states of the radiation field, namely the *squeezed states*, provided measurements are made of a single one of a pair of conjugate field variables. A field mode prepared in such a state exhibits reduced quantum noise in one of the variables at the expense of increased noise in the other, so that the uncertainty relation involving their variances is still satisfied. The potential of such states for increasing the obtainable precision of optical measurements is self-evident. However, the practical realization of the potential benefits requires delicate experimental techniques that we describe only briefly here.[1]

5A.1 Squeezed states: definition and properties

5A.1.1 Definition

In this complement we consider a single mode of the radiation field, as in Section 5.3 of Chapter 5. We thus omit the subscript ℓ corresponding to this mode. We start by introducing a new operator $\hat{A}_R$, a function of the real variable R, and defined by

$$\hat{A}_R = \hat{a}\cosh R + \hat{a}^\dagger \sinh R. \tag{5A.1}$$

It is easy to show that this operator satisfies

$$\left[\hat{A}_R, \hat{A}_R^\dagger\right] = \cosh^2 R - \sinh^2 R = 1. \tag{5A.2}$$

[1] For a more detailed discussion see, H. A. Bachor and T. Ralph, *A Guide to Experiments in Quantum Optics*, Wiley (2004).

Thus the operator $\hat{A}_R$ and its Hermitian conjugate $\hat{A}_R^\dagger$ obey the same commutation relation as the photon annihilation and creation operators $\hat{a}$ and $\hat{a}^\dagger$ of the mode under consideration. They therefore share the same properties. In particular, by employing the procedure we adopted in introducing the quasi-classical states $|\alpha\rangle$, which we defined as being the eigenstates of the photon annihilation operator, we can introduce the equivalent states $|\alpha, R\rangle$, the eigenstates of $\hat{A}_R$:

$$\hat{A}_R |\alpha, R\rangle = \alpha |\alpha, R\rangle . \tag{5A.3}$$

These states are known as *squeezed states*. Notice that when $R = 0$ they reduce to the usual quasi-classical states, and that they are not stationary states, like the quasi-classical states.

5A.1.2 Expectation values of field observables for a squeezed state

In order to calculate the expectation value of the electric field for a squeezed state and its variance we must first express the electric field operator in terms of the operators $\hat{A}_R$ and $\hat{A}_R^\dagger$. The annihilation operator may be written

$$\hat{a} = \hat{A}_R \cosh R - \hat{A}_R^\dagger \sinh R, \tag{5A.4}$$

so that the electric field operator (Equation (4.109)) becomes

$$\begin{aligned}\hat{E}(\mathbf{r}) =& \mathrm{i}\,\mathcal{E}^{(1)} \Big\{\hat{A}_R \left(\mathrm{e}^R \cos \mathbf{k}\cdot\mathbf{r} + \mathrm{i}\mathrm{e}^{-R} \sin \mathbf{k}\cdot\mathbf{r}\right) \\ &- \hat{A}_R^\dagger \left(\mathrm{e}^R \cos \mathbf{k}\cdot\mathbf{r} - \mathrm{i}\mathrm{e}^{-R} \sin \mathbf{k}\cdot\mathbf{r}\right)\Big\} .\end{aligned} \tag{5A.5}$$

The expectation value of this operator is easily evaluated using (5A.3). We obtain

$$\langle \alpha, R|\, \hat{E}(\mathbf{r})\, |\alpha, R\rangle = \mathrm{i}\,\mathcal{E}^{(1)} \Big\{\left(\alpha \cosh R - \alpha^* \sinh R\right) \mathrm{e}^{\mathrm{i}\mathbf{k}\cdot\mathbf{r}} - \text{c.c.}\Big\} . \tag{5A.6}$$

This mean value for a squeezed state is indistinguishable from that for a quasi-classical state $|\alpha'\rangle$ with

$$\alpha' = \alpha \cosh R - \alpha^* \sinh R. \tag{5A.7}$$

The same is not true of the field variance. Using expression (5A.5) and the commutation relation (5A.2), we find

$$\langle \alpha, R|\, \hat{E}^2(\mathbf{r})\, |\alpha, R\rangle = \left(\langle \alpha, R|\, \hat{E}(\mathbf{r})\, |\alpha, R\rangle\right)^2 + \left|\mathcal{E}^{(1)}\left(\mathrm{e}^R \cos \mathbf{k}\cdot\mathbf{r} + \mathrm{i}\mathrm{e}^{-R} \sin \mathbf{k}\cdot\mathbf{r}\right)\right|^2 , \tag{5A.8}$$

so that

$$\Delta^2 E(\mathbf{r}) = \left[\mathcal{E}^{(1)}\right]^2 \left(\mathrm{e}^{2R} \cos^2 \mathbf{k}\cdot\mathbf{r} + \mathrm{e}^{-2R} \sin^2 \mathbf{k}\cdot\mathbf{r}\right) . \tag{5A.9}$$

This expression must be compared with the value $\left[\mathcal{E}^{(1)}\right]^2$ obtained for a quasi-classical state (see Equation (5.87)).

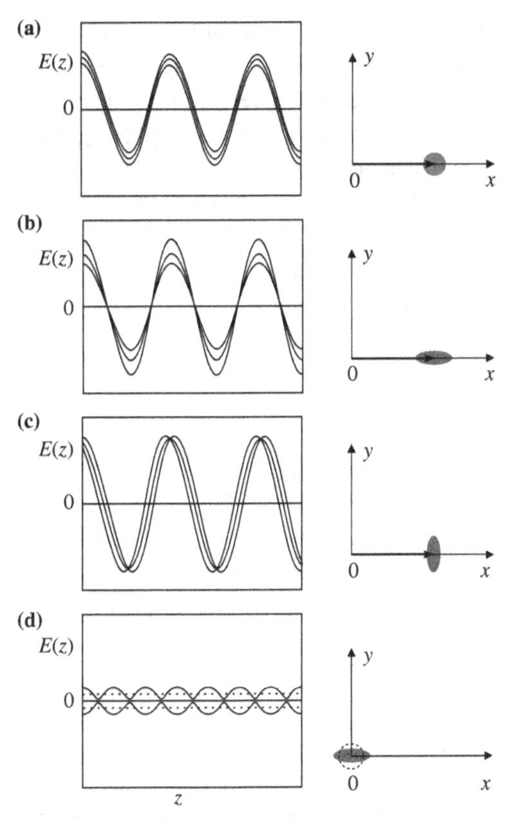

Figure 5A.1 Form of the spatial dependence of the electric field and the phasor representation for various states of the field: (a) a quasi-classical state ($R = 0$), characterized by a band of possible field values of width $2\mathcal{E}^{(1)}$ and by a circular zone in the phasor diagram; (b) squeezed state with $R < 0$, giving a reduced field uncertainty at points where this field passes through zero and an increased uncertainty at points where the field is large; (c) squeezed state with $R > 0$, distinguished by a small uncertainty at field maxima and maximal uncertainty at zero crossings; (d) squeezed vacuum. The vacuum-state field variance is shown by the dotted lines of constant separation $2\mathcal{E}^{(1)}$.

Thus, for a squeezed state, the variance of the field depends on the position coordinate $\mathbf{r}$ and, moreover, oscillates between a value smaller than the variance of the vacuum fluctuations $\left[\mathcal{E}^{(1)}\right]^2$ and one that is larger, with a period equal to half the optical wavelength.

We show in Figure 5A.1, for various values of α and R, the variation of the instantaneous electric field for a squeezed state as a function of the distance z in the direction of propagation. The central solid line depicts the expectation value given by (5A.6) and the vertical distance between the two outer lines is the square-root of the variance (5A.9). We have sketched in the same figure the corresponding phasor representation of the field in each case. Figure 5A.1(a) corresponds to a quasi-classical state ($R = 0$) whilst figures 5A.1(b) and 5A.1(c) correspond to α real and R, respectively, negative and positive (the value of α' defined above being held constant). Figure 5A.1(d) corresponds to $\alpha = 0$ and $R < 0$

('squeezed vacuum'). It is apparent from (5A.9) that when $kz = p\pi + \pi/2$ (where p is an integer) for Figures 5A.1(b) and 5A.1(d), or when $kz = p\pi$ in the case of Figure 5A.1(c), the electric field is defined more accurately than the electric field of a quasi-classical state: the variance is reduced by a factor of $\exp\{2|R|\}$, which is known as the *squeezing parameter* and which can be arbitrarily large, at least in theory. The price to pay is that a quarter of a wavelength later the uncertainty in the field is increased over that of a quasi-classical state by the same factor.

Additionally, we have the following expressions for the quadrature operators as functions of the operators $\hat{A}_R$ and $\hat{A}_R^\dagger$:

$$\hat{E}_Q = \mathcal{E}^{(1)}\,\mathrm{e}^R(\hat{A}_R + \hat{A}_R^\dagger) \tag{5A.10}$$

$$\hat{E}_P = -\mathrm{i}\,\mathcal{E}^{(1)}\,\mathrm{e}^{-R}(\hat{A}_R - \hat{A}_R^\dagger). \tag{5A.11}$$

This straightforwardly implies

$$(\Delta E_Q)^2 = \left[\mathcal{E}^{(1)}\right]^2 \mathrm{e}^{2R} \tag{5A.12}$$

$$(\Delta E_P)^2 = \left[\mathcal{E}^{(1)}\right]^2 \mathrm{e}^{-2R} \tag{5A.13}$$

$$\Delta E_P \Delta E_Q = \left[\mathcal{E}^{(1)}\right]^2. \tag{5A.14}$$

Thus, the squeezed states are *minimum uncertainty states* for which one quadrature fluctuates less than in the vacuum state.

Finally, we derive, for a squeezed state, the expectation value and the variance of the photon number. The photon number operator $\hat{N}$ is given in terms of the operators $\hat{A}_R$ and $\hat{A}_R^\dagger$ by

$$\hat{N} = \hat{A}_R^\dagger \hat{A}_R \cosh^2 R + \hat{A}_R \hat{A}_R^\dagger \sinh^2 R - \sinh R \cosh R\left(\hat{A}_R^2 + \hat{A}_R^{\dagger 2}\right), \tag{5A.15}$$

leading easily to the expectation number of photons in the state $|\alpha, R\rangle$,

$$\langle \hat{N}\rangle = \langle \alpha, R|\,\hat{N}\,|\alpha, R\rangle = |\alpha'|^2 + \sinh^2 R, \tag{5A.16}$$

where α' is given by (5A.7). This expression shows that, even for the squeezed state with $\alpha = \alpha' = 0$ (known as the *squeezed vacuum*), the mean number of photons is no longer zero if $R \neq 0$.

In calculating the variance of the photon number, we restrict our attention to the simple case of a real value of α. A tedious, but straightforward manipulation (using the commutator (5A.2) and expression (5A.3)) leads to the result,

$$(\Delta N)^2 = \alpha^2 \mathrm{e}^{-4R} + 2\sinh^2 R \cosh^2 R = \alpha'^2 \mathrm{e}^{-2R} + 2\sinh^2 R \cosh^2 R. \tag{5A.17}$$

Consider the case of positive R (that of Figure 5A.1(c)) for which $\alpha'^2 \gg \exp 2R$. The second term in (5A.17) is then negligible and the variance in $\hat{N}$ differs little from $\langle \hat{N}\rangle \exp(-2R)$, that is, the uncertainty in the intensity is reduced from that expected for a quasi-classical state by a factor of the squeezing parameter $\exp(-2R)$. This is clearly shown in the phasor representation of the field in Figure 5A.1(c), which shows that the fluctuations of the amplitude of the phasor are smaller than those for the representation of a quasi-classical state. The light field in this state is said to have *sub-Poissonian* statistics, because the photon number variance is less than that found for a Poissonian distribution for which $(\Delta N)^2 = \langle \hat{N} \rangle$. Measurements of the conjugate quantity, in this case the phase of the wave, of course exhibit larger fluctuations than expected for a quasi-classical state, as can be deduced from Figure 5A.1(c) from the greater extent of the shaded area in a direction perpendicular to that of the phasor. Obviously the above conclusions are reversed for the case of R negative (that of Figure 5A.1(b)).

The time-dependence of the expectation values of the field variables can be found by generalizing the treatment of Section 5.3. This leads to results analogous to (5A.6) and (5A.9) in which the phase $\mathbf{k}.\mathbf{r}$ is replaced by $\mathbf{k}.\mathbf{r} - \omega t$. Thus the above discussion of the spatial dependence of the expectation values of the field variables can be extended into the temporal domain: the field variance at a given point in space oscillates between values smaller and larger than the vacuum fluctuations with a period equal to half the optical period. The temporal dependence of the field in such a state (the case of α real, $R < 0$) then corresponds to that depicted in Figure 5.7 of Chapter 5, which is analogous to Figure 5A.1(b).

5A.1.3 The squeezing operator

Any squeezed state can be derived from a quasi-classical state by a unitary transformation known as a *squeezing transformation*:

$$|\alpha, R\rangle = \hat{S}(R)\,|\alpha\rangle\,, \tag{5A.18}$$

with

$$\hat{S}(R) = \exp\left[R(\hat{a}^2 - \hat{a}^{\dagger 2})/2\right]. \tag{5A.19}$$

In order to justify (5A.18) we employ the identity:[2]

$$\mathrm{e}^{\hat{B}}\hat{A}\,\mathrm{e}^{-\hat{B}} = \hat{A} + \frac{1}{1!}\left[\hat{B},\hat{A}\right] + \frac{1}{2!}\left[\hat{B},\left[\hat{B},\hat{A}\right]\right] + \frac{1}{3!}\left[\hat{B},\left[\hat{B},\left[\hat{B},\hat{A}\right]\right]\right] + \ldots, \tag{5A.20}$$

which is valid for any pair of operators $\hat{A}$ and $\hat{B}$. We consider the case for which $\hat{B} = R(\hat{a}^2 - \hat{a}^{\dagger 2})/2$ and $\hat{A} = \hat{a}$. The commutators $[\hat{B}, \hat{a}]$ and $[\hat{B}, \hat{a}^\dagger]$ are given by

$$\left[\hat{B}, \hat{a}\right] = R\hat{a}^\dagger \tag{5A.21}$$

$$\left[\hat{B}, \hat{a}^\dagger\right] = R\hat{a}, \tag{5A.22}$$

from which we deduce,

$$\begin{aligned} \hat{S}(R)\hat{a}\hat{S}^{-1}(R) &= \hat{a}\left(1 + \tfrac{R^2}{2!} + \ldots\right) + \hat{a}^\dagger\left(R + \tfrac{R^3}{3!} + \ldots\right) \\ &= \hat{a}\cosh R + \hat{a}^\dagger \sinh R = \hat{A}_R. \end{aligned} \tag{5A.23}$$

[2] See, for example, W. Louisell, *Quantum Statistical Properties of Radiation*, p. 136, Wiley (1973).

Applying this operator to the state vector $\hat{S}(R)|\alpha\rangle$, we obtain

$$\hat{S}(R)\alpha\,|\alpha\rangle = \hat{A}_R\hat{S}(R)\,|\alpha\rangle\,. \tag{5A.24}$$

Thus $\hat{S}(R)|\alpha\rangle$ is an eigenvector of $\hat{A}_R$ corresponding to eigenvalue α, and therefore satisfies Equation (5A.3), which defines the squeezed state $|\alpha, R\rangle$.

5A.1.4 Transmission of a squeezed state by a beamsplitter

We now consider the effect of a partially reflecting beamsplitter on incident light described by a squeezed state. The beamsplitter, similar to the one in Figure 5.1, has amplitude reflection and transmission coefficients r and t, which are assumed to satisfy $r^2 + t^2 = 1$.

We use here the results of Section 5.1.2 concerning partially reflecting beamsplitters. Let $\hat{E}_1^{(+)}$ and $\hat{E}_2^{(+)}$ be the complex field operators associated with the input modes (1) and (2). Generalizing (5.7), the transmitted complex field operator is given by

$$\hat{E}_4^{(+)} = t\hat{E}_1^{(+)} - r\hat{E}_2^{(+)}. \tag{5A.25}$$

For simplicity, we restrict our discussion to the consideration of a single field mode of frequency ω in each port. The operators $\hat{E}_i^{(+)}$ are then simply proportional to the photon annihilation operators $\hat{a}_i$. Thus, we have

$$\hat{a}_4 = t\hat{a}_1 - r\hat{a}_2. \tag{5A.26}$$

From this expression, we can derive one for the quadrature operator $\hat{E}_Q = \mathcal{E}^{(1)}(\hat{a} + \hat{a}^\dagger)$ of the transmitted field:

$$\hat{E}_{Q4} = t\hat{E}_{Q1} - r\hat{E}_{Q2}, \tag{5A.27}$$

assuming real transmission and reflection coefficients t and r.

We can now calculate the quadrature fluctuations in the transmitted field for the situation in which a squeezed state $|\alpha, R\rangle$, which we suppose to be of the *sub-Poissonian* variety for which α is real and positive, is incident at port (1). The initial state of the field $|\psi\rangle$ is therefore the tensor product of the squeezed state at port (1) and the vacuum state at port (2):

$$|\psi\rangle = |\alpha, R\rangle \otimes |0\rangle\,. \tag{5A.28}$$

The action of the quadrature operator $\hat{E}_{Q4}$ on this state is described by

$$\hat{E}_{Q4}\,|\psi\rangle = t\hat{E}_{Q1}\,|\alpha, R\rangle \otimes |0\rangle - \mathcal{E}^{(1)}\,r\,|\alpha, R\rangle \otimes |1\rangle\,, \tag{5A.29}$$

since $\hat{E}_{Q2}|0\rangle = \mathcal{E}^{(1)}|1\rangle$. From this expression we can derive the expectation value of $\hat{E}_{Q4}$ and its variance. For the former quantity we find

$$\left\langle\psi\left|\hat{E}_{Q4}\right|\psi\right\rangle = t\left\langle\alpha, R\left|\hat{E}_{Q1}\right|\alpha, R\right\rangle, \tag{5A.30}$$

which is identical to the result expected classically; the mean field quadrature is simply reduced by the transmission factor t. For the variance, $(\Delta E_{Q4})^2 = \|\hat{E}_{Q4} \mid \psi\rangle\|^2 - (\langle\psi \mid \hat{E}_{Q4} \mid \psi\rangle)^2$, we obtain

$$(\Delta E_{Q4})^2 = t^2(\Delta E_{Q1})^2 + r^2\left[\mathcal{E}^{(1)}\right]^2 = \left[\mathcal{E}^{(1)}\right]^2(t^2\,\mathrm{e}^{-2R} + r^2). \tag{5A.31}$$

When the beamsplitter has a transmission coefficient close to one, $r \approx 0$ and the fluctuations on the transmitted field are close to those on the incident field and therefore remain squeezed ($(\Delta E_{Q4})^2 < \left[\mathcal{E}^{(1)}\right]^2$). However, if t is small, r approaches unity and the fluctuations on the output field arise principally from the second term in (5A.31), that is from the vacuum fluctuations entering the unused input port. These fluctuations have the magnitude expected for a quasi-classical state, even if the input field is very strongly squeezed. Note that (5A.31) implies that if the input field is a quasi-classical state, then so is the output field, whatever the value of the reflection coefficient r.

5A.1.5 Effect of losses

From the point of view of the transmitted beam, the coefficient r^2 represents a loss from the transmitted intensity. Using an argument analogous to that of the preceding paragraph, expression (5A.31) can be generalized to the case of a system of several beamsplitters encountered successively by the transmitted beam: it is sufficient to replace r^2 in (5A.31) by the sum of the intensity reflection coefficients of the various beamsplitters. More generally, any global loss of light which is proportional to the incident intensity can be considered as the scattering of a portion of the light into a collection of unused loss modes. Formula (5A.31) then applies if r^2 is replaced by the intensity loss coefficients A, and t^2 by $1 - A$. From this point of view, *absorption* has the same effect as *partial reflection*.

Any linear loss therefore has the effect of reducing the squeezing of the quantum fluctuations (but also of reducing any initial excess noise) so that the state of the output field is closer to a quasi-classical one than the input field. The quasi-classical states are, therefore, in a sense, privileged in that they are the only states that are stable with respect to dissipative losses. Expression (5A.31) shows in addition that the introduction even of very small losses suffices to disrupt the suppression of fluctuations that occurs in the squeezed state. The reduction of such losses in order to conserve the squeezing, obtained with much effort from certain non-linear interactions, is one of the principal technical challenges in experiments aimed at the production and manipulation of squeezed light.

Comment This property is not only true of the quadrature component of the output beam, but also applies to other observables such as the intensity noise. A sub-Poissonian beam becomes less sub-Poissonian when it is transmitted through a beamsplitter.

An alternative argument, in terms of photons, can be given to explain the fact that the existence of a loss tends to increase the Poissonian character of the light field. This relies on the fact that a Poisson distribution is associated with an ensemble of statistically independent photons. Thus for a sub-Poissonian distribution the photons are not randomly distributed, but have a regular order. The effect of a loss, being simply to remove photons at random from this regular array, is to cause the photon distribution in the output field to be more random than that in the input field, so that the distribution is made more Poissonian.

5A.2 Generation of squeezed light

5A.2.1 Generation by parametric processes

Expressions (5A.18) and (5A.19) can be put in the form:

$$|\alpha, R\rangle = \exp(-\mathrm{i}\hat{H}_{\mathrm{I}}T/\hbar)\,|\alpha\rangle\,, \tag{5A.32}$$

with

$$\hat{H}_{\mathrm{I}} = \mathrm{i}\frac{\hbar R}{2T}(\hat{a}^{+2} - \hat{a}^2), \tag{5A.33}$$

which shows that a squeezed state results from the temporal evolution over a period T of a quasi-classical state under the influence of a Hamiltonian $\hat{H}_{\mathrm{I}}$. The form of this Hamiltonian is characteristic of processes in which there is the *simultaneous* creation or annihilation of two photons of the mode under consideration. Interactions of this type are encountered in non-linear optics, for example in *degenerate parametric mixing* (see Chapter 7) in which the annihilation of a pump photon at frequency 2ω is associated with the simultaneous creation of two photons at frequency ω. Figure 5A.2 shows an experimental setup using parametric mixing to generate squeezed light. It consists of a suitable non-linear medium enclosed in a Fabry–Perot cavity whose two mirrors are transparent at the pump frequency 2ω, whilst the first, M, is totally reflecting at frequency ω and the second M' is partially transmitting at the same frequency. The cavity is resonant at frequency ω, so that the degenerate down-conversion $\omega_1 = \omega_2 = \omega$ is preferred over non-degenerate down-conversion ($\omega_1 + \omega_2 = 2\omega$ with $\omega_1 \neq \omega_2$), the frequencies ω_1 and ω_2 in the latter case not being resonant with the cavity. The system is pumped by a laser

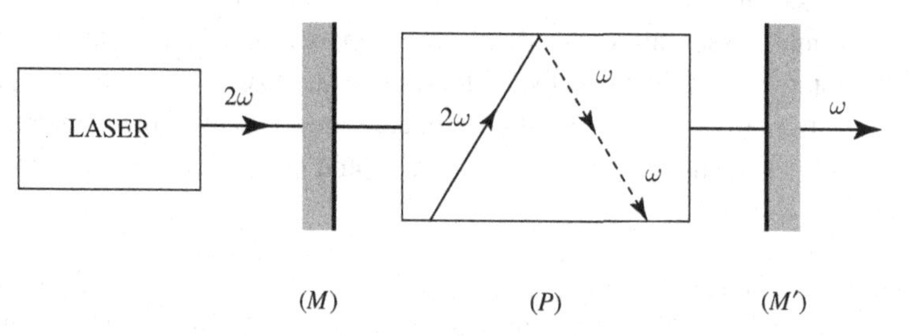

Figure 5A.2 Schematic representation of a setup employing degenerate parametric down-conversion for the production of squeezed light. The two mirrors M and M' are totally transmitting at the pump frequency 2ω, whilst they form a resonant cavity at frequency ω.

emitting at frequency 2ω. Thus the initial state of the ω-frequency mode is the vacuum state. Below the threshold for parametric oscillation as we shall see in Complement 7A, this field state is transformed under the effect of the Hamiltonian for the parametric down-conversion, which is of the form (5A.33), into a *squeezed vacuum* state. Because of the finite transmission of mirror M' at frequency ω, a steady regime is established in which the field leaving the cavity via M' is of the type described by Figure 5A.1(d). This state of the radiation field, which, as we pointed out previously, has no classical counterpart, was first demonstrated in 1985. Later experiments have led to the production of squeezed light characterized by squeezing parameters as large as 10. Squeezed states are also produced using degenerate parametric down-conversion without the help of a resonant cavity, but in a pulsed regime, using for example mode-locked lasers as a pump.

5A.2.2 Other methods

Numerous non-linear optical processes are capable of producing squeezed states of light for suitably chosen experimental configurations and provided losses are kept to a minimum.[3] These include *frequency doubling*, which is also described by a Hamiltonian of the form (5A.33) and which can produce a finite-amplitude squeezed state in the transmitted pump light as well as in the generated second-harmonic beam. A medium characterized by a *Kerr non-linearity* (see Complement 7B) is also capable of producing squeezed states of light from an incident quasi-classical field propagating through it. In contrast to the use of parametric interactions, in this case it is not necessary to pump the medium with light of a frequency different from that of the squeezed mode. The non-linear refractive index of silica in particular is sufficiently large that the propagation of an initially quasi-classical field in a single-mode fibre made of this material just a few metres long transforms its state into a squeezed one.

Finally, we point out that a laser can also emit a sub-Poissonian field. We saw in Chapter 3 that the fluctuations of the phase of the laser output, resulting from the phase diffusion driven by spontaneous emission that leads to the Schawlow–Townes limit for its linewidth, are intrinsically very large. In the laser there is no mechanism for locking the phase of the output wave, so that it is free to undergo unlimited diffusion (see Complement 3D). Owing to the resulting large value of the variance of the phase, $\Delta\phi$, the uncertainty relation concerning the phase and the photon number, which can be written,

$$\Delta N \geq \frac{1}{2\Delta\phi}, \tag{5A.34}$$

implies that the output field can be very strongly sub-Poissonian. In practice, for common laser systems, losses and the noise introduced by the mechanism that drives the population inversion (pumping noise) ensure that the output field is characterized by a Poissonian distribution. However, a semiconductor laser pumped by a current with strongly suppressed

[3] See Footnote 1, this complement.

fluctuations produces a squeezed field with reduced intensity fluctuations (under optimum conditions it is possible to obtain a squeezing parameter of the order of 10). The simplicity of squeezed-light generators of this type, if they become easily available, should ensure that squeezed states of the radiation field will cease to be just laboratory curiosities and will find practical applications. It is to some of these potential uses that we turn our attention in the following section.

5A.3 Applications of squeezed states

The judicious use of squeezed states can permit the resolution of measurements on an optical field, which generally concern a single field variable, for instance its intensity, to be improved beyond the 'shot-noise' limit that applies to quasi-classical fields. In what follows we give two examples in which the manipulation of squeezed states does yield such a potential improvement in precision.

5A.3.1 Measurement of small absorption coefficients

In its simplest form a direct absorption measurement is achieved by passing a light beam through the absorbing medium and measuring the proportion transmitted, the absorption A being given by the difference between the incident intensity $\langle \hat{N}_1 \rangle$ and the intensity transmitted, $\langle \hat{N}_4 \rangle = (1 - A)\langle \hat{N}_1 \rangle$. The absorption is therefore given by $\langle \hat{N}_1 - \hat{N}_4 \rangle = A\langle \hat{N}_1 \rangle$. The noise on this signal is of magnitude $((\Delta N_1)^2 + (\Delta N_4)^2)^{1/2}$: the noise on the two measurements that must be carried out in order to determine the absorption are added quadratically. Now $(\Delta N_4)^2$, which is given by (5A.31) with $r^2 = A \ll 1$, is very close to $(\Delta N_1)^2$. The minimum possible value of A that can be measured, $A_{\min}$, corresponds to a signal to noise ratio of 1. It is given by

$$A_{\min} = \sqrt{2}\frac{1}{2}\frac{\Delta N_1}{\langle \hat{N}_1 \rangle}. \tag{5A.35}$$

In the case of a quasi-classical field $A_{\min} = \sqrt{2/\langle \hat{N}_1 \rangle}$. In order to increase the sensitivity of the measurement it is necessary either to:

- increase the mean number of photons detected $\langle \hat{N}_1 \rangle$, that is, increase the intensity of the incident beam or the measurement time; or
- employ a sub-Poissonian, squeezed field which reduces $\Delta N_1 / \langle \hat{N}_1 \rangle$ and hence $A_{\min}$ by a factor of e^R.

In the ideal case of a field perfectly stabilized in intensity, described by a number state (see Section 5.3.3), the limiting value of $A_{\min}$ corresponds to the detection of the absorption of a single photon in the relevant field mode and is therefore $1/\langle \hat{N}_1 \rangle$.

If we take the example of a light beam of wavelength 1 μm of power 1 μW, the incident photon current is about 5×10^{19} per second. For a quasi-classical state, the minimum absorption that can be detected in an integration time of 1 ms is, according to (5A.35), $A_{\text{min}} \approx 6 \times 10^{-9}$. This is the so-called standard quantum limit. For a number state the corresponding ideal result is $A_{\text{min}} \approx 2 \times 10^{-17}$. Real experiments are far from that ideal, but measurements below the standard quantum limit have been demonstrated.

5A.3.2 Interferometric measurements

Interferometric optical measurements are amongst the most precise made. For example, a gravitational wave detector described in Complement 3F, designed to be sensitive to the gravity waves emitted by a supernova, is a Michelson interferometer with an arm length of 3 km. The magnitude of the path-length shift that is expected to be caused by such an event is minute: of the order of 10^{-18} m. In fact it was the need to optimize the accuracy of such measuring devices that motivated much of the research on squeezed states.

Consider, then, an interferometer of the kind depicted in Figure 5.8(a), into which is injected at input port (1) an intense, monochromatic light beam at frequency ω. It can be shown that the minimum discernible phase-shift, $\Delta\phi_{\text{min}}$, of the light in one interferometer arm compared to that in the other, that for which the signal to noise ratio is one, is[4]

$$\Delta\phi_{\text{min}} = \frac{1}{\sqrt{\langle \hat{N}_1 \rangle}} \frac{\Delta E_{P2}}{\mathcal{E}^{(1)}}, \tag{5A.36}$$

where ΔE_{P2} is the variance of that quadrature component of the field entering at port (2) that is in phase with the input field at port (1). This quantity is therefore independent of the fluctuations of the field injected at port (1) of the interferometer. Usually, no field is injected at port (2), so that ΔE_{P2} is just the amplitude of the vacuum fluctuations, $\mathcal{E}^{(1)}$. In this case we find that $\Delta\phi_{\text{min}}$ is equal to $1/\sqrt{\langle \hat{N}_1 \rangle}$, which is the same as the minimum absorption that can be measured using a quasi-classical field (standard quantum limit). To attain the sensitivity need to detect gravitational waves for a measurement integration time of 1 ms would require a laser power of the order of 100 kW. However, the sensitivity can be increased by injecting a squeezed vacuum state into the second input port of the interferometer, (2). Expressions (5A.13) and (5A.36) indicate that this gives rise to an increase in sensitivity by a factor of $e^{|R|}$. Squeezing factors $e^{|R|}$ of the order of four can be obtained for integration times of a millisecond. This might seem to be a small improvement considering the effort expended and the increased complexity of the apparatus. However, since the expected rate of supernova detections is of the order of one per year, at the standard quantum limit, even such a small increase is very welcome!

[4] C. M. Caves, Quantum Mechanical Noise in an Interferometer, *Physical Review D* **23**, 1693 (1981).

5B Complement 5B One-photon wave packet

One-photon sources are important elements in quantum optics. The archetypal example is an atom raised to an excited state at time $t = 0$, then de-exciting with emission of a single photon. The development of this kind of source depends on progress with experimental techniques, e.g. the possibility of isolating a single atom, molecule or quantum well. In this complement, we present the formalism for describing the corresponding radiation, and use it to discuss some spectacular experiments which bring out properties quite incompatible with a classical description of the electromagnetic field. We begin in Section 5B.2 by describing the anti-correlation between detections on either side of a semi-reflecting mirror, establishing the quantitative difference with a classical field. Section 5B.3 discusses a quantum optical effect that was only demonstrated at the beginning of the twenty-first century, namely the quantum coalescence of two one-photon wave packets on a semi-reflecting mirror, which occurs even when the two photons were emitted by independent atoms. An analogous effect, the Hong–Hou–Mandel effect, is discussed in Chapter 7. These effects exemplify quantum interference involving two photons. Finally, Section 5B.4 is concerned with quantum calculations involving quasi-classical states. As we now know, this leads to results that are identical to the predictions of semi-classical theory.

5B.1 One-photon wave packet

5B.1.1 Definition and single photodetection probability

Consider a one-photon state of the form (5.132), i.e.

$$|1\rangle = \sum_{\ell} c_\ell |0,\ldots,n_\ell = 1,0,\ldots\rangle = \sum_{\ell} c_\ell |1_\ell\rangle, \tag{5B.1}$$

with $\sum_\ell |c_\ell|^2 = 1$. It is an eigenstate of $\hat{N} = \sum_\ell \hat{N}_\ell$ with eigenvalue 1, but it is not an eigenstate of $\hat{N}_\ell$, and nor is it an eigenstate of the Hamiltonian $\hat{H}_R = \sum_\ell \hbar\omega_\ell(\hat{N}_\ell + 1/2)$, whenever there are two non-zero coefficients c_ℓ corresponding to two modes of different frequencies. It changes in time and the state is given at time t by

$$|1(t)\rangle = \sum_{\ell} c_\ell \, \mathrm{e}^{-\mathrm{i}\omega_\ell t} |1_\ell\rangle. \tag{5B.2}$$

Using (5.4), the photodetection signal at time t and point $\mathbf{r}$ is given by

$$w(\mathbf{r},t) = s\left\|\hat{\mathbf{E}}^{(+)}(\mathbf{r})|1(t)\rangle\right\|^2 = s\left|\sum_\ell c_\ell \mathscr{E}_\ell^{(1)} \boldsymbol{\varepsilon}_\ell \mathrm{e}^{\mathrm{i}(\mathbf{k}_\ell.\mathbf{r}-\omega_\ell t)}\right|^2 = s|\mathbf{E}^{(+)}(\mathbf{r},t)|^2, \tag{5B.3}$$

with

$$\mathbf{E}^{(+)}(\mathbf{r},t) = \sum_\ell c_\ell \mathscr{E}_\ell^{(1)} \boldsymbol{\varepsilon}_\ell \mathrm{e}^{\mathrm{i}(\mathbf{k}_\ell.\mathbf{r}-\omega_\ell t)}. \tag{5B.4}$$

Note that this single photodetection probability is exactly the same as would be found for the classical field (5B.4), using the semi-classical model of photodetection (Equation 5.2).

5B.1.2 One-dimensional wave packet

In classical electromagnetism, we know how to construct wave packets occupying a bounded region of spacetime. We consider a set of coefficients c_ℓ different from zero for values of $\mathbf{k}_\ell$ distributed over a volume of $\mathbf{k}$-space of extent δk_x, δk_y, δk_z about a value $\mathbf{k}_0$. We thus obtain, at time $t = 0$, a wave packet localized in a volume of real space with dimensions of the order of $(\delta k_x)^{-1}$, $(\delta k_y)^{-1}$, $(\delta k_z)^{-1}$. When the same set of coefficients c_ℓ is substituted into (5B.1), we thus obtain a photodetection probability (5B.3) that differs from zero only within some bounded region.

The volume of this region generally increases without limit as time goes by, and this in each space dimension, but there are specific forms for which the spreading effect does not occur. An example is the one-dimensional wave packet we are about to discuss.

Consider the case in which the wavevectors $\mathbf{k}_\ell$ associated with the non-zero coefficients c_ℓ are all parallel to the same unit vector $\mathbf{u}$, i.e.

$$\mathbf{k}_\ell = \frac{\omega_\ell}{c}\mathbf{u} = \ell\frac{2\pi}{L}\mathbf{u}, \tag{5B.5}$$

where L is an arbitrary quantization length. The function (5B.4) then takes the form

$$\mathbf{E}^{(+)}(\mathbf{r},t) = \sum_\ell c_\ell \mathscr{E}_\ell^{(1)} \boldsymbol{\varepsilon}_\ell\, \mathrm{e}^{\mathrm{i}\omega_\ell(\mathbf{r}\cdot\mathbf{u}/c-t)}, \tag{5B.6}$$

and the photodetection probability (5B.3) depends on space and time only through the quantity

$$\tau = t - \mathbf{r}\cdot\mathbf{u}/c. \tag{5B.7}$$

The photodetection signal thus propagates without distortion at speed c in the direction specified by $\mathbf{u}$.

This kind of wave packet is not very realistic, in the sense that it extends infinitely in the plane perpendicular to $\mathbf{u}$. One can imagine a cylindrical beam of cross-section $S_\perp$, propagating in the direction $\mathbf{u}$. If the transverse dimensions are much larger than the wavelength, diffraction will be negligible, and the cylindrical beam is an approximate solution for the

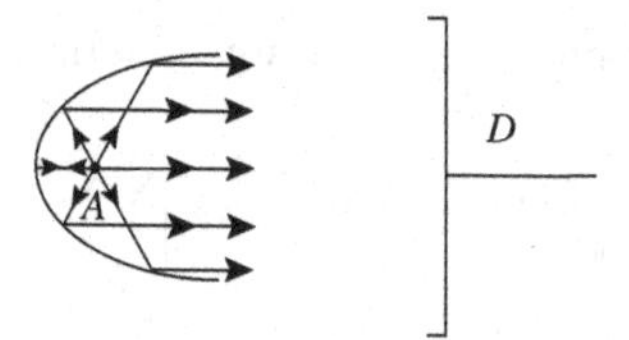

Figure 5B.1 A single atom A, placed at the focus of a parabolic mirror, emits a collimated one-photon wave packet, well modelled by the one-dimensional wave packet of Section 5B.1.2. The detector D is wide enough to detect the whole wave packet.

equations of electromagnetism. The beam obtained by placing a very small source, e.g. an excited atom at the focus of a parabolic mirror, is adequately described by this model (see Figure 5B.1).

Let us consider the case where all the modes have the same polarization $\boldsymbol{\varepsilon}$. The coefficients c_ℓ then depend only on the frequency ω_ℓ, and a wave packet can be formed by considering a distribution peaking at some ω_0, described by

$$c(\omega_\ell) = f(\omega_\ell - \omega_0), \tag{5B.8}$$

where $f(\Omega)$ is a function centred on 0 and having a typical half-width $\delta\omega$ that is small compared with ω_0. The function (5B.6) will then be proportional to the Fourier transform $\tilde{f}(\tau)$ of $f(\Omega)$, yielding a wave packet with width of the order of $1/\delta\omega$. To carry out the calculation explicitly, the sum $\sum_\ell$ in (5B.6) is replaced by an integral, introducing the one-dimensional mode density deduced from (5B.5):

$$\frac{\mathrm{d}\ell}{\mathrm{d}\omega_\ell} = \frac{L}{2\pi c}. \tag{5B.9}$$

The quantization volume is $S_\perp L$, and the constant $\mathscr{E}_\ell^{(1)}$ is thus,

$$\mathscr{E}_\ell^{(1)} = \sqrt{\frac{\hbar\omega_\ell}{2\varepsilon_0 L S_\perp}}. \tag{5B.10}$$

As it varies little over the interval $\delta\omega$, it can be replaced by its value at ω_0 and brought outside the integral. The final result is

$$\begin{aligned} \mathbf{E}^{(+)}(\mathbf{r},t) &= \boldsymbol{\varepsilon}\sqrt{\frac{\hbar\omega_0}{2\varepsilon_0 L S_\perp}}\frac{L}{2\pi c}\int_{-\infty}^{+\infty}\mathrm{d}\omega_\ell\, c(\omega_\ell)\mathrm{e}^{-\mathrm{i}\omega_\ell\tau} \\ &= \boldsymbol{\varepsilon}\sqrt{\frac{\hbar\omega_0 L}{4\pi\varepsilon_0 c^2 S_\perp}}\mathrm{e}^{-\mathrm{i}\omega_0(t-\frac{\mathbf{u}\cdot\mathbf{r}}{c})}\tilde{f}\left(t-\frac{\mathbf{u}\cdot\mathbf{r}}{c}\right), \end{aligned} \tag{5B.11}$$

with

$$\tilde{f}(\tau) = \frac{1}{\sqrt{2\pi}}\int_{-\infty}^{+\infty}\mathrm{d}\Omega f(\Omega)\,\mathrm{e}^{-\mathrm{i}\Omega\tau}. \tag{5B.12}$$

The photodetection probability per unit time and per unit detector area is, therefore,

$$w(\mathbf{r}, t) = s\frac{\hbar\omega_0 L}{4\pi\varepsilon_0 c^2 S_\perp}\left|\tilde{f}\left(t - \frac{\mathbf{u}\cdot\mathbf{r}}{c}\right)\right|^2, \tag{5B.13}$$

and the detection probability per unit time over the whole detector is

$$\frac{\mathrm{d}\mathcal{P}}{\mathrm{d}t} = s\frac{\hbar\omega_0 L}{4\pi\varepsilon_0 c^2}\left|\tilde{f}\left(t - \frac{\mathbf{u}\cdot\mathbf{r}}{c}\right)\right|^2. \tag{5B.14}$$

The total probability of photodetection, integrated over a time interval covering the whole wave packet, is obtained by integrating (5B.14). With the form (5B.12) for the Fourier transform, the Parseval–Plancherel relation gives

$$\int \mathrm{d}\tau|\tilde{f}(\tau)|^2 = \int \mathrm{d}\Omega|f(\Omega)|^2. \tag{5B.15}$$

The right-hand side is easily evaluated by writing down the normalization condition for the $|c_\ell|^2$, given the density of modes (5B.9):

$$\sum_\ell |c_\ell|^2 = \frac{L}{2\pi c}\int \mathrm{d}\Omega|f(\Omega)|^2 = 1. \tag{5B.16}$$

The final result is

$$\mathcal{P} = s\frac{\hbar\omega_0}{2\varepsilon_0 c}. \tag{5B.17}$$

As one would hope, the arbitrary length L disappears from the final expression.

For a single photon, a perfect detector must give $\mathcal{P} = 1$, and the sensitivity s is, therefore,

$$s_{\text{perfect}} = \frac{2\varepsilon_0 c}{\hbar\omega_0}. \tag{5B.18}$$

A real detector generally has lower sensitivity than this, by a factor (less than unity) called the quantum efficiency or quantum yield of the detector. For certain spectral ranges, there are detectors with quantum efficiencies very close to unity.

5B.1.3 Spontaneous emission photon

Spontaneous emission by a *single* atom in an excited state gives a one-photon wave packet. Indeed, consider a two-level atom in the excited state $|b\rangle$ in the radiation vacuum at time $t = 0$ (situation studied in Section 6.4 of Chapter 6). The initial state $|b; 0\rangle$ is coupled by the quantum interaction Hamiltonian with all states of the form $|a; 1_\ell\rangle$, where $|a\rangle$ is the ground state of the atom, and $|1_\ell\rangle$ represents a one-photon state of mode ℓ. From the detailed investigation of the coupling between a discrete state and a continuum, discussed in Chapter 1, we can calculate the final state of the system after a long lapse of time compared with the lifetime of the excited state. It is found that

$$\begin{aligned}|\psi\rangle &= \sum_\ell \gamma_\ell(t\to\infty)|a; 1_\ell\rangle \\ &= |a\rangle \otimes \sum_\ell \gamma_\ell(t\to\infty)|1_\ell\rangle.\end{aligned} \tag{5B.19}$$

The radiation part of this state does indeed have the form (5B.1).

Suppose now that the atom is placed at the focus of an optical system, e.g. a parabolic mirror, which transforms all emitted spherical waves into plane waves propagating along the same direction. We then obtain a one-dimensional one-photon wave packet.

Generalizing the result (1.88) to the case of spontaneous emission (see Section 6.4), we obtain the coefficients

$$c_\ell = \frac{K}{\omega_\ell - \omega_0 + \mathrm{i}\Gamma_{\mathrm{sp}}/2}, \tag{5B.20}$$

where

$$K = \left(\Gamma_{\mathrm{sp}}\frac{c}{L}\right)^{\frac{1}{2}}, \tag{5B.21}$$

to ensure normalization of c_ℓ, with the mode density (5B.9). Note that the emitted light spectrum is described by a Lorentzian line centred at ω_0, with width Γ_{sp} at half-maximum:

$$|c(\omega_\ell)|^2 = \frac{K^2}{(\omega_\ell - \omega_0)^2 + \frac{\Gamma_{\mathrm{sp}}^2}{4}}. \tag{5B.22}$$

We now write $\mathbf{E}^{(+)}(\mathbf{r}, t)$ in the form (5B.11), taking the Fourier transform of

$$f(\Omega) = \frac{K}{\Omega + \mathrm{i}\frac{\Gamma_{\mathrm{sp}}}{2}}. \tag{5B.23}$$

It can be shown that this yields

$$\tilde{f}(\tau) = K\sqrt{2\pi}\, H(\tau)\, \mathrm{e}^{-\frac{\Gamma_{\mathrm{sp}}}{2}\tau}, \tag{5B.24}$$

where $H(\tau)$ is the Heaviside step function, equal to 0 for $\tau < 0$ and 1 for $\tau \geq 0$. We thus obtain a wave packet that begins suddenly at $\tau = 0$ and dies off exponentially. Taking a perfect photodetector that covers the whole cross-section $S_\perp$ of the beam (see Figure 5B.1), the detection probability per unit time (5B.14) is

$$\frac{\mathrm{d}\mathcal{P}}{\mathrm{d}t} = \Gamma_{\mathrm{sp}}\, H\left(t - \frac{\mathbf{u}\cdot\mathbf{r}}{c}\right)\, \mathrm{e}^{-\Gamma_{\mathrm{sp}}\left(t - \frac{\mathbf{u}\cdot\mathbf{r}}{c}\right)}. \tag{5B.25}$$

Figure 5B.2 shows this probability. The curve can be obtained experimentally by carrying out many repeats of the following experiment: the atom is excited at time t_{exc} and we then measure the time t_{det} at which a photon is detected. The histogram giving the distribution of intervals $t_{\mathrm{det}} - t_{\mathrm{exc}}$ looks like the distribution (5B.25), apart from statistical fluctuations.

Comment We have considered the case of a polarized beam to simplify the treatment. In the general case, each polarization is handled separately.

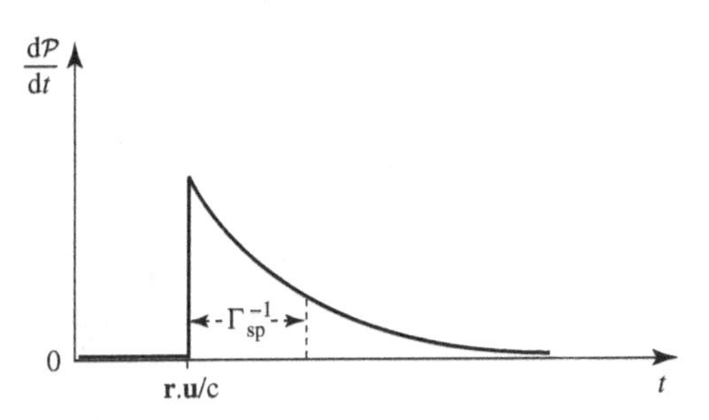

Figure 5B.2 Probability of photodetection at point r and time t for a one-photon wave packet emitted at $\mathbf{r} = 0$ and $t = 0$. The signal begins at time $t = \mathbf{r}\,.\,\mathbf{u}/c$, then dies down exponentially. Note that, for each individual measurement, only one photodetection can be observed. The above signal can be obtained by repeating the experiment many times and measuring the time interval between excitation of the atom and detection of a photon by the detector.

5B.2 Absence of double detection and difference with a classical field

In Section 5B.1.1, as in Section 5.4.3 of Chapter 5, we have seen that for a one-photon wave packet the detection probability $w(\mathbf{r}, t)$ at a point has the same form as would be obtained for a classical field. Moreover, we showed in Section 5.4.3 that, for this kind of one-photon state, the probability of double detection at two points $\mathbf{r}_1$ and $\mathbf{r}_2$ is strictly zero:

$$w^{(2)}(\mathbf{r}_1, \mathbf{r}_2, t) = 0. \tag{5B.26}$$

This corresponds perfectly with the intuitive picture one would have of a single photon. We also indicated in Section 5.5.1 that the probability of a double detection in the output ports of a beamsplitter is zero in the case of a one-photon state. We give here the detailed calculation leading to this result, as well as the calculation in the case of a classical wave packet.

5B.2.1 Semi-reflecting mirror

The one-photon wave packet is now sent through the input port (1) of a semi-reflecting mirror and two detectors are placed at the output ports (3) and (4) (see Figure 5B.3).

We express the field $\hat{E}_3^{(+)}(D_3)$ in terms of the incoming field at ports (1) and (2). At port (1), the origin 0_1 is chosen symmetrically to the focus with respect to the apex of the paraboloid. Using (5.8) and including the propagation in the input and output space, we obtain

$$\hat{E}_3^{(+)}(D_3) = \frac{\mathrm{i}}{\sqrt{2}} \sum_{\ell_1} \mathcal{E}_{\ell_1}^{(1)} \mathrm{e}^{\mathrm{i}k_{\ell_1}(r_1+r_3)} \hat{a}_{\ell_1} + \frac{\mathrm{i}}{\sqrt{2}} \sum_{\ell_2} \mathcal{E}_{\ell_2}^{(1)} \mathrm{e}^{\mathrm{i}k_{\ell_2}(r_2+r_3)} \hat{a}_{\ell_2}, \tag{5B.27}$$

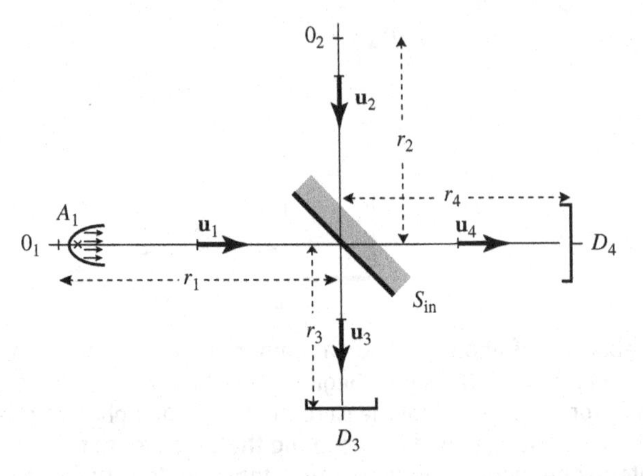

Figure 5B.3 **Photodetection on either side of the semi-reflecting mirror for a one-photon wave packet emitted by an atom whose image is at distance r_1 from the mirror. An electronic system counts detection events at D_3 and D_4, as well as the number of joint detections at D_3 and D_4.**

where ℓ_1 and ℓ_2 characterize the incoming modes at ports (1) and (2), respectively. Likewise, for the fields at detector D_4,

$$\hat{E}_4^{(+)}(D_4) = \frac{i}{\sqrt{2}} \sum_{\ell_1} \mathcal{E}_{\ell_1}^{(1)} e^{ik_{\ell_1}(r_1+r_4)} \hat{a}_{\ell_1} - \frac{i}{\sqrt{2}} \sum_{\ell_2} \mathcal{E}_{\ell_2}^{(1)} e^{ik_{\ell_2}(r_2+r_4)} \hat{a}_{\ell_2}. \tag{5B.28}$$

The single detection probabilities per unit time are obtained as in Section 5.5, by taking an input state of the form $|1_1\rangle \otimes |0_2\rangle$. Only the terms of (5B.28) related to modes ℓ_1 contribute, and the rest of the calculation is as in Section 5B.1.3. For perfect detectors, we find

$$\frac{d\mathcal{P}_3}{dt} = \frac{1}{2} \Gamma_{sp} H(\tau_3) e^{-\Gamma_{sp}\tau_3}, \tag{5B.29}$$

with

$$\tau_3 = t - \frac{r_1 + r_3}{c}, \tag{5B.30}$$

and

$$\frac{d\mathcal{P}_4}{dt} = \frac{1}{2} \Gamma_{sp} H(\tau_4) e^{-\Gamma_{sp}\tau_4}, \tag{5B.31}$$

with

$$\tau_4 = t - \frac{r_1 + r_4}{c}. \tag{5B.32}$$

The joint detection probability is

$$w^{(2)}(r_3, r_4, t) = s^2 \frac{1}{4} \langle 1(t)| \hat{E}_3^{(-)}(r_3) \hat{E}_4^{(-)}(r_4) \hat{E}_4^{(+)}(r_4) \hat{E}_3^{(+)}(r_3) |1(t)\rangle. \tag{5B.33}$$

When $\hat{E}_3^{(+)}$ and $\hat{E}_4^{(+)}$ are replaced by their expressions (5B.27–5B.28) in terms of the incoming field, one obtains, once again, terms

$$\sum_{\ell_1'}\sum_{\ell_1''} \hat{a}_{\ell_1'}\hat{a}_{\ell_1''}|1_{\ell_1}\rangle = 0. \tag{5B.34}$$

The joint detection probability is zero, even if $r_3 = r_4$, i.e. even if the single detection probabilities are simultaneously non-zero. We recover here the result discussed above in the absence of the mirror (see also Section 5.4.3).

Comment In the above calculations, and remembering that port (2) is empty, the second terms in expressions (5B.27) and (5B.28) contribute nothing. However, it is nevertheless important to include them because these terms play a role in certain fluctuation calculations, even when port (2) is empty.

5B.2.2 Double detection with a classical wave packet

In classical electromagnetism, one can imagine a polarized one-dimensional wave packet described by the analytic signal,

$$E_{\text{cl}}^{(+)}(r,t) = \sum_{\ell} \mathcal{E}_{\ell}^{(1)} c_{\ell} \mathrm{e}^{-\mathrm{i}\omega_{\ell}(t-\frac{r}{c})}, \tag{5B.35}$$

with the same coefficients c_ℓ as in Section 5B.1.3, and the form (5B.10) for $\mathcal{E}_{\ell}^{(1)}$. In the sense of classical electromagnetism, the energy contained in this wave packet is $\hbar\omega_0$. Calculations to find the single photodetection probabilities are strictly analogous to those presented above, and expressions (5B.29) and (5B.31) are obtained behind the semi-reflecting mirror.

Let us now find the double photodetection probability. In the semi-classical model, there is a wave packet at each detector, and the photodetections at the two detectors are independent random events. The joint detection probability is thus the product of the single detection probabilities, i.e.

$$w^{(2)}(r_1,t_1;r_2,t_2) = w(r_1,t_1)\, w(r_2,t_2) \tag{5B.36}$$

and

$$\frac{\mathrm{d}^2\mathcal{P}}{\mathrm{d}t_3\mathrm{d}t_4} = \frac{\mathrm{d}\mathcal{P}_3}{\mathrm{d}t}\cdot\frac{\mathrm{d}\mathcal{P}_4}{\mathrm{d}t} = \frac{1}{4}\Gamma_{\text{sp}}^2 H(\tau_3)H(\tau_4)\mathrm{e}^{-\Gamma_{\text{sp}}(\tau_3+\tau_4)}. \tag{5B.37}$$

There are clearly some values of τ_3 and τ_4 for which this joint detection probability differs from zero. To compare with the calculation in Section 5B.2.1, consider the case $t_3 = t_4$. The simultaneous photodetection probability differs from zero if the two wave packets overlap, i.e. if $|r_3 - r_4|$ is not large compared with c/Γ_{sp}.

In contrast with the one-photon wave packet of quantum optics, a classical wave packet gives rise to a certain probability of joint detection on either side of the semi-reflecting mirror. A lower bound for the simultaneous detection probability can in fact be found for the semi-classical model of photodetection.

To this end, consider the single detection probabilities for detectors D_3 and D_4, assumed to be at the same distance from the emitter ($r_3 = r_4$). Measurements thus refer to the classical field taken at the same instant of time, which we shall write as $E^{(+)}(t)$, omitting the retardation to simplify the notation. The single detection probabilities are

$$w(D_3, t) = \frac{s}{\sqrt{2}} |E^{(+)}(t)|^2 \tag{5B.38}$$

$$w(D_4, t) = \frac{s}{\sqrt{2}} |E^{(+)}(t)|^2, \tag{5B.39}$$

and the simultaneous detection probability is

$$w^{(2)}(D_3, D_4, t) = w(D_3, t) w(D_4, t) = \frac{s^2}{2} |E^{(+)}(t)|^4. \tag{5B.40}$$

In a real experiment, the measurement is repeated many times, which amounts to averaging over possibly different values of $E(t)$. The average photodetection probabilities are, therefore,

$$\overline{w(D_3)} = \frac{s}{\sqrt{2}} \overline{|E^{(+)}(t)|^2} \tag{5B.41}$$

$$\overline{w(D_4)} = \frac{s}{\sqrt{2}} \overline{|E^{(+)}(t)|^2} \tag{5B.42}$$

$$\overline{w^{(2)}(D_3, D_4)} = \frac{s^2}{2} \overline{|E^{(+)}(t)|^4}. \tag{5B.43}$$

Now any real quantity $f(t)$ satisfies the Cauchy–Schwartz inequality:

$$\overline{[f(t)]^2} \geq [\overline{f(t)}]^2. \tag{5B.44}$$

In the experiment, measurements of the single and simultaneous detection probabilities refer to the same time intervals, and the averages are the same. It follows that

$$\overline{w^{(2)}(D_3, D_4)} \geq \overline{w(D_3)} \cdot \overline{w(D_4)}. \tag{5B.45}$$

In the case where the classical wave packets all have the same amplitude, equality holds in (5B.45). In any case, for a classical field and the semi-classical photodetection model, the simultaneous detection probability cannot be less than the product of the single detection probabilities.

If we find $\overline{w^{(2)}(D_3, D_4)} < \overline{w(D_3)} \cdot \overline{w(D_4)}$ in an experiment, we can be sure that the radiation under investigation cannot be correctly described by classical electromagnetism. The violation of inequality (5B.45) is a criterion used to test single-photon light sources, which play an important role in quantum optics.

The above photodetection probabilities are instantaneous detection probabilities for pointlike detectors, and one may wonder whether there is an analogous inequality when one considers photodetection probabilities integrated over a finite observation time window θ and over detectors of surface area $S_\perp$. Consider first the case of point detectors, but measurements integrated over a time lapse θ. We define the single photodetection probabilities over θ by

$$\pi_\theta(D_3,t) = \int_t^{t+\theta} w(D_3,t)\mathrm{d}t \tag{5B.46}$$

$$\pi_\theta(D_4,t) = \int_t^{t+\theta} w(D_4,t)\mathrm{d}t. \tag{5B.47}$$

The probability of a coincidence detection during θ is obtained by double integration of $w^{(2)}(D_3,D_4;t_3,t_4)$, which is the double detection probability at D_3 at time t_3 and at D_4 at time t_4 :

$$w^{(2)}(D_3,D_4;t_3,t_4) = w(D_3,t_3)\,w(D_4,t_4) = \frac{s^2}{2}|E^{(+)}(t_3)|^2\,|E^{(+)}(t_4)|^2. \tag{5B.48}$$

Given (5B.47–5B.48), this yields

$$\begin{aligned}\pi_\theta^{(2)} &= \int_t^{t+\theta} \mathrm{d}t_3 \int_t^{t+\theta} \mathrm{d}t_4\; w^{(2)}(D_3,D_4;t_3,t_4)\\ &= \pi_\theta(D_3,t)\,\pi_\theta(D_4,t).\end{aligned} \tag{5B.49}$$

When the experiment is repeated many times, the average is taken over a large number of different samples of $E(t)$ taken during θ. Consider the sample i, between t_i and $t_i+\theta$, and define

$$\lambda_i = \int_{t_i}^{t_i+\theta} \mathrm{d}t|E^{(+)}(t)|^2. \tag{5B.50}$$

The probabilities for this sample are

$$\{\pi_\theta(D_3)\}_i = \frac{s}{\sqrt{2}}\,\lambda_i \tag{5B.51}$$

$$\{\pi_\theta(D_4)\}_i = \frac{s}{\sqrt{2}}\,\lambda_i \tag{5B.52}$$

and

$$\{\pi_\theta^{(2)}(D_3,D_4)\}_i = \frac{s^2}{2}\,\lambda_i^2. \tag{5B.53}$$

Averaging over all the samples i, we obtain the average probabilities $P_\theta(D_3)$, $P_\theta(D_4)$ and $P_\theta^{(2)}(D_3,D_4)$, which are the quantities obtained at the end of the experiment. Now there is once again a Cauchy–Schwartz inequality for the numbers λ_i:

$$\overline{\lambda_i^2} \geq (\overline{\lambda_i})^2, \tag{5B.54}$$

from which it follows that

$$P_\theta^{(2)}(D_3,D_4) \geq P_\theta(D_3)\cdot P_\theta(D_4). \tag{5B.55}$$

An analogous argument can be made when considering integration over the whole surface of the detectors, and the inequality (5B.55) applies to the probabilities obtained by carrying out measurements with finite detectors and finite time windows. In practice it is this inequality that is used to check whether a source does in fact emit one-photon pulses. For example, for the one-photon interference illustrated in Figure 5.9, it was found that

$$\frac{P_\theta^{(2)}(D_3, D_4)}{P_\theta(D_3)\,P_\theta(D_4)} = 0.18, \tag{5B.56}$$

well below the critical value of one corresponding to the limits of (5B.55). The fact that this number is not zero is easy to explain. For one thing, the detectors give a residual signal even in the absence of any light (thermal noise), and for another, there is a non-zero probability of exciting two individual emitters during the same time window θ. But the result (5B.56) nevertheless violates the inequality (5B.55), whatever corrections might be made to it. There can be no doubt that this radiation is non-classical, and has a clear single-photon character.

5B.3 Two one-photon wave packets on a semi-reflecting mirror

5B.3.1 Single detections

A one-photon wave packet now enters each of the input ports (see Figure 5B.4). The two wave packets are identical, i.e. the incoming radiation has the form

$$|\psi\rangle = |1_1\rangle \otimes |1_2\rangle, \tag{5B.57}$$

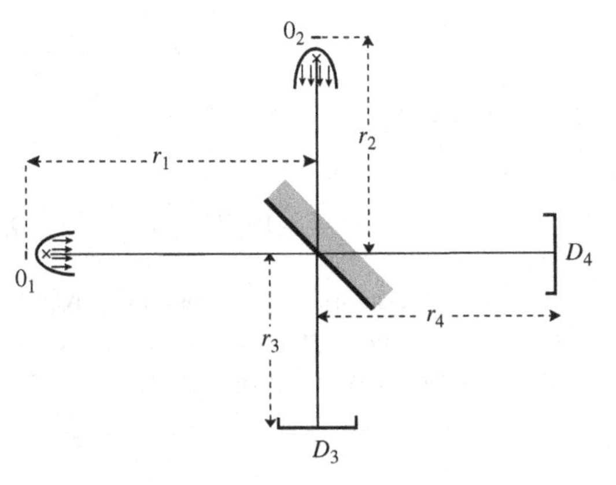

Figure 5B.4 Two one-photon wave packets on a semi-reflecting mirror. The two emitters are excited at the same time, and the two wave packets arrive at the mirror at the same time if $r_1 = r_2$. Electronic circuits record single and joint detections at D_3 and D_4.

where the coefficients c_{ℓ_1} and c_{ℓ_2} (Equation 5B.1) are the same. To be specific, we shall use the form (5B.20).

The expression (5B.27) for $\hat{E}_3^{(+)}(D_3)$ yields,

$$\begin{aligned}\hat{E}_3^{(+)}(D_3)|\psi(t)\rangle &= \frac{\mathrm{i}}{\sqrt{2}}\sum_{\ell_1}\mathcal{E}_{\ell_1}^{(1)}\,c_{\ell_1}\mathrm{e}^{-\mathrm{i}\omega_{\ell_1}\left(t-\frac{r_1+r_3}{c}\right)}|0_1\rangle\otimes|1_2\rangle \\ &+\frac{\mathrm{i}}{\sqrt{2}}\sum_{\ell_2}\mathcal{E}_{\ell_2}^{(1)}\,c_{\ell_2}\mathrm{e}^{-\mathrm{i}\omega_{\ell_2}\left(t-\frac{r_2+r_3}{c}\right)}|1_1\rangle\otimes|0_2\rangle,\end{aligned} \tag{5B.58}$$

where $|0_1\rangle$ and $|0_2\rangle$ denote the vacuum at ports (1) and (2). Taking the square of the norm, a calculation as in Section 5B.1.3 yields the single detection probability over the whole detector as

$$\begin{aligned}\frac{\mathrm{d}\mathcal{P}_3}{\mathrm{d}t}(t) &= \frac{1}{2}H\left(t-\frac{r_1+r_3}{c}\right)\Gamma_{\mathrm{sp}}\mathrm{e}^{-\Gamma_{\mathrm{sp}}\left(t-\frac{r_1+r_3}{c}\right)} \\ &+\frac{1}{2}H\left(t-\frac{r_2+r_3}{c}\right)\Gamma_{\mathrm{sp}}\mathrm{e}^{-\Gamma_{\mathrm{sp}}\left(t-\frac{r_2+r_3}{c}\right)}.\end{aligned} \tag{5B.59}$$

An analogous calculation for detector D_4 yields

$$\begin{aligned}\hat{E}_4^{(+)}(D_4)|\psi(t)\rangle &= \frac{\mathrm{i}}{\sqrt{2}}\sum_{\ell_1}\mathcal{E}_{\ell_1}^{(1)}\,c_{\ell_1}\mathrm{e}^{-\mathrm{i}\omega_{\ell_1}\left(t-\frac{r_1+r_4}{c}\right)}|0_1\rangle\otimes|1_2\rangle \\ &-\frac{\mathrm{i}}{\sqrt{2}}\sum_{\ell_2}\mathcal{E}_{\ell_2}^{(1)}\,c_{\ell_2}\mathrm{e}^{-\mathrm{i}\omega_{\ell_2}\left(t-\frac{r_2+r_4}{c}\right)}|1_1\rangle\otimes|0_2\rangle\end{aligned} \tag{5B.60}$$

and

$$\begin{aligned}\frac{\mathrm{d}\mathcal{P}_4}{\mathrm{d}t}(t) &= \frac{1}{2}H\left(t-\frac{r_1+r_4}{c}\right)\Gamma_{\mathrm{sp}}\mathrm{e}^{-\Gamma_{\mathrm{sp}}\left(t-\frac{r_1+r_4}{c}\right)} \\ &+\frac{1}{2}H\left(t-\frac{r_2+r_4}{c}\right)\Gamma_{\mathrm{sp}}\mathrm{e}^{-\Gamma_{\mathrm{sp}}\left(t-\frac{r_2+r_4}{c}\right)}.\end{aligned} \tag{5B.61}$$

At each detector, we find the sum of the probabilities associated with each wave packet, with no interference term. The effects are additive in terms of probabilities, as for independent classical particles distributed randomly by the beamsplitter between the output ports.

5B.3.2 Joint detections

To calculate the joint detection probability $w^{(2)}(D_3, D_4)$ on either side of the mirror, we must take the square of the norm of $\hat{E}_3^{(+)}(D_3)\,\hat{E}_4^{(+)}(D_4)|\psi(t)\rangle$. Referring to the incoming space and using (5B.60), this yields

$$\hat{E}_3^{(+)}(D_3)\,\hat{E}_4^{(+)}(D_4)|\psi(t)\rangle =$$

$$\frac{1}{2}\left\{\left[\sum_{\ell_1} c_{\ell_1}\mathcal{E}_{\ell_1}^{(1)}\mathrm{e}^{-\mathrm{i}\omega_{\ell_1}\left(t-\frac{r_1+r_4}{c}\right)}\right]\left[\sum_{\ell_2} c_{\ell_2}\mathcal{E}_{\ell_2}^{(1)}c_{\ell_2}\mathrm{e}^{-\mathrm{i}\omega_{\ell_2}\left(t-\frac{r_2+r_3}{c}\right)}\right]\right.$$

$$\left.-\left[\sum_{\ell_1} c_{\ell_1}\mathcal{E}_{\ell_1}^{(1)}\mathrm{e}^{-\mathrm{i}\omega_{\ell_1}\left(t-\frac{r_1+r_3}{c}\right)}\right]\left[\sum_{\ell_2} c_{\ell_2}\mathcal{E}_{\ell_2}^{(1)}c_{\ell_2}\mathrm{e}^{-\mathrm{i}\omega_{\ell_2}\left(t-\frac{r_2+r_4}{c}\right)}\right]\right\}|0_1\rangle\otimes|0_2\rangle. \tag{5B.62}$$

Since c_{ℓ_1} and c_{ℓ_2} are the same, the subscripts 1 and 2 can be interchanged in the above expression and it turns out that, when $r_1 = r_2$, the expressions in square brackets are the same and exactly cancel. *The probability of joint detection on either side of the mirror is zero.* This result contradicts the simple picture developed in Section 5B.3.1, according to which the single photons are distributed randomly between the output ports of the semi-reflecting mirror, like independent classical particles. The calculation just carried out shows that, when they perfectly overlap on the beamsplitter ($r_1 = r_2$), the two photons never leave by different ports, but leave together either by port (3) or by port (4). This is referred to as 'quantum coalescence'.

This phenomenon is the result of a quantum interference effect between the quantum amplitudes associated with the two symbolic diagrams shown in Figure 5B.5. The negative sign responsible for the cancellation results from the opposite signs for the reflection terms in the input–output relations for a semi-reflecting mirror.

This purely quantum effect should not be confused with an interference effect between classical fields. What we have here is interference between quantum amplitudes associated with the diagrams of Figure 5B.5, each one involving both photons. For interference to occur, the two processes must be indistinguishable, and this requires the two wave packets to be identical, i.e. they must have the same distribution of c_ℓ and the same emission time, and r_1 must be equal to r_2. But we should stress that there is no coherence, in the classical sense, between the wave packets, since they are produced by independent emitters.

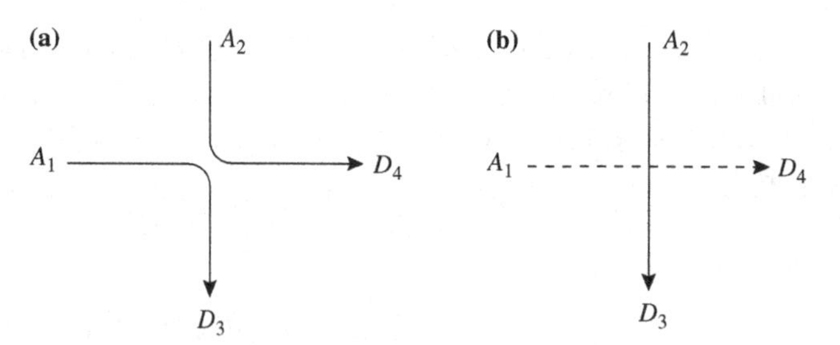

Figure 5B.5 Schematic diagrams representing the two two-photon processes whose amplitudes interfere destructively. **(a)** The photon emitted by A_1 is detected by D_3 and the photon emitted by A_2 is detected by D_4. **(b)** The photon emitted by A_1 is detected by D_4 and the photon emitted by A_2 is detected by D_3.

The experiment has indeed been carried out in this form, using two distinct atoms or ions excited independently and re-emitting spontaneous photons.[1]

Comments

(i) Insofar as two strictly indistinguishable photons are picked up at detector D_3 or detector D_4, the effect is related to the bosonic nature of photons. Two fermions with the same characteristics would never be detected simultaneously at the same detector.

(ii) Quantitative significance can be given to the indistinguishability of the two photons by calculating the joint detection probability when the distances r_1 and r_2 are not the same. The calculation is straightforward but tedious, starting from (5B.62). After integration, it is found that the total joint probability of a detection at D_3 and a detection at D_4 is given by

$$\mathcal{P}^{(2)}(D_3, D_4) = \frac{1}{2}\left(1 - \mathrm{e}^{-\Gamma_{\mathrm{sp}}\frac{|r_2 - r_1|}{c}}\right), \tag{5B.63}$$

whatever the detection times (within the wave packet). If $r_2 = r_1$, $\mathcal{P}^{(2)} = 0$. But if $|r_2 - r_1| > c/\Gamma_{\mathrm{sp}}$, which means that one can tell whether the photon came from A_1 or from A_2 by the detection time (the excitation times are known), the result is $\mathcal{P}^{(2)}(D_3, D_4) = 1/2$. This is just what would be obtained by considering two classical particles, each of which has probability 1/2 of being reflected and probability 1/2 of being transmitted. Indeed, in this case, there are four possible final situations, two of which correspond to joint detection on either side of the semi-reflecting mirror.

5B.4 Quasi-classical wave packet

Consider a multimode quasi-classical state of the form (5.122):

$$|\psi_{\mathrm{qc}}(t)\rangle = \Pi_\ell |\alpha_\ell \mathrm{e}^{-\mathrm{i}\omega_\ell t}\rangle, \tag{5B.64}$$

where all modes have the same direction of propagation $\mathbf{k}_\ell / k_\ell$ and the same polarization $\boldsymbol{\varepsilon}$. Each state $|\alpha_\ell \mathrm{e}^{-\mathrm{i}\omega_\ell t}\rangle$ is an eigenstate of $\hat{a}_\ell$, and the α_ℓ are taken in the form (5B.20), but without imposing the normalization condition (5B.21). The expected number of photons in this state is equal to

$$N = \sum_\ell \langle \psi_{\mathrm{qc}} | \hat{a}_\ell^\dagger \hat{a}_\ell | \psi_{\mathrm{qc}} \rangle = \sum_\ell |\alpha_\ell|^2. \tag{5B.65}$$

We saw in Section 5.4.2 that $|\psi_{\mathrm{qc}}\rangle$ is an eigenstate of $\hat{\mathbf{E}}^{(+)}(\mathbf{r})$, with eigenvalue

$$\begin{aligned} \mathbf{E}_{\mathrm{cl}}^{(+)}(r,t) &= \mathrm{i}\boldsymbol{\varepsilon}_\ell \sum_\ell \mathcal{E}_\ell^{(1)} c_\ell \mathrm{e}^{-\mathrm{i}\omega_\ell (t - \frac{r}{c})} \\ &= \boldsymbol{\varepsilon}_\ell E_{\mathrm{cl}}^{(+)}(r,t), \end{aligned} \tag{5B.66}$$

where $E_{\mathrm{cl}}^{(+)}(\mathbf{r}, t)$ is precisely the analytic signal introduced in (5B.35). It can be calculated explicitly using (5B.11), (5B.20), (5B.24) and (5B.65), and we obtain

[1] J. Beugnon *et al.*, Quantum Interference Between Two Single Photons Emitted by Independently Trapped Atoms, *Nature* **440**, 779 (2006).

$$E_{\text{cl}}^{(+)}(r,t) = \sqrt{N}\sqrt{\frac{\hbar\omega_0\Gamma_{\text{sp}}}{2\varepsilon_0 c S_\perp}}\, H\left(t - \frac{r}{c}\right) \mathrm{e}^{-\left(\frac{\Gamma_{\text{sp}}}{2} + i\omega_\ell\right)\left(t - \frac{r}{c}\right)}. \tag{5B.67}$$

Such a classical wave packet has cross-section $S_\perp$. Its amplitude is zero for $t < \frac{r}{c}$, and becomes negligible again after a time of the order of Γ_{sp}^{-1}. The classical electromagnetic energy contained in this packet is $N\hbar\omega_0$.

The single photodetection probability for the quasi-classical wave packet is

$$\begin{aligned} w(r,t) &= s\|\hat{\mathbf{E}}^{(+)}(\mathbf{r})|\psi_{1\text{qc}}(t)\rangle\|^2, \\ &= s|E_{\text{cl}}^{(+)}(r,t)|^2, \end{aligned} \tag{5B.68}$$

where $E_{\text{cl}}^{(+)}(r,t)$ is the classical wave packet (5B.67). If we take $N = 1$, this result is the same as the one obtained in the semi-classical photodetection model with the classical field (5B.67), which is identical to the probability calculated for the quantum one-photon wave packet of Section 5B.1.3.

The double photodetection probability is

$$w^{(2)}(r_1, r_2, t) = s^2\|\hat{E}^{(+)}(r_2)\hat{E}^{(+)}(r_1)|\psi_{1\text{qc}}(t)\rangle\|^2. \tag{5B.69}$$

As $|\psi_{1\text{qc}}\rangle$ is an eigenstate of $\hat{E}^{(+)}(r_1)$ and $\hat{E}^{(+)}(r_2)$, we have immediately,

$$w^{(2)}(r_1, r_2, t) = w(r_1, t) \cdot w(r_2, t), \tag{5B.70}$$

which is precisely the result of the semi-classical photodetection model for the classical field (5B.67).

Note that the above result remains true even if the expected number of photons N is well below unity. Now it can be shown that a light pulse that has been highly attenuated by passing through an absorbent material gives precisely a multimode quasi-classical state, where N can assume a value much less than unity. Such a state is radically different from a one-photon state, since the inequalities (5B.45) or (5B.55) are not violated. It has been checked that the inequality (5B.55) remains true even for an expected number of photons per pulse equal to $N = 10^{-2}$.[2] It is therefore incorrect to claim that a highly attenuated beam is made up of single photons.

The properties of a quasi-classical wave packet can thus be interpreted using a classical model of the electromagnetic field. The above properties can also be interpreted in terms of photons by recalling that, for a quasi-classical state, the probabilities of having $\mathcal{N}$ photons are not zero when $\mathcal{N} > 1$, but are in fact equal to $[\mathcal{P}(\mathcal{N} = 1)]^{\mathcal{N}}/\mathcal{N}!$. In a quasi-classical pulse, one thus has a certain probability of having 2 photons, 3 photons, N photons. It is these groups of photons that give rise to a sufficient number of coincidences on either side of the semi-reflecting mirror to avoid violation of the inequality (5B.45).

[2] The experiment was carried out with the same detection system as the one used to violate the inequality (5B.55) in the case of a one-photon wave packet (see Section 5B.2.2 and in particular 5B.55). The same experimental setup was used to observe the interference patterns in Figure 5.9 with one-photon wave packets. P. Grangier, G. Roger and A. Aspect, Experimental Evidence for a Photon Anticorrelation Effect on a Beam Splitter: a New Light on Single-Photon Interferences, *Europhysics Letters* **1**, 173 (1986).

Complement 5C Polarization-entangled photons and violation of Bell's inequalities[1]

5C.1 From the Bohr–Einstein debate to the Bell inequalities and quantum information: a brief history of entanglement

Entanglement is one of the most surprising features of quantum mechanics. However, it was not until the last decades of the twentieth century that its full importance was understood and it was realized that it could lead to revolutionary applications in the area of quantum information. It was A. Einstein who discovered the extraordinary properties of non-factorizable two-particle states, when seeking to demonstrate that the formalism of quantum mechanics is incomplete. He presented his findings in 1935 in his famous article published jointly with B. Podolsky and N. Rosen, now referred to as the 'EPR' paper. Soon afterwards, Schrödinger coined the term 'entangled states' to emphasize the fact that the properties of the two particles are inextricably bound together.

In the EPR article, Einstein and his colleagues used quantum predictions to conclude that the formalism of quantum mechanics was incomplete, in the sense that it did not account for the whole of physical reality, and that the task of physics was therefore to find a more complete theory. They did not contest the validity of the quantum formalism, but suggested that a further, more detailed level of description would have to be introduced, in which each particle of the EPR pair would have well-defined properties that were not taken into account in the quantum formalism. For Einstein, a complete description of the physical reality of each particle required the introduction of further parameters, the aim then being to recover the predictions of standard quantum mechanics by taking a statistical average over these extra 'hidden' variables. This approach is analogous to what is done in the kinetic theory of gases, where each molecule of a gas can, in principle, be described by specifying its trajectory, even though, in practice, a statistical description (for instance the Maxwell–Boltzmann distribution) is generally considered to be adequate to account for the properties of the gas which interest the physicist.

Apparently, Niels Bohr was very impressed by the EPR argument, which exploits the quantum formalism itself in order to contest the completeness of that same theory. He thus set about refuting the argument, asserting that in an entangled state, one can no longer speak of the individual properties of each particle.

[1] For more details and a more complete bibliography, the reader is referred to A. Aspect, in *Quantum [Un]speakables – From Bell to Quantum Information*, edited by R. A. Bertelmann and A. Zeilinger, Springer (2002). This text is also available on the Web at http://arXiv.org under the reference quant-ph/04 02001.

It might be thought that this controversy between two of the great men of twentieth-century physics would pass like a shock wave through the physics community. In actual fact, at the time that the EPR paper was published in 1935, quantum mechanics was going from one success to another and most physicists paid little attention to the debate, considering it academic. It seemed that the decision to support one or the other position was more a question of personal taste (or epistemological position) than anything else, without consequence for the way the quantum formalism was put into practice, as Einstein himself seemed to accept.

It was thirty years before this general consensus was called into question in a short paper published in 1964 by John Bell. The situation was radically altered by this publication. In just a few lines of calculation, the article shows that, if the EPR argument is taken seriously and hidden variables are introduced explicitly to understand the strong correlations between the two particles, then one ends up in contradiction with the predictions of the quantum theory. As a consequence, Bell's theorem took the Bohr–Einstein debate out of the domain of pure epistemology (the interpretation of physical theories) and into the arena of experimental physics. It suddenly became possible to decide the debate by measuring correlations and seeing whether they violated Bell's inequalities, as predicted by quantum theory, or whether they satisfied these inequalities, as followed from Einstein's position.

The importance of Bell's theorem (the incompatibility between quantum mechanics and any attempt to complete it in the way suggested by Einstein) was only gradually recognized. In 1969, a paper by J. Clauser, M. Horne, A. Shimony and R. Holt suggested using photon pairs with entangled polarizations emitted in certain atomic radiative cascades. This triggered a series of pioneering experiments at the beginning of the 1970s. However, the technology of the day was limited and the first experiments – still some way from the ideal setup – produced contradictory results, although there was a general tendency to favour quantum mechanics. But with the technical advances of laser physics, it became possible at the end of the 1970s to build a more efficient source of polarization-entangled photon pairs. The question was dealt with quite unambiguously: Bell's inequalities were violated in experimental setups that came ever closer to the ideal thought experiment or Gedanken experiment so dear to Einstein.

More and more physicists then realized the extraordinary nature of entanglement as revealed by the violation of Bell's inequalities.[2] A third generation of sources for entangled photon pairs would soon be developed, based on nonlinear optical effects, but in anisotropic

[2] An example of this change of mind is provided by two quotations from R. Feynman concerning the EPR correlations:

- 'This point was never accepted by Einstein... It became known as the Einstein–Podolsky–Rosen paradox. But when the situation is described as we have done it here, there doesn't seem to be any paradox at all...' *The Feynman Lectures on Physics*, Vol. III, Chapter 18, Addison-Wesley (1965).
- 'I have entertained myself always by squeezing the difficulty of quantum mechanics into a smaller and smaller place, so as to get more and more worried about this particular item. It seems to be almost ridiculous that you can squeeze it to a numerical question that one thing is bigger than another. But there you are – it is bigger...' R. P. Feynman, Simulating Physics with Computers, *Intl. Journal of Theoretical Physics* **21**, 467 (1982).

It is interesting to note that it was in this article that Feynman introduced the idea that a quantum computer might be fundamentally more powerful than an ordinary one.

crystals rather than in atoms (see Chapter 7). One major step forward which resulted from this development was a way of controlling the direction in which the entangled photons are emitted. This makes it possible to inject the two members of each pair into two optical fibres arranged in opposite directions. In this way, experiments were run with source–detector distances of several hundred metres, or several tens of kilometres, as in an experiment using the commercial fibre optic network of the Swiss telecommunications company.

All these experiments confirm the violation of Bell's inequalities, and highlight the extraordinary properties of entanglement. It was at this point, that a new idea emerged, according to which entanglement might open up novel possibilities for transmitting and processing information. The field of quantum information was born, based on the state-of-the-art resource provided by entangled states. A new world had opened up for physicists and mathematicians, for computer scientists and engineers alike (see Complement 5E).

5C.2 Photons with correlated polarization: EPR pairs

5C.2.1 Measuring the polarization of a single photon

Consider two modes ℓ' and ℓ'' of the electromagnetic field characterized by the same wavevector $\mathbf{k}$ (and hence the same frequency) parallel to the Oz-axis, but with polarizations $\boldsymbol{\varepsilon}'$ and $\boldsymbol{\varepsilon}''$ along the Ox- and Oy-axes, respectively. These two modes generate the space $\mathcal{E}_{\mathbf{k}}$ of one-photon states with wavevector $\mathbf{k}$, defined by the basis $\{|1_x\rangle; |1_y\rangle\}$, which we shall write $\{|x\rangle; |y\rangle\}$ to simplify the notation. This space of polarization states of a $\mathbf{k}$ photon is two dimensional.

The basis $\{|x\rangle\ ;\ |y\rangle\}$ is in fact associated with an observable, namely, the polarization in the Ox direction, which can be measured using a polarization analyser, or polarizer for short (see Figure 5C.1). This device has two output ports labelled $+1$ and -1, such that photons in the state $|x\rangle$ are certain to go into port $+1$, while photons in $|y\rangle$ will go into port -1. To describe this type of measurement, we introduce the observable $\hat{A}(0)$ which has eigenvectors $|x\rangle$ and $|y\rangle$ with eigenvalues $+1$ and -1, respectively. Relative to the basis $\{|x\rangle\ ;\ |y\rangle\}$, the corresponding operator can be written

$$\hat{A}(0) = \begin{pmatrix} +1 & 0 \\ 0 & -1 \end{pmatrix}. \qquad (5C.1)$$

The polarizer can in fact rotate about the Oz-axis and its orientation is indicated by a unit vector $\mathbf{u}$, or by the angle $\theta = (Ox, \mathbf{u})$ between $\mathbf{u}$ and Ox. A measurement of the polarization in the orientation specified by θ is associated with linear polarizations along θ or $\theta + \dfrac{\pi}{2}$.

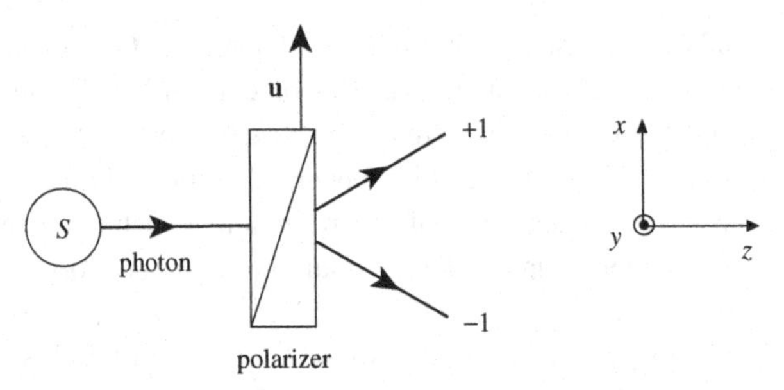

Figure 5C.1 **Light polarization measurement.** The polarizer, which can rotate about the Oz-axis along which the light propagates, measures the polarization along **u**, perpendicular to Oz. The figure shows the particular case in which **u** lies along Ox [$\theta = (Ox, \mathbf{u}) = 0$]. Light polarized in the u direction leaves via port $+1$, whereas light polarized in the direction perpendicular to **u** leaves via port -1. In the general case, for a classical beam, a certain fraction leaves by each port. A single photon, which cannot divide itself between the two ports, leaves either by port $+1$ or by port -1, with probabilities that depend on the quantum state. We then say that a polarization measurement along **u** has been carried out on the photon, in which the results may be either $+1$ or -1.

The eigenvectors of the observable $\hat{A}(\theta)$ are obtained by rotation:

$$|+_\theta\rangle = \cos\theta\,|x\rangle + \sin\theta\,|y\rangle, \tag{5C.2}$$

$$|-_\theta\rangle = -\sin\theta\,|x\rangle + \cos\theta\,|y\rangle, \tag{5C.3}$$

and $\hat{A}(\theta)$ is expressed relative to the basis $\{|x\rangle\,,\,|y\rangle\}$ as,

$$\hat{A}(\theta) = \begin{pmatrix} \cos 2\theta & \sin 2\theta \\ \sin 2\theta & -\cos 2\theta \end{pmatrix}. \tag{5C.4}$$

It is easy to check that $\hat{A}|\pm_\theta\rangle = \pm|\pm_\theta\rangle$.

If we consider an incident photon linearly polarized along a direction making an angle λ with Ox, its state can be written

$$|\psi\rangle = |+_\lambda\rangle = \cos\lambda\,|x\rangle + \sin\lambda\,|y\rangle. \tag{5C.5}$$

A measurement made by the polarizer oriented in the θ direction will give results $+1$ or -1 with probabilities

$$P_+(\theta,\lambda) = |\langle +_\theta|+_\lambda\rangle|^2 = \cos^2(\theta-\lambda) \tag{5C.6}$$

$$P_-(\theta,\lambda) = |\langle -_\theta|+_\lambda\rangle|^2 = \sin^2(\theta-\lambda). \tag{5C.7}$$

Equations (5C.6) and (5C.7) express in probabilistic terms and for a single photon the classical result known as Malus' law, which gives the intensities transmitted to the polarizer ports $+1$ and -1 for an incident beam polarized in the direction specified by λ.

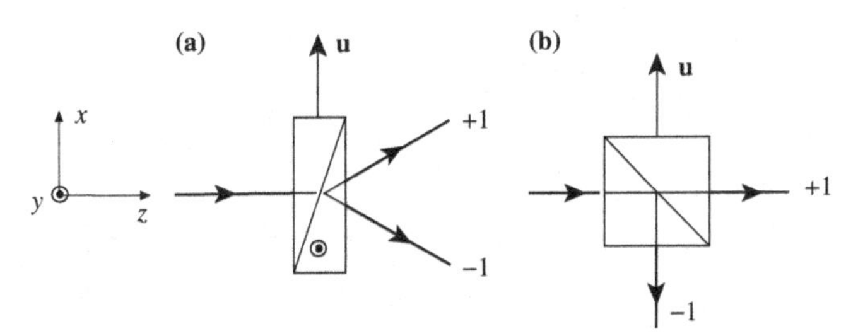

Figure 5C.2 (a) Birefringent polarizer and (b) dielectric layer polarizer. Both are oriented along $\theta = 0$. Polarization measurements are made by placing photon counters at ports +1 and −1.

Comments

(i) Polarizers are commonplace devices in the visible region of the spectrum, usually made either from birefringent anisotropic crystals, or by joining together two prisms along their hypotenuses and covering with a judiciously calculated stack of dielectric layers. In the first case **u** corresponds to the optical axis of the birefringent crystal. In the second, **u** lies in the plane of incidence on the hypotenuses (see Figure 5C.2).

(ii) There is a clear analogy between a polarization measurement made on a photon by a polarizer and measurement of a spin component on a spin-1/2 particle using a Stern–Gerlach setup. Care must nevertheless be taken over the factor of 2 which accompanies angles characterizing orientations. This arises because orthogonal spin states $|\uparrow\rangle$ and $|\downarrow\rangle$ in the Hilbert space correspond to orientations separated by an angle of 180° in physical space, whereas for photon polarizations, orthogonality in Hilbert space corresponds to orthogonal orientations of the polarizers. These properties reflect the fact that a rotation through an angle θ about the Oz-axis is described by an operator $\exp\{-\mathrm{i}\theta\hat{J}_z\}$ and that the angular momentum of the photon is $J = 1$ whilst the spin of the electron is $J = 1/2$ (CDL, Complement B.VI).

(iii) Measurements associated with the observable $\hat{\mathbf{A}}(\mathbf{u})$ are made by placing devices capable of detecting a single photon at the output ports +1 and −1, i.e. photomultipliers or avalanche photodiodes.

(iv) The above approach can be generalized to cover the case of circular polarizations:

$$|\boldsymbol{\varepsilon}_+\rangle = -\frac{1}{\sqrt{2}}(|x\rangle + \mathrm{i}\,|y\rangle) \tag{5C.8}$$

$$|\boldsymbol{\varepsilon}_-\rangle = \frac{1}{\sqrt{2}}(|x\rangle - \mathrm{i}\,|y\rangle) \tag{5C.9}$$

The kets $|\boldsymbol{\varepsilon}_+\rangle$ and $|\boldsymbol{\varepsilon}_-\rangle$ clearly form an orthonormal basis for the space $\mathcal{E}_\mathbf{k}$. The overall sign, which is a priori arbitrary, has been chosen to be consistent with the classical definitions (see, for example, 2B.25–2B.26).

5C.2.2 Photon pairs and joint polarization measurements

We now consider a pair of photons ν_1 and ν_2 with frequencies ω_1 and ω_2, respectively, emitted simultaneously along $-\mathbf{Oz}$ and $+\mathbf{Oz}$, respectively (see Figure 5C.3). The only unspecified degree of freedom is the polarization of each photon. The polarization state of the pair is described by a ket in the space

$$\mathcal{E} = \mathcal{E}_1 \otimes \mathcal{E}_2, \tag{5C.10}$$

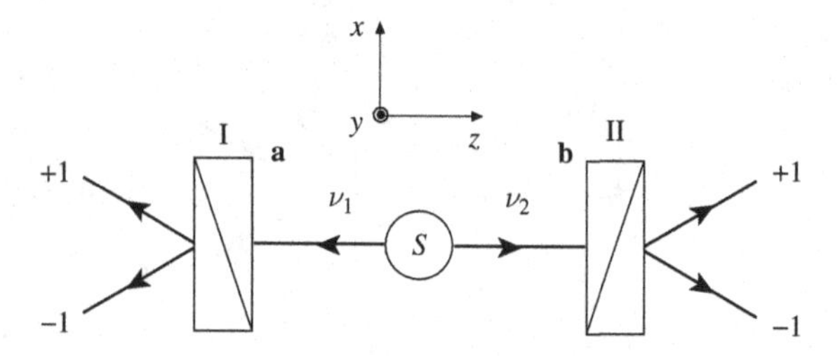

Figure 5C.3 **EPR thought experiment using photon pairs with correlated polarizations.** The polarizations of photons ν_1 and ν_2 from the same pair are analysed in directions a and b by polarizers I and II, where the vectors a and **b** characterizing the polarizer orientations are perpendicular to the *Oz*-axis. The measurement results reveal the existence of polarization correlations.

which is the tensor product of the two-dimensional spaces $\mathcal{E}_1$ and $\mathcal{E}_2$ describing the polarizations of ν_1 and ν_2, respectively.

The space $\mathcal{E}$ is four-dimensional. A basis for this space is given by the four kets,

$$\mathcal{E} = \{|x_1, x_2\rangle\,;\ |x_1, y_2\rangle\,;\ |y_1, x_2\rangle\,;\ |y_1, y_2\rangle\}. \tag{5C.11}$$

The polarization properties of a pair are described by a vector $|\psi\rangle$ in this space.

Using polarizers I and II, oriented in the chosen directions **a** and **b** (making angles $\theta_\mathbf{a}$ and $\theta_\mathbf{b}$ with the Ox-axis), polarization measurements can be carried out on each photon. A joint measurement on two photons of the same pair can yield one of the four results $(+1, +1)$, $(+1, -1)$, $(-1, +1)$, or $(-1, -1)$. The corresponding probabilities are

$$P_{++}(\mathbf{a}, \mathbf{b}) = |\langle +_\mathbf{a}, +_\mathbf{b} \mid \psi\rangle|^2 \tag{5C.12}$$

$$P_{+-}(\mathbf{a}, \mathbf{b}) = |\langle +_\mathbf{a}, -_\mathbf{b} \mid \psi\rangle|^2 \tag{5C.13}$$

and so on.

We can also obtain the probabilities for measurements on one photon. These are related to the joint probabilities. For example, the probability of obtaining $+1$ for the photon ν_1 is

$$P_{+}(\mathbf{a}) = P_{++}(\mathbf{a}, \mathbf{b}) + P_{+-}(\mathbf{a}, \mathbf{b}). \tag{5C.14}$$

Comments

(i) These probabilities are measured by placing photon detectors at the corresponding ports of the polarizers and using coincidence circuits capable of identifying detections that are simultaneous to within a few nanoseconds.

(ii) In order to keep the notation as transparent as possible, the indices 1 and 2 will often be omitted, as has been done above from (5C.12) onwards, with the understanding that the first quantity refers to the photon ν_1 and the second to photon ν_2.

5C.2.3 EPR pairs with correlated polarizations

We consider photon pairs in the state

$$|\psi_{\text{EPR}}\rangle = \frac{1}{\sqrt{2}}(|x,x\rangle + |y,y\rangle). \tag{5C.15}$$

Note immediately the special nature of this state, in which a specific polarization cannot be attributed either to photon ν_1 or to photon ν_2. Indeed, this state cannot be factorized into a tensor product of two terms, one associated with ν_1 and the other with ν_2, in sharp contrast to the states $|x,x\rangle$ or $|y,y\rangle$ taken separately. (The state $|x,x\rangle$ represents a photon ν_1 polarized in the Ox direction and a photon ν_2 polarized in the Ox direction.) This impossibility of factorizing a state like $|\psi_{\text{EPR}}\rangle$ lies at the heart of entanglement.

It is easy to calculate the probabilities of joint detections expected for the state $|\psi_{\text{EPR}}\rangle$, when the polarizers I and II are oriented along **a** and **b**, respectively, making angles $\theta_{\mathbf{a}}$ and $\theta_{\mathbf{b}}$ to the Ox-axis. For example, using (5C.12) and (5C.2), we obtain

$$P_{++}(\mathbf{a},\mathbf{b}) = |\langle +_{\mathbf{a}}, +_{\mathbf{b}} \mid \psi_{\text{EPR}}\rangle|^2 = \frac{1}{2}\cos^2(\theta_{\mathbf{a}} - \theta_{\mathbf{b}}) = \frac{1}{2}\cos^2(\mathbf{a},\mathbf{b}). \tag{5C.16}$$

In a like manner, for the three other joint detection probabilities, we find,

$$P_{--}(\mathbf{a},\mathbf{b}) = \frac{1}{2}\cos^2(\mathbf{a},\mathbf{b}), \tag{5C.17}$$

$$P_{+-}(\mathbf{a},\mathbf{b}) = P_{-+}(\mathbf{a},\mathbf{b}) = \frac{1}{2}\sin^2(\mathbf{a},\mathbf{b}). \tag{5C.18}$$

Note that these probabilities depend only on the angle $(\mathbf{a},\mathbf{b}) = \theta_{\mathbf{b}} - \theta_{\mathbf{a}}$ between the polarizers and not on their absolute orientations. The result is invariant under rotation about Oz.

The probability of obtaining $+1$ for the photon ν_1, regardless of the result for ν_2, is

$$P_{+}(\mathbf{a}) = P_{++}(\mathbf{a},\mathbf{b}) + P_{+-}(\mathbf{a},\mathbf{b}) = \frac{1}{2}. \tag{5C.19}$$

Likewise, we obtain the other single detection probabilities:

$$P_{-}(\mathbf{a}) = \frac{1}{2}, \tag{5C.20}$$

$$P_{+}(\mathbf{b}) = P_{-}(\mathbf{b}) = \frac{1}{2}. \tag{5C.21}$$

We first observe from (5C.19–5C.21) that each measurement yields results which, taken separately, appear to be random. If we represent the measurement result from polarizer I (oriented along **a**) by means of a classical random variable $\mathcal{A}(\mathbf{a})$ which can only assume the values $+1$ or -1, then (5C.19) and (5C.20) imply that

$$\mathcal{P}[\mathcal{A}(\mathbf{a}) = +1] = \mathcal{P}[\mathcal{A}(\mathbf{a}) = -1] = \frac{1}{2}. \tag{5C.22}$$

This result is therefore perfectly random, and the statistical average (denoted by a bar over the top) of $\mathcal{A}(\mathbf{a})$ is

$$\overline{\mathcal{A}(\mathbf{a})} = 0. \tag{5C.23}$$

Likewise, the result of a polarization measurement on ν_2 by polarizer II oriented in direction $\mathbf{b}$ is a random variable $\mathcal{B}(\mathbf{b})$ which can only assume the values $+1$ and -1 and has average zero, i.e.

$$\overline{\mathcal{B}(\mathbf{b})} = 0. \tag{5C.24}$$

Hence, in the EPR state, each photon taken separately appears to be unpolarized. However, we shall now see that the polarizations of ν_1 and ν_2 are in fact *correlated.* To do this, we consider the correlation coefficient between the random variables $\mathcal{A}(\mathbf{a})$ and $\mathcal{B}(\mathbf{b})$, defined by

$$E(\mathbf{a},\mathbf{b}) = \frac{\overline{\mathcal{A}(\mathbf{a})\cdot\mathcal{B}(\mathbf{b})} - \overline{\mathcal{A}(\mathbf{a})}\cdot\overline{\mathcal{B}(\mathbf{b})}}{\left(\overline{|\mathcal{A}(\mathbf{a})|^2}\right)^{1/2}\cdot\left(\overline{|\mathcal{B}(\mathbf{b})|^2}\right)^{1/2}}. \tag{5C.25}$$

The above definition is relative to the random variables $\mathcal{A}(\mathbf{a})$ and $\mathcal{B}(\mathbf{b})$, which represent the measurement results. If these measurement results are given by the quantum theoretical predictions regarding the EPR pair, as specified in (5C.16–5C.18), we have

$$\begin{aligned}\overline{\mathcal{A}(\mathbf{a})\cdot\mathcal{B}(\mathbf{b})} &= P_{++}(\mathbf{a},\mathbf{b}) + P_{--}(\mathbf{a},\mathbf{b}) - P_{+-}(\mathbf{a},\mathbf{b}) - P_{-+}(\mathbf{a},\mathbf{b}) \\ &= \cos 2(\mathbf{a},\mathbf{b}).\end{aligned} \tag{5C.26}$$

Taking into account (5C.23) and (5C.24) and the fact that

$$\mathcal{A}(\mathbf{a})^2 = \mathcal{B}(\mathbf{b})^2 = 1, \tag{5C.27}$$

we find that quantum mechanics predicts the following correlation coefficient for the polarizations in the EPR state:

$$E_{\mathrm{QM}}(\mathbf{a},\mathbf{b}) = \overline{\mathcal{A}(\mathbf{a})\cdot\mathcal{B}(\mathbf{b})} = \cos\, 2(\mathbf{a},\mathbf{b}). \tag{5C.28}$$

If polarizers I and II are oriented in the same direction, i.e. $(\mathbf{a},\mathbf{b}) = 0$, the correlation coefficient predicted by quantum mechanics is unity. In other words, we have perfect correlation.

This perfect correlation can be seen directly by considering the value of the joint probabilities when $(\mathbf{a},\mathbf{b}) = 0$. For example, we have $P_{++}(\mathbf{a},\mathbf{a}) = 1/2$. Recalling that $P_{+}(\mathbf{a}) = 1/2$, we deduce that the conditional probability of finding $+1$ for ν_2 in the direction $\mathbf{b} = \mathbf{a}$,[3] having found $+1$ for ν_1 in the direction $\mathbf{a}$, is just

$$\begin{aligned}P\{\mathcal{B}(\mathbf{a}) = +1 \mid \mathcal{A}(\mathbf{a}) = +1\} &= \frac{\mathcal{P}\{\mathcal{B}(\mathbf{a}) = +1 \text{ AND } \mathcal{A}(\mathbf{a}) = +1\}}{\mathcal{P}\{\mathcal{A}(\mathbf{a}) = +1\}} \\ &= \frac{P_{++}(\mathbf{a},\mathbf{a})}{P_{+}(\mathbf{a})} = 1.\end{aligned} \tag{5C.29}$$

[3] P. Réfrégier, *Noise Theory and Applications to Physics, from Fluctuations to Information*, Springer, NY (2004).

We can thus be certain to find $+1$ for ν_2 if we have found $+1$ for ν_1, when the polarizers have the same orientation. It can likewise be shown that if we find -1 for ν_1, we then find -1 for ν_2. This perfect correlation is confirmed by the fact that $P_{+-}(\mathbf{a},\mathbf{a}) = P_{-+}(\mathbf{a},\mathbf{a}) = 0$, i.e. if we find $+1$ for ν_1, we never then find -1 for ν_2, and vice versa.

The level of correlation (5C.28) depends on the angle between the polarizers. For $(\mathbf{a},\mathbf{b}) = \pi/4$ it is zero, but for $(\mathbf{a},\mathbf{b}) = \pi/2$ it is -1, which corresponds once again to a perfect correlation (the minus sign indicating that if we find $+1$ on one side, we are certain to find -1 on the other).

The quantum theoretical prediction of perfect correlation between certain measurements made on two distant particles, spatially separated but described by an entangled state, was discovered by Einstein, Podolsky and Rosen. They concluded that quantum mechanics was an incomplete theory. This is the question we shall now investigate.

Comment The expression of $E_{QM}(\mathbf{a},\mathbf{b})$ in (5C.28) can be found directly by substituting the quantum observables $\hat{A}(\mathbf{a})$ and $\hat{B}(\mathbf{b})$ introduced in Section 5C.2.1, for the random variables $\mathcal{A}(\mathbf{a})$ and $\mathcal{B}(\mathbf{b})$ into the definition (5C.25), and then taking the quantum expectation value (indicated by angle brackets) rather than the statistical average (indicated by the overbar). The reason why we have carried out the calculation in detail is to make a clear distinction in the argument between what is quantum theoretical (the calculation of the probabilities of different measurement results) and what is part of standard probability theory (the definition of the correlation coefficient). The measurement results (the occurrence of a 'click' at the photomultiplier) are macroscopic events, to which the ordinary notions of probability apply without difficulty.

5C.2.4 The search for a picture to interpret the correlations between widely separated measurements

Picture based on quantum calculations

As we have presented them, the calculations were carried out in a configuration space in which the two particles are described globally, and it is difficult to represent what is happening in real space. We may thus ask whether there is a way to interpret the calculation which led to the perfect correlation prediction (5C.28) using pictures in the real space of the laboratory. To attempt to do this, let us imagine that polarizer I is a little closer to the source than polarizer II, in such a way that the measurement takes place in two stages.

The first stage, in which the polarization of ν_1 is measured in the **a** direction, can give results $+1$ or -1 with equal probability, according to (5C.19–5C.20). To continue the calculation, we apply the 'postulate of wave packet reduction':[4] just after the measurement, the state vector of the system is the (normalized) projection of the initial state onto the eigenspace associated with the obtained result. To simplify, we shall assume that polarizer I is oriented along the Ox-axis. If we obtain $+1$, the corresponding eigenspace, which has two dimensions, is generated by $\{|x,x\rangle\ ;\ |x,y\rangle\}$. The projection of $|\psi_{\mathrm{EPR}}\rangle$ of (5C.15) onto

[4] See, for example, Chapter III of CDL, or Chapter 5 of BD.

this sub-space gives $|\psi'_+\rangle = |x, x\rangle$. Likewise, if we obtain -1 for the photon ν_1 (eigenvalue associated with $|y\rangle$), the corresponding eigenspace is generated by $\{|y, x\rangle\ ;\ |y, y\rangle\}$, and after projecting $|\psi_{\text{EPR}}\rangle$, the state of the pair becomes $|\psi'_-\rangle = |y, y\rangle$.

Quite generally, it is a straightforward matter to show that, if polarizer I is oriented along arbitrary $\mathbf{a}$, and if the measurement result on ν_1 is $+1$, we obtain just after the measurement,

$$|\psi'_{+\mathbf{a}}\rangle = |+_{\mathbf{a}}, +_{\mathbf{a}}\rangle. \tag{5C.30}$$

For a result -1, the 'reduced' state vector is

$$|\psi'_{-\mathbf{a}}\rangle = |-_{\mathbf{a}}, -_{\mathbf{a}}\rangle. \tag{5C.31}$$

The states (5C.30) and (5C.31) are factorizable: the vector ν_2 now has a well-defined polarization, parallel to $\mathbf{a}$ in the case $+1$, and perpendicular to $\mathbf{a}$ in the case -1. The probabilities of obtaining $+1$ or -1 during the measurement by polarizer II, oriented along $\mathbf{b}$, are $\cos^2(\mathbf{a}, \mathbf{b})$ and $\sin^2(\mathbf{a}, \mathbf{b})$, respectively. Multiplying by the probabilities 1/2 of finding $+1$ or -1 in the first measurement (on ν_1), we recover the expressions (5C.16–5C.18) for the joint probabilities.

The two-stage calculation thus yields the same results as the global calculation. It is more complicated, but it has the advantage of suggesting a picture of what happens in real space. This picture is as follows. As long as no measurement has been made, there is as much chance of finding $+1$ or -1 with each photon. But as soon as a first measurement has been made and has given a result, e.g. $+1$ along $\mathbf{a}$ for ν_1, the second photon is projected into an identical polarization state to the one found for the first, i.e. the state $|+_{\mathbf{a}}\rangle$ in the above example.

It is easy to understand why such a picture was unacceptable to Einstein, inventor of the relativistic causality principle, which forbids any influence from propagating faster than the speed of light. Indeed, the picture we have developed suggests an instantaneous influence of the first measurement on the state of the second photon, which may be located a macroscopic distance away. One might well imagine that the projection of the wave packet could propagate at the speed of light, and that the correlations (5C.16–5C.18) could only be observed if the detections were separated by a timelike interval in the relativistic sense.[5] But then what happens if the two detections are separated by a spacelike interval? In principle, the result of the quantum calculation remains valid, and if we reject the idea of an instantaneous influence violating relativistic causality – also referred to as a 'nonlocal influence' – we must reject the picture we have just constructed and look for a better one. This approach, put forward in the EPR paper, has indeed been pursued by several authors, and in particular David Bohm, who developed specific models, until John Bell came up with a general argument leading to his famous inequalities. It is this argument that we shall discuss next.

[5] The relativistic interval between two events $(\mathbf{r}_1, t_1)$ and $(\mathbf{r}_2, t_2)$ is timelike if and only if the distance $|r_1 - r_2|$ is less than the distance $c|t_1 - t_2|$ travelled by light during the time interval $|t_1 - t_2|$. In the opposite case, i.e. $|\mathbf{r}_1 - \mathbf{r}_2| > c|t_1 - t_2|$, the interval is said to be spacelike, and the special theory of relativity implies that no causal influence can then connect the two events.

Classical picture via shared parameters

There is a priori a simple picture for understanding the correlations between the two photons: if there is an identical parameter for two photons belonging to the same pair, and if the measurement results on each particle depend on this common parameter, then it is easy to account for the correlations. For example, we may imagine that half of the pairs are emitted at the outset with a common polarization in the Ox direction, and the other half with a polarization in the Oy direction. If the polarizers are oriented in the Ox direction, we will indeed recover the results predicted by quantum mechanics for this orientation, i.e.

$$P_{++} = P_{--} = \frac{1}{2}, \tag{5C.32}$$

$$P_{+-} = P_{-+} = 0. \tag{5C.33}$$

Before further examining the possibility of understanding the whole set of results due to quantum theory in this way, let us say immediately that such a description must necessarily go beyond the formalism of quantum mechanics. This is because we introduce different types of pair at the emission stage. In the above example, we have pairs polarized along Ox and other pairs polarized along Oy. In contrast, the quantum state (5C.15) is the same for all of the pairs. Introducing a parameter which distinguishes different types of pair thus amounts to supplementing the quantum formalism, thereby admitting that this formalism provides only an incomplete picture of reality, along the lines of the argument in the EPR paper. These quantities, which are intended to complete the quantum formalism in this way, are referred to as hidden variables.

The model presented above is too simplistic to account for the quantum predictions for all orientations of the polarizers, but we may attempt to improve it, following John Bell's line of reasoning. Suppose, for example, that the two photons of a given pair have from the outset a well-defined polarization $\mathbf{p}$ specified by the angle λ between this polarization (perpendicular to Oz) and the Ox-axis. The orientation of $\mathbf{p}$ varies from one pair to another, and we take λ to be a random variable evenly distributed between 0 and 2π, and hence characterized by the constant probability density function

$$\rho(\lambda) = \frac{1}{2\pi}. \tag{5C.34}$$

We model the measurement by polarizer I oriented along $\mathbf{a}$ (making angle $\theta_{\mathbf{a}}$ with Ox) by introducing a function,

$$A(\lambda, \mathbf{a}) = \text{sign}\left\{\cos 2(\theta_{\mathbf{a}} - \lambda)\right\}, \tag{5C.35}$$

which is equal to $+1$ if the absolute value of the angle between the polarization $\mathbf{p}$ and the measurement direction $\mathbf{a}$ (modulo π) is less than $\pi/4$, and -1 if this angle lies between $\pi/4$ and $\pi/2$ (see Figure 5C.4). An identical model is adopted for polarizer II oriented along $\theta_{\mathbf{b}}$:

$$B(\lambda, \mathbf{b}) = \text{sign}\left\{\cos 2(\theta_{\mathbf{b}} - \lambda)\right\}. \tag{5C.36}$$

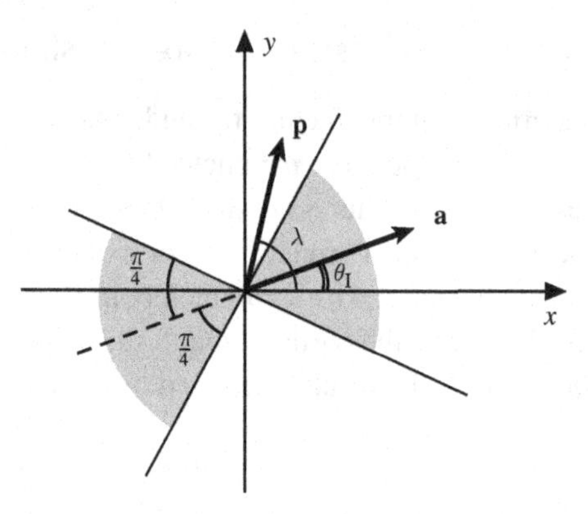

Figure 5C.4 **A particular hidden variable model.** Each photon has a well-determined polarization **p**, characterized by the angle λ with Ox. A polarization measurement along a (angle θ with Ox) gives $+1$ if $0 \leq |\theta - \lambda| \leq \pi/4$ modulo π (shaded region), and -1 if $\pi/4 \leq |\theta - \lambda| \leq \pi/2$ modulo π (unshaded region).

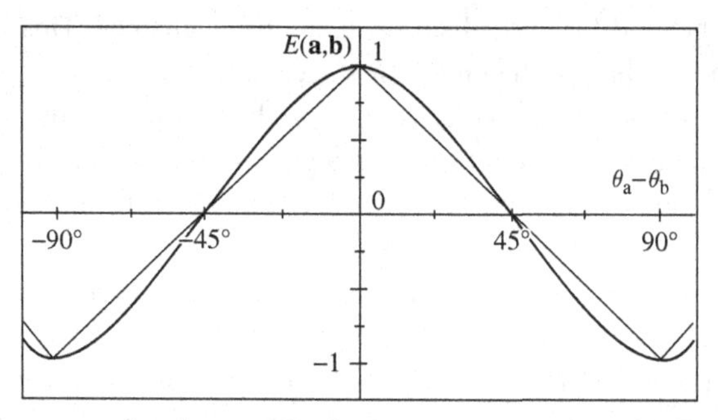

Figure 5C.5 **Polarization correlation coefficient as a function of the angle between the polarizers.** Comparison between the results of the quantum theoretical calculation (thick curve) and those of the hidden variable model described in the text (thin curve). The predictions of the two models are identical when the polarizers are parallel, perpendicular, or at 45° to one another, and remain very close at intermediate angles.

It is easy to show that this model predicts many results that are identical or close to the quantum predictions for the EPR pairs. For example, the single detection probabilities $P_+(\mathbf{a})$, $P_-(\mathbf{a})$, $P_+(\mathbf{b})$, and $P_-(\mathbf{b})$ are all equal to 1/2 (compare with (5C.19–5C.21)). Moreover, the correlation coefficient has period π, and it is easy to show that

$$E(\mathbf{a}, \mathbf{b}) = \int_0^{2\pi} d\lambda\, \rho(\lambda) A(\lambda, \mathbf{a}) \cdot B(\lambda, \mathbf{b}) = 1 - 4\frac{|\theta_\mathbf{a} - \theta_\mathbf{b}|}{\pi} \tag{5C.37}$$

for

$$-\frac{\pi}{2} \leq \theta_\mathbf{a} - \theta_\mathbf{b} \leq \frac{\pi}{2}. \tag{5C.38}$$

Figure 5C.5 compares the results (5C.37–5C.38) with the quantum theoretical prediction (5C.28). We observe that the hidden variable model exactly reproduces the perfect correlations predicted by quantum mechanics when the polarizers are parallel or perpendicular, and gives very similar values for the other orientations. One may well wonder whether the model cannot be refined, e.g. by choosing less simplistic forms for the response $A(\lambda,\theta)$ of the polarizers, in such a way as to obtain results that agree exactly with the quantum predictions for all orientations of the polarizers. The answer to this question is provided by Bell's theorem. And it is negative!

5C.3 Bell's theorem

5C.3.1 Bell inequalities

Bell's argument is perfectly general. It applies to any local hidden variable theory (LHVT) and not only to the model that we have just given as an example. In the spirit of the EPR argument, we wish to account for the polarization correlations between two photons of the same pair by introducing a parameter λ that is common to the two photons of the pair and which changes randomly from one pair to another. It is described as a random variable characterized by a positive definite probability density $\rho(\lambda)$, i.e.

$$\rho(\lambda) \geq 0, \tag{5C.39}$$

$$\int \mathrm{d}\lambda\, \rho(\lambda) = 1. \tag{5C.40}$$

In addition, we describe the polarization measurements I and II on photons carrying the parameter λ by means of functions $A(\lambda,\mathbf{a})$ and $B(\lambda,\mathbf{b})$ which can only assume the values $+1$ or -1:

$$|A(\lambda,\mathbf{a})| = |B(\lambda,\mathbf{b})| = 1. \tag{5C.41}$$

We assume that there is as much chance of obtaining $+1$ as -1 for each polarization measurement, i.e.

$$\int \mathrm{d}\lambda\, \rho(\lambda) A(\lambda,\mathbf{a}) = \int \mathrm{d}\lambda\, \rho(\lambda) B(\lambda,\mathbf{b}) = 0. \tag{5C.42}$$

It is then clear that the polarization correlation coefficient, defined in the usual way by (5C.25), can be written in this model as

$$E_{\mathrm{LHVT}}(\mathbf{a},\mathbf{b}) = \overline{A(\lambda,\mathbf{a}) \cdot B(\lambda,\mathbf{b})} = \int \mathrm{d}\lambda\, \rho(\lambda) A(\lambda,\mathbf{a})\, B(\lambda,\mathbf{b}). \tag{5C.43}$$

The Bell inequalities will apply to any hidden variable theory of the kind described here, whatever the specific form of the functions $A(\lambda,\mathbf{a}), B(\lambda,\mathbf{b})$, or $\rho(\lambda)$, provided that they satisfy properties (5C.39–5C.42). To prove these inequalities, consider the quantity

$$\begin{aligned} s(\lambda,\mathbf{a},\mathbf{a}',\mathbf{b},\mathbf{b}') = {} & A(\lambda,\mathbf{a}) \cdot B(\lambda,\mathbf{b}) - A(\lambda,\mathbf{a}) \cdot B(\lambda,\mathbf{b}') \\ & + A(\lambda,\mathbf{a}') \cdot B(\lambda,\mathbf{b}) + A(\lambda,\mathbf{a}') \cdot B(\lambda,\mathbf{b}'). \end{aligned} \tag{5C.44}$$

It can be factorized to give

$$s(\lambda, \mathbf{a}, \mathbf{a}', \mathbf{b}, \mathbf{b}') = A(\lambda, \mathbf{a})\Big(B(\lambda, \mathbf{b}) - B(\lambda, \mathbf{b}')\Big) + A(\lambda, \mathbf{a}')\Big(B(\lambda, \mathbf{b}) + B(\lambda, \mathbf{b}')\Big). \quad (5C.45)$$

Using (5C.41) and (5C.45), it is easy to see that

$$s(\lambda, \mathbf{a}, \mathbf{a}', \mathbf{b}, \mathbf{b}') = \pm 2, \quad (5C.46)$$

whatever the value of λ, by noting that either $B(\lambda, \mathbf{b}) = B(\lambda, \mathbf{b}')$ or $B(\lambda, \mathbf{b}) = -B(\lambda, \mathbf{b}')$. If we average the relation (5C.46) over λ, we thus obtain a quantity lying between -2 and $+2$:

$$-2 \leq \int \mathrm{d}\lambda\, \rho(\lambda)\, s(\lambda, \mathbf{a}, \mathbf{a}', \mathbf{b}, \mathbf{b}') \leq +2. \quad (5C.47)$$

Using (5C.43), four values of the correlation coefficient arise, taken for the four orientations $(\mathbf{a}, \mathbf{b})$, $(\mathbf{a}, \mathbf{b}')$, $(\mathbf{a}', \mathbf{b})$ and $(\mathbf{a}', \mathbf{b}')$, constrained by the inequalities

$$-2 \leq S(\mathbf{a}, \mathbf{a}', \mathbf{b}, \mathbf{b}') \leq 2, \quad (5C.48)$$

where S is defined by

$$S = E(\mathbf{a}, \mathbf{b}) - E(\mathbf{a}, \mathbf{b}') + E(\mathbf{a}', \mathbf{b}) + E(\mathbf{a}', \mathbf{b}'). \quad (5C.49)$$

We have just established a particularly useful form of Bell's inequalities, known as the Bell–Clauser–Horn–Shimony–Holt inequalities. These restrict any correlation that can be written in the form (5C.43), where the quantities $A(\lambda, \mathbf{a})$, $B(\lambda, \mathbf{b})$, and $\rho(\lambda)$ can assume any form compatible with (5C.39–5C.42).

Comments

(i) The above demonstration is not limited to the case of scalar variables λ. The mathematical nature of λ is of little importance as long as (5C.39–5C.41) are satisfied.

(ii) The original Bell inequalities were only valid in the case where $E(\mathbf{a}, \mathbf{a}) = 1$. The great advantage of the inequalities in the form (5C.48–5C.49), referred to as the Bell–Clauser–Horne–Shimony–Holt inequalities after their discoverers, is that they can be directly applied in an experimental test in which $E(\mathbf{a}, \mathbf{a}) = 1$ does not hold exactly.

(iii) The demonstration of the Bell inequalities is not restricted to the case where the relationship between λ and the measurement result is given by a deterministic function $A(\lambda, \mathbf{a})$. The inequalities also apply to the case where the relationship between λ and the measurement result is probabilistic, i.e. one can define probabilities $p_{\pm}(\lambda, \mathbf{a})$ of obtaining the results ± 1, once $\mathbf{a}$ and λ are fixed.

5C.3.2 Conflict with quantum mechanics

The Bell inequalities are very general: they apply to any model in which polarization correlations are accounted for by introducing extra variables in the same spirit as suggested by Einstein's ideas. But it turns out that the polarization correlations (5C.28) predicted by

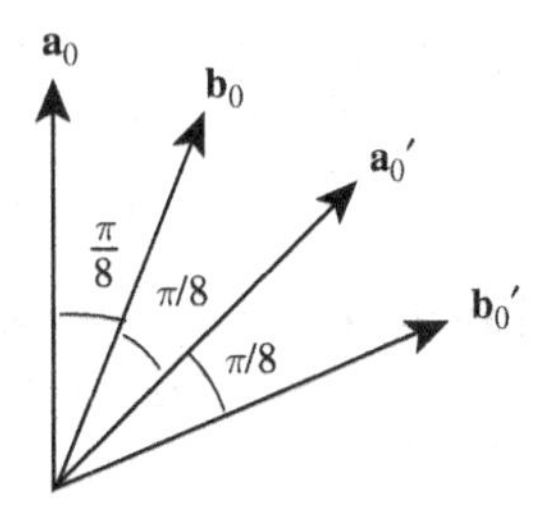

Figure 5C.6 **Maximal violation of the Bell inequalities.** For the set of orientations $\{a_0, a_0', b_0, b_0'\}$ of polarizers I and II obeying $(a_0, b_0) = (b_0, a_0') = (a_0', b_0') = \pi/8$, the quantity S_{QM} predicted by quantum mechanics is $2\sqrt{2}$, well above the upper bound of the inequalities (5C.48).

quantum mechanics for photons in the EPR state given by (5C.15) violate these inequalities for certain orientations of the polarizers. This violation can amount to a considerable discrepancy, as we shall now show.

Indeed, consider the orientations $(\mathbf{a}_0, \mathbf{a}_0', \mathbf{b}_0, \mathbf{b}_0')$ shown in Figure 5C.6, where

$$(\mathbf{a}_0, \mathbf{b}_0) = (\mathbf{b}_0, \mathbf{a}_0') = (\mathbf{a}_0', \mathbf{b}_0') = \frac{\pi}{8}, \tag{5C.50}$$

and hence,

$$(\mathbf{a}_0, \mathbf{b}_0') = \frac{3\pi}{8}. \tag{5C.51}$$

Using (5C.28), we then find that, for the EPR state, quantum mechanics predicts a value for S equal to

$$S_{\mathrm{QM}}(\mathbf{a}_0, \mathbf{a}_0', \mathbf{b}_0, \mathbf{b}_0') = 2\sqrt{2} = 2.828\ldots, \tag{5C.52}$$

clearly contradicting (5C.48), i.e. exceeding the upper bound of (5C.48) by a considerable amount.

The conceptual impact of this result cannot be overestimated. The violation of the Bell inequalities by the quantum theoretical predictions shows explicitly that *the EPR quantum correlations cannot be reduced to the classical concepts* which lead to those inequalities. These concepts seem so natural when we are dealing with correlations between classical objects that their failure is particularly surprising. It was, in fact, by identical reasoning that biologists were led to introduce extra parameters when they observed that certain twins had identical features such as hair colour, eye colour, blood group, histocompatibility types, and so on, concluding in that case that these characteristics had to be determined by chromosomes which were identical for the two twins in a given pair. What is more, this deduction was made well before it became possible to observe such chromosomes with the electron microscope. In the light of Bell's theorem, we are compelled to admit that this kind of argument does not apply to EPR-type quantum correlations.

Another equally important consequence of Bell's theorem is that *the debate between Einstein and Bohr can be decided by experiment.* In principle, one only has to measure the correlations in a situation where quantum mechanics predicts a violation of Bell's inequalities to find out whether it is the 'Einstein-type' interpretation that should be rejected, or whether one has identified a situation in which quantum mechanics fails, given that it is the shared fate of all physical theories to be brought, in the end, to the limits of their validity.

5C.3.3 Locality condition and relativistic causality. Experiment with variable polarizers

Having identified a contradiction between the predictions of quantum mechanics and the predictions of any hidden variable theory attempting to complete quantum mechanics according to the EPR programme, one is compelled to seek the deeper reasons for this conflict. What exactly are the hypotheses, either explicit or implicit, that are required to obtain Bell's inequalities? With forty years of hindsight and after hundreds of papers devoted to this question, it would seem that two hypotheses are sufficient:

(1) the introduction of extra, shared variables to explain the correlations between measurements on the separate systems;
(2) the locality hypothesis, emphasized by Bell in his first paper, and which we shall now make explicit. In the context of the thought experiment in Figure 5C.3, the *locality hypothesis* requires that the result of the measurement by one polarizer, e.g. polarizer I, cannot depend on the orientation (**b**) of the other polarizer II – and conversely, a measurement at II cannot be affected by the orientation **a** of I. Likewise, the state of the photons at emission cannot depend on the orientations **a** and **b** of the polarizers which will later carry out the measurements on these same photons.

We have implicitly assumed this hypothesis by using the forms $\rho(\lambda)$, $A(\lambda, \mathbf{a})$, and $B(\lambda, \mathbf{b})$ to describe the emission of photon pairs, or the response of the polarizers. Hence, it is clear that the function $A(\lambda, \mathbf{a})$ describing the measurement by polarizer I does not depend on the orientation **b** of the distant polarizer II, and $B(\lambda, \mathbf{b})$ does not depend on **a**. Likewise, the probability distribution $\rho(\lambda)$ of the variables over the pairs at the instant of emission does not depend on the orientations **a** or $\mathbf{a}'$, **b** or $\mathbf{b}'$, of the polarizers which will carry out the measurement. These locality conditions are essential if we are to obtain the Bell inequalities. Indeed, it is easy to check by going through the proof that if the response of polarizer I were allowed, for example, to depend on **b**, by expressing it in the form $A(\lambda, \mathbf{a}, \mathbf{b})$, one would not then be able to prove that the quantity $s(\lambda, \mathbf{a}, \mathbf{a}', \mathbf{b}, \mathbf{b}')$ is equal to ± 2.

However natural it may seem, the locality condition does not follow a priori from any fundamental law of physics. As Bell himself observed, there is nothing to forbid some unknown interaction from allowing the orientation of polarizer II to influence polarizer I. But Bell also remarked that, if an experiment could be carried out in which *the orientations of the polarizers are changed quickly enough* as the photons are propagating between the source and the polarizers, then Einstein's *relativistic causality* would step in to forbid the existence of such an interaction (see Figure 5C.7). Indeed, as no interaction can propagate faster than light, the information concerning the orientation of polarizer II at the instant of measurement would not arrive at I in time to influence the measurement. It is even more obvious that, in this kind of thought experiment, the initial state of the photons at the time

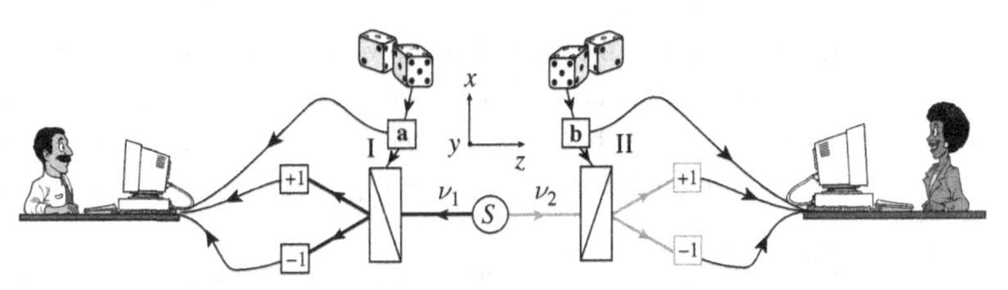

Figure 5C.7 **Thought experiment with variable polarizers. If the orientations a and b of polarizers I and II can be randomly modified during the propagation time of the photons between the source and the polarizers, then Bell's locality condition becomes a consequence of Einstein's relativistic causality. To emphasize the separation between the measurement operations, we have indicated that it is possible to record the following data separately by each measurement device: the result of the polarization measurement, the orientation of the polarizer at the instant when it gave this result, and the time of this result. The correlation is then determined later by comparing the results obtained at the same time by the two measurement devices.**

of their preparation in the source cannot depend on orientations that the polarizers have not yet adopted, but will have at the instant of measurement. In such a setup, the *locality condition* (condition 2 above) is no longer a new hypothesis, but it becomes *a consequence of Einstein's relativistic causality*.

Hence, the thought experience of Figure 5C.7 tests the whole range of ideas that Einstein liked to put forward in the context of the EPR correlations: on the one hand, the possibility (or for Einstein, even the necessity) of completing the quantum formalism by introducing parameters completely characterizing the state of each photon, which Einstein called the physical reality of each photon, and on the other hand, the impossibility of a direct interaction between events separated by a spacelike interval, in the relativistic jargon, i.e. an interval such that no signal can propagate from one event to the other at a speed less than or equal to the speed of light. *So it is indeed the conflict between quantum mechanics and the world view defended by Einstein, associated with the above assumptions 1 and 2, sometimes called local realism, that will be decided in experiments to test the Bell inequalities.*

5C.4 The experimental verdict and violation of the Bell inequalities[6]

It is astonishing that in 1964, after decades of experimental observations in agreement with the predictions of quantum theory, there were no results allowing a test of the Bell inequalities. It became clear that situations in which the theory predicted a violation of the Bell inequalities were actually extremely rare, and that such 'sensitive'

[6] References to a series of relevant experiments can be found in: A. Aspect, Bell's Inequality Test: More Ideal than Ever, *Nature* **398**, 189 (1999). See also footnote 1, page 413.

arrangements would therefore have to be devised using the experimental methods of the day. In 1969, the paper by J. Clauser, M. Horne, A. Shimony and R. Holt showed that the pairs of visible photons produced when an atom decays by cascading down from level to level might be good candidates, provided that the levels were suitably chosen (see Complement 6C).

Inspired by this paper, pioneering experiments were set up in the early 1970s, showing that entangled photon pairs could be generated, even though the very low efficiency of the light sources made these experiments extremely difficult. It should therefore come as no surprise that the first tests of the Bell inequalities yielded somewhat contradictory results. Although they eventually did give the advantage to quantum mechanics, the experimental setups were still far-removed from the ideal thought experiment. But it was not until the end of the 1970s that advances in laser physics made it possible to build a source of entangled photons with unprecedented intensity (see Section 6C.4 of Complement 6C). It then became possible to set up experimental arrangements that were much closer to the ideal thought experiment (see Figure 5C.7). For example, using polarization analysers with two output ports, it became possible to measure the polarization correlation coefficient $E(\mathbf{a}, \mathbf{b})$ of (5C.25) for various orientations of the polarizers, and this directly, without auxiliary calibration. The Bell inequalities (5C.48) were violated by more than 40 standard deviations. For the orientations (5C.50), a value of $S_{\text{exp}} = 2.697 \pm 0.015$ was found. Furthermore, the results were in excellent agreement with the quantum predictions, given the experimental imperfections ($S_{\text{QM}} = 2.70 \pm 0.05$). It also became possible to move the polarizers further and further away from the source, and to carry out measurements separated by a spacelike interval. Finally, it became possible to achieve the setup suggested by Bell, in which the orientations of the polarizers are modified while the photon travels from source to detector. In fact, the orientation of the polarizers was changed every 10 ns, compared with the 20 ns required by light to cover the source–detector separation. A clear violation of Bell's inequalities was observed in this situation where the measurements were spacelike-separated, in the relativistic sense, from the choice of polarizer orientation.

Towards the end of the 1980s, a third generation of sources for entangled photon pairs was developed, based on nonlinear optical effects that were no longer provided by atoms (see Complement 6C and Figure 5C.8), but by nonlinear crystals (see Chapter 7). The major step forward here is the control over the directions in which the entangled photons are emitted, making it possible to inject the two members of each pair into two optical fibres pointing in opposite directions. Experiments could then be carried out with source–detector separations of several hundred metres, or even several tens of kilometres in one experiment, which used the commercial fibre optic network of the Swiss telecommunications company. With such distances between source and measurement devices, it became possible to choose the orientation of each polarizer in a strictly random way during the propagation time of the photons from the source, which was not completely the case in the 1982 experiment. Such an experiment, exactly reproducing the setup in Figure 5C.7, was carried out in 1998, unambiguously confirming the violation of the Bell inequalities in a situation where the locality condition is a consequence of relativistic causality.

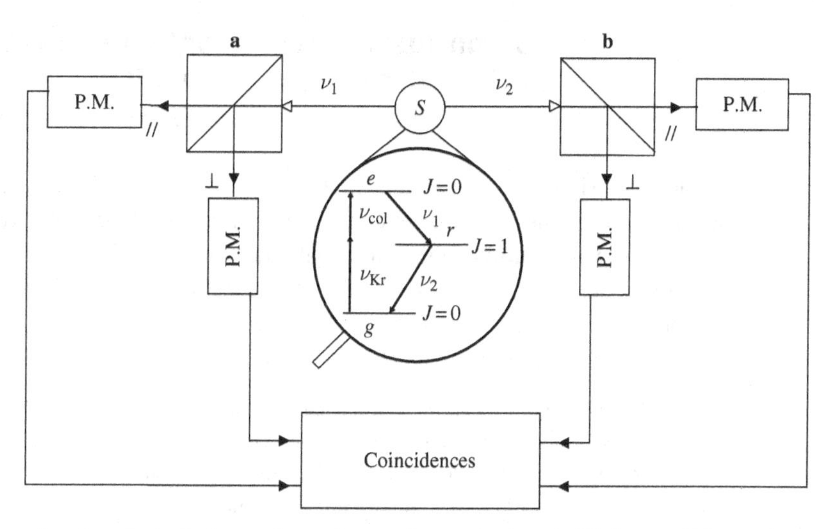

Figure 5C.8 **Source of entangled photon pairs and experiment to test Bell's inequalities.** Using two lasers with strictly controlled frequencies, a calcium atom is efficiently excited into level e, and a pair of polarization-entangled photons (ν_1, ν_2) is subsequently obtained (see Complement 6C). This source was used at the beginning of the 1980s to carry out polarization correlation measurements with an accuracy of 1% in less than 2 minutes, making it possible to set up experimental arrangements that come ever closer to the ideal thought experiment. For example, the above diagram shows how to use electronic circuitry and polarization analysers with two ports, oriented along **a** and **b**, to measure the polarization correlation coefficient $E(\mathbf{a}, \mathbf{b})$ directly (note the similarity with Figure 5C.3).

With growing interest in the entanglement phenomenon, new entangled systems began to appear, on which various tests of the Bell inequalities with different features could be carried out. For example, an experiment where the internal states of two ions are entangled led to a test in which the quantum efficiencies of the detectors were good enough to refute an objection known as the 'sensitivity loophole'.

Today there is an impressive stockpile of experimental data which unequivocally violate the Bell inequalities in a wide range of different setups, each one serving to deal with one of the various loopholes that have been imagined.[7] Moreover, it should be noted that, not only do these experiments lead to unambiguous violation of the Bell inequalities, but they also provide an arena for testing the quantitative predictions of quantum mechanics to a very high level of accuracy, revealing them to be perfectly precise whenever experimental conditions are sufficiently well controlled. The observed agreement is indeed impressive. Today we may consider that the Bell inequalities have been violated, and that the astonishing predictions of quantum theory with regard to entangled systems have been confirmed, leaving us with the task of drawing our conclusions.

[7] Insofar as no experiment has yet dealt with the sensitivity loophole and at the same time achieved spacelike-separated measurements, it is in fact possible, by pushing logical reasoning to its very limits, to uphold the view that all the investigations carried out so far still deviate from the ideal experiment. See, for example: R. Garcia-Patron, J. Fiurášek, N. J. Cerf, J. Wenger, R. Tualle-Brouri and Ph. Grangier, Proposal for a Loophole-Free Bell Test Using Homodyne Detection, *Physical Review Letters* **93**, 130409 (2004).

5C.5 Conclusion: from quantum nonlocality to quantum information

What conclusions can be drawn from the violation of the Bell inequalities? To begin with, we must accept the idea that an entangled system cannot be conceived of as being composed of separate sub-systems with locally defined physical properties that are unable to influence one another across relativistically spacelike intervals. This notion of 'separability' nevertheless seemed so fundamental to Einstein that he made it a cornerstone in his demonstration that quantum mechanics would need to be completed:[8]

> 'One may avoid this conclusion [that quantum mechanics is incomplete] only by accepting that measurement of S1 changes the real situation of S2 (by telepathy) or by denying the independence of the real situations of such spatially separated objects. Both alternatives seem to me equally unacceptable.'

Now that the violation of the Bell inequalities has been established, we can no longer reject these alternatives. We must abandon the world view advocated by Einstein, sometimes referred to as 'local realism'.

One might ask which of the ideas leading to the Bell inequalities should be abandoned: locality or realism? For our part, it seems difficult to understand these two notions as being independent. How could one conceive of independent physical realities for two spatially separated systems but which were nevertheless able to remain in contact via an instantaneous, superluminal interaction? *It is our view that the nonlocality of quantum mechanics, often presented as the conclusion to be drawn from the violation of the Bell inequalities, represents a negation of the whole local realist view of the world.* Twin entangled photons are not just two distinct systems carrying identical copies of the same set of parameters. A pair of entangled photons must be considered as a single, inseparable system, described by a global quantum state, which cannot be decomposed into two states, one for each photon. The properties of the pair cannot be reduced to the sum of the properties of the individual photons.

We stress here that careless use of the expression 'nonlocality' might lead one to conclude, wrongly, that a usable signal of some kind could be sent at a speed greater than the speed of light. Returning to the setup shown in Figure 5C.7, it might indeed be thought that a measurement by polarizer II of the polarization of photon ν_2 instantaneously informs as to the direction of polarizer I, which has just made a measurement on photon ν_1. For example, imagine that a measurement on ν_1, made in the direction $\mathbf{a}$, gives the result $+1$. As we have seen, we may then say that photon ν_2 instantaneously adopts a linear polarization parallel to $\mathbf{a}$, at the instant when ν_1 is analysed. Are we not therefore instantaneously aware at II of the value of the orientation $\mathbf{a}$ which could be chosen at the last moment at I? We would thereby have transmitted a piece of information, namely the orientation $\mathbf{a}$, at a speed faster than light. In fact, looking closely at (5C.16–5C.21), we find that it is not the results of single measurements at II that depend on the orientation $\mathbf{a}$ of I, but rather the results of joint measurements. In order to detect the change in direction of I, the observer at II must

[8] A. Einstein in *A. Einstein Philosopher Scientist*, Open Court and Cambridge University Press (1949).

compare his own measurements with those made at I. Now the result of the latter measurements is a classical piece of data that can only be communicated to the observer at II by a classical channel connecting I to II, in which information does not propagate faster than light. It is only after receiving these data, which in themselves do not supply the desired information, and then comparing them with his own, that the observer at II can determine the value of **a**.[9]

In fact, the above argument rests on the fact that the quantum state of a single system, in this case the polarization of photon ν_2, cannot be completely determined by measurement. In order to fully determine the state of a system, we need to be able to prepare a large number of copies, all in exactly the same state, and to make a large number of measurements of different observables, in this case with a series of polarizers having different orientations. But it is in fact impossible to duplicate a quantum system, i.e. obtain two systems that are in strictly the same initial state. This is the no-cloning theorem.[10] It is not therefore possible to know the orientation of the polarization state of ν_2 and one must abandon the dream so dear to science fiction authors of a superluminal telegraph.

At the end of a journey that has taken us from the EPR paper in 1935, via the Bell inequalities in 1964, to the experiments they have inspired, one might be left with the frustrating feeling of a series of negative conclusions: the established properties of quantum entanglement leave no choice but to abandon the local realist view, and the no-cloning theorem forbids the superluminal telegraph. In fact, these discoveries pave the way to quite remarkable new prospects. We are beginning to understand how we may take advantage of the quantum properties that we have just discussed in new concepts of data processing and transmission, where quantum entanglement and the no-cloning theorem play a central role: this is the field of quantum information, which has come into its own since the end of the 1990s, and which we shall review in Complement 5E.

[9] A scheme of the same kind is at work in the process known as quantum teleportation, in which a quantum state of a system is transferred from one system to another (C. H. Bennett *et al.*, Teleporting an Unknown Quantum State Via Dual Classical and Einstein–Podolsky–Rosen Channels, *Physical Review Letters* **70**, 1895–1899 (1993)). Although one stage of the process seems to be instantaneous, the operation as a whole is only accomplished when a piece of information has been transmitted by a classical channel.

[10] W. K. Wooters and W. H. Zurek, A Single Quantum Cannot be Cloned, *Nature* **299**, 802 (1982).

Complement 5D Entangled two-mode states

In this complement, we discuss another example of entangled states of the electromagnetic field: those that can be constructed in the Hilbert space describing two modes of the field. These states give rise to very similar correlations to those discovered by Einstein, Podolsky and Rosen, used by these authors to bring out some of the weirdest features of the quantum world, as demonstrated by John Bell (see Complement 5C). We shall give a brief outline of certain properties of these states and describe various methods for producing them.

5D.1 General description of a two-mode state

5D.1.1 General considerations

According to (4.131), the most general two-mode state of the electromagnetic field can be written,

$$|\psi\rangle = \sum_{n_\ell=0}^{+\infty} \sum_{n_{\ell'}=0}^{+\infty} c_{n_\ell n_{\ell'}} |0, ..., n_\ell, 0..., n_{\ell'}, 0...\rangle. \tag{5D.1}$$

Only two particular modes, indexed by ℓ and ℓ', are not in the vacuum state. We use the abbreviated notation,

$$|\psi\rangle = \sum_{n_\ell=0}^{+\infty} \sum_{n_{\ell'}=0}^{+\infty} c_{n_\ell n_{\ell'}} |n_\ell, n_{\ell'}\rangle \tag{5D.2}$$

for such a state. The two modes may differ by the polarization, the direction of propagation or the frequency. Apart from the higher dimension of the state space, the new feature compared with the single-mode case investigated in Section 5.3 of Chapter 5 is that *the two modes can be physically separated* by means of a polarization splitter, a diaphragm or a prism, respectively. It is then possible to carry out independent measurements on each of them. And it is *correlations between these measurements* that we shall be concerned with through most of this complement.

Comment The situation considered here is different from the one described in Complement 5C, where four different modes were considered (two for each photon ν_1 and ν_2). But here, we will consider the possibility of states with more than one photon in a mode, in contrast with the state (5C.15) when $|x\rangle$ or $|y\rangle$ refers to a single photon (see Section 5C.2.1).

5D.1.2 Schmidt decomposition

There is a theorem that any two-mode state of the form (5D.2) can be expressed in the following form, known as the 'Schmidt decomposition':

$$|\psi\rangle = \sum_{i=1}^{S} \sqrt{p_i}|u_i\rangle \otimes |u'_i\rangle, \tag{5D.3}$$

where $|u_i\rangle$ and $|u'_i\rangle$ are single-mode quantum states in modes ℓ and ℓ', respectively, which form two orthonormal sets of vectors, and the p_i are strictly positive real numbers such that $\sum_{i=1}^{S} p_i = 1$. The Schmidt decomposition is not unique, but the number of terms in the sum, denoted here by S and called the 'Schmidt number', is fixed once the vector $|\psi\rangle$ is given. There are then two possibilities:

- $S = 1$, in which case $|\psi\rangle$ can be expressed in the form $|u_1\rangle \otimes |u'_1\rangle$. This is a *factorized state*. The physical system is described in each of the two modes by a single-mode state of the same kind as those studied in Section 5.3 of this chapter.
- $S > 1$, in which case $|\psi\rangle$ is not factorizable. This is an entangled state. Physically, this means that it is impossible to describe the radiation in the modes ℓ and ℓ' by separate state vectors. The system forms an inseparable whole, even if the modes are physically distinct and measurements are carried out on each of them by means of measurement devices a very great distance apart, such that they cannot be connected by a physical interaction.

Comment These considerations can be extended to the case where the system is described, not by a state vector $|\psi\rangle$, but by a density matrix ρ. The generalization of the factorized state is what we shall call a separable quantum state. This is a state described by a density matrix that can be written in the form,

$$\rho = \sum_{j=1}^{n} q_j \rho_j \otimes \rho'_j, \tag{5D.4}$$

where ρ_j and ρ'_j are single-mode density matrices in modes ℓ and ℓ', and q_j are strictly positive real numbers such that $\sum_{j=1}^{n} q_j = 1$. This is therefore a statistical superposition of factorized states, in which the mode ℓ, for example, has probability q_j of being in the single-mode state described by a density matrix ρ_j. The generalization of the entangled state is the non-separable state, which cannot be written in the form (5D.4). The characterization of non-separability for a statistical mixture is significantly more involved than for a pure case.[1]

[1] See, for example, D. Bruss and G. Leuchs, editors, *Lectures on Quantum Information*, Wiley-VCH (2006).

5D.1.3 Correlations between measurements carried out on the two modes

Let $\hat{A}$ and $\hat{A}'$ be observables acting, respectively, in the Hilbert spaces $\mathcal{F}_\ell$ and $\mathcal{F}_{\ell'}$ associated with modes ℓ and ℓ'. These may be field quadrature operators, for example, or photon number operators. As these operators act in different spaces, they necessarily commute. The correlation coefficient $E(A, A')$ between measurements of these two quantities lies between -1 and $+1$ and is given by

$$E(A, A') = \frac{\langle\psi|\hat{A}\hat{A}'|\psi\rangle - \langle\psi|\hat{A}|\psi\rangle\langle\psi|\hat{A}'|\psi\rangle}{\Delta A \Delta A'}. \tag{5D.5}$$

It generalizes to arbitrary measurements the coefficient $E(\mathbf{a}, \mathbf{b})$ introduced for correlations between polarization measurements in (5C.25) of Complement 5C. When $E(A, A') = 0$, the measurements are not correlated, i.e. measurement of $\hat{A}$ provides no information about the value of $\hat{A}'$ in the same state. On the other hand, if $E(A, A') = 1$ (or $E(A, A') = -1$), the measurements are perfectly correlated (or anticorrelated), and the result of a measurement of $\hat{A}$ for mode ℓ tells us the result of a measurement of $\hat{A}'$ for mode ℓ' without our having to actually carry out this second measurement. The existence of strong correlations thus provides a way of carrying out measurements 'at a distance', also called 'non-destructive' measurements, because the measurement of $\hat{A}'$ carried out in this way cannot physically affect mode ℓ'.

Comment Note that very strong, even perfect correlations can exist between classical quantities. The specifically quantum nature of the correlations appears when they arise for several quantities associated with observables that do not commute. This is what was discussed in great detail in Complement 5C for the example of photon polarization correlations. We shall be discussing another example at the end of this complement for measurements of the field quadrature components.

Suppose now that the system is in a factorized state $|\psi\rangle = |u_1\rangle \otimes |u'_1\rangle$. Then,

$$\langle\psi|\hat{A}\hat{A}'|\psi\rangle = \langle u_1|\hat{A}|u_1\rangle\langle u'_1|\hat{A}'|u'_1\rangle = \langle\psi|\hat{A}|\psi\rangle\langle\psi|\hat{A}'|\psi\rangle. \tag{5D.6}$$

The correlation coefficient $E(A, A')$ is therefore zero. When the system is described by a state vector ('a pure state'), *there can only therefore be a correlation between measurements if the state* $|\psi\rangle$ *is entangled*, i.e. the Schmidt number S of the decomposition (5D.3) is larger than 1.

The form of the observables for which this correlation is perfect in the entangled state $|\psi\rangle$ can be determined. Indeed, consider the observables with eigenvectors $|u_i\rangle$ and $|u'_i\rangle$ appearing in (5D.3). They can be written in the form

$$\hat{A} = \sum_{i=1}^{S} a_i |u_i\rangle\langle u_i| \quad ; \quad \hat{A}' = \sum_{i=1}^{S} a'_i |u'_i\rangle\langle u'_i|, \tag{5D.7}$$

where the a_i and a'_i are real numbers, assumed to be all different. If the system is in the state (5D.3) and measurement of $\hat{A}$ gives the result $|u_{i_0}\rangle$, then after this measurement, the state is projected onto the corresponding eigenstate $|u_{i_0}\rangle \otimes |u'_{i_0}\rangle$, and one can thus be certain that measurement of $\hat{A}'$ will give the result $|u'_{i_0}\rangle$. The correlation between the measurements is then perfect.

5D.2 Twin photon states

5D.2.1 Definition and properties

Consider the two-mode state

$$|\psi\rangle = \sum_{n=0}^{+\infty} c_n |n_\ell = n, n_{\ell'} = n\rangle = \sum_{n=0}^{+\infty} c_n |n, n\rangle, \tag{5D.8}$$

where the coefficients c_n are all positive. It has already been expressed in the Schmidt form (5D.3). The Schmidt number of the state is thus the number of non-zero coefficients c_n. It is entangled whenever at least two of the coefficients are non-zero.

Now consider the photon number operators $\hat{N}_\ell = \hat{a}_\ell^\dagger \hat{a}_\ell$ and $\hat{N}_{\ell'} = \hat{a}_{\ell'}^\dagger \hat{a}_{\ell'}$. They are diagonal relative to the number state basis, of the form (5D.7). There will therefore be *a perfect correlation between measurements of the number of photons in each of these modes*, whence the name attributed to such states. Furthermore, there will be a perfect correlation over all observables that are proportional to the photon number operators, e.g. the momentum operators for twin states like (5D.8) involving two modes of a travelling plane wave, and the orbital angular momentum operators for twin states like (5D.8) involving two Laguerre–Gauss modes.

Note that we also have

$$(\hat{N}_\ell - \hat{N}_{\ell'})|\psi\rangle = \sum_{n=0}^{+\infty} c_n (n-n)|n, n\rangle = 0. \tag{5D.9}$$

The state $|\psi\rangle$ is therefore an eigenstate of the difference in the number of photons, with eigenvalue zero. We thus have $\Delta(N_\ell - N_{\ell'}) = 0$, and so there will be strictly no fluctuations in the signal obtained by taking the difference in the instantaneous values of these two photodetection signals, whereas one would record fluctuations in the numbers of photons at photodetectors measuring the intensity of each mode separately.

Let us now consider the correlations between the quadrature components of each mode, defined in (5.58) and (5.59). A short calculation using (5D.5) and (5D.8) leads to the correlation coefficient,

$$E(E_{Q\ell}, E_{Q\ell'}) = -E(E_{P\ell}, E_{P\ell'}) = \frac{\sum_n n(c_n c_{n-1}^* + c_n^* c_{n-1})}{\sum_n (2n+1)|c_n|^2}. \tag{5D.10}$$

There is therefore a non-zero correlation between the quadrature components Q of the two modes (and an equal anticorrelation between the quadrature components P) whenever two consecutive coefficients in the expansion (5D.8) are non-zero. The magnitude of the correlation depends on the precise dependence of c_n on n. For example, if the c_n form a real geometric series with ratio r, so that $c_n = Ar^n$ for some A, then it follows from (5D.10) that $E(E_{Q\ell}, E_{Q\ell'}) = 2r/(1+r^2)$. The correlation between the quadrature components Q, and the anticorrelation between the quadrature components P, become simultaneously perfect when the ratio r of the series tends to 1.

5D.2.2 Production

Consider the Hamiltonian

$$\hat{H}_{\mathrm{I}} = \mathrm{i}\hbar\xi \left(\hat{a}_\ell^\dagger \hat{a}_{\ell'}^\dagger - \hat{a}_\ell \hat{a}_{\ell'} \right). \tag{5D.11}$$

It describes a non-degenerate parametric interaction, studied in more detail in Chapter 7 on nonlinear optics, in which there is simultaneous generation or destruction of a photon called the signal photon and a photon called the idler photon. When it acts over a time T on the vacuum, this interaction produces the state,

$$|\psi(T)\rangle = \exp(-\frac{\mathrm{i}T\hat{H}_{\mathrm{I}}}{\hbar})|0\rangle, \tag{5D.12}$$

whose exact expression is obtained by expanding the exponential as a series. Since photons are always created in pairs by the interaction Hamiltonian (5D.11), the state thereby produced will necessarily be a twin-photon state of the form (5D.8). A detailed calculation shows that[2]

$$|\psi(T)\rangle = \frac{1}{\cosh R} \sum_n (\tanh R)^n |n, n\rangle, \tag{5D.13}$$

with $R = \xi T$. The number of non-negligible terms in this expansion depends on the value of R, and hence on the efficiency of the parametric interaction, which is in turn largely dependent on the intensity of the pump beam. It is generally only a few units. Note finally that the coefficients c_n do indeed form a geometric series in this case, leading therefore to a very simple expression for the quadrature correlation coefficients:

$$E(E_{Q\ell}, E_{Q\ell'}) = -E(E_{P\ell}, E_{P\ell'}) = \tanh 2R. \tag{5D.14}$$

The correlation tends to 1 when the pump power becomes large.

We shall see in Complement 7A that one can create a device analogous in many ways to a laser by enclosing the parametric medium in a Fabry–Perot cavity for the two modes. In this device, called an optical parametric oscillator, photons are always created in pairs. Above the oscillation threshold, it will produce signal and idler light beams described by a twin-photon state, but this time containing a very high average number of photons.

[2] See, for example, D. F. Walls and G. J. Milburn, *Quantum Optics*, p. 84, Springer (1995).

5D.3 Relation between squeezing and entanglement

5D.3.1 General considerations

The Hamiltonian (5D.11) used to produce the twin-photon state is very similar to the Hamiltonian (5A.33) producing a squeezed state. If we transform (5D.11) by changing the mode:

$$\hat{a}_+ = \frac{1}{\sqrt{2}}(\hat{a}_\ell + \hat{a}_{\ell'}) \tag{5D.15}$$

$$\hat{a}_- = \frac{1}{\sqrt{2}}(\hat{a}_\ell - \hat{a}_{\ell'}), \tag{5D.16}$$

which is the input–output relation for a semi-reflecting mirror, we obtain

$$\hat{H}_\text{I} = \text{i}\frac{\hbar\xi}{2}\left(\hat{a}_+^{\dagger 2} - \hat{a}_-^{\dagger 2} - \hat{a}_+^2 + \hat{a}_-^2\right), \tag{5D.17}$$

i.e. the sum of two squeezing Hamiltonians, acting on each of the new modes defined by (5D.15) and (5D.16). Squeezing and entanglement arise here as two aspects of a single physical phenomenon, and the transformation (5D.17) associated with a semi-reflecting mirror tells us how to switch from one to the other. We shall spell this out by using the input–output relations directly on the operators in order to calculate the correlations.

5D.3.2 Mixing two squeezed states on a semi-reflecting mirror

Consider now two squeezed vacuum states, $|\alpha = 0, R\rangle$ and $|\alpha = 0, -R\rangle$ with $R > 0$ (see Complement 5A), sent to the two input ports $\ell = 1$ and $\ell' = 2$ of a semi-reflecting mirror. The input–output relations of the semi-reflecting mirror are the same for the annihilation operators and for the quadrature operators. Hence, at ports 3 and 4 of Figure 5D.1, we have

$$\hat{E}_{Q3} = \frac{1}{\sqrt{2}}(\hat{E}_{Q1} + \hat{E}_{Q2}) \tag{5D.18}$$

$$\hat{E}_{Q4} = \frac{1}{\sqrt{2}}(\hat{E}_{Q1} - \hat{E}_{Q2}), \tag{5D.19}$$

with analogous relations for the quadrature components P. Since the incoming states are squeezed vacuum states, the expectation values of the quadratures of the incoming and outgoing modes are zero. Using the expressions (5A.10–5A.11) for the quadrature operators in terms of the operators $\hat{A}_R$ and $\hat{A}_R^\dagger$ and the commutation relation (5A.2), we find their variances to be

$$\Delta(E_{Q3})^2 = \Delta(E_{Q4})^2 = \Delta(E_{P3})^2 = \Delta(E_{P4})^2 = \left[\mathcal{E}_\ell^{(1)}\right]^2 \cosh 2R. \tag{5D.20}$$

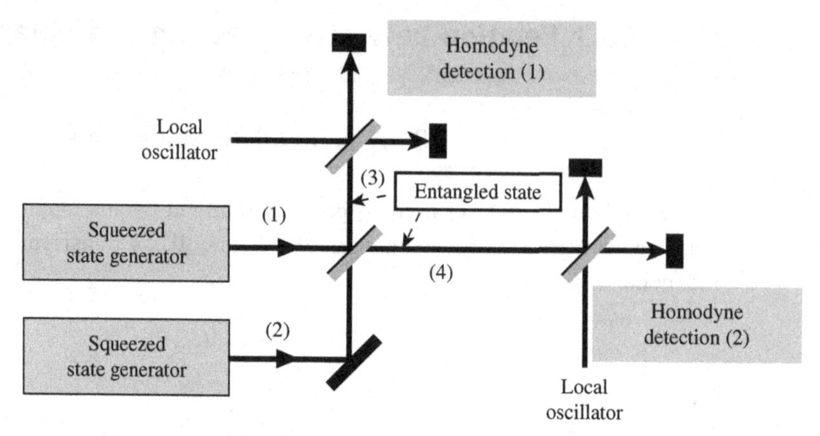

Figure 5D.1 **Generating and detecting entangled two-mode states.**

The mixing of two squeezed states thus produces two beams with greater fluctuations than the vacuum in all its quadrature components, because the outgoing fluctuations depend mainly on the noisy, 'anti-squeezed' quadrature components $E_{P\ell}$ and $E_{Q\ell'}$. In the phasor representation, they are both described by circles centred at the origin and of ever greater diameter as the incoming states are more squeezed. However, these noisy quadrature components are in fact highly correlated because the state describing the two outgoing modes is, as we know, an entangled state. In order to measure the quadrature correlations, experimenters generally use a *double homodyne detection*, as shown schematically in Figure 5D.1. One can then simultaneously measure the quadrature fluctuations E_{Q3} and E_{Q4}, or indeed E_{P3} and E_{P4}, depending on the phase of the local oscillators. By taking the sum and the difference of the instantaneous outgoing photocurrents from the homodyne detections, one obtains the standard deviations $\Delta(E_{Q3} - E_{Q4})$ and $\Delta(E_{P3} + E_{P4})$. An analogous calculation to the last shows that these quantities are given by

$$\Delta(E_{Q3} - E_{Q4}) = \Delta(E_{P3} + E_{P4}) = 2\left[\mathcal{E}_\ell^{(1)}\right]^2 \mathrm{e}^{-2R}. \tag{5D.21}$$

They become very close to zero whenever R is of the order of a few units. This calculation clearly shows that, at the two output ports of the semi-reflecting mirror, there are simultaneously fields with identical quadratures Q and opposite quadratures P.

One might wonder whether the variances found in (5D.21) are compatible with the Heisenberg inequality, because they involve quadrature observables $(\hat{E}_{Q3}, \hat{E}_{P3})$ and $(\hat{E}_{Q4}, \hat{E}_{P4})$ which are not mutually commuting. From the commutators $[\hat{E}_{Q3}, \hat{E}_{P3}] = [\hat{E}_{Q4}, \hat{E}_{P4}] = \left[\mathcal{E}_\ell^{(1)}\right]^2$, it follows that

$$[\hat{E}_{Q3} - \hat{E}_{Q4}, \hat{E}_{P3} + \hat{E}_{P4}] = 0. \tag{5D.22}$$

There is therefore no Heisenberg inequality to forbid $\Delta(E_{Q3} - E_{Q4})$ and $\Delta(E_{P3} + E_{P4})$ from being simultaneously very small, or even zero.

Comments

(i) It is the simultaneous presence of correlation and anticorrelation between two non-commuting observables that is the signature of the entanglement of the state for the given system. In the general case where the system is not necessarily in a pure state, it has been shown that,[3] if the inequality

$$\Delta(E_{Q3} - E_{Q4}) + \Delta(E_{P3} + E_{P4}) < 4\left[\mathcal{E}_\ell^{(1)}\right]^2 \tag{5D.23}$$

holds, then the state of the system is non-separable in the sense of the comment at the end of Section 5D.1.2 of this complement. This is indeed the case in the example of this section whenever R is positive.

(ii) There is a simpler way of producing an entangled two-mode state. The idea is to divide a squeezed state into two on a semi-reflecting mirror, a situation in many ways analogous to the one in which a one-photon state encounters a semi-reflecting mirror and a one-photon entangled state $(|0,1\rangle + |1,0\rangle)/\sqrt{2}$ is produced. The correlation calculation is as before, replacing the incoming squeezed state at the first input port by the vacuum. The correlation coefficients are found to be $E(E_{Q3}, E_{Q4}) = -E(E_{P3}, E_{P4}) = \tanh R$. Consequently, there is still a correlation for one quadrature and an anticorrelation for the other, but less than when two squeezed states are mixed.

5D.3.3 Non-destructive measurement of two complementary variables: the 'EPR paradox'

As mentioned at the beginning of this complement, any correlation between quantities measured on different modes can be used to carry out a non-destructive measurement. In the setup of Figure 5D.1, with two highly squeezed incoming states, a measurement of $\hat{E}_{Q4}$, or of $\hat{E}_{P4}$, can be used to determine the quadrature $\hat{E}_{Q3}$, or $\hat{E}_{P3}$, of mode 3 exactly, without having to insert a detector for this mode, and hence without disturbing it, and a fortiori, without destroying it. By carrying out measurements for mode 4, one thus obtains perfect knowledge, without uncertainty, of two mode 3 physical quantities that happen to be complementary in Bohr's sense, i.e. associated with non-commuting operators, for which there is a Heisenberg inequality:

$$\Delta E_{Q3} \Delta E_{P3} \geq \left[\mathcal{E}_\ell^{(1)}\right]^2. \tag{5D.24}$$

It would therefore seem that, in the situation we are considering here, it should be possible to measure with great accuracy two quantities which do not commute, and that there will be a violation of (5D.24). This paradoxical situation is perfectly analogous to the one considered by Einstein, Podolsky and Rosen ('EPR') in their famous paper, up to replacement of the position and momentum operators of two particles in an entangled state by the quadrature operators of two field modes.

In fact the vanishing variances are those relating to the operators $[\hat{E}_{Q3} - \hat{E}_{Q4}]$ and $[\hat{E}_{P3} + \hat{E}_{P4}]$ which commute, and for which there is no Heisenberg inequality, whereas the 'raw'

[3] L. M. Duan, G. Giedke, I. Cirac and P. Zoller, Inseparability Criterion for Continuous Variable Systems, *Physical Review Letters* **84**, 2722 (2000).

variances ΔE_{Q3} and ΔE_{P3} are very large, as we saw in Section 5D.3.2. When the mode 3 quadratures are measured through information obtained for mode 4, *a selection is made among the set of measurements carried out for this mode*, retaining only those for which a precise value of a mode 4 quadrature has been measured. These are therefore *conditional measurements*. The variances calculated from these selected data are known in statistics as *conditional variances*, or 'scedastic functions', and they are different statistical quantities from the raw variances.

Many experiments have been carried out in this type of situation.[4] Conditional variances have been measured for mode 3 given mode 4, to yield experimental values of the product of the conditional variances of the two quadratures that are much smaller than the Heisenberg 'limit' (5D.24). As in the last complement, we have once again a situation where a thought experiment conceived of by the founders of quantum mechanics has become a real experiment that can be carried out in the laboratory.

[4] See, for example, M. D. Reid, P. D. Drummond, E. G. Cavalcanti, W. P. Bowen, P. K. Lam, H. A. Bachor, U. L. Andersen and G. Leuchs, The Einstein–Podolsky–Rosen paradox: from concepts to applications, *Review of Modern Physics* **81**, 1727 (2009).

5E Complement 5E Quantum information

The work of John Bell in the mid 1960s and experiments carried out to test his famous inequalities in the following decades have led to a detailed re-examination of the concepts of quantum mechanics, and revealed the full importance of the notion of entanglement. This reconsideration helped to generate the new and extremely rich field of research known as quantum information in the 1980s.[1] The guiding idea behind this field of activity is that, by exploiting the specific rules of quantum physics, one can conceive of new ways of calculating and communicating, in which the rules of play are no longer the well-known classical rules.

One can thus develop new methods of cryptography in which the message is protected by the basic principles of quantum mechanics, and new computation methods that can be exponentially more efficient than classical algorithms. Quantum information is not therefore a mere sideline for physics, but concerns information theory, algorithmics and the mathematics of complexity theory. This research has already led to proposals for new algorithms and new computation architectures based on quantum logic gates with no classical equivalent. Still on the fundamental level, the meeting of information theory and quantum mechanics which lies at the heart of quantum information has led to a very stimulating regeneration of the theoretical tools used on both sides. One can thus envisage new approaches to the fundamental principles of quantum theory, associated with new ways of defining and processing information. Recalling the well-known remark due to Rolf Landauer, according to which information is of its very nature physical, it is no surprise to find that quantum mechanics, which underlies the whole of modern physics, should reveal such close ties with information theory.

5E.1 Quantum cryptography

5E.1.1 From classical to quantum cryptography

A first example of the use of these ideas is quantum cryptography. Quite generally, the aim of cryptography is to transmit a secret message from a sender (Alice) to a recipient (Bob),

[1] For more details, the reader may consult M. A. Nielsen and I. L. Chuang, *Quantum Computation and Quantum Information*, Cambridge University Press (2000).

while minimizing the risks of a spy (named Eve, by reference to eavesdropping) being able to intercept and decode the message. Cryptography has long played an essential role in the protection of commercial and military secrets, but it has recently become relevant to the public in general, with the development of electronic message systems, from credit cards to internet shopping. Classical cryptography usually uses sophisticated coding systems that cannot be broken in any reasonable length of time given currently available means of computation. The level of security is thus acceptable, but not absolute, because it does depend on the means available to one's opponents. Moreover, it cannot generally be demonstrated mathematically.

However, there is a simple method of cryptography that is 'unconditionally certain' from the mathematical standpoint, based on the idea that Alice and Bob may have previously exchanged a secret key, i.e. a long sequence of random characters known only to them. If this key is as long as the message, and if it is used only once, then the absolute safety of the code follows from a mathematical theorem proved by Claude Shannon in 1948. By virtue of this theorem, the level of security of the message thus reduces to the care with which the key has been communicated. And this is where quantum cryptography comes in, for it allows Alice and Bob to exchange a secret key with absolute security guaranteed by the very principles of quantum physics!

5E.1.2 Quantum cryptography with entangled photons

There are already many quantum cryptography 'protocols', and we shall discuss here a method due to Artur Ekert, who uses entangled photon pairs as in the experimental tests of the Bell inequalities.[2] This will show that what is transmitted in such an experiment is not a message, but a sequence of correlated random numbers, i.e. precisely what one means by a secret key! According to the principle discussed above, this key could then be used to encode the 'true' message, with mathematically demonstrated security.

Suppose therefore that Alice and Bob share polarization-entangled photon pairs (see Complement 5C). Alice and Bob can arbitrarily choose the measurements they will carry out on their photon, but the spy (Eve) cannot yet know these measurements when the photon is transmitted through the line. In addition, as we shall see in the next section, Eve cannot 'clone' the photon which reaches her, i.e. she cannot make an identical copy. In fact, we shall see that any attempt by Eve to intercept the photon will perturb its state and create transmission errors that Alice and Bob will be able to detect. On the other hand, when there is no transmission error, Alice and Bob will know that there is no spy on the line.

More precisely, Alice and Bob agree to carry out their measurements for four linear polarization states, oriented along a horizontal axis denoted by $|h\rangle$, a vertical axis $|v\rangle$, an axis at 45° to the right $|d\rangle$, and an axis at 45° to the left $|g\rangle$. The orthogonal states $|h\rangle$ and $|v\rangle$ are easy to distinguish because they give results + and − for the same orientation of the polarizer, called the *hv* basis. Likewise, the states $|d\rangle$ and $|g\rangle$ give results $+$ and $-$ when

[2] A. K. Ekert, Quantum Cryptography Based on Bell's Theorem, *Physical Review Letters* **67**, 661 (1991).

the polarizer is oriented at 45° to hv, in the orientation called the dg basis. On the other hand, the bases hv and dg are said to be 'incompatible', meaning that, if the polarization is known in one of the bases, it is completely random in the other, since we have

$$|d\rangle = \frac{1}{\sqrt{2}}(|h\rangle + |v\rangle), \quad |g\rangle = \frac{1}{\sqrt{2}}(|h\rangle - |v\rangle), \tag{5E.1}$$

$$|h\rangle = \frac{1}{\sqrt{2}}(|d\rangle + |g\rangle), \quad |v\rangle = \frac{1}{\sqrt{2}}(|d\rangle - |g\rangle). \tag{5E.2}$$

Suppose that an experimental device produces the entangled state[3]

$$|\psi_{\mathrm{AB}}\rangle = \frac{1}{\sqrt{2}}(|hh\rangle + |vv\rangle). \tag{5E.3}$$

The part A of this state is sent to Alice and the part B to Bob. It is easy to see using (5E.1) that this state can also be written,

$$|\psi_{\mathrm{AB}}\rangle = \frac{1}{\sqrt{2}}(|dd\rangle + |gg\rangle). \tag{5E.4}$$

The form (5E.3) of the state shows that, if Alice obtains the measurement result h (or v), then Bob will also obtain the result h (or v) without uncertainty. The form (5E.4) for the same state shows that this will also be true using the dg basis. On the other hand, suppose that Alice uses the hv basis and Bob the dg basis. Then if Alice measures h, for example, the state received by Bob will be $|h\rangle = 1/\sqrt{2}(|d\rangle + |g\rangle)$ by reduction of the wave packet, and Bob will have a 50% chance of measuring d and a 50% chance of measuring g. The same will happen if Alice measures g. Their measurements will then be totally decorrelated.

After receiving the photons, Bob discloses the full set of choices he has made for the measurement axes, hv or dg, as well as a fraction of the results, + or −. When Alice examines these results, she can detect the presence of a spy if there is one, by the following reasoning. The spy Eve does not know any better than Bob what orientation hv or dg she has chosen to measure the polarization of each photon she has received. Suppose therefore that Eve aligns her polarizer arbitrarily with hv or dg and, at each detection, re-emits a photon in exactly the same polarization state as the one she has just measured. For example, if she chooses hv and measures +, she re-emits a photon polarized along h to Bob. But this intervention by Eve can in fact be detected, because she thereby introduces errors into the detection by Bob.

Consider for example the case depicted in the second column of Figure 5E.1, where Alice has detected a d photon, and Bob has also aligned his polarizer with the dg basis, but where Eve has aligned hers with the hv basis. Eve will then measure + with probability 1/2 and − with probability 1/2. Depending on her result, she then sends Bob a photon in the state h or v. In either case, when his polarizer is aligned with the dg basis, Bob can measure + (d) with probability 1/2 and − (g) with probability 1/2. On the other hand, if

[3] This is the same state as (5C.15), written in different notation.

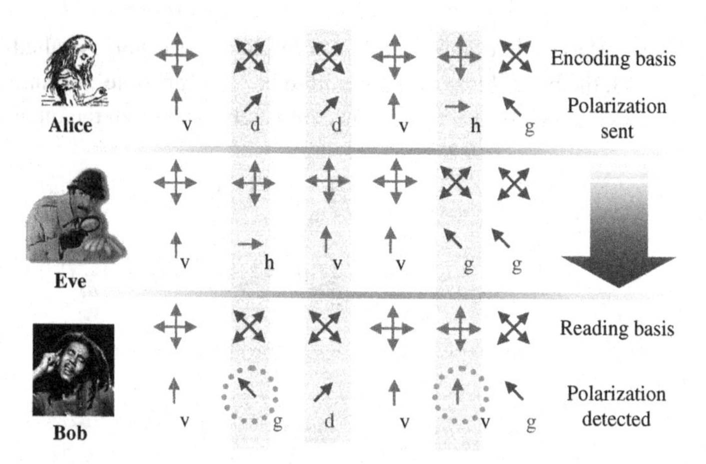

Figure 5E.1 **Quantum cryptography protocol. Only those situations where Alice and Bob choose the same basis are illustrated. This happens half the time on average. In the second column, Alice has detected a photon *d*, but Bob has detected a photon *g*. This error has been caused by the eavesdropper Eve, who detected and re-emitted a photon *h*.**

Eve had not been interfering, Bob would have measured d with unit probability. *The spy's interference thus introduces errors in 25% of the cases*, whence Alice and Bob can detect it and stop transmission.

5E.1.3 From theory to practice

There will always be errors in the transmission line, usually due to technical imperfections. Being cautious, Alice and Bob ought to attribute all errors to a potential spy. One might then conclude that the line will be unusable, but this is not so. In fact, Alice and Bob will first assess the level of errors in the line, using 'test' data that Bob makes public. It can then be shown that, knowing this error level, quantitative bounds can be obtained for the amount of information extracted by Eve.[4] The smaller the error level, the smaller the amount of information available for Eve from the transmitted photons.

The sequence of bits exchanged and not divulged by Bob will then be used to form the secret cryptographic key. In order to use such a key, Alice and Bob must eliminate the errors it may have contained, using standard error detection protocols. Then, depending on the level of errors they have ascertained, they will further reduce the number of useful bits to obtain a final key, which is smaller, but totally unknown to Eve. A higher and higher error level on the line will not endanger the security of this final key, but merely reduce its length. It can be shown that, if the error level in the photons measured is greater than 11%, the correction protocol will in the end yield no secret bit. On the other hand, for an error level below 11%, Alice and Bob will be able to produce a perfectly safe key, with no

[4] N. Gisin *et al.*, Quantum Cryptography, *Review of Modern Physics* **74**, 145 (2002).

errors. The existence of technical errors, typically producing error levels of a few per cent, would not therefore prevent transmission of a secret key.

5E.1.4 The no-cloning theorem

We have assumed that Eve could choose the orientation of her polarizer in an arbitrary manner for each photon, then re-emit to Bob a photon polarized in accordance with her measurement result. One might wonder whether this is the best strategy for avoiding detection. In particular, if she could 'clone' (or duplicate) the photon without modifying its polarization state, she could then send one of the two clones to Bob, while keeping the other in order to carry out her own measurement. This kind of espionage would, indeed, be undetectable. However, the cloning of an unknown state is (fortunately for Alice and Bob) impossible in quantum mechanics, as demonstrated by W. K. Wooters and W. H. Zurek.[5] Indeed, there is no reliable way to produce one or more copies of a quantum state, unless that state is at least partly known beforehand.

To prove this result, let $|\alpha_1\rangle$ be the original quantum state to be duplicated. The system on which the copy must be 'imprinted', so to speak, is initially in a known state $|\phi\rangle$ (the reader may think of it as a sheet of white paper in a photocopier). The evolution of the combined original/copy system during the cloning operation must, therefore, be

$$\text{Cloning} : |\text{original} : \alpha_1\rangle|\text{copy} : \phi\rangle \longrightarrow |\text{original} : \alpha_1\rangle|\text{copy} : \alpha_1\rangle. \tag{5E.5}$$

This evolution is governed by a Hamiltonian that we shall not need to specify here, but which does not depend on $|\alpha_1\rangle$, since this is assumed to be unknown. For another state of the original $|\alpha_2\rangle$ (orthogonal to $|\alpha_1\rangle$), we must also have,

$$\text{Cloning} : |\text{original} : \alpha_2\rangle|\text{copy} : \phi\rangle \longrightarrow |\text{original} : \alpha_2\rangle|\text{copy} : \alpha_2\rangle. \tag{5E.6}$$

The impossibility of cloning then arises for the initial state,

$$|\alpha_3\rangle = \frac{1}{\sqrt{2}}\left(|\alpha_1\rangle + |\alpha_2\rangle\right). \tag{5E.7}$$

If the copying operation works for this state, we should find

$$\text{Cloning} : |\text{original} : \alpha_3\rangle|\text{copy} : \phi\rangle \longrightarrow |\text{original} : \alpha_3\rangle|\text{copy} : \alpha_3\rangle. \tag{5E.8}$$

However, linearity of the Schrödinger equation requires, by linear combination of (5E.5) and (5E.6),

$$\begin{aligned} |\text{original} : \alpha_3\rangle|\text{copy} : \phi\rangle \longrightarrow (1/\sqrt{2})\Big(&|\text{original} : \alpha_1\rangle|\text{copy} : \alpha_1\rangle \\ &+ |\text{original} : \alpha_2\rangle|\text{copy} : \alpha_2\rangle\Big). \end{aligned} \tag{5E.9}$$

But this final entangled state is not the one desired, i.e. (5E.8). By examining this proof, we can establish exactly what quantum mechanics brings to cryptography. If we restrict

[5] W. K. Wooters and W. H. Zurek, A Single Quantum Cannot be Cloned, *Nature* **299**, 802 (1982).

transmission to two orthogonal states $|\alpha_1\rangle = |h\rangle$ and $|\alpha_2\rangle = |v\rangle$, a spy can remain undetected, as we have seen in the previous sections. The operations (5E.5) and (5E.6) are possible, simply by measuring the polarization of the photon in the hv basis and then re-emitting it in the same state. It is the fact of using at the same time both the states $|\alpha_1\rangle$ and $|\alpha_2\rangle$ and linear combinations of them, such as $|\alpha_3\rangle = |d\rangle$ and $|\alpha_3\rangle = |g\rangle$, that sets quantum cryptography apart, preventing any reliable duplication of a message intercepted by a spy.

5E.1.5 And if there were no entangled states? The BB84 protocol

The above arguments are based on the idea of sharing entangled photon pairs between Alice and Bob. In fact the source of the photon pairs can be located with Alice, so that Eve and Bob only have access to the second photon of the pair. It can then be shown that nothing changes either for Eve or for Bob if Alice does not use this source, but simply sends Bob a single photon polarized in one of the four directions h, v, d or g chosen at random. In such a scheme, entanglement seems to play no role but, in fact, it can be shown that the security of the cryptographic protocol depends on the ability of the channel to transmit entanglement. The error threshold that prevents successful production of a secret key is the same as that at which the entanglement of the photon pair would be destroyed by Eve's intervention. In a situation where Alice simply sends a polarized photon to Bob, the equivalent of the protocol we have just discussed was proposed by Charles Bennett and Gilles Brassard in 1984, and is known by the name of 'protocol BB84'.

Initially, security proofs used true single-photon wave packets (see Section 5.4.3 and Complement 5B). However, it was shown later that one can be more tolerant of the state sent by Alice, and that weak quasi-classical pulses (see Section 5.4.2) are also acceptable. But with such pulses, the maximum bit rate for secure transmission is much lower, since one must have a small fraction of pulses with more than one photon, and many of the cases then have no photon at all (Equation 5.78 with $|\alpha|^2 \ll 1$). In fact, this can be improved by using weak pulses with randomly modulated average numbers of photons. This is known as the decoy state protocol.

One can also show that photon counting techniques are not necessarily required, and that one can instead apply homodyne detection techniques, which use a 'local oscillator' as in coherent optical telecommunications. Such techniques, known as 'continuous variable quantum key distribution', have been developing rapidly, and now reach levels of secrecy and bit rates comparable to their photon counting analogues.

So what is the exact role of entanglement in quantum cryptography? Entanglement plays a crucial conceptual role in security proofs, due to the equivalence already quoted above. In addition, it has been suggested that an ideal test of Bell's inequalities, i.e. without loophole (see Complement 5C), would be an ultimate qualifying test for cryptographic security.

5E.1.6 Experimental results

Many laboratories around the world have carried out experiments demonstrating the efficiency of quantum cryptography over propagation distances up to a hundred kilometres or so. These experiments were done either in the visible range by propagation through the atmosphere, or at the commercial telecommunications wavelength of 1.55 μm using optical fibres. One can even find commercial quantum cryptography systems.

5E.2 Quantum computing

5E.2.1 Quantum bits or 'qubits'

We saw in the last section that a bit of information (0 or 1) can be coded by two orthogonal states of a polarized photon. But what happens in terms of information content if the photon is placed in a quantum linear combination of these two states? Heuristically, one might say that the bit is equal to neither 0 nor 1, but rather some linear superposition of these two values. In order to account for this possibility, one introduces the notion of 'qubit', short for quantum bit. In contrast to the classical bit, this allows for the possibility of such linear superpositions. We shall see that this idea has important implications when we come to consider quantum computers, based on the manipulation of large numbers of qubits.

We shall take a highly simplified definition of a computer, considering it as a machine able to carry out operations on sets of N bits called registers. The content of a register is a binary word, representing a number memorized by the computer. For $N = 3$, there are therefore $2^3 = 8$ possible words, i.e. the triplets 111, 110, 101, 011, 100, 010, 001 and 000. Consider now a q-register, made up of a set of N qubits. The 2^N possible states of the classical register will define a basis for the state space of the q-register, which can be placed in an arbitrary linear combination of all the states in the basis,

$$\begin{aligned} |\psi\rangle = c_0\,|000\rangle + c_1\,|001\rangle + c_2\,|010\rangle + c_3\,|011\rangle \\ + c_4\,|100\rangle + c_5\,|101\rangle + c_6\,|110\rangle + c_7\,|111\rangle. \end{aligned} \tag{5E.10}$$

Suppose now that the computer calculates something, i.e. carries out an operation on the state of the q-register. Since the operation is carried out on a linear superposition of states, one can consider it as being done in 'parallel' on the 2^N classical numbers. In contrast, a classical computer would have to carry out the 2^N operations one by one. This notion of quantum parallelism implies a gain in the efficiency of the computer, which can in principle be exponential if the 2^N calculations corresponding to N qubits are indeed carried out simultaneously.

This immediately raises several questions. On the fundamental level, what type of calculations and what type of algorithms can be carried out with such a device? And on the practical level, how could such a device actually be implemented?

5E.2.2 The Shor factorization algorithm

In the last section, we alluded to non-quantum cryptographic systems, often called algorithm protocols. One of these protocols exploits the fact that certain mathematical operations are very easy to carry out in one direction, but much harder to carry out in the other. For example, it is easy and fast for a computer to find the product of two numbers, but it is generally much more difficult to decompose a number into its prime factors. For example, if we consider the product P of two large prime numbers, approximately $\sqrt{P}$ divisions must be carried out in order to identify the factors. The computation time is then of the order of $e^{a \log P}$, i.e. it increases exponentially with the number of figures (or bits) in P, and this soon becomes dissuasive. The 'RSA' cryptographic method inspired by this observation and first proposed by Rivest, Shamir and Adelman, is currently very widely used for credit cards, internet transactions and so on, being considered particularly safe.

It is not difficult therefore to imagine the impact of a paper published in 1994 by Peter Shor, claiming that a quantum computer could factorize the product P of two prime numbers in a time of the order of $(\log P)^3$, i.e. an exponential factor shorter than could be achieved by a classical computer! Now that the initial storm has blown over, the situation seems to be as follows. The algorithm put forward by Shor is correct in principle, and it would indeed achieve the stated gain in efficiency. However, it seems that the construction of a competitive quantum computer remains beyond the reach of current technology, even though it is not disallowed by the laws of physics. The design of a quantum computer would thus appear to be a long-term aim for science, rather than an immediate threat to algorithmic cryptographic systems. In the meantime, this topic has led to a deeper understanding of the differences between classical and quantum calculations.

Calculations are carried out by means of algorithms, which mathematicians classify in terms of difficulty. This is the subject of complexity theory. This classification suggests that the difficulty of a problem is a property of the problem, rather than a property of the machine that is doing the calculation. And this is indeed the case, as long as the physics governing the calculator remains unchanged. For example, factorization is a 'difficult' problem for all classical computers. This just means that the time required to do the calculation rises very quickly (in fact, exponentially) when one needs to factorize bigger and bigger numbers. But Peter Shor showed in 1994 that this problem becomes an 'easy' one for the quantum computer. The simple fact of changing the basic physics governing the computer allows one to design a new algorithm, which is exponentially faster in this particular case.

Shor's algorithm exploits a subtle method. The naive algorithm which simply carries out divisions is in any case extremely inefficient, and there are far superior classical algorithms. For example, a 155-figure number can be factorized in just a few months of distributed computation using classical methods. But Shor's algorithm uses a different method, based on number theory. Indeed, there is a theorem which says that, to factorize a number P, one can construct a simple function F of P and the integer variable n which is periodic in n. More precisely, one has $F(P, n) = a^n \bmod P$, where a is an integer coprime with P. If the period of F is known, call it R, it is easy to obtain the factors of P, which are given by the greatest common divisor of n and $a^{R/2} + 1$ and of n and $a^{R/2} - 1$. This argument is classical, but in order to find R, the function F must be evaluated many times over, and this makes the calculation rather inefficient, hence 'difficult' for the classical computer.

On the other hand, the quantum computer evaluates all the values taken by F in parallel, for all values of n which the registers can contain. One then carries out several operations (projective measurement, Fourier transform, etc.) to obtain a random number, the 'result' of the calculation. By repeating the calculation, this produces a sequence of numbers, which are not in fact completely random. By analysing the regularities in the sequence, the period R can be extracted and the original number factorized. It is clear from this that the quantum computer functions here in a rather particular way. It does not really produce a 'result' in the usual sense, but rather a clue that can be used to obtain the required result exponentially faster.

5E.2.3 Working principle of a quantum computer

We shall attempt here to give some intuitive idea of how a quantum computer might carry out a calculation, taking the example of the Shor algorithm. The basic idea is that the calculation must be reducible to a quantum evolution of some initial entangled state, followed by a 'measurement' which determines the state of the q-register but at the same time interrupts its evolution. In accordance with the principles of quantum mechanics, the value obtained will be associated with one of the eigenstates of the measured observable, which corresponds here to a classical state of the register, i.e. a binary word. On the other hand, the evolution of the computer during the calculation itself will have involved, in parallel, the 2^N states corresponding to all the numbers the register can contain.

To be able to carry out successive operations, the system of qubits must be made to evolve in a controlled way, under the action of a clock which determines the rate of the calculation. At first sight, the determination of this evolution seems to be an inextricable problem if one hopes to carry out a non-trivial calculation. But in fact, it can be shown that the construction can be rather simply achieved, because any calculation can be decomposed into a series of simple operations affecting only one or two qubits. As in classical information theory, these simple operations are carried out by 'logic gates', the classically well-known examples being NOT, AND and OR gates, and so on. However, the *quantum logic gates* required by Shor's algorithm have certain peculiar features:

(1) They must be 'reversible', in order to be compatible with the quantum evolution of the q-register.
(2) They must manipulate qubits, on which it must be possible to carry out certain logical operations that would be inconceivable classically.

Simple examples of quantum logic gates are $\sqrt{1}$ and $\sqrt{\text{NOT}}$, which must be applied twice to obtain the identity (1 gate), or to swap 0 and 1 (NOT gate). These two gates in fact prepare a qubit in a linear superposition of 0 and 1 with equal weights. For example, two applications of the gate $\sqrt{1}$, also called the Hadamard gate, carry out the successive transformations:

$$|0\rangle \longrightarrow \frac{1}{\sqrt{2}}(|0\rangle + |1\rangle) \longrightarrow \frac{1}{2}((|0\rangle + |1\rangle) + (|0\rangle - |1\rangle)) = |0\rangle$$

$$|1\rangle \longrightarrow \frac{1}{\sqrt{2}}(|0\rangle - |1\rangle) \longrightarrow \frac{1}{2}((|0\rangle + |1\rangle) - (|0\rangle - |1\rangle)) = |1\rangle.$$

Another very important gate is the 'C–NOT' (controlled NOT) gate, which is a two-qubit gate carrying out the following operation:

$$|0\rangle\ |0\rangle \longrightarrow |0\rangle\ |0\rangle$$

$$|0\rangle\ |1\rangle \longrightarrow |0\rangle\ |1\rangle$$

$$|1\rangle\ |0\rangle \longrightarrow |1\rangle\ |1\rangle$$

$$|1\rangle\ |1\rangle \longrightarrow |1\rangle\ |0\rangle.$$

This gate keeps the first qubit unchanged and produces an 'exclusive OR' on the second qubit. One can also say that the second qubit is switched (NOT gate) if the first is in state $|1\rangle$, so that this is indeed a controlled NOT gate. Now consider the effect of applying a Hadamard gate to the first qubit, followed by a C–NOT gate:

$$|0\rangle\ |0\rangle \longrightarrow \frac{1}{\sqrt{2}}(|0\rangle\ |0\rangle + |1\rangle\ |0\rangle) \longrightarrow \frac{1}{\sqrt{2}}(|0\rangle\ |0\rangle + |1\rangle\ |1\rangle)$$

$$|0\rangle\ |1\rangle \longrightarrow \frac{1}{\sqrt{2}}(|0\rangle\ |1\rangle + |1\rangle\ |1\rangle) \longrightarrow \frac{1}{\sqrt{2}}(|0\rangle\ |1\rangle + |1\rangle\ |0\rangle)$$

$$|1\rangle\ |0\rangle \longrightarrow \frac{1}{\sqrt{2}}(|0\rangle\ |0\rangle - |1\rangle\ |0\rangle) \longrightarrow \frac{1}{\sqrt{2}}(|0\rangle\ |0\rangle - |1\rangle\ |1\rangle)$$

$$|1\rangle\ |1\rangle \longrightarrow \frac{1}{\sqrt{2}}(|0\rangle\ |1\rangle - |1\rangle\ |1\rangle) \longrightarrow \frac{1}{\sqrt{2}}(|0\rangle\ |1\rangle - |1\rangle\ |0\rangle).$$

The two qubits, which were initially in one of the four possible factorized states, end up in one of the four entangled states! These final states are called 'Bell states', and form a 'maximally entangled' basis for the state space of the two qubits. It is interesting to note that the opposite operation, i.e. applying the C–NOT gate and then the Hadamard gate, 'disentangles' the two qubits and thereby allows one to identify the four 'Bell states'. This inverse operation is called a Bell measurement. A simple but spectacular demonstration of these tools is quantum teleportation of a qubit, as described below.

As we have already mentioned, an arbitrary calculation can be decomposed into a sequence of applications of one-qubit and two-qubit gates. It might be thought that it would suffice to have the computer evolve into a single-component state, which would be the result of the calculation. Unfortunately, very few algorithms are accessible to such a simple manipulation. The final state of the computer is usually a linear superposition, and the result obtained is thus random. For example, if we consider Shor's algorithm, the result should rather be considered as a clue pointing the way to factorization. It is easy to check by conventional methods whether the result is the right one, and repeat the calculation in case of failure. Peter Shor showed that this trial and error procedure leads to the right answer with a probability arbitrarily close to unity, by carrying out a number of trials that increases linearly – rather than exponentially – with the number of figures in the number to be factorized.

5E.2.4 Practical matters

The idea of the quantum computer is thus compatible with the laws of physics, and such a computer does seem to be feasible, at least as long as one considers simple calculations, involving only a small number of gates. But in the case of large calculations, the global state of the computer will be a linear superposition of a very large number of states, i.e. an entangled state, whose evolution must be controlled while preserving all the properties of the linear superposition. It is not obvious that this type of system could actually be achieved in practice. Current research aims in particular at the following points:

(1) On the one hand, the evolving q-register must be extremely well isolated from the outside environment. Any coupling with the environment would otherwise induce '*decoherence*', likely to scramble the quantum superpositions. This requirement is only compatible with the idea of controlling the evolution of the calculation for extremely well-understood systems.

(2) On the other hand, in order to withstand the effects of residual perturbations, 'correction codes' must be designed to reset the computer in the state it had before the perturbation. It has been shown theoretically that such quantum correction codes can be devised in principle, and could ensure that the computer continues to operate without error, provided that the level of error per operation is not too high and that the calculation has a high enough level of redundancy.

These two lines of research, i.e. design of a system with very low decoherence, and preparation of suitable quantum correction codes, have stimulated both experimental quantum physics and algorithmics. It is difficult to predict the outcome of such research, but it seems perfectly plausible to think that simple quantum algorithms will find applications, e.g. in quantum cryptography.

There are many experimental studies aiming to implement and manipulate qubits in a wide range of physical systems. It is easy to manipulate photons, and to get them to propagate, as we have seen in the example of quantum cryptography, but they do not lend themselves well to the construction of logic gates. Systems that have been widely studied are ions, trapped atoms, cavity quantum electrodynamics (see Complement 6B), superconducting junctions, quantum dots and others. Experiments have been concerned with q-registers involving only a very limited number of qubits, because decoherence, which destroys the quantum superpositions required for the operation of the quantum computer, increases extremely fast with the number of qubits. Only experience will tell whether these difficulties can be overcome.

5E.3 Quantum teleportation

While quantum computing remains a long-term objective, there are already several extremely interesting devices for *quantum information processing*, which are less difficult to achieve experimentally. The most emblematic of these is known as 'quantum teleportation', which we shall discuss briefly here.

As we have already seen, it is impossible to clone or perfectly determine the state of a single qubit if nothing is known about its state. But can this unknown state be 'transferred over a large distance'? Interestingly, the answer here is affirmative, and this is the subject of quantum teleportation. The reader should note that it is not a superluminal transfer of matter of the kind so often invoked in *Star Trek*, but rather the transfer through classical and quantum communication channels of the unknown quantum state of a quantum object to another quantum object similar to the first and located some distance away. In order to satisfy the no-cloning theorem, the state of the initial quantum object is necessarily destroyed in the operation.

The teleportation of the state of some qubit A to a target qubit C can be considered as a quantum algorithm involving the following steps:

(1) Alice starts with 3 qubits: qubit A in an arbitrary state $|\psi_A\rangle = \alpha|0_A\rangle + \beta|1_A\rangle$ which Alice does not know, the target qubit C and an auxiliary qubit B. The qubits B and C are initially in state $|0\rangle$.

(2) Alice entangles qubits B and C by the method discussed above, using a Hadamard gate and a C–NOT gate, then keeps qubit B and sends qubit C to Bob.

(3) Alice then carries out a Bell measurement on qubits A and B (see above). This measurement 'projects' the pair (A, B) into one of the four Bell states, which is the result of the measurement. The state of qubit A is destroyed in this operation.

(4) Alice transmits the result of her measurement to Bob, that is, she tells him which of the four Bell states she has obtained. Bob thus has qubit C and two classical bits of information ($m = 0, 1, 2$ or 3). Bob then applies a transformation to C which depends on the value of m, and reconstitutes the state of qubit A by the calculation specified below.

The sequence of operations carried out on the state $|\psi_{ABC}\rangle$ of the three qubits is thus (the calculation carried out in stage 2 is essential here!):

$$\text{Stage 1:} \quad |\psi_{ABC}\rangle_{\text{initial}} = (\alpha\ |0_A\rangle + \beta\ |1_A\rangle)\ |0_B\rangle\ |0_C\rangle$$

$$\text{Stage 2:} \quad (\alpha\ |0_A\rangle + \beta\ |1_A\rangle)\ \frac{1}{\sqrt{2}}(|0_B 0_C\rangle + |1_B 1_C\rangle) =$$

$$\frac{1}{\sqrt{2}}(\alpha|0_A 0_B 0_C\rangle + \alpha|0_A 1_B 1_C\rangle + \beta|1_A 0_B 0_C\rangle + \beta|1_A 1_B 1\rangle) =$$

$$\frac{1}{2^{3/2}}\left(|0_A 0_B\rangle + |1_A 1_B\rangle\right)\left(\alpha\,|0_C\rangle + \beta\,|1_C\rangle\right) +$$

$$\frac{1}{2^{3/2}}\left(|0_A 0_B\rangle - |1_A 1_B\rangle\right)\left(\alpha\,|0_C\rangle - \beta\,|1_C\rangle\right) +$$

$$\frac{1}{2^{3/2}}\left(|0_A 1_B\rangle + |1_A 0_B\rangle\right)\left(\alpha\,|1_C\rangle + \beta\,|0_C\rangle\right) +$$

$$\frac{1}{2^{3/2}}\left(|0_A 1_B\rangle - |1_A 0_B\rangle\right)\left(\alpha\,|1_C\rangle - \beta\,|0_C\rangle\right)$$

Stage 3: State measured by Alice	Action taken by Bob on qubit C
$\frac{1}{\sqrt{2}}(\lvert 0_A 0_B\rangle + \lvert 1_A 1_B\rangle)$	Nothing
$\frac{1}{\sqrt{2}}(\lvert 0_A 0_B\rangle - \lvert 1_A 1_B\rangle)$	PI gate
$\frac{1}{\sqrt{2}}(\lvert 0_A 1_B\rangle + \lvert 1_A 0_B\rangle)$	NOT gate
$\frac{1}{\sqrt{2}}(\lvert 0_A 1_B\rangle - \lvert 1_A 0_B\rangle)$	NOT and PI gates.

The NOT gate (bit inversion) has already been defined, and the PI gate (phase inversion) achieves the operation $|0\rangle \to |0\rangle$, $|1\rangle \to -|1\rangle$.

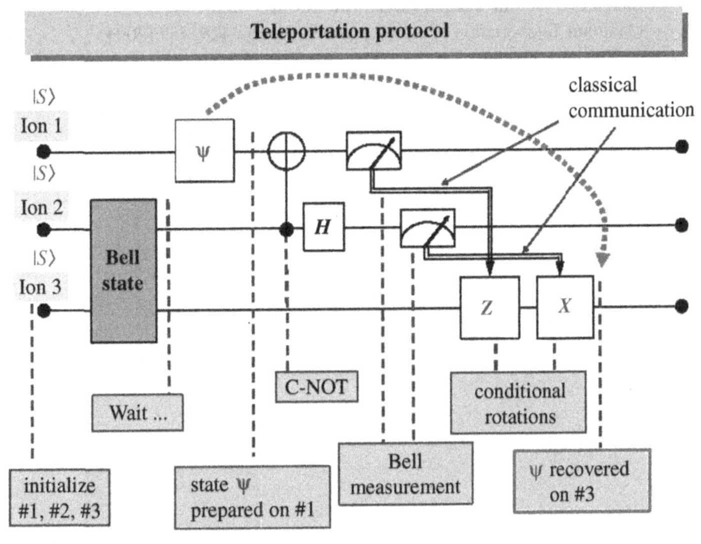

Figure 5E.2 Quantum teleportation protocol using trapped ions. The figure shows the various steps in the algorithm, implemented one by one by acting on three qubits encoded in terms of the quantum state of individual ions a few micrometres apart in a trap.

By examining the expression for the state given at the end of stage 2, we find that, in every case, qubit C ends up in the initial state $|\psi_A\rangle$ of qubit A. It is very important to observe that neither Bob nor Alice has access to $|\psi_A\rangle$. All the operations carried out above are done blind, on a qubit whose state remains unknown to them. This possibility of acting on a qubit without knowing its state plays a fundamental role in quantum calculation, and it underlies the design of quantum error correction codes.

Quantum teleportation has been achieved experimentally with entangled states of photons and trapped ions (Figure 5E.2).[6]

5E.4 Conclusion

This complement has given an overview of some examples in which the laws of quantum mechanics open up radically new prospects for data processing and transmission. The study of these possibilities began with theoretical proposals, and proceeded with experiments to demonstrate them on model systems. The next step, which is the manufacture of practical and commercially viable devices, faces major technological challenges. Practical realizations exist in the case of quantum cryptography, but they look a long way off as far as the quantum computer and quantum teleportation of complex systems are concerned. Whatever the issue, concepts and techniques developed in this type of research lead to innovative physical ideas and applications, of an unprecedented kind.

[6] D. Bouwmeester *et al.*, Experimental Quantum Teleportation, *Nature* **390**, 575 (1997); M. Riebe *et al.*, Deterministic Quantum Teleportation with Atoms, *Nature* **429**, 734 (2004); M. D. Barrett *et al.*, Deterministic Quantum Teleportation of Atomic Qubits, *Nature* **429**, 737 (2004).

6 Interaction of an atom with the quantized electromagnetic field

In this chapter, we discuss a purely quantum approach to the interaction between an atom and the electromagnetic field. In this treatment, the atom and electromagnetic field form a single quantum system, whose evolution is handled globally within a unified formalism. This will thus have the merit of being fully consistent from the theoretical standpoint. But the main advantage in an entirely quantum approach is that it can treat the full range of matter–radiation interaction phenomena. In particular, it provides a rigorous description of the *spontaneous emission of light by an excited atom*, something that falls outside the scope of the semi-classical framework applied in Chapters 2 and 3, where the lifetime of an excited atomic state had to be fed in phenomenologically. It can describe other phenomena of the same type, such as *parametric fluorescence* by a nonlinear crystal subjected to pumping radiation (see Chapter 7), which underlies many recent developments in quantum optics. The fully quantum approach also has the merit of allowing a *simple interpretation in terms of photons* for the various matter–radiation interaction processes, such as absorption, stimulated emission, scattering, and also the basic processes of nonlinear optics. Indeed, it provides a unified framework for both stimulated and spontaneous processes. Finally, it can be used to tackle completely new situations where matter and radiation interact, which lie outside the scope of any semi-classical description, such as *cavity quantum electrodynamics* or the *production of single-photon or entangled-photons states*.

Section 6.1 deals with classical electrodynamics, i.e. the classical description of a system comprising charged particles interacting with the electromagnetic field they produce. The aim is to identify a set of pairs of conjugate canonical variables that can be used with the Hamiltonian formalism to recover the dynamical equations of the field and charges, i.e. the Maxwell–Lorentz equations. We decompose the electromagnetic field into a longitudinal part, controlled by the dynamics of the charges, and a transverse part with autonomous dynamics called radiation. Then by imposing the Coulomb gauge, we can identify conjugate canonical variables for the radiation by a simple generalization of the approach described in Chapter 4. After introducing the subtle notion of generalized momentum for particles in the presence of radiation, we obtain a total Hamiltonian describing the system of interacting particles and radiation, and this can then be used to carry out the canonical quantization procedure. This rather formal Section 6.1 can be skipped if the reader is in a hurry to see applications of the quantum formalism, as presented in Section 6.2. However, it should be stressed that these are key ideas for anyone who wishes to acquire a full understanding of this formalism. For our part, we have done everything in our power to simplify the formalism as far as possible, leaving certain of the more tedious calculations to Complement 6A so that the reader may focus on the main features of the method.

Section 6.2 presents the basic elements of the quantum formalism, which describe the system of interacting charges and radiation. It is shown in particular that the quantum Hamiltonian can be decomposed into three parts: the first describes the radiation alone; the second describes a set of charged particles subject to their mutual Coulomb interactions, which constitutes an elementary model for the atom; and the third describes the interaction between the quantized radiation and the quantized atom. In Section 6.3, the main interaction processes between matter and radiation are described briefly from this new standpoint. Some of these, such as stimulated emission and absorption, could be described using the semi-classical formalism (Chapter 2), but what we get now is a description of spontaneous emission, a very important phenomenon that earns its own section (Section 6.4). In contrast, in Section 6.5, we discuss a process that can be just as well described using the semi-classical formalism, namely elastic scattering. This section can be considered as an exercise in the application of the quantum formalism.

As indicated, Complement 6A treats the Hamiltonian formalism of classical electrodynamics, justifying some results applied in Section 6.1. Complement 6B discusses an active field of research in quantum optics, namely cavity quantum electrodynamics, used to produce a certain number of fascinating quantum physical phenomena: entangled states of atoms and radiation, modification of spontaneous emission by the cavity, quantum revival and others. These phenomena play a key role in the field of quantum information (Complement 5E). Complement 6C presents a rudimentary calculation showing how an atom excited to a well chosen level, can emit by two successive spontaneous emissions 'in cascade', pairs of photons entangled in polarization. Such photon pairs have been considered in Complement 5C as an example of entangled systems lending themselves to a test of Bell's inequalities.

6.1 Classical electrodynamics and interacting fields and charges

6.1.1 The Maxwell–Lorentz equations

The equations describing the dynamics of a system of charged particles interacting with fields $\mathbf{E}(\mathbf{r},t)$ and $\mathbf{B}(\mathbf{r},t)$ include Maxwell's equations, which relate the electromagnetic fields with the charge and current densities $\rho(\mathbf{r},t)$ and $\mathbf{J}(\mathbf{r},t)$, respectively:

$$\nabla \cdot \mathbf{E}(\mathbf{r},t) = \frac{1}{\varepsilon_0}\rho(\mathbf{r},t) \tag{6.1}$$

$$\nabla \times \mathbf{B}(\mathbf{r},t) = \frac{1}{c^2}\frac{\partial}{\partial t}\mathbf{E}(\mathbf{r},t) + \frac{1}{\varepsilon_0 c^2}\mathbf{J}(\mathbf{r},t) \tag{6.2}$$

$$\nabla \cdot \mathbf{B}(\mathbf{r},t) = 0 \tag{6.3}$$

$$\nabla \times \mathbf{E}(\mathbf{r},t) = -\frac{\partial}{\partial t}\mathbf{B}(\mathbf{r},t). \tag{6.4}$$

For point-charged particles with positions $\mathbf{r}_\mu$, charges q_μ, and masses m_μ, the charge and current densities are given as a function of the positions $\mathbf{r}_\mu(t)$ and velocities $\mathbf{v}_\mu(t) = \mathrm{d}\mathbf{r}_\mu/\mathrm{d}t$ by,

$$\rho(\mathbf{r},t) = \sum_\mu q_\mu \delta[\mathbf{r} - \mathbf{r}_\mu(t)] \tag{6.5}$$

$$\mathbf{J}(\mathbf{r},t) = \sum_\mu q_\mu \mathbf{v}_\mu \delta[\mathbf{r} - \mathbf{r}_\mu(t)]. \tag{6.6}$$

The particle dynamics is determined by the Newton–Lorentz equation which governs the motions of the point charges under the effect of the electromagnetic field in the non-relativistic limit:

$$m_\mu \frac{\mathrm{d}}{\mathrm{d}t}\mathbf{v}_\mu = q_\mu \Big[\mathbf{E}(\mathbf{r}_\mu,t) + \mathbf{v}_\mu \times \mathbf{B}(\mathbf{r}_\mu,t)\Big]. \tag{6.7}$$

The above equations, referred to together as the Maxwell–Lorentz equations, fully determine the dynamics of the particle–field system provided that the initial state of the system has been sufficiently specified. The initial conditions required at some initial time t_0 in order to determine the whole future evolution of the system in an unambiguous way are the values of the electric and magnetic fields at all points $\mathbf{r}$ of space, and the position and velocity of each particle. These quantities fully determine the state of the system at any later time, and thus constitute a complete set of dynamical variables. We have already seen in Section 4.2.2 of Chapter 4 that, when we restrict to a system confined within a cube of volume L^3, the electric and magnetic fields can be described using a discrete set of complex functions of time, i.e. the spatial Fourier components $\tilde{\mathbf{E}}_\mathbf{n}(t)$ and $\tilde{\mathbf{B}}_\mathbf{n}(t)$. In terms of these Fourier components, Maxwell's equations take the form

$$\mathrm{i}\mathbf{k}_\mathbf{n} \cdot \tilde{\mathbf{E}}_\mathbf{n} = \frac{1}{\varepsilon_0}\tilde{\rho}_\mathbf{n} \tag{6.8}$$

$$\mathrm{i}\mathbf{k}_\mathbf{n} \cdot \tilde{\mathbf{B}}_\mathbf{n} = 0 \tag{6.9}$$

$$\mathrm{i}\mathbf{k}_\mathbf{n} \times \tilde{\mathbf{E}}_\mathbf{n} = -\frac{\mathrm{d}}{\mathrm{d}t}\tilde{\mathbf{B}}_\mathbf{n} \tag{6.10}$$

$$\mathrm{i}\mathbf{k}_\mathbf{n} \times \tilde{\mathbf{B}}_\mathbf{n} = \frac{1}{c^2}\frac{\mathrm{d}}{\mathrm{d}t}\tilde{\mathbf{E}}_\mathbf{n} + \frac{1}{\varepsilon_0 c^2}\tilde{\mathbf{J}}_\mathbf{n}, \tag{6.11}$$

where $\tilde{\rho}_\mathbf{n}$ and $\tilde{\mathbf{J}}_\mathbf{n}$ are the spatial Fourier components of $\rho(\mathbf{r},t)$ and $\mathbf{J}(\mathbf{r},t)$, respectively. The transition to Fourier components raises no difficulty because the Maxwell equations are linear in all these quantities.

As in Chapter 4, the switch to the reciprocal (momentum) space yields a system of ordinary differential equations for a set of purely time-dependent functions. What is more, these equations are local in the reciprocal space, in the sense that (6.8–6.11) only couple quantities with the same index $\mathbf{n}$. However, it should not be concluded that the quantities associated with different values of $\mathbf{n}$, i.e. with different field modes, are in fact decoupled. Indeed, the motion of the charges is governed by the Lorentz equation (6.7), which is nonlinear (hence nonlocal in the reciprocal space), and the quantities $\tilde{\rho}_\mathbf{n}$ and $\tilde{\mathbf{J}}_\mathbf{n}$ may depend on field components for values $\mathbf{n}'$ different from $\mathbf{n}$. The situation here thus differs from the one considered in Chapter 4, where the space was empty of charges and currents, i.e. $\rho = 0$ and $\mathbf{J} = 0$, and the equations for different plane travelling wave modes were decoupled.

6.1.2 Decomposition of the electromagnetic field into transverse and longitudinal components. Radiation

Transverse and longitudinal components of the electric field

Consider the electric field, described in reciprocal space by its spatial Fourier components $\tilde{\mathbf{E}}_\mathbf{n}$. These vector quantities can always be expressed in terms of an orthonormal basis comprising a unit vector $\mathbf{e_n}$ in the direction $\mathbf{k_n}$ and unit vectors $\boldsymbol{\varepsilon}_{\mathbf{n},1}$ and $\boldsymbol{\varepsilon}_{\mathbf{n},2}$ in the plane orthogonal to $\mathbf{k_n}$ (see Figure 4.1). One can then write

$$\tilde{\mathbf{E}}_\mathbf{n} = \tilde{E}_{//\mathbf{n}}\mathbf{e_n} + \tilde{E}_{\perp\mathbf{n},1}\boldsymbol{\varepsilon}_{\mathbf{n},1} + \tilde{E}_{\perp\mathbf{n},2}\boldsymbol{\varepsilon}_{\mathbf{n},2}. \tag{6.12}$$

We shall call the vectors $\tilde{\mathbf{E}}_{//\mathbf{n}} = \tilde{E}_{//\mathbf{n}}\mathbf{e_n}$ and $\tilde{\mathbf{E}}_{\perp\mathbf{n}} = \tilde{E}_{\perp\mathbf{n},1}\boldsymbol{\varepsilon}_{\mathbf{n},1} + \tilde{E}_{\perp\mathbf{n},2}\boldsymbol{\varepsilon}_{\mathbf{n},2}$ the longitudinal and transverse components of $\tilde{\mathbf{E}}_\mathbf{n}$, respectively. Summing the Fourier series and collecting together longitudinal and transverse terms in the manner of (4.18), we obtain two fields in real space, denoted in the obvious way by $\mathbf{E}_{//}(\mathbf{r}, t)$ and $\mathbf{E}_\perp(\mathbf{r}, t)$, such that

$$\mathbf{E}(\mathbf{r}, t) = \mathbf{E}_{//}(\mathbf{r}, t) + \mathbf{E}_\perp(\mathbf{r}, t). \tag{6.13}$$

A necessary consequence of (6.12) is

$$\nabla \cdot \mathbf{E}_\perp(\mathbf{r}, t) = 0 \tag{6.14}$$

$$\nabla \times \mathbf{E}_{//}(\mathbf{r}, t) = \mathbf{0}. \tag{6.15}$$

By this procedure, we have *decomposed the electric field into its longitudinal part* $\mathbf{E}_{//}(\mathbf{r}, t)$, *with zero curl, and its transverse part* $\mathbf{E}_\perp(\mathbf{r}, t)$, *with zero divergence*. Naturally, one could do this for any vector field, e.g. the magnetic field or the vector potential. Maxwell's equation (6.9) expresses the fact that $\tilde{\mathbf{B}}_\mathbf{n}$ is perpendicular to $\mathbf{k_n}$, whence *the magnetic field is purely transverse*.

Longitudinal electric field

Equation (6.8) can also be written

$$i\mathbf{k_n} \cdot \tilde{\mathbf{E}}_{//\mathbf{n}} = \frac{1}{\varepsilon_0}\tilde{\rho}_\mathbf{n}. \tag{6.16}$$

It provides a purely algebraic relation between the longitudinal field $\tilde{E}_{//\mathbf{n}}$ and the charge density at the same time t. Returning to the direct (position) space and real fields, we obtain the equation

$$\nabla \cdot \mathbf{E}_{//}(\mathbf{r}, t) = \frac{1}{\varepsilon_0}\rho(\mathbf{r}, t). \tag{6.17}$$

This is the *fundamental equation of electrostatics*, written here for a charge distribution that may not be static. In the case of a point-charge distribution, we then have the well-known solution,

$$\mathbf{E}_{//}(\mathbf{r},t) = \frac{1}{4\pi\varepsilon_0}\sum_\mu q_\mu \frac{\mathbf{r}-\mathbf{r}_\mu(t)}{|\mathbf{r}-\mathbf{r}_\mu(t)|^3}. \tag{6.18}$$

This shows that *the longitudinal electric field at time t is the instantaneous Coulomb field associated with* ρ, and calculated as though the charge distribution were static, frozen at its value at time t. Equation (6.18) shows that $\mathbf{E}_{//}$ is then an explicit function of the particle positions. The way it evolves is completely controlled by the particle dynamics, i.e. the values of $\mathbf{E}_{//}$ *are not independent dynamical variables of the system.*

Comment

The fact that the longitudinal electric field instantaneously follows the evolution of the charge distribution does not imply the existence of electrical phenomena propagating faster than light. Indeed, only the total electric field can be measured, and has a physical meaning. It can be shown that the transverse field $\mathbf{E}_\perp$ also has an instantaneous component which exactly balances that of $\mathbf{E}_{//}$ in such a way that the total field remains purely retarded.[1]

Transverse electric and magnetic fields. Radiation

Since field $\mathbf{B}$ is transverse according to (6.9), Maxwell's equations (6.10) and (6.11) apply to $\tilde{\mathbf{B}}_\mathbf{n}$ and the transverse part $\tilde{\mathbf{E}}_{\perp\mathbf{n}}$ of $\mathbf{E}$, yielding,

$$\frac{\mathrm{d}}{\mathrm{d}t}\tilde{\mathbf{B}}_\mathbf{n} = -\mathrm{i}\mathbf{k}_\mathbf{n} \times \tilde{\mathbf{E}}_{\perp\mathbf{n}}, \tag{6.19}$$

$$\frac{\mathrm{d}}{\mathrm{d}t}\tilde{\mathbf{E}}_{\perp\mathbf{n}} = \mathrm{i}c^2\mathbf{k}_\mathbf{n} \times \tilde{\mathbf{B}}_{\perp\mathbf{n}} - \frac{1}{\varepsilon_0}\tilde{\mathbf{J}}_{\perp\mathbf{n}}, \tag{6.20}$$

where $\tilde{\mathbf{J}}_{\perp\mathbf{n}}$ is the transverse Fourier component of the current density.

Equations (6.19) and (6.20) determine the coupled dynamics of the transverse electric field and the magnetic field, which depends on the sources solely through the transverse part of the current density. As we have seen in Chapter 4, these fields may differ from zero even if the source current is zero. Unlike the longitudinal components, *the transverse components of the fields are thus genuine dynamical variables.*

The transition to reciprocal space has provided a way of distinguishing two parts of the electromagnetic field when it interacts with a charge distribution: the longitudinal components, zero for $\mathbf{B}$, and a simple function of the instantaneous charge distribution for $\mathbf{E}$; the transverse components, with their own independent dynamics, obeying (6.19) and (6.20), which correspond to the 'radiated' part of the field. *In the presence of charges, radiation can thus be identified as the transverse part of the electromagnetic field.*

[1] See CDG I Complement C_I, Exercise 3.

6.1.3 Polarized Fourier components of the radiation and the vector potential in the Coulomb gauge

As in Chapter 4, we introduce the vector potential $\mathbf{A}(\mathbf{r}, t)$ such that $\mathbf{B} = \nabla \times \mathbf{A}$, and impose the Coulomb gauge condition $\nabla \cdot \mathbf{A} = 0$. In this gauge, the vector potential $\mathbf{A}(\mathbf{r}, t)$ is transverse, like $\mathbf{E}_\perp(\mathbf{r}, t)$ and $\mathbf{B}(\mathbf{r}, t)$. As in Section 4.2.3, the transverse Fourier components $\tilde{\mathbf{E}}_{\perp\mathbf{n}}$, $\tilde{\mathbf{A}}_{\mathbf{n}}$, and $\tilde{\mathbf{J}}_{\perp\mathbf{n}}$ can be decomposed relative to the polarizations $\boldsymbol{\varepsilon}_{\mathbf{n},1}$ and $\boldsymbol{\varepsilon}_{\mathbf{n},2}$, thereby introducing the scalar components $\tilde{A}_\ell$ and $\tilde{E}_{\perp\ell}$. Using the polarizations $\boldsymbol{\varepsilon}'_{\mathbf{n}} = \mathbf{k}_\ell \times \boldsymbol{\varepsilon}_\ell / k_\ell$ for $\tilde{\mathbf{B}}_{\mathbf{n}}$ (see (4.33)), Maxwell's equations ((6.19–6.20)) can be expressed in the following form generalizing (4.44–4.45):

$$\frac{\mathrm{d}}{\mathrm{d}t}\tilde{B}_\ell(t) = -\mathrm{i}k_\ell \tilde{E}_{\perp\ell}(t) \tag{6.21}$$

$$\frac{\mathrm{d}}{\mathrm{d}t}\tilde{E}_{\perp\ell}(t) = -\mathrm{i}c^2 k_\ell \tilde{B}_\ell(t) - \frac{1}{\varepsilon_0}\tilde{J}_{\perp\ell}. \tag{6.22}$$

Once again, $\tilde{B}_\ell$ can be replaced by $\mathrm{i}k_\ell \tilde{A}_\ell$ to give

$$\frac{\mathrm{d}}{\mathrm{d}t}\tilde{A}_\ell(t) = -\tilde{E}_\ell(t) \tag{6.23}$$

$$\frac{\mathrm{d}}{\mathrm{d}t}\tilde{E}_\ell(t) = \omega_\ell^2 \tilde{A}_\ell(t) - \frac{1}{\varepsilon_0}\tilde{J}_{\perp\ell}(t). \tag{6.24}$$

The above system determines the dynamics of the radiation. It is analogous to (4.47–4.48), which characterizes the dynamics of a free harmonic oscillator, except that here we also have a source term. What we find will thus be the dynamics of a forced harmonic oscillator.

6.1.4 Normal variables for radiation and expansion in polarized, travelling plane waves

As in Section 4.3.2, Equations (6.23) and (6.24) can be decoupled by introducing normal variables α_ℓ and β_ℓ (see (4.51) and (4.52)), which gives

$$\frac{\mathrm{d}\alpha_\ell}{\mathrm{d}t} + \mathrm{i}\,\omega_\ell \alpha_\ell = \frac{\mathrm{i}}{2\varepsilon_0 \mathcal{E}_\ell^{(1)}}\tilde{J}_{\perp\ell} \tag{6.25}$$

$$\frac{\mathrm{d}\beta_\ell}{\mathrm{d}t} - \mathrm{i}\,\omega_\ell \beta_\ell = -\frac{\mathrm{i}}{2\varepsilon_0 \mathcal{E}_\ell^{(1)}}\tilde{J}_{\perp\ell}. \tag{6.26}$$

These are indeed equations for two harmonic oscillators with proper frequencies ω_ℓ and $-\omega_\ell$, but the right-hand sides are not necessarily harmonic functions varying with these frequencies. However, this need not prevent us from inverting the defining relations (4.51–4.52) to express the various transverse fields in terms of $\alpha_\ell(t)$ and $\beta_\ell(t)$. Since the fields are real, we have once again $\beta_\ell^*(t) = \alpha_{-\ell}(t)$ (see (4.61)), and by exactly the same argument as in Section 4.3.3, the radiation can be expressed in the form:

$$\mathbf{A}(\mathbf{r},t) = \sum_\ell \boldsymbol{\varepsilon}_\ell \frac{\mathcal{E}_\ell^{(1)}}{\omega_\ell} \left[\alpha_\ell(t)\mathrm{e}^{\mathrm{i}\mathbf{k}_\ell\cdot\mathbf{r}} + \alpha_\ell^*(t)\mathrm{e}^{-\mathrm{i}\mathbf{k}_\ell\cdot\mathbf{r}}\right] \tag{6.27}$$

$$\mathbf{E}_\perp(\mathbf{r},t) = \sum_\ell \boldsymbol{\varepsilon}_\ell \mathcal{E}_\ell^{(1)} \left[\mathrm{i}\alpha_\ell(t)\mathrm{e}^{\mathrm{i}\mathbf{k}_\ell\cdot\mathbf{r}} - \mathrm{i}\alpha_\ell^*(t)\mathrm{e}^{-\mathrm{i}\mathbf{k}_\ell\cdot\mathbf{r}}\right] \tag{6.28}$$

$$\mathbf{B}(\mathbf{r},t) = \sum_\ell \boldsymbol{\varepsilon}'_\ell \frac{\mathcal{E}_\ell^{(1)}}{c} \left[\mathrm{i}\alpha_\ell(t)\mathrm{e}^{\mathrm{i}\mathbf{k}_\ell\cdot\mathbf{r}} - \mathrm{i}\alpha_\ell^*(t)\mathrm{e}^{-\mathrm{i}\mathbf{k}_\ell\cdot\mathbf{r}}\right] \tag{6.29}$$

with

$$\mathcal{E}_\ell^{(1)} = \sqrt{\frac{\hbar\omega_\ell}{2\varepsilon_0 L^3}} \tag{6.30}$$

and

$$\boldsymbol{\varepsilon}'_\ell = \frac{1}{k_\ell}\mathbf{k}_\ell \times \boldsymbol{\varepsilon}_\ell. \tag{6.31}$$

6.1.5 Generalized particle momentum. Radiation momentum

The momentum of the electromagnetic field in volume V is proportional to the integral of the Poynting vector over this volume (see Complement 4B):

$$\mathbf{P}^{\mathrm{em}} = \varepsilon_0 \int_V \mathrm{d}^3 r\, \mathbf{E}(\mathbf{r},t) \times \mathbf{B}(\mathbf{r},t). \tag{6.32}$$

Separating out the longitudinal part of the electric field, we define the longitudinal momentum:

$$\mathbf{P}^{\mathrm{em}}_{\mathrm{long}} = \varepsilon_0 \int_V \mathrm{d}^3 r\, \mathbf{E}_{/\!/}(\mathbf{r},t) \times \mathbf{B}(\mathbf{r},t). \tag{6.33}$$

From the Fourier decomposition of fields $\mathbf{E}$ and $\mathbf{B}$ (see (4.18)) and the reality condition (4.20), $\mathbf{P}^{\mathrm{em}}_{\mathrm{long}}$ takes the form

$$\mathbf{P}^{\mathrm{em}}_{\mathrm{long}} = \varepsilon_0 L^3 \sum_{\mathbf{n}} \tilde{\mathbf{E}}^*_{/\!/\mathbf{n}} \times \tilde{\mathbf{B}}_{\mathbf{n}}. \tag{6.34}$$

In (6.34), the longitudinal electric field can be expressed in the form deduced from (6.16):

$$\tilde{\mathbf{E}}_{/\!/\mathbf{n}} = -\frac{\mathrm{i}}{\varepsilon_0}\frac{\mathbf{k}_{\mathbf{n}}}{k_n^2}\tilde{\rho}_{\mathbf{n}}, \tag{6.35}$$

while $\tilde{\mathbf{B}}_{\mathbf{n}}$ can be replaced by $\mathbf{i}\mathbf{k}_{\mathbf{n}} \times \tilde{\mathbf{A}}_{\mathbf{n}}$. Recalling that $\tilde{\mathbf{A}}_{\mathbf{n}}$ is transverse when we impose the Coulomb gauge, this leads in the end to

$$\mathbf{P}^{\mathrm{em}}_{\mathrm{long}} = L^3 \sum_{\mathbf{n}} \tilde{\rho}^*_{\mathbf{n}} \tilde{\mathbf{A}}_{\mathbf{n}}. \tag{6.36}$$

Going back to real space, it then follows that

$$\mathbf{P}^{\mathrm{em}}_{\mathrm{long}} = \int \mathrm{d}^3 r\, \rho(\mathbf{r},t)\mathbf{A}(\mathbf{r},t) = \sum_\mu q_\mu \mathbf{A}(\mathbf{r}_\mu,t). \tag{6.37}$$

Equation (6.37) shows that the longitudinal momentum of the electromagnetic field is closely related to the charges, since it depends only on **A** at the location of the charges, and it is zero if $q_\mu = 0$. This suggests grouping this momentum with the momentum of the charges, and defining the *generalized momentum* of the charge immersed in the electromagnetic radiation field by the sum,

$$\mathbf{p}_\mu = m_\mu \mathbf{v}_\mu + q_\mu \mathbf{A}(\mathbf{r}_\mu, t). \tag{6.38}$$

This new dynamical variable associated with the particle is in fact *canonically conjugate* to $\mathbf{r}_\alpha$ in the presence of the electromagnetic field, as is shown in Complement 6A.

Given (6.32), to which we add the momentum $\sum_\mu m_\mu \mathbf{v}_\mu$ of the charges, the total momentum of the system comprising charges and electromagnetic field becomes

$$\mathbf{P} = \sum_\mu \mathbf{p}_\mu + \varepsilon_0 \int_V \mathrm{d}^3 r \, \mathbf{E}_\perp(\mathbf{r}, t) \times \mathbf{B}(\mathbf{r}, t). \tag{6.39}$$

The first term is the sum of the generalized momenta of the particles. The second, which can be expressed solely in terms of the transverse fields, is the *momentum of the radiation.*

Comments

The total angular momentum **J** of the system is another constant of the motion. The angular momentum about the coordinate origin is given by

$$\mathbf{J} = \sum_\mu \mathbf{r}_\mu(t) \times m_\mu \mathbf{v}_\mu(t) + \varepsilon_0 \int \mathrm{d}^3 r \, \mathbf{r} \times \left(\mathbf{E}(\mathbf{r}, t) \times \mathbf{B}(\mathbf{r}, t)\right). \tag{6.40}$$

It can be shown that, for an isolated system of fields and particles, $\mathrm{d}\mathbf{J}/\mathrm{d}t = 0$. It can also be shown that the angular momentum is given in terms of the generalized momentum by

$$\mathbf{J} = \sum_\mu \mathbf{r}_\mu \times \mathbf{p}_\mu + \varepsilon_0 \int \mathrm{d}^3 r \, \mathbf{r} \times \left[\mathbf{E}_\perp(\mathbf{r}, t) \times \mathbf{B}(\mathbf{r}, t)\right]. \tag{6.41}$$

The total angular momentum **J** is thus given simply as the sum of the moments of the generalized momenta $\mathbf{p}_\mu$ and the moment about the origin of the momentum density $\varepsilon_0 \mathbf{E}_\perp \times \mathbf{B}$ of the transverse field (see Complement 4B). As for the momentum, we have thus been able to separate the contributions of the particles and the radiation in the angular momentum.

6.1.6 Hamiltonian in the Coulomb gauge

Total energy of the charge–field system

The expression for the energy of a set of charges interacting with an electromagnetic field can be found in any standard textbook on classical electrodynamics:[2]

$$H = \sum_\mu \frac{1}{2} m_\mu \mathbf{v}_\mu^2 + \frac{\varepsilon_0}{2} \int \mathrm{d}^3 r \left(\mathbf{E}^2(\mathbf{r}, t) + c^2 \mathbf{B}^2(\mathbf{r}, t)\right). \tag{6.42}$$

One justification for this expression is that it can be shown using the Maxwell–Lorentz equations that $\mathrm{d}H/\mathrm{d}t = 0$, which expresses conservation of energy for the isolated

[2] See, for example, J. D. Jackson, *Classical Electrodynamics*, Wiley (1998).

field–particle system. A more complete justification comes from the fact that the Maxwell–Lorentz equations can be recovered from H using the Hamiltonian formalism (see Complement 6A). It may be somewhat surprising to observe that the explicit interaction term between charges and field does not appear in (6.42). However, this term will turn up when the electromagnetic field is decomposed into longitudinal and transverse parts, and the kinetic energy of the particles expressed in terms of the generalized momenta, i.e.

$$H_{\text{kin}} = \sum_\mu \frac{1}{2} m \, \mathbf{v}_\mu^2 = \sum_\mu \frac{1}{2m_\mu} \left(\mathbf{p}_\mu - q\mathbf{A}(\mathbf{r}_\mu, t) \right)^2. \tag{6.43}$$

Transverse and longitudinal electromagnetic energy

When we go to the Fourier space, where the longitudinal and transverse fields are orthogonal, the electromagnetic energy can be separated into longitudinal and transverse contributions according to

$$H_{\text{em}} = \frac{\varepsilon_0}{2} \int \mathrm{d}^3 r \left(\mathbf{E}^2(\mathbf{r}, t) + c^2 \mathbf{B}^2(\mathbf{r}, t) \right) = H_{\text{trans}}^{\text{em}} + H_{\text{long}}^{\text{em}}. \tag{6.44}$$

The transverse part is given by

$$H_{\text{trans}}^{\text{em}} = \frac{\varepsilon_0}{2} \int \mathrm{d}^3 r \left(\mathbf{E}_\perp^2(\mathbf{r}, t) + c^2 \mathbf{B}^2(\mathbf{r}, t) \right). \tag{6.45}$$

Using expansions (6.28) and (6.29) for the transverse fields in terms of the modes ℓ, and arguing as in Section 4.4.1, we obtain

$$H_{\text{trans}}^{\text{em}} = \sum_\ell \hbar \omega_\ell |\alpha_\ell|^2. \tag{6.46}$$

This is the energy of the radiation, denoted in what follows by H_{R}, as in Chapter 4.

Regarding the longitudinal part, it is given in terms of the Fourier components by

$$H_{\text{long}}^{\text{em}} = \frac{\varepsilon_0}{2L^3} \sum_{\mathbf{n}} |\tilde{\mathbf{E}}_{//\mathbf{n}}|^2 = \frac{1}{2\varepsilon_0 L^3} \sum_{\mathbf{n}} \frac{|\tilde{\rho}_{\mathbf{n}}|^2}{k_{\mathbf{n}}^2}. \tag{6.47}$$

The charge density is given in terms of the particle positions (see (6.5)), and $\tilde{\rho}_{\mathbf{n}}$ takes the form

$$\tilde{\rho}_{\mathbf{n}} = \frac{1}{L^3} \sum_\mu q_\mu \, \mathrm{e}^{-\mathrm{i}\mathbf{k}_{\mathbf{n}} \cdot \mathbf{r}_\mu}, \tag{6.48}$$

which can be substituted into (6.47) to obtain

$$H_{\text{long}} = \frac{1}{2\varepsilon_0 L^3} \sum_{\mathbf{n}} \sum_\mu \sum_\nu q_\mu q_\nu \, \mathrm{e}^{-\mathrm{i}\mathbf{k}_{\mathbf{n}} \cdot (\mathbf{r}_\mu - \mathbf{r}_\nu)}. \tag{6.49}$$

Consider, to begin with, the term corresponding to a single particle in the sum over μ and ν ($\mu = \nu$). Replacing the discrete sum by an integral and taking the limit $L \to \infty$, this becomes

$$\frac{q_\alpha^2}{2\varepsilon_0 L^3} \sum_{\mathbf{n}} \frac{1}{k_{\mathbf{n}}^2} \xrightarrow[L \to \infty]{} \frac{q_\alpha^2}{2\varepsilon_0 (2\pi)^3} \int \frac{\mathrm{d}^3 k}{k^2}. \tag{6.50}$$

It is easy to see that this integral is divergent. The divergence of the 'self-energy' term is well known in electrostatics. It arises because the Coulomb energy of a point particle is infinite. But since this term does not vary in time, it can be considered as a simple shift in the zero energy. Removing the self-energy terms from the total energy, the energy of the longitudinal electric field in (6.47) reduces to a sum in which μ and ν are different. Now taking the limit $L \to \infty$ and using the fact that the Fourier transform of $1/k^2$ is $1/r$, we find

$$\begin{aligned} H_{\text{long}}^{\text{em}} &= \frac{1}{2\varepsilon_0(2\pi)^3}\sum_{\mu}\sum_{\nu\neq\mu} q_\mu q_\nu \int \mathrm{d}^3k \frac{\mathrm{e}^{\mathrm{i}\mathbf{k}(\mathbf{r}_\mu-\mathbf{r}_\nu)}}{k^2} = \sum_{\mu}\sum_{\nu\neq\mu}\frac{q_\mu q_\nu}{8\pi\varepsilon_0|\mathbf{r}_\mu-\mathbf{r}_\nu|} \\ &= V_{\text{coul}}(\mathbf{r}_1,\ldots,\mathbf{r}_\mu\ldots). \end{aligned} \tag{6.51}$$

It is found that $H_{\text{long}}^{\text{em}}$ is just the instantaneous Coulomb interaction energy V_{Coul} between the different point particles with charges q_μ located at points $\mathbf{r}_\mu$.

Conjugate canonical variables for the field and charges

The total energy (6.42) of the system now has the form,

$$H = \sum_\mu \frac{1}{2} m_\mu \mathbf{v}_\mu^2 + V_{\text{coul}} + H_{\text{R}}. \tag{6.52}$$

It is the sum of the kinetic energy of the particles, their Coulomb interaction energy, and the energy H_{trans} of the transverse field or radiation field, which is given by the same expression H_{R} as the free radiation discussed in Chapter 4. To get (6.52) into the Hamiltonian form, we must use conjugate canonical variables, which we identify by checking that Hamilton's equations do indeed return the Maxwell–Lorentz equations. In Complement 6A, we show that this is indeed the case when H_{R} is taken as the decomposition of the radiation into normal modes obtained in Section 6.1.3, and when we introduce the same canonical variables Q_ℓ and P_ℓ as in Chapter 4 for the free electromagnetic field. Regarding the first term, it has to be expressed in terms of the generalized momenta $\mathbf{p}_\mu$, which are the canonical variables conjugate to the particle positions $\mathbf{r}_\mu$. The final result is, therefore,

$$H = \sum_\mu \frac{1}{2m_\mu}\left(\mathbf{p}_\mu - q_\mu \mathbf{A}(\mathbf{r}_\mu, t)\right)^2 + V_{\text{coul}} + H_{\text{R}}, \tag{6.53}$$

where V_{Coul} is given as a function of $\mathbf{r}_\mu$ (see (6.51)) and the radiation Hamiltonian H_{R} is expressed in terms of the variables Q_ℓ and P_ℓ by (4.105):

$$H_{\text{R}} = \sum_\ell \frac{\omega_\ell}{2}\left(Q_\ell^2 + P_\ell^2\right). \tag{6.54}$$

Hamiltonian for the charges

We expand the square in (6.53) and gather the terms into the form

$$H = H_{\text{R}} + H_{\text{P}} + H_{\text{I}}. \tag{6.55}$$

The second contribution H_{P} refers to the particles alone:

$$H_{\mathrm{P}} = \sum_{\mu} \frac{\mathbf{p}_{\mu}^{2}}{2m_{\mu}} + V_{\mathrm{coul}} = \sum_{\mu} \frac{\mathbf{p}_{\mu}^{2}}{2m_{\mu}} + \sum_{\mu} \sum_{\nu<\mu} \frac{q_{\mu} q_{\nu}}{4\pi \varepsilon_0 |\mathbf{r}_{\mu} - \mathbf{r}_{\nu}|}. \tag{6.56}$$

It describes the charges interacting via the Coulomb interaction in the absence of radiation. This corresponds to the simplest possible model of an atom, in which only the instantaneous Coulomb interaction between the charges is taken into account and all relativistic effects are ignored, as are the magnetic moments due to the spins of the different particles (electrons and nucleus). The Hamiltonian H_{P} depends only on the conjugate canonical variables $\mathbf{r}_{\mu}$ and $\mathbf{p}_{\mu}$ of the particles.

Interaction Hamiltonian

The term H_{I} involves both particles and radiation, since it has the form

$$H_{\mathrm{I}} = \sum_{\mu} \left(-\frac{q_{\mu}}{m_{\mu}} \mathbf{p}_{\mu} \cdot \mathbf{A}(\mathbf{r}_{\mu}, t) + \frac{q_{\mu}^{2}}{2m_{\mu}} \mathbf{A}^{2}(\mathbf{r}_{\mu}, t) \right). \tag{6.57}$$

This is the term that describes the interaction between particles and radiation, and it is indeed an interaction Hamiltonian when $\mathbf{A}(\mathbf{r}, t)$ (the vector potential in the Coulomb gauge) is expressed in terms of the conjugate canonical variables (Q_{ℓ}, P_{ℓ}) of the radiation.

6.2 Interacting fields and charges and quantum description in the Coulomb gauge

6.2.1 Canonical quantization

In Section 6.1, the problem has been set in the Hamiltonian form by expressing the total energy (6.55) of the system comprising charges and electromagnetic field in terms of the pairs of conjugate canonical variables $(\mathbf{r}_{\mu}, \mathbf{p}_{\mu})$ and (Q_{ℓ}, P_{ℓ}). Canonical quantization consists in replacing these pairs of canonical variables by pairs of Hermitian operators with commutators set equal to $\mathrm{i}\hbar$.

We thereby obtain the quantum Hamiltonian for the system, which we shall return to in Section 6.2.2, and also the observables relating to the particles, and those relating to the fields, the latter being given by

$$\hat{\mathbf{A}}(\mathbf{r}) = \sum_{\ell} \boldsymbol{\varepsilon}_{\ell} \frac{\mathcal{E}_{\ell}^{(1)}}{\omega_{\ell}} \left(\hat{a}_{\ell}\, \mathrm{e}^{\mathrm{i}\mathbf{k}_{\ell}\cdot\mathbf{r}} + \hat{a}_{\ell}^{\dagger}\, \mathrm{e}^{-\mathrm{i}\mathbf{k}_{\ell}\cdot\mathbf{r}} \right) \tag{6.58}$$

$$\hat{\mathbf{E}}_{\perp}(\mathbf{r}) = \sum_{\ell} \boldsymbol{\varepsilon}_{\ell}\, \mathcal{E}_{\ell}^{(1)} \left(\mathrm{i}\,\hat{a}_{\ell}\, \mathrm{e}^{\mathrm{i}\mathbf{k}_{\ell}\cdot\mathbf{r}} - \mathrm{i}\,\hat{a}_{\ell}^{\dagger}\, \mathrm{e}^{-\mathrm{i}\mathbf{k}_{\ell}\cdot\mathbf{r}} \right) \tag{6.59}$$

$$\hat{\mathbf{B}}(\mathbf{r}) = \sum_{\ell} \frac{\mathbf{k}_{\ell}}{\omega_{\ell}} \times \boldsymbol{\varepsilon}_{\ell}\, \mathcal{E}_{\ell}^{(1)} \left(\mathrm{i}\,\hat{a}_{\ell}\, \mathrm{e}^{\mathrm{i}\mathbf{k}_{\ell}\cdot\mathbf{r}} - \mathrm{i}\,\hat{a}_{\ell}^{\dagger}\, \mathrm{e}^{-\mathrm{i}\mathbf{k}_{\ell}\cdot\mathbf{r}} \right). \tag{6.60}$$

As for free radiation, the non-Hermitian operators $\hat{a}_\ell$ and $\hat{a}_\ell^\dagger$ are defined in terms of the Hermitian operators $\hat{Q}_\ell$ and $\hat{P}_\ell$ by the relations (4.102–4.103). They satisfy the commutation relations (4.100–4.101):

$$[\hat{a}_\ell, \hat{a}_{\ell'}^\dagger] = \delta_{\ell\ell'} \tag{6.61}$$

$$[\hat{a}_\ell, \hat{a}_{\ell'}] = 0. \tag{6.62}$$

Note also the 'electric field for one photon',

$$\mathcal{E}_\ell^{(1)} = \sqrt{\frac{\hbar\omega_\ell}{2\varepsilon_0 L^3}}. \tag{6.63}$$

6.2.2 Hamiltonian and state space

The classical Hamiltonian (6.55) becomes the operator $\hat{H}$, given by

$$\hat{H} = \hat{H}_\text{P} + \hat{H}_\text{R} + \hat{H}_\text{I}. \tag{6.64}$$

The first term depends only on the particle observables, while the second only involves the radiation observables. We can thus consider the eigenvalues and eigenstates of $\hat{H}_\text{P}$ and $\hat{H}_\text{R}$ separately, and thereby construct a basis for the state space of the whole system. This basis, called the *decoupled basis*, can then be used to express the third term $\hat{H}_\text{I}$, the interaction term, which involves both particle and radiation observables.

The Hamiltonian $\hat{H}_\text{P}$:

$$\hat{H}_\text{P} = \sum_\mu \frac{\hat{\mathbf{p}}_\mu^2}{2m_\mu} + V_\text{coul}(\hat{\mathbf{r}}_1, \ldots, \hat{\mathbf{r}}_\mu, \ldots), \tag{6.65}$$

describes a set of charges undergoing Coulomb interaction and with kinetic energies $\hat{\mathbf{p}}_\mu^2/2m_\mu$. It corresponds to the simplest model of an atom, encountered in elementary quantum mechanics textbooks. Its eigenstates $|i\rangle$ are associated with energies E_i by

$$\hat{H}_\text{P}|i\rangle = E_i\,|i\rangle. \tag{6.66}$$

These energy levels form a discrete sequence for bound states, and a continuum for ionized states. The set of states $|i\rangle$ constitutes a basis for the state space describing the particles.

The second term,

$$\hat{H}_\text{R} = \sum_\ell \hbar\omega_\ell\left(\hat{a}_\ell^\dagger\hat{a}_\ell + \frac{1}{2}\right), \tag{6.67}$$

is the free radiation Hamiltonian. Its set of eigenstates $|n_1, \ldots, n_\ell, \ldots\rangle$ (see Section 4.6.1) generates the state space of the radiation.

The state space of the full system field–particles is the tensor product of the particle state space and the radiation state space. A basis for this global space can be obtained by taking the tensor products of particle and radiation basis states. This basis thus contains all states of the form $|i\rangle \otimes |n_1, \ldots, n_\ell, \ldots\rangle$, which we shall denote by $|i; n_1, \ldots, n_\ell, \ldots\rangle$. Such a state represents an atom in state $|i\rangle$ in the presence of n_1 photons in mode $1, \ldots, n_\ell$ photons in mode ℓ. It is an eigenstate of $\hat{H}_\text{P} + \hat{H}_\text{R}$ with

$$(\hat{H}_P + \hat{H}_R)|i; n_1, \dots n_\ell, \dots\rangle = \left(E_i + \sum_\ell n_\ell \hbar\omega_\ell + E_V\right)|i; n_1, \dots n_\ell \dots\rangle, \tag{6.68}$$

where E_V is the energy of the vacuum given by (4.121).

The third term of (6.64),

$$\hat{H}_I = -\sum_\mu \frac{q_\mu}{m_\mu}\hat{\mathbf{p}}_\mu \cdot \hat{\mathbf{A}}(\hat{\mathbf{r}}_\mu) + \sum_\mu \frac{q_\mu^2}{2m_\mu}\left(\hat{\mathbf{A}}(\hat{\mathbf{r}}_\mu)\right)^2, \tag{6.69}$$

describes the interaction between particles and radiation. It acts on the state space describing the full system, generated by the basis states $|i; n_1, \dots, n_\ell, \dots\rangle$. But when we take into account the interaction Hamiltonian $\hat{H}_I$, the state $|i; n_1, \dots, n_\ell, \dots\rangle$ is no longer an eigenstate of the total Hamiltonian (6.64), and this state will thus evolve into a new state, which is a superposition of states of the form $|f; n'_1, \dots, n'_\ell, \dots\rangle$: both the atom and the radiation have changed state. The *interaction Hamiltonian* $\hat{H}_I$ is thus responsible for *transitions in which the atom changes state, while photons are absorbed or re-emitted*, although the total energy of the system will remain the same since the total Hamiltonian is time independent.

Note a crucial difference with the semi-classical approach described in Chapter 2: the interaction term is time independent here. There are therefore stationary states and well-defined values of the energy of the total system comprising interacting particles and radiation. These are the eigenstates of the total Hamiltonian (see, for example, Complement 6B).

Comments

(i) The three terms in the Hamiltonian $\hat{H}$ all act in the particle–field tensor product space. Strictly speaking, the first term should be written in the form, $\hat{H}_P \otimes \hat{1}_R$, where $\hat{1}_R$ is the identity operator on the radiation state space. Likewise, we should write $\hat{1}_P \otimes \hat{H}_R$ for the second term. We shall omit these identity operators, according to common practice, thereby simplifying the notation.

(ii) To write (6.69), we had to use the fact that $\hat{\mathbf{A}}(\hat{\mathbf{r}}_\mu)$ commutes with $\hat{\mathbf{p}}_\mu$, which follows from the choice of the Coulomb gauge. This can be checked directly using expression (6.58) for $\hat{\mathbf{A}}$, replacing $\mathbf{r}$ by $\hat{\mathbf{r}}_\mu$ and using the $\mathbf{r}$ representation, where $\hat{\mathbf{r}}_\mu$ is simply $\mathbf{r}_\mu$, while $\hat{\mathbf{p}}_\mu$ becomes $\frac{\hbar}{i}\nabla_{\mathbf{r}_\mu}$, and remembering that $\mathbf{k}_\ell \cdot \boldsymbol{\varepsilon}_\ell = 0$.

(iii) All the considerations discussed above also apply to the case where the potential V_{Coul} in the particle Hamiltonian (6.51) results from an externally imposed electrostatic field. This is exploited in the model for an atom in which an electron is immersed in the Coulomb potential of the nucleus, the latter being located at the coordinate origin.

6.2.3 Interaction Hamiltonian

Long-wavelength approximation

As in Section 2.2.4 of Chapter 2, we shall simplify the form of the interaction Hamiltonian by assuming that the spatial variation of the electromagnetic field can be neglected on the atomic length scale. We thus replace the vector potential $\hat{\mathbf{A}}(\hat{\mathbf{r}}_\mu)$ at the positions of the electrons by its value at the position $\mathbf{r}_0$ of the nucleus, and furthermore, we treat the latter as a classical quantity. Restricting to the case of an atom with a single electron to simplify the notation, the interaction Hamiltonian then becomes[3]

$$\hat{H}_{\mathrm{I}} = -\frac{q}{m}\hat{\mathbf{p}}\,.\,\hat{\mathbf{A}}(\mathbf{r}_0) + \frac{q^2}{2m}\left(\hat{\mathbf{A}}(\mathbf{r}_0)\right)^2. \tag{6.70}$$

Comments

(i) Application of the long-wavelength approximation requires a little more care here than it did in Chapter 2. Indeed, we must be sure that the relevant physical processes really do only involve modes whose wavelengths are much longer than the atomic length scale. Now we shall find throughout this chapter that vacuum modes can play a role. So it is not enough to check that the incident radiation satisfies the long-wavelength condition.[4]

(ii) It is interesting to ask whether there is a transformation generalizing the Göppert–Mayer transformation introduced in Chapter 2 (see Section 2.2.4), which led to an electric dipole interaction Hamiltonian, of the form

$$\hat{H}'_{\mathrm{I}} = -\hat{\mathrm{D}}\,.\,\hat{\mathrm{E}}(r_0), \tag{6.71}$$

in the long-wavelength approximation. Such a transformation does in fact exist (see CDG I, Complement $\mathrm{A_{IV}}$), but the total Hamiltonian is not simply the sum of $\hat{H}_{\mathrm{P}}$, $\hat{H}_{\mathrm{R}}$ and $\hat{H}'_{\mathrm{I}}$. Indeed, as a result of the transformation, an extra term appears in the particle Hamiltonian. This term plays an important role when calculating the radiative corrections to the atomic energy levels (Lamb shift). However, it can often be ignored in absorption and emission calculations, which can then be treated using the electric dipole Hamiltonian $\hat{H}'_{\mathrm{I}}$.

The choice between the forms (6.71) or (6.70) is often dictated purely by calculational convenience. Note that (6.70) is quadratic in the field, whereas the electric dipole Hamiltonian is linear, which sometimes simplifies calculations. However, there is no general rule here, as can be seen from the calculation of Thomson scattering discussed later in this chapter, where (6.70) leads to the simpler treatment.

Decomposition of the interaction Hamiltonian

The interaction Hamiltonian (6.70) is a sum of two terms, one linear and one quadratic in $\hat{\mathbf{A}}_\perp$. As in Chapter 2, we shall denote these two terms by $\hat{H}_{\mathrm{I1}}$ and $\hat{H}_{\mathrm{I2}}$. To simplify the notation, we set $\mathbf{r}_0 = 0$ in (6.70), i.e. we take the atomic nucleus to remain fixed at the coordinate origin. We then obtain

$$\hat{H}_{\mathrm{I1}} = -\frac{q}{m}\sum_\ell \sqrt{\frac{\hbar}{2\varepsilon_0\omega_\ell L^3}}\;\hat{\mathbf{p}}\,.\,\boldsymbol{\varepsilon}_\ell(\hat{a}_\ell + \hat{a}_\ell^\dagger) \tag{6.72}$$

$$\hat{H}_{\mathrm{I2}} = \frac{q^2}{2m}\frac{\hbar}{2\varepsilon_0 L^3}\sum_j\sum_\ell \frac{\boldsymbol{\varepsilon}_j\,.\,\boldsymbol{\varepsilon}_\ell}{\sqrt{\omega_j\,.\,\omega_\ell}}(\hat{a}_j\hat{a}_\ell^\dagger + \hat{a}_j^\dagger\hat{a}_\ell + \hat{a}_j\hat{a}_\ell + \hat{a}_j^\dagger\hat{a}_\ell^\dagger). \tag{6.73}$$

It should be borne in mind that this form has been established in the context of the long-wavelength approximation.

[3] We neglect terms relative to the nucleus, smaller by a factor m/M, where M is the mass of the nucleus.

[4] See CDG I, Chapter IV.

Comment As noted in Chapter 2 (Comment (iv) in Section 2.2.4), there are physical situations where the position of the nucleus cannot be made to disappear. For example, when the atom is moving, the position of the nucleus must be expressed as explicitly depending on time, and this amounts to taking into account the Doppler effect. One can go even further, considering that the motion of the nucleus itself (or more exactly, the motion of the atomic centre of mass) must be quantized. One must then return to expression (6.58) for the radiation at position $\mathbf{r}_0$, and replace $\mathbf{r}_0$ by the position operator $\hat{\mathbf{r}}_0$ of the atomic centre of mass. In this situation, the Hamiltonian for the atom must contain a kinetic energy term $\hat{\mathbf{p}}_0^2/2M$, where $\hat{\mathbf{p}}_0$ is the momentum operator of the atomic centre of mass, conjugate to $\hat{\mathbf{r}}_0$, and M is the mass of the atom (see, for example, Complement 8A).

6.3 Interaction processes

6.3.1 The Hamiltonian $\hat{H}_{I1}$

The interaction Hamiltonian $\hat{H}_{I1}$ is linear in $\hat{a}_\ell$ and $\hat{a}_\ell^\dagger$ (see Equation (6.72)). In first-order perturbation theory (see Chapter 1), it can only therefore induce transitions between states in which *the number of photons differs by unity*. If the final state has one photon more than the initial state, we have an *emission process*. If the final state has one photon less than the initial state, we have an *absorption process*.

The Hamiltonian $\hat{H}_{I1}$ also contains the atomic operator $\hat{\mathbf{p}}$. Since this operator is odd, the diagonal elements in its matrix representation are zero (because atomic states have definite parity). Transitions induced by $\hat{H}_{I1}$ necessarily involve *a change of atomic state between two levels of opposite parity*.

6.3.2 Absorption

Consider the atom in a state $|a\rangle$, in the presence of n_j photons in some specific mode j, assuming that all other modes are empty. We use the simplified notation

$$|\phi_i\rangle = |a; n_1 = 0, \ldots n_j \ldots 0\rangle = |a; n_j\rangle \tag{6.74}$$

for this initial state, which is an eigenstate of the unperturbed Hamiltonian $\hat{H}_P + \hat{H}_R$ (see Section 6.2.2). Under the effect of $\hat{H}_{I1}$, the system can move into a final state containing one photon less:

$$|\phi_f\rangle = |b; n'_j = n_j - 1\rangle,$$

which is also an eigenstate of $\hat{H}_P + \hat{H}_R$.

Using (4.115),

$$\hat{a}_j|n_j\rangle = \sqrt{n_j}|n_j - 1\rangle, \tag{6.75}$$

we obtain

$$\langle b; n_j - 1|\hat{H}_{I1}|a; n_j\rangle = -\frac{q}{m}\sqrt{\frac{\hbar}{2\varepsilon_0\omega_j L^3}}\,\langle b|\hat{\mathbf{p}}\,.\,\boldsymbol{\varepsilon}_j|a\rangle\sqrt{n_j}. \tag{6.76}$$

Here we may apply the general results of Chapter 1 for the evolution of a system under the influence of a constant perturbation. In particular, the transition between eigenstates of the unperturbed Hamiltonian only occurs with a significant probability between two states of the same energy. This condition concerns the total system, and is expressed here by

$$E_a + n_j\hbar\omega_j = E_b + (n_j - 1)\hbar\omega_j, \tag{6.77}$$

whence,

$$E_b = E_a + \hbar\omega_j. \tag{6.78}$$

The final state of the atom has a higher energy than the initial state, and the process has maximal probability if the energy $\hbar\omega_j$ of the photon that disappears is exactly equal to the energy gained by the atom. Here we have the *resonance condition*, already encountered in the semi-classical treatment of Chapter 2. Moreover, *we have now justified the interpretation of the absorption process in terms of photons*, as shown schematically in Figure 6.1.

The results from perturbation theory provide other information. For example, we know that, in the perturbative limit, the transition probability is proportional to the squared modulus of the matrix element (6.76), i.e. to the number of photons n_j, which is itself proportional to the wave intensity (see Section 5.3.7).

Comment We have made very different use of perturbation theory here to investigate the absorption process, as compared with what was done in the semi-classical context. In Chapter 2, the system under consideration was an atom which evolved between two different energy levels under the effect of a perturbation depending sinusoidally on time. Here, the quantum system is the full system comprising the atom and

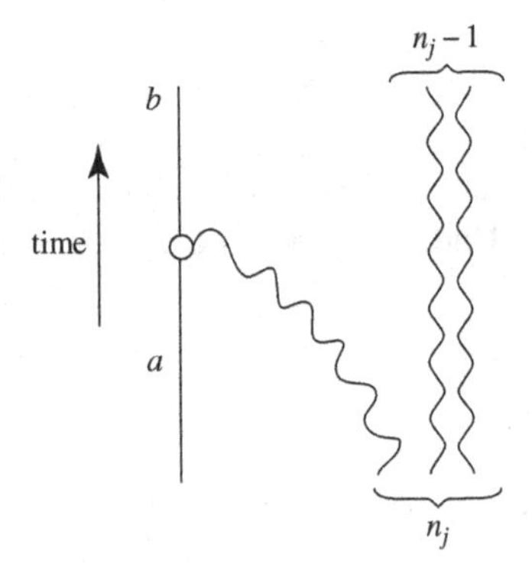

Figure 6.1 The absorption process. The diagram should be read from bottom to top. A photon disappears and the atom goes from state $|a\rangle$ to state $|b\rangle$.

photons. It evolves between two states of the same energy under the effect of a constant perturbation considered to be switched on during the interaction time.

6.3.3 Emission

Now assume that the atom is initially in a state $|b\rangle$, in the presence of n_j photons, and that it evolves to a final state $|a; n_j + 1\rangle$ containing one more photon. Using the relation

$$\hat{a}_j^\dagger |n_j\rangle = \sqrt{n_j + 1}\, |n_j + 1\rangle, \tag{6.79}$$

we can calculate the matrix element,

$$\langle a; n_j + 1|\hat{H}_{I1}|b; n_j\rangle = -\frac{q}{m}\sqrt{\frac{\hbar}{2\varepsilon_0\omega_j L^3}}\langle a|\hat{\mathbf{p}}\,.\,\boldsymbol{\varepsilon}_j|b\rangle\sqrt{n_j + 1}. \tag{6.80}$$

Conservation of energy between the initial and final states is expressed by

$$E_a = E_b - \hbar\omega_j. \tag{6.81}$$

The process described here is the *emission* of a photon. The atom goes from level b to level a of a lower energy, by emitting a photon. Figure 6.2 shows this process schematically.

If the number of photons n_j is much greater than 1, the probability of the process is, in the perturbative approximation, proportional to n_j, i.e. to the intensity of the wave. Here we recognize stimulated emission, a process that becomes more probable as the intensity of the incident radiation increases.

But (6.79) reveals another phenomenon. *Even when there are no incident photons* ($n_j = 0$), the transition amplitude is non-zero, given by

$$\langle a; n_j = 1|\hat{H}_{I1}|b; 0\rangle = -\frac{q}{m}\sqrt{\frac{\hbar}{2\varepsilon_0\omega_j L^3}}\langle a|\hat{\mathbf{p}}\,.\,\boldsymbol{\varepsilon}_j|b\rangle. \tag{6.82}$$

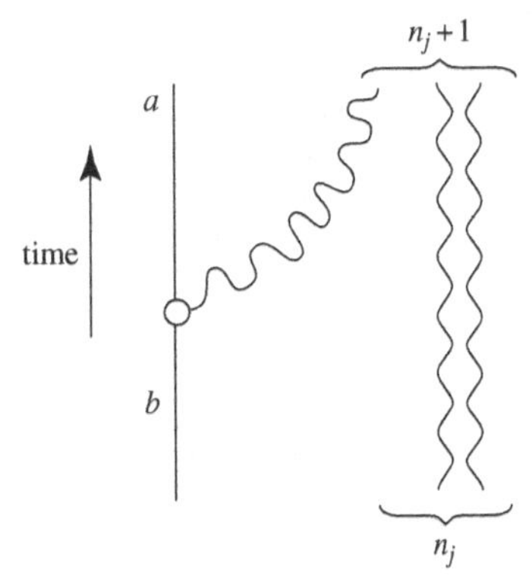

Figure 6.2 An emission process. The atom goes from state $|b\rangle$ to state $|a\rangle$ by emitting a photon. If $n_j = 0$, this is spontaneous emission.

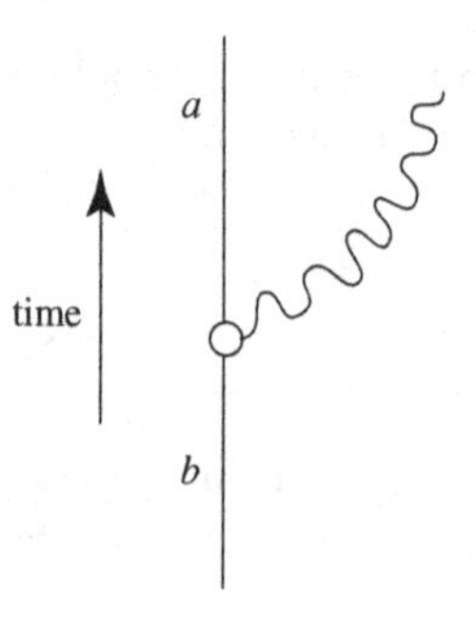

Figure 6.3 **Spontaneous emission. This is nothing other than Figure 6.2, with no photon in the initial state.**

An isolated atom, placed in a vacuum, is thus able to emit a photon, provided that there is a lower energy state for which the matrix element (6.82) is non-zero. The energy of the emitted photon is equal to the energy lost by the atom (see (6.81)). The process described here is *spontaneous emission*. It will be discussed in more detail in Section 6.4.

Comments

(i) Equation (6.79) describes both stimulated and spontaneous emission. In the expression $n_j + 1$, term n_j is usually associated with stimulated emission and term 1 with spontaneous emission. Comparing with (6.75), the symmetry between stimulated emission and absorption is clearly apparent.

(ii) If mode j is initially empty, while mode ℓ contains n_ℓ photons, the matrix element responsible for emission of a photon in mode j remains the same, i.e.

$$\langle a; n_j = 1, n_\ell | \hat{H}_{I1} | b; n_j = 0, n_\ell \rangle = -\frac{q}{m}\sqrt{\frac{\hbar}{2\varepsilon_0 \omega_j L^3}} \langle a | \hat{\mathbf{p}} \,.\, \boldsymbol{\varepsilon}_j | b \rangle. \tag{6.83}$$

The spontaneous emission rate in a given mode does not therefore depend on the presence or otherwise of photons in other modes.

It should not be thought, however, that the presence of photons in mode ℓ has no effect on the situation. If we consider an excited atom in state $|b\rangle$ in the presence of n_ℓ photons in mode ℓ, it will have a higher probability of emitting in mode ℓ than in an initially empty mode. A set of excited atoms will therefore emit preferentially in a mode that is already populated, not because the spontaneous emission rate in the empty modes gets smaller, but because the induced emission in the populated mode ℓ occurs before spontaneous emission in the initially empty mode j has had time to take place. This property is one of the working principles of the laser amplifier.

6.3.4 Rabi oscillation

In Chapter 1 (see Section 1.2.6), we obtained the exact solution for the dynamical equation governing a two-level system subject to a constant interaction coupling the two available states. If the system is initially in one of the two states, the probability of finding it in the other state is a sinusoidal function of time. This is known as Rabi oscillation.

These results can be applied to the total system comprising the atom and radiation in mode j. Let $|\phi_i\rangle = |a; n_j\rangle$ be the initial state, where the atom is in state $|a\rangle$ and the initial number of photons in mode j is n_j (see Equation (6.74)). Under the effect of $\hat{H}_{I1}$, state $|\phi_i\rangle$ is coupled to $|\phi_f\rangle = |b; n_j - 1\rangle$. *Neglecting the possibility of spontaneous emission in other modes*, we may restrict attention to the two states $|\phi_i\rangle$ and $|\phi_f\rangle$, between which the Rabi oscillation occurs. Coupling between the two levels can be written in the form

$$\langle\phi_f|\hat{H}_{I1}|\phi_i\rangle = \frac{\hbar\Omega_1}{2}, \tag{6.84}$$

where Ω_1 is the Rabi angular frequency, equal to (see (6.72) and (6.75))

$$\Omega_1 = -\frac{q}{m}\frac{2}{\sqrt{2\varepsilon_0\hbar\omega_j L^3}}\langle b|\hat{\mathbf{p}}\,.\,\boldsymbol{\varepsilon}_j|a\rangle\sqrt{n_j}. \tag{6.85}$$

The results of Chapter 1, and in particular (1.63), then give the probability of finding the system in state $|\phi_f\rangle$ after a time T:

$$P_{i\to f}(T) = \frac{\Omega_1^2}{\delta^2 + \Omega_1^2}\sin^2\left(\sqrt{\Omega_1^2 + \delta^2}\,\frac{T}{2}\right), \tag{6.86}$$

where we have introduced the detuning,

$$\delta = \omega_j - \frac{E_b - E_a}{\hbar}. \tag{6.87}$$

We thus obtain the Rabi oscillation under the effect of quasi-resonant radiation for the probability of finding the atom in state $|a\rangle$ or state $|b\rangle$.

Comments

(i) The Rabi oscillation under the effect of radiation was discussed in Chapter 2. We described it there as the oscillation of a quantum system (the atom) between two states ($|a\rangle$ and $|b\rangle$) with different energies, under the effect of a sinusoidal coupling with frequency $\omega \cong (E_b - E_a)/\hbar$. Here the system is considered as an atom–photon ensemble making transitions between two states of the same energy, i.e. the atomic ground state with n_j photons, and the excited atomic state with one photon less. The coupling is constant and applied for a time interval T.

(ii) The different processes described in Sections 6.3.2, 6.3.3 and 6.3.4 depend on the value of the matrix element $\langle a|\hat{\mathbf{p}}\cdot\boldsymbol{\varepsilon}_j|b\rangle$. Let us recall here that we have shown in Chapter 2 that this quantity is proportional to the matrix element $\langle a|\hat{\mathbf{r}}\cdot\boldsymbol{\varepsilon}_j|b\rangle$ of the electron position operator, and therefore of the atomic dipole $\hat{\mathbf{D}} = q\hat{\mathbf{r}}$. From (2.96), we obtain more precisely that

$$\langle a|\hat{\mathbf{p}}\cdot\boldsymbol{\varepsilon}_j|b\rangle = -\mathrm{i}m\omega_0\langle a|\hat{\mathbf{r}}\cdot\boldsymbol{\varepsilon}_j|b\rangle = -\mathrm{i}\frac{m\omega_0}{q}\langle a|\hat{\mathbf{D}}\cdot\boldsymbol{\varepsilon}_j|b\rangle. \tag{6.88}$$

6.3.5 The Hamiltonian $\hat{H}_{I2}$ and elastic scattering

As can be seen from (6.73), in the long-wavelength approximation, the Hamiltonian $\hat{H}_{I2}$ only acts on radiation. As far as the atom is concerned, it behaves just like the identity operator. The transitions caused by $\hat{H}_{I2}$ thus occur *without changing the atomic state*, affecting only the radiation.

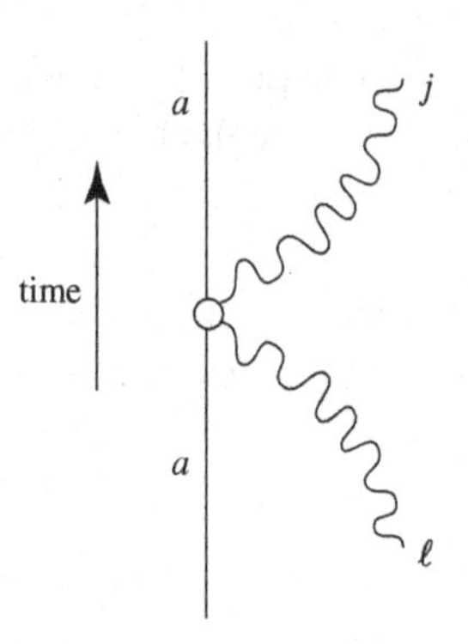

Figure 6.4 **Scattering of a mode ℓ photon to mode j without change in the atomic state. This process conserves energy for $\omega_j = \omega_\ell$. We then have elastic scattering.**

The various processes associated with the four terms in $\hat{H}_{12}$ do not all occur with the same probability. The term $\hat{a}_j^\dagger \hat{a}_\ell^\dagger$ describes the appearance of two photons. It does not respect equality of the initial and final energies, and the associated probability is thus very small. The same goes for $\hat{a}_j \hat{a}_\ell$, which describes the disappearance of two photons. These two process are thus very unlikely to occur.

On the other hand, a term like $\hat{a}_j^\dagger \hat{a}_\ell$ describes the disappearance of one photon in mode ℓ and the appearance of a photon in mode j. This process can conserve energy if the frequencies ω_j and ω_ℓ are equal. The photons j and ℓ then *differ in their direction of propagation and their polarization*. This is an *elastic scattering process*, without change of frequency, depicted in the diagram of Figure 6.4.

The matrix element for such a process is obtained from (6.73). If mode ℓ initially contains n_ℓ photons, with mode j being initially empty, we have

$$\langle a; n_j = 1, n_\ell - 1|\hat{H}_{12}|a; n_j = 0, n_\ell\rangle = \frac{q^2}{m} \frac{\hbar}{2\varepsilon_0 L^3} \frac{\boldsymbol{\varepsilon}_j \cdot \boldsymbol{\varepsilon}_\ell}{\omega_j} \sqrt{n_\ell}. \tag{6.89}$$

In this expression, we have used the equality of the incident and scattered frequencies,

$$\omega_\ell = \omega_j, \tag{6.90}$$

and the fact that there are *two terms* ($\hat{a}_j^\dagger \hat{a}_\ell$ and $\hat{a}_\ell \hat{a}_j^\dagger$) associated with the same process in the sum of (6.73). Equation (6.89) will be used later (Section 6.5) to calculate the *scattering cross-section*. Note immediately that the probability of the scattering process described here is proportional to the squared modulus of (6.89) and hence to the number n_ℓ of photons, or indeed the intensity of the radiation.

Comment Equation (6.89) describes scattering from an occupied mode ℓ to an initially unoccupied mode j. If mode j were also occupied in the initial state, the matrix element (6.89) would be multiplied by a factor $\sqrt{n_j + 1}$, and the process would be much more likely. This would be *stimulated scattering* from mode ℓ to mode j. This is a *nonlinear process*, and its probability is proportional to the *product of the light intensities* in modes j and ℓ.

6.4 Spontaneous emission

6.4.1 Principle of the calculation

Discrete level coupled to a continuum

We have seen in Section 6.3.3 that an atom in an excited state $|b\rangle$ can spontaneously de-excite to a state $|a\rangle$ with lower energy E_a by emitting a spontaneous photon in mode j. In this process, the atom–radiation system goes from state $|b, 0\rangle$ (excited atom in vacuum) to state $|a, 1_j\rangle$ (atom in its ground state and one photon in mode j). We know that such a transition, induced by a time-independent coupling, will only occur if the initial and final states have very similar energies, and the emitted photon will then have energy $\hbar\omega_j$ almost equal to $E_b - E_a$, denoted by $\hbar\omega_0$:

$$\hbar\omega_j \simeq E_b - E_a = \hbar\omega_0. \tag{6.91}$$

This process can thus be represented as in Figure 6.5, which shows the same phenomenon as in Figure 6.3 in another guise. To simplify the discussion, we assume here that there is only one state with energy lower than b to which the atom can de-excite by emitting a photon.

As the modes of the electromagnetic field have frequencies forming a quasi-continuum, the final states $|a; 1_j\rangle$ of the process have energies that also form a quasi-continuum, and we may apply the results obtained in Chapter 1 for the coupling between a discrete level and a continuum. Recall that this formalism can be used to carry out a perturbative calculation, via the Fermi golden rule, of a *departure rate* Γ_{sp} from the discrete level. Here any departure leads to the appearance of a spontaneous photon and the departure rate can thus be interpreted as a *spontaneous emission rate*, i.e. a *rate of photon production*. This rate will be calculated in the following sections. Furthermore, we can use the method of Wigner and Weisskopf (see Section 1.3.2d) to obtain the key result in this section, namely, that the probability of the atom remaining in the excited state $|b\rangle$ *decreases exponentially* with time constant Γ_{sp}^{-1}, which is interpreted as the *radiative lifetime* of state $|b\rangle$,

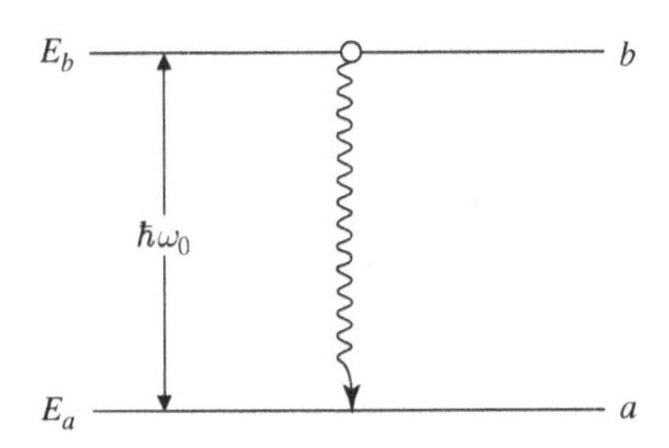

Figure 6.5 Spontaneous emission between an excited state $|b\rangle$ and a state $|a\rangle$ of lower energy. The emitted photon has energy close to $\hbar\omega_0 = E_b - E_a$. The direction of emission and the polarization are random variables. Their probability distribution gives the radiation pattern of the atom, depending on the characteristics of states $|a\rangle$ and $|b\rangle$.

$$\tau_{sp} = \frac{1}{\Gamma_{sp}}. \tag{6.92}$$

In fact the application of Fermi's golden rule, spelt out in detail in Sections 6.4.3 and 6.4.4, gives the *differential spontaneous emission rate of a photon with given direction and polarization*, the energy being fixed by (6.91) (the modes of the electromagnetic field still form a quasi-continuum versus frequency, even at a fixed direction of emission and polarization). The integral of this quantity over all emission directions and polarizations then gives the total spontaneous emission rate Γ_{sp}. The ratio of the differential rate and the total rate is used to find the *radiation pattern* (or radiation diagram) for the transition $|b\rangle \to |a\rangle$, i.e. the angular distribution of the emitted radiation, for each polarization. We shall see that, in free space, this pattern is determined (see Equation (6.109)), by the matrix element,

$$\langle a|\hat{\mathbf{p}}\,.\,\boldsymbol{\varepsilon}_\ell|b\rangle, \tag{6.93}$$

which, up to a multiplicative constant, depends only on the angular momenta of states $|a\rangle$ and $|b\rangle$. In order to see a complete calculation, we consider a particular transition that we shall now specify, and then we shall be able to give an explicit demonstration of the procedure just presented.

Transition $(l = 1, m = 0) \to (l = 0, m = 0)$

We consider the particular case of an atom with just one electron, whose ground state $|a\rangle$ has zero orbital angular momentum ($l = 0$, and hence $m_l = 0$, where m_l relates to the component $\hat{L}_z = \mathbf{e}_z\,.\,\hat{\mathbf{L}}$ of the angular momentum in the Oz direction). For the excited state, we take a state $|b\rangle$ with orbital angular momentum ($l = 1, m_l = 0$). It is then easy to show that

$$\langle a|\hat{p}_x|b\rangle = \langle a|\hat{p}_y|b\rangle = 0, \tag{6.94}$$

so that the matrix element (6.93) can be expressed simply as a function of $\langle a|\hat{p}_z|b\rangle$, i.e.

$$\langle a|\hat{\mathbf{p}}\,.\,\boldsymbol{\varepsilon}_j|b\rangle = \langle a|\hat{p}_z|b\rangle\left(\mathbf{e}_z\,.\,\boldsymbol{\varepsilon}_j\right). \tag{6.95}$$

The fact that the matrix elements in (6.94) are zero can be shown by recalling that $\langle a|\hat{p}_x|b\rangle$ and $\langle a|\hat{p}_y|b\rangle$ are proportional to $\langle a|\hat{x}|b\rangle$ and $\langle a|\hat{y}|b\rangle$, respectively (see Equation (2.96)). The fact that the matrix elements of $\hat{x}$ and $\hat{y}$ are zero in a transition $(l = 1, m = 0) \to (l = 0, m = 0)$ follows from the angular dependence of the corresponding wavefunctions (see Complement 2B). For example, we have

$$\langle a|\hat{x}|b\rangle = \int r^2 \sin\theta\, \mathrm{d}r\, \mathrm{d}\theta\, \mathrm{d}\varphi\, R_a^*(r)\, Y_0^{0*}(\theta,\varphi)\, r\sin\theta\cos\varphi\, R_b(r)\, Y_1^0(\theta,\varphi), \tag{6.96}$$

which is zero, because the spherical harmonics Y_0^0 and Y_1^0 do not depend on angle φ. The same goes for $\langle a|\hat{y}|b\rangle$, which is given as above but replacing

$$x = r\sin\theta\cos\varphi \tag{6.97}$$

by

$$y = r \sin\theta \sin\varphi. \tag{6.98}$$

However, remembering that

$$z = r \cos\theta, \tag{6.99}$$

we see that there is in principle no reason why the matrix element $\langle a|\hat{z}|b\rangle$, and hence $\langle a|\hat{p}_z|b\rangle$, should be zero.

6.4.2 Quasi-continuum of one-photon states and density of states

To apply the Fermi golden rule, one must know the final *density of states* $|a; 1_j\rangle$, which is the same as the *density of one-photon states* $|1_j\rangle$. In fact, we only need to study the *density of modes j* over which the radiation has been quantized. We assume that the modes are defined as in Chapter 4, using the periodic boundary conditions for a cubic discretization volume of side L. The modes are then plane waves with wavevectors,

$$\mathbf{k}_j = \frac{2\pi}{L}\left(n_x\mathbf{e}_x + n_y\mathbf{e}_y + n_z\mathbf{e}_z\right) \tag{6.100}$$

and frequencies,

$$\omega_j = ck_j. \tag{6.101}$$

In these relations, $\mathbf{e}_x$, $\mathbf{e}_y$ and $\mathbf{e}_z$ are unit vectors in an orthogonal triad and the numbers n_x, n_y and n_z are integers.

Two modes are associated with each wavevector $\mathbf{k}_j$, corresponding to two polarizations $\boldsymbol{\varepsilon}$ and $\boldsymbol{\varepsilon}'$, forming a triad with $\mathbf{k}_j$ (see Figure 4.1 in Chapter 4). If $\boldsymbol{\varepsilon}$ is chosen in the plane $(\mathbf{e}_z, \mathbf{k}_j)$, the vector $\boldsymbol{\varepsilon}'$ is perpendicular to $\mathbf{e}_z$, and the matrix element (6.95) is zero for the transition considered here. We thus ignore modes polarized perpendicularly to $\mathbf{e}_z$, and in the density of relevant states, count a single mode for each wavevector $\mathbf{k}_j$, namely the one polarized in plane $(\mathbf{e}_z, \mathbf{k}_j)$. If θ is the angle between $\mathbf{e}_z$ and $\mathbf{k}_j$, the matrix element (6.95) is proportional to

$$\mathbf{e}_z \cdot \boldsymbol{\varepsilon}_j = \sin\theta. \tag{6.102}$$

A single wavevector $\mathbf{k}_j$, and hence a single point in the reciprocal space, corresponds to each of these modes (see Figure 6.6). Equation (6.100) then shows that the full set of these points constitutes a cubic lattice, in which the unit cell has volume $(2\pi/L)^3$. The density of states in the reciprocal space is the reciprocal of this volume, and is thus equal to $(L/2\pi)^3$. But this is not yet the quantity we need, because in order to apply Fermi's golden rule, we require the density of states as a function of the energy.

The energy of state $|1_j\rangle$ is

$$E_j = \hbar\omega_j = \hbar ck_j. \tag{6.103}$$

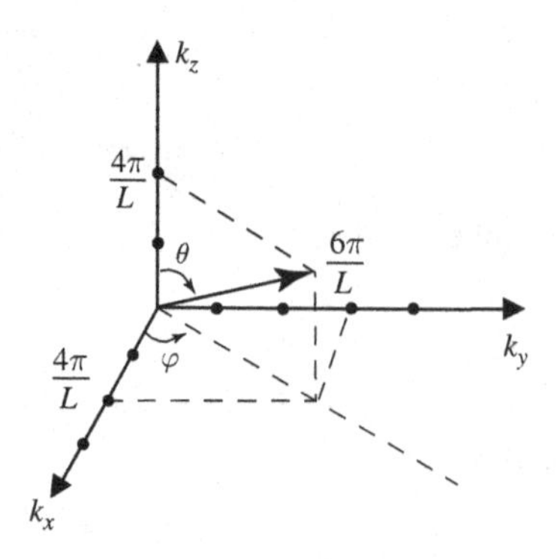

Figure 6.6 Representing the modes in reciprocal space. As an example, mode ($n_x = 2, n_y = 3, n_z = 2$) has been represented by its wavevector. Direction $\mathbf{k}_j$ is specified by the angles θ and φ. The set of endpoints of vectors $\mathbf{k}_j$ associated with the modes defined by (6.100) constitutes a cubic lattice with side $2\pi/L$.

As L tends to infinity, Equation (6.100) tells us that the energies E_j become more and more closely spaced,[5] whereupon we do indeed have a quasi-continuum of states.

If we specify a state $|1_j\rangle$ by its energy E_j and by the direction (θ, φ) of the wavevector $\mathbf{k}_j$ (Figure 6.6), the density of states to be determined is $\rho(\theta, \varphi; E)$, defined by

$$\mathrm{d}N = \rho(\theta, \varphi; E)\, \mathrm{d}E\, \mathrm{d}\Omega, \tag{6.104}$$

where $\mathrm{d}N$ is the number of states with energy in the range between E and $E + \mathrm{d}E$ and whose wavevector points into the solid angle $\mathrm{d}\Omega$ about the direction (θ, φ). The endpoints of the corresponding wavevectors lie between spheres of radii k and $k + \mathrm{d}k$, corresponding to energies $E/\hbar c$ and $(E + \mathrm{d}E)/\hbar c$, and they point into the solid angle $\mathrm{d}\Omega$. Given that $E = \hbar ck$, the corresponding volume is

$$\mathrm{d}^3k = k^2 \mathrm{d}k \mathrm{d}\Omega = \frac{E^2}{(\hbar c)^3}\, \mathrm{d}E\, \mathrm{d}\Omega. \tag{6.105}$$

Dividing this volume by the volume $(2\pi/L)^3$ associated with each mode, we obtain the number of states in the elementary volume (6.105) as

$$\mathrm{d}N = \left(\frac{L}{2\pi}\right)^3 \frac{E^2}{(\hbar c)^3}\, \mathrm{d}E\, \mathrm{d}\Omega, \tag{6.106}$$

whence the required density of states is

$$\rho(\theta, \varphi; E) = \left(\frac{L}{2\pi}\right)^3 \frac{E^2}{(\hbar c)^3}. \tag{6.107}$$

[5] Recall that the whole procedure of letting L tend to infinity is only valid in free space. In a cavity, the quantization volume has a physical meaning, and the results obtained here are no longer necessarily valid (see Complement 6B).

Comments

(i) The density of states (6.107) is isotropic, i.e. it is independent of the direction of propagation. This is to be expected, since space is isotropic.

(ii) Expression (6.107) for the density of states is only completely unambiguous if it is associated with the defining relation (6.104). We could have defined another density of states $\rho'(\theta,\varphi)$ by the relation $\mathrm{d}N = \rho'(\theta,\varphi)\mathrm{d}E\mathrm{d}\theta\mathrm{d}\varphi$, and this would clearly lead to the same physical predictions. The advantage of definition (6.104) based on the solid angle is that the resulting density (6.107) is explicitly isotropic, whereas $\rho'(\theta,\varphi)$ depends on θ, and this might misleadingly suggest that there is no isotropy.

6.4.3 Spontaneous emission rate in a given direction

To apply Fermi's golden rule, we must take the final density of states of the total system $|a;1_j\rangle$ around the energy value E_a+E_j equal to the energy E_b of the initial state $|b;0\rangle$. This density is equal to the density of the radiation modes around the value $E_j = E_b - E_a = \hbar\omega_0$ (see Equation (6.91)). This leads to the spontaneous emission rate per unit solid angle in the direction (θ,φ) (see Equation (1.97)):

$$\frac{\mathrm{d}\Gamma_{\mathrm{sp}}}{\mathrm{d}\Omega} = \frac{2\pi}{\hbar}\left|\langle a;1_j|\hat{H}_{I1}|b;0\rangle\right|^2 \rho(\theta,\varphi;E_j=\hbar\omega_0). \tag{6.108}$$

Replacing the matrix element of $\hat{H}_{I1}$ by the expression in (6.72) and the density of states by (6.107), and taking into account (6.95) and (6.102), we arrive at an expression that is independent of the side L of the discretization box. In the particular case of a transition $(l=1,m=0)\to(l=0,m=0)$, we obtain

$$\frac{\mathrm{d}\Gamma_{\mathrm{sp}}}{\mathrm{d}\Omega}(\theta,\varphi) = \frac{q^2}{8\pi^2\varepsilon_0}\frac{\omega_0}{m^2\hbar c^3}\left|\langle a|\hat{p}_z|b\rangle\right|^2\sin^2\theta. \tag{6.109}$$

The term $\sin^2\theta$ describes the spontaneous emission pattern for a transition $(l=1,m=0)\to(l=0,m=0)$. The emission is zero in the Oz direction ($\theta=0$), and maximal in the directions perpendicular to Oz. Recall also that, for such a transition, the emitted light has polarization $\boldsymbol{\varepsilon}$ in the plane $(Oz,\mathbf{k})$ and perpendicular to the emission direction $\mathbf{k}$.

Comments

(i) The radiation diagram obtained here is the same as the one obtained by the classical calculation of radiation from a dipole oscillating in the Oz direction (see Complement 2A). This similarity can be attributed to the particular choice of transition $(l=1,m=0)\to(l=0,m=0)$.

Had we considered a transition $(l=1,m=\pm1)\to(l=0,m=0)$, we would also have found a radiation pattern like the one for a classical dipole rotating in the (xOy) plane. These similarities are due to the vectorial nature of the operator $\hat{\mathbf{p}}$ arising in the Hamiltonian $\hat{H}_{I1}$, and the particular choice of a transition $(\ell=1)\to(\ell=0)$.

(ii) For an arbitrary transition, the spontaneous emission rate in a given direction must be calculated for two mutually orthogonal polarizations $\boldsymbol{\varepsilon}$ and $\boldsymbol{\varepsilon}'$ (both perpendicular to the emission direction), which gives two different rates:

$$\frac{\mathrm{d}\Gamma_{\mathrm{sp}}}{\mathrm{d}\Omega}(\theta,\varphi,\boldsymbol{\varepsilon}) \quad\propto\quad |\langle a|\mathbf{p}.\boldsymbol{\varepsilon}|b\rangle|^2 \tag{6.110}$$

and

$$\frac{\mathrm{d}\Gamma_{\mathrm{sp}}}{\mathrm{d}\Omega}(\theta,\varphi,\boldsymbol{\varepsilon}') \quad \propto \quad \left|\langle a|\mathbf{p}\,.\boldsymbol{\varepsilon}'|b\rangle\right|^2. \tag{6.111}$$

The total spontaneous emission rate in direction (θ,φ) is the sum of the two rates in (6.110) and (6.111).

The two polarizations $\boldsymbol{\varepsilon}$ and $\boldsymbol{\varepsilon}'$ can be complex, in order to describe circular or elliptical polarizations. For a given transition and emission direction, one can then find a pair of mutually orthogonal polarizations (in the sense of Hilbert space, i.e. $\boldsymbol{\varepsilon}'\,.\,\boldsymbol{\varepsilon}^* = 0$), such that one of the two corresponds to a zero matrix element (6.111). This method is a generalization of the calculation carried out for the transition $(l = 1, m = 0) \rightarrow (l = 0, m = 0)$.

Apart from simplifying the calculation, the approach provides useful indications concerning the polarization of the emitted light. For example, for the transition $(l = 1, m = 0) \rightarrow (l = 0, m = 0)$, we found a linear polarization in the plane containing the Oz-axis. For a transition $(l = 1, m = \pm 1) \rightarrow (l = 0, m = 0)$, we would find that the light emitted in an arbitrary direction is elliptically polarized. But for emission in the Oz direction, the light is circularly polarized, while for emission perpendicular to Oz, the light is linearly polarized, in the plane perpendicular to Oz. Once again, the result is analogous to the one obtained using classical electromagnetism for a dipole rotating in the plane (xOy).

In general, for a transition $(l = 1) \rightarrow (l = 0)$, the polarization of the emitted light is as would be predicted by classical electromagnetism for a classical dipole suitably chosen with regard to the magnetic sub-level m of the atom in the initial state.

6.4.4 Lifetime of the excited state and natural width

To find the total spontaneous emission rate, one simply integrates the differential rate (6.109) over all emission directions:

$$\Gamma_{\mathrm{sp}} = \int \frac{\mathrm{d}\Gamma_{\mathrm{sp}}}{\mathrm{d}\Omega}\mathrm{d}\Omega. \tag{6.112}$$

In spherical coordinates, the element of solid angle is given by

$$\mathrm{d}\Omega = \sin\theta\,\mathrm{d}\theta\,\mathrm{d}\varphi. \tag{6.113}$$

Given (6.112), the angle integral in (6.109) is thus,

$$\int_0^{2\pi}\mathrm{d}\varphi\int_0^{\pi}\mathrm{d}\theta\,\sin^3\theta = \frac{8\pi}{3}. \tag{6.114}$$

So finally, for the transition $(l = 1, m = 0) \rightarrow (l = 0, m = 0)$,

$$\Gamma_{\mathrm{sp}} = \frac{q^2}{3\pi\varepsilon_0}\frac{\omega_0}{m^2\hbar c^3}\left|\langle a|\hat{p}_z|b\rangle\right|^2. \tag{6.115}$$

This rate corresponds to the *total spontaneous emission probability per unit time*. The *radiative lifetime* of the state $|b\rangle$ is the reciprocal of this quantity (see Equation (6.92)).

Comment If state $|b\rangle$ can de-excite by spontaneous emission to several lower energy states $|a_1\rangle, |a_2\rangle \ldots$, the total probability of spontaneous emission is the sum of the probabilities for each transition, and we thus have a rate,

$$\Gamma_{sp} = \Gamma_{sp}^{(1)} + \Gamma_{sp}^{(2)} + \ldots, \tag{6.116}$$

which implies the radiative lifetime,

$$\frac{1}{\tau} = \frac{1}{\tau_1} + \frac{1}{\tau_2} + \ldots \tag{6.117}$$

This situation often occurs for highly excited states. It also arises for transitions to the ground state when the latter is degenerate. In this case, one must sum the transition rates for all the sub-levels of the ground state to obtain the radiative lifetime of the excited state.

As indicated in Chapter 1 (see Equation (1.91)), the energy distribution in the final state of the process is Lorentzian with full-width at half-maximum equal to the transition rate. Spontaneously emitted photons thus have a frequency spectrum centred on ω_0 and a full-width at half-maximum equal to Γ_{sp}. More precisely, the probability per unit frequency band of observing a spontaneously emitted photon near the frequency ω is given by

$$\frac{dP}{d\omega} = \frac{\Gamma_{sp}}{2\pi} \frac{1}{(\omega - \omega_0)^2 + \frac{\Gamma_{sp}^2}{4}}. \tag{6.118}$$

The full-width at half-maximum of this curve, called the *natural width* of the radiative transition between the two states, is therefore equal here to the spontaneous emission rate Γ_{sp} of the excited state.[6]

Using (6.88), we may also write,

$$\Gamma_{sp} = \frac{1}{3\pi\varepsilon_0} \frac{\omega_0^3}{\hbar c^3} \left| \langle a|\hat{D}_z|b\rangle \right|^2. \tag{6.119}$$

This expression lends itself well to lifetime calculations, e.g. in the case of the hydrogen atom, where the wavefunctions are known. For example, in the case of the *Lyman α transition*, from level $2p$ to level $1s$, the calculation gives a rate $\Gamma_{sp} = 6 \times 10^8\ \text{s}^{-1}$, which corresponds to the measured lifetime of 1.6 ns.

It may also be useful to express Γ_{sp} in terms of the oscillator strength of the transition, a dimensionless quantity defined by

$$f_{ab} = \frac{2m\omega_0 \left| \langle a|\hat{z}|b\rangle \right|^2}{\hbar}. \tag{6.120}$$

One then has

$$\Gamma_{sp} = f_{ab} \Gamma_{cl}, \tag{6.121}$$

where Γ_{cl} is defined by

$$\Gamma_{cl} = \frac{q^2}{6\pi\varepsilon_0} \frac{\omega_0^2}{mc^3}, \tag{6.122}$$

which is the reciprocal of the classical radiative damping of an elastically bound electron (see Complement 2A).

[6] The natural width of the transition only coincides with the reciprocal of the lifetime of the excited state if the lower energy state in the transition is stable, i.e. has zero width.

For transitions between the ground state and the first excited level of a one-electron atom, the oscillator strength is of order 1, making (6.120) particularly useful. For example, for the Lyman α transition of hydrogen, we have $f_{ab} = 0.42$, and for the transition $3S_{1/2} \rightarrow 3P_{1/2}$ of sodium, $f_{ab} = 0.33$.

Comments

(i) When the matrix elements of $\hat{p}_x$ and $\hat{p}_y$ between $|a\rangle$ and $|b\rangle$ are non-zero, and for a state $|b\rangle$ with zero orbital angular momentum l, Equation (6.121) remains true, provided that the definition (6.120) is generalized to

$$f_{ab} = \frac{2m\omega_0}{\hbar} \left|\langle a|\hat{\mathbf{r}}|b\rangle\right|^2 . \tag{6.123}$$

(ii) As can be seen from the above expressions, the spontaneous emission rate increases with the transition frequency (as ω_0^2 for constant oscillator strength). It is thus much bigger for a transition in the ultra-violet than for a transition in the infra-red. Typically, radiative lifetimes are of nanosecond order in the ultra-violet, ten nanoseconds in the visible and microsecond order in the near infra-red. In the far infra-red, and a fortiori in the microwave range, spontaneous emission is a very weak process. (It nevertheless plays a role in astrophysics, e.g. for the 21 cm emission by interstellar hydrogen.)

(iii) Spontaneous emission increases faster with frequency than induced emission. This property is due to the fact that induced emission occurs in a specific mode, whereas spontaneous emission leads to photon emission in a number of modes that increases as ω_0^2, owing to the frequency dependence (6.107) of the density of states. This is the reason why it is generally easier to obtain a laser effect at longer wavelengths (see Comment (i) of 3.2.1).

(iv) For reasons of symmetry, it may happen that the matrix element (6.119) of the transition is zero. Spontaneous emission cannot then occur. This is called a *forbidden line*. This is the case between two levels of the same parity, since the position or electric dipole operators are odd operators. An interesting example is the level $2S$ ($n = 2$, $\ell = 0$) of hydrogen, which cannot decay by electric dipole transition to the ground state $1S$ ($n = 1$, $\ell = 0$), of the same parity. The only other energy level lower than $2S$ is the level $2P_{1/2}$, but the corresponding transition has frequency 1 GHz, in the microwave range, and the corresponding spontaneous emission rate is therefore negligible. The $2S$ level is thus *metastable*, i.e. it is stable despite the fact that it is not the minimal energy. In fact, this level does have a small probability of decaying to the $1S$ state by spontaneous emission of two photons. The rate for this decay is calculated as above but using the second-order perturbative interaction Hamiltonian. The lifetime associated with this higher-order process is of the order of one second.

(v) It can be shown that all Zeeman sub-levels of a given level (n, l) have the same radiative lifetime.

6.4.5 Spontaneous emission: a joint property of the atom and the vacuum

The calculation of the rate of spontaneous emission is an example of a calculation that can be done only with the fully quantum formalism of atom–radiation interaction. This is sufficient to justify its importance. In addition, this calculation has allowed physicists to realize that spontaneous emission is not an intrinsic property of the atom alone, it is in fact a property of the atom coupled to the quantized vacuum.

This statement is more than an academic remark, if one notices that it is possible to change the properties of the quantized vacuum by adding boundary conditions. The result is to modify the density of states of the modes into which the spontaneous photon can be emitted, which entails a modification of the spontaneous emission rate (see for instance Equation (6.108)) compared to the case of spontaneous emission in free space.[7]

This has important consequences. It is for instance possible to inhibit spontaneous emission if the density of states is null around ω_0. On the contrary, it is possible to enhance spontaneous emission, or to favour a specific direction of emission. This effect was first observed in the microwave domain, by placing in a wave guide an atom in a highly excited state able to de-excite by microwave emission only.[8] Inhibition or enhancement of spontaneous emission were observed, depending on the value of the density of states in the wave guide available for spontaneous emission. A similar effect has been demonstrated at optical wavelengths.[9] Many more examples can be found in experiments of 'cavity quantum electrodynamics' (Complement 6B), and in particular the famous Purcell effect (Section 6B.4). Other means have been considered to control spontaneous emission, such as use of 'photonic crystals' or 'disordered superlattices' (footnote 5 in Section 6B.2.3), and these may lead to important applications.

6.5 Photon scattering by an atom

6.5.1 Scattering matrix elements

A scattering process is one in which, in the presence of an atom, a photon disappears from one mode to reappear in another. The final atomic state may be the same as or different from the initial atomic state. In the first case, owing to energy conservation, the scattered photon has the same frequency as the incident photon and the scattering is *elastic*. However, if the internal atomic energy is altered during the process, we have *inelastic scattering*, with a change in the frequency of the radiation. This is *Raman scattering*.[10]

We showed in Section 6.3.5 that the interaction Hamiltonian $\hat{H}_{I2}$ can induce scattering without change in the atomic state (see Figure 6.4). This is therefore elastic scattering. Since $\hat{H}_{I2}$ is quadratic in the vector potential, the scattering appears in first-order perturbation theory and is given in terms of the transition matrix element by (6.89).

Another mechanism can lead to a scattering process, namely the one described by the action of $\hat{H}_{I1}$ when we carry out a second-order perturbation calculation as in Section 1.2.5.

[7] D. Kleppner, Inhibited Spontaneous Emission, *Physical Review Letters* **47**, 233 (1981).

[8] R. G. Hulet, E. S. Hilfer and D. Kleppner, Inhibited Spontaneous Emission by a Rydberg Atom, *Physical Review Letters* **55**, 2137 (1985).

[9] W. Jhe *et al.*, Suppression of Spontaneous Decay at Optical Frequencies: Tests of Vacuum-Field Anisotropy in Confined Space, *Physical Review Letters* **58**, 666 (1987) and **58**, 1497 (1987).

[10] Here we assume implicitly that the initial atomic level is non-degenerate. If this is not so, scattering may occur in such a way that the atom changes state between two sub-levels but there is nevertheless no change in the energy of the level.

We have seen how a transition can be described in second-order perturbation theory, by introducing an effective Hamiltonian whose matrix element between the initial and final states is given in the case of a single incident photon ($n_\ell = 1$) scattered into mode j, by

$$\langle a'; 1_j|\hat{H}_{\mathrm{I}}^{\mathrm{eff}}|a; 1_\ell\rangle = \sum_b \frac{\langle a'; 1_j\left|\hat{H}_{\mathrm{I1}}\right|b; 0\rangle\langle b; 0\left|\hat{H}_{\mathrm{I1}}\right|a; 1_\ell\rangle}{E_a - E_b + \hbar\omega_\ell} + \sum_b \frac{\langle a'; 1_j\left|\hat{H}_{\mathrm{I1}}\right|b; 1_j, 1_\ell\rangle\langle b; 1_j, 1_\ell\left|\hat{H}_{\mathrm{I1}}\right|a; 1_\ell\rangle}{E_a - E_b - \hbar\omega_j}. \tag{6.124}$$

The first sum in (6.124) describes a process for which the intermediate scattering state is $|b; 0\rangle$ (atom in state $|b\rangle$ and no photons). This first term is illustrated in Figure 6.7. The second sum in (6.124) corresponds to the diagram in Figure 6.8, since the intermediate scattering state $|b; 1_j, 1_\ell\rangle$ contains two photons (one in mode j and the other in mode ℓ).

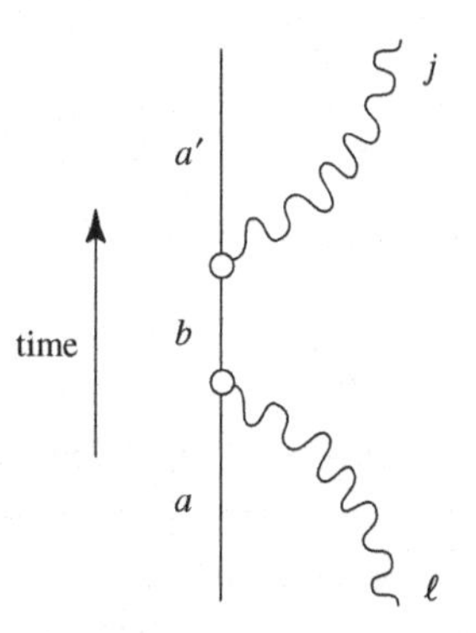

Figure 6.7 Scattering of a mode ℓ photon into mode j under the action of the Hamiltonian $\hat{H}_{\mathrm{I1}}$, to second order in perturbation theory.

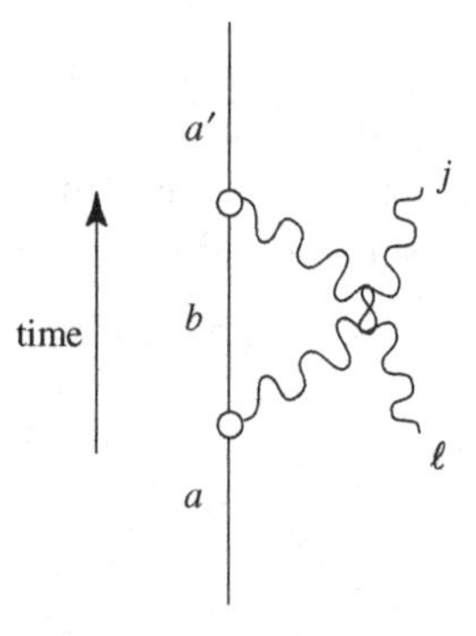

Figure 6.8 Second process for scattering of a mode ℓ photon into a mode j photon under the action of $\hat{H}_{\mathrm{I1}}$, to second order in perturbation theory. The intermediate state in which the atom and the field have gained in energy is highly non-resonant.

Since the interaction Hamiltonian is independent of time, the transition can only occur if the initial state $|a; 1_\ell\rangle$ and the final state $|a'; 1_j\rangle$ have the same energy:

$$E_a + \hbar\omega_\ell = E_{a'} + \hbar\omega_j. \tag{6.125}$$

However, no energy conservation condition is imposed on the intermediate stage in the process, and E_b is generally different from $E_a + \hbar\omega_\ell$. The excitation of the state $|b\rangle$ is said to be *virtual*. It should also be noted that, while the process shown in Figure 6.7 is the more intuitive of the two, the diagram in Figure 6.8 corresponds to the, in principle, much less obvious process in which the atom emits a photon in mode j before absorbing a photon in mode ℓ. In this case, the intermediate state in the scattering process is $|b; 1_j, 1_\ell\rangle$. This state, where the atom is in the excited state $|b\rangle$ and the field has an extra photon, clearly has a much higher energy than does the initial state. It must nevertheless be taken into account.

Using (6.72), Equation (6.124) may be rewritten in the form

$$\begin{aligned}\langle a'; 1_j \left|\hat{H}_{\mathrm{I}}^{\mathrm{eff}}\right| a; 1_\ell\rangle = \frac{q^2\hbar}{m^2\varepsilon_0 L^3\sqrt{\omega_j\omega_\ell}} \sum_b \Bigg[&\frac{\langle a'|\mathbf{p}\,.\,\boldsymbol{\varepsilon}_j|b\rangle\langle b|\mathbf{p}\,.\,\boldsymbol{\varepsilon}_\ell|a\rangle}{E_a - E_b + \hbar\omega_\ell} \\ &+ \frac{\langle a'|\mathbf{p}\,.\,\boldsymbol{\varepsilon}_\ell|b\rangle\langle b|\mathbf{p}\,.\,\boldsymbol{\varepsilon}_j|a\rangle}{E_a - E_b - \hbar\omega_j}\Bigg].\end{aligned} \tag{6.126}$$

If the incident mode ℓ contains n_ℓ photons, the coupling matrix element is then,

$$\langle a'; 1_j, n_\ell - 1|\hat{H}_{\mathrm{I}}^{\mathrm{eff}}|a; 0_j, n_\ell\rangle = \sqrt{n_\ell}\langle a'; 1_j, 0_\ell|\hat{H}_{\mathrm{I}}^{\mathrm{eff}}|a; 0_j, 1_\ell\rangle, \tag{6.127}$$

which is equal to the matrix element (6.126) associated with a single incident photon, multiplied by $\sqrt{n_\ell}$. As for the scattering term associated with $\hat{H}_{12}$ (see (6.90)), the scattering amplitude is proportional to the square root of the number of incident photons.

6.5.2 Scattering cross-section

As for spontaneous emission, the scattering process couples a discrete state with a quasi-continuum of final states, since the modes j form such a continuum (see Section 6.4). The Fermi golden rule can thus be applied to calculate a scattering probability per unit time (transition rate) from mode ℓ to the quasi-continuum, proportional to the squared modulus of the transition matrix element. This scattering rate is therefore proportional to the number n_ℓ of photons present in the incident mode.

By analogy with (6.108), we write the scattering rate from mode ℓ (containing n_ℓ photons) to an initially empty mode j, whose wavevector $\mathbf{k}_j$ points into the solid angle $\mathrm{d}\Omega$ about the direction (θ, φ), in the form,

$$\frac{\mathrm{d}\Gamma}{\mathrm{d}\Omega}(\boldsymbol{\varepsilon}_j, \theta, \varphi) = \frac{2\pi}{\hbar} n_\ell \left|\langle a'; 1_j|\hat{H}_{\mathrm{I}}|a; 1_\ell\rangle\right|^2 \rho(\theta, \varphi; \hbar\omega_j = \hbar\omega_\ell + E_a - E_{a'}). \tag{6.128}$$

In this equation, the interaction Hamiltonian $\hat{H}_{\mathrm{I}}$ is either $\hat{H}_{12}$ or $\hat{H}_{\mathrm{I}}^{\mathrm{eff}}$ (see Equation (6.124)), or more generally the sum of the two. Using the expressions (6.89) or (6.126) for the matrix elements, and (6.107) for the density of states, we find that the right-hand side of (6.128) is proportional to L^{-3}. On the face of things, this looks rather surprising, since the volume L^3, chosen arbitrarily, should drop out of the final result.

In fact, as explained in Section 5.3.7 of Chapter 5, the *intensity* of a travelling wave associated with mode ℓ is not characterized by the number n_ℓ of photons in mode ℓ, but rather by the *photon current* Π_ℓ^{phot} (photon flux per unit area and per unit time), which is related to the irradiance Π_ℓ (incident power per unit area) by

$$\Pi_\ell^{\text{phot}} = \frac{\Pi_\ell}{\hbar\omega_\ell}. \tag{6.129}$$

The photon current Π_ℓ^{phot} can be related to the number n_ℓ of photons in the mode by remembering that there are n_ℓ/L^3 photons per unit volume, travelling at speed c:

$$\Pi_\ell^{\text{phot}} = c\frac{n_\ell}{L^3}. \tag{6.130}$$

Replacing n_ℓ by $\Pi_\ell^{\text{phot}}L^3/c$ in (6.128), we obtain a relation that no longer involves L^3, whence we may define the *differential scattering cross-section* $\frac{\mathrm{d}\sigma}{\mathrm{d}\Omega}(\boldsymbol{\varepsilon}_j,\theta,\varphi)$ by

$$\frac{\mathrm{d}\Gamma}{\mathrm{d}\Omega}(\boldsymbol{\varepsilon}_j,\theta,\varphi) = c\frac{n_\ell}{L^3}\frac{\mathrm{d}\sigma}{\mathrm{d}\Omega}(\boldsymbol{\varepsilon}_j,\theta,\varphi) = \Pi_\ell^{\text{phot}}\frac{\mathrm{d}\sigma}{\mathrm{d}\Omega}(\boldsymbol{\varepsilon}_j,\theta,\varphi). \tag{6.131}$$

Integrating over all directions (θ,φ) and summing over the two orthogonal polarizations $\boldsymbol{\varepsilon}_j'$ and $\boldsymbol{\varepsilon}_j''$ of two modes associated with each direction, we then obtain the *total scattering cross-section* for a mode ℓ photon:

$$\sigma = \int \mathrm{d}\Omega\left(\frac{\mathrm{d}\sigma}{\mathrm{d}\Omega}(\boldsymbol{\varepsilon}_j',\theta,\varphi) + \frac{\mathrm{d}\sigma}{\mathrm{d}\Omega}(\boldsymbol{\varepsilon}_j'',\theta,\varphi)\right). \tag{6.132}$$

The cross-section has the physical dimensions of an area. Heuristically, it can be thought of as the effective area intercepting photons in the incident flux in such a way as to scatter them.

6.5.3 Qualitative description of some scattering processes

Rayleigh scattering

Rayleigh scattering is a low-energy elastic scattering process. By low-energy process, we mean that the energy of the photon is small compared with the energies required to actually raise the atom to an excited state. The Rayleigh scattering process can be represented by the diagram in Figure 6.9.

Starting from (6.89) and (6.126), it can be shown that,[11] for a two-level atom (Bohr frequency ω_0), the total cross-section for Rayleigh scattering is equal to

$$\sigma_{\text{R}} = \frac{8\pi}{3}r_0^2\frac{\omega^4}{\omega_0^4}. \tag{6.133}$$

[11] See CDG II, Exercise 3.

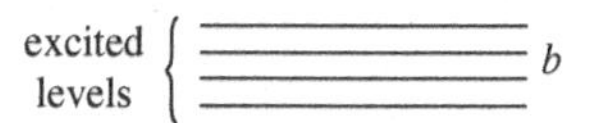

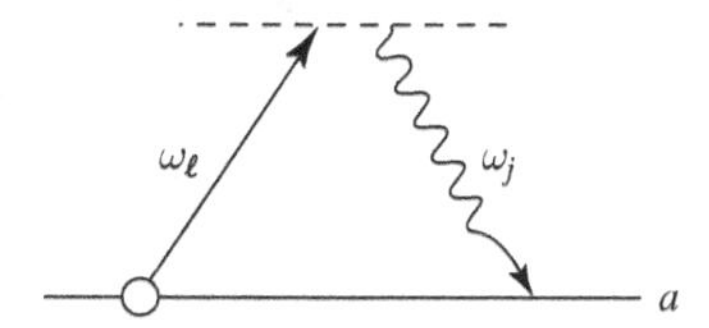

Figure 6.9 **Rayleigh scattering. The energy of the photon is smaller than the energies of the excited states.**

In this formula, ω is the frequency of the incident (and scattered) photons, and r_0 is the so-called *classical electron radius* (see Complement 2A):

$$r_0 = \frac{q^2}{4\pi\varepsilon_0 mc^2} \tag{6.134}$$

equal to 2.8×10^{-15}m.

Note that the Rayleigh scattering cross-section σ_R varies as the fourth power of the angular frequency ω, i.e. short wavelength radiation is more scattered than long wavelength radiation. Hence, in the visible range, scattering is greater for blue light than for red light, which explains the blue colour of the sky, since the scattering of the sun's light by molecules in the atmosphere is largely Rayleigh scattering, associated with electronic molecular resonances in the ultra-violet.

Comment Far from resonance, the model provided by the two-level atom is not very realistic. However, it can be shown that (6.133) remains valid, provided that we replace ω_0 by an effective atomic resonance frequency, defined by

$$\frac{1}{(\omega_0^{\text{eff}})^2} = \hbar^2 \sum_{b \neq a} \frac{f_{\text{ab}}}{(E_b - E_a)^2}, \tag{6.135}$$

where f_{ab} is the oscillator strength for the transition $a \leftrightarrow b$ defined by (6.123).

Thomson scattering

Thomson scattering is a high-energy elastic scattering process, shown schematically in Figure 6.10. The term 'high energy' means that the photon energy is large compared with the ionization energy E_{I} of the atom.

It will be shown in Section 6.5.4 that the Thomson scattering cross-section is equal to

$$\sigma_{\text{T}} = \frac{8\pi}{3} r_0^2, \tag{6.136}$$

where r_0 is the length defined in (6.134). This cross-section is independent of the frequency of the incident photons. Thomson scattering causes the attenuation of X rays when they

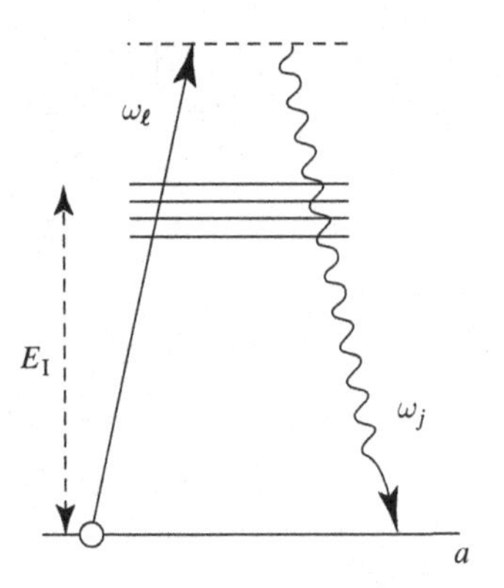

Figure 6.10 Thomson scattering. The energy of the incident photon is greater than the energy of all the excited atomic states.

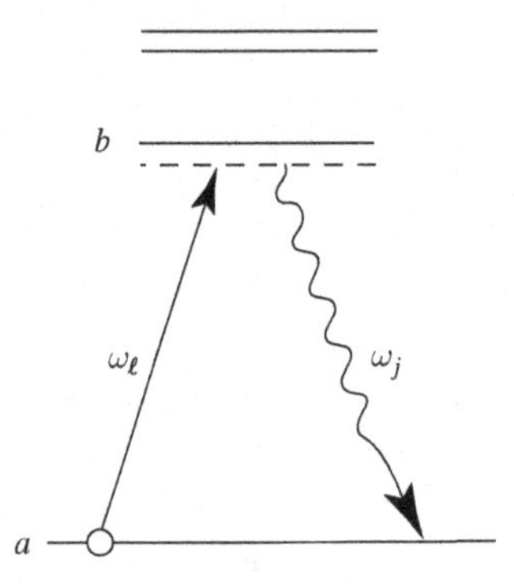

Figure 6.11 Resonant scattering. The energy of the incident photon is very close to the energy of one of the excited atomic states.

propagate through matter. It increases with the number of electrons. The electron density of a solid can be determined by measuring this attenuation. Measurements of this kind played an important role in determining the atomic numbers of certain elements, still under investigation at the beginning of the twentieth century. It is also this phenomenon that gives rise to the different X-ray transparencies of biological tissues, well known in medical applications. So-called 'heavy' elements, i.e. with large numbers of electrons Z, are not very transparent to X-rays because they tend to scatter more.

Resonant scattering

When the angular frequency of the incident photon is close to the atomic angular frequency ω_0, the situation is intermediate between the two cases considered above. This is resonant scattering, shown schematically in Figure 6.11.

In this process, absorption of a mode ℓ photon in the ground state is practically resonant, since the transition of the atom into the excited state $|b\rangle$ is a step in the scattering process that conserves energy. If we analyse the three possible scattering diagrams (Figures 6.4, 6.7

and 6.8), we observe that only the diagram in Figure 6.7 has an intermediate step in which the incident photon is absorbed and the atom is in the excited state. It follows that this diagram makes the dominant contribution to the resonant scattering calculation and that it describes the actual physical course taken by the process. To be more precise, note that when $\hbar\omega_\ell$ is very close to $E_b - E_a$, a divergence appears in the first term of (6.126). This divergence arises because we are using second-order perturbation theory. A more complete treatment, bringing in higher-order terms in the perturbation series, eliminates the problem. The result is to add an imaginary term $i\hbar\Gamma_{\text{sp}}/2$ to the resonant denominator of (6.126),[12] where Γ_{sp} is the natural width of level $|b\rangle$ given by (6.115). If the quasi-resonant scattering cross-section (i.e. for $\hbar\omega_\ell \approx E_b - E_a = \hbar\omega_0$) is calculated in this way, the result is[13]

$$\sigma_{\text{res}} = \frac{3\lambda_0^2}{2\pi} \frac{1}{1 + 4\frac{(\omega-\omega_0)^2}{\Gamma_{\text{sp}}^2}}, \tag{6.137}$$

where ω is the frequency of the incident (and scattered) photons and $\lambda_0 = 2\pi c/\omega_0$ is the resonance wavelength. For exact resonance, we obtain the value

$$\sigma_{\text{res}} = \frac{3\lambda_0^2}{2\pi}. \tag{6.138}$$

Note that the scattering cross-section at exact resonance only depends on the resonance frequency and not on the specific characteristics of the atomic levels (although Γ_{sp} does depend on those characteristics). For frequencies associated with visible light, the numerical value of this resonant cross-section is much higher than the values of the Rayleigh and Thomson scattering cross-sections given by (6.133) and (6.136), respectively. Indeed, a simple calculation shows that, in the visible range, σ_{res} is some 16 orders of magnitude greater than σ_{T} (λ_0 is of the order of 10^{-6} m, while r_0 is 2.8×10^{-15} m), and the Rayleigh scattering cross-section is smaller than the Thomson scattering cross-section. This resonant scattering cross-section, corresponding to dimensions of the order of the light wavelength, is considerably bigger than what one might consider to be atomic dimensions (the Bohr radius being 0.5×10^{-10} m).

The high numerical value makes it easy to observe resonant scattering with the naked eye on an atomic vapour (*optical resonance*). It has even been possible to observe light scattered by a single trapped atom or ion, illuminated by a resonant laser. This surprising fact is easy to understand when one remembers that a 0.1 milliwatt laser beam carries 10^{15} photons per second. If the beam has a diameter of 1 centimetre, the ion will scatter some 10^8 photons per second, and an observer placed 0.3 m away will pick up about 10^4 photons per second, which is detectable by the human eye.

Comments

(i) The classical calculation of the scattering of an electromagnetic wave by an elastically bound electron (see Complement 2A) yields scattering cross-sections with the same form as (6.133),

[12] See CDG II, Chapter III.
[13] See CDG II, Exercise 5.

(6.136) and (6.137). While the quantum theoretical value (6.135) of the Rayleigh scattering cross-section involves the oscillator strength of the transitions, i.e. the matrix elements depending on the atomic wavefunctions, it is remarkable to observe that the Thomson and resonant scattering cross-sections, (6.136) and 6.138), respectively, obtained by the two approaches are absolutely identical.

It can also be shown that the angular and polarization dependence of the differential cross-sections obtained by classical calculation are the same as those obtained by quantum theoretical calculation when the atom is in a state $|a\rangle$ with angular momentum $l = 0$.

(ii) Equation (6.137) was established to lowest order with regard to incident field intensity and is thus only applicable when the incident intensity is small compared with a characteristic intensity, $\Pi_{\text{sat}} = \dfrac{\pi}{3}\dfrac{hc\Gamma_{\text{sp}}}{\lambda_0^3}$ called the *saturation intensity* (typically of the order of a few milliwatts per square centimetre). When the intensity increases, two new phenomena arise. One is that the intensity of light scattered without frequency change is no longer proportional to the incident intensity. This is the saturation phenomenon. The other is that nonlinear processes occur, so that photons appear with different frequencies from those of the incident radiation (see Figure 6.13).

Raman scattering

Raman scattering is a low-energy inelastic scattering process. In contrast to the processes considered so far, the final atomic state $|a'\rangle$ differs from the initial atomic state $|a\rangle$, and the frequency of the scattered photon differs from the frequency of the incident photon. The change in frequency of the radiation is, according to (6.125), directly related to the energy interval between levels a and a':

$$\hbar(\omega_\ell - \omega_j) = E_{a'} - E_a. \tag{6.139}$$

Raman scattering is shown schematically in Figure 6.12.

Raman scattering is particularly useful in molecular physics, because it can be used to measure energy intervals between vibrational or rotational levels belonging to the same electronic ground state. For such levels, direct absorption with $|a\rangle$ going to $|a'\rangle$ is often forbidden, while Raman scattering is allowed.

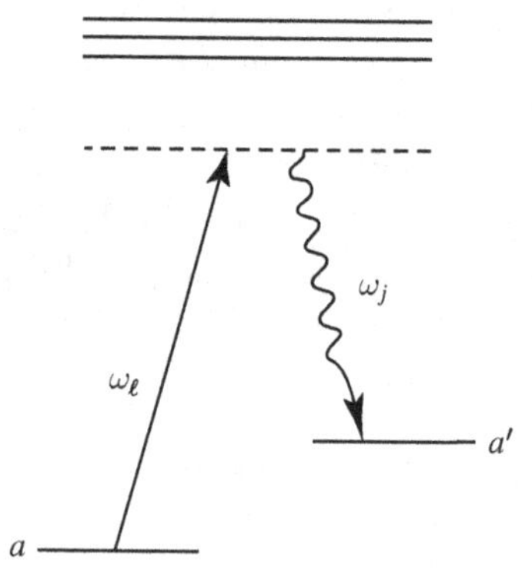

Figure 6.12 Raman scattering. The energies of the scattered and incident photons are not the same. This is inelastic scattering.

6.5.4 Thomson scattering cross-section

In this section, we shall exemplify the calculation of scattering cross-sections by working out the result for Thomson scattering in some detail. As mentioned previously, in order to calculate a scattering cross-section, one must add the first-order contribution of $\hat{H}_{I2}$ to the second-order contribution of $\hat{H}_{I1}$. The case of Thomson scattering is particularly simple, because the second-order effects of $\hat{H}_{I1}$ can be neglected, as we shall see.

Second-order contribution of $\hat{H}_{I1}$

The importance of this contribution can be assessed using (6.126) for the matrix element,

$$\langle a;1_j|\hat{H}_{\mathrm{I}}^{\mathrm{eff}}|a;1_\ell\rangle = \frac{q^2}{m^2}\frac{\hbar}{2\varepsilon_0 L^3\omega}\sum_b\left(\frac{\langle a|\hat{\mathbf{p}}\,.\,\boldsymbol{\varepsilon}_j|b\rangle\langle b|\hat{\mathbf{p}}\,.\,\boldsymbol{\varepsilon}_\ell|a\rangle}{E_a-E_b+\hbar\omega}\right.$$
$$\left.+\frac{\langle a|\hat{\mathbf{p}}\,.\,\boldsymbol{\varepsilon}_\ell|b\rangle\langle b|\hat{\mathbf{p}}\,.\,\boldsymbol{\varepsilon}_j|a\rangle}{E_a-E_b-\hbar\omega}\right), \tag{6.140}$$

where we have set $\omega=\omega_\ell=\omega_j$, since Thomson scattering is elastic.

In the Thomson scattering scenario, the energy $\hbar\omega$ of the photons is very high compared with the difference E_b-E_a between the atomic energies. This means that E_b-E_a can be neglected in comparison with $\hbar\omega$ in the denominators of the fractions appearing in (6.140). Using the closure relation,

$$\sum_b |b\rangle\langle b| = 1,$$

Equation (6.140) can be rewritten in the form

$$\langle a;1_j|\hat{H}_{\mathrm{I}}^{\mathrm{eff}}|a;1_\ell\rangle = \frac{q^2}{m^2 2\varepsilon_0 L^3\omega^2}\langle a|[(\hat{\mathbf{p}}\,.\,\boldsymbol{\varepsilon}_j)(\hat{\mathbf{p}}\,.\,\boldsymbol{\varepsilon}_\ell)-(\hat{\mathbf{p}}\,.\,\boldsymbol{\varepsilon}_\ell)(\hat{\mathbf{p}}\,.\,\boldsymbol{\varepsilon}_j)]|a\rangle. \tag{6.141}$$

As any two components of $\hat{\mathbf{p}}$ commute, we obtain

$$\langle a;1_j|\hat{H}_{\mathrm{I}}^{\mathrm{eff}}|a;1_\ell\rangle = 0. \tag{6.142}$$

The second-order contribution of $\hat{H}_{I1}$ is thus negligible under the conditions of Thomson scattering.[14]

Frequency domain of Thomson scattering

Before calculating the Thomson scattering cross-section, it is useful to specify the frequency range over which the calculation can be justified. Indeed, we have made assumptions which limit this frequency range. For one thing, we have assumed that $\hbar\omega$ is very big compared with the atomic transition energies. This implies that

[14] A more accurate calculation can be carried out, taking into account further terms in the power series expansion of $(E_b-E_a)/\hbar\omega$. It can then be shown (see CDG II, Exercise 4) that the second-order contribution of $\hat{H}_{I1}$ is smaller than that of $\hat{H}_{I2}$ by a factor of the order of $(E_I/\hbar\omega)^2$.

$$\hbar\omega \gg E_{\mathrm{I}}, \tag{6.143}$$

where E_{I} is the ionization energy of the atom. As an example, recall that the ionization energy of the hydrogen atom is[15]

$$E_{\mathrm{I}} = \frac{1}{2}\alpha^2 mc^2, \tag{6.144}$$

where $\alpha = 1/137$ is the fine structure constant.

Furthermore, we have used the long-wavelength approximation in the interaction Hamiltonian, which means that

$$a_0 \ll \frac{c}{\omega}, \tag{6.145}$$

where a_0 is the Bohr radius, characteristic of atomic dimensions, given by[15]

$$a_0 = \frac{1}{\alpha}\frac{\hbar}{mc}. \tag{6.146}$$

We conclude that the energy range for Thomson scattering obtained from (6.132) and (6.144) is

$$\frac{\alpha^2}{2}mc^2 \ll \hbar\omega \ll \alpha mc^2. \tag{6.147}$$

For the hydrogen atom, this implies energies in the range from 14 eV to 3.6 keV, or wavelengths in the range from 90 nm (far UV) to 0.3 nm (X-rays).

Differential cross-section

It is now straightforward to calculate the differential scattering rate using the form (6.128) in which the interaction Hamiltonian $\hat{H}_{\mathrm{I}}$ is taken to be just $\hat{H}_{\mathrm{I2}}$ of (6.73), and the density of states was calculated in (6.107). Using the definition (6.131), we obtain the *differential cross-section for Thomson scattering* from mode ℓ to mode j:

$$\frac{\mathrm{d}\sigma_{\mathrm{T}}}{\mathrm{d}\Omega} = r_0^2(\boldsymbol{\varepsilon}_j.\boldsymbol{\varepsilon}_\ell)^2, \tag{6.148}$$

a remarkably simple formula, involving only the scalar product of the incident and scattered polarizations, and in which r_0 is the classical electron radius (6.134).

Total cross-section

This is obtained by summing (6.148) over all polarizations, then over the emission directions of the scattered photon. The calculation is the same as the one for spontaneous emission in Section 6.4.

Since the polarization $\boldsymbol{\varepsilon}_\ell$ of the incident photon is given, we consider a scattering direction $\mathbf{k}_j$ making an angle θ with $\boldsymbol{\varepsilon}_\ell$. The two possible polarizations $\boldsymbol{\varepsilon}_\ell$ and $\boldsymbol{\varepsilon}'_\ell$ of the scattered

[15] BD, Chapter 10.

photon can be chosen in the plane $(\boldsymbol{\varepsilon}_\ell, \mathbf{k}_j)$ and orthogonal to this plane. With this choice, the second contribution is zero:

$$\boldsymbol{\varepsilon}'_j \, . \, \boldsymbol{\varepsilon}_\ell = 0, \tag{6.149}$$

while the first is

$$\boldsymbol{\varepsilon}_j \, . \, \boldsymbol{\varepsilon}_\ell = \sin\theta. \tag{6.150}$$

The differential scattering cross-section in a direction making an angle θ with the incident polarization is thus equal to

$$\frac{\mathrm{d}\sigma}{\mathrm{d}\Omega} = r_0^2 \sin^2\theta. \tag{6.151}$$

Integrating over the solid angle (see Section 6.4.4), we obtain the total Thomson scattering cross-section,

$$\sigma_\mathrm{T} = \frac{8\pi}{3} r_0^2, \tag{6.152}$$

which is the expression given in Section 6.5.3.

6.6 Conclusion. From the semi-classical to the quantum treatment of atom–light interaction

In this chapter we have given the formalism to treat the interaction of quantized radiation with atoms, and we have applied it to first examples. As indicated on several occasions, the first triumph of the quantum theory is to give a consistent and quantitative treatment of spontaneous emission by an excited atom. More generally, processes in which photons are emitted in an initially empty mode cannot in general be described correctly using the semi-classical description, and must be treated using this approach. Although in this chapter detailed calculations have been limited to the simplest cases – spontaneous emission from a discrete atomic state, Thomson scattering – nothing further is required to treat more complicated processes, such as Raman scattering (Figure 6.12).

At this point in this book, it is tempting to compare the quantum treatment that we now know, with the semi-classical model, which was developed in Chapters 2 and 3. In particular, in Complement 2A, the Lorentz model of the elastically bound electron allowed us to obtain quantitative results about the radiative damping of the electron oscillations (Section 2A.3), and about scattering when the electron is driven by an incident classical field (Section 2A.4). A comparison shows that the results of the two treatments are quite similar. As already noted, the spontaneous emission rate, as calculated in this chapter (Section 6.4) is the same as the classical damping rate, within a factor of order 1, the oscillator strength. The scattering cross-sections in the various regimes are also similar, exactly equal in the case of Thomson scattering, and within the same correcting factor in the case of Rayleigh and resonant scatterings. Such similarities are crying out for comment.

Let us first address the case of spontaneous emission. Here, there is no question that the fully quantum model is the only consistent model, since an atom initially in an excited state, i.e. a stationary state where nothing oscillates, bears no resemblance with a damped oscillating electron. The difference is even more striking when we consider an excited state that can decay spontaneously towards several lower-lying states, a situation without any classical analogue in the case of a single atom. In addition, the fact that the quantum model yields a precise value of the spontaneous rate in quantitative agreement with experimental observations, confirms that the full quantum treatment is the correct one in the case of spontaneous emission. Moreover, as indicated in Section 6.4.5, by emphasizing the fact that spontaneous emission is a joint property of the atom and vacuum fluctuations, the quantum description has led to modern quantum optics developments in which spontaneous emission is controlled by changing the environment of the atom. So, apart from convenient images, e.g. to describe the polarization of the emitted photons in various types (σ_+, σ_- or π) of transitions, there is not much to gain from the semi-classical model in the case of spontaneous emission.

A similar conclusion might be drawn in the case of light scattering towards initially empty modes, as for instance Raman scattering (Figure 6.12). However, the case of elastic scattering where the initial and final states of the atom are identical (Rayleigh, resonant or Thomson scattering), deserves a special comment. In that case, indeed, the Lorentz model of the elastically bound electron not only yields results very similar to the quantum treatment, but it also points out interesting properties of the process. Take for instance the case of resonant scattering, as described in Figure 6.11. This figure might suggest that the scattered radiation, in analogy with spontaneous emission as described in Section 6.4, has no well-defined phase. But on the other hand, the semi-classical treatment of Section 2A.4, which gives the correct result provided that one replaces Γ_{cl} by Γ_{sp}, suggests that the scattered radiation has a phase perfectly related to that of the incident radiation. Actually, this property (which can be checked experimentally) is also predicted by the fully quantum treatment, provided that one considers an incident field consisting of a quasi-classical state with a well-defined phase (Section 5.3.4). The scattered field is then also a quasi-classical state, with a phase related to the phase of the incident field, as in the semi-classical treatment. This is a case where it is fruitful to consider both the semi-classical and the fully quantum model of atom–light interaction.

This too schematic discussion then suggests that the quantum treatment of atom–light interaction is particularly important when there is creation of new frequencies, not present in the initial radiation. Actually, this can happen even in the case of quasi-resonant scattering (Figure 6.11), where new frequencies, different from the incident radiation frequency, appear in the scattered light when the incident radiation is sufficiently intense. This can easily be understood by considering higher-order perturbative terms, in the quantum treatment. One can then have several photons created in the same elementary process, with frequencies different from the initial one, the only constraint being total energy conservation for the whole process. Figure 6.13 presents an example of such a process, leading to the appearance of two lines centred at ω_0 and $2\omega - \omega_0$ when an atom resonant at ω_0 is irradiated with intense light at frequency ω close to but different from ω_0. The diagram of Figure 6.13 describes a process where two photons at ω disappear, while two photons are

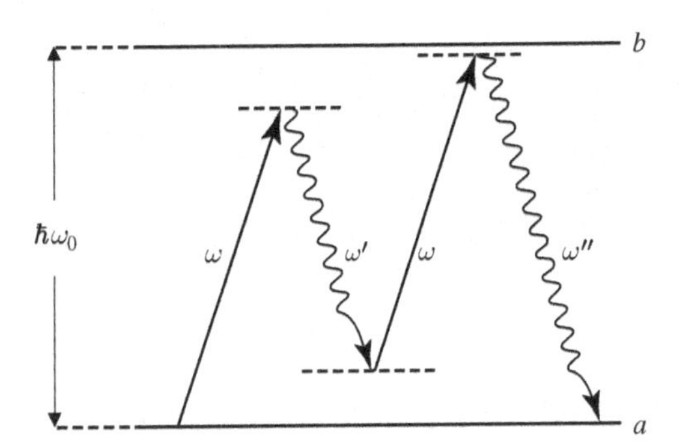

Figure 6.13 Higher-order scattering process in which two incident photons disappear and two photons are created with different frequencies. This process produces the sidebands of the resonance fluorescence triplet.

created at ω' and ω'', such that $2\hbar\omega = \hbar\omega' + \hbar\omega''$. This higher-order process is resonant when ω'' is equal to ω_0 within Γ_{sp} (the spontaneous emission rate from b). Such a process yields what is known as the 'sidebands of the resonance fluorescence triplet'[16] at frequencies ω_0 and $2\omega - \omega_0$ (since there is also a component at ω due to resonant scattering, there are three different components in the resonance fluorescence light).

As the incident radiation gets more and more intense, higher and higher-order interaction terms need to be taken into account. It is then often useful to adopt a more powerful method, such as the model of *an atom dressed by photons*.[17] Here the idea of treating the *interacting atom and radiation* as a *unique global quantum system* is pushed to its natural conclusion. This approach provides simpler calculational methods and clearer pictures of what is happening. It is thus a natural extension of the description of the interacting atom and radiation discussed in this chapter.

The material presented here should thus be viewed as more than just the elaboration of a consistent theory. It is in fact the starting point for understanding many modern developments of quantum optics and of the physics of the matter–radiation interaction.

[16] B. R. Mollow, Power Spectrum of Light Scattered by Two-Level Systems, *Physical Review* **188**, 1969 (1969).

[17] The dressed atom approach, developed by C. Cohen-Tannoudji and his collaborators, is discussed in CDG II, Chapter VI.

6A Complement 6A Hamiltonian formalism for interacting fields and charges

6A.1 Hamiltonian formalism and canonical quantization

In this book, we use the classical Hamiltonian formalism to write the energy of a system in a form that lends itself to canonical quantization. For this to work, we must be sure to identify pairs of conjugate canonical variables which, after quantization, will give pairs of operators with commutators equal to $i\hbar$. To achieve this, we write down Hamilton's equations for the relevant energy expression and check that they do indeed return the previously established dynamical equations. This is what was done in Chapter 4 for the electromagnetic field in vacuum. We expressed the electromagnetic energy in terms of the real and imaginary parts of normal variables, and showed that Hamilton's equations were equivalent to Maxwell's equations. In Chapter 6, we used an analogous approach for an interacting system of charges and fields, but we did not supply all the details of the calculations.

The aim in this complement is to show explicitly that the Hamiltonian for classical electrodynamics, expressed in the form given at the end of Section 6.1, does indeed lead back to the equations of motion of the particles under the effect of the fields, but also to the dynamical equations of the fields themselves in the presence of the charges, i.e. the Maxwell–Lorentz equations as given in Section 6.1.1. This is then used to confirm that we have indeed identified pairs of conjugate canonical variables.

6A.2 Hamilton's equations for particles and radiation

6A.2.1 Classical Hamiltonian for the charge–field system

It was shown in Section 6.2 that the total classical Hamiltonian describing the interacting system of field and charges could be written

$$H = \sum_{\mu} \frac{1}{2m_{\mu}} \left[\mathbf{p}_{\mu} - q_{\mu}\mathbf{A}(\mathbf{r}_{\mu}, t)\right]^2 + V_{\text{coul}}(\mathbf{r}_1, \ldots \mathbf{r}_{\mu} \ldots) + \sum_{\ell} \hbar\omega_{\ell} |\alpha_{\ell}|^2. \qquad (6A.1)$$

Let us now show that the pairs $(\mathbf{r}_{\mu}, \mathbf{p}_{\mu})$ and (Q_{ℓ}, P_{ℓ}) are each pairs of conjugate canonical variables. The real variables Q_{ℓ} and P_{ℓ} are proportional to the real and imaginary parts of α_{ℓ}, according to the definition,

$$\alpha_\ell = \frac{1}{\sqrt{2\hbar}}(Q_\ell + \mathrm{i}P_\ell). \tag{6A.2}$$

We must therefore express the fields in (6A.1) in terms of Q_ℓ and P_ℓ.

6A.2.2 Hamilton's equations for the charges

Hamilton's equations for a Cartesian component of the pair $(\mathbf{r}_\mu, \mathbf{p}_\mu)$ can be written, e.g. for the x component,

$$\frac{\mathrm{d}}{\mathrm{d}t}x_\mu = \frac{\partial H}{\partial p_{x\mu}} \tag{6A.3}$$

$$\frac{\mathrm{d}}{\mathrm{d}t}p_{x\mu} = -\frac{\partial H}{\partial x_\mu}. \tag{6A.4}$$

As we have already seen, e.g. in Complement 4A, the first equation gives, using expression (6A.1) for the Hamiltonian,

$$\frac{\mathrm{d}}{\mathrm{d}t}x_\mu = v_{x\mu} = \frac{1}{m}\left[p_{x\mu} - q_\mu A_x(\mathbf{r}_\mu, t)\right], \tag{6A.5}$$

which is (6.38). Equation (6A.4) requires more work, but the calculation for the first term in the Hamiltonian (6A.1) has already been carried out in Complement 4A. It shows that we obtain the equation of motion of charge μ subjected to electric and magnetic fields deriving from the vector potential $\mathbf{A}(\mathbf{r}, t)$. The second term in (6A.1) adds to the force the term due to the Coulomb interaction between charge μ and all the other charges. As regards the third term, it is not a function of the $\mathbf{p}_\mu$ and $\mathbf{r}_\mu$ and so plays no role in the particle motions.

We thus conclude that, if the total energy is expressed in the form (6A.1), we do indeed obtain the dynamical equations of the charges. The pairs $r_{i\mu}$, $p_{i\mu}$ $(i = x, y, z)$ are thus conjugate canonical variables.

6A.2.3 Hamilton's equations for the radiation

Hamilton's equations for Q_ℓ and P_ℓ are

$$\frac{\mathrm{d}}{\mathrm{d}t}Q_\ell = \frac{\partial H}{\partial P_\ell} \tag{6A.6}$$

$$\frac{\mathrm{d}}{\mathrm{d}t}P_\ell = -\frac{\partial H}{\partial Q_\ell}. \tag{6A.7}$$

Consider first the last term of (6A.1), which refers to the radiation alone. Writing it in the form

$$H_\mathrm{R} = \sum_\ell \frac{\omega_\ell}{2}\left(Q_\ell^2 + P_\ell^2\right), \tag{6A.8}$$

we obtain

$$\frac{d}{dt}Q_\ell = \omega_\ell P_\ell \tag{6A.9}$$

$$\frac{d}{dt}P_\ell = -\omega_\ell Q_\ell \tag{6A.10}$$

and hence,

$$\frac{d}{dt}(Q_\ell + iP_\ell) = -i\omega_\ell(Q_\ell + iP_\ell). \tag{6A.11}$$

We do therefore recover

$$\frac{d}{dt}\alpha_\ell = -i\omega_\ell\alpha_\ell, \tag{6A.12}$$

that is, the dynamical equation for α_ℓ in the absence of source currents.

Now consider the first term of (6A.1), H_{kin}, in which we express $\mathbf{A}(\mathbf{r}_\mu, t)$ as a function of the Q_ℓ and P_ℓ. The first Hamilton equation is

$$\frac{dQ_\ell}{dt} = \frac{\partial H_{\text{cin}}}{\partial P_\ell} = \sum_\mu \frac{1}{m_\mu}\left[\mathbf{p}_\mu - q_\mu\mathbf{A}(\mathbf{r}_\mu, t)\right]\frac{\partial}{\partial P_\ell}\left[\mathbf{p}_\mu - q_\mu\mathbf{A}(\mathbf{r}_\mu, t)\right]. \tag{6A.13}$$

Replacing the first square bracket by $m_\mu\mathbf{v}_\mu$ and expanding $\mathbf{A}(\mathbf{r}_\mu, t)$ in the second square bracket using (6.27) and (6A.2), we obtain

$$\frac{dQ_\ell}{dt} = -\sum_\mu q_\mu\mathbf{v}_\mu\cdot\boldsymbol{\varepsilon}_\ell\frac{\mathcal{E}_\ell^{(1)}}{\omega_\ell\sqrt{2\hbar}}\left(ie^{i\mathbf{k}_\ell\cdot\mathbf{r}_\mu} - ie^{-i\mathbf{k}_\ell\cdot\mathbf{r}_\mu}\right). \tag{6A.14}$$

In a similar way, the second Hamilton equation becomes

$$\frac{dP_\ell}{dt} = -\frac{\partial H}{\partial Q_\ell} = \sum_\mu q_\mu\mathbf{v}_\mu\cdot\boldsymbol{\varepsilon}_\ell\frac{\mathcal{E}_\ell^{(1)}}{\omega_\ell\sqrt{2\hbar}}\left(e^{i\mathbf{k}_\ell\cdot\mathbf{r}_\mu} + e^{-i\mathbf{k}_\ell\cdot\mathbf{r}_\mu}\right). \tag{6A.15}$$

It follows that

$$\frac{d\alpha_\ell}{dt} = i\sum_\mu q_\mu\mathbf{v}_\mu\cdot\boldsymbol{\varepsilon}_\ell\frac{\mathcal{E}_\ell^{(1)}}{\hbar\omega_\ell}e^{-i\mathbf{k}_\ell\cdot\mathbf{r}_\mu}. \tag{6A.16}$$

The right-hand side of the above equation is proportional to the transverse component $\tilde{J}_{\perp\ell}$ of the current. Indeed, starting with (6.6), we have

$$\begin{aligned}\tilde{J}_{\perp\ell} &= \frac{1}{L^3}\int d^3r\,\boldsymbol{\varepsilon}_\ell\cdot e^{-i\mathbf{k}_\ell\cdot\mathbf{r}}\sum_\mu q_\mu\mathbf{v}_\mu\delta(\mathbf{r}-\mathbf{r}_\mu)\\ &= \frac{1}{L^3}\sum_\mu q_\mu\mathbf{v}_\mu\cdot\boldsymbol{\varepsilon}_\ell\,e^{-i\mathbf{k}_\ell\cdot\mathbf{r}_\mu}\end{aligned} \tag{6A.17}$$

and then, taking into account the expression (4.99) for $\mathcal{E}_\ell^{(1)}$, Equation (6A.16) implies that

$$\frac{d\alpha_\ell}{dt} = \frac{i}{2\varepsilon_0\mathcal{E}_\ell^{(1)}}\tilde{J}_{\perp\ell}. \tag{6A.18}$$

We thus find the source term of (6.25).

6A.2.4 Conclusion

We have shown in this complement how the energy of a system of interacting fields and charges can be written in terms of pairs of variables relating to the charges (the components $r_{i\mu}$ and $p_{i\mu}$) and pairs of variables relating to the radiation (the real and imaginary parts of the normal variables α_ℓ used to express the transverse fields). We have checked explicitly that the Hamilton equations for these pairs of variables do indeed return the Maxwell–Lorentz equations describing the dynamics of the interacting fields and charges. These variables are therefore conjugate canonical variables, and we may use the canonical quantization procedure in which these variables are replaced by pairs of operators with commutators equal to $i\hbar$. The resulting expression for the energy is then the quantum Hamiltonian for the problem, and we obtain expressions for the relevant quantum observables, in particular the fields, in the same manner.

6B Complement 6B Cavity quantum electrodynamics

Up to now in this work we have not been concerned with the environment in which an atom and the radiation field interact. We have implicitly assumed that the radiation propagates in free space and that there are no boundaries to reflect radiation emitted by the atom. We shall show in this complement that when such boundaries do exist and are sufficiently reflecting, or more especially when the atom is enclosed in a resonant cavity, its radiative properties, such as its absorption spectrum and the rate of spontaneous emission, are drastically altered. This can be the case even if the cavity boundaries themselves are very far from the atom, on the atomic scale of distances.

The conditions under which these *cavity quantum electrodynamic* effects can be observed are actually quite difficult to reach and this is why, usually, it is possible to assume that the radiative properties of a system are independent of the enclosure surrounding it. Nevertheless, thanks to some outstanding technical achievements, these effects can be observed in remarkable experiments. Atoms coupled to cavities then appear as a promising system in quantum information either for quantum processing or for single photon sources (see Complement 5E).[1]

6B.1 Presentation of the problem

Consider the system sketched in Figure 6B.1, in which an atom at rest at the origin of coordinates is enclosed in a cavity of volume V, with perfectly reflecting walls. Because of the boundary conditions on the field imposed by the presence of the cavity, the field can only exist as a superposition of a discrete set of modes, determined by the cavity geometry, each associated with a definite oscillation frequency. For the rhombic geometry of Figure 6B.1 the modes correspond to plane standing waves parallel to the Cartesian axes and the allowed frequencies are determined by the requirement that the field should vanish at the boundary walls.

In this complement we consider only the simple case in which all the cavity modes have frequencies very different from the Bohr frequencies of the atom, except for one mode which is quasi-resonant with the transition from the ground state $|a\rangle$ to an excited state $|b\rangle$,

[1] For a detailed description of cavity QED physics, see. S. Haroche and J.-M. Raimond, *Exploring the Quantum*, Oxford University Press (2007).

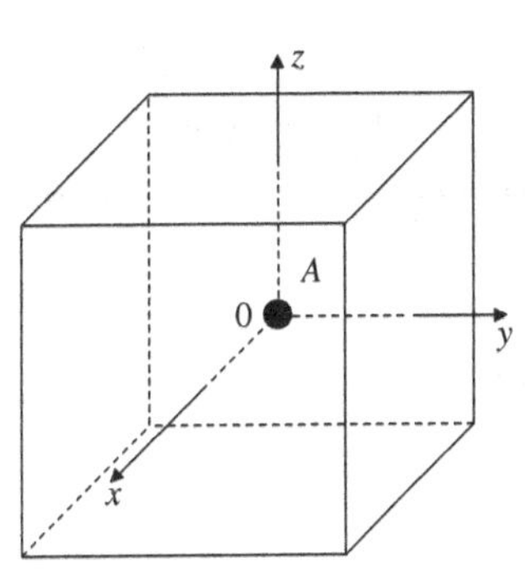

Figure 6B.1 **System composed of a two-level atom at the origin of coordinates enclosed in a rhombic cavity with perfectly reflecting walls.**

which is characterized by a Bohr frequency ω_0. To determine the evolution of the system we therefore need only to consider the sub-system composed of the two-level atom with states $|a\rangle$ and $|b\rangle$ coupled to a single mode of the radiation field. We suppose, in addition, that the long-wavelength approximation is valid. These assumptions make it much easier to obtain a solution for the quantum mechanical system.

Thus, in the situation we describe here the fictitious quantization volume, used in our earlier treatment of the quantized electromagnetic field, is replaced by a real cavity and field operators such as $\hat{N}$, the operator for the total number of photons in the cavity, now have a real physical significance.

Comment There are different experimental realizations of the model we present here:

- One can use a cavity with dimensions of the order of 1 cm formed of superconducting metal (see Figure 6B.2a). Its eigenmodes have frequencies not less than a few tens of GHz. Small holes in the cavity walls allow the passage of a beam of atoms prepared in a highly excited (Rydberg) state. Such states closely approximate hydrogenic eigenstates with very large principal quantum number. A cavity mode is then quasi-resonant with a transition between levels with principal quantum number n and $n+1$ with n of the order of 50, corresponding to a Bohr frequency of about 50 GHz.
- Alternatively, one can use a Fabry–Perot cavity of length less than 1 mm formed by two highly reflecting mirrors at optical wavelengths (see Figure 6B.2b). A single atom from an ultra-cold sample (Chapter 8) can be put almost at rest in the cavity. The cavity is tuned to have a mode at resonance with one atomic transition in the optical range. It can be shown that it is possible to neglect the coupling of the atom with all the other modes if the separation of the atomic levels is sufficiently large (this is the case for the transition between the ground state and the first excited state for an alkali-metal atom such as caesium) and if the cavity finesse is sufficiently large (finesses up to 10^6 are obtainable).
- Another option, made possible by recent developments in nanotechnology, is to use epitaxial techniques to make a microcavity with Bragg mirrors a few microns in length, containing quantum wells or quantum dots, which provide a good approximation to two-level systems (see Figure 6B.2c).

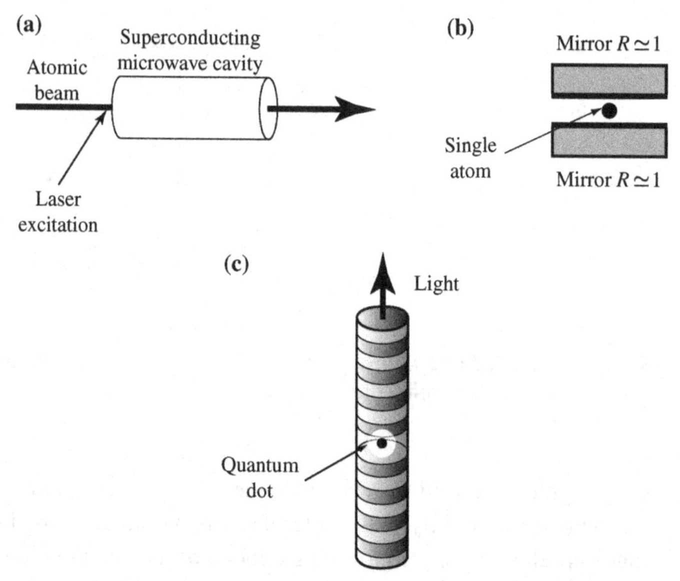

Figure 6B.2 Three experimental configurations for demonstrating cavity QED effects. **(a)** Interaction of a dilute atomic beam of atoms, prepared by laser excitation in a Rydberg state, with a resonant mode, in the microwave spectral region, of a superconducting cavity of centimetric dimensions; **(b)** interaction of a single atom in its ground state with a resonant mode, in the visible domain, of a high-finesse Fabry–Perot cavity of millimetric dimensions; **(c)** interaction of a solid-state quantum two-level system (such as a semiconductor quantum well or quantum dot) with the fundamental mode of a Fabry–Perot cavity of nanometric dimensions. The mirrors of the cavity are constituted by alternate layers of dielectrics of different indices acting as Bragg reflectors. The quantum dot lies at an antinode of the cavity standing wave.

6B.2 Eigenmodes of the coupled atom–cavity system

6B.2.1 Jaynes–Cummings model

The state space in which the evolution of the system can be described consists of the tensor product space of the two states $|a\rangle$ and $|b\rangle$ of the atom and the state space of a single mode of the field, for which the photon number states constitute a natural basis. One possible basis of the tensor product space is the tensor product of the bases of the atom and of the field:

$$\{|i,n\rangle = |a\rangle \otimes |n\rangle \text{ or } |b\rangle \otimes |n\rangle, \text{ with } i = a \text{ or } b, \text{ and } n = 0,1,...\}. \tag{6B.1}$$

We restrict the Hamiltonian of the atom–radiation system to this sub-space by writing

$$\hat{H} = \hat{H}_{\text{at}} + \hat{H}_{\text{R}} + \hat{H}_{\text{I}}, \tag{6B.2}$$

with

$$\hat{H}_{\text{at}} = \hbar\omega_0 |b\rangle\langle b| \tag{6B.3}$$

and

$$\hat{H}_{\mathrm{R}} = \hbar\omega\hat{a}^{\dagger}\hat{a}. \tag{6B.4}$$

Here we have taken the zero of energy such that it coincides with the energy of the state $|a, 0\rangle$. The interaction term $\hat{H}_{\mathrm{I}}$ is given for the general case by expression (6.69). Making the long-wavelength approximation and taking the atom at the origin of coordinates we use expressions (6.72) and (6.73), and we obtain

$$\hat{H}_{\mathrm{I}} = -\frac{q}{m}\sqrt{\frac{\hbar}{2\varepsilon_0\omega V}}\hat{\mathbf{p}}.\boldsymbol{\varepsilon}(\hat{a}^{\dagger} + \hat{a}) + \frac{q^2}{2m}\frac{\hbar}{2\varepsilon_0\omega V}(\hat{a}^{\dagger} + \hat{a})^2, \tag{6B.5}$$

where V is the cavity volume. The second term of the right-hand side of the expression above, arising from the term in $\hat{\mathbf{A}}^2$ in the interaction Hamiltonian (see Equation (6.70)), is an operator that acts only on field variables. It modifies the energy eigenvalues and eigenstates, but does not appear in the atom–field coupling. We assume low enough radiation intensity to be able to neglect effects due to this term. Additionally, we use the fact that the diagonal matrix elements of the momentum operator $\hat{\mathbf{p}}$ in the basis of the atomic states are zero to write $\hat{H}_{\mathrm{I}}$ in the form,

$$\hat{H}_{\mathrm{I}} = \frac{\hbar\Omega_1^{(1)}}{2}\left(|a\rangle\langle b| + |b\rangle\langle a|\right)(\hat{a}^{\dagger} + \hat{a}), \tag{6B.6}$$

where we have assumed that $\langle a|\boldsymbol{\varepsilon}.\hat{\mathbf{p}}|b\rangle$ is real, and introduced the *single photon Rabi frequency*,

$$\Omega_1^{(1)} = -\frac{2q}{m}\frac{1}{\sqrt{2\hbar\varepsilon_0\omega V}}\langle a|\hat{\mathbf{p}}.\boldsymbol{\varepsilon}|b\rangle. \tag{6B.7}$$

It is identical to (6.85) to within a factor of $\sqrt{n_j}$. For the situation described in Figure 6B.2a, that is for $V \approx 1\ \mathrm{cm}^3$, a transition frequency of 50 GHz and for a matrix element of $\hat{\mathbf{p}}$ equal to $-i\omega m\langle\hat{z}\rangle$ (see expression (6.88)) with $\langle\hat{z}\rangle \approx n^2 a_0 \approx 100$ nm ($n \approx 50$ is the principal quantum number), the single photon Rabi frequency $\Omega_1^{(1)}/2\pi$ is of the order of 100 kHz.

Comment

This simplified model of the matter–light interaction, in which one considers the interaction of a two-level atom with a single mode of the radiation field, is known as the *Jaynes–Cummings* model.[2]

6B.2.2 Diagonalization of the Hamiltonian

The first step towards a calculation of the evolution of the system consists in finding the eigenvalues of the Hamiltonian and the associated eigenstates. We start by considering the Hamiltonian $\hat{H}_{\mathrm{at}} + \hat{H}_{\mathrm{R}}$, which does not take into account the atom–field coupling. The

[2] E. T. Jaynes and F. W. Cummings, Comparison of Quantum and Semi-Classical Radiation Theories with Application to the Beam Maser, *Proceedings IEEE* **51**, 89 (1963).

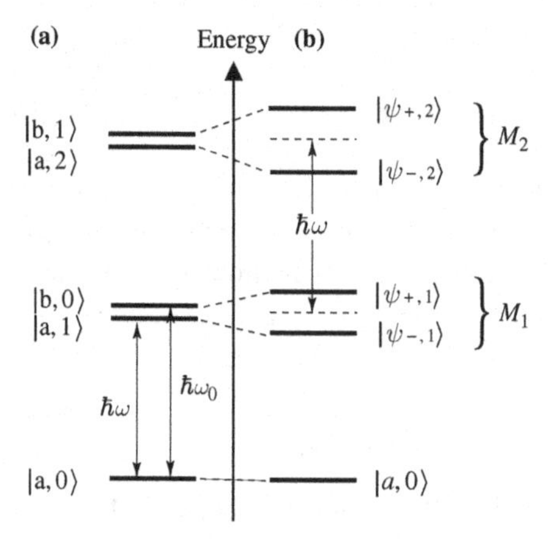

Figure 6B.3 **Energy levels (a) in the absence and (b) in the presence of the atom–field coupling.**

eigenstates of this part of the Hamiltonian are the *decoupled* states $|i, n\rangle$ (Equation 6B.1), with eigenenergies given by

$$\left(\hat{H}_{\text{at}} + \hat{H}_{\text{R}}\right)|a, n\rangle = \hbar n\omega |a, n\rangle, \tag{6B.8}$$

$$\left(\hat{H}_{\text{at}} + \hat{H}_{\text{R}}\right)|b, n\rangle = \hbar(\omega_0 + n\omega)|b, n\rangle. \tag{6B.9}$$

The ensemble of states $\{|i, n\rangle; i = a, b; n = 0, 1, 2, \ldots\}$ constitutes a basis for the state space of the system comprising the atom and a single cavity mode. It can be used even if the interaction term is taken into account.

Figure 6B.3(a) illustrates the eigenenergies of the decoupled Hamiltonian $\hat{H}_{\text{at}} + \hat{H}_{\text{R}}$. The ground state is the state $|a, 0\rangle$, which we take to have zero energy. Since $\omega \approx \omega_0$ the remaining levels are arranged in closely spaced doublets M_n, the energies of the two levels of a doublet differing by $\hbar\delta = \hbar(\omega - \omega_0)$ and successive doublets being separated by an energy of the order of $\hbar\omega$.

We consider now the effect of the interaction term. First of all we evaluate the matrix elements of $\hat{H}_{\text{I}}$ (Equation 6B.6), which are given by

$$\langle i, n|\hat{H}_{\text{I}}|i', n'\rangle = \frac{\hbar\Omega_1^{(1)}}{2}\left\{\langle i|a\rangle\langle b|i'\rangle + \langle i|b\rangle\langle a|i'\rangle\right\}\langle n|(\hat{a} + \hat{a}^{\dagger})|n'\rangle. \tag{6B.10}$$

These matrix elements are non-zero if and only if $i = a$ and $i' = b$ or $i = b$ and $i' = a$, with $n' = n \pm 1$. Consequently, this interaction Hamiltonian only has non-zero elements off the diagonal (i.e. all diagonal elements are zero); it connects, on the one hand, the two states $|a, n\rangle$ and $|b, n - 1\rangle$ within a given doublet M_n, the corresponding matrix element having the value $\langle a, n|\hat{H}_{\text{I}}|b, n - 1\rangle = \sqrt{n}\hbar\Omega_1^{(1)}/2$. On the other hand, it couples states in doublets M_n and $M_{n\pm 2}$, the corresponding matrix elements being of the same order of magnitude. For instance, one has $\langle a, n|\hat{H}_{\text{I}}|b, n + 1\rangle = \sqrt{n + 1}\hbar\Omega_1^{(1)}/2$.

An exact diagonalization of the total Hamiltonian matrix is not possible. However, given that the levels are grouped in tightly spaced doublets with a considerably greater spacing between successive doublets, the effect of the non-resonant couplings between different doublets can be neglected, to a good approximation: we make use of the fact that a time-independent coupling has an appreciable effect only on states of similar energies. It can be shown that this approximation gives results that are correct to within a factor of $\langle \hat{H}_{\mathrm{I}} \rangle / \hbar\omega \approx \sqrt{n}\, \Omega_1^{(1)}/2\omega$. It is equivalent to the quasi-resonant approximation made in Chapter 1 (Section 1.2.4).

The restriction of the interaction Hamiltonian to a given doublet M_n can be written in the operator form:

$$\hat{H}_{\mathrm{I}} = \frac{\hbar\Omega_1^{(1)}}{2}\{|a\rangle\langle b|\, \hat{a}^{\dagger} + |b\rangle\langle a|\, \hat{a}\}. \tag{6B.11}$$

In this sub-space the total Hamiltonian takes the form

$$\hat{H} = \hbar \begin{bmatrix} n\omega & \dfrac{\Omega_1^{(1)}\sqrt{n}}{2} \\ \dfrac{\Omega_1^{(1)}\sqrt{n}}{2} & n\omega - \delta \end{bmatrix}. \tag{6B.12}$$

The eigenvalues and eigenvectors resulting from the diagonalization of such a 2×2 matrix were given in Chapter 1 (Section 1.2.6). The energy eigenvalues are

$$E_{\pm,n} = \hbar\left(n\omega - \frac{\delta}{2} \pm \frac{1}{2}\sqrt{n\left[\Omega_1^{(1)}\right]^2 + \delta^2}\right), \tag{6B.13}$$

corresponding to eigenvectors $|\psi_{\pm n}\rangle$,

$$|\psi_{+,n}\rangle = \cos\theta_n |a, n\rangle + \sin\theta_n |b, n-1\rangle \tag{6B.14}$$

$$|\psi_{-,n}\rangle = -\sin\theta_n |a, n\rangle + \cos\theta_n |b, n-1\rangle, \tag{6B.15}$$

with

$$\tan 2\theta_n = \frac{\Omega_1^{(1)}\sqrt{n}}{\delta}. \tag{6B.16}$$

Thus the effect of the coupling is to repel the energy levels within each doublet (see Figure 6B.3(b)). The ground state $|a, 0\rangle$ does not belong to a doublet and is unaffected by the interaction Hamiltonian, at least in the approximation adopted here.

Figure 6B.4 shows the form of the energy levels of the coupled atom–cavity system as a function of the detuning δ (Equation 6B.13). In each doublet an *avoided crossing* occurs around $\delta = 0$. On resonance ($\delta = 0$) the energy separation of the two levels is $\sqrt{n}\, \hbar\Omega_1^{(1)}$, and the parameter θ_n has the value $\pi/4$, so that the eigenstates $|\psi_{\pm n}\rangle$ take the simple form,

$$|\psi_{\pm,n}(\delta = 0)\rangle = \frac{1}{\sqrt{2}}(\pm|a, n\rangle + |b, n-1\rangle). \tag{6B.17}$$

These states are usually known as the *dressed states* of the atom (i.e. dressed by the photons).[3] They are *entangled states* of the atom–field system. They cannot be factorized

[3] See CDG II, Chapter VI.

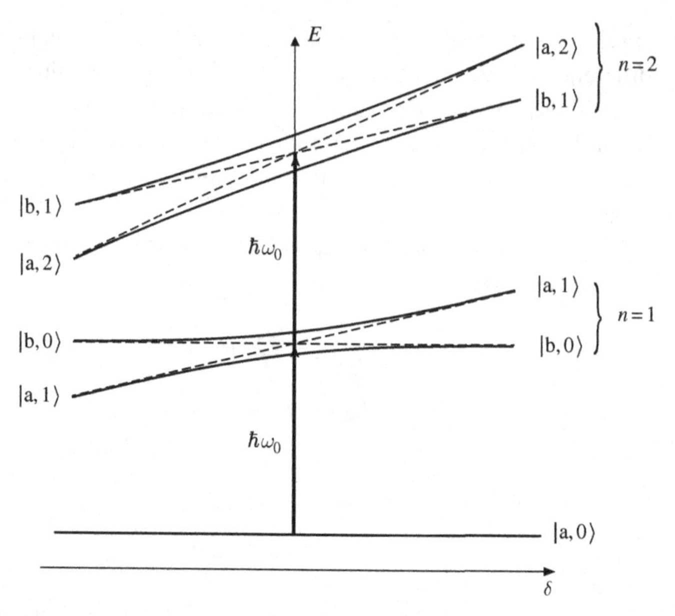

Figure 6B.4 Lower part of the energy-level diagram for the coupled atom–cavity system, as a function of the detuning of the cavity mode from the free-atom resonance. The dotted lines correspond to the energy levels in the absence of the interaction term $\hat{H}_{\mathrm{I}}$.

into a part involving only the atom and another involving only the field. This is evidence of a strong correlation of the state of the atom with that of the field. As the modulus of the detuning $|\delta|$ increases, the parameter θ_n tends to 0 or $\pi/2$, and the two eigenstates (6B.14), (6B.15), tend to the factorizable, uncorrelated states $|a,n\rangle$ and $|b,n-1\rangle$.

Figure 6B.4 shows only the two lowest energy doublets. At higher energies the diagram has a similar form but the on-resonance separation of the levels of a doublet increases in proportion to $\sqrt{n}$. Note that when n is a large number, it is more convenient to plot $E_{\pm,n} - n\hbar\omega$ rather than $E_{\pm,n}$.

6B.2.3 Spontaneous emission of an excited atom placed in the empty cavity

In the following we shall consider the evolution of the system from an initial state in which an excited atom is placed in the cavity empty of radiation. The cavity mode with which the atom interacts is assumed to be resonant with the atomic transition ($\delta = 0$). The initial state of the system can therefore be written $|\psi(0)\rangle = |b,0\rangle$. This is not an eigenstate of the total Hamiltonian and so is not stationary. However, it can be expanded in terms of the eigenstates of the doublet M_1, $|\psi_{\pm 1}\rangle$:

$$|\psi(0)\rangle = |b,0\rangle = \frac{1}{\sqrt{2}}(|\psi_{+,1}\rangle + |\psi_{-,1}\rangle). \tag{6B.18}$$

Thus at a later time t the system is in the state,

$$|\psi(t)\rangle = \frac{1}{\sqrt{2}}\left(|\psi_{+,1}\rangle e^{-iE_{+,1}t/\hbar} + |\psi_{-,1}\rangle e^{-iE_{-,1}t/\hbar}\right)$$
$$= e^{-i\omega t}\left(-i|a,1\rangle \sin\frac{\Omega_1^{(1)}}{2}t + |b,0\rangle \cos\frac{\Omega_1^{(1)}}{2}t\right). \tag{6B.19}$$

The probability, $P_b(t)$, of finding the atom in the excited state is then,

$$P_b(t) = \sum_n |\langle b,n|\psi(t)\rangle|^2 = |\langle b,0|\psi(t)\rangle|^2 = \cos^2\left(\frac{\Omega_1^{(1)}t}{2}\right). \tag{6B.20}$$

Thus spontaneous emission of an excited 2-level atom placed in a resonant cavity manifests itself as an oscillatory evolution between the two atomic levels. This is obviously very different from the monotonic decay observed in free space. This arises because in the present case the coupling is between a pair of discrete states, which gives rise to Rabi oscillations, whereas in free space spontaneous emission couples an initial discrete state with a continuum, leading to an exponential decay. Note that the system oscillates between the two states $|b,0\rangle$ and $|a,1\rangle$, where the former is the initial state of the system, so the evolution can be considered to arise from the periodic exchange of energy between atom and cavity. The atom goes into the ground state when a photon is emitted in the cavity mode, and returns to the excited state when it reabsorbs the photon circulating in the lossless cavity. Such a system is said to undergo *one-photon Rabi oscillations*.[4]

Consider now the situation where the cavity is no longer exactly resonant with the atomic transition. An identical calculation to that presented above reveals that the probability of finding the atom in the excited state at time t is now,

$$P_b(t) = 1 - \frac{\left[\Omega_1^{(1)}\right]^2}{\delta^2 + \left[\Omega_1^{(1)}\right]^2}\sin^2\left(\sqrt{\delta^2 + \left[\Omega_1^{(1)}\right]^2}\,\frac{t}{2}\right). \tag{6B.21}$$

This probability remains close to 1 for $\delta \gg \Omega_1^{(1)}$. The atom therefore remains in the excited state with high probability.

In the two cases considered above the behaviour of the excited atom is dramatically different from what it would be in free space: in a resonant cavity it oscillates between the ground and excited states instead of undergoing an irreversible decay to the ground state. If the cavity is far from resonance with the atomic transition, no spontaneous decay occurs because no quasi-continuum of modes exists over the range of frequencies at which spontaneous emission could occur. Thus *spontaneous emission is not an intrinsic property of an atom*; it arises from the coupling of an atom to its environment.

Other means can be envisaged for inhibiting spontaneous emission, for example by placing the atom in a medium with a *photonic band-gap* at its resonance frequency.[5] Such a

[4] Sometimes referred to in the literature as *vacuum Rabi oscillations*.

[5] E. Yablonovich, Inhibited Spontaneous Emission in Solid-State Physics and Electronics, *Physical Review Letters* **58**, 2059 (1987). Similarly, spontaneous emission can be inhibited by placing the atom in a disordered

medium, which is characterized by a periodic spatial variation, for example of its refractive index, is the analogue for photons of a periodic crystalline structure for electrons:[6] it exhibits forbidden bands of energies at which photons cannot propagate.

Comment We should stress that we have been considering a highly idealized situation in which the atom is coupled to only one mode of a cavity, and the cavity itself is treated as ideal. In real experiments, cavities have losses and there is some residual coupling of the atom to other modes. These extra effects must be made small enough to be able to control spontaneous emission. A considerable effort is invested in achieving such situations, since control of spontaneous emission would be a major advantage in light sources.

6B.3 Evolution in the presence of an intracavity field

In this section we still consider the simple situation in which the cavity supports a mode exactly resonant with the atomic transition, the effect of all other cavity modes being neglected, i.e. the case $\delta = 0$. We then determine the evolution of the system as a function of the state of the radiation field initially stored in the cavity.

6B.3.1 Field initially in a number state

We have already performed this calculation in Section 6.3.4 of Chapter 6. We find, in the case in which the field is a number state $|n\rangle$, an oscillatory evolution characterized by a Rabi frequency, which is now $\Omega_1 = \sqrt{n}\,\Omega_1^{(1)}$. The time evolution in this case is similar to that found for a classical electromagnetic wave (see Section 2.3.2), but there is a remarkable feature here: the state of the system at a given time is a non-factorizable, *entangled state*. It is interesting to consider the case of a $\pi/2$ pulsed interaction, for which the interaction, switched on at time $t = 0$, is switched off at time $t = T$, such that $\Omega_1 T = \pi/2$ (in the example of Figure 6B.2(a), T is the time taken for the atoms to cross the cavity). The system, initially in state $|b, n\rangle$, is then found in the final state,

$$|\psi(T)\rangle = \frac{1}{\sqrt{2}}(-\mathrm{i}|a, n+1\rangle + |b, n\rangle). \tag{6B.22}$$

For this state the results of measurements performed on the atom and on the field are perfectly correlated: if the atom is found to be in state $|a\rangle$ (respectively, $|b\rangle$), the cavity field contains $n + 1$ (respectively, n) photons with certainty, even if the state of the atom is recorded arbitrarily far from the cavity.

dielectric: S. John, Strong Localization of Photons in Certain Disordered Dielectric Superlattices, *Physical Review Letters* **58**, 2486 (1987).

[6] See, for example, N. W. Ashcroft and N. D. Mermin, *Solid State Physics*, Saunders College Publishing (1976).

Let us note another specific feature of the case of a number state as radiation field: the mean values of the atomic dipole moment $\langle\psi(t)|\hat{\mathbf{D}}|\psi(t)\rangle$ and of the electric field $\langle\psi(t)|\hat{\mathbf{E}}|\psi(t)\rangle$ are zero at all times, whereas the mean value of the dipole has an oscillatory behaviour in the semi-classical approach (Equation (2.126)).

6B.3.2 Field initially in an 'intense' quasi-classical state: semi-classical limit

Consider now the case in which the atom in the cavity is initially in its excited state $|b\rangle$ and the field in the mode with which it interacts is in a quasi-classical state $|\alpha\rangle$. We consider firstly the limit $|\alpha| \gg 1$. Recall that, expressed in the basis of the number states, the quasi-classical state is given by (see Section 5.3.4)

$$|\alpha\rangle = \sum_{n=0}^{\infty} c_n|n\rangle = \sum_{n=0}^{\infty} \mathrm{e}^{-|\alpha|^2/2}\frac{\alpha^n}{\sqrt{n!}}|n\rangle. \tag{6B.23}$$

The photon number probabilities $|c_n|^2$ form a Poisson distribution, centred on $\bar{n} = |\alpha|^2$, with a width $\Delta n = \sqrt{\bar{n}} = |\alpha|$. To obtain the time evolution of the system we re-express the initial state $|\psi(0)\rangle = |b\rangle \otimes |\alpha\rangle$ in the basis of the eigenstates of the atom–cavity system:

$$|\psi(0)\rangle = \frac{1}{\sqrt{2}}\sum_{n\geq 1} c_{n-1}(|\psi_{+,n}\rangle + \psi_{-,n}\rangle). \tag{6B.24}$$

The state vector at subsequent times is, therefore, given by

$$\begin{aligned}|\psi(t)\rangle &= \frac{1}{\sqrt{2}}\sum_{n\geq 1} c_{n-1}\mathrm{e}^{-\mathrm{i}n\omega t}\left(|\psi_{+,n}\rangle\mathrm{e}^{-\mathrm{i}\Omega_1^{(1)}\sqrt{n}t/2} + |\psi_{-,n}\rangle\mathrm{e}^{\mathrm{i}\Omega_1^{(1)}\sqrt{n}t/2}\right)\\ &= \sum_{n\geq 1} c_{n-1}\mathrm{e}^{-\mathrm{i}n\omega t}\left[-\mathrm{i}|a,n\rangle\sin\left(\frac{\Omega_1^{(1)}\sqrt{n}}{2}t\right) + |b,n-1\rangle\cos\left(\frac{\Omega_1^{(1)}\sqrt{n}}{2}t\right)\right].\end{aligned} \tag{6B.25}$$

To simplify this expression, we expand $\sqrt{n}$ around its value for $n = \bar{n}$:

$$\sqrt{n} \approx \sqrt{\bar{n}} + (n - \bar{n})\frac{1}{2\sqrt{\bar{n}}}. \tag{6B.26}$$

For values of n distant from $\bar{n}$ by less than $\sqrt{\bar{n}}$, the second term is negligible compared to the first. Thus, to a good approximation, the Rabi frequency is constant over the range of n for which the c_n are non-negligible. We can therefore write

$$\begin{aligned}|\psi(t)\rangle \approx &-\mathrm{i}\left(\sum_n c_{n-1}\mathrm{e}^{-\mathrm{i}n\omega t}|a,n\rangle\right)\sin\frac{\sqrt{\bar{n}}\,\Omega_1^{(1)}}{2}t\\ &+\left(\sum_n c_{n-1}\mathrm{e}^{-\mathrm{i}n\omega t}|b,n-1\rangle\right)\cos\frac{\sqrt{\bar{n}}\,\Omega_1^{(1)}}{2}t.\end{aligned} \tag{6B.27}$$

We shall make a further approximation, consisting in replacing c_{n-1} by c_n in the first sum of the expression above; according to the recurrence relation (5.71), the ratio of these two coefficients is $\alpha/\sqrt{n}$, which is very close to $\alpha/\sqrt{\bar{n}} = 1$ for all the non-negligible terms of the sum. We obtain, finally,

$$|\psi(t)\rangle \approx \left(-\mathrm{i}|a\rangle \sin \frac{\sqrt{\bar{n}}\,\Omega_1^{(1)}}{2} t + |b\rangle \mathrm{e}^{-\mathrm{i}\omega t} \cos \frac{\sqrt{\bar{n}}\,\Omega_1^{(1)}}{2} t \right) \otimes |\alpha \mathrm{e}^{-\mathrm{i}\omega t}\rangle. \tag{6B.28}$$

Thus to within terms of the order of $1/\sqrt{\bar{n}} = 1/\alpha$, the state of the system is the tensor product of a state involving only the atom and a quasi-classical state of the field. This quasi-classical state is exactly what would have resulted from the free evolution of the field without the atoms being present. So in contrast to the previous case, a measurement of the state of either the atom or the field in this case yields no information at all about the state of the field or the atom, respectively. A further distinction is that for the state $|\psi(t)\rangle$ the atomic dipole moment has a non-zero expectation value.

In fact the atomic state appearing in expression (6B.28) is none other than that obtained in the interaction of a two-level atom initially in its excited state with a classical, resonant electromagnetic wave (the situation discussed in Chapter 2). Thus the calculation presented here justifies the semi-classical treatment that we presented in Chapter 2 (Section 2.3.2). It shows that the semi-classical evolution is simply a special case of a purely quantum mechanical evolution, which applies to the interaction of an atom with a quasi-classical state with a large amplitude (corresponding to a large value of $|\alpha|$). This is in contrast to the strikingly non-classical behaviour of the field in a number state.

6B.3.3 Field initially in a quasi-classical state with a small number of photons

If α is of the order of unity, expression (6B.25) remains valid, but not the simplified form of (6B.28). The probability $P_b(t)$ of finding the atom in the state $|b\rangle$, whatever the state of the field, is

$$\begin{aligned} P_b(t) &= \sum_n |\langle b, n|\psi(t)\rangle|^2 = \sum_{n\geq 1} |c_{n-1}|^2 \cos^2\left(\frac{\Omega_1^{(1)}}{2}\sqrt{n}\,t\right) \\ &= \mathrm{e}^{-|\alpha|^2} \sum_{n'=0}^{+\infty} \frac{|\alpha|^{2n'}}{n'!} \cos^2\left(\frac{\Omega_1^{(1)}}{2}\sqrt{n'+1}\,t\right). \end{aligned} \tag{6B.29}$$

Figure 6B.5 illustrates the evolution of $P_b(t)$ for the case of $\alpha = 4$. The complex evolution of the system exhibits three principal characteristics, each occurring over a different timescale:

- The atom undergoes Rabi oscillations, the Rabi frequency being determined by the mean number of photons $|\alpha|^2$, as in the semi-classical case.
- The Rabi oscillations are damped over a timescale of a few periods leading to their eventual disappearance and the equalizing of the populations of the ground and excited

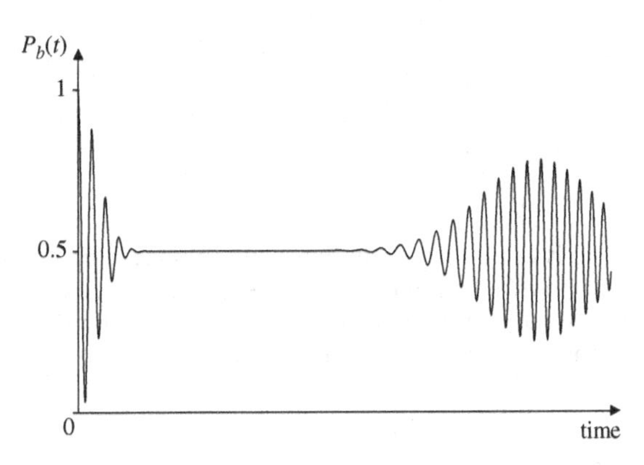

Figure 6B.5 Probability of finding the atom in its excited state when it interacts resonantly with a cavity mode in a quasi-classical state with $\alpha = 4$.

atomic levels, $P_b(t) = 1/2$. This happens because the various sine waves contributing to expression (6B.29) have slightly differing periods. Thus, although they start in phase, they rapidly become out of phase leading eventually to their mutual cancellation.

- The Rabi oscillation eventually reappears on a longer timescale. This occurs because the number of sine waves contributing to expression (6B.29) is *finite*. Thus the time taken for the various oscillations to return to their initial phase relationship is also finite. This phenomenon is known as quantum revival. Of these three phenomena, only the first has a classical counterpart. Rabi oscillation damping and revival are purely quantum phenomena which have both been observed experimentally.

6B.4 Effect of cavity losses: the Purcell effect

In realistic situations, cavities are not perfect: the cavity walls inevitably exhibit losses or non-zero transmission. As a result, the electromagnetic energy inside the empty cavity decreases exponentially in time, with a decay rate that we will call Γ_{cav}.[7] The atom–cavity mode system is no longer isolated, and can no longer be described by a state vector, only by a density matrix. A detailed analysis, which can be found for instance in reference 1, shows that there exist two qualitatively different regimes for the evolution of the system:

- $\Gamma_{\mathrm{cav}}/2 < \Omega_1^{(1)}$: **strong coupling regime**
  When the cavity has small enough losses, the evolution is somewhat similar to that of the perfect cavity case: the probability of finding the atom in the excited state oscillates in time, but the oscillation is damped with the decay constant Γ_{cav}. When $\Gamma_{\mathrm{cav}} \ll \Omega_1^{(1)}$,

[7] If the cavity is a two-mirror Fabry–Perot cavity, Γ_{cav} is related to the cavity finesse $\mathcal{F}$ (see Complement 3A) by $\Gamma_{\mathrm{cav}} = \dfrac{c}{2L}\dfrac{2\pi}{\mathcal{F}}$, where L is the cavity length. It is also related to the so-called quality factor Q of the resonator by $\Gamma_{\mathrm{cav}} = \omega/Q$.

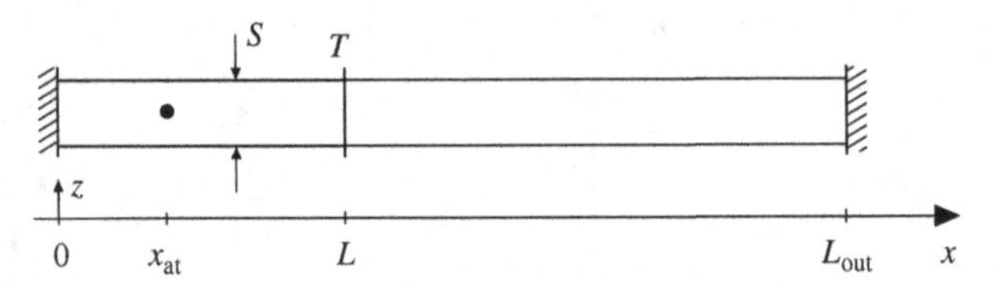

Figure 6B.6 The atom is at position x_{at} in the cavity composed of a perfect mirror at $x = 0$, and a partially transmitting mirror (T) at $x = L$. The field is quantized by adding a perfect mirror at position $x = L_{out}$, with $L_{out} \to \infty$ at the end of the calculation. We consider modes with a field transversely homogeneous over an area S.

the oscillation frequency is close to that of a perfect cavity, but it decreases when Γ_{cav} approaches $\Omega_1^{(1)}/2$.

- $\Gamma_{cav}/2 > \Omega_1^{(1)}$: **weak coupling regime**
  The Rabi oscillation disappears, and for the case of an atom initially in the excited state and an empty cavity, the excited state population decreases monotonically in time. However, the decay rate is different from the decay rate of the atom in free space Γ_{sp}: a cavity, even not perfect, modifies spontaneous emission and may enhance it when it is resonant with the atomic transition. This effect was mentioned for the first time by Purcell, and is called the Purcell effect.[8]

We give here a derivation of the Purcell effect, where the damping of the cavity is due to a partially transmitting output coupler. We consider a two-mirror Fabry–Perot cavity (Figure 6B.6) of length L (the length of a round trip in this cavity is $L_{cav} = 2L$) with a perfect output coupler of transmission and reflection factors T and R for intensity ($R + T = 1$).

In order to quantize the field, we consider a fictitious cavity with a perfect mirror at position $x = L_{out}$, with $L_{out} \gg L$ ($L_{out} \to \infty$ at the end of the calculation). We consider modes propagating along x, with a constant amplitude over a section S perpendicular to x. As in the calculation of spontaneous emission in Section 6.4, we consider an atomic transition such that the atom emits radiation polarized in the plane containing the z-axis and the axis of propagation. Here it is thus polarized along z. The field is of the form (compare Complement 4C)

$$\hat{\mathbf{A}}(\mathbf{r}) = \mathbf{e}_z \sum_m \frac{\mathcal{E}_m^{(1)}}{\omega_m} v_m(x)\,\hat{b}_m + \text{h.c.} \tag{6B.30}$$

In this expression, $\hat{b}_m$ and $\hat{b}_m^\dagger$ obey the standard commutation relations for destruction and creation operators, and v_m is a normalized mode of the field such that

$$\int_V \mathrm{d}^3r |v_m(r)|^2 = V \tag{6B.31}$$

[8] E. M. Purcell, Spontaneous Emission Probabilities at Radio Frequencies, *Physical Review* **69**, 681 (1946).

and

$$\mathcal{E}_m^{(1)} = \sqrt{\frac{\hbar\omega_m}{2\varepsilon_0 V}}, \tag{6B.32}$$

where V is the volume of quantization. To determine $v_m(z)$, we consider a plane wave incident from the right onto the beamsplitter at L, and we use the standard transmission and reflection relations for a perfect beamsplitter. We then find a solution of the form

$$v_m(x) = i\sqrt{2}\mu_m \sin k_m x \qquad \text{for} \quad 0 < x < L \tag{6B.33}$$

$$v_m(x) = i\sqrt{2}\mu'_m \sin(k_m x + \theta_m) \qquad \text{for} \quad L < x < L_{\text{out}}. \tag{6B.34}$$

The angle θ_m is determined by

$$\operatorname{tg}\theta_m = \frac{\sqrt{R}\sin 2k_m L}{1 - \sqrt{R}\cos 2k_m L}, \tag{6B.35}$$

and the ratio of μ_m to μ'_m is

$$\left|\frac{\mu_m}{\mu'_m}\right|^2 = \frac{T}{(1-\sqrt{R})^2}\frac{1}{1+\frac{4\sqrt{R}}{(1-\sqrt{R})^2}\sin^2 k_m L}. \tag{6B.36}$$

This expression corresponds to the result (3A.42) of Complement 3A:

$$\left|\frac{\mu_m}{\mu'_m}\right|^2 = \frac{2\mathcal{F}}{\pi}\frac{1}{1+\frac{4}{\pi^2}\sin^2 k_m L}, \tag{6B.37}$$

where we have introduced the finesse $\mathcal{F}$, and assumed $T \ll 1$ (large finesse cavity), so that

$$\mathcal{F} = \frac{\pi R^{1/4}}{1-\sqrt{R}} \simeq \frac{2\pi}{T}. \tag{6B.38}$$

To determine the absolute values of μ_m and μ'_m, we use the normalization (6B.31). When $L_{\text{out}} \to \infty$, this condition is totally dominated by the interval $L < x < L_{\text{out}}$, and we find

$$|\mu'_m|^2 \simeq 1. \tag{6B.39}$$

The amplitude $|\mu_m|$ in the cavity is then determined by (6B.37). It can be much larger than 1 at resonance.

The quantized values k_m are determined by the condition that the field is null on the fictitious perfect mirror at L_{out}, i.e. from (6B.34),

$$k_m L_{\text{out}} + \theta_m = m\,\pi. \tag{6B.40}$$

In order to use the Fermi golden rule, we need to know the density of modes per unit energy. Neglecting the variation of θ_m in (6B.40), we find

$$\frac{\mathrm{d}m}{\mathrm{d}(\hbar c k_m)} = \frac{L_{\text{out}}}{\pi\hbar c}. \tag{6B.41}$$

As in Section 6.4, we use expression (6.72) of the interaction Hamiltonian to calculate the coupling matrix element between the mode m and the atom at x_{at}:

$$W_m = \left\langle a, 1_m | \hat{H}_{I1} | b, 0 \right\rangle = \frac{q_e}{m_e} \langle a | p_z | b \rangle \frac{\mathcal{E}_m^{(1)}}{\omega_m} v_m(x_{\text{at}}), \tag{6B.42}$$

where q_e and m_e are the charge and mass of the electron. Since the atom is in the cavity, we take expression (6B.33) of $v_m(x)$ and find

$$|W_m|^2 = \left| \left\langle a, 1_m | \hat{H}_{I1} | b, 0 \right\rangle \right|^2 = \frac{q_e^2}{m_e^2} |\langle a | p_z | b \rangle|^2 \left(\frac{\mathcal{E}_m^{(1)}}{\omega_m} \right)^2 2 |\mu_m|^2 \sin^2(k_m x_{\text{at}}). \tag{6B.43}$$

We can use (6B.37) and (6B.39) to express $|\mu_m|^2$, and the Fermi golden rule yields the rate of spontaneous emission in the cavity as

$$\Gamma_{\text{sp,cav}} = \frac{2\pi}{\hbar} \left[\frac{\text{d}m}{\text{d}(\hbar\omega_m)} |W_m|^2 \right]_{\omega_m = \omega_0}, \tag{6B.44}$$

whence we find

$$\Gamma_{\text{sp,cav}} = \frac{q_e^2}{\pi \varepsilon_0} \frac{|\langle a | p_z | b \rangle|^2}{m_e^2 \hbar c S \omega_0} \frac{4\mathcal{F}}{1 + \frac{4}{\pi^2} \mathcal{F}^2 \sin^2 k_0 L} \sin^2(k_0 x_{\text{at}}). \tag{6B.45}$$

The size L_{out} of the fictitious cavity has disappeared from the final result, as it should.

Comparing this with the rate Γ_{sp} of spontaneous emission in free space (6.115), we find

$$\frac{\Gamma_{\text{sp,cav}}}{\Gamma_{\text{sp}}} = \frac{3\mathcal{F}\lambda^2}{\pi^2 S} \sin^2 k_0 x_{\text{at}} \frac{1}{1 + \frac{4\mathcal{F}^2}{\pi^2} \sin^2 k_0 L}. \tag{6B.46}$$

Equation (6B.46) shows that the rate of spontaneous emission in the cavity is a maximum when the cavity is at exact resonance with the atom ($\sin k_0 L = 0$), and when the atom lies at an antinode of the cavity mode ($\sin k_0 x_{\text{at}} = 1$). When these two conditions are fulfilled, we are left with

$$\frac{\Gamma_{\text{sp,cav}}}{\Gamma_{\text{sp}}} = \frac{3\mathcal{F}\lambda^2}{\pi^2 S}. \tag{6B.47}$$

The ratio $\Gamma_{\text{sp,cav}} / \Gamma_{\text{sp}}$ is called the Purcell factor, and it is usually expressed in terms of the cavity Q factor $Q = 2L\mathcal{F}/\lambda$ and effective mode volume V_{mode} defined by,

$$V_{\text{mode}} = \iiint_{\text{mode}} \text{d}x \, \text{d}y \, \text{d}z \, \sin^2 kx = \frac{1}{2} LS, \tag{6B.48}$$

leading to

$$\frac{\Gamma_{\text{sp,cav}}}{\Gamma_{\text{sp}}} = \frac{3}{4\pi^2} \frac{Q\lambda^3}{V_{\text{mode}}}. \tag{6B.49}$$

When this factor is large compared to 1, spontaneous emission in the cavity dominates spontaneous emission in all other directions. As a result, the spontaneous emission rate is increased, and light is predominantly emitted along the cavity axis, which can be very useful for applications.

In order to get a large Purcell factor, one needs a cavity having a very high Q factor and a very small volume. Enhancement factors of order 10–100 have been observed in semiconductor microcavities.[9]

Comment The use of the Fermi golden rule is valid only if the coupling matrix element (6B.43) is constant over an energy interval significantly larger than $\hbar\Gamma_{\text{sp,cav}}$. At a resonance of the cavity, this demands that $\Gamma_{\text{sp,cav}}$ should be small compared to the width Γ_{cav} of the resonance. One can show that this is equivalent to the condition $\Omega_1^{(1)} \ll \Gamma_{\text{cav}}$.

6B.5 Conclusion

In this complement we have studied a simple application of the formalism developed in Chapter 6. We have shown that the interaction of an atom with a single mode of the electromagnetic field leads to a coupled global system known as the *dressed atom*. In fact, this concept is quite general, and can be applied profitably to the case of matter–light interactions in free space, for example, to the case of an atom interacting with a resonant or quasi-resonant single-mode radiation source (whether in the visible or the radio-frequency spectral range), all other field modes being empty. In this case, the energy-level diagram of the dressed atom, with its doublets of levels separated by the Rabi frequency $\sqrt{\bar{n}}\,\Omega_1^{(1)}$ easily accounts for many phenomena, such as:

- When the two-level atom is dressed by an intense electromagnetic wave, one can observe a doublet spectrum by probing this system with a weak auxiliary light source. This splitting of the atomic line into a doublet is called the Autler–Townes effect.
- When the two-level atom is coupled to an intense monochromatic field, its spontaneous emission spectrum is modified and exhibits three peaks separated by quantities related to the Rabi frequency of the field. This effect is known as the triplet spectrum of resonance fluorescence.

We have also given an example of how a cavity can modify the rate of spontaneous emission of the atom coupled to a mode of the cavity. This effect, the Purcell effect, is finding ever more applications in the domain of light sources.

Once again, subtle concepts of quantum optics play a role in phenomena extending from basic physics to applications.

[9] J.-M. Gérard and B. Gayral, Strong Purcell Effect for InAs Quantum Boxes in Three-Dimensional Solid-State Microcavities, *Journal of Lightwave Technology* **17**, 2089 (1999).

6C

Complement 6C Polarization-entangled photon pairs emitted in an atomic radiative cascade

6C.1 Introduction. Entangled photon pairs for real experiments

In Complement 5C, we discussed the argument of Einstein, Podolsky and Rosen and Bell's theorem concerning polarization-entangled photons described by the state (5C.15):

$$|\Psi_{\mathrm{EPR}}\rangle = \frac{1}{\sqrt{2}}\Big(|x,x\rangle + |y,y\rangle\Big). \tag{6C.1}$$

To make the transition from thought experiment to actual experiment, one must first be able to produce a pair of photons with the following properties:

- Their frequencies must lie in the visible (or near infra-red or ultra-violet), the only frequency range in which we have polarizers able to carry out a genuine polarization measurement in the sense specified in Section 5C.2.1. This condition excludes higher-energy photons, such as X or γ, for which two-channel polarizers do not exist.
- The emission process must be able to occur via two distinct channels starting from the same initial state and leading to the same final state, in such a way that one obtains a superposition of two terms.

As has been shown by Clauser, Horne, Shimony and Holt, photon pairs emitted in certain atomic radiative cascades satisfy these requirements.[1] A radiative cascade is a process in which an atom successively emits several photons, usually at different frequencies, by de-exciting in several successive steps from an excited state to lower energy states. An atom starting in a level e, going through an intermediate level r, and ending up in the ground state g, emits a pair of photons of energy:

$$\hbar\omega_1 = E_{\mathrm{e}} - E_{\mathrm{r}} \tag{6C.2}$$

and

$$\hbar\omega_2 = E_{\mathrm{r}} - E_{\mathrm{g}}. \tag{6C.3}$$

We shall see that, if level r is degenerate, we can indeed obtain a pair of entangled photons.

[1] J. Clauser, M. Horne, A. Shimony and R. Holt, Proposed Experiments to Test Local Hidden Variable Theories, *Physical Review Letters* **23**, 880 (1969).

To make this more concrete, calcium provides an example of a radiative cascade in which one photon is emitted in the violet (423 nm) and another is emitted in the green (513 nm). The photon pairs emitted by this cascade have been used to carry out very convincing tests of the Bell inequalities.

In this complement, we shall use the formalism set up in this chapter to show that this spontaneous decay process does indeed produce a pair of photons with polarizations described by the state (6C.1), if the atomic levels have angular momenta $J = 0, J = 1$ and $J = 0$, in that order, and if the photons are filtered by direction.

6C.2 Photon pair emitted in an atomic radiative cascade $J = 0 \rightarrow J = 1 \rightarrow J = 0$. Elementary process

6C.2.1 Description of the system

Consider three atomic levels g, r and e, with total angular momenta $J = 0, J = 1$ and $J = 0$, respectively (see Figure 6C.1). Level e is not degenerate, and the wavefunction associated with state $|\mathrm{e}\rangle$ is even. Likewise for $|\mathrm{g}\rangle$. However, the intermediate level r is degenerate, since the quantum number m associated with the projection $\hat{J}_z$ of the angular momentum in the Oz direction can take values $m_{\mathrm{r}} = -1, 0$ or $+1$. The three corresponding states have odd wavefunctions.

When excited into the upper level e, the atom can thus de-excite spontaneously, by electric dipole transition, into any of the intermediate states $|m_{\mathrm{r}}\rangle$ by emitting a photon ν_1. From there, it can once again spontaneously de-excite by electric dipole transition into the ground state $|\mathrm{g}\rangle$ by emitting a photon ν_2. However, it cannot go directly from $|\mathrm{e}\rangle$ to $|\mathrm{g}\rangle$ by emitting a single photon since an electric dipole transition can only occur between two levels of

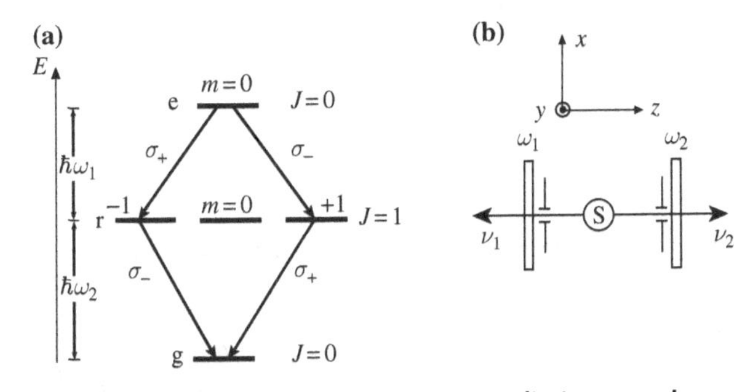

Figure 6C.1 Source of entangled photon pairs emitted in an atomic radiative cascade $J = 0 \rightarrow J = 1 \rightarrow J = 0$. In the source S, an atom excited to level e decays spontaneously by emitting a photon ν_1 of frequency ω_1, and a photon ν_2 of frequency ω_2. By means of diaphragms and interference filters, we isolate photons ν_1 propagating along the negative z direction and photons ν_2 propagating along the positive z direction. Note that the photons do not have the same frequency, and can thus be distinguished by the filters.

opposite parity (see Complement 2B). We shall be concerned with the state describing a pair of photons (ν_1, ν_2) emitted in certain directions. As can be seen from the figure, we select a photon ν_1 propagating along the negative z direction, and a photon ν_2 propagating along the positive z direction.

6C.2.2 Emission of photon ν_1 and entangled atom–radiation state

To determine the state of the field obtained after the transition from state $|e\rangle$ to state $|g\rangle$, and after filtering the direction and frequency, we need to specify the matrix elements of the interaction Hamiltonian. Note first that, since the atom changes level, $\hat{H}_{I2}$ does not contribute, as explained in Section 6.3.5. Regarding $\hat{H}_{I1}$, it can couple two atomic levels, even when there is no initial photon, causing the spontaneous emission of a photon (see Section 6.3.3). More precisely, starting from the initial state with 0 photons:

$$|i\rangle = |e\,;\,0\rangle, \tag{6C.4}$$

the Hamiltonian $\hat{H}_{I1}$ can in principle cause a transition to any state:

$$|j\rangle = |m_r\,;\,n_\ell = 1\rangle, \tag{6C.5}$$

where $|m_r\rangle$ denotes the atom in the state $|r, m_r\rangle$, and ℓ is a field mode.

The corresponding matrix element is (see (6.89))

$$\langle m_r\,;\,n_\ell = 1\,|\,\hat{H}_{I1}\,|\,e\,;\,0\rangle = -\frac{q}{m}\sqrt{\frac{\hbar}{2\varepsilon_0\omega_\ell L^3}}\langle m_r\,|\,\hat{\mathbf{p}}\cdot\boldsymbol{\varepsilon}_\ell\,|\,e\rangle. \tag{6C.6}$$

In this calculation, we are only concerned with the state of the field after filtering, and we thus consider only those modes ℓ propagating in the negative z direction. For a given frequency ω', there are then two possible modes corresponding to two orthogonal polarizations. We choose circular polarizations σ_+ and σ_- around Oz, writing them in the form (see Complement 2A)

$$\boldsymbol{\varepsilon}_+ = -\frac{\mathbf{e}_x + \mathrm{i}\mathbf{e}_y}{\sqrt{2}}, \tag{6C.7}$$

$$\boldsymbol{\varepsilon}_- = \frac{\mathbf{e}_x - \mathrm{i}\mathbf{e}_y}{\sqrt{2}}. \tag{6C.8}$$

Given the selection rules discussed in Complement 2A, the only non-zero matrix elements are

$$\langle m_r = -1\,;\,n_{\sigma_+} = 1\,|\,\hat{H}_{I1}\,|\,e\,;\,0\rangle = -\frac{q}{m}\sqrt{\frac{\hbar}{2\varepsilon_0\omega' L^3}}\langle m_r = -1\,|\,\hat{\mathbf{p}}\cdot\boldsymbol{\varepsilon}_+\,|\,e\rangle \tag{6C.9}$$

and

$$\langle m_r = 1\,;\,n_{\sigma_-} = 1\,|\,\mathbf{H}_{I1}\,|\,e\,;\,0\rangle = -\frac{q}{m}\sqrt{\frac{\hbar}{2\varepsilon_0\omega' L^3}}\langle m_r = 1\,|\,\hat{\mathbf{p}}\cdot\boldsymbol{\varepsilon}_-\,|\,e\rangle. \tag{6C.10}$$

It is easy to show that, for the two modes considered (polarizations $\boldsymbol{\varepsilon}_+$ and $\boldsymbol{\varepsilon}_-$, same direction of propagation along $-Oz$), the two matrix elements are equal. The projection onto the space associated with these modes of the state obtained after de-excitation thus has the form (after normalization),

$$|\psi'\rangle = \frac{1}{\sqrt{2}}\Big(|m_{\rm r} = -1\,;\, n_{\sigma_+} = 1\rangle + |m_{\rm r} = +1\,;\, n_{\sigma_-} = 1\rangle\Big). \tag{6C.11}$$

This is an entangled (non-factorizable) state of the atom and the radiation, a superposition of two distinct states of the system:

- The atom is in $|{\rm r}, m_{\rm r} = -1\rangle$, and a photon with polarization σ_+ and frequency ω' propagates in the negative $-Oz$ direction.
- The atom is in $|{\rm r}, m_{\rm r} = +1\rangle$, and a photon with polarization σ_- and frequency ω' propagates in the negative $-Oz$ direction.

6C.2.3 Emission of photon ν_2 and elementary EPR pair

Level r is not stable and can de-excite to the ground state $|{\rm g}\rangle$. More precisely, under the effect of $\hat{H}_{\rm I1}$, a state $|{\rm r}, m_{\rm r}\,;\,0\rangle$ is coupled with $|{\rm g}\,;\,n_k = 1\rangle$, where k denotes a radiation mode with frequency ω_k and polarization $\boldsymbol{\varepsilon}_k$. The coupling matrix element is

$$\langle {\rm g}\,;\, n_k = 1\,|\,\hat{H}_{\rm I1}\,|\,{\rm r}, m_{\rm r}\,;\,0\rangle = -\frac{q}{m}\sqrt{\frac{\hbar}{2\varepsilon_0\omega_k L^3}}\langle {\rm g}\,|\,\hat{\mathbf{p}}\cdot\boldsymbol{\varepsilon}_k\,|\,{\rm r}\rangle. \tag{6C.12}$$

Once again, we take into account the directional filtering (in the positive $+Oz$ direction), and we consider a single frequency ω''. The only degeneracy in the modes is thus the degeneracy in the polarization. According to the selection rules, for photons propagating in the positive $+Oz$ direction, the only non-zero matrix elements are

$$\langle {\rm g}\,;\, n_{\sigma_-} = 1\,|\,\hat{H}_{\rm I1}\,|\,m_{\rm r} = -1\,;\,0\rangle = -\frac{q}{m}\sqrt{\frac{\hbar}{2\varepsilon_0\omega'' L^3}}\langle {\rm g}\,|\,\hat{\mathbf{p}}\cdot\boldsymbol{\varepsilon}_-\,|\,m_{\rm r} = -1\rangle \tag{6C.13}$$

and

$$\langle {\rm g}\,;\, n_{\sigma_+} = 1\,|\,\hat{H}_{\rm I1}\,|\,m_{\rm r} = 1\,;\,0\rangle = -\frac{q}{m}\sqrt{\frac{\hbar}{2\varepsilon_0\omega'' L^3}}\langle {\rm g}\,|\,\hat{\mathbf{p}}\cdot\boldsymbol{\varepsilon}_+\,|\,m_{\rm r} = 1\rangle. \tag{6C.14}$$

These matrix elements are equal.

Under the effect of $\hat{H}_{\rm I1}$, the intermediate state (6C.11) thus evolves after long times to the final state,

$$|\psi_{\rm f}\rangle = \frac{1}{\sqrt{2}}\Big(|{\rm g}\,;\, n_{\sigma_+} = 1\,,\, n_{\sigma_-} = 1\rangle + |{\rm g}\,;\, n_{\sigma_-} = 1\,,\, n_{\sigma_+} = 1\rangle\Big). \tag{6C.15}$$

In this expression, it is understood that the first photon propagates in the negative z direction with a frequency ω' close to ω_1, while the second propagates in the positive z direction with a frequency ω'' close to ω_2.

We may extract the atomic part $|g\rangle$ as a factor, whence the radiation part is

$$|\psi_R\rangle = \frac{1}{\sqrt{2}}\Big(|n_{\sigma_+} = 1\,,\, n_{\sigma_-} = 1\rangle + |n_{\sigma_-} = 1\,,\, n_{\sigma_+} = 1\rangle\Big). \tag{6C.16}$$

We now have an entangled state of the radiation alone, a superposition of two distinct states of the pairs (ν_1, ν_2):

- A photon with frequency ω' and polarization σ_+ propagating in the negative $-Oz$ direction and a photon with frequency ω'' and polarization σ_- propagating in the positive $+Oz$ direction.
- A photon with frequency ω' and polarization σ_- propagating in the negative $-Oz$ direction and a photon with frequency ω'' and polarization σ_+ propagating in the positive $+Oz$ direction.

At this stage, we may simplify the notation as we did in Complement 5C, taking it as understood that we only have one-photon states. Letting $|\boldsymbol{\varepsilon}_+\rangle$ and $|\boldsymbol{\varepsilon}_-\rangle$ denote the one-photon states with circular polarizations σ_+ and σ_-, respectively, relative to the Oz-axis, the radiation state (6C.16) can be written in the form

$$|\psi_R\rangle = \frac{1}{\sqrt{2}}\Big(|\boldsymbol{\varepsilon}_+\,,\,\boldsymbol{\varepsilon}_-\rangle + |\boldsymbol{\varepsilon}_-\,,\,\boldsymbol{\varepsilon}_+\rangle\Big). \tag{6C.17}$$

In each tensor product $|\boldsymbol{\varepsilon}_+, \boldsymbol{\varepsilon}_-\rangle$ or $|\boldsymbol{\varepsilon}_-, \boldsymbol{\varepsilon}_+\rangle$, it is understood that the first term refers to one photon propagating in the negative z direction, while the second term refers to one photon propagating in the positive z direction.

The polarization basis can be changed by introducing orthogonal linear polarizations, e.g. along Ox or Oy. Then, letting $|x\rangle$ and $|y\rangle$ denote one-photon states polarized in the x and y directions, we may write, as in (5C.8) and (5C.9),

$$|\boldsymbol{\varepsilon}_+\rangle = -\frac{1}{\sqrt{2}}\Big(|x\rangle + \mathrm{i}|y\rangle\Big), \tag{6C.18}$$

$$|\boldsymbol{\varepsilon}_-\rangle = \frac{1}{\sqrt{2}}\Big(|x\rangle - \mathrm{i}|y\rangle\Big). \tag{6C.19}$$

Substituting into (6C.17), we obtain the state of the photon pair:

$$|\psi_R\rangle = -\frac{1}{\sqrt{2}}\Big(|x, x\rangle + |y, y\rangle\Big), \tag{6C.20}$$

Up to a minus sign which is of no physical significance, this is the same as the state

$$|\psi_{EPR}\rangle = \frac{1}{\sqrt{2}}\Big(|x, x\rangle + |y, y\rangle\Big), \tag{6C.21}$$

introduced in Complement 5C without justification.

Comments (i) Note that the pair of orthogonal axes Ox and Oy can be taken with any orientation in the plane perpendicular to Oz. This is related to the fact that the state (6C.17) is invariant under rotation about Oz, as can be checked directly by making a change of basis.

(ii) A priori the state of the system comprises photon pairs propagating in all space directions. The spatial filtering device uses diaphragms to eliminate those that are not emitted in the Oz direction. Strictly speaking, to describe this system, we should use a density matrix formalism and take a partial trace over all modes absorbed by the diaphragms. However, we are only interested in the state of the field when the photodetectors effectively observe the presence of photons emitted in the Oz direction. There is then a 'reduction of the wave packet' to the state (6C.17) we have identified in this section.

(iii) An analogous calculation can be done by filtering the photons ν_1 and ν_2 in any specific direction. It is then found that the probability of joint detection behind two polarizers carrying out a polarization analysis in directions **a** and **b** can be written as in Complement 2C, i.e.

$$P_{++}(\mathbf{a}, \mathbf{b}) = \frac{1}{2}\cos^2(\mathbf{a}, \mathbf{b}), \tag{6C.22}$$

and this even if the directions of propagation are not along the same axis Oz. By integration, this formula can be used to treat the important practical case where the photons are captured by a lens occupying a certain finite solid angle.

6C.3 Generalization and sum over frequencies

The calculation just described reveals the mechanism leading to entanglement of the two emitted photons: transition to an intermediate state in which the first photon and the atom are entangled, then emission of a second photon leading to a state in which the two photons are entangled. But it is unrealistic in the sense that it assumes that we can carry out an infinitely narrow filtering of the frequencies ω' and ω'' of the photons propagating in the negative and positive z directions, respectively. In reality, we use interference filters centred on the resonance frequencies ω_1 and ω_2, respectively, with passbands that are wider than the natural widths of these two lines. The final state of the field after filtering is then a superposition of terms of the form (6C.21):

$$|\psi\rangle = \frac{1}{\sqrt{2}}\iint \mathrm{d}\omega_\ell\, \mathrm{d}\omega_k\, C(\omega_\ell\,,\, \omega_k)\Big(|x_\ell\,,\, x_k\rangle + |y_\ell\,,\, y_k\rangle\Big), \tag{6C.23}$$

where $|x_\ell, x_k\rangle$ denotes a pair of photons of frequencies ω_ℓ and ω_k propagating in the negative and positive z directions, respectively, and linearly polarized along Ox; and likewise for $|y_\ell, y_k\rangle$. Normalization of this state then requires,

$$\iint \mathrm{d}\omega_\ell\, \mathrm{d}\omega_k \mid C(\omega_\ell\,,\, \omega_k)|^2 = 1. \tag{6C.24}$$

The total photodetection probabilities are then simply obtained by summing the elementary probabilities associated with each elementary state $(|x_\ell, x_k\rangle + |y_\ell, y_k\rangle)/\sqrt{2}$. Indeed, one can in principle distinguish the results by the energy of the emitted photoelectrons. The integral associated with the sum over frequencies can be extracted as a factor, and the single or coincidence detection probabilities have the same values as those found for the elementary pairs. As an example, we may write the total joint photodetection probability in the two channels $+1$ and $+1$ of polarizers I and II oriented along **a** and **b** in the form,

$$P_{++}(\mathbf{a},\mathbf{b}) = \frac{1}{2}\iint \mathrm{d}\omega_\ell\,\mathrm{d}\omega_k\,|C(\omega_\ell\,,\,\omega_k)|^2\Big|\langle +_{\mathbf{a}},+_{\mathbf{b}}|(|x_\ell\,,\,x_k\rangle + |y_\ell\,,\,y_k\rangle)\Big|^2$$

$$= \frac{1}{2}\Big|\langle +_{\mathbf{a}}\,,\,+_{\mathbf{b}}|(|x\,,\,x\rangle + |y\,,\,y\rangle)\Big|^2 = \frac{1}{2}\cos^2(\mathbf{a}\,\mathbf{b}).$$

We thus obtain the same result as in Complement 5C.

6C.4 Two-photon excitations

To obtain an efficient source of entangled photon pairs by the atomic radiative cascade mechanism, it must be possible to excite the atom selectively into state $|\mathrm{e}\rangle$. Since states $|\mathrm{g}\rangle$ and $|\mathrm{e}\rangle$ have the same parity, this precludes the simple absorption of a photon with frequency $(E_\mathrm{e} - E_\mathrm{g}/\hbar)$. However, state $|\mathrm{e}\rangle$ can be selectively excited through a nonlinear process in which two waves of frequencies ω_Kr and ω_D cause a transition between the two states $|\mathrm{e}\rangle$ and $|\mathrm{g}\rangle$, provided that

$$\hbar(\omega_\mathrm{Kr} + \omega_\mathrm{D}) = E_\mathrm{e} - E_\mathrm{g}, \tag{6C.25}$$

and there exists a non-zero coupling matrix element between the atom and each of the two waves (see Sections 1.2.5 and 2.3.3). Figure 6C.2 shows this situation schematically.

Generalizing the calculation of Section 2.3.3 (see Equation (2.127)) to the case of two unequal frequencies, we find a transition probability after a time T equal to

$$P_{\mathrm{g}\to\mathrm{e}}(T) = T\frac{2\pi}{\hbar}\left|\frac{1}{4}\frac{\hbar\Omega_\mathrm{Kr}\Omega_\mathrm{D}}{\omega_\mathrm{Kr} - \omega_2}\right|^2 \delta_T\left[E_\mathrm{e} - E_\mathrm{g} - \hbar(\omega_\mathrm{Kr} + \omega_\mathrm{D})\right], \tag{6C.26}$$

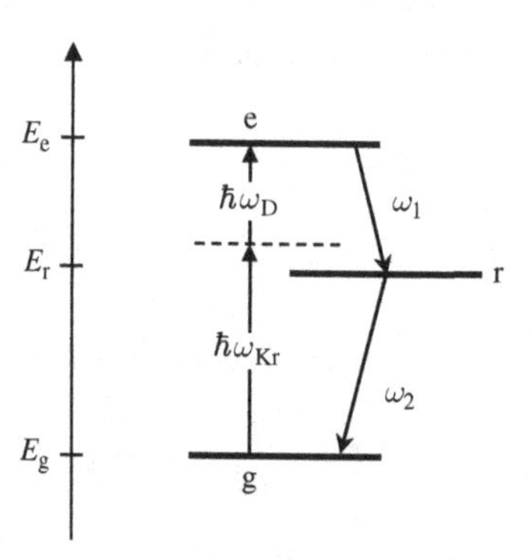

Figure 6C.2 Two-photon excitation of the radiative cascade e → r → g. The atom can be excited to state e by a nonlinear process involving two lasers of frequencies ω_Kr and ω_D, such that $\hbar(\omega_\mathrm{Kr} + \omega_\mathrm{D}) = E_\mathrm{e} - E_\mathrm{g}$. In order to satisfy this 'two-photon' resonance condition exactly in the calcium cascade used to test Bell's inequalities (wavelengths $\lambda_1 = 551$ nm and $\lambda_2 = 423$ nm), we use an ionized krypton laser emitting at 406 nm, and a tunable laser emitting at 581 nm, whose frequency can be controlled to the nearest MHz, i.e. with a relative accuracy of a few 10^{-9}, in order to meet the condition (6C.25).

where Ω_{Kr} and Ω_{D} are the Rabi angular frequencies characterizing the coupling of each wave with the atom. We can take into account the finite width Γ_{e} of level e by replacing $\delta_T(E)$ by the Lorentzian of unit area and full-width at half-maximum $\hbar\Gamma_{\mathrm{e}}$, so that finally,

$$P_{\mathrm{g}\to\mathrm{e}}(T) = \frac{T}{\Gamma_{\mathrm{e}}}\left|\frac{1}{4}\frac{\hbar\Omega_{\mathrm{Kr}}\Omega_{\mathrm{D}}}{\omega_{\mathrm{Kr}}-\omega_2}\right|^2 \frac{1}{1+\left(\frac{\omega_{\mathrm{Kr}}+\omega_{\mathrm{D}}-\omega_1-\omega_2}{\Gamma_{\mathrm{e}}/2}\right)^2}. \tag{6C.27}$$

At the resonance of the two-photon transition, the excitation rate for the cascade is

$$\frac{\mathrm{d}}{\mathrm{d}t}P_{\mathrm{g}\to\mathrm{e}} = \frac{1}{\Gamma_{\mathrm{e}}}\left|\frac{1}{4}\frac{\Omega_{\mathrm{Kr}}\Omega_{\mathrm{D}}}{\omega_{\mathrm{Kr}}-\omega_2}\right|^2. \tag{6C.28}$$

It is easy to obtain Rabi angular frequencies of $10^{10}\ \mathrm{s}^{-1}$ for electric dipole transitions by focusing laser beams of a few tens of milliwatts on a fraction of a square millimetre. In the example of Figure 6C.2, Γ_{e} is of the order of $10^7\ \mathrm{s}^{-1}$, and the detuning $\omega_{\mathrm{Kr}}-\omega_2$ of the order of $10^{14}\ \mathrm{s}^{-1}$, which gives an excitation rate of the order of $10^4\ \mathrm{s}^{-1}$. It has thus been possible to obtain an intense enough entangled photon source to carry out sophisticated tests of the Bell inequalities (see Complement 5C).

PART III

APPLYING BOTH APPROACHES

7 Nonlinear optics. From the semi-classical approach to quantum effects

7.1 Introduction

In 1961, just a few months after Maiman invented the ruby laser, Franken focused the pulses emitted from such a laser, of wavelength 694 nm, on a quartz plate, and examined the spectrum of the light transmitted using a simple prism (see Figure 7.1). He thus discovered that ultra-violet light of wavelength 347 nm was emerging from the quartz plate. Clearly, as it propagated through the quartz, the light of frequency ω had generated the second harmonic, of frequency 2ω.

It thus transpires that in optics, as in any other part of physics, a system subjected to a strong enough sinusoidal excitation will leave the linear response regime. Nonlinearities cause harmonics of the excitation frequency to appear.

But what intensity is needed before nonlinear effects will appear? One might think that a natural scale would be the electric field of the nucleus at the location of an atomic electron. In the case of the hydrogen atom in its ground state, this field is about $e/4\pi\varepsilon_0 a_0^2$, or 3×10^{11} V.m^{-1}. (Here e is the charge of the electron, and a_0 the Bohr radius, of the order of 5×10^{-11} m). In fact, experiment shows that, in the transparency zone of a dielectric material like quartz, a field of just 10^7 V.m^{-1} (corresponding to a light intensity of 2.5 kW.cm^{-2}) is sufficient for nonlinear effects to appear perturbatively. Close to electron resonances, even weaker intensities, of the order of mW.cm^{-2}, can cause a system to go into saturation, which is a nonlinear regime (see Chapter 2 and Complement 7B).

This chapter and the following will provide an opportunity to apply the various tools set up in previous chapters. To begin with, we shall treat the radiation classically, with the matter properties expressed through the susceptibility coefficients, which can be obtained by quantum calculations. This is the semi-classical approach of Chapter 2. We shall use it to obtain the equation of propagation of the fields in a nonlinear medium (see Section 7.2), and to study in some detail, as an example, the phenomenon of three-wave mixing (see Section 7.3). However, there are very important phenomena, e.g. parametric fluorescence, that cannot be described by this approach. We then have recourse to the purely quantum theoretical approach introduced from Chapter 4 onwards. This will be the subject of Section 7.4, which describes in particular one of the most fascinating phenomena of quantum optics, namely, the creation of twin photon pairs.

Two complements will provide the reader with a deeper grasp of the ideas introduced in the chapter. Complement 7A details the classical and quantum theoretical features of the phenomenon known as parametric amplification, and also the oscillators that can be based on this amplification effect ('optical parametric oscillators'). Complement 7B

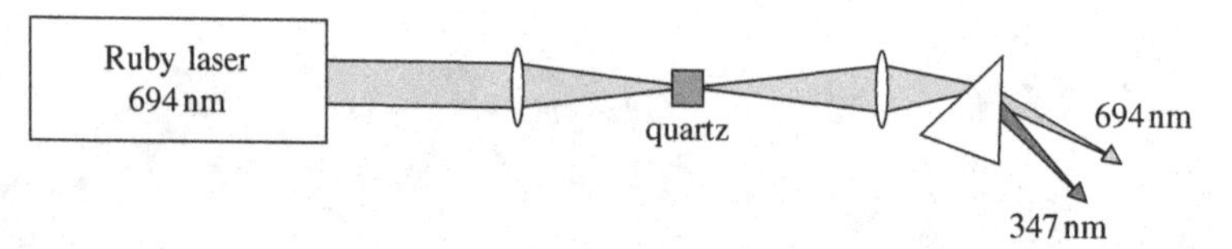

Figure 7.1 Franken's experiment. The laser pulse focused in the quartz gives rise to light at twice the frequency.

presents several examples of higher order nonlinear phenomena, in which the refractive index depends on the light intensity (optical Kerr effect).

7.2 Electromagnetic field in a nonlinear medium. Semi-classical treatment

7.2.1 Linear susceptibility

We know that the propagation of an electromagnetic wave in matter is affected by the dielectric polarization induced by the electric field of the wave (see, for example, Chapter 2). The dielectric polarization $\mathbf{P}(\mathbf{r}, t)$ near a point $\mathbf{r}$ (induced dipole per unit volume) is clearly related to the electric field at the point $\mathbf{r}$, and it is the dielectric susceptibility that characterizes this relation. In order to calculate this, one must determine the motions of the charges in the matter under the effect of the monochromatic classical field:

$$\mathbf{E}(\mathbf{r}, t) = \boldsymbol{\varepsilon}\mathcal{E}(\mathbf{r}, \omega)\,\mathrm{e}^{-i\omega t} + \mathrm{c.c.} \tag{7.1}$$

This calculation involves a quantum theoretical description of matter. If we consider a polarizable object (atom in a vapour, ion, atom or molecule in a solid), characterized by an electric dipole operator $\hat{\mathbf{D}}$, the effect of the field $\mathbf{E}(\mathbf{r}, t)$ is to establish a forced regime characterized by a density matrix $\hat{\sigma}_{\mathrm{st}}(\mathbf{r}, t)$, and its dipole moment assumes the average value (see Complement 2C):

$$\langle\hat{\mathbf{D}}\rangle(\mathbf{r}, t) = \mathrm{Tr}\left[\hat{\sigma}_{\mathrm{st}}(\mathbf{r}, t)\hat{\mathbf{D}}\right]. \tag{7.2}$$

For a density N/V (number of atoms per unit volume), the dielectric polarization at point $\mathbf{r}$ is given by

$$\mathbf{P}(\mathbf{r}, t) = \frac{N}{V}\langle\hat{\mathbf{D}}\rangle(\mathbf{r}, t). \tag{7.3}$$

Under the effect of a sinusoidal excitation of frequency ω, the forced regime is periodic with period $2\pi/\omega$. If the electric field is weak enough, the system is in the linear response regime, $\mathbf{P}(\mathbf{r}, t)$ is a sinusoidal function of time, and it can be written in an analogous form to (7.1) by introducing the complex amplitude $\mathcal{P}(\mathbf{r}, \omega)$. We thus define the complex linear susceptibility $\chi^{(1)}(\omega)$ such that

$$\mathcal{P}(\mathbf{r}, \omega) = \varepsilon_0\chi^{(1)}(\omega)\mathcal{E}(\mathbf{r}, \omega). \tag{7.4}$$

For a transparent medium, $\chi^{(1)}(\omega)$ is real, and to a first approximation, we can neglect the frequency dependence (dispersion of the refractive index). We then have proportionality between the polarization and real fields, even when they are not monochromatic:

$$\mathbf{P}(\mathbf{r}, t) = \varepsilon_0 \chi^{(1)} \mathbf{E}(\mathbf{r}, t). \tag{7.5}$$

If the medium is not isotropic, the dielectric polarization is not necessarily parallel to the electric field. In this case, the susceptibility is a rank 2 tensor, expressed in the form of a 3×3 matrix, and (7.5) is replaced by

$$P_i(\mathbf{r}, t) = \varepsilon_0 \sum_j \chi_{ij}^{(1)} E_j(\mathbf{r}, t), \tag{7.6}$$

where P_i and E_j are the Cartesian components of the vectors $\mathbf{P}$ and $\mathbf{E}$, and the indices i and j take the values (x, y, z).

7.2.2 Nonlinear susceptibility

When the electric field becomes stronger, we enter the nonlinear regime, which can be described by generalizing the relations in Section 7.2.1. For example, for a transparent material, neglecting the frequency dependence (small dispersion), (7.6) generalizes to

$$P_i = \varepsilon_0 \sum_j \chi_{ij}^{(1)} E_j + \varepsilon_0 \sum_j \sum_k \chi_{ijk}^{(2)} E_j E_k + \varepsilon_0 \sum_j \sum_k \sum_\ell \chi_{ijk\ell}^{(3)} E_j E_k E_\ell + \dots, \tag{7.7}$$

where $[\chi_{ijk}^{(2)}]$ is a real rank 3 tensor (a 3-index matrix with 27 components), which characterizes the second-order nonlinear susceptibility, while $[\chi_{ijkl}^{(3)}]$ is a rank 4 tensor which characterizes the third-order nonlinear susceptibility, and so on. For intensities accessible using typical lasers, successive terms in the expansion of (7.7) usually drop off quickly, and it suffices to keep only up to the second-order term, unless $\chi^{(2)}$ happens to be zero, in which case the third-order term in $\chi^{(3)}$ must also be taken into account (see below).

The tensorial nature of the nonlinear polarization, illustrated by (7.7), can give rise to a wide range of phenomena. However, in this chapter, we shall generally be concerned with only one Cartesian component of the nonlinear polarization, itself induced by a linearly polarized field. This simplification can be a consequence of the symmetry of the problem, or of the phase matching condition (see below). We shall thus omit the Cartesian indices of the electric field or the induced polarization, as well as the tensor indices of the single relevant component of the nonlinear susceptibility. Moreover, as we did in (7.7), we shall not specify the position $\mathbf{r}$ explicitly, in order to simplify the notation, unless it seems important for a clearer understanding.

Consider then the case in which the point $\mathbf{r}$ in the material is subjected to a superposition of two monochromatic fields of frequencies ω_1 and ω_2. The total field is therefore $E(t) = E_1(t) + E_2(t)$, where

$$E_1(t) = \mathcal{E}_1 \, \mathrm{e}^{-\mathrm{i}\omega_1 t} + \text{c.c.} \tag{7.8}$$

$$E_2(t) = \mathcal{E}_2 \, \mathrm{e}^{-\mathrm{i}\omega_2 t} + \text{c.c.} \tag{7.9}$$

Table 7.1 The different terms than can arise when two beams of frequencies ω_1 and ω_2 propagate in a second-order nonlinear medium.

$\chi^{(2)}(-2\omega_1\,;\,\omega_1,\omega_1)$	$\left(\mathcal{E}_1^2\,\mathrm{e}^{-2\mathrm{i}\omega_1 t}+\mathrm{c.c.}\right)$	Frequency doubling
$\chi^{(2)}(-2\omega_2\,;\,\omega_2,\omega_2)$	$\left(\mathcal{E}_2^2\,\mathrm{e}^{-2\mathrm{i}\omega_2 t}+\mathrm{c.c.}\right)$	Frequency doubling
$2\chi^{(2)}(0\,;\,\omega_1,-\omega_1)$	$\lvert\mathcal{E}_1\rvert^2$	Optical rectification
$2\chi^{(2)}(0\,;\,\omega_2,-\omega_2)$	$\lvert\mathcal{E}_2\rvert^2$	Optical rectification
$2\chi^{(2)}(-\omega_1-\omega_2\,;\,\omega_1,\omega_2)$	$\left(\mathcal{E}_1\mathcal{E}_2\,\mathrm{e}^{-\mathrm{i}(\omega_1+\omega_2)t}+\mathrm{c.c.}\right)$	Frequency addition
$2\chi^{(2)}(-\omega_1+\omega_2\,;\,\omega_1,-\omega_2)$	$\left(\mathcal{E}_1\mathcal{E}_2^*\,\mathrm{e}^{-\mathrm{i}(\omega_1-\omega_2)t}+\mathrm{c.c.}\right)$	Frequency difference

The second-order nonlinear polarization is then,

$$\begin{aligned}P^{(2)}(t)&=\varepsilon_0\chi^{(2)}E(t)^2\\&=\varepsilon_0\chi^{(2)}\left(\mathcal{E}_1\,\mathrm{e}^{-\mathrm{i}\omega_1 t}+\mathrm{c.c.}+\mathcal{E}_2\,\mathrm{e}^{-\mathrm{i}\omega_2 t}+\mathrm{c.c.}\right)^2,\end{aligned}\tag{7.10}$$

or

$$\begin{aligned}P^{(2)}(t)=\varepsilon_0\chi^{(2)}\Big(&\mathcal{E}_1^2\,\mathrm{e}^{-2\mathrm{i}\omega_1 t}+\mathrm{c.c.}+\mathcal{E}_2^2\,\mathrm{e}^{-2\mathrm{i}\omega_2 t}+\mathrm{c.c.}+2\lvert\mathcal{E}_1\rvert^2+2\lvert\mathcal{E}_2\rvert^2\\&+2\mathcal{E}_1\mathcal{E}_2\,\mathrm{e}^{-\mathrm{i}(\omega_1+\omega_2)t}+\mathrm{c.c.}+2\mathcal{E}_1\mathcal{E}_2^*\,\mathrm{e}^{-\mathrm{i}(\omega_1-\omega_2)t}+\mathrm{c.c.}\Big)\end{aligned}\tag{7.11}$$

The above equation underlies a great many nonlinear processes, such as second harmonic generation (terms oscillating at twice the frequency of the applied fields), or frequency addition, where the wave of frequency ω_1 mixes with the wave of frequency ω_2 to produce a polarization at the frequency $\omega_1+\omega_2$.

When one approaches the resonances of the system, neither the frequency dependence of the susceptibility nor the phase offset between the nonlinear polarization and the electric field can continue to be ignored, and this leads to a complex-valued, frequency-dependent susceptibility. For example, for the addition term between the frequencies ω_1 and ω_2, we shall use the notation $\chi^{(2)}(-\omega_1-\omega_2;\omega_1,\omega_2)$ for $\chi^{(2)}$. By convention, the first of the three arguments is always equal to the negative of the sum of the other two arguments, and so corresponds, up to a sign, to the frequency produced in the nonlinear polarization. The various terms contributing to the nonlinear polarization are displayed in Table 7.1.

Note that some terms, such as the frequency addition term, involve a factor of 2. This arises simply from the doubled cross-product appearing in the expansion of the square of the total electric field (see (7.11)). This factor of 2 is a source of confusion in the literature, because many authors incorporate it into the definition of $\chi^{(2)}$. This definition, which we shall not be using here, has the disadvantage that it gives rise to a singularity in the susceptibility when two frequencies are equal, a purely mathematical singularity with no physical counterpart.[1]

Nonlinear media are often crystals, and symmetries play a fundamental role because they frequently show at the outset that many components of the nonlinear susceptibility tensors will be zero. Without going into the details of this question, although it is essential in practice, we shall give an important example, namely the centrally symmetric media. These are invariant under inversion through the origin. This symmetry implies that second-order terms like (7.10–7.11) are necessarily zero. Indeed, inversion with respect to the origin changes the sign of the components E_1 and E_2, but also of the dielectric polarization and hence $P^{(2)}$, which is incompatible with a relation like (7.10). We may conclude that, in a centrally symmetric medium, only the odd terms differ from zero, and the first nonlinear term is of order 3. We also see that, in order to have second-order nonlinearities, which are in principle much bigger, we must either use a crystal that is not centrally symmetric, or apply some external stress that breaks the symmetry, e.g. a mechanical stress, a static electric field, an interface with a different material. These matters are essential when considering applications of nonlinear optics.

Comment If instead of adopting the convention of (7.8) and (7.9) for the complex representation, we introduce a factor of 1/2 (we write $E = \mathrm{Re}\,\{\mathcal{E}\exp(-\mathrm{i}\omega t)\}$), a factor of $1/2^n$ will appear in the expression for the order n nonlinear polarization. This adds to the confusion mentioned above, and one must therefore take great care when making quantitative comparisons between susceptibility values published by different authors.

7.2.3 Propagation in a nonlinear medium

We know that the propagation of plane monochromatic waves in a dielectric medium can be handled simply by means of the equation of propagation that is a direct consequence of Maxwell's equations:

$$\Delta\mathbf{E} - \frac{1}{c^2}\frac{\partial^2}{\partial t^2}\left(\mathbf{E} + \frac{\mathbf{P}}{\varepsilon_0}\right) = 0. \tag{7.12}$$

This equation is also valid for complex fields, or analytic signals, $\mathbf{E}^{(+)}(\mathbf{r}, t)$ and $\mathbf{P}^{(+)}(\mathbf{r}, t)$. In a linear dielectric, neglecting dispersion, we have

$$\mathbf{P}^{(+)}(\mathbf{r}, t) = \mathbf{P}_{\mathrm{L}}^{(+)}(\mathbf{r}, t) = \varepsilon_0 \chi^{(1)} \mathbf{E}^{(+)}(\mathbf{r}, t), \tag{7.13}$$

and the equation of propagation becomes

$$\Delta\mathbf{E}^{(+)}(\mathbf{r}, t) - \frac{n^2}{c^2}\frac{\partial^2 \mathbf{E}^{(+)}(\mathbf{r}, t)}{\partial t^2} = 0, \tag{7.14}$$

[1] There is a clear discussion in the book by Y. R. Shen (Section 9 of Chapter 2): Y. R. Shen, *Principles of Nonlinear Optics*, Wiley Interscience (1984). Another good reference is P. N. Butcher and D. Cotter (Section 2.3.3): *The Elements of Nonlinear Optics*, Cambridge Studies in Modern Physics (1990).

where n is the refractive index:

$$n^2 = 1 + \chi^{(1)} = \varepsilon_r. \tag{7.15}$$

When there is dispersion, the susceptibility $\chi^{(1)}$, and hence the index n, depend on the frequency ω. However, in this case Equation (7.13) must be replaced by a relation between complex amplitudes, such as (7.4). If the medium is absorbing or amplifying, $\chi^{(1)}$ (and hence the wavevector $\mathbf{k}_\ell$) has an imaginary part, which corresponds to attenuated or amplified waves (see Sections 2.4.4 and 2.5.2 in Chapter 2). We expand the field $\mathbf{E}^{(+)}(\mathbf{r}, t)$ in the form of a series of plane monochromatic waves:

$$\mathbf{E}^{(+)}(\mathbf{r}, t) = \sum_\ell \boldsymbol{\varepsilon}_\ell \mathcal{E}_\ell \exp\{\mathrm{i}(\mathbf{k}_\ell \cdot \mathbf{r} - \omega_\ell t)\}, \tag{7.16}$$

with

$$\mathbf{k}_\ell^2 = n_\ell^2 \frac{\omega_\ell^2}{c^2} \tag{7.17}$$

and

$$\boldsymbol{\varepsilon}_\ell \cdot \mathbf{k}_\ell = 0. \tag{7.18}$$

The parameter n_ℓ is the index of refraction of the medium at frequency ω_ℓ.

In a nonlinear medium, Equation (7.14) remains true, but it is useful to show the nonlinear polarization explicitly:

$$\Delta \mathbf{E}^{(+)}(\mathbf{r}, t) - \frac{1}{c^2} \frac{\partial^2}{\partial t^2} \left(\mathbf{E}^{(+)}(\mathbf{r}, t) + \frac{\mathbf{P}_{\mathrm{L}}^{(+)}(\mathbf{r}, t) + \mathbf{P}_{\mathrm{NL}}^{(+)}(\mathbf{r}, t)}{\varepsilon_0} \right) = 0. \tag{7.19}$$

To simplify, we shall only consider the case in which all the waves propagate in the same direction Oz. We seek a solution of (7.19) in the form (7.16), but with an amplitude, or envelope, $\mathcal{E}_\ell(z)$ which now depends on the z-coordinate. We shall also make the *slowly varying envelope approximation*, assuming that the amplitude of the waves varies slowly on the scale of the wavelength $\lambda_\ell = 2\pi/k_\ell$, i.e. it satisfies

$$\left| \frac{\mathrm{d}\mathcal{E}_\ell(z)}{\mathrm{d}z} \right| \ll k_\ell \, |\mathcal{E}_\ell(z)|, \qquad \left| \frac{\mathrm{d}^2\mathcal{E}_\ell(z)}{\mathrm{d}z^2} \right| \ll k_\ell \left| \frac{\mathrm{d}\mathcal{E}_\ell(z)}{\mathrm{d}z} \right|. \tag{7.20}$$

This condition allows us to neglect terms in $\mathrm{d}^2\mathcal{E}_\ell/\mathrm{d}z^2$ in the Laplacian of (7.19), which generally has the form

$$\Delta \mathbf{E}^{(+)}(z, t) = \sum_\ell \left[-k_\ell^2 \mathcal{E}_\ell(z) + \mathrm{i}2k_\ell \frac{\mathrm{d}\mathcal{E}_\ell(z)}{\mathrm{d}z} + \frac{\mathrm{d}^2\mathcal{E}_\ell(z)}{\mathrm{d}z^2} \right] \boldsymbol{\varepsilon}_\ell \exp\{\mathrm{i}(k_\ell z - \omega_\ell t)\}. \tag{7.21}$$

Then, taking into account (7.15) and (7.17), the equation of propagation takes the form

$$\sum_\ell \mathrm{i}2k_\ell \frac{\mathrm{d}\mathcal{E}_\ell(z)}{\mathrm{d}z} \boldsymbol{\varepsilon}_\ell \exp\{\mathrm{i}(k_\ell z - \omega_\ell t)\} = \frac{1}{\varepsilon_0 c^2} \frac{\partial^2}{\partial t^2} \mathbf{P}_{\mathrm{NL}}^{(+)}(z, t). \tag{7.22}$$

$\mathbf{P}_{\mathrm{NL}}^{(+)}(z,t)$ can be decomposed into monochromatic components,

$$\mathbf{P}_{\mathrm{NL}}^{(+)}(z,t) = \sum_{\ell} \boldsymbol{\varepsilon}_{\ell} \mathcal{P}_{\mathrm{NL}}^{\omega_\ell}(z) \exp\{-\mathrm{i}\omega_\ell t\}, \tag{7.23}$$

and we obtain a set of equations relating the amplitudes $\mathcal{E}_\ell$ to the components $\mathcal{P}_{\mathrm{NL}}^{\omega_\ell}$:

$$\frac{\mathrm{d}\mathcal{E}_\ell(z,t)}{\mathrm{d}z} \exp\{\mathrm{i}k_\ell z\} = \mathrm{i}\frac{\omega_\ell}{2\varepsilon_0 n_\ell c} \mathcal{P}_{\mathrm{NL}}^{\omega_\ell}(z). \tag{7.24}$$

Then all we need to do is to express $\mathcal{P}_{\mathrm{NL}}^{\omega_\ell}(z)$ in terms of the amplitudes $\mathcal{E}_m(z)$ at the point z using the nonlinear susceptibilities, to obtain a system of coupled first-order differential equations that can be solved to yield the amplitudes $\mathcal{E}_\ell(z)$ of the different waves.

Rather than try to develop this formalism in all its generality, which would take up a lot of space, *we shall confine our attention in this chapter to the simplest case, in which a second-order nonlinearity couples three waves together.* The case of 4-wave mixing, induced by third-order nonlinearities, is discussed in Complement 7B.

7.3 Three-wave mixing. Semi-classical treatment

7.3.1 Frequency addition

We consider two intense waves, called pump waves, propagating in the Oz direction in a second-order nonlinear medium. As we have just seen, there are many terms associated with the physical processes of frequency doubling, addition or difference. We shall consider frequency addition, associated with the polarization oscillating at frequency $\omega_3 = \omega_1 + \omega_2$. The corresponding complex amplitude, which arises in (7.11), is

$$\mathcal{P}_{\mathrm{NL}}^{\omega_3}(z) = 2\varepsilon_0\, \chi^{(2)} \mathcal{E}_1(z)\mathcal{E}_2(z) \exp\{\mathrm{i}(k_1+k_2)z\}, \tag{7.25}$$

where $\chi^{(2)}$ in fact stands for $\chi^{(2)}(-\omega_3\,;\,\omega_1,\omega_2)$. This nonlinear polarization produces a field $\mathcal{E}_3(z)$ at frequency ω_3, which evolves according to (7.24):

$$\frac{\mathrm{d}\mathcal{E}_3}{\mathrm{d}z} \exp\{ik_3 z\} = \mathrm{i}\frac{\omega_3}{2\varepsilon_0 n_3 c} \mathcal{P}_{NL}^{\omega_3}(z). \tag{7.26}$$

Substituting (7.25) into (7.26), we obtain

$$\frac{\mathrm{d}\mathcal{E}_3}{\mathrm{d}z} = \mathrm{i}\frac{\omega_3 \chi^{(2)}}{n_3 c} \mathcal{E}_1(z)\mathcal{E}_2(z) \exp\{\mathrm{i}\Delta k \cdot z\}, \tag{7.27}$$

with

$$\Delta k = k_1 + k_2 - k_3. \tag{7.28}$$

We now express the fact that, at $z = 0$ (the entry face of the plate of nonlinear material, see Figure 7.2), the amplitude of the wave at ω_3 is zero:

$$\mathcal{E}_3(0) = 0. \tag{7.29}$$

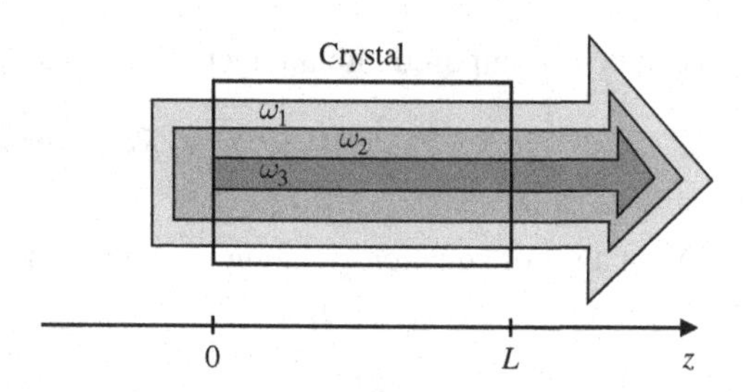

Figure 7.2 Frequency addition in a plate of thickness L, with second-order nonlinearity $\chi^{(2)} \neq 0$. The two intense pump beams at ω_1 and ω_2 combine to give rise to a wave of frequency $\omega_3 = \omega_1 + \omega_2$. Note that the beam diameters have been represented as different for clarity, while in reality they are the same.

The nonlinear effect is often weak so, to a first approximation, we may assume that the intensity of the pump beams does not vary over the length of the nonlinear material, and we can then take $\mathcal{E}_1$ and $\mathcal{E}_2$ to be constant. Provided that z remains small compared with Δk^{-1}, Equation (7.27) can be simplified to

$$\frac{\mathrm{d}\mathcal{E}_3}{\mathrm{d}z} \simeq \mathrm{i}\frac{\omega_3 \chi^{(2)}}{n_3 c}\mathcal{E}_1\mathcal{E}_2, \tag{7.30}$$

and we observe that the amplitude of the wave produced at $\omega_3 = \omega_1 + \omega_2$ increases linearly with z. If the thickness L of the nonlinear plate is smaller than Δk^{-1}, we thus have an output intensity (Poynting vector Π_3, see Equation (7.53)) equal to

$$\Pi_3 = C L^2\, \Pi_1 \Pi_2, \tag{7.31}$$

with

$$C = \frac{\left[\omega_3 \chi^{(2)}\right]^2}{2\varepsilon_0 c^3\, n_1 n_2 n_3}. \tag{7.32}$$

It is useful to introduce the powers Φ_1, Φ_2 and Φ_3 which, in the simplified model where the intensity is constant over the beam profile of area S, can be written

$$\Phi_i = S\, \Pi_i. \tag{7.33}$$

The output power is then,

$$\Phi_3 = C L^2\, \frac{\Phi_1 \Phi_2}{S}, \tag{7.34}$$

which increases as the cross-sectional area S decreases. *The nonlinear optical processes are more pronounced when the beams are more highly focused.*

To give an order of magnitude of the effect in the case of a typical nonlinear material, consider KTP, for which $\chi^{(2)}$ is equal to 5×10^{-12} m.V^{-1}. For a crystal of length $L = 1$ cm, illuminated by two waves ($\lambda \simeq 1\ \mu$m), of the same intensity, focused on an optimal spot ($S \simeq \lambda L = 10^{-2}$ mm^2), we find that the frequency addition wave has power 0.2 mW for input powers $\Phi_1 = \Phi_2 = 1$ W. This is a very low value, but quite enough for many applications in which one requires a coherent monochromatic wave at frequency $\omega_1 + \omega_2$.

Note that, if we use pulsed lasers, whose peak power can easily exceed a kilowatt, the nonlinear process becomes remarkably efficient. However, care must be taken not to exceed the threshold at which the material is damaged (10^{13} W.m^{-2} for KTP).

Comments

(i) The constant pump approximation, i.e. Φ_1 and Φ_2 constant, implies that Φ_3 remains small compared with Φ_1 and Φ_2.

(ii) In the more realistic case of a laser with Gaussian profile, the area S in (7.34) can be replaced by the square w_0^2 of the radius w_0 at the waist, provided that the constant C is modified by a factor very different from unity to account for the transverse intensity variation.

This is only true if the nonlinear interaction occurs near the waist, over a length L that is smaller than the Rayleigh length $z_R = \pi w_0^2/\lambda$ over which the Gaussian beam remains sufficiently cylindrical (see Complement 3B). If length L of the nonlinear material is greater than z_R, the cross-sectional area S arising in (7.34) can no longer be treated as constant, and the 3-wave mixing process is only efficient over an effective length L_{eff} of order z_R, for which area S keeps its minimal value of order w_0^2. The power Φ_3, which varies as L_{eff}^2/S, is then proportional to w_0^2 and not w_0^{-2}. The optimum situation is obtained when z_R is equal to length L of the material, i.e. $L = \pi w_0^2/\lambda$.

(iii) Note that the argument in comment (ii) is only true for free propagation. In a wave guide, e.g. plane wave guide or optical fibre, transverse confinement is maintained over great distances. One can then obtain very efficient conversion, provided that the phase-matching condition is satisfied.

(iv) The above considerations apply to the case where there is only one incident field of frequency ω_1, and where the second harmonic $\omega_3 = 2\omega_1$ is generated. The above formulas can be used, replacing $\mathbf{k}_2$ by $\mathbf{k}_1$ and ω_2 by ω_1, and dividing the coefficient $\chi^{(2)}$ by 2 (see Section 7.2.2).

7.3.2 Phase matching

If the thickness L of the nonlinear plate is not small compared with Δk^{-1}, we should recover (7.27), with the initial condition (7.29). In the constant pump approximation, i.e. $\mathcal{E}_1$ and $\mathcal{E}_2$ constant, the integration yields

$$\mathcal{E}_3(z) = \frac{\omega_3 \chi^{(2)}}{n_3 c} \mathcal{E}_1 \mathcal{E}_2 \frac{\exp\{\mathrm{i}\Delta k \cdot z\} - 1}{\Delta k}. \tag{7.35}$$

The intensity of the ω_3 wave leaving the nonlinear plate is thus proportional to

$$|\mathcal{E}_3(L)|^2 = \left(2\frac{\omega_3 \chi^{(2)}}{n_3 c}\right)^2 |\mathcal{E}_1|^2 |\mathcal{E}_2|^2 \left(\frac{\sin \Delta k \cdot L/2}{\Delta k}\right)^2. \tag{7.36}$$

This equation shows that, for a given mismatch Δk, frequency addition will have maximal efficiency for a plate of thickness

$$L_{\text{opt}} = \frac{\pi}{|\Delta k|}, \tag{7.37}$$

and the optimal intensity of the ω_3 wave leaving the plate will then be proportional to

$$|\mathcal{E}_3(L_{\text{opt}})|^2 = \left(2\frac{\omega_3\chi^{(2)}}{n_3 c}\right)^2 |\mathcal{E}_1|^2\,|\mathcal{E}_2|^2 \frac{1}{\Delta k^2}. \tag{7.38}$$

Clearly, the aim will be to make $|\Delta k|$ as small as possible, in order to be able to use an optimal nonlinear plate that is as long as possible.

The perfect phase-matching condition:

$$\Delta k = 0, \tag{7.39}$$

yields a field $\mathcal{E}_z(z)$ increasing linearly with z for all z. This condition can be written

$$k_3 = k_1 + k_2. \tag{7.40}$$

Since we must also have

$$\omega_3 = \omega_1 + \omega_2,$$

perfect phase matching would appear in principle to be impossible to achieve in a dispersive material, where

$$k_i = n(\omega_i)\,\frac{\omega_i}{c}, \tag{7.41}$$

and where the refractive index $n(\omega)$ is a monotonic (increasing) function of ω. We can get around this difficulty by using a birefringent material, where the refractive index depends on the polarization.

Consider for instance the case of a type I nonlinear crystal, in which the two waves at ω_1 and ω_2, polarized along Ox, give rise to a wave at ω_3 polarized along Oy. (The waves propagate along Oz, as shown in Figure 7.3.) Owing to the birefringence, the refractive index is different for the waves polarized along Ox (ordinary index n_o) and for the wave polarized along Oy (extraordinary index n_e). When $n_e(\omega) < n_o(\omega)$, exact phase matching can be achieved, as shown in Figure 7.4, which is plotted for the special case where $\omega_1 = \omega_2 = \omega_3/2$ to simplify.

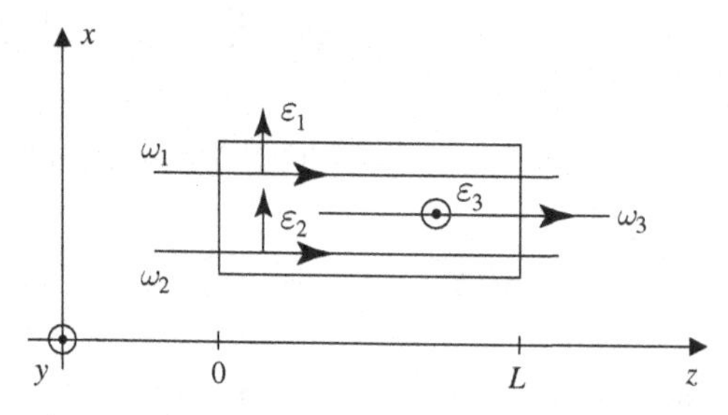

Figure 7.3 Frequency addition in a birefringent nonlinear crystal of type I. The pump waves at ω_1 and ω_2, polarized along Ox, propagate with the ordinary refractive index n_o, whereas the sum wave at $\omega_3 = \omega_1 + \omega_2$, polarized along Oy, propagates with the extraordinary index n_e. If we have $n_e(\omega) < n_o(\omega)$, phase matching can be achieved (see also Figure 7.4).

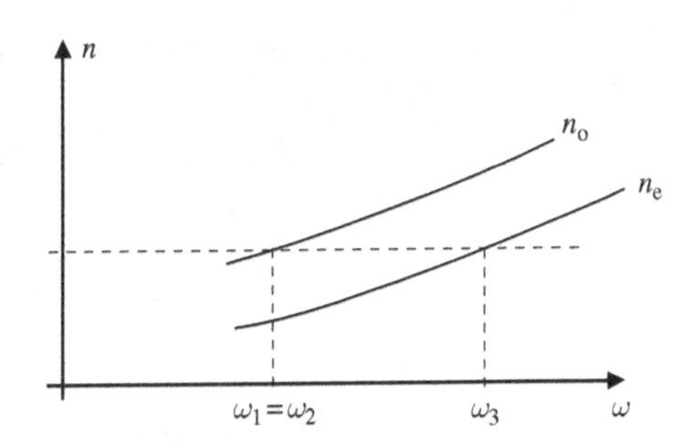

Figure 7.4 Perfect phase matching for frequency doubling in a type I crystal. The extraordinary index at $\omega_3 = 2\omega_1$ is equal to the ordinary index at ω_1. This situation is obtained by judicious choice of the crystal orientation and temperature.

Figure 7.5 Nonlinear interaction between waves with non-collinear propagation, and the vectorial phase-matching condition $\mathbf{k}_3 = \mathbf{k}_1 + \mathbf{k}_2$.

Exact matching can be obtained for each set of frequencies $(\omega_1, \omega_2, \omega_3 = \omega_1 + \omega_2)$ by judicious choice of the direction of propagation in the crystal (which is not isotropic). To modify the frequencies where matching is achieved, one simply changes the orientation of the crystal. It is also possible to change the temperature of the crystal, since this shifts the dispersion curves in Figure 7.4.

The argument is easily generalized to the case of non-collinear waves (see Figure 7.5). Two waves propagating in different directions (with wavevectors $\mathbf{k}_1$ and $\mathbf{k}_2$) will be phase matched with a wave at frequency $\omega_3 = \omega_1 + \omega_2$, propagating with wavevector

$$\mathbf{k}_3 = \mathbf{k}_1 + \mathbf{k}_2. \tag{7.42}$$

This type of situation has the disadvantage of poor beam overlap, which limits the interaction length L (see Figure 7.5). However, the transverse directions offer extra degrees of freedom which can be extremely useful in certain setups.

Comment For very long interaction lengths, e.g. in guided propagation, it is often impossible to obtain a good enough phase matching to actually benefit from the whole interaction length. An elegant solution to this problem is the technique known as quasi-phase matching, in which several pieces of nonlinear medium are set end-to-end, each piece being of length $\pi/|\Delta k|$, but with the signs of the coefficients $\chi^{(2)}$ alternating from one piece to the next. One can show, by integrating (7.27) in this situation, that there will be a compensation of the phase mismatches in adjacent sections of the nonlinear material. One material that lends itself well to this method is lithium niobate, and the technique is then called periodically poled lithium niobate (PPLN).

7.3.3 Coupled dynamics of three-wave mixing

Consider a situation with perfect phase matching. Within the nonlinear material, there are three collinear coupled waves such that,

$$\omega_1 + \omega_2 = \omega_3 \tag{7.43}$$

$$k_1 + k_2 = k_3. \tag{7.44}$$

Let us now relax the assumption of constant pumps, i.e. let us seek the dynamical equations of the three amplitudes $\mathcal{E}_1(z)$, $\mathcal{E}_2(z)$ and $\mathcal{E}_3(z)$.

One might wonder why one should consider only process (7.43), shown schematically in Figure 7.6(a). For example, one could envisage the process in Figure 7.6(b), which gives rise to a new wave at frequency

$$\omega_4 = \omega_1 + \omega_3. \tag{7.45}$$

In fact it is easy to see that, if the phase-matching condition is satisfied for the process (7.43) of Figure 7.6(a), it is not for (7.45), which we may therefore neglect. However, the process in Figure 7.6(c), in which the waves at ω_1 and ω_3 combine to give

$$\omega_2 = \omega_3 - \omega_1, \tag{7.46}$$

is favoured by the phase-matching condition

$$k_2 = k_3 - k_1, \tag{7.47}$$

which is the same as (7.44). This frequency-difference process shows up when we write down an analogous equation to (7.25):

$$P_{\rm NL}(z,t) = 2\varepsilon_0 \chi^{(2)} [\mathcal{E}_3(z) \exp\{i(k_3 z - \omega_3 t)\} + \text{c.c.}] \times [\mathcal{E}_1(z) \exp\{i(k_1 z - \omega_1 t)\} + \text{c.c.}]. \tag{7.48}$$

(a)	(b)
ω_1, ω_2 → $\omega_3 = \omega_1 + \omega_2$, $\chi^{(2)}\mathcal{E}_1\mathcal{E}_2$	ω_1, ω_3 → $\omega_4 = \omega_1 + \omega_3$, $\chi^{(2)}\mathcal{E}_1\mathcal{E}_3$
(c)	**(d)**
ω_1, ω_3 → $\omega_2 = \omega_3 - \omega_1$, $\chi^{(2)}\mathcal{E}_3\mathcal{E}_1^*$	ω_2, ω_3 → $\omega_1 = \omega_3 - \omega_2$, $\chi^{(2)}\mathcal{E}_3\mathcal{E}_2^*$

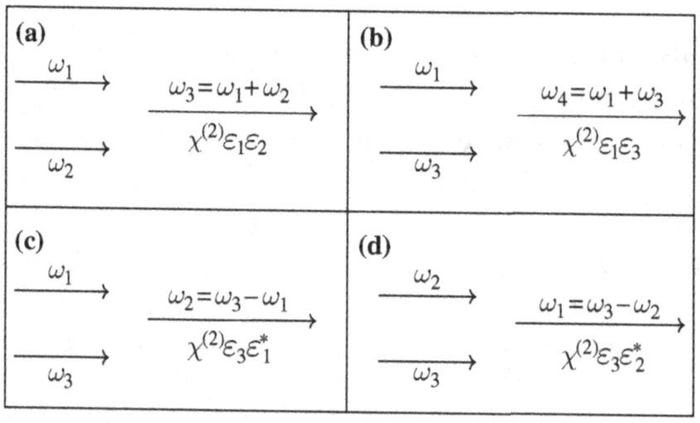

Figure 7.6 Some 3-wave nonlinear processes. Processes (a), (c) and (d) are favoured by the same phase-matching condition $k_3 = k_1 + k_2$, whereas process (b) is in this case negligible. There is essentially an interaction between ω_1, ω_2 and ω_3.

Arguing as in Section 7.3.1, we obtain the equation,

$$\frac{d\mathcal{E}_2}{dz} = i\frac{\omega_2\chi^{(2)}}{n_2c}\mathcal{E}_3(z)\mathcal{E}_1^*(z), \tag{7.49}$$

where we have kept only the term favoured by the phase-matching relation (7.47) (the term in $\mathcal{E}_3\mathcal{E}_1$ contains a factor $\exp\{i(k_3 + k_1 - k_2)z\}$ which oscillates during the propagation).

A similar analysis shows that the process in Figure 7.6(d) is also favoured by the phase-matching condition (7.44).

So it turns out that, if we obtain the phase-matching condition (7.44) for the three waves at ω_1, ω_2, and $\omega_3 = \omega_1 + \omega_2$, there are three efficient processes which couple the three waves, while bringing in no other frequencies. We end up with a *closed* system of coupled differential equations:

$$\frac{d\mathcal{E}_3}{dz} = i\frac{\omega_3}{n_3c}\chi^{(2)}\mathcal{E}_1(z)\mathcal{E}_2(z) \tag{7.50}$$

$$\frac{d\mathcal{E}_1}{dz} = i\frac{\omega_1}{n_1c}\chi^{(2)}\mathcal{E}_3(z)\mathcal{E}_2^*(z) \tag{7.51}$$

$$\frac{d\mathcal{E}_2}{dz} = i\frac{\omega_2}{n_2c}\chi^{(2)}\mathcal{E}_3(z)\mathcal{E}_1^*(z). \tag{7.52}$$

This nonlinear system is not only able to treat the frequency addition scenario in cases where the constant pump condition is relaxed, but it also applies to the parametric amplification phenomenon to be discussed below.

There is a completely general solution for (7.50–7.52) in terms of elliptic functions.[2] However, it is extremely complex and barely amenable to physical interpretation. On the other hand, it is instructive to determine the invariants of this system of equations, i.e. those functions of the three interacting fields that remain constant during the propagation.

The coupled system (7.50–7.52) was established for a transparent nonlinear medium, far from any absorbing region, i.e. $\chi^{(2)}$ is real. One thus expects to find conservation of energy. To check this, we consider the average Poynting vectors (average power per unit area transverse to the beam). Within a dielectric of index n, the Poynting vector is

$$\mathbf{\Pi} = \frac{\overline{\mathbf{E}\times\mathbf{B}}}{\mu_0} = \frac{\overline{E^2}}{\mu_0\omega}\mathbf{k} = n\varepsilon_0 c\,\overline{E^2}\mathbf{e}_z = \Pi\mathbf{e}_z$$

$$= 2n\varepsilon_0 c\,\mathcal{E}^*\mathcal{E}\mathbf{e}_z, \tag{7.53}$$

where the overbar denotes a time average over some long lapse of time compared with the optical period. Then, for each wave,

$$\frac{d\Pi}{dz} = 2n\varepsilon_0 c\left(\frac{d\mathcal{E}}{dz}\mathcal{E}^* + \mathcal{E}\frac{d\mathcal{E}^*}{dz}\right), \tag{7.54}$$

[2] J. A. Armstrong, N. Bloembergen, J. Ducuing and P. S. Pershan, Interactions between Light Waves in a Nonlinear Dielectric, *Physical Review* **127**, 1918 (1962).

whence Equations (7.50–7.52) imply

$$\frac{\mathrm{d}\Pi_1}{\mathrm{d}z} = \mathrm{i}\omega_1\, 2\varepsilon_0\chi^{(2)}(\mathcal{E}_1^*\mathcal{E}_2^*\mathcal{E}_3 - \text{c.c.}) \tag{7.55}$$

$$\frac{\mathrm{d}\Pi_2}{\mathrm{d}z} = \mathrm{i}\omega_2\, 2\varepsilon_0\chi^{(2)}(\mathcal{E}_1^*\mathcal{E}_2^*\mathcal{E}_3 - \text{c.c.}) \tag{7.56}$$

$$\frac{\mathrm{d}\Pi_3}{\mathrm{d}z} = -\mathrm{i}\omega_3\, 2\varepsilon_0\chi^{(2)}(\mathcal{E}_1^*\mathcal{E}_2^*\mathcal{E}_3 - \text{c.c.}). \tag{7.57}$$

It thus follows that

$$\frac{d}{dz}(\Pi_1 + \Pi_2 + \Pi_3) = \mathrm{i}(\omega_1 + \omega_2 - \omega_3)\, 2\varepsilon_0\chi^{(2)}(\mathcal{E}_1^*\mathcal{E}_2^*\mathcal{E}_3 - \text{c.c.}) = 0, \tag{7.58}$$

which expresses energy conservation. No energy is therefore transferred to the nonlinear medium, which serves only to facilitate the coupling process. The situation here is very different from the propagation of a wave through an ensemble of quasi-resonant atoms, where the medium can remove or supply energy to the waves crossing it (see Section 2.5 of Chapter 2).

The second invariant is also obtained from (7.55–7.57), which imply

$$\frac{1}{\omega_1}\frac{\mathrm{d}\Pi_1}{\mathrm{d}z} = \frac{1}{\omega_2}\frac{\mathrm{d}\Pi_2}{\mathrm{d}z} = -\frac{1}{\omega_3}\frac{\mathrm{d}\Pi_3}{\mathrm{d}z}. \tag{7.59}$$

It follows that

$$\frac{\mathrm{d}}{\mathrm{d}z}\left(\frac{\Pi_1}{\omega_1} - \frac{\Pi_2}{\omega_2}\right) = 0. \tag{7.60}$$

This is the Manley–Rowe relation, which has a simple interpretation in the quantum theoretical description, in terms of photons, as we shall see later (see Section 7.4.1).

Comment As can be seen from (7.58), the intensity Π_3 of the wave produced by frequency addition can only grow at the expense of the pump waves Π_1 and Π_2. This is the pump depletion phenomenon, which limits the growth of Π_3. The exact form of the dependence of Π_1, Π_2 and Π_3 on z is obtained by integrating the coupled system of differential equations (7.50–7.52).

7.3.4 Parametric amplification

The nonlinear crystal is now illuminated by an intense wave (the pump) at frequency ω_3 and a weak wave (the signal) at ω_1 (see Figure 7.7). We assume that the phase-matching condition (7.44) is satisfied for $\omega_2 = \omega_3 - \omega_1$.

If we now make the constant pump approximation, i.e. $\mathcal{E}_3$ constant, Equations (7.51) and (7.52) imply that

$$\frac{\mathrm{d}^2\mathcal{E}_2}{\mathrm{d}z^2} = \left(\frac{\chi^{(2)}|\mathcal{E}_3|}{c}\right)^2 \frac{\omega_1\omega_2}{n_1 n_2}\mathcal{E}_2 = \gamma^2\mathcal{E}_2. \tag{7.61}$$

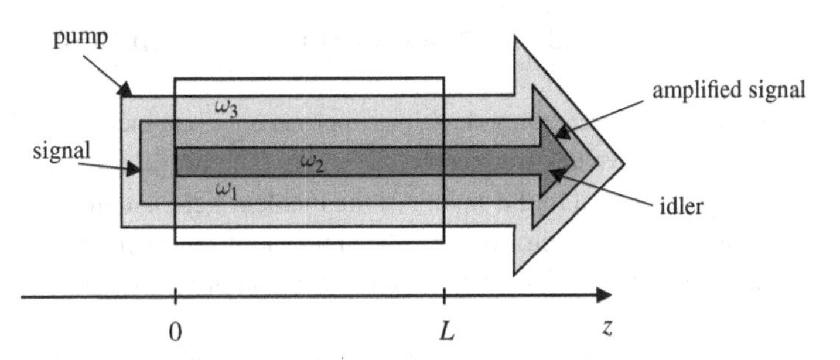

Figure 7.7 Parametric amplification. By combining the intense pump ω_3 with the weak wave at ω_1, the latter (the signal) can be amplified. Another wave, the idler, appears at $\omega_2 = \omega_3 - \omega_1$. The beams are shown here as separate for clarity but, in reality, they are actually superposed.

For the initial conditions,

$$\mathcal{E}_2(z = 0) = 0 \tag{7.62}$$

$$\mathcal{E}_1(z = 0) = \mathcal{E}_1(0), \tag{7.63}$$

the solution is

$$\mathcal{E}_2(z) = \mathrm{i}\sqrt{\frac{\omega_2 n_1}{\omega_1 n_2}}\frac{\mathcal{E}_3}{|\mathcal{E}_3|}\mathcal{E}_1^*(0)\sinh(\gamma z) \tag{7.64}$$

$$\mathcal{E}_1(z) = \mathcal{E}_1(0)\cosh(\gamma z), \tag{7.65}$$

with

$$\gamma = \frac{\chi^{(2)}|\mathcal{E}_3|}{c}\left(\frac{\omega_1\omega_2}{n_1 n_2}\right)^{1/2}. \tag{7.66}$$

We observe that the signal wave $\mathcal{E}_1$ is amplified, while a complementary wave known as the idler also appears. This process of light wave amplification, called *parametric amplification*, is useful because it can be tuned over a rather broad frequency range, by adjusting the phase-matching condition using the methods discussed in Section 7.3.2 (altering the orientation and/or the temperature of the nonlinear crystal). If it is placed inside a Fabry–Perot resonant cavity, an optical parametric amplifier (OPA) gives a tunable laser source called an optical parametric oscillator (OPO), which is useful in many applications, and which we shall discuss in detail in Complement 7A.

Comment Consider once again the 1 cm KTP crystal (see Section 7.3.1), optimally pumped by a focused laser beam ($\lambda L \approx S$). The intensity gain $(\cosh \gamma L)^2$ is then equal to 1.13 for a power of 1 W. This is a modest gain, similar to a gas laser. If the incident power is greater than 100 W, the gain $(\cosh \gamma L)^2$ is greater than 200. According to (7.58), the power transferred to the signal and the idler at the expense of the pump is then significant, and the constant pump approximation is not generally valid in this case. Parametric amplification can be extremely efficient if one uses high instantaneous powers delivered by pulsed lasers, such as mode-locked lasers.

7.3.5 Frequency doubling with pump depletion

In the case of frequency doubling, there is only one incident field of frequency ω_1, and the generated wave has frequency $\omega_3 = 2\omega_1$. The useful term in the nonlinear polarization is then proportional to the square of the incident field, and not the double product $\mathcal{E}_1\mathcal{E}_2$ of the fields. Once this factor of 2 has been taken into account, the coupled equations for the two fields $\mathcal{E}_1(z)$ and $\mathcal{E}_3(z)$ can be written in a similar way to (7.50–7.52):

$$\frac{d\mathcal{E}_1}{dz} = i\frac{\omega_1}{n_1 c}\chi^{(2)}\mathcal{E}_3(z)\mathcal{E}_1^*(z)\, e^{i(k_3-2k_1)z} \tag{7.67}$$

$$\frac{d\mathcal{E}_3}{dz} = i\frac{\omega_3}{2n_3 c}\chi^{(2)}\mathcal{E}_1^2(z)\, e^{i(2k_1-k_3)z}. \tag{7.68}$$

Conversion is maximal if the phase-matching condition $k_3 = 2k_1$ is satisfied. The two coupled equations can then be solved exactly. With the initial conditions $\mathcal{E}_1(0) = E_1$ (E_1 real) and $\mathcal{E}_3(0) = 0$, the solution is

$$\mathcal{E}_1(z) = \frac{E_1}{\cosh \gamma' z} \tag{7.69}$$

$$\mathcal{E}_3(z) = iE_1 \tanh \gamma' z, \tag{7.70}$$

with $\gamma' = \dfrac{\omega_1}{n_1 c}\chi^{(2)}E_1$.

For z small compared with γ'^{-1}, the power of the second harmonic wave grows as the square of the incident power and as the square of the interaction length, in conformity with the constant pump approximation. When $\gamma' z$ is of order 1, the process saturates at the value $\mathcal{E}_3 = iE_1$, while the pump field tends to zero. There is then *total conversion* of the pump into the second harmonic when the medium is thick enough. In fact this is a highly simplified model, because it neglects the transverse variation of the pump field and the associated diffraction. In reality, one can nevertheless achieve extremely high conversion efficiencies using pulsed lasers with very high peak powers. For example, frequency doubling of the Megajoule laser (see Section 3E.4 of Complement 3E) occurs with an efficiency of around 80%. One can also achieve conversion efficiencies of more than 50% with continuous-wave lasers of power around 1 watt, provided that one inserts the nonlinear crystal in a Fabry–Perot cavity that is resonant for the wave of frequency ω_1. One then benefits from the increase in the intra-cavity light intensity, proportional to the finesse (see Complement 3A).

7.3.6 Parametric fluorescence

Consider once again the above situation with an intense pump at ω_3 and the phase-matching condition established for $\omega_1 = \omega_3/2$, but this time sending no field at frequency ω_1 into the nonlinear crystal. Equations (7.67) and (7.68) with $\mathcal{E}_1 = 0$ imply that the system no longer evolves. Likewise in the more general case of three-wave mixing (see (7.50–7.52))

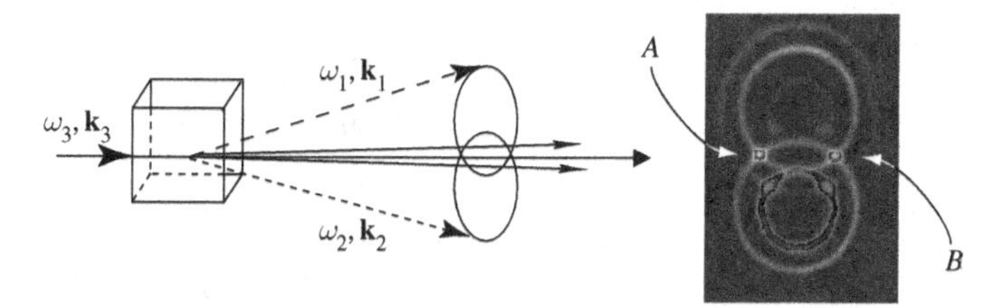

Figure 7.8 Parametric fluorescence. The intense pump (ω_3, $\mathbf{k}_3$) generates an ensemble of pairs of conjugate waves (ω_1, $\mathbf{k}_1$; ω_2, $\mathbf{k}_2$) such that $\omega_1 + \omega_2 = \omega_3$ and $\mathbf{k}_1 + \mathbf{k}_2 = \mathbf{k}_3$. A double cone is observed, with a cross-section in the form of coloured rings. If two diaphragms *A* and *B* are placed at the intersections of the middle cones, pairs of polarization-correlated photons can be obtained (see comment (iii) in Section 7.4.6).

when only the pump wave is non-zero ($\mathcal{E}_1 = \mathcal{E}_2 = 0$). According to the equations, no non-zero field except for the pump field appears in the solution. And yet, experience shows that new waves do indeed appear. The phenomenon is called parametric fluorescence, shown schematically in Figure 7.8. If the crystal is subjected to an intense pump wave $\mathcal{E}_3$ of wavevector $\mathbf{k}_3$, a double series of coloured rings is output from the crystal. They are formed by a set of conjugate pairs of waves coupled to the same pump field $\mathcal{E}_3$, of frequency ω_1 and wavevector $\mathbf{k}_1$ and of frequency ω_2 and wavevector $\mathbf{k}_2$, satisfying the relations,

$$\omega_1 + \omega_2 = \omega_3 \tag{7.71}$$

$$\mathbf{k}_1 + \mathbf{k}_2 = \mathbf{k}_3. \tag{7.72}$$

This process *creates photons from the vacuum, and is analogous to spontaneous emission by an excited atom. Like the latter, it can only be handled by a purely quantum theoretical approach. This is what we shall undertake in the next section.*

7.4 Quantum treatment of parametric fluorescence

7.4.1 Unavoidability and advantages of the quantum treatment

The relation $\omega_3 = \omega_1 + \omega_2$ suggests describing the parametric fluorescence process as a kind of 'reaction' in which a photon $\hbar\omega_3$ disappears to give rise to two photons $\hbar\omega_1$ and $\hbar\omega_2$, with conservation of energy (transparent medium):

$$\hbar\omega_3 \longrightarrow \hbar\omega_1 + \hbar\omega_2. \tag{7.73}$$

This picture agrees with the Manley–Rowe relations, which follow directly from the classical equations (7.55–7.57), and which can be written,

$$\frac{1}{\hbar\omega_1}\frac{d\Pi_1}{dz} = \frac{1}{\hbar\omega_2}\frac{d\Pi_2}{dz} = -\frac{1}{\hbar\omega_3}\frac{d\Pi_3}{dz}. \tag{7.74}$$

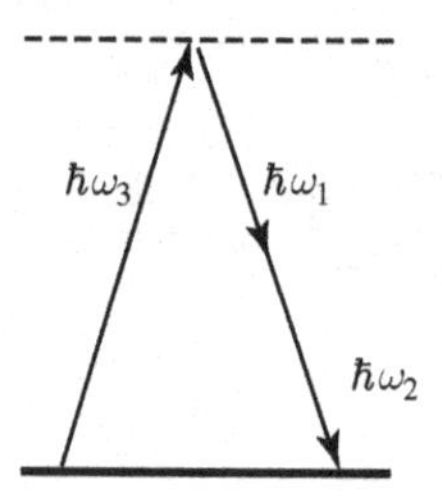

Figure 7.9 **Elementary process in the parametric interaction. A pump photon is destroyed and one signal photon and one idler photon are simultaneously created.**

Interpreting $\Pi/\hbar\omega$ as a photon current, we see that one photon $\hbar\omega_1$ and one photon $\hbar\omega_2$ will appear for each photon $\hbar\omega_3$ that gets destroyed. Clearly, the phase-matching condition (7.42) can be interpreted as conservation of momentum in the elementary process (7.73), since it can be rewritten in the form

$$\hbar\mathbf{k}_3 = \hbar\mathbf{k}_1 + \hbar\mathbf{k}_2. \tag{7.75}$$

The picture associated with the process (7.73) suggests that the photons $\hbar\omega_1$ and $\hbar\omega_2$ are emitted in pairs by an elementary process that could be depicted schematically as in Figure 7.9. These *twin-photon pairs* give rise to typically quantum phenomena, quite inconceivable in the classical world, and we shall give several examples. Indeed, the quantum description is much more than just a formalism for carrying out quantitative calculations of the parametric fluorescence rate. It is utterly indispensable for predicting and describing the properties of the photon pairs produced in this process.

Comment Note that the condition (7.75) does not need to be rigorously fulfilled. For one thing, in a crystal of finite thickness L, it actually has the form $\Delta k \leq \pi/L$, with $\Delta\mathbf{k} = \mathbf{k}_3 - \mathbf{k}_1 - \mathbf{k}_2$, and for another, the mixing process has a non-zero amplitude even if the condition is not satisfied. It thus has a quite different status to the relation $\omega_3 = \omega_1 + \omega_2$, which holds exactly when continuous beams are used. (The situation is quite different with ultra-short pulses, where the frequency dispersion is high.)

7.4.2 Quantum treatment of three-wave mixing

Consider modes 1, 2 and 3 such that,

$$\omega_3 = \omega_1 + \omega_2, \tag{7.76}$$

$$\mathbf{k}_3 = \mathbf{k}_1 + \mathbf{k}_2. \tag{7.77}$$

For a transparent nonlinear medium, there is no energy exchange with the matter in the medium, and we assume that the nonlinear process can be described using an *effective Hamiltonian* responsible for the disappearance or appearance of photons in waves 1, 2 and 3. We shall thus describe the radiation by a Hamiltonian containing one freely evolving term, and a coupling term:

$$\hat{H} = \sum_{i=1}^{3} \hbar\omega_i \left(\hat{a}_i^+ \hat{a}_i + \frac{1}{2}\right) + \hbar g(\hat{a}_1^+ \hat{a}_2^+ \hat{a}_3 + \hat{a}_1 \hat{a}_2 \hat{a}_3^+)$$

$$= \hat{H}_0 + \hat{H}_{\mathrm{I}}. \tag{7.78}$$

The first coupling term $(\hat{a}_1^+ \hat{a}_2^+ \hat{a}_3)$ corresponds to the process (7.73), in which one photon $\hbar\omega_3$ disappears, and one pair $(\hbar\omega_1, \hbar\omega_2)$ appears. The Hermitian conjugate term corresponds to the opposite process, in which a pair $(\hbar\omega_1, \hbar\omega_2)$ disappears, and a photon $\hbar\omega_3$ is created. We can check some conservation laws regarding the number of photons in these processes. To do this, let us calculate the commutators of the photon number operators $\hat{N}_i = \hat{a}_i^+ \hat{a}_i$ with the Hamiltonian. We have,

$$[\hat{N}_1, \hat{H}] = \hbar g[\hat{a}_1^+ \hat{a}_1, \hat{a}_1^+ \hat{a}_2^+ \hat{a}_3 + \hat{a}_1 \hat{a}_2 \hat{a}_3^+]$$

$$= \hbar g[\hat{a}_1^+ \hat{a}_2^+ \hat{a}_3 - \hat{a}_1 \hat{a}_2 \hat{a}_3^+] = \hat{F} \tag{7.79}$$

$$[\hat{N}_2, \hat{H}] = \hat{F} \tag{7.80}$$

$$[\hat{N}_3, \hat{H}] = -\hat{F}. \tag{7.81}$$

We observe that $\hat{N}_1 - \hat{N}_2$, $\hat{N}_1 + \hat{N}_3$, $\hat{N}_2 + \hat{N}_3$, and $\hat{N}_1 + \hat{N}_2 - 2\hat{N}_3$ all commute with the total Hamiltonian. We say that these are *good quantum numbers*, meaning that these observables are conserved during the evolution of the system. For example, an initial state $|\psi_{\mathrm{i}}\rangle = |N_1, N_2, N_3\rangle$ will lead to a final state $|\psi_{\mathrm{f}}\rangle = |N_1 + 1, N_2 + 1, N_3 - 1\rangle$, and it is easy to check that $|\psi_{\mathrm{i}}\rangle$ and $|\psi_{\mathrm{f}}\rangle$ are eigenvectors of the above observables $(\hat{N}_1 + \hat{N}_3, \hat{N}_2 + \hat{N}_3,$ etc.) associated with the same eigenvalue.

7.4.3 Perturbative treatment of parametric fluorescence

Consider an initial state in which there are only photons in the pump mode, i.e.

$$|\psi(t = 0)\rangle = |\psi_i\rangle = |0, 0, N_3\rangle. \tag{7.82}$$

Under the effect of the interaction Hamiltonian $\hat{H}_{\mathrm{I}}$ of (7.78), in the usual perturbative approximation,[3] $|\psi_{\mathrm{i}}\rangle$ becomes, for sufficiently small times t,

$$|\psi(t)\rangle = \mathrm{e}^{-\mathrm{i}\omega_3\left(N_3+\frac{1}{2}\right)t}\left(|0, 0, N_3\rangle - \mathrm{i}g\sqrt{N_3}\,t\,|1, 1, N_3 - 1\rangle\right). \tag{7.83}$$

Consider the situation shown in Figure 7.8, where the light propagates inside the crystal, and we seek the stationary quantum state $|\psi\rangle$ output from the crystal as a function of the input quantum state. The interaction time t is thus taken as the propagation time L/c for the light to cross the crystal of length L. Up to a global phase factor of no physical significance, we then have

$$|\psi_f\rangle = |0, 0, N_3\rangle + C|1, 1, N_3 - 1\rangle,$$

[3] See Section 1.2 of Chapter 1.

where $|C| = \frac{g\sqrt{N_3}L}{c} \ll 1$. In fact there are many pairs of modes (i,j) which satisfy the condition (7.76) exactly, and condition (7.77) approximately. The final state (after the crystal) can thus be written, to first order in perturbation theory,

$$|\psi_f\rangle = |0, 0, N_3\rangle + \left(\sum_{i,j} C_{ij}|1_i, 1_j\rangle\right) \otimes |N_3 - 1\rangle, \tag{7.84}$$

where modes i and j satisfy the condition (7.76), and $|C_{ij}|$ increases the more closely satisfied (7.77) is.

The state vector,

$$|\psi'\rangle = \sum_{i,j} C_{ij}|1_i, 1_j\rangle, \tag{7.85}$$

describes a state containing twin photon pairs. It is an eigenstate of the total photon number operator $\hat{N}$ with eigenvalue 2, and an eigenstate of the Hamiltonian with energy $\hbar\omega_3$, and hence a stationary state. It is also an entangled state as soon as two of the coefficients C_{ij} are non-zero.

7.4.4 Change of picture: the Heisenberg representation

For a better description of the purely quantum effects that can be observed with twin photon pairs, it is convenient to change the representation of the quantum fields. Up to now (Chapters 4 to 6), we have used the Schrödinger picture, in which the field is described by a time-dependent state $|\psi(t)\rangle$, while the observables, such as the electric field, vector potential, momentum, etc., are described by time-independent operators acting in the Hilbert space of states $|\psi(t)\rangle$. The evolution of the system as time goes by is then specified by the Schrödinger equation:

$$i\hbar\frac{d}{dt}|\psi(t)\rangle = \hat{H}|\psi(t)\rangle, \tag{7.86}$$

where $\hat{H}$ is the Hamiltonian of the system.

In quantum physics, we also use the so-called Heisenberg picture,[4] in which the states $|\psi(t)\rangle$ do not depend on time. The time dependence is then carried by the observables, and the Schrödinger equation is replaced by the Heisenberg equation, which specifies the temporal evolution of the operator $\hat{A}(t)$:

$$i\hbar\frac{d\hat{A}(t)}{dt} = [\hat{A}(t), \hat{H}]. \tag{7.87}$$

In quantum optics, the Heisenberg representation can be used to give the quantum fields a very similar form to the one we know so well for the classical fields. Let us consider

[4] See, for example, BD, Chapter 5, Exercise 3, and CDL, Complement G_{III}.

the evolution of the free field (when there are no charges) under the influence of the Hamiltonian,

$$\hat{H} = \sum_{\ell} \hbar\omega_{\ell} \left(\hat{a}_{\ell}^{\dagger}\hat{a}_{\ell} + \frac{1}{2} \right). \tag{7.88}$$

The field observables are expressed in terms of operators $\hat{a}_{\ell}$ and $\hat{a}_{\ell}^{\dagger}$. So, for example, recall that the electric field is given by (see (4.109))

$$\hat{\mathbf{E}}(\mathbf{r}) = \sum_{\ell} \mathrm{i}\boldsymbol{\varepsilon}_{\ell} E_{\ell}^{(1)} \left(\mathrm{e}^{\mathrm{i}\mathbf{k}_{\ell}\cdot\mathbf{r}}\hat{a}_{\ell} - \mathrm{e}^{-\mathrm{i}\mathbf{k}_{\ell}\cdot\mathbf{r}}\hat{a}_{\ell}^{\dagger} \right). \tag{7.89}$$

In the Heisenberg representation, the time dependence of the operator $\hat{a}_{\ell}(t)$ is obtained from (7.87) and (7.88):

$$\mathrm{i}\hbar\frac{\mathrm{d}\hat{a}_{\ell}}{\mathrm{d}t} = [\hat{a}_{\ell}(t), \hat{H}] = [\hat{a}_{\ell}, \hbar\omega_{\ell}\hat{a}_{\ell}^{\dagger}\hat{a}_{\ell}] = \hbar\omega_{\ell}(\hat{a}_{\ell}\hat{a}_{\ell}^{\dagger} - \hat{a}_{\ell}^{\dagger}\hat{a}_{\ell})\hat{a}_{\ell} = \hbar\omega_{\ell}\,\hat{a}_{\ell}. \tag{7.90}$$

This equation can be integrated immediately to give

$$\hat{a}_{\ell}(t) = \hat{a}_{\ell}(0)\,\mathrm{e}^{-\mathrm{i}\omega_{\ell}t}. \tag{7.91}$$

It is thus very easy to go from the Schrödinger picture to the Heisenberg picture by multiplying the annihilation operators $\hat{a}_{\ell}$ by $\mathrm{e}^{-\mathrm{i}\omega_{\ell}t}$ and the creation operators $\hat{a}_{\ell}^{\dagger}$ by $\mathrm{e}^{\mathrm{i}\omega_{\ell}t}$. For example, the free electric field (7.89) becomes, in the Heisenberg picture,

$$\hat{\mathbf{E}}(\mathbf{r},t) = \sum_{\ell} \mathrm{i}\boldsymbol{\varepsilon}_{\ell} E_{\ell}^{(1)} \left(\mathrm{e}^{\mathrm{i}(\mathbf{k}_{\ell}\cdot\mathbf{r}-\omega_{\ell}t)}\hat{a}_{\ell} - \mathrm{e}^{-\mathrm{i}(\mathbf{k}_{\ell}\cdot\mathbf{r}-\omega_{\ell}t)}\hat{a}_{\ell}^{\dagger} \right) = \hat{\mathbf{E}}^{(+)}(\mathbf{r},t) + \hat{\mathbf{E}}^{(-)}(\mathbf{r},t), \tag{7.92}$$

which has the form of a superposition of plane waves of type $\mathrm{e}^{\mathrm{i}(\mathbf{k}_{\ell}\cdot\mathbf{r}-\omega_{\ell}t)}$, just as in classical electromagnetism.

In general, quantities involving both operators and state vectors take the same form in either the Schrödinger or the Heisenberg representation. For example, the single photodetection probability is given in the Schrödinger representation by (see (5.4))

$$w^{(1)}(\mathbf{r},t) = s\,\langle\psi(t)|\hat{\mathbf{E}}^{(-)}(\mathbf{r})\cdot\hat{\mathbf{E}}^{(+)}(\mathbf{r})|\psi(t)\rangle. \tag{7.93}$$

In the Heisenberg picture, this becomes,

$$w^{(1)}(\mathbf{r},t) = s\langle\psi|\hat{\mathbf{E}}^{(-)}(\mathbf{r},t)\cdot\hat{\mathbf{E}}^{(+)}(\mathbf{r},t)|\psi\rangle. \tag{7.94}$$

When we consider joint photodetection rates (5.5), the Heisenberg picture is more general, because it allows one to express the rate of joint detections at two different times t and t':

$$w^{(2)}(\mathbf{r}_1,t;\mathbf{r}_2,t') = \langle\psi|\hat{\mathbf{E}}^{(-)}(\mathbf{r}_1,t)\cdot\hat{\mathbf{E}}^{(-)}(\mathbf{r}_2,t')\cdot\hat{\mathbf{E}}^{(+)}(\mathbf{r}_2,t')\cdot\hat{\mathbf{E}}^{(+)}(\mathbf{r}_1,t)|\psi\rangle. \tag{7.95}$$

In this expression, the scalar products are between $\hat{\mathbf{E}}^{(-)}(\mathbf{r}_1,t)$ and $\hat{\mathbf{E}}^{(+)}(\mathbf{r}_1,t)$, and $\hat{\mathbf{E}}^{(-)}(\mathbf{r}_2,t')$ and $\hat{\mathbf{E}}^{(+)}(\mathbf{r}_2,t')$, but the operators have to be kept in the order in which they appear in (7.95), because they do not commute in general. We shall use the Heisenberg picture throughout the rest of this chapter.

7.4.5 Simultaneous emission of parametric fluorescence photons

Starting from the situation described in Section 7.4.3, we place two filter diaphragms after the nonlinear crystal, defining two directions $\mathbf{u}_1$ and $\mathbf{u}_2$ such that the phase-matching condition is strictly satisfied for a particular pair of frequencies $\omega_1^{(0)}$ and $\omega_2^{(0)}$ (see Figure 7.10). In fact there are many conjugate pairs of modes (ω_1, ω_2) propagating along $\mathbf{u}_1$ and $\mathbf{u}_2$ and satisfying the phase-matching condition approximately. Thanks to (7.76), their frequencies are

$$\omega_1 = \omega_1^{(0)} + \delta \tag{7.96}$$

and

$$\omega_2 = \omega_2^{(0)} - \delta, \tag{7.97}$$

where the detuning δ is much smaller than $\omega_3/2$. Their wavevectors are

$$\mathbf{k}_1 = \omega_1 \frac{\mathbf{u}_1}{c} \tag{7.98}$$

$$\mathbf{k}_2 = \omega_2 \frac{\mathbf{u}_2}{c}. \tag{7.99}$$

In the Heisenberg picture, the radiation state restricted to the modes transmitted through the filtering diaphragms is then

$$\begin{aligned} |\psi''\rangle &= \sum_\delta C(\delta)|1_{\mathbf{u}_1,\omega_1^{(0)}+\delta}, 1_{\mathbf{u}_2,\omega_2^{(0)}-\delta}\rangle \\ &= \sum_\delta C(\delta)|1_{\mathbf{u}_1,\delta}, 1_{\mathbf{u}_2,-\delta}\rangle. \end{aligned} \tag{7.100}$$

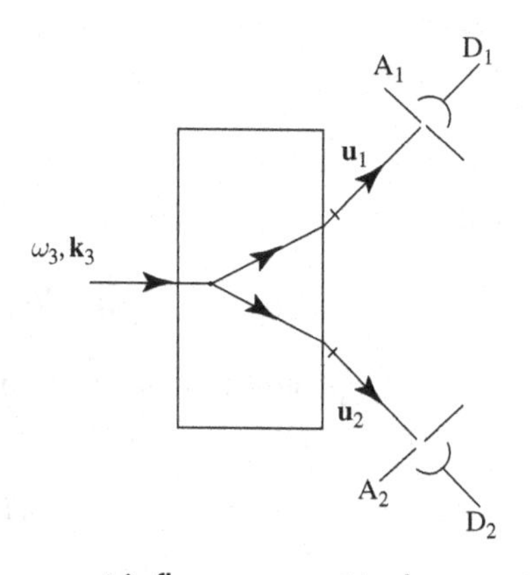

Figure 7.10 Twin photons produced by parametric fluorescence. Diaphragms A_1 and A_2 select conjugate directions $\mathbf{u}_1$ and $\mathbf{u}_2$ such that the phase-matching condition is strictly satisfied for frequencies $\omega_1^{(0)}$ and $\omega_2^{(0)}$. The detectors D_1 and D_2 measure single or coincidence photodetection signals. In this way, we check that the photons are produced in pairs.

The second, simpler expression is not ambiguous, when we recall that the frequencies of the two photons are given by (7.96) and (7.97). Insofar as δ characterizes the deviation from perfect phase matching, the coefficient $|C(\delta)|$ goes through a maximum for $\delta = 0$, and is described by a bell-shaped curve of typical width 10^{13} Hz $(10^{-2}\omega_3)$.

Photodetectors D_1 and D_2 are placed behind each diaphragm. We begin by calculating the single photodetection rate at D_1 (see (7.93)):

$$w^{(1)}(\mathbf{r}_1, t) = s\|\hat{E}_1^{(+)}(\mathbf{r}_1, t)|\psi''\rangle\|^2 \tag{7.101}$$

with

$$\hat{E}_1^{(+)}(\mathbf{r}_1, t) = \mathrm{i}\sum_{\ell_1} \mathcal{E}_{\ell_1}^{(1)} \exp\{\mathrm{i}\omega_{\ell_1}(\tau_1 - t)\}\hat{a}_{\ell_1}. \tag{7.102}$$

We introduce the delay,

$$\tau_1 = \frac{\mathbf{u}_1 \cdot \mathbf{r}_1}{c}, \tag{7.103}$$

where $\mathbf{r}_1$ is the position of detector D_1, taken with respect to the crystal output. The index 1 appearing in the expression for the electric field $\hat{\mathbf{E}}_1$ and in the notation ℓ_1 for the modes reminds us that we only consider modes propagating along $\mathbf{u}_1$. We then find,

$$\hat{E}_1^{(+)}(\mathbf{r}_1, t)|\psi''\rangle = \mathrm{i}\sum_{\delta} C(\delta)\,\mathcal{E}_{\omega_1}^{(1)}\,\mathrm{e}^{\mathrm{i}\omega_1(\tau_1 - t)}|0_{\mathbf{u}_1,\delta}, 1_{\mathbf{u}_2,-\delta}\rangle, \tag{7.104}$$

where we have defined $\mathcal{E}_{\omega_1}^{(1)} = \sqrt{\dfrac{\hbar\omega_1}{2\varepsilon_0 L^3}}$.

In the above expression, it is understood that ω_1 depends on δ through (7.96). Insofar as this expression is an expansion in terms of a sequence of orthogonal kets, the single detection rate (7.101), which is just the squared modulus of (7.104), is given simply by

$$w^{(1)}(\mathbf{r}_1, t) = s\sum_{\delta} |C(\delta)|^2 \left[\mathcal{E}_{\omega_1}^{(1)}\right]^2. \tag{7.105}$$

This quantity is time independent. The single photodetection probability is constant at D_1. Of course, one has an analogous result for D_2.

We now calculate the joint detection rate at D_1 and D_2 (see (7.95)):

$$w^{(2)}(\mathbf{r}_1, t; \mathbf{r}_2, t') = s^2\|\hat{E}_2^{(+)}(\mathbf{r}_2, t')\hat{E}_1^{(+)}(\mathbf{r}_1, t)|\psi''\rangle\|^2. \tag{7.106}$$

Then, proceeding as above, we find

$$\hat{E}_1^{(+)}(\mathbf{r}_1, t)\hat{E}_2^{(+)}(\mathbf{r}_2, t')|\psi''\rangle = -\sum_{\delta} C(\delta)\mathcal{E}_{\omega_1}^{(1)}\mathcal{E}_{\omega_2}^{(1)}\,\mathrm{e}^{\mathrm{i}\omega_1(\tau_1 - t)}\,\mathrm{e}^{\mathrm{i}\omega_2(\tau_2 - t')}|0, 0\rangle, \tag{7.107}$$

where the quantities ω_1 and ω_2 depend on δ according to (7.96) and (7.97), and τ_2 is the light propagation time from the crystal to detector D_2, given by

$$\tau_2 = \frac{\mathbf{u}_2 \cdot \mathbf{r}_2}{c}. \tag{7.108}$$

Equation (7.107) is radically different from (7.104), in the sense that the vector $|0, 0\rangle$ (the vacuum) is a factor of the sum of complex amplitudes, which must therefore be summed before taking the squared modulus. Replacing ω_1 and ω_2 by their expressions in terms of δ, we obtain

$$w^{(2)}(\mathbf{r}_1, t; \mathbf{r}_2, t') = s^2 \left| \sum_{\delta} C(\delta) \mathcal{E}^{(1)}_{\omega_1} \mathcal{E}^{(1)}_{\omega_2} \exp\{i\delta(\tau_1 - \tau_2 - t + t')\} \right|^2 . \tag{7.109}$$

To calculate the above expression, we replace the discrete sum by an integral, introducing the density of states (see, for example, Section 6.4.2). This density of states varies slowly with δ. The same is true of $\mathcal{E}^{(1)}_{\omega_1}$ and $\mathcal{E}^{(1)}_{\omega_2}$, and they can be brought outside the integral, which is thus simply proportional to the Fourier transform $\tilde{C}(\tau)$ of $C(\delta)$, taken at $\tau = \tau_1 - \tau_2 - t + t'$. Since $C(\delta)$ is bell-shaped and of typical width $\Delta = 10^{13}\ \mathrm{s}^{-1}$, its transform $\tilde{C}(\tau)$ is a very narrow bell-shaped function, of typical width 10^{-13} s.

The joint detection probability at times t and t' is thus a function of $t - t'$ that is very strongly peaked around

$$t - t' = \tau_1 - \tau_2. \tag{7.110}$$

Recalling that τ_1 and τ_2 are the light propagation times from the crystal to the detectors D_1 and D_2, respectively, this result can be interpreted as meaning that two photons are emitted at exactly the same time in the crystal, then propagate, respectively, to D_1 and D_2.

Progress with methods of coincidence photon detection in the 1970s and 1980s made it possible to demonstrate the simultaneity of these emissions experimentally (see Figure 7.11). Combined with experiments of the kind described in Chapter 5, showing that each photon is indeed a single photon, these experiments constitute a convincing proof of the need to quantize the electromagnetic field. There is no way to interpret them in the framework of the semi-classical model, where the atom is quantized but the

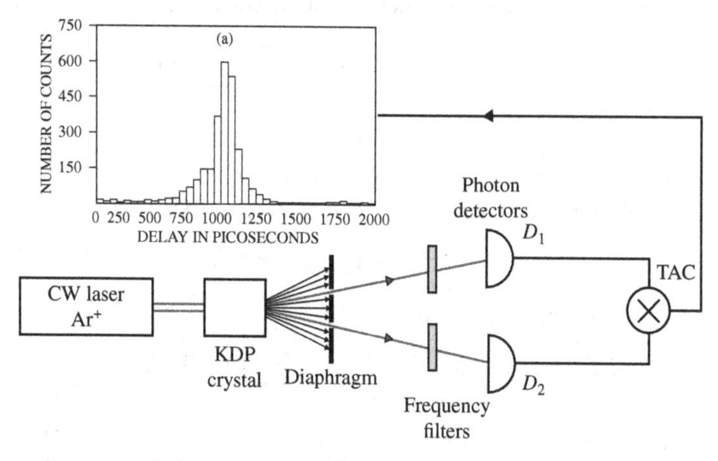

Figure 7.11 Simultaneous detection of photons emitted by parametric fluorescence. From S. Friberg, C. Hong and L. Mandel, Measurement of time delays in the parametric production of photon pairs, *Physical Review Letters* 54, 2011 (1985). The experimental setup includes a time–amplitude converter (TAC) which provides a histogram of the time intervals between detections. The half-width of this histogram (170 picoseconds) is of the same order as the time resolution of the system, and the intrinsic width of the process is therefore much smaller. The expected theoretical width is $\Delta^{-1} \simeq 0.1$ picosecond.

electromagnetic field is not. In particular, the semi-classical model, in which the fluorescence fields E_1 and E_2 have constant amplitudes, cannot explain the peak of the correlation function $w^{(2)}(\mathbf{r}_1, t; \mathbf{r}_2; t')$.

7.4.6 Two-photon interference

The Hong–Ou–Mandel experiment

The experiment we have just described (see Figure 7.11) can be given a perfectly clear interpretation in the particle view, where the photons are emitted in pairs. One might wonder whether it is really useful to appeal to such an elaborate analysis as the one using quantum electrodynamics presented above. The answer is in fact that this is the only formalism capable of describing both particle-like behaviour, e.g. pair emission, and wave-like behaviour. For example, with each photon of the pair, one can set up an interference experiment of the type discussed in Section 5.5 (see Figure 5.8).

The ability of the quantum optics formalism to describe both particle and wave behaviour is quite remarkable, and in itself justifies the introduction of such a formalism. But quantum optics does much more than that, because it predicts effects that would be quite inconceivable and, indeed, impossible to describe in the context of classical particle or wave models. This is particularly true of the two-photon interference phenomenon that we shall now discuss.

To demonstrate the two-photon interference phenomenon, let us go back to the parametric fluorescence setup of Figure 7.11, but instead of detecting photons emitted in two directions $\mathbf{u}_1$ and $\mathbf{u}_2$, let us recombine them on a beamsplitter S, and consider the photodetection signals at D_3 and D_4 (see Figure 7.12).

As in Section 7.4.5, the directions $\mathbf{u}_1$ and $\mathbf{u}_2$ correspond to exact phase matching for frequencies $\omega_1^{(0)}$ and $\omega_2^{(0)}$, and a pair of photons emitted in the directions $\mathbf{u}_1$ and $\mathbf{u}_2$ with frequencies ω_1 and ω_2, respectively, are characterized by the deviation δ from $\omega_1^{(0)}$ and $\omega_2^{(0)}$ (see (7.96–7.97)). Given the directional filtering, the outgoing radiation state from the nonlinear crystal is therefore still described by (7.100).

To calculate the photodetection probabilities at D_3 and D_4, we use once again the Heisenberg picture, expressing the fields $\hat{E}_3^{(+)}$ and $\hat{E}_4^{(+)}$ output from the beamsplitter in terms of the fields $\hat{E}_1^{(+)}$ and $\hat{E}_2^{(+)}$ output from the crystal. We first write, just after the beamsplitter, taking the transmission and reflection coefficients to be equal (see Chapter 5),

$$\hat{E}_3^{(+)}(\mathbf{r}_S, t) = \frac{1}{\sqrt{2}}\left(\hat{E}_1^{(+)}(\mathbf{r}_S, t) + \hat{E}_2^{(+)}(\mathbf{r}_S, t)\right) \tag{7.111}$$

$$\hat{E}_4^{(+)}(\mathbf{r}_S, t) = \frac{1}{\sqrt{2}}\left(\hat{E}_1^{(+)}(\mathbf{r}_S, t) - \hat{E}_2^{(+)}(\mathbf{r}_S, t)\right). \tag{7.112}$$

To refer everything to the crystal output, we express $\hat{E}_1^{(+)}(\mathbf{r}_S, t)$ and $\hat{E}_2^{(+)}(\mathbf{r}_S, t)$, as in Section 7.4.5, taking into account the propagation times τ_1 and τ_2 from leaving the crystal to arriving at the beamsplitter, following paths 1 and 2, respectively (see (7.103) and (7.108)):

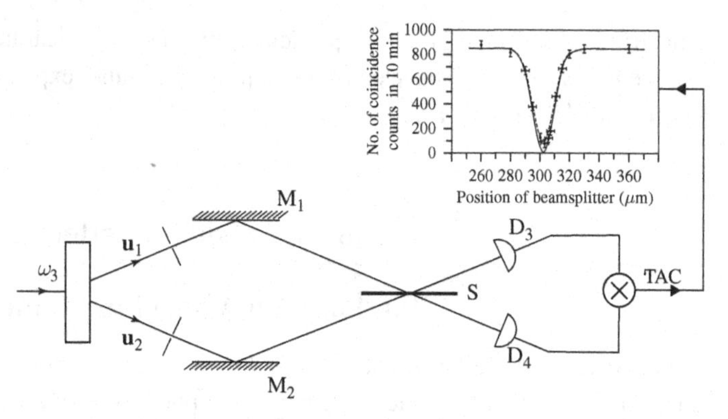

Figure 7.12 Two-photon interference. The photons in each pair are recombined on the beamsplitter S, and we consider the photodetection signals at D_1 and D_2, and in particular the double photodetection (coincidence signals). The insert shows the coincidence rate measured by Hong, Ou and Mandel for different values of the delay between the times of arrival of the photons on the beamsplitter S.[5] This delay can be modified by displacing the beamsplitter by means of a micrometer screw. When the photons arrive simultaneously, the coincidence rate falls to zero, which shows that the two photons leave the beamsplitter from the same side. The width of the dip in the count rate corresponds to a displacement of 30 μm, i.e. a delay that varies by 100 femtoseconds.

$$\hat{E}_1^{(+)}(\mathbf{r}_\mathrm{S}, t) = \mathrm{i} \sum_{\ell_1} \mathcal{E}_{\ell_1}^{(1)} \exp\{\mathrm{i}\omega_{\ell_1}(\tau_1 - t)\}\, \hat{a}_{\ell_1} \tag{7.113}$$

$$\hat{E}_2^{(+)}(\mathbf{r}_\mathrm{S}, t) = \mathrm{i} \sum_{\ell_2} \mathcal{E}_{\ell_2}^{(1)} \exp\{\mathrm{i}\omega_{\ell_2}(\tau_2 - t)\}\, \hat{a}_{\ell_2}. \tag{7.114}$$

As in Section 7.4.5, the modes ℓ_1 and ℓ_2 appearing in the expansion of the state (7.100) leaving the crystal are completely determined by the parameter δ, and we may replace ℓ_1 by $\mathbf{u}_1$, δ and ℓ_2 by $\mathbf{u}_2$, $-\delta$ in the expressions involving the fields $\hat{E}_3^{(+)}$ and $\hat{E}_4^{(+)}$ and the state $|\psi''\rangle$.

We begin by calculating the single detection rate at D_3, assuming it to be placed immediately after the output of the beamsplitter S:

$$w^{(1)}(D_3, t) = s \left\| \hat{E}_3^{(+)}(\mathbf{r}_\mathrm{S}, t)|\psi''\rangle \right\|^2. \tag{7.115}$$

When we calculate $\hat{E}_3^{(+)}(\mathbf{r}_\mathrm{S}, t)|\psi''\rangle$, we obtain once again an expansion in terms of orthogonal vectors:

$$\begin{aligned}\hat{E}_3^{(+)}(\mathbf{r}_\mathrm{S}, t)|\psi''\rangle = {} & \frac{\mathrm{i}}{\sqrt{2}} \sum_\delta C(\delta)\, \mathcal{E}_{\omega_1}^{(1)}\, \mathrm{e}^{\mathrm{i}\omega_1(\tau_1 - t)} |0_{\mathbf{u}_1,\delta}, 1_{\mathbf{u}_2,-\delta}\rangle \\ & + \frac{\mathrm{i}}{\sqrt{2}} \sum_\delta C(\delta)\, \mathcal{E}_{\omega_2}^{(1)}\, \mathrm{e}^{\mathrm{i}\omega_2(\tau_2 - t)} |1_{\mathbf{u}_1,\delta}, 0_{\mathbf{u}_2,-\delta}\rangle\end{aligned} \tag{7.116}$$

[5] C. K. Hong, Z. Y. Ou and L. Mandel, Measurement of Subpicosecond Time Intervals between Two Photons by Interference, *Physical Review Letters* **59**, 2044 (1987)

and we find

$$w^{(1)}(\mathrm{D}_3, t) = \frac{s}{2} \sum_{\delta} |C(\delta)|^2 \left(\left[\mathcal{E}_{\omega_1}^{(1)} \right]^2 + \left[\mathcal{E}_{\omega_2}^{(1)} \right]^2 \right), \tag{7.117}$$

where the frequencies ω_1 and ω_2 are related to δ by (7.96) and (7.97). The detection rate at D_3 is therefore independent of time. This is clearly true also for $w^{(1)}(\mathrm{D}_4, t)$. As in Section 7.4.5, we find constant single photodetection rates, implying that the photons arrive at the detectors at a steady average rate.

We now calculate the double photodetection rate just behind S:

$$w^{(2)}(\mathrm{D}_3, t; \mathrm{D}_4, t') = s^2 \| \hat{E}_3^{(+)}(\mathbf{r}_\mathrm{s}, t)\, \hat{E}_4^{(+)}(\mathbf{r}_\mathrm{s}, t') |\psi''\rangle \|^2. \tag{7.118}$$

Using (7.111–7.112) and (7.113–7.114), we obtain

$$\begin{aligned} &\hat{E}_3^{(+)}(\mathbf{r}_\mathrm{s}, t)\, \hat{E}_4^{(+)}(\mathbf{r}_\mathrm{s}, t')\, |\psi''\rangle \\ &\quad = -\frac{1}{2} \sum_{\delta} C(\delta) \mathcal{E}_{\omega_1}^{(1)} \mathcal{E}_{\omega_2}^{(1)} \left(\mathrm{e}^{\mathrm{i}\omega_1(\tau_1 - t')}\, \mathrm{e}^{\mathrm{i}\omega_2(\tau_2 - t)} - \mathrm{e}^{\mathrm{i}\omega_1(\tau_1 - t)}\, \mathrm{e}^{\mathrm{i}\omega_2(\tau_2 - t')} \right) |0, 0\rangle. \end{aligned} \tag{7.119}$$

As in Section 7.4.5, when we calculate the coincidence count rate, the state vector $|0, 0\rangle$ of the vacuum appears as a factor, and we must carry out a sum of complex amplitudes before taking the square of the modulus.

To simplify the calculation, we shall assume from here on that $\omega_1^{(0)} = \omega_2^{(0)} = \omega_3/2$. Using (7.96) and (7.97), the sum of the amplitudes in (7.119) can be expressed in the form,

$$A = -\frac{1}{2}\, \mathrm{e}^{\mathrm{i}\frac{\omega_3}{2}(\tau_1 + \tau_2 - t - t')} \sum_{\delta} C(\delta)\, \mathcal{E}_{\omega_1}^{(1)} \mathcal{E}_{\omega_2}^{(1)} \left(\mathrm{e}^{\mathrm{i}\delta(\tau_1 - \tau_2 + t - t')} - \mathrm{e}^{\mathrm{i}\delta(\tau_1 - \tau_2 - t + t')} \right). \tag{7.120}$$

Recall that $C(\delta)$ is a bell-shaped function, with a maximum at $\delta = 0$ and typical width $\Delta \simeq 10^{13}\ \mathrm{s}^{-1}$. In order to carry out the calculation explicitly, we take a Gaussian of half-width Δ (at $\mathrm{e}^{-1/2}$):

$$C(\delta) = \frac{C_0}{\sqrt{2\pi}\, \Delta} \exp\left\{ -\frac{\delta^2}{2\Delta^2} \right\}. \tag{7.121}$$

We replace the sum $\sum_{\delta}$ by an integral, and we bring $\mathcal{E}_{\omega_1}^{(1)} \mathcal{E}_{\omega_2}^{(1)}$ outside the integral, since it varies only slightly over the width Δ. Integration then yields $A_1 - A_2$, where

$$\begin{aligned} A_1 &= \int d\delta\, C(\delta)\, \mathrm{e}^{\mathrm{i}\delta(\tau_1 - \tau_2 + t - t')} \\ &= C_0\, \mathrm{e}^{-\frac{\Delta^2}{2}(\tau_1 - \tau_2 + t - t')^2} \end{aligned} \tag{7.122}$$

and

$$\begin{aligned} A_2 &= \int d\delta\, C(\delta)\, \mathrm{e}^{\mathrm{i}\delta(\tau_1 - \tau_2 - t + t')} \\ &= C_0\, \mathrm{e}^{-\frac{\Delta^2}{2}(\tau_1 - \tau_2 - t + t')^2}. \end{aligned} \tag{7.123}$$

The joint detection rate is given, finally, by

$$w^{(2)}(\mathrm{D}_3, t; \mathrm{D}_4, t') = \frac{s^2}{4}\left[\mathcal{E}^{(1)}_{\omega_{3/2}}\right]^4 \left|A_1(t-t') - A_2(t-t')\right|^2. \tag{7.124}$$

The functions A_1 and A_2 are extremely narrow (Δ^{-1} is typically equal to 10^{-13} s), and there is no detector or electronic system with such a time resolution. We shall thus assume that in fact we detect all coincidences characterized by a time interval $t - t'$ lying in the window centred on zero and with width much greater than Δ^{-1}. We also assume that we have adjusted the paths between the crystal and the beamsplitter in such a way that $|\tau_1 - \tau_2|$ is less than the width of the coincidence window. The global coincidence rate $w_\mathrm{c}(\mathrm{D}_3; \mathrm{D}_4)$ is calculated by integrating (7.124) over $\theta = t - t'$, which involves the integrals,

$$\int A_1^2 \, d\theta = \int A_2^2 \, d\theta = C_0^2 \sqrt{\pi}\, \Delta^{-1}, \tag{7.125}$$

$$\int A_1 A_2 \, d\theta = C_0^2 \sqrt{\pi}\, \Delta^{-1}\, \mathrm{e}^{-\Delta^2(\tau_1-\tau_2)^2}. \tag{7.126}$$

The final result is

$$w_\mathrm{c}(\mathrm{D}_3; \mathrm{D}_4) = \frac{s^2}{2}\left[\mathcal{E}^{(1)}_{\omega_{3/2}}\right]^4 C_0^2 \sqrt{\pi}\, \Delta^{-1}\left(1 - \mathrm{e}^{-\Delta^2(\tau_1-\tau_2)^2}\right). \tag{7.127}$$

Note that the coincidence detection rate is constant, except around $\tau_1 = \tau_2$, where it vanishes exactly (the light propagation times from the crystal to the edges of the beamsplitter are exactly equal). The width of the dip in the count rate is of the order of Δ^{-1}, i.e. typically 10^{-13} s^{-1}. This is indeed what is observed experimentally (see the insert of Figure 7.12).

Comment To keep the above calculation simple, we have assumed that the detectors D_3 and D_4 are positioned immediately after the beamsplitter S, which is clearly unrealistic. Going back through the calculation, it is easy to check that taking into account the propagation times τ_3 and τ_4 between the beamsplitter S and the detectors D_3 and D_4, respectively, leads to the same result, provided that one replaces $\theta = t' - t$ by $\theta = t' - t + \tau_3 - \tau_4$. On the practical level, this means that the coincidence window must be centred on $\tau_3 - \tau_4$, and not on zero, which was obvious anyway.

Discussion

This physical effect, observed for the first time by Hong, Ou and Mandel,[6] is remarkable for several reasons. To begin with, note that the coincidence rate w_c can be obtained as a function of $\tau_1 - \tau_2$ with a much better time resolution than one of the photodetectors themselves, since it depends only on the accuracy with which one checks the difference in the propagation times to the beamsplitter S, i.e. the position of the beamsplitter S. By displacing it with piezoelectric transducers, it is very easy to obtain a position resolution

[6] See Footnote 5, this chapter.

better than the micrometer, which means a (light travel) time resolution of a few femtoseconds (10^{-15} m). The insert in Figure 7.12 shows the results of the experiment, confirming that the width of the dip in the count rate is indeed of the order of Δ^{-1}.

On the conceptual level, this result reveals some of the most surprising properties of quantum physics. As explained above, experimental facts force us to admit that we have photons emitted in pairs. But then we also have the fact that, if the two photons arrive at exactly the same moment on the beamsplitter, then rather than being distributed randomly and independently over the two output channels of the beamsplitter, they are compelled to leave through the same channel, and never on either side of the beamsplitter. Clearly, this property could never be understood in terms of classical particles randomly distributed on either side of the beamsplitter S, because one would have coincidence in 50% of cases. Returning to the calculations, e.g. Equation (7.120), we observe that the vanishing of the coincidence term results from *destructive interference between two probability amplitudes*.

More precisely, the destructive interference occurs (for $\tau_1 = \tau_2$) between the amplitudes $\exp\{i\delta\theta\}$ and $\exp\{i(-\delta)\theta\}$ associated with the following two processes involving two photons (see Figure 7.13): (a) a photon at $\omega_3/2 + \delta$ is emitted in channel 1, then detected at D_3, while a photon at $\omega_3/2 - \delta$ is emitted in channel 2, then detected at D_4; (b) a photon at $\omega_3/2 - \delta$ is emitted in channel 1 then detected at D_4, while a photon at $\omega_3/2 + \delta$ is emitted in channel 2, then detected at D_3. There is no way of distinguishing these two processes, which correspond to the same final state in which a photon at $\omega_3/2 + \delta$ is detected at D_3, while a photon at $\omega_3/2 - \delta$ is detected at D_4. The amplitudes can thus be legitimately added together. If on the other hand the travel times τ_1 and τ_2 are not strictly equal, it is in principle possible to find out whether a detection is associated with process (a) or process (b), and then the amplitudes should not be added together. There is no longer any destructive interference. Note that the argument does not require the use of detectors able to resolve such short timescales. It is enough that the distinction should be possible in principle, should one have such perfect equipment at one's disposal, limited only by the fundamental rules of quantum physics.

We may wonder whether the semi-classical model can render an account of the effect described above. The answer is negative. A model with two classical waves impinging on the beamsplitter cannot mimic both facts that joint detection events never happen on opposite sides, and that the side of double detection is random. Once again, we find a situation that can be interpreted neither by a classical particle model, nor by a classical

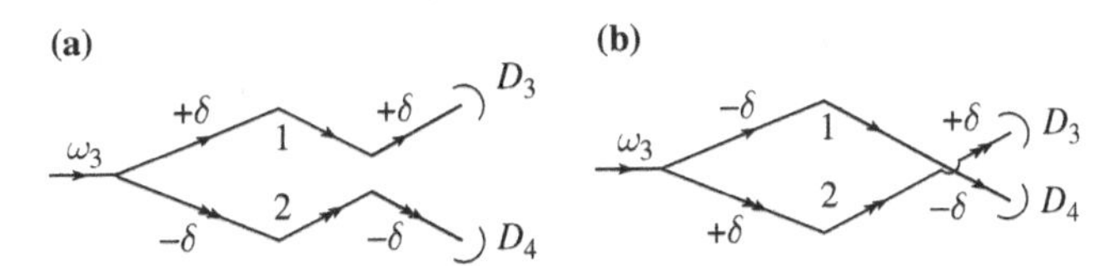

Figure 7.13 Two-photon processes whose amplitudes interfere, because they cannot be distinguished, when the distances between the source and the beamsplitter are exactly equal (see Figure 7.12 and text).

wave model, and where only the full quantum optics treatment can render an account of the observed facts.

Comments

(i) The Hong–Ou–Mandel experiment we have just described is reminiscent of the experiment of coalescence of indistinguishable photons emitted by two independent sources, described in Section 5B.3. In both situations, the cancellation of the joint detection probability at zero delay is due to a destructive interference between quantum amplitudes associated with indistinguishable processes. Note however that the timescales involved in the two experiments are different. In the Hong–Ou–Mandel experiment the width of the dip in Figure 7.13 is given by the reciprocal of the phase-matching spectral bandwidth Δ. In the coalescence experiment, the width of the dip (see Equation (5B.63)) is given by the coherence time of the photons, and it is in general much broader.

(ii) Among the various optical phenomena that cannot be described by classical models, of which we have now described several examples, we can distinguish different classes. A first one is related to wave–particle duality for *a single photon*, and has been commented on in detail in Chapter 5 (Section 5.5, or 5B.1 and 5B.2). In such situations, one has interference effects associated with different paths, and it is reminiscent of interference of ordinary waves following different paths. What is non-classical in such situations is the fact that distinct paths are followed by a single particle. Here we have a situation with an interference between *quantum amplitudes associated with two photons*. Such an effect belongs to the same class as, for instance, the quantum correlations observed with entangled photons (Complement 5C), which are also described as an interference between amplitudes associated with two photons. A similar comment can be made about the experiment of coalescence of indistinguishable photons of Section 5B.3.

(iii) The elementary analysis of Section 7.4.6a leads us to consider the radiation after the two diaphragms as an entangled state of two photons,

$$|\psi_{\text{ent}}\rangle = \frac{1}{\sqrt{2}}\Big[|1_{\mathbf{u}_1,+\delta}\,;\,1_{\mathbf{u}_2,-\delta}\rangle + |1_{\mathbf{u}_1,-\delta}\,;\,1_{\mathbf{u}_2,+\delta}\rangle\Big], \tag{7.128}$$

a superposition of two two-photon states;

- The state $|1_{\mathbf{u}_1,+\delta}; 1_{\mathbf{u}_2,-\delta}\rangle$, where a photon of frequency $\omega_3/2+\delta$ is emitted in channel $\mathbf{u}_1$, and a photon at $\omega_3/2-\delta$ is emitted in channel $\mathbf{u}_2$.
- The state $|1_{\mathbf{u}_1,-\delta}; 1_{\mathbf{u}_2,+\delta}\rangle$, where a photon of frequency $\omega_3/2-\delta$ is emitted in channel $\mathbf{u}_1$, and a photon at $\omega_3/2+\delta$ is emitted in channel $\mathbf{u}_2$.

The state (7.128) is a frequency-entangled state. One can also obtain polarization-entangled photons from the parametric fluorescence phenomenon. Consider, for example, the experimental situation shown in Figure 7.8. This uses a so-called ‘type II’ nonlinear crystal, in which the twin photons have orthogonal polarizations $\boldsymbol{\varepsilon}_1$ and $\boldsymbol{\varepsilon}_2$. As the crystal is birefringent, the phase-matching conditions then differ for the two polarizations. The upper circle in Figure 7.8 is made up of photons of polarization $\boldsymbol{\varepsilon}_1$, while the lower circle comprises photons of polarization $\boldsymbol{\varepsilon}_2$. We position the diaphragms for filtering the modes at the points of intersection A and B of the two circles. Both polarizations are then transmitted, and the entangled state is given by

$$|\psi_{\text{ent}}\rangle = \frac{1}{\sqrt{2}}\Big[|1_{\mathbf{u}_1,+\delta,\boldsymbol{\varepsilon}_1}\,;\,1_{\mathbf{u}_2,-\delta,\boldsymbol{\varepsilon}_2}\rangle + |1_{\mathbf{u}_1,-\delta,\boldsymbol{\varepsilon}_2}\,;\,1_{\mathbf{u}_2,+\delta,\boldsymbol{\varepsilon}_1}\rangle\Big]. \tag{7.129}$$

This is a polarization-entangled state, similar to the one produced by an atomic cascade (see Complements 5C and 6C), used to carry out many experiments demonstrating and/or using the entanglement phenomenon.

7.5 Conclusion

Nonlinear optics is above all an extremely powerful technique, upon which a great many applications of modern optics are now based. For example, in laser technology, nonlinear optics is used to produce coherent beams at new frequencies, starting out from frequencies for which lasers already exist. Furthermore, nonlinear effects such as the Kerr effect (see Complement 7B) can be used to control a light signal by means of another light signal, and thus play an important role in optical data processing. The optical Kerr effect also allows the propagation of ultrashort soliton-type pulses without spreading, or optical bistability.

The applications mentioned above can often be understood using the semi-classical formalism of nonlinear optics. However, we have also shown in Section 7.4 that there is a whole class of phenomena that can only be described using the framework provided by quantum optics, a spectacular example being the creation of entangled twin-photon pairs. While these things remain in the domain of fundamental research, there are also some prospects for applications to quantum information. But that is another story (see Complement 5E).

7A Complement 7A Parametric amplification and oscillation. Semi-classical and quantum properties

In this complement, we consider one of the most important properties of the parametric interaction, already mentioned in Section 7.3.4, namely the possibility of amplifying an incident field, and hence constructing an optical oscillator by inserting the parametric medium inside a resonant cavity, usually a Fabry–Perot cavity. To begin with, we shall examine the classical characteristics of parametric amplification and oscillation, emphasizing the similarities and differences compared with amplification by induced emission, studied in Chapter 3. Then we shall consider the main quantum properties of the same system: amplification without added noise, squeezing of quantum fluctuations below threshold by an OPO and the production of twin beams.

7A.1 Classical description of parametric amplification

7A.1.1 Non-degenerate case

This configuration has already been mentioned in Section 7.3.4. An intense pump wave of frequency ω_3 and a weak signal wave of frequency $\omega_1 < \omega_3$ are directed into a parametric crystal (see Figure 7.7). We have seen that an idler wave of frequency $\omega_2 = \omega_3 - \omega_1$ appears in the crystal when the phase-matching condition $\mathbf{k}_2 \simeq \mathbf{k}_3 - \mathbf{k}_1$ is satisfied. We shall assume in this section that $\omega_2 \neq \omega_1$ (non-degenerate case), that the pump is constant, and that the phase-matching condition holds exactly. The solution of the coupled propagation equations (7.50–7.52) (for a pump of pure imaginary amplitude) is then given by (7.64–7.65). There is amplification of the signal wave, with a gain in intensity G equal to

$$G = \cosh^2(\gamma L), \tag{7A.1}$$

where L is the length of the crystal and $\gamma = \dfrac{\chi^{(2)}|\mathcal{E}_3|}{c}\sqrt{\dfrac{\omega_1\omega_2}{n_1 n_2}}$. In addition, the phase of the signal wave output from the crystal is equal to its value on entry into the crystal increased by the propagation term $k_1 L$. This is indeed a coherent amplification phenomenon, parametric amplification, accompanied by the appearance of a new wave, the idler wave. In Complement 7B, we shall see another arrangement leading to coherent amplification of an incident wave, called the 'phase conjugate mirror', due to a third-order nonlinear optical phenomenon.

7A.1.2 Degenerate case

Consider the arrangement in the last paragraph, but assuming now that the phase-matching condition is fulfilled for $\omega_2 = \omega_1 = \omega_3/2$, and for a crystal of type I, where the signal and idler modes have the same polarization, and with all the fields propagating in the same direction. In this particular case, called the degenerate case, the phase-matching condition is simply $n_1 = n_3$.

Under such conditions, the signal and idler modes cannot be distinguished, and we find ourselves in the opposite situation to the frequency doubling scenario discussed in Section 7.3.5. We must then use (7.67) and (7.68). In the constant pump approximation, we have the following propagation equation for the signal field:

$$\frac{\mathrm{d}\mathcal{E}_1(z)}{\mathrm{d}z} = \mathrm{i}\frac{\omega_1}{n_1 c}\chi^{(2)}\mathcal{E}_3(0)\mathcal{E}_1^*(z). \tag{7A.2}$$

By suitable choice of phase origin, consider a purely imaginary pump, $\mathcal{E}_3(0) = -\mathrm{i}E_3$, with E_3 real and positive. The solution of (7A.2) is then simply,

$$\mathcal{E}_1(z) = \mathcal{E}_1(0)\cosh\gamma' z + \mathcal{E}_1^*(0)\sinh\gamma' z, \tag{7A.3}$$

with $\gamma' = \chi^{(2)}E_3\omega_1/n_1 c$. This solution is very different from the non-degenerate case (7.65), because the behaviour of the system now depends on the *initial phase of the incoming field*:

- If $\mathcal{E}_1(0)$ is real, then $\mathcal{E}_1(z) = \mathcal{E}_1(0)\mathrm{e}^{\gamma' z}$, and the incident field is amplified.
- If $\mathcal{E}_1(0)$ is imaginary, then $\mathcal{E}_1(z) = \mathcal{E}_1(0)\mathrm{e}^{-\gamma' z}$, and the incident field is attenuated. One often says that the field is deamplified, to distinguish this process from a simple loss effect during propagation. The energy of the signal field is not in fact dissipated but, rather, it is wholly transferred to the pump field.

As we have made a particular choice of phase for the pump field $\mathcal{E}_3(0)$, the gain in fact depends on the relative phase between the incoming signal field $\mathcal{E}_1(0)$ and the pump field. It is thus important to control the optical paths of these two waves very carefully, as one would in an interferometer setup, if a stable gain is to be obtained. A new kind of behaviour is thus observed here, quite different from amplification by stimulated emission. In fact, other branches of physics are full of examples of such phase-dependent amplifiers. For example, in mechanics, the amplitude of oscillation of a pendulum in which one parameter, such as the length or the moment of inertia, is modulated at twice its natural frequency will either increase or decrease depending on the relative phase between the oscillation and the modulation of the parameter. Any child playing on a swing will know how sensitive its oscillations are to the phase when attempts are made to increase the amplitude of swing by periodic movements of the body. Indeed, the term 'parametric amplifier', and by extension, 'parametric interaction', comes from this physical situation, in which one of the system parameters is 'pumped' at twice its natural frequency. We find the same behaviour in hydrodynamics (the Faraday instability) and in nonlinear electronic oscillators.

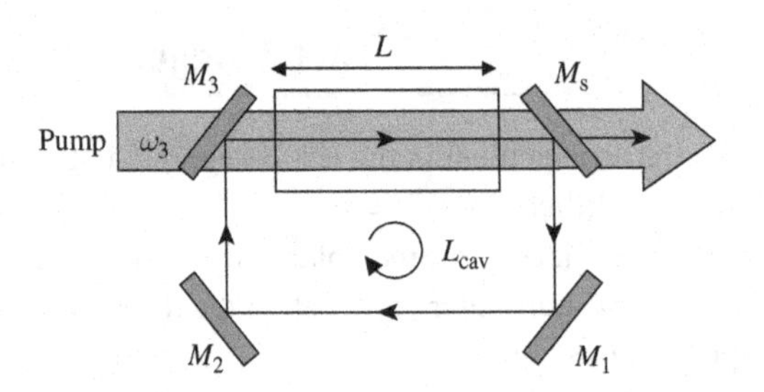

Figure 7A.1 Optical parametric oscillator (OPO). The optical cavity can be resonant for just the signal beam (singly resonant OPO) or for both the signal and idler beams (doubly resonant OPO).

7A.2 The optical parametric oscillator (OPO)

We saw in Chapter 3 that, by inserting a stimulated emission light amplifier in a resonant optical cavity, we could obtain an *optical oscillator*, the laser, able to produce a coherent light wave, i.e. a highly directional and highly monochromatic light wave. We shall see that the same situation arises when a parametric amplifier is placed inside a resonant cavity, to obtain what is called an *optical parametric oscillator* (OPO). Here we shall consider only the non-degenerate case, where the parametric interaction can amplify a signal field of any phase.

7A.2.1 Description of the system

The OPO is shown schematically in Figure 7A.1. It comprises a nonlinear crystal of length L, in which the parametric interaction occurs, inserted within an optical cavity of length L_{cav}, assumed to be a ring cavity to simplify the discussion. This cavity is totally transparent at the wavelength of the pump beam. We investigate two configurations:

- The *singly resonant OPO*, where the cavity, transparent at the wavelength of the pump beam and idler beams, comprises perfectly reflecting mirrors M_1, M_2 and M_3 and an output mirror M_s that is transparent for the pump and idler, but low transmission at the wavelength of the signal beam. Let R and T be the reflection and transmission coefficients for the intensity at this mirror.
- The *doubly resonant OPO*, where the cavity, still transparent at the wavelength of the pump beam, is now resonant for both the signal and idler beam. The mirrors M_1, M_2 and M_3 are perfectly reflecting, and the output mirror M_s weakly transmitting at frequencies ω_1 and ω_2. To simplify this, we shall assume that the transmission and reflection coefficients are the same at the two frequencies.

As for a laser, an oscillation is only set up if, for each round trip, the parametric gain dominates losses due to transmission at the output mirror. So, just as happens for the laser, there will be an oscillation threshold, governed here by the intensity of the pump beam.

7A.2.2 Singly resonant OPO

Recalling that the complex field is equal to the slowly varying amplitude $\mathcal{E}_1(z)$ multiplied by the propagation phase factor $e^{ik_1 z}$, we obtain from (7.65) the following value for the signal field leaving the crystal:

$$E_1^{(+)}(z=L) = \mathcal{E}_1(0)\,e^{in_1\omega_1 L/c}\cosh\gamma L. \tag{7A.4}$$

We deduce the value of the signal field $E_1^{(+)}(L_{cav})$ when it has reflected off the output mirror and propagated once around the cavity:

$$E_1^{(+)}(L_{cav}) = \sqrt{R}\mathcal{E}_1(0)\,e^{i\phi_1}\cosh\gamma L, \tag{7A.5}$$

where $\phi_1 = (L_{cav} + (n_1 - 1)L)\omega_1/c$ is the phase accumulated by the signal as it propagates over a complete round trip of the optical cavity.

The conditions for steady-state oscillation of the system are obtained, as for the laser, by expressing the fact that, after one trip around the cavity, the field is absolutely identical to what it was before, in both phase and amplitude, hence in complex value, i.e. $E_1^{(+)}(L_{cav}) = E_1^{(+)}(0) = \mathcal{E}_1(0)$. It then follows that

$$\mathcal{E}_1(0)\left(1 - \sqrt{R}\,e^{i\phi_1}\cosh\gamma L\right) = 0. \tag{7A.6}$$

As for the laser, we obtain a solution in which the oscillator is 'off', i.e. $\mathcal{E}_1(0) = 0$, and another solution in which the value of the signal field is non-zero, when

$$\sqrt{R}\,e^{i\phi_1}\cosh\gamma L = 1, \tag{7A.7}$$

which implies two real conditions,

$$\cosh\gamma L = \frac{1}{\sqrt{R}} \tag{7A.8}$$

$$(n_1 - 1)L + L_{cav} = m\lambda_1, \tag{7A.9}$$

with m an integer. The first condition reflects the equality of gain and loss, while the second expresses the resonant nature of the cavity for the signal wave.

For a low-intensity pump ($\gamma L \ll 1$) and a highly resonant cavity ($\sqrt{R} \simeq 1 - T/2$), the first condition implies that

$$|\mathcal{E}_3|^2 = \frac{n_1 n_2 c^2}{\omega_1\omega_2}\frac{2\pi}{\mathcal{F}(\chi^{(2)})^2 L^2}. \tag{7A.10}$$

This gives the intensity of the pump beam at the oscillation threshold, which is inversely proportional to both the square of the nonlinearity and to the finesse of the cavity $\mathcal{F} = 2\pi/T$. As in the frequency doubling setup, for optimal focusing conditions and under typical experimental conditions, the value of the pump power threshold is of the order of a few watts.

When the pump field exceeds the threshold value, the signal field is amplified on each trip around the cavity, thereby taking ever higher values. The energy transferred from the pump to the signal and idler waves is no longer negligible. The amplitude of the pump wave diminishes during propagation, and this leads to a drop in the parametric gain. This is a *gain saturation* phenomenon, similar to what is observed with stimulated emission, and which provides a way of finding the signal intensity under steady-state conditions. We shall not give the details of the calculation here, and the reader is referred to the specialist literature.[1]

There are several differences between the OPO and the laser:

- Gain saturation is due to very different physical effects in the two systems: depletion of the pump by energy transfer in parametric amplification, and reduction of population inversion in amplification by stimulated emission.
- The dynamics is very different: existence of an excited level able to store energy for an appreciable length of time in the laser, and practically instantaneous energy transfer from the pump mode to the signal and idler modes in the OPO. It follows that an OPO pumped by a pulsed laser could follow the temporal variation of the pump wave, even if it is very fast.

Due to the very high value of the threshold intensity, the singly resonant OPO can only operate with intense continuous wave lasers, or with pulsed lasers. Singly resonant OPOs are very useful sources of coherent radiation, because the emission wavelength is not related to the level structure of the active medium, as it is with a laser, but rather to the phase-matching conditions. Since the latter can be modified by changing the temperature or orientation of the crystal, an OPO can be tuned over a very broad spectral band, essentially only limited by the transparency zone of the nonlinear medium. For a given adjustment of the phase matching, the spectral width of the emitted light is related to that of the pump laser, and can thus be very narrow. The conversion efficiency of the pump to the sum of the signal and idler waves can reach values of around 90%. OPOs are very widely used in spectroscopy, e.g. in the infra-red, to detect extremely low concentrations of toxic gases or pollutants.

7A.2.3 Doubly resonant OPO

In this scenario, the cavity recycles the pump and idler waves, and these two fields are now, in principle, non-zero at the crystal input. To simplify this, we shall consider the case where the pump laser has a relatively low power. Under such conditions $\gamma' L \ll 1$, and (7.50–7.52) can be solved to first order in $\gamma' L$. With the constant pump approximation, the amplitudes of the fields output from the crystal are

$$\mathcal{E}_1(L) = \mathcal{E}_1(0) + \mathrm{i} g_1 \mathcal{E}_3 \mathcal{E}_2^*(0) \tag{7A.11}$$

$$\mathcal{E}_2(L) = \mathcal{E}_2(0) + \mathrm{i} g_2 \mathcal{E}_3 \mathcal{E}_1^*(0), \tag{7A.12}$$

[1] R. W. Boyd, *Nonlinear Optics*, Academic Press (2008).

with

$$g_i = \frac{\omega_i}{n_i c}\chi^{(2)}L, \quad i = 1,2. \tag{7A.13}$$

This is a 'cross-gain' phenomenon, the signal field amplifying the idler field, and vice versa. By a similar argument to the one given in the last section, we deduce the values of the complex signal and idler fields after one round trip:

$$E_1^{(+)}(L_{\text{cav}}) = \sqrt{R}\,\mathrm{e}^{\mathrm{i}\phi_1}\left[\mathcal{E}_1(0) + \mathrm{i}g_1\mathcal{E}_3\mathcal{E}_2^*(0)\right] \tag{7A.14}$$

$$E_2^{(+)}(L_{\text{cav}}) = \sqrt{R}\,\mathrm{e}^{\mathrm{i}\phi_2}\left[\mathcal{E}_2(0) + \mathrm{i}g_2\mathcal{E}_3\mathcal{E}_1^*(0)\right], \tag{7A.15}$$

where $\phi_i = (L_{\text{cav}} + (n_i - 1)L)\omega_i/c$ for $i = 1,2$. Steady-state conditions are obtained by requiring the fields to equal their values at the crystal input. Multiplying both sides of the last two equations by $\mathrm{e}^{-\mathrm{i}\phi_1}$ and $\mathrm{e}^{-\mathrm{i}\phi_2}$, respectively, and taking the complex conjugate of the second equation, this implies

$$(\mathrm{e}^{-\mathrm{i}\phi_1} - \sqrt{R})\mathcal{E}_1(0) - \mathrm{i}g_1\sqrt{R}\mathcal{E}_3\mathcal{E}_2^*(0) = 0 \tag{7A.16}$$

$$\mathrm{i}g_2\sqrt{R}\mathcal{E}_3^*\mathcal{E}_1(0) + (\mathrm{e}^{\mathrm{i}\phi_2} - \sqrt{R})\mathcal{E}_2^*(0) = 0. \tag{7A.17}$$

This yields a system of two homogeneous linear equations, the zero fields constituting one solution (when the OPO is off). A non-zero solution only exists if the determinant of the system is zero, whence,

$$(\mathrm{e}^{-\mathrm{i}\phi_1} - \sqrt{R})(\mathrm{e}^{\mathrm{i}\phi_2} - \sqrt{R}) = Rg_1g_2|\mathcal{E}_3|^2. \tag{7A.18}$$

Since the right-hand side of this equation is real, so must the left-hand side also be real, and this can only be true if

$$\phi_1 = \phi_2 + 2p\pi; \quad p \text{ integer.} \tag{7A.19}$$

The vanishing of the real part implies the following value for the pump intensity:

$$|\mathcal{E}_3|^2 = \frac{1 + R - 2\sqrt{R}\cos\phi_1}{Rg_1g_2}. \tag{7A.20}$$

This is the threshold value of the pump intensity. When the output mirror has low transmission, then to lowest order in T,

$$|\mathcal{E}_3|^2 = \frac{n_1n_2c^2}{4\omega_1\omega_2}\,\frac{16\sin^2(\phi_1/2) + T^2}{(\chi^{(2)})^2L^2}. \tag{7A.21}$$

When ϕ_1 is not a multiple of 2π, the first term in the numerator dominates, and the threshold is extremely high. The minimal threshold value is thus obtained for

$$\phi_1 = 2p_1\pi; \quad \phi_2 = 2p_2\pi \qquad p_1, p_2 \text{ integers,} \tag{7A.22}$$

i.e. when the cavity is resonant for both the signal mode and the idler mode. Introducing the finesse of the cavity, it is then equal to

$$|\mathcal{E}_3|^2 = \frac{n_1n_2c^2}{\omega_1\omega_2}\,\frac{\pi^2}{\mathcal{F}^2(\chi^{(2)})^2L^2}. \tag{7A.23}$$

As might be expected, for a doubly resonant system, the threshold is significantly smaller than for a singly resonant OPO (dependence on $1/\mathcal{F}^2$ rather than $1/\mathcal{F}$ in (7A.10)). It is typically of the order of 100 mW, hence easily accessible to low-power continuous wave lasers.

This reduction in the threshold is accompanied by the double resonance condition (7A.22), which represents an extra constraint as compared with the singly resonant OPO. If we take into account the condition $\omega_1 + \omega_2 = \omega_3$, double resonance implies that we must have

$$\frac{\omega_3}{2\pi c} = \frac{p_1}{(n_1 - 1)L + L_{\text{cav}}} + \frac{p_2}{(n_2 - 1)L + L_{\text{cav}}}. \tag{7A.24}$$

For a given pump frequency, the values of the cavity length L_{cav} for which oscillation occurs are determined by the integers p_1 and p_2. It follows that the doubly resonant OPO can only oscillate stably for a *discrete set of cavity lengths*, in contrast to the laser or singly resonant OPO.

To simplify, let us consider exact double resonance. When the oscillation condition (7A.18) is satisfied, Equation (7A.16) is satisfied with $e^{i\phi_1} = 1$. If we also assume that $T \ll 1$, the signal and idler fields are then related by

$$\frac{T}{2}\mathcal{E}_1 = ig_1\mathcal{E}_3\mathcal{E}_2^*. \tag{7A.25}$$

Note first that, if $\mathcal{E}_1$ and $\mathcal{E}_2$ are solutions, then $\mathcal{E}_1 e^{i\theta}$ and $\mathcal{E}_2 e^{-i\theta}$ are also solutions for any value of θ. We found an analogous property for the laser, where the oscillation condition fixed the intensity, but not the phase of the intra-cavity field (see Section 3.1.2). As we saw in Complement 3D, this freedom implied the existence of a *phase diffusion* phenomenon for the laser field due to effects leading to fluctuations, such as spontaneous emission. The OPO has a similar property, but now it concerns the *phase difference between the signal and idler fields*. On the other hand, the sum of the phases of these two fields is related to the phase of the pump field.

Finally, Equation (7A.25) implies that

$$\frac{T^2}{4}|\mathcal{E}_1|^2 = g_1^2|\mathcal{E}_3|^2|\mathcal{E}_2|^2. \tag{7A.26}$$

Given (7A.13) and (7A.23), this leads to

$$\frac{n_1}{\omega_1}|\mathcal{E}_1|^2 = \frac{n_2}{\omega_2}|\mathcal{E}_2|^2. \tag{7A.27}$$

Under parametric oscillation conditions, we thus recover the Manley–Rowe relation (7.60), which was originally obtained from the propagation equations of the semi-classical theory, but which can be simply interpreted in the quantum context, because it says that the photon number fluxes of the signal and idler waves are equal. The quantum features of the amplification, and then of parametric oscillation, will be tackled in Section 7A.3.

Let us end by mentioning that, as in the laser, the transverse properties of the signal and idler beams are imposed by the resonant cavity, which must necessarily involve one or more concave mirrors to compensate for diffraction. In this case, the beams output from the OPO are Gaussian modes (see Complement 3B). They thus have the same coherence properties as for the laser.

7A.3 Quantum features of parametric amplification

7A.3.1 Quantum description of attenuation and amplification processes

Classically, attenuation and amplification processes are described by a linear input–output relation for the complex amplitudes:

$$E_{\text{out}}^{(+)} = gE_{\text{in}}^{(+)}, \tag{7A.28}$$

where the gain g is a complex number with modulus greater than unity for amplification, and less than unity for attenuation. If we wish to describe the same process in a quantum framework, it is tempting to write the same relation for the field operators, as we did in the case of the semi-reflecting mirror in Section 5.1.2:

$$\hat{E}_{\ell,\text{out}}^{(+)} = g\hat{E}_{\ell,\text{in}}^{(+)}. \tag{7A.29}$$

In (7A.29), it was assumed that each mode with index (ℓ, in) at the amplifier (or attenuator) input is related to a single mode with index (ℓ, out) of the free field at the amplifier (or attenuator) output. However, there is a strong constraint on what constitutes an acceptable transformation in the quantum context: *such a transformation must preserve the value of the commutators*, i.e. we must have, both at the input and the output, and for all modes affected by the transformation,

$$\left[\hat{E}_{\ell}^{(+)}, \left(\hat{E}_{\ell'}^{(+)}\right)^{\dagger}\right] = \delta_{\ell\ell'}\left[\mathcal{E}_{\ell}^{(1)}\right]^2, \tag{7A.30}$$

where ℓ and ℓ' denote any two modes of the field. Indeed, the incoming and outgoing fields are free fields, and their commutator has a value that does not depend on the way the relevant field has been produced. Such transformations, respecting the commutation rules (7A.30), are said to be 'canonical'. It can be checked that this is effectively the case for the example of the semi-reflecting mirror. However, for the transformation (7A.29), one has

$$\left[\hat{E}_{\ell,\text{out}}^{(+)}, \left(\hat{E}_{\ell,\text{out}}^{(+)}\right)^{\dagger}\right] = |g|^2\left[\hat{E}_{\ell,\text{in}}^{(+)}, \left(\hat{E}_{\ell,\text{in}}^{(+)}\right)^{\dagger}\right]. \tag{7A.31}$$

The two commutators can thus only be equal if g is a simple phase factor of unit modulus, as for propagation in vacuum. In the case of attenuation or amplification, one must therefore write

$$\hat{E}_{\ell,\text{out}}^{(+)} = g\hat{E}_{\ell,\text{in}}^{(+)} + \hat{B}_{\ell}, \tag{7A.32}$$

where $\hat{B}_{\ell}$ is a *noise operator*, with zero expected value and commutation relations ensuring that the transformation thereby defined is canonical. We shall see in the following that the presence of this further term significantly increases the quantum fluctuations of the beam as it crosses such systems. We shall take g real to simplify. Consider first the attenuation case $g < 1$. It is easy to see that one possible form of canonical transformation is

$$\hat{E}^{(+)}_{\ell,\text{out}} = g\hat{E}^{(+)}_{\ell,\text{in}} + \sqrt{1-g^2}\hat{E}^{(+)}_{\ell,\text{b}}, \tag{7A.33}$$

where $\hat{E}^{(+)}_{\ell,\text{b}}$ is a complex field operator in a given mode, analogous to $\hat{E}^{(+)}_{\ell,\text{in}}$ and $\hat{E}^{(+)}_{\ell,\text{out}}$, and which commutes with them. This was the transformation we wrote down in the case of the semi-reflecting mirror (which corresponds to the special case $g = 1/\sqrt{2}$). In the present case, $\hat{E}^{(+)}_{\ell,\text{b}}$ corresponds to the field entering via the other channel of the mirror, which is here in the vacuum state. We have generalized this here to the case of a linear attenuation of arbitrary origin. The effect of the term $\hat{E}^{(+)}_{\ell,\text{b}}$ on the fluctuations of the beam has already been examined in Section 5A.1.4 of Complement 5A. In particular, we saw that an attenuated coherent state was transformed into another coherent state of smaller expected value.

Consider now the amplification case $g > 1$. One possible form of canonical transformation is then,

$$\hat{E}^{(+)}_{\ell,\text{out}} = g\hat{E}^{(+)}_{\ell,\text{in}} + \sqrt{g^2-1}\left(\hat{E}^{(+)}_{\ell,\text{b}}\right)^{\dagger}, \tag{7A.34}$$

where $\left(\hat{E}^{(+)}_{\ell,\text{b}}\right)^{\dagger}$ has replaced $\hat{E}^{(+)}_{\ell,\text{b}}$. We deduce the following relations for the field quadrature operators:

$$\hat{E}_{P\ell,\text{out}} = g\hat{E}_{P\ell,\text{in}} - \sqrt{g^2-1}\,\hat{E}_{P\ell,\text{b}}; \quad \hat{E}_{Q\ell,\text{out}} = g\hat{E}_{Q\ell,\text{in}} + \sqrt{g^2-1}\hat{E}_{Q\ell,\text{b}}. \tag{7A.35}$$

These can be used to determine the effect of amplification on the variances of the two field quadratures. The incoming state of the system is the tensor product of an arbitrary state in the amplifier mode and of the vacuum in the noise mode. We thus have,

$$(\Delta E_{P\ell,\text{out}})^2 = G(\Delta E_{P\ell,\text{in}})^2 + (G-1)\left[\mathcal{E}^{(1)}\right]^2 \tag{7A.36}$$

$$(\Delta E_{Q\ell,\text{out}})^2 = G(\Delta E_{Q\ell,\text{in}})^2 + (G-1)\left[\mathcal{E}^{(1)}\right]^2, \tag{7A.37}$$

where $G = g^2$ is the gain in the intensity of the system.

We see that, due to the second term of (7A.36) and (7A.37), *the amplification process necessarily introduces noise*. If we begin with a state that is highly squeezed in the quadrature P, for example $\Delta E_{P\ell,\text{in}} \simeq 0$, the amplified state will have fluctuations in this quadrature greater than $\left[\mathcal{E}^{(1)}\right]^2$, i.e. greater than the fluctuations of the vacuum, as soon as $G > 2$. If we begin with a coherent state $\left((\Delta E_{P\ell,\text{in}})^2 = \left[\mathcal{E}^{(1)}\right]^2\right)$, the quantum noise at the output is multiplied by $2G-1$, whereas the expected value of the signal is only multiplied by G. We define the *noise factor* of the amplifier, which characterizes the degradation of the signal-to-noise ratio between input and output, as the energy amplification factor of the noise divided by that of the signal, i.e. the noise added by the amplifier relative to the input. In the present case, the noise factor is given by

$$F = \frac{2G-1}{G}. \tag{7A.38}$$

For a very high gain amplifier, this factor is equal to 2, or '3 dB' on the logarithmic scale, whereas a theoretical amplifier adding no noise would have a noise factor of 1. We thus see that very simple and fundamental quantum considerations regarding the preservation of commutation relations allow us to predict a practical limitation for any optical amplifier operating on quasi-classical states: an ideal amplifier, which adds no noise to the incoming signal, is forbidden by the laws of quantum mechanics.

7A.3.2 Non-degenerate parametric amplification

When the signal and idler fields are non-zero at the crystal input, the semi-classical propagation equations ((7.50–7.52)) have the following solution, making the constant pump approximation with $\mathcal{E}_3$ pure imaginary and with negative imaginary part:

$$\mathcal{E}_1(L) = \cosh(\gamma L)\,\mathcal{E}_1(0) + \sqrt{\frac{\omega_1 n_2}{\omega_2 n_1}}\,\mathcal{E}_2^*(0)\sinh(\gamma L) \tag{7A.39}$$

$$\mathcal{E}_2(L) = \cosh(\gamma L)\,\mathcal{E}_2(0) + \sqrt{\frac{\omega_2 n_1}{\omega_1 n_2}}\,\mathcal{E}_1^*(0)\sinh(\gamma L), \tag{7A.40}$$

where γ is as defined in (7.66).[2] Here we seek once again to extend these relations by replacing the complex signal and idler fields by the corresponding quantum operators:

$$\hat{E}_1^{(+)}(L) = \left[\cosh(\gamma L)\,\hat{E}_1^{(+)}(0) + \sqrt{\frac{\omega_2 n_1}{\omega_1 n_2}}\left(\hat{E}_2^{(+)}(0)\right)^\dagger \sinh(\gamma L)\right] \mathrm{e}^{\mathrm{i}n_1\omega_1 L/c} \tag{7A.41}$$

$$\hat{E}_2^{(+)}(L) = \left[\cosh(\gamma L)\,\hat{E}_2^{(+)}(0) + \sqrt{\frac{\omega_1 n_2}{\omega_2 n_1}}\left(\hat{E}_1^{(+)}(0)\right)^\dagger \sinh(\gamma L)\right] \mathrm{e}^{\mathrm{i}n_2\omega_2 L/c}. \tag{7A.42}$$

The reader may check that these linear input–output relations do preserve the commutation relations, which are given in a dielectric medium of index n_ℓ by,

$$\left[\hat{E}_\ell^{(+)}, \left(\hat{E}_{\ell'}^{(+)}\right)^\dagger\right] = \frac{\hbar\omega_\ell}{2n_\ell\varepsilon_0 L^3}\,\delta_{\ell\ell'}. \tag{7A.43}$$

The appearance of the factor $1/n_\ell$ in the commutator arises because the field amplitudes are divided by $\sqrt{n_\ell}$ as the fields pass from the vacuum into a medium of index n_ℓ. These relations are thus acceptable from a quantum standpoint, and it can be shown that they follow from the Hamiltonian approach to the parametric interaction (Section 7.4.2). They have the form (7A.34) required for a quantum amplifier, provided we set $g = \cosh(\gamma L)$, and up to the square root factor, which is necessary to preserve the commutators, because the latter have values that differ by the factor $\omega_1 n_2/\omega_2 n_1$ for the signal and idler fields.

Comparing (7A.41) with (7A.34), we can identify the 'noise mode' that adds noise when the signal mode is amplified. It is quite simply the idler mode. One can thus envisage reducing the noise introduced by the idler mode by injecting a squeezed state in this mode

[2] The solutions (7A.39) and (7A.40) used for the doubly resonant OPO are the same as (7A.11) and (7A.12) to first order in γL.

rather than the vacuum. In this case, even if the amplifier enhances any incoming signal field in the same manner, whatever its phase, the noise factor will only be improved for a signal field whose expected value is in phase with the squeezed quadrature of the idler mode. On the other hand, there will be a degradation of the orthogonal quadrature in the same proportions.

7A.3.3 Degenerate parametric amplification

Finally, consider the degenerate case $\omega_1 = \omega_2$, for which there is a phase sensitive amplification of the incoming field (see Section 7A.1.2). Setting $\hat{E}_1^{(+)} \equiv \hat{E}_2^{(+)}$ in (7A.41) or (7A.42), we then obtain

$$\hat{E}_1^{(+)}(L) = \left[\cosh(\gamma' L)\hat{E}_1^{(+)}(0) + \left(\hat{E}_1^{(+)}(0)\right)^\dagger \sinh(\gamma' L)\right] \mathrm{e}^{\mathrm{i}n_1\omega_1 L/c}, \tag{7A.44}$$

the quantum theoretical extension of the classical relation (7A.3), where γ' is defined after that relation. We deduce the following relations for the quadrature operators:

$$\hat{E}_{P1}(L) = \mathrm{e}^{\gamma' L}\hat{E}_{P1}(0), \tag{7A.45}$$

$$\hat{E}_{Q1}(L) = \mathrm{e}^{-\gamma' L}\hat{E}_{Q1}(0). \tag{7A.46}$$

These relations preserve the commutator $\left[\hat{E}_{P1}, \hat{E}_{Q1}\right]$ without the need for additional 'noise' terms. Regarding the standard deviations of the quantum fluctuations, they imply

$$\Delta E_{P1}(L) = \mathrm{e}^{\gamma' L}\Delta E_{P1}(0), \tag{7A.47}$$

$$\Delta E_{Q1}(L) = \mathrm{e}^{-\gamma' L}\Delta E_{Q1}(0). \tag{7A.48}$$

The amplification (or deamplification) factor is therefore the same for the signal (7A.44) and for the noise (7A.45). This means that *the degenerate parametric amplifier has noise factor equal to unity*. This does not contradict the general analysis in Section 7A.2.1, because it is a phase-sensitive amplifier, for which the initial expression does not have the simple form (7A.34). So there are indeed 'perfect' amplifiers that do not increase the signal-to-noise ratio, but in order to achieve such a performance, one must restrict to incident optical signals of well-determined phase.

Note that, if we send in 'nothing', i.e. the vacuum, at the input of the degenerate amplifier, we obtain a state such that

$$\Delta E_{P1}(L) = \mathrm{e}^{\gamma' L}\left[\mathcal{E}^{(1)}\right]^2, \tag{7A.49}$$

$$\Delta E_{Q1}(L) = \mathrm{e}^{-\gamma' L}\left[\mathcal{E}^{(1)}\right]^2. \tag{7A.50}$$

This is a *squeezed state* with zero expected value (the squeezed vacuum), described in Complement 5A.

7A.4 Quantum fluctuations in the fields produced by a doubly resonant OPO

We have just seen that the parametric interaction in a nonlinear crystal could produce non-classical states, such as squeezed states, with a compression factor of $e^{\gamma' L}$ depending on the pump laser power, hence rather modest when using continuous-wave lasers, and only really significant with pulsed lasers involving high instantaneous power outputs. We shall show that, by inserting the crystal in a resonant cavity, we can obtain a theoretical compression factor that tends to infinity as we approach the oscillation threshold of the system, hence even for a rather moderate pump power. An OPO can also be used to produce highly correlated and entangled beams.

To determine the quantum properties of the fields emitted by the OPO, we shall use the tools set up in the previous chapters in a different way from the situation where the modes contained only a few photons. Indeed, an OPO emitting a light intensity of mW order at around $1\,\mu\text{m}$ produces some 10^{16} photons per second, and no existing photodetector today would be fast enough to detect them separately. We are no longer under photon counting conditions but, rather, under 'continuous variable' conditions, where the photodetector produces an electric current proportional to the light flux it receives and which varies continuously. Quantum effects are then manifested through the fluctuations of the fields and their mutual correlations.

7A.4.1 The small quantum fluctuation limit

Consider a 'macroscopic' light beam of mW order. During the measurement time, which we may take to be around $100\,\mu\text{s}$ for concreteness, the number of photons arriving at the photodetector is $N = 10^{12}$. Imagine for the moment that it is in a coherent or quasi-classical state. The quantum fluctuations in the photon number around this value will be of the order of $\sqrt{N}$, or 10^6, a significant and measurable value, yet nevertheless much smaller than N itself. This is therefore a situation where *the quantum fluctuations are small compared with the average value*, and one we have already considered in Section 5.3.6, in order to be able to define a phase operator. To determine the dynamics of these fluctuations, it will be justified to consider them as infinitesimal, and treat the problem to lowest order, i.e. *linearize* the dynamical quantum equations around the average value. In particular, we shall express the complex field operator,

$$\hat{E}^{(+)} = \langle \hat{E} \rangle + \delta \hat{E}^{(+)}, \tag{7A.51}$$

where $\langle \hat{E} \rangle$ is the expected value of the field, determined using classical equations, like those established for the OPO in this complement, and $\delta \hat{E}^{(+)}$ is a *fluctuation operator* with zero expected value, which will be treated to first order.

We saw in Section 7A.3.2 that the propagation equations for the quantum fields in a parametric medium were the same as the classical equations. When they are solved to first order, we obtain analogous relations to (7A.11), up to the propagation phase factor:

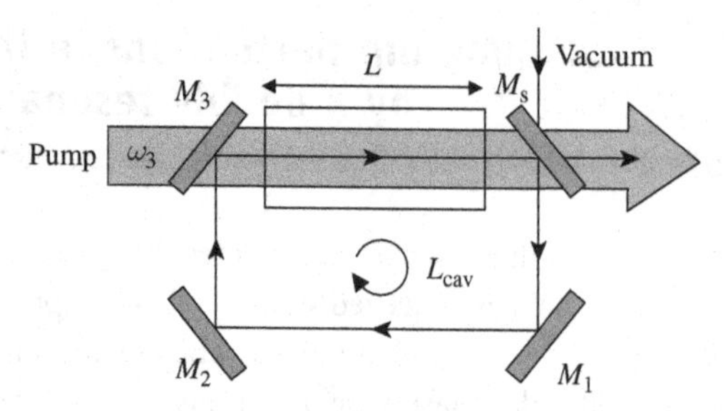

Figure 7A.2 **Diagram of the OPO showing the incoming and outgoing quantum fields.**

$$\hat{E}_1^{(+)}(L) = \left(\hat{E}_1^{(+)}(0) + \mathrm{i}g_1\hat{E}_3^{(+)}(0)\,[\hat{E}_2^{(+)}(0)]^\dagger\right)\mathrm{e}^{\mathrm{i}n_1\omega_1 L/c} \tag{7A.52}$$

$$\hat{E}_2^{(+)}(L) = \left(\hat{E}_2^{(+)}(0) + \mathrm{i}g_2\hat{E}_3^{(+)}(0)\,[\hat{E}_1^{(+)}(0)]^\dagger\right)\mathrm{e}^{\mathrm{i}n_2\omega_2 L_c/c}, \tag{7A.53}$$

where the coordinate origin along the direction of propagation of the light is taken at the point of entry into the crystal of length L. The propagation in the remaining part of the cavity in Figure 7A.2 involves a partial reflection at the output mirror and propagation through empty space, for both of which we know the quantum expression. For thesc operators, they are

$$\hat{E}_i^{(+)}(L_{\mathrm{cav}}) = \sqrt{R}\hat{E}_i^{(+)}(L)\,\mathrm{e}^{\mathrm{i}\frac{\omega_i}{c}(L_{\mathrm{cav}}-L)} + \sqrt{T}\mathrm{e}^{\mathrm{i}\psi_i}\hat{E}_{i,\mathrm{in}}^{(+)}, \tag{7A.54}$$

where $i = 1, 2$ and L_{cav} is the cavity length.

This differs from the classical relation (7A.14) through the term in $\hat{E}_{i,\mathrm{in}}^{(+)}$. Indeed, the output mirror M_S has to be treated like the semi-reflecting mirror of Section 5.1.2, taking into account the fact that it couples the intra-cavity signal and idler field operators with the field operators incident on the mirror and coming from the outside (see Figure 7A.2). The factor $\mathrm{e}^{\mathrm{i}\psi_i}$ is simply a phase factor due to the propagation of the field $\hat{E}_{i,\mathrm{in}}^{(+)}$ from the mirror to the crystal entry point. Combining (7A.51) and (7A.52) with $\hat{E}_i^{(+)}(L_{\mathrm{cav}}) = \hat{E}_i^{(+)}(0)$ for steady-state conditions, taking into account the fact that we assume the double resonance conditions to hold in the cavity for the signal and idler fields (see (7A.22)), and writing $\sqrt{R} \simeq 1 - T/2$, we obtain the following operator relations for the fields $\hat{E}_i^{(+)}(0) = \hat{E}_i^{(+)}$, $i = 1, 2, 3$:

$$\frac{T}{2}\hat{E}_1^{(+)} = \mathrm{i}g_1\hat{E}_3^{(+)}(\hat{E}_2^{(+)})^\dagger + \sqrt{T}\hat{E}_{1,\mathrm{in}}^{(+)}\,\mathrm{e}^{\mathrm{i}\psi_1} \tag{7A.55}$$

$$\frac{T}{2}\hat{E}_2^{(+)} = \mathrm{i}g_2\hat{E}_3^{(+)}(\hat{E}_1^{(+)})^\dagger + \sqrt{T}\hat{E}_{2,\mathrm{in}}^{(+)}\,\mathrm{e}^{\mathrm{i}\psi_2}. \tag{7A.56}$$

Linearizing with respect to the fluctuations, these imply

$$\frac{T}{2}\delta\hat{E}_1^{(+)} = \mathrm{i}g_1\delta\hat{E}_3^{(+)}\langle(\hat{E}_2^{(+)})^\dagger\rangle + \mathrm{i}g_1\langle\hat{E}_3^{(+)}\rangle(\delta\hat{E}_2^{(+)})^\dagger + \sqrt{T}\delta\hat{E}_{1,\mathrm{in}}^{(+)}\mathrm{e}^{\mathrm{i}\psi_1} \tag{7A.57}$$

$$\frac{T}{2}\delta\hat{E}_2^{(+)} = \mathrm{i}g_2\delta\hat{E}_3^{(+)}\langle(\hat{E}_1^{(+)})^\dagger\rangle + \mathrm{i}g_2\langle\hat{E}_3^{(+)}\rangle(\delta\hat{E}_1^{(+)})^\dagger + \sqrt{T}\delta\hat{E}_{2,\mathrm{in}}^{(+)}\mathrm{e}^{\mathrm{i}\psi_2}. \tag{7A.58}$$

Now if we wish to determine the fields leaving the OPO (and these are in fact the only quantities available for measurement), we must use the second coupling equation for the fields on the mirror M_S:

$$\hat{E}^{(+)}_{i,\text{out}} = -\sqrt{R}\hat{E}^{(+)}_{i,\text{in}} + \sqrt{T}\,\hat{E}^{(+)}_i(L)\mathrm{e}^{\mathrm{i}\theta_i}, \tag{7A.59}$$

where $i = 1, 2$, and where we have taken care to change the sign of the amplitude of the reflection factor from one side of the mirror to the other. Note that $\mathrm{e}^{\mathrm{i}\theta_i}$ is the propagation phase factor from the crystal exit to the output mirror. This relation simplifies because T, g_1 and g_2 are all small. This means that we may neglect all terms of second order in these quantities. So to a first approximation, we have

$$\hat{E}^{(+)}_{i,\text{out}} \simeq -\hat{E}^{(+)}_{i,\text{in}} + \sqrt{T}\,\hat{E}^{(+)}_i\mathrm{e}^{-\mathrm{i}\psi_i}, \tag{7A.60}$$

for $i = 1, 2$. In this relation, we have used the fact that the total phase ϕ_i for propagation through one round trip is $\phi_i = \psi_i + n_i\omega_i L/c + \theta_i$, and this is a multiple of 2π for a resonant cavity. In the following, we shall assume, in order to simplify the notation, that θ_i is also a multiple of 2π.

We shall not solve these equations for the general case here but, rather, give two examples of remarkable quantum phenomena that can be obtained with such a system.

7A.4.2 Frequency-degenerate OPO below threshold: producing squeezed states of the field

We now consider the degenerate case as discussed in Sections 7A.1.2 and 7A.3.3, where the signal and idler modes are the same. We must then take $\delta\hat{E}_1(0) \equiv \delta\hat{E}_2(0)$ in (7A.55). Furthermore, since we are below threshold, the expected value of the signal and idler fields is zero. Hence,

$$\frac{T}{2}\delta\hat{E}^{(+)}_1 = \mathrm{i}g_1\langle\hat{E}^{(+)}_3\rangle(\delta\hat{E}^{(+)}_1)^\dagger + \sqrt{T}\delta\hat{E}^{(+)}_{1,\text{in}}. \tag{7A.61}$$

In the rest of this section, we shall drop the indices 1 in the field expressions. By suitable choice of phase origin, we take the expected value of the pump field to be pure imaginary: $\langle\hat{E}^{(+)}_3(0)\rangle = -\mathrm{i}E_3$, with E_3 real. We then have the following solution for the fluctuations of the quadrature components at the crystal input:

$$\delta\hat{E}_Q = \sqrt{T}\frac{\delta\hat{E}_{Q,\text{in}}}{T/2 - g_1E_3} \tag{7A.62}$$

$$\delta\hat{E}_P = \sqrt{T}\frac{\delta\hat{E}_{P,\text{in}}}{T/2 + g_1E_3}. \tag{7A.63}$$

Using (7A.60), the fluctuations in the quadrature components of the outgoing signal field are found to be

$$\delta\hat{E}_{Q,\text{out}} = \frac{T/2 + g_1 E_3}{T/2 - g_1 E_3}\delta\hat{E}_{Q,\text{in}}, \tag{7A.64}$$

$$\delta\hat{E}_{P,\text{out}} = \frac{T/2 - g_1 E_3}{T/2 + g_1 E_3}\delta\hat{E}_{P,\text{in}}. \tag{7A.65}$$

These can be used to calculate the standard deviation of the fluctuations in the quadrature components of the outgoing field as a function of those of the incoming field, in this case, the vacuum:

$$\Delta E_{Q,\text{out}} = \frac{T/2 + g_1 E_3}{T/2 - g_1 E_3}\mathcal{E}^{(1)}, \tag{7A.66}$$

$$\Delta E_{P,\text{out}} = \frac{T/2 - g_1 E_3}{T/2 + g_1 E_3}\mathcal{E}^{(1)}, \tag{7A.67}$$

where $\mathcal{E}^{(1)}$ gives the standard deviation of the vacuum fluctuations at the frequency of the signal mode. Note that the fluctuations in the quadrature E_P are smaller than those of the vacuum, while the fluctuations in E_Q are greater than this value, in such a way as to conserve the product of the two at its minimal value allowed by the Heisenberg inequality (5.32). Below threshold, the degenerate OPO thus produces a *squeezed state*, in the same way as the degenerate parametric amplifier (see Section 7A.3.3). The difference with the case of simple propagation in the parametric crystal is that the quadrature E_P has a quantum noise that tends to zero when the pump field E_3 approaches the finite value $T/2g_1$, whereas an infinite pump power is required in the case of single passage. This finite value is none other than the *oscillation threshold* of the doubly resonant OPO (see (7A.26)), and it is thus easy to reach, even with low-power pump lasers.[3] Reductions in the variance up to 90% below the vacuum fluctuations are now obtained experimentally in this kind of device.

7A.4.3 Non-frequency-degenerate OPO above threshold: producing twin beams

Consider now a non-degenerate OPO. To simplify this analysis, we shall assume that the two frequencies ω_1 and ω_2 are different, but close enough to one another for the quantities g_1 and g_2 to be approximately equal:

$$g_1 \simeq g_2 = g. \tag{7A.68}$$

The OPO is taken above threshold, so that the pump field intensity is fixed by the oscillation condition (7A.26) at exact resonance, i.e. $|\mathcal{E}_3|^2 = T^2/4g^2$ with the approximations made

[3] L. A. Wu, H. J. Kimble, J. Hall and H. Wu, Generation of Squeezed States by Parametric Down Conversion, *Physical Review Letters* **57**, 2520 (1986).

here. By suitable choice of phase origin, we shall take $\langle \hat{E}_3(0)\rangle = \mathcal{E}_3 = -\mathrm{i}T/2g$. Equations (7A.57) then become,

$$\frac{T}{2}\left(\delta\hat{E}_1^{(+)} - (\delta\hat{E}_2^{(+)})^\dagger\right) = \mathrm{i}g\delta\hat{E}_3^{(+)}\langle(\hat{E}_2^{(+)})^\dagger\rangle + \sqrt{T}\delta\hat{E}_{1,\mathrm{in}}^{(+)}, \tag{7A.69}$$

$$\frac{T}{2}\left(\delta\hat{E}_2^{(+)} - (\delta\hat{E}_1^{(+)})^\dagger\right) = \mathrm{i}g\delta\hat{E}_3^{(+)}\langle(\hat{E}_1^{(+)})^\dagger\rangle + \sqrt{T}\delta\hat{E}_{2,\mathrm{in}}^{(+)}. \tag{7A.70}$$

Moreover, above threshold, the expected values of the signal and idler fields are non-zero, and related by the following equation derived from (7A.25):

$$\frac{T}{2}\langle\hat{E}_1^{(+)}\rangle = \mathrm{i}g\langle\hat{E}_3^{(+)}\rangle\langle(\hat{E}_2^{(+)})^\dagger\rangle = \frac{T}{2}\langle\hat{E}_2^{(+)}\rangle^*. \tag{7A.71}$$

The possibility of changing the phase of the signal and idler fields as discussed in Section 7A.2.3 means that we may simplify by choosing the solution where $\langle\hat{E}_1(0)\rangle$ and $\langle\hat{E}_2(0)\rangle$ are real, hence equal according to (7A.71). Subtracting one equation from the other, we may then eliminate the term depending on the fluctuations of the pump field in favour of the quadrature operators of the signal and idler fields. This yields

$$\delta\hat{E}_{Q1} - \delta\hat{E}_{Q2} = \frac{2}{\sqrt{T}}\left(\delta\hat{E}_{1,\mathrm{in}}^{(+)} - \delta\hat{E}_{2,\mathrm{in}}^{(+)}\right). \tag{7A.72}$$

The left-hand side of this equation is Hermitian, and hence also equal to the average of the right-hand side and its Hermitian conjugate. It follows that

$$\begin{aligned}\delta\hat{E}_{Q1} - \delta\hat{E}_{Q2} &= \frac{1}{\sqrt{T}}\left(\delta\hat{E}_{1,\mathrm{in}}^{(+)} + (\delta\hat{E}_{1,\mathrm{in}}^{(+)})^\dagger - \delta\hat{E}_{2,\mathrm{in}}^{(+)} - (\delta\hat{E}_{2,\mathrm{in}}^{(+)})^\dagger\right)\\ &= \frac{1}{\sqrt{T}}\left(\delta\hat{E}_{Q1,\mathrm{in}} - \delta\hat{E}_{Q2,\mathrm{in}}\right).\end{aligned} \tag{7A.73}$$

If we use relation (7A.59) at the output mirror to calculate the same combination for the outgoing fields, we finally obtain

$$\delta\hat{E}_{Q1,\mathrm{out}} - \delta\hat{E}_{Q2,\mathrm{out}} = 0. \tag{7A.74}$$

The quantum fluctuations in this quantity thus cancel completely, and we shall show that it is in fact proportional to the fluctuations in the difference in intensity between the outgoing signal and idler beams. Indeed, the intensity of the signal field $I_{1,\mathrm{out}}$, for example, is given by the expected value of the observable $(\hat{E}_{1,\mathrm{out}}^{(+)})^\dagger\hat{E}_{1,\mathrm{out}}^{(+)}$. Using the linearized fluctuation approximation, the intensity fluctuation operator is therefore given by

$$\delta\hat{I}_{1,\mathrm{out}} = \langle(\hat{E}_{1,\mathrm{out}}^{(+)})^\dagger\rangle\delta\hat{E}_{1,\mathrm{out}}^{(+)} + (\delta\hat{E}_{1,\mathrm{out}}^{(+)})^\dagger\langle\hat{E}_{1,\mathrm{out}}^{(+)}\rangle. \tag{7A.75}$$

The expected value $\langle\hat{E}_{1,\mathrm{out}}^{(+)}\rangle$ is equal to $\sqrt{T}\langle\hat{E}_1^{(+)}\rangle$, which is real. Hence,

$$\delta\hat{I}_1^{\mathrm{out}} = \sqrt{T}\langle\hat{E}_1^{(+)}\rangle\left(\delta\hat{E}_{1,\mathrm{out}}^{(+)} + (\delta\hat{E}_{1,\mathrm{out}}^{(+)})^\dagger\right) = \sqrt{T}\langle\hat{E}_1^{(+)}\rangle\delta\hat{E}_{Q1,\mathrm{out}}. \tag{7A.76}$$

So we do indeed have $\delta\hat{I}_{1,\mathrm{out}} - \delta\hat{I}_{2,\mathrm{out}} = 0$. Even if the intensities of the signal and idler beams are affected by non-zero fluctuations, these fluctuations are perfectly correlated, hence equal at any moment of time, in such a way that they cancel exactly when one is subtracted from the other. The beams leaving the OPO are twin beams. It can be shown that the resulting quantum state is a twin-photon state of the kind studied in Section 5D.2.

When two beams of the same intensity are formed by dividing an incident beam into two equal parts using a semi-reflecting mirror, we do not obtain this type of twin beam. In this case, the expected values are equal, but the fluctuations in the two output channels are not correlated. A simple calculation using (5.8) and (5.9) relating the fields on a semi-reflecting mirror shows that, in this case,

$$\Delta(I_{1,\text{out}} - I_{2,\text{out}})^2 = \left[\mathcal{E}^{(1)}\right]^2 (\langle I_{1,\text{out}}\rangle + \langle I_{2,\text{out}}\rangle). \tag{7A.77}$$

The fact that the OPO produces twin beams, in contrast to the semi-reflecting mirror, may seem natural enough in retrospect, since the parametric emission phenomenon gives rise to *twin photons* in the cavity, and these eventually leave the cavity and contribute to creating identical intensity fluctuations in these two beams. The semi-reflecting mirror for its part does not 'cut the incident photons in two', but rather directs them randomly, and with equal probabilities, into its two output channels. It is this random process that underlies the noise given in (7A.77).

This reduction in fluctuations in the above-threshold, non-degenerate OPO has been observed experimentally.[4] Because there are non-zero losses, which withdraw photons in the signal and idler modes in a non-correlated way, the fluctuations do not cancel perfectly in the difference. Reductions in the fluctuations of this quantity by as much as 90%, compared with the value given in (7A.77), have been obtained and used to reduce the quantum fluctuations in *differential measurements*, e.g. in the measurement of very low absorptions. For example, the absorbing medium intercepts the signal beam and its presence causes the appearance of a signal in the intensity difference between the two beams.

[4] A. Heidmann, R. Horowicz, S. Reynaud, E. Giacobino, C. Fabre and G. Camy, Observation of Quantum Noise Reduction on Twin Laser Beams, *Physical Review Letters* **59**, 2555 (1987).

Complement 7B Nonlinear optics in optical Kerr media

In this complement we discuss several examples of optical phenomena in media where *the refractive index depends nonlinearly on the intensity*, known as optical Kerr media. This nonlinear effect exists in all materials, even isotropic ones, like glass or fused silica, but it is particularly marked in certain physical systems to be exemplified in Section 7B.1. After investigating the propagation of light through such media in Section 7B.2, we shall discuss three applications of the optical Kerr effect (which can be studied in any order). We begin by describing a *bistable optical system*, when this nonlinear medium is inserted in a Fabry–Perot cavity (Section 7B.3). We then study *phase conjugate mirrors* and examine their potential applications in *adaptive optics* (Section 7B.4). Finally, we discuss certain effects occurring during the propagation of an isolated wave, bounded either transversely or temporally, in a Kerr medium, and describe *self-focusing effects* (Section 7B.5) and *self-phase-modulation* effects (Section 7B.6). In particular, we shall show that nonlinear effects and dispersion effects can compensate to produce stable structures known as *solitons*, which maintain their shape during propagation.

7B.1 Examples of third-order nonlinearities

7B.1.1 Nonlinear response of two-level atoms

We begin by studying a simple case of a nonlinear interaction, namely a two-level quantum system under the effects of a plane wave. Because the calculations are straightforward, this model will allow us to identify certain key features of these phenomena that can then be generalized to more complicated systems. Note that the field is treated here as a classical variable.

We have seen in Chapter 2 that, when an electric field is applied to an ensemble of atoms or molecules, displacements of the electric charges will create an *induced electric dipole moment*. For weak electric fields, and assuming the medium to be isotropic, these induced dipoles will be proportional to the electric field and aligned with it. However, as the electric field strength is increased, the dipoles cease to grow in linear proportion to the applied field and saturation occurs. In Chapter 2 it was shown that, for two-level atoms, the average dipole moment per unit volume (or *polarization*) is related to the incident electric field by a relation of the form,

$$\mathbf{P}^{(+)}(\mathbf{r},t) = \varepsilon_0 \chi\, \mathbf{E}^{(+)}(\mathbf{r},t), \tag{7B.1}$$

where $\mathbf{P}^{(+)}$ and $\mathbf{E}^{(+)}$ are the complex polarization and field at the point $\mathbf{r}$, and χ is the complex susceptibility at the same point, i.e.

$$\chi = \chi' + i\chi''. \tag{7B.2}$$

The susceptibility χ is found by seeking the steady-state solution of the optical Bloch equations. In the case of a monochromatic field $\mathbf{E}^{(+)}(\mathbf{r})\mathrm{e}^{-i\omega t}$ and purely radiative relaxation by spontaneous emission, we then obtain (see (2.188)),

$$\chi' = \frac{N}{V}\frac{d^2}{\varepsilon_0\hbar}\frac{\omega_0 - \omega}{\frac{\Gamma_{\mathrm{sp}}^2}{4} + \frac{\Omega_1^2}{2} + (\omega_0 - \omega)^2} \tag{7B.3}$$

$$\chi'' = \frac{N}{V}\frac{d^2}{\varepsilon_0\hbar}\frac{\frac{\Gamma_{\mathrm{sp}}}{2}}{\frac{\Gamma_{\mathrm{sp}}^2}{4} + \frac{\Omega_1^2}{2} + (\omega_0 - \omega)^2}. \tag{7B.4}$$

In these relations, N/V is the number of atoms per unit volume near the point $\mathbf{r}$, d is the matrix element of the electric dipole between the two levels, $\omega - \omega_0$ is the detuning from resonance, Γ_{sp} is the reciprocal of the radiative lifetime of the excited level, and $\Omega_1 = -dE_0/\hbar$ is the Rabi angular frequency at resonance. Saturation of the optical dipole shows up through the presence of the term $(\Omega_1)^2$ in the denominators of (7B.3–7B.4), this being proportional to the light intensity at point $\mathbf{r}$. Recall that (7B.3–7B.4) were deduced in the *quasi-resonance approximation*, i.e. assuming $|\omega_0 - \omega| \ll \omega_0$. We shall assume here that this condition is satisfied. Furthermore, we shall only consider situations in which dispersive effects, related to χ', dominate over absorption effects, related to χ'', which will be valid far enough away from resonance, when

$$|\omega_0 - \omega| \gg \Gamma_{\mathrm{sp}}. \tag{7B.5}$$

When (7B.5) is satisfied, χ'' is very small compared with χ', and Γ_{sp} can be neglected in the denominator of the formula giving χ', whence,

$$\chi' = \frac{N}{V}\frac{d^2}{\varepsilon_0\hbar}\frac{\omega_0 - \omega}{\frac{\Omega_1^2}{2} + (\omega_0 - \omega)^2}. \tag{7B.6}$$

If in addition the detuning $|\omega_0 - \omega|$ is much bigger than the Rabi angular frequency Ω_1 at resonance, then one can carry out a perturbative expansion of the formula in (7B.6) in powers of $\Omega_1/|\omega_0 - \omega|$. Keeping only the first two terms in the expansion, we have

$$\chi' = \chi_1' + \chi_3' I + \ldots, \tag{7B.7}$$

where χ_1' is the linear susceptibility, χ_3' is the third-order nonlinear susceptibility, and here $I = \frac{1}{2}E_0^2$ is the intensity of the incident field. Equations (7B.6) and (7B.7) then imply the following values for χ_1' and χ_3':

$$\chi_1' = \frac{N}{V} \frac{d^2}{\varepsilon_0 \hbar(\omega_0 - \omega)} \tag{7B.8}$$

$$\chi_3' = -\frac{N}{V} \frac{d^4}{\varepsilon_0 \hbar^3(\omega_0 - \omega)^3}. \tag{7B.9}$$

The quantities χ_1' and χ_3' have *opposite* signs, as one would expect for a *saturation effect*. Note that the sign of χ_3' is different on either side of the resonance. We shall see in Section 7B.4 that very different physical effects occur depending on the sign of χ_3'. These effects can thus be explored by considering the system on one side of the resonance or the other.

If the polarization $\mathbf{P}$ is decomposed into a linear part $\mathbf{P}_L$ and a nonlinear part $\mathbf{P}_{NL}$ as in Section 7.2, then using (7B.7) and complex notation, we find,

$$\mathbf{P}_L^{(+)} = \varepsilon_0 \chi_1' \, \mathbf{E}^{(+)} \tag{7B.10}$$

$$\mathbf{P}_{NL}^{(+)} = \varepsilon_0 \chi_3' \, I \, \mathbf{E}^{(+)}, \tag{7B.11}$$

with $I = \dfrac{1}{2} E_0^2 = 2|\mathbf{E}^{(+)}|^2$ independent of time for a monochromatic field.

Comments

(i) When the relaxation is not purely radiative, we have

$$\chi_3' = -\frac{2\Gamma_D}{\Gamma_{sp}} \frac{N}{V} \frac{d^4}{\varepsilon_0 \hbar^3(\omega_0 - \omega)^3} \tag{7B.12}$$

where Γ_D is the dipole relaxation rate.

(ii) Equation (7B.11) relating $\mathbf{E}^{(+)}(\mathbf{r}, t)$ and $\mathbf{P}_{NL}^{(+)}(\mathbf{r}, t)$ depends in principle on the chosen field point, owing to spatial variations in the light intensity $I(\mathbf{r})$. The coefficients χ_1' and χ_3' may also depend on $\mathbf{r}$ if the density N/V is not constant.

7B.1.2 Nonlinearity by optical pumping

Optical pumping, as described in Complement 2B, also leads to a nonlinear susceptibility under steady-state conditions.

Consider, for example, a transition linking a ground state of angular momentum $J_a=1/2$ with an excited level of angular momentum $J_b=1/2$ (see Figure 7B.1). Under the action of a wave with polarization σ_+ and amplitude E_+ (see Section 2B.3.1 of Complement 2B), the atoms are optically pumped from level $m_a = -1/2$ to level $m_a = +1/2$. This can be expressed mathematically by the following pumping equations for the populations N_- and N_+ of the Zeeman sub-levels $m_a = -1/2$ and $m_a = +1/2$ of the ground state:

$$\left(\frac{dN_-}{dt}\right)_{\text{pump}} = -K_p E_+^2 N_- \tag{7B.13}$$

$$\left(\frac{dN_+}{dt}\right)_{\text{pump}} = K_p E_+^2 N_-, \tag{7B.14}$$

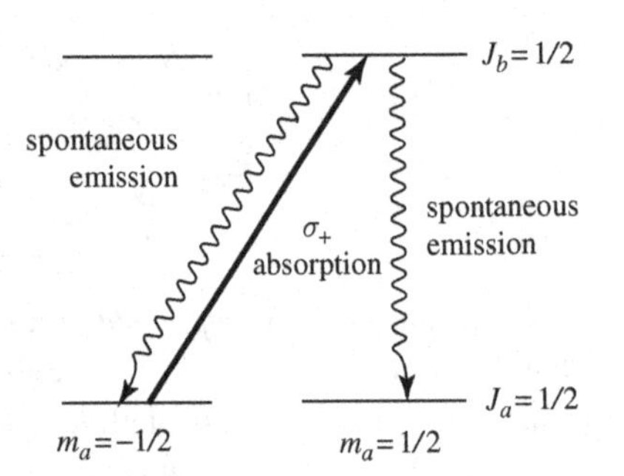

Figure 7B.1 Population transfer from the sub-level $m_a = -1/2$ to the sub-level $m_a = +1/2$ under the action of a wave with circular polarization σ_+.

where K_p is given by the following expression, valid for a non-resonant excitation:

$$K_p = \frac{\Gamma_{\text{sp}}}{12} \frac{d^2}{\hbar^2(\omega_0 - \omega)^2}, \tag{7B.15}$$

with $d = \langle b, +1/2|\hat{\mathbf{D}} \cdot \boldsymbol{\varepsilon}_+|a, -1/2\rangle$. This expression is obtained by a similar method to the one presented in Section 2B.3.2.[1]

The optical pumping has to compete with *relaxation phenomena* in the medium, e.g. collisions between atoms, collisions with the walls, fluctuating magnetic field, which tend to equalize the populations of the Zeeman sub-levels. The effect of this relaxation is described via damping terms in the Bloch equations. If the relaxation rate of the coherence is very fast, this leads to rate equations for the populations of the different levels (see Section 2C.5.1 of Complement 2C). In particular, the time dependence of the population difference under the effect of relaxation will satisfy

$$\left[\frac{\mathrm{d}(N_+ - N_-)}{\mathrm{d}t}\right]_{\text{relax}} = -\gamma_{\text{R}}(N_+ - N_-), \tag{7B.16}$$

where γ_{R} is the relaxation rate of the populations (assumed the same in the two levels). In the steady-state regime, optical pumping and relaxation effects balance one another, which implies the equation,

$$\left[\frac{\mathrm{d}(N_+ - N_-)}{\mathrm{d}t}\right]_{\text{pump}} + \left[\frac{\mathrm{d}(N_+ - N_-)}{\mathrm{d}t}\right]_{\text{relax}} = 2K_pE_+^2N_- - \gamma_R(N_+ - N_-) = 0. \tag{7B.17}$$

When the transition coupling the ground state and excited state is not saturated (negligible population in the excited levels), using the notation $N = N_+ + N_-$ for the total population, we find

[1] The numerical coefficient appearing in (7B.15) comes from the value of the Clebsch–Gordan coefficients for the transition $1/2 \to 1/2$. For another transition liable to give rise to optical pumping, the numerical coefficient would generally be different, but the functional dependence on the physical parameters, e.g. detuning etc., would be the same.

$$N_- = \frac{N}{2\left(1 + \frac{K_p E_+^2}{\gamma_R}\right)}. \tag{7B.18}$$

According to (7B.8), the susceptibility of the medium for a monochromatic incident wave with polarization σ_+ is equal to

$$\chi' = \frac{N_-}{V} \frac{d^2}{\varepsilon_0 \hbar(\omega_0 - \omega)}, \tag{7B.19}$$

whence, by (7B.18),

$$\chi' = \frac{N}{2V} \frac{d^2}{\varepsilon_0 \hbar(\omega_0 - \omega)} \frac{1}{1 + \frac{K_p E_+^2}{\gamma_R}}. \tag{7B.20}$$

We recover a susceptibility depending on the intensity, but the origin of the nonlinearity is associated here with population transfers between Zeeman sub-levels of the lower level involved in the transition. In practice, these nonlinearities due to optical pumping are observed with intensities *well below* those required to saturate an atomic transition. On the other hand, they are slower to get established.

Apart from the situations we have just described, there are many other processes that can lead to an intensity-dependent susceptibility. We could mention effects related to the orientation of anisotropic polarizable molecules under the effect of the electric field of the applied wave, multiphoton effects, necessarily nonlinear or the spectacular photorefractive effects related to charge transport in electro-optical media.[2]

7B.2 Field propagation in Kerr media

We shall use the results of Section 7.2, and in particular the propagation equation (7.24). Equation (7B.11) for the nonlinear polarization shows that it only oscillates at frequency ω of the incident wave. In contrast to second-order nonlinear media, there are no waves with new frequencies in the nonlinear interaction. The optical Kerr effect modifies the propagation of the incident wave, or produces waves going in different directions, but at the same frequency.

7B.2.1 Single incident wave

Suppose to begin with that the incident wave is a single plane wave propagating in the direction Oz. In the slowly varying wave approximation and using (7B.10) and (7.24), the dynamical equation for its envelope $\mathcal{E}(z)$ is

$$\frac{d\mathcal{E}}{dz} = i\frac{\omega}{2n_0 c}\chi_3' I\mathcal{E}, \tag{7B.21}$$

[2] For all these effects, see R. W. Boyd, *Nonlinear Optics*, Academic Press (2008).

where $I = 2|\mathcal{E}|^2$. Since χ'_3 is real, it follows immediately from this equation that

$$\frac{\mathrm{d}I}{\mathrm{d}z} = 0. \tag{7B.22}$$

The propagation of a plane wave in a Kerr medium thus occurs at constant intensity. If I_0 is the initial intensity of the wave, we have

$$\mathcal{E}(z) = \sqrt{\frac{I_0}{2}}\,\mathrm{e}^{\mathrm{i}\varphi(z)}. \tag{7B.23}$$

Equation (7B.21) implies the following expression for $\varphi(z)$:

$$\varphi(z) = \varphi_0 + \frac{\omega}{2n_0 c}\chi'_3 I_0 z, \tag{7B.24}$$

and also the equation for the wave propagating in the medium:

$$E^{(+)}(z) = E^{(+)}(0)\,\mathrm{e}^{\mathrm{i}\omega/c(n_0+\chi'_3 I_0/2n_0)z}. \tag{7B.25}$$

It is just as though the medium had refractive index n given by

$$n = n_0 + n_2 I_0, \tag{7B.26}$$

with

$$n_2 = \frac{\chi'_3}{2n_0}. \tag{7B.27}$$

The term $n_2 I_0$, called the 'nonlinear index', thus corrects the usual index by an amount proportional to the intensity of the incident field. This is called the *optical Kerr effect*, because it is analogous to the usual Kerr effect in the case of oscillating electric fields, i.e. a modification of the index of a medium under the effect of an electrostatic field. We shall see in the following some effects arising due to the dependence of the index on the local power of the wave.

The relative magnitude of the nonlinear term $n_2 I_0$ and the linear term n_0 depends a lot on the material (and the light intensity). It may be as much as 10^{-2} for a gaseous medium in the immediate vicinity of the resonance and for a continuous-wave laser of moderate power, but it is generally much smaller. For instance, n_2 is equal to 2.7×10^{-20} m^2.W^{-1} for silica. This is a rather low value compared with other media, such as semiconductors, but due to the very low absorption by silica (0.2 dB.km^{-1}, or 5×10^{-5} m^{-1}), a significant nonlinear phase offset can be obtained if the light propagates in a silica optical fibre several kilometres long.

7B.2.2 Two travelling waves propagating in opposite directions

Consider now the case of two waves of the same frequency travelling in opposite directions along the Oz-axis. The complex electric field is given by

$$E^{(+)}(\mathbf{r}, t) = \left(\mathcal{E}_1\,\mathrm{e}^{\mathrm{i}kz} + \mathcal{E}_2\,\mathrm{e}^{-\mathrm{i}kz}\right)\mathrm{e}^{-\mathrm{i}\omega t}. \tag{7B.28}$$

The nonlinear polarization is then given by

$$P^{(+)}_{\mathrm{NL}}(\mathbf{r},t) = \varepsilon_0 \chi'_3 \left|\mathcal{E}_1 \mathrm{e}^{ikz} + \mathcal{E}_2 \mathrm{e}^{-ikz}\right|^2 \left(\mathcal{E}_1 \mathrm{e}^{ikz} + \mathcal{E}_2 \mathrm{e}^{-ikz}\right) \mathrm{e}^{-i\omega t}. \quad (7B.29)$$

There are terms in $\mathrm{e}^{\pm ikz}$ whose phase is matched with the incident waves and which will significantly alter their amplitude. Let $P^{(1)}_{\mathrm{NL}}$ be the term in e^{ikz} in $P^{(+)}_{\mathrm{NL}}$. It is given by

$$P^{(1)}_{\mathrm{NL}}(\mathbf{r}) = \varepsilon_0 \chi'_3 \left[|\mathcal{E}_1|^2 + 2|\mathcal{E}_2|^2\right] \mathcal{E}_1 \, \mathrm{e}^{ikz}. \quad (7B.30)$$

The second term in this expression is responsible for a nonlinear coupling between the two counter-propagating waves called the *cross-Kerr effect*. Its nonlinear coefficient is twice that of the self-Kerr effect, and it leads to a nonlinear index for wave (1):

$$n^{(1)} = n_0 + n_2(I_1 + 2I_2). \quad (7B.31)$$

Comment Note that the factor of 2 between the cross- and 'direct' Kerr effects appearing in (7B.31) is not universal. Even for two-level atoms, a different factor (between 1 and 2) can be found for a relaxation that is not purely radiative, and when the motion of the atoms is taken into account.[3] In particular, if averaging processes due to the motion are significant, it can be shown that the formula for $n^{(1)}$ becomes symmetric in I_1 and I_2:

$$n^{(1)} = n_0 + n_2(I_1 + I_2). \quad (7B.32)$$

7B.3 Optical bistability

Consider a Fabry–Perot cavity (see Complement 3A) with identical input and output mirrors, each with transmission coefficient T, containing a Kerr medium of length ℓ. To avoid complications due to the cross-Kerr effect, we shall assume the cavity to be a ring of total length L_{cav}, so that a single wave (treated as a plane wave) crosses the medium (see Figure 7B.2).

As we saw in Complement 3A, the transmission T_{cav} of this cavity has resonant peaks when the optical length L_{cav} lies near values of L that are integer multiples $p\lambda$ of the wavelength λ of the incident field, according to the formula derived from (3A.12) and (3A.34):

$$T_{\mathrm{cav}} = \frac{1}{1 + \frac{4\mathcal{F}^2}{\pi^2} \sin^2 \frac{kL_{\mathrm{cav}}}{2}}, \quad (7B.33)$$

where $\mathcal{F}$ is the finesse of the cavity. The dependence of T_{cav} on L_{cav} is given by the resonance curve (Airy function) of Figure 7B.3.

[3] See G. Grynberg and M. Pinard, Inelastic and Adiabatic Contributions to Atomic Polarizability, *Physical Review A* **32**, 3772 (1985).

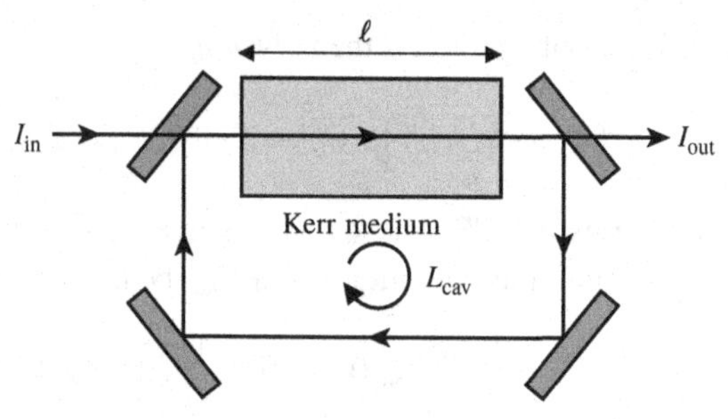

Figure 7B.2 Fabry–Perot ring cavity containing a Kerr medium.

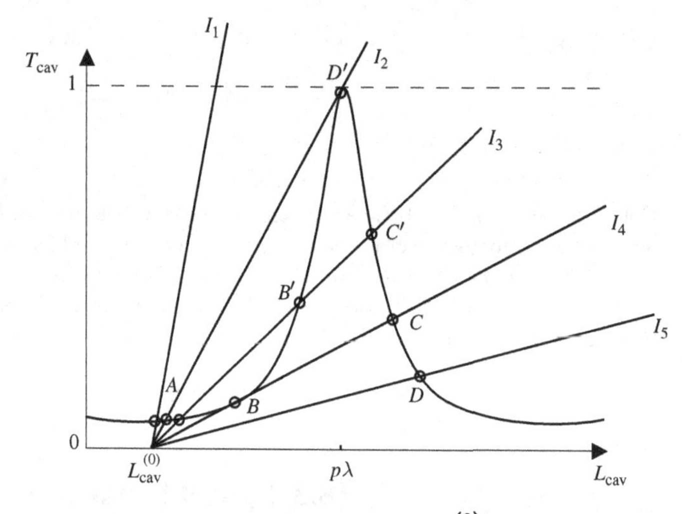

Figure 7B.3 Cavity transmission for given geometry (determined by $L_{cav}^{(0)}$) and intensity. The operating point of the device lies at the intersection of the straight line and the resonance curve of the Fabry–Perot cavity. One solution is found when the intensity of the incident wave is less than I_2 or greater than I_4, and three solutions when it lies between these values.

The optical length L_{cav} of the cavity containing the Kerr medium is

$$L_{cav} = L + (n_0 - 1)\ell + n_2 I_{cav}\ell, \tag{7B.34}$$

where L is the geometric length and I_{cav} is the light intensity in the cavity. We may then deduce the following expression for the transmission coefficient T_{cav} of the cavity containing the Kerr medium:

$$T_{cav} = \frac{I_{out}}{I_{in}} = T\frac{I_{cav}}{I_{in}} = \frac{T}{n_2\ell}\frac{L_{cav} - L_{cav}^{(0)}}{I_{in}}, \tag{7B.35}$$

where $L_{cav}^{(0)} = L + (n_0 - 1)\ell$ is the optical length of the cavity when $I_{cav} \approx 0$. This linear relation between T_{cav} and L_{cav} is shown in Figure 7B.3 for different values of I_{in}. The operating point of the device thus lies at the intersection of the curve (7B.33) and the straight line (7B.35).

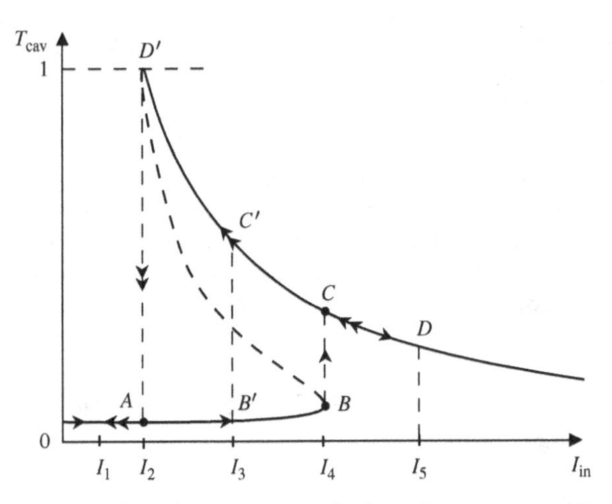

Figure 7B.4 Transmission as a function of incident intensity. The branch *BD* corresponds to an unstable solution. *Bistability* is observed between I_2 and I_4, and this gives rise to the hysteresis curve *ABCD'*.

Consider the case illustrated in Figure 7B.3, where L is not a multiple of $p\lambda$. At very low incident intensities, the straight line of (7B.35) is very steep ($I_{in} = I_1$ in the figure), and the transmission T_{cav} barely differs from $T_{cav}(L_{cav}^{(0)})$, i.e. it is very low in value. When I_{in} reaches the value I_2, where the straight line is tangent to the Airy curve, a second solution appears, with transmission close to unity. Then, as I_{in} continues to increase, there are three possible solutions for the system of two equations ($I_{in} = I_3$ in the figure). Finally, when I_{in} is greater than I_4, corresponding to the second situation where the straight line is tangent to the Airy function, there is once again only one solution ($I_{in} = I_5$).

Figure 7B.4 gives the values of T_{cav} as a function of I_{in}. It can be shown that the part of the curve with negative gradient (between B and D') corresponds to an unstable configuration of the system, which cannot therefore be observed experimentally. For $I_2 < I_{in} < I_4$, *two stable solutions are thus simultaneously available to the system*, whence the name *optical bistability* attributed to this phenomenon.

The solution actually observed will depend on *the past history of the system*. Suppose that the incident intensity is initially very low and increases gradually. The system will move continuously along the lower branch of the bistability curve until the intensity I_{in} goes above the value I_4. The system is then compelled to jump to the upper branch, whence the operating point moves suddenly from B to C (see Figure 7B.4).

When the incident intensity decreases from the operating point C, the transmitted intensity moves gradually along the upper branch of the bistability curve until I_{in} reaches the value I_2, whereupon the system has no option but to jump from the operating point D' to the point A.

The Fabry–Perot cavity can thus be switched from low to high transmission by pushing I_{in} temporarily above the value I_4, and from high to low transmission by reducing I_{in} temporarily below I_2. If I_{in} is then held at some value between I_2 and I_4, it will remain in the given state. The device thus behaves like an *optical memory*, capable of storing one bit of information.

Comment

Would it be possible to build an 'all-optical computer' using this type of element, in which light signals would replace electrical currents? This question was actively investigated in the 1990s, the main advantage being the possibility of very short switching times. However, it never led to any large-scale development, owing to the considerable technical obstacles, e.g. high switching energy and difficulty miniaturizing the optical components and integrating them on a large scale.

7B.4 Phase conjugate mirror

Nonlinear optics provides a way of making mirrors for which the reflection laws are quite different from those of Descartes. In this section, we shall see that the phase conjugate mirror reflects a light ray in exactly the opposite direction to the incident ray, rather than in the symmetric direction relative to the normal to the mirror, as would happen for ordinary reflection governed by the Snell–Descartes law.

7B.4.1 Degenerate four-wave mixing

Consider a Kerr medium, interacting with three incident waves of amplitudes $\mathcal{E}_\mathrm{p}$, $\mathcal{E}'_\mathrm{p}$ and $\mathcal{E}_\mathrm{s}$ and the same angular frequency ω (see Figure 7B.5). Waves $\mathcal{E}_\mathrm{p}$ and $\mathcal{E}'_\mathrm{p}$, which propagate in *opposite directions*, have much bigger amplitude than wave $\mathcal{E}_\mathrm{s}$, which propagates in the Oz direction. $\mathcal{E}_\mathrm{p}$ and $\mathcal{E}'_\mathrm{p}$ will be called the pump waves and $\mathcal{E}_\mathrm{s}$ the signal (or probe) wave.

We assume that the three incident electric fields have the same polarization $\boldsymbol{\varepsilon}$ perpendicular to the plane of the figure. The total complex field is then,

$$\mathbf{E}^{(+)}(\mathbf{r},t) = \boldsymbol{\varepsilon}\left[\mathcal{E}_\mathrm{p}\mathrm{e}^{\mathrm{i}\mathbf{k}\cdot\mathbf{r}} + \mathcal{E}'_\mathrm{p}\mathrm{e}^{-\mathrm{i}\mathbf{k}\cdot\mathbf{r}} + \mathcal{E}_\mathrm{s}\mathrm{e}^{\mathrm{i}kz}\right]\mathrm{e}^{-\mathrm{i}\omega t}. \tag{7B.36}$$

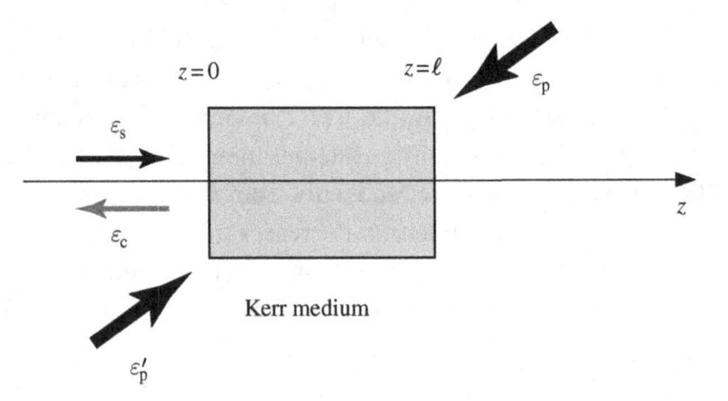

Figure 7B.5 Arrangement of incident waves in a phase conjugation experiment by four-wave mixing. A so-called conjugate wave $\mathcal{E}_\mathrm{c}$ is produced, propagating in the opposite direction to $\mathcal{E}_\mathrm{s}$.

The third-order nonlinear polarization is then given at any point of the nonlinear medium by:

$$\mathbf{P}_{\mathrm{NL}}^{(+)}(\mathbf{r}) = 2\varepsilon_0 \chi_3' |E^{(+)}(\mathbf{r}, t)|^2 \, E^{(+)}(\mathbf{r}, t). \tag{7B.37}$$

It involves many terms with a range of spatial dependences, each of which will serve as a source term for the propagation of waves in the medium. There is one term in particular varying like e^{-ikz}, denoted by $\mathbf{P}_{\mathrm{c}}^{(+)}$, given by

$$\mathbf{P}_{\mathrm{c}}^{(+)}(\mathbf{r}, t) = 4\varepsilon_0 \chi_3' \, \mathcal{E}_{\mathrm{p}} \mathcal{E}_{\mathrm{p}}' \mathcal{E}_{\mathrm{s}}^* \, \mathrm{e}^{-ikz} \, \mathrm{e}^{-\mathrm{i}\omega t}. \tag{7B.38}$$

This term is phase matched with a wave $\mathcal{E}_{\mathrm{c}}(z)\mathrm{e}^{-ikz}\,\mathrm{e}^{-\mathrm{i}\omega t}$ in the medium called the 'conjugate wave', whose amplitude satisfies

$$\frac{\mathrm{d}\mathcal{E}_{\mathrm{c}}}{\mathrm{d}z} = -\mathrm{i}\frac{2\omega}{n_0 c} \, \chi_3' \, \mathcal{E}_{\mathrm{p}} \mathcal{E}_{\mathrm{p}}' \mathcal{E}_{\mathrm{s}}^*. \tag{7B.39}$$

The minus sign comes from the fact that $k = -n_0 \dfrac{\omega}{c}$ for the conjugate wave in (7.24). The process we have just identified involves four waves of the same frequency. This is why we speak of *degenerate four-wave mixing*. Note also that the equation of propagation involves $\mathcal{E}_{\mathrm{s}}^*$ rather than $\mathcal{E}_{\mathrm{s}}$, whence the term *phase conjugation* used to describe such a situation. This will be considered in more detail in the next section.

The appearance of a wave propagating in the opposite direction to the incident wave can be understood qualitatively in the following way. The superposition of the pump wave $\mathcal{E}_{\mathrm{p}}$ and the wave $\mathcal{E}_{\mathrm{s}}$ creates a stationary wave structure with intensity modulation in the direction parallel to $k\mathbf{e}_z - \mathbf{k}$. This intensity modulation in the Kerr medium leads to an index modulation, which then behaves like a thick periodic grating. The diffraction of the second pump wave $\mathcal{E}_{\mathrm{p}}'$, with wavevector $-\mathbf{k}$, on this grating leads to the appearance of a wave with wavevector $-\mathbf{k} - (k\mathbf{e}_z - \mathbf{k}) = -k\mathbf{e}_z$.

7B.4.2 Phase conjugation

We write the complex signal field in the form

$$E_{\mathrm{s}}^{(+)}(z, t) = \mathcal{E}_{\mathrm{s}}(z)\,\mathrm{e}^{-\mathrm{i}\omega t} = A\,\mathrm{e}^{\mathrm{i}\varphi_{\mathrm{s}}}\,\mathrm{e}^{ikz}\,\mathrm{e}^{-\mathrm{i}\omega t}. \tag{7B.40}$$

The corresponding real field,

$$E_{\mathrm{s}}(z, t) = 2A \cos\left(\omega t - kz - \varphi_{\mathrm{s}}\right), \tag{7B.41}$$

describes a wave propagating in the positive z direction. Equation (7B.39) shows that the conjugate field has the form

$$E_{\mathrm{c}}^{(+)}(z, t) = \mathcal{E}_{\mathrm{c}}(z)\,\mathrm{e}^{-\mathrm{i}\omega t} = A'\,\mathrm{e}^{-\mathrm{i}\varphi_{\mathrm{s}}}\,\mathrm{e}^{-ikz}\,\mathrm{e}^{-\mathrm{i}\omega t}, \tag{7B.42}$$

i.e. it is a real field:

$$E_{\mathrm{c}}(z, t) = 2A' \cos\left(\omega t + kz + \varphi_z\right). \tag{7B.43}$$

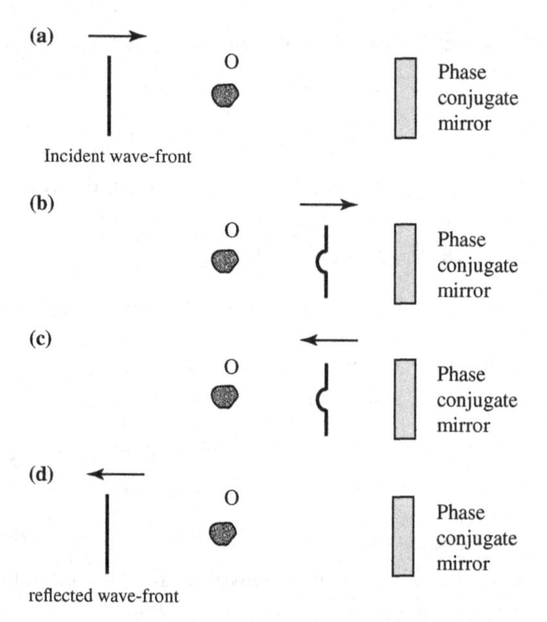

Figure 7B.6 Distortion of an initially plane wave-front (a) after crossing a defect of higher index than the surrounding medium (b). The effect of the phase conjugate mirror is to reverse the direction of propagation of the rays while conserving the wave-front (c). According to the law of reversed return, the reflected beam has a wave-front that is once again plane after crossing the non-uniform optical medium (d).

This is a wave propagating in the negative z direction. Note that it can also be written,

$$E_c(z,t) = 2A' \cos\left(- \omega t - kz - \varphi_z \right) = \frac{A'}{A} E_s(z, -t). \tag{7B.44}$$

It can be thought of as the signal wave going 'back in time'.

Since the propagation equations are invariant under combined time reversal and reversal of the direction of propagation, *the conjugate wave will follow exactly the same path* as the incident wave, *but in the opposite direction*, no matter how complex the medium in which the wave is propagating.

To illustrate this property, let us follow an initially plane incident wave (see Figure 7B.6(a)), which first crosses a distorting transparent medium (the object O in figure 7B.6), and is then reflected on a *phase conjugate mirror*. After crossing the medium O, the wave-front is distorted, since the optical paths followed by the various light rays in the medium O are different (Figure 7B.6(b)). Since the action of the phase conjugate mirror is analogous to a time reversal, after reflection (Figure 7B.6(c)), we obtain *an identical wave-front to the incident wave-front* of Figure 7B.6(b). Applying the above rule about the *exact reversal of the propagation*, we find that the reflected wave-front is once again a *plane wave* after crossing back through the distorting medium (Figure 7B.6(d)).

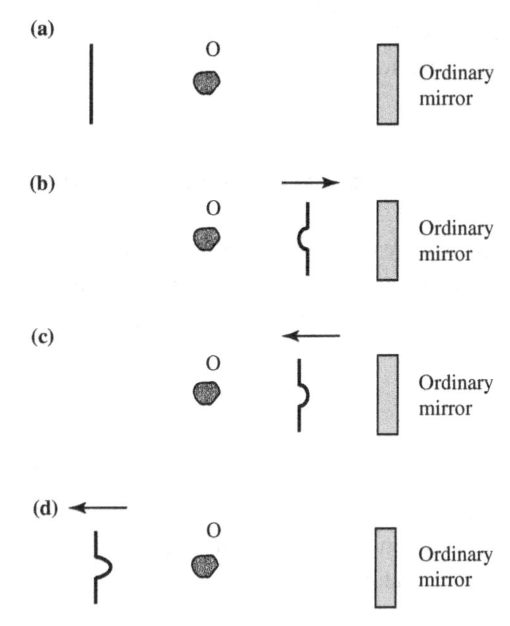

Figure 7B.7 An ordinary mirror reflects the rays and the distortion of the wave-front. After going through the distorting medium a second time, the distortion is twofold (compare with Figure 7B.6).

If the phase conjugating medium of Figure 7B.6 is replaced by a *normal mirror*, there is no rectification of the wave-front. Quite the opposite, in fact, because *the distortion is double* after a second crossing of the distorting medium (Figure 7B.7).

There are natural applications for this kind of wave-front correction when electromagnetic radiation has to travel great distances through the atmosphere. Indeed, owing to *fluctuations in the air density*, the atmosphere behaves as a distorting medium. (What is more, distortions of the wave-front also fluctuate in time, as attested by the scintillation of the stars.) So a laser beam with Gaussian transverse intensity distribution and cylindrical symmetry when it leaves the laser will gradually lose these properties as it travels through the atmosphere. It follows that the quality of the beam can be rapidly degraded. If the beam is reflected back on itself by a normal mirror, the deterioration in quality will be further increased on the return journey, and the emitter will receive a highly perturbed return beam. This is detrimental for the quality of communications between remote stations. One way to solve the problem would be to replace the normal mirror by a phase conjugate mirror. In this case, provided that the fluctuations of the atmosphere are not too fast compared with the timescale for the return trip of the wave, the reflected light rays will follow the same path as the incident rays, and the return beam will be *automatically targeted* on the emitting station. (Information can be transferred between the two stations by modulating the intensity of the reflected beam, for example.)

However, it should be mentioned that, even though some demonstration experiments have been successful, practical applications of the phase conjugate mirror remain few and far between. On the other hand, mirrors based on phase conjugation have been commercialized for high-gain amplifiers.[4]

7B.4.3 Calculating the reflection coefficient

For simplicity, we shall assume that the two pump waves are real, with the same intensity, and that the intensity remains constant in the medium. According to (7B.39), the equation of propagation for the conjugate field is then,

$$\frac{\mathrm{d}\mathcal{E}_\mathrm{c}}{\mathrm{d}z} = -\mathrm{i}\kappa\mathcal{E}_\mathrm{s}^*, \tag{7B.45}$$

where $\kappa = \dfrac{2\omega}{n_0 c}\chi_3'\mathcal{E}_\mathrm{p}\mathcal{E}_\mathrm{p}'$. Under the effect of the signal wave $\mathcal{E}_\mathrm{s}$, a conjugate wave appears. In the presence of the two pump waves, this conjugate wave in its turn produces a nonlinear polarization that will modify the amplitude $\mathcal{E}_\mathrm{s}$ of the signal field, which obeys the propagation equation,

$$\frac{\mathrm{d}\mathcal{E}_\mathrm{s}}{\mathrm{d}z} = \mathrm{i}\kappa\mathcal{E}_\mathrm{c}^*. \tag{7B.46}$$

The sign is positive on the right-hand side, because k is positive for the pump wave. As in the case of parametric mixing, studied in Chapter 7, we thus have several coupled propagation equations. The initial conditions are $\mathcal{E}_\mathrm{s}(0) = \mathcal{E}_0$ and $\mathcal{E}_\mathrm{c}(\ell) = 0$. Equations (7B.45) and (7B.46) then have solutions 0 and ℓ :

$$\mathcal{E}_\mathrm{s}(z) = \mathcal{E}_0 \frac{\cos|\kappa|(z-\ell)}{\cos|\kappa|\ell} \tag{7B.47}$$

$$\mathcal{E}_\mathrm{c}(z) = -\mathrm{i}\frac{\kappa}{|\kappa|}\mathcal{E}_0^* \frac{\sin|\kappa|(z-\ell)}{\cos|\kappa|\ell}. \tag{7B.48}$$

The reflection coefficient is therefore given by

$$R_\mathrm{c} = \left|\frac{\mathcal{E}_\mathrm{c}(0)}{\mathcal{E}_\mathrm{s}(0)}\right|^2 = \tan^2|\kappa\ell|. \tag{7B.49}$$

This coefficient is greater than 1 when $|\kappa\ell|$ lies between $\pi/4$ and $\pi/2$. *The reflected beam is then more intense than the incident beam.* When $|\kappa\ell|$ approaches the value $\pi/2$, the reflection coefficient tends to infinity. This (unphysical) result can be put down to our assumption that the amplitudes of the pump waves could be constant in the medium. When R_c becomes large, the probe and conjugate waves can have amplitudes comparable with the pump waves, and thereby modify their intensity. To find the solution to the problem in this case, one must solve a system of four coupled differential equations, since the complex amplitude of the pump waves is then a function of z. It is found that R_c can indeed

[4] For further detail, see R. W. Boyd, *Nonlinear Optics*, Academic Press (2008) and A. Brignon and J. P. Huignard (Eds.), *Phase Conjugate Laser Optics*, Wiley (2003).

become very big when $|\kappa\ell|$ is close to $\pi/2$, and values greater than 100 have been observed experimentally.

It is thus possible to obtain an optical oscillator, like the laser or the OPO discussed in the last complement, by placing a high-gain phase conjugate mirror opposite a mirror.

Note that the appearance of the conjugate wave is accompanied by *an amplification of the probe wave* transmitted by the medium. Indeed, according to (7B.47), at the output of the nonlinear medium (in the plane $z = \ell$), the probe wave has amplitude

$$\mathcal{E}_s(\ell) = \mathcal{E}_0 \frac{1}{\cos|\kappa\ell|}. \tag{7B.50}$$

The intensity transmission coefficient T_c for the probe wave is thus,

$$T_c = \frac{1}{\cos^2|\kappa\ell|}, \tag{7B.51}$$

which is bigger than 1.

It is interesting to note that, combining (7B.49) and (7B.51), we have

$$T_c - R_c = 1. \tag{7B.52}$$

This relation tells us that *the difference between the intensities of the probe and conjugate waves is independent of* ℓ and coincides with the intensity of the probe wave where it enters the medium. The increase in intensity of the probe wave is thus equal to the intensity of the conjugate wave. In other words, *the physical process produces as many photons in the probe wave as in the conjugate wave.*

This property can be understood by seeking the origin of the energy supplied to the waves E_s and E_c. This energy comes from each of the pump waves which, in an elementary process, each lose one photon in favour of the probe and conjugate waves (see Figure 7B.8). What we have here is a quantum process creating twin photons in the probe and conjugate waves, analogous to the one studied in detail in Chapter 7 in the case of the parametric interaction in a second-order non-linear medium. The differences between the two processes lie in the fact that here the four interacting waves all have the same frequency, and phase matching is automatic.

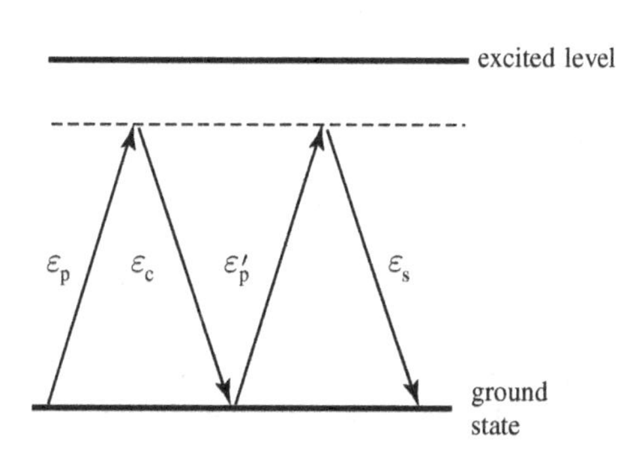

Figure 7B.8 Elementary process in four-wave mixing. There is absorption of one photon in each pump wave and emission of a photon in the probe wave and conjugate wave.

7B.5 Propagation of a spatially non-uniform wave in a Kerr medium

In this section and the next, we shall investigate effects associated with the propagation of a wave in a Kerr medium. In contrast to the simple situation of the plane wave discussed up to now, we shall consider a wave whose amplitude may depend on space and time. We shall study the simplest cases, i.e. *a wave with time-independent power but spatially varying amplitude* (self-focusing and self-defocusing), and *a plane wave with time-dependent amplitude* (self-phase modulation).

7B.5.1 Self-focusing

Consider a *Gaussian beam* TEM_{00} (see Complement 3B), for which the complex amplitude at the waist is (see (3B.2))

$$E^{(+)}(x, y, z = 0) = E_0 \exp\left(-\frac{r^2}{w_0^2}\right), \tag{7B.53}$$

where $r = \sqrt{x^2 + y^2}$. To simplify, we consider the case where the Rayleigh length is much larger than the thickness ℓ of the Kerr medium. The amplitude profile remains equal to (7B.53) right through the medium. Since the intensity of the wave is greater at $r = 0$, the optical thickness $n\ell = (n_0 + n_2 I)\ell$ will be different at the centre and the edge of the beam, just as happens when a light beam goes through a lens. If $n_2 > 0$, the optical thickness is greater at the centre and we have a convergent lens effect created by the light itself, whence the name self-focusing given to this phenomenon (see Figure 7B.9). If $n_2 < 0$, the optical thickness is lower at the centre and the equivalent lens is divergent. This is known as self-defocusing.

More quantitatively, the Kerr medium induces a modification of the phase φ of the wave that depends on the transverse coordinate r, and is given by

$$\varphi(r) = \omega\frac{\ell}{c}(n_0 + n_2 I) = \omega_0\frac{\ell}{c}\left[n_0 + 2n_2 E_0^2\, e^{-2r^2/w_0^2}\right]. \tag{7B.54}$$

To obtain this expression, we assume that the Kerr medium is thin enough to justify neglecting the local dependence $I(r)$ of the intensity on r through the medium. For small transverse distances r compared with w_0, we have

$$\varphi(r) \approx \omega_0\frac{\ell}{c}\left[n_0 + 2n_2 E_0^2 - \frac{4n_2 E_0^2}{w_0^2} r^2\right]. \tag{7B.55}$$

There is a quadratic dependence of the phase on the transverse position variable r, which is the same as for a Gaussian wave with radius of curvature R (see (3B.2)), where

$$R = -\frac{w_0^2}{8n_2 E_0^2 \ell} = -\frac{w_0^2}{4n_2 I_0 \ell}. \tag{7B.56}$$

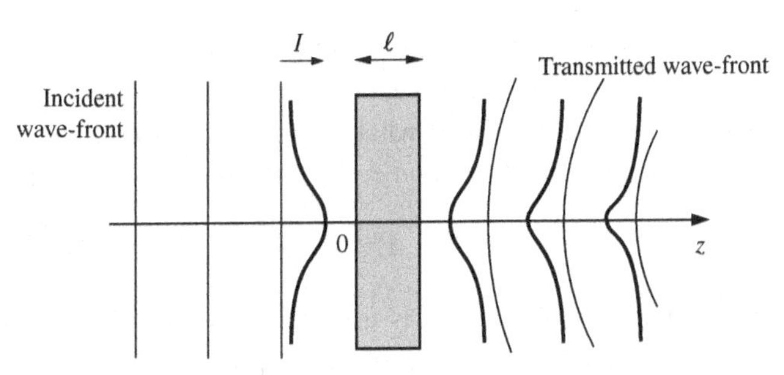

Figure 7B.9 Self-focusing through a thin Kerr medium.

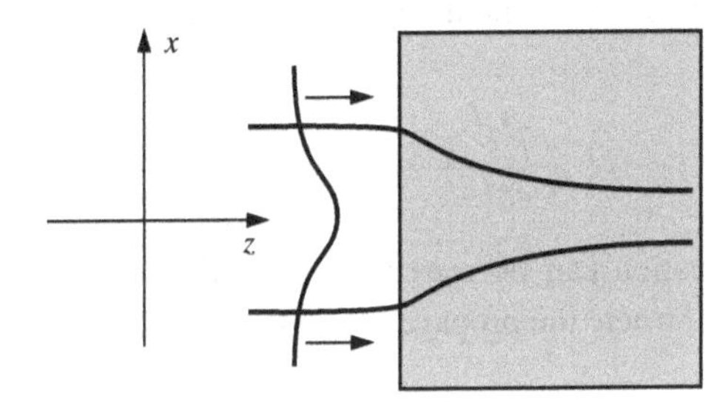

Figure 7B.10 Self-focusing inside a Kerr medium.

We thus obtain a convergent Gaussian wave leaving the medium if $n_2 > 0$, and a divergent Gaussian wave if $n_2 < 0$. The absolute value of the focal length of the equivalent thin lens is given precisely by (7B.56). To give an order of magnitude, consider a laser pulse produced by a YAG laser ($\lambda = 1\,\mu$m, energy 10 mJ, width 10 ns, peak power 1 MW) in the form of a Gaussian beam with waist $w_0 = 100\,\mu$m. According to (7B.56), a glass plate ($n_2 \approx 3 \times 10^{-12}$ m^2 W^{-1}) of thickness 1 cm will have the same effect as a lens with a focal length of about 10 cm. The effect is therefore far from being negligible. It can be observed under moderate intensity conditions and plays a very important role in certain pulsed lasers with self-phase-locked modes.

The above reasoning is only valid if the medium is thin enough for the self-focusing effect to manifest itself *outside* the medium at a great enough distance compared with its thickness ℓ. The validity condition of (7B.56) is therefore,

$$\ell \ll \frac{w_0}{\sqrt{n_2 I_0}}. \tag{7B.57}$$

7B.5.2 Spatial soliton and self-focusing

Suppose now that the Kerr medium is thick. The beam focusing effect leads to increased intensity on its axis, and the self-focusing effect will be further accentuated during propagation. In other words, the beam will converge more and more quickly. However, another effect will oppose the convergence induced by the Kerr effect. This is *diffraction*, which causes the beam to diverge, and all the more so as its transverse cross-section is reduced.

The self-focusing phenomenon will thus stabilize if diffraction is able to exactly balance the induced lensing effect. We then obtain a beam of constant transverse dimensions, called a spatial soliton, and this is what we shall now investigate. Here we limit the discussion to *a single transverse dimension* x. We thus consider the complex field,

$$\mathbf{E}^{(+)}(x,z,t) = \boldsymbol{\varepsilon}\,\mathcal{E}(x,z)\,\mathrm{e}^{\mathrm{i}(kz-\omega t)}, \tag{7B.58}$$

where the envelope $\mathcal{E}(x,z)$ is assumed to vary slowly on the length scale of λ in both the z and x directions. The equation of propagation of this envelope is obtained from the general equation (7.19) by a similar argument to the one used in Section 7.2.3, but now taking into account the transverse x dependence in the expression for the Laplacian. This leads to

$$\frac{\partial^2\mathcal{E}}{\partial x^2} + 2\mathrm{i}\omega\frac{n_0}{c}\frac{\partial\mathcal{E}}{\partial z} = -\frac{\omega^2}{\varepsilon_0 c^2}\mathcal{P}^{(+)}_{\mathrm{NL}} = -\frac{2\omega^2}{c^2}\chi_3'|\mathcal{E}|^2\,\mathcal{E}. \tag{7B.59}$$

This equation can be rewritten in a mathematical form close to that of Schrödinger's equation, where the propagation in the z direction replaces the time variation:

$$\mathrm{i}\frac{\partial\mathcal{E}}{\partial z} = -\frac{c}{2n_0\omega}\frac{\partial^2\mathcal{E}}{\partial x^2} - \frac{\omega}{n_0 c}\chi_3'|\mathcal{E}|^2\,\mathcal{E}. \tag{7B.60}$$

This equation, often encountered in nonlinear physics, is called the nonlinear Schrödinger equation. It has the following analytical solution, called a spatial soliton, valid for any value of E_0,

$$\mathcal{E} = E_0\,\mathrm{e}^{\mathrm{i}\delta kz}\frac{1}{\cosh(x/x_0)}, \tag{7B.61}$$

where x_0, the transverse extent of the soliton, and δk, the additional term in the wavevector of the soliton wave, are given by

$$x_0 = \frac{c}{\omega}\frac{1}{\sqrt{\chi_3' E_0^2}}, \tag{7B.62}$$

$$\delta k = \frac{\omega}{2n_0 c}\chi_3'\,E_0^2. \tag{7B.63}$$

Note that the soliton gets narrower when it becomes more intense. It should be noted that the analytical solution presented here does not generalize to two transverse dimensions. Indeed, the focusing effect increases the intensity more quickly along the axis when there are two dimensions, than when there is only one, and diffraction cannot compensate for this extra growth. There is then no stable solution. Experimentally, however, it is observed in the two-dimensional case that the light propagates in the Kerr medium in the form of one or more filaments. The steady-state transverse structure is then due to a *saturation* of the nonlinearity, hence to terms of higher order in the expansion in powers of I. The existence of several filaments propagating parallel to one another, often observed, is due to the amplification of very slight initial perturbations in the wave-front.

7B.6 Propagation of a pulse in a Kerr medium

7B.6.1 Self-phase modulation

Up to now, we have considered *steady-state* phenomena and we have investigated the *spatial* manifestations of nonlinear processes. But interesting effects are also observed when the *temporal* aspects of these phenomena are taken into account.

Suppose an intense pulse of variable intensity $I(t)$ is directed onto a nonlinear medium of thickness ℓ. The phase of the output wave,

$$\varphi = \left[\omega t - n\frac{\omega}{c}\ell\right], \tag{7B.64}$$

will depend on the '*instantaneous*' intensity of the wave like the index $n = n_0 + n_2 I(t)$. The phase-shift due to propagation will thus be time-dependent. This is the phenomenon known as self-phase modulation. The *frequency* ω_{ins} of the wave, which is nothing other than the *time derivative of the phase*, will thus be modified:

$$\omega_{\text{ins}} = \omega - \frac{\omega}{c} n_2 \ell \frac{\mathrm{d}I}{\mathrm{d}t}. \tag{7B.65}$$

If the incident wave has a well-defined frequency, the transmitted wave will have a *broadened frequency spectrum*. Moreover, we may note that for $n_2 > 0$, the frequency is rather shifted toward the red when the intensity increases and toward the blue when it decreases. The pulse is said to be chirped. This spectral broadening of the pulse can be exploited in many devices to obtain very short pulses. Indeed, a short pulse in the time domain must have a broad frequency spectrum. By sending a picosecond pulse into a Kerr medium such as an optical fibre, the output from the medium is a pulse that is broadened in the frequency domain. We may thus squeeze the pulse temporally into the femtosecond range.

A second consequence is that the propagation time in the medium (equal to $n\ell/c$ for a medium with non-dispersive index n) is then itself modulated, since n is a function of time, leading to a temporal distortion of the incident pulse. If n_2 is negative, for example, the top of the pulse will propagate more quickly than its sides. The result will be a *distorted pulse shape*, with a steeper and steeper rise side.

7B.6.2 Propagation in a dispersive linear medium

There is another phenomenon that can distort a temporal pulse, namely, a frequency-dependent (or dispersive) linear index n_0. This will be the subject of the present section.

We write the amplitude of the complex field of the pulse in the form

$$E^{(+)}(z,t) = \int \tilde{E}(z,\omega)\,\mathrm{e}^{-\mathrm{i}\omega t}\frac{\mathrm{d}\omega}{\sqrt{2\pi}}, \tag{7B.66}$$

assuming it to have infinite transverse extent. If the pulse propagates in a linear dielectric medium, Maxwell's equations are easily solved and we have,

$$E^{(+)}(z,t) = \int \tilde{E}(0,\omega)\, \mathrm{e}^{\mathrm{i}(k(\omega)z-\omega t)} \frac{\mathrm{d}\omega}{\sqrt{2\pi}}, \tag{7B.67}$$

where $k(\omega)$ is a function of the linear susceptibility $\chi(\omega)$. We assume that the pulse has Fourier components $\tilde{E}(0,\omega)$ that are non-zero only in a narrow band $\left[\omega_0 - \Delta\omega/2, \omega_0 + \Delta\omega/2\right]$ around a central frequency ω_0, which implies that its duration, of order $1/\Delta\omega$, is long compared with the optical period. We can then use a truncated expansion of $k(\omega)$:

$$k(\omega) = k_0 + k_0'\,\delta\omega + \frac{1}{2}k_0''\,\delta\omega^2, \tag{7B.68}$$

where

$$k_0 = k(\omega_0), \tag{7B.69}$$

$$k_0' = \left.\frac{\mathrm{d}k}{\mathrm{d}\omega}\right|_{\omega=\omega_0}, \tag{7B.70}$$

$$k_0'' = \left.\frac{\mathrm{d}^2k}{\mathrm{d}\omega^2}\right|_{\omega=\omega_0}, \tag{7B.71}$$

$$\delta\omega = \omega - \omega_0. \tag{7B.72}$$

Writing the pulse in the form of a 'carrier' of frequency ω_0, multiplied by an envelope that varies on a slow timescale, of the order of $\dfrac{1}{\Delta\omega}$,

$$E^{(+)}(z,t) = \mathrm{e}^{\mathrm{i}(k_0 z-\omega_0 t)}\mathcal{E}(z,t), \tag{7B.73}$$

we find from (7B.68) that $\mathcal{E}(z,t)$ is given by

$$\mathcal{E}(z,t) = \int \tilde{E}(0,\omega)\, \exp\left\{-\mathrm{i}\delta\omega t + \mathrm{i}z\left(k_0'\delta\omega + \frac{k_0''}{2}\delta\omega^2\right)\right\} \frac{\mathrm{d}(\delta\omega)}{\sqrt{2\pi}}. \tag{7B.74}$$

We then deduce the following expression for $\dfrac{\partial\mathcal{E}}{\partial z}$:

$$\frac{\partial\mathcal{E}}{\partial z} = \mathrm{i}\int \tilde{E}(0,\omega)\,\mathrm{e}^{-\mathrm{i}\delta\omega t}(k_0'\delta\omega + \frac{k_0''}{2}\delta\omega^2)\frac{\mathrm{d}(\delta\omega)}{\sqrt{2\pi}}. \tag{7B.75}$$

The terms with $\delta\omega$ and $(\delta\omega)^2$ under the integral sign are expressed in terms of the time derivatives $\dfrac{\partial\mathcal{E}}{\partial t}$ and $\dfrac{\partial^2\mathcal{E}}{\partial t^2}$. We thus have

$$\frac{\partial\mathcal{E}}{\partial z} = -k_0'\frac{\partial\mathcal{E}}{\partial t} - \mathrm{i}\frac{k_0''}{2}\frac{\partial^2\mathcal{E}}{\partial t^2}. \tag{7B.76}$$

If the medium is such that $k_0'' = 0$, the solution of this equation is extremely simple:

$$\mathcal{E}(z,t) = \mathcal{E}(0, t - k_0' z). \tag{7B.77}$$

The envelope of the pulse at any z is equal to its value at the input to the medium $z = 0$ a time $k_0' z$ earlier, whence there is propagation without distortion at the *group velocity* $v_g = 1/k_0'$.

Now consider the case $k_0'' \neq 0$, and set $\tau = t - k_0' z$. This is the time measured from the top of the pulse. Equation (7B.76) can then be written,

$$\mathrm{i}\frac{\partial \mathcal{E}}{\partial z}(z,\tau) = \frac{k_0''}{2}\frac{\partial^2 \mathcal{E}}{\partial \tau^2}(z,\tau). \tag{7B.78}$$

This equation is formally equivalent to the Schrödinger equation for a free particle (where z replaces the time variable). We know that, in this case, there will be something like spreading of the wave packet, i.e. spreading of the light pulse, due to the phenomenon known as group velocity dispersion.

7B.6.3 Propagation in a dispersive Kerr medium. Temporal soliton

In the Kerr medium, this dispersion-related pulse spreading phenomenon will compete with shortening of the pulse due to self-phase modulation. One thus expects to find solutions in which the two effects balance one another, that remain stable under propagation.

The existence of a nonlinear polarization $P_{\mathrm{NL}}^{(+)}$ will add a term in the equation of propagation (7B.78). In an analogous way to (7B.60), we can thus write

$$\mathrm{i}\frac{\partial \mathcal{E}}{\partial z} - \frac{k_0''}{2}\frac{\partial^2 \mathcal{E}}{\partial \tau^2} = -\frac{\omega_0}{2n_0 c}\chi_3' I \mathcal{E}, \tag{7B.79}$$

where $I = 2|\mathcal{E}|^2$. We obtain a nonlinear Schrödinger equation, analogous to the one found in the last section (see (7B.60)). There is therefore a temporal soliton, which does not deform during propagation, analogue of the spatial soliton, hence of the form,

$$\mathcal{E}(z,\tau) = E_0\, \mathrm{e}^{\mathrm{i}\delta k' z}\frac{1}{\cosh(\tau/T_{\mathrm{sol}})}, \tag{7B.80}$$

with

$$T_{\mathrm{sol}} = \left(-\frac{k_0'' n_0 c}{\omega_0 \chi_3' E_0^2}\right)^{1/2}, \tag{7B.81}$$

$$\delta k' = \frac{\omega_0}{2n_0 c}\chi_3' E_0^2. \tag{7B.82}$$

Note that soliton pulses can only exist if $k_0''/\chi_3' < 0$. In optical fibres, this happens if the wavelength is greater than $1.3\,\mu\mathrm{m}$. It is thus possible to obtain solitons at the wavelength

1.5μm, the one used by optical telecommunications, which propagate without distortion over hundreds of kilometres of fibre and can be regenerated by passing through erbium-doped fibre amplifiers (see Complement 3F.4). These pulses may be elements of a digital data transfer where the rate is not restricted by pulse spreading over long propagation distances.

8 Laser manipulation of atoms. From incoherent atom optics to atom lasers

When the Nobel Prize for Physics was awarded to Claude Cohen-Tannoudji, William D. Philipps and Steven Chu in 1997 for the development of methods for cooling and trapping atoms using lasers, this was the reward for two decades of investigations not only touching upon the fundamental aspects of the light–matter interaction, but also leading to a range of applications.[1] On the theoretical level, the advent of this field of research in the late 1970s stimulated the development of various theoretical methods for describing the radiative forces by which light affects the motion of atoms. On the experimental level, judicious use of the radiative forces exerted by lasers made it possible to obtain a drastic reduction of atomic velocities – in other words, to cool an atomic vapour. These advances were characterized by a remarkable cross-fertilization of theoretical and experimental innovations. It was not long before applications began to appear in the field of high-resolution spectroscopy, since ultracold, i.e. slow, atoms can be observed for longer periods, and this allows greater accuracy when measuring atomic resonance frequencies than could be obtained for atoms at room temperature. Based on such resonances, the most accurate clocks in the world use laser-cooled atoms or ions. Atom interferometry is another application of laser manipulation of atoms, which can be applied to measure inertial effects, due for example to the rotation of the Earth or the motion of a vehicle, with an accuracy that already exceeds traditional methods. In a different domain, atom nanolithography may provide elegant solutions to the problems raised by the manufacture of ever smaller microelectronic components; and the list continues. All these applications, realized with samples of cold atoms, are reminiscent of classical optics, based on incoherent sources of light. Indeed, even when we need to use a quantum description of the atomic motion, the wavefunctions of the different atoms of a sample are mutually incoherent, as the wavelets emitted by different points of an incoherent light source. We therefore call 'incoherent atom optics' the domain to which these applications belong.

On the fundamental side, one remarkable application of laser-cooled atoms is the emergence of a new area of research, namely, the investigation of gaseous Bose–Einstein condensates, in turn rewarded by the Nobel Prize, obtained by E. Cornell, C. Wieman and W. Ketterle in 2001.[2] Indeed, it was by starting with samples of laser-cooled and trapped atoms that it became possible to observe, under almost ideal conditions, the quantum phase

[1] See the Nobel lectures, reprinted in *Reviews of Modern Physics* **70** (July 1998).

[2] See the Nobel lectures of E. Cornell, C. Wieman and W. Ketterle, reprinted in *Review of Modern Physics* **74** (2002).

transition predicted by Einstein in 1924,[3] characterized by the formation of a macroscopic atomic population (i.e. involving a large number of atoms) in the same quantum state. The phenomenon is reminiscent of the laser transition, characterized by the appearance of a macroscopic field (a large number of photons) in the same mode of the electromagnetic field. The analogy is so striking that it is even called an atom laser. One could not dream of a better example as a conclusion for this book.

In this chapter, we shall emphasize the basic features of the laser manipulation of atoms, which will appeal to everything we have seen so far regarding the interaction between an atom or an ensemble of atoms and light. Absorption, stimulated emission and spontaneous emission are the basic mechanisms underlying radiative forces, so they provide a good application of the material in this book. But the interest in discussing radiative forces is further enhanced by the fact that, depending of the phenomenon studied, it is easier to supply a simple physical picture – and the corresponding theoretical treatment – in one or other of the light–matter interaction models we have already discussed, i.e. the semi-classical model, where the light is treated as a classical electromagnetic wave, or the quantum model, where the light is quantized and treated as composed of photons carrying momentum. We begin with a first section (8.1), where we use both models to show that momentum as well as energy exchanges between light and atoms are quantized. In Section 8.2, we use the semi-classical model to introduce the two main types of radiative forces: the resonance-radiation pressure and the dipole force. However, we shall find that, while the semi-classical model provides a simple and fruitful picture of the dipole force, it is the quantum model for light that leads to the simplest interpretation of resonance-radiation pressure. The quantum model will also be required to understand the limits of radiative cooling, presented in Section 8.3. Indeed, fluctuations due to spontaneous emission, a phenomenon fundamentally related to the quantum aspects of radiation, as well as the quantization of momentum exchanges, lead to heating, and this places a basic limit on the quest for low temperatures, at values that are already spectacularly low, since the nK range is accessible. In Section 8.4, we shall introduce and discuss elementary notions on Bose–Einstein condensates and atom lasers. The chapter is supplemented with a complement presenting a method for cooling atoms to temperatures in the nanokelvin range. This method employs a subtle interference effect between quantum amplitudes associated with atom–laser interactions, and appeals to unusual statistics known as 'Lévy statistics'. It allows one to obtain atom samples with a residual velocity less than the recoil velocity associated with absorption or emission of a single photon, a surprising result when one recalls that the velocity of the atom changes by multiples of this recoil velocity.

8.1 Energy and momentum exchanges in the atom–light interaction

Radiative forces allow one to modify the motion of an atom, i.e. its momentum as well as its kinetic energy. In this section, we will use the quantum formalism to describe the

[3] A. Einstein, *Quanten theorie des einatomigen idealen Gases*, Zweite Abhandlung, Preussische Akademie der Wissenschaften, Physikalisch-Mathematische Klasse, Sitzungsberichte, 3–14 (1925).

atomic motion, and show that the changes in energy or momentum proceed by quantized quantities, $\hbar\omega$ and $\hbar\mathbf{k}$, respectively.

This result can be obtained with the semi-classical model of atom–light interaction, but an enlightening image stems from the fully quantized formalism of Chapter 6, which allows us to interpret these quantized changes in terms of absorption or emission of photons. This will lead us to show that the resonance condition in the atom–light interaction has to be corrected, not only by the usual Doppler shift associated with the atomic velocity, but also by a supplementary term known as 'the recoil shift'.

8.1.1 Quantum description of the external degrees of freedom of the atom

To describe the quantum state of an atom in motion, we can no longer simply define its internal state $|\psi_{\text{int}}\rangle = \sum_i c_i|i\rangle$, where the states $|i\rangle$ are the eigenstates of the atomic Hamiltonian $\hat{H}_0$ which describes the motion of the electrons around the nucleus. We must also describe the state of the centre of mass of the atom taken as a whole. In the case where the atom is in free space, the 'external' Hamiltonian $\hat{H}_{\text{ext}}$ that describes the motion of the atom is simply $\hat{\mathbf{P}}^2/2M$, where $\hat{\mathbf{P}}$ is the momentum operator for the centre of mass of the atom, and M is its total mass. Joint eigenstates of $\hat{H}_{\text{ext}}$ and $\hat{\mathbf{P}}$ are plane waves $|\mathbf{K}\rangle$ such that

$$\langle\hat{\mathbf{r}}|\mathbf{K}\rangle = \frac{1}{L^{3/2}}\,\mathrm{e}^{\mathrm{i}\mathbf{K}\cdot\mathbf{r}}, \tag{8.1}$$

where $\hat{\mathbf{r}}$ describes the position of the centre of mass, and L^3 is the discretization volume. For such an eigenstate,

$$\hat{\mathbf{P}}|\mathbf{K}\rangle = \hbar\mathbf{K}|\mathbf{K}\rangle = \mathbf{P}|\mathbf{K}\rangle, \tag{8.2}$$

so the velocity is perfectly known, and equal to $\dfrac{\hbar\mathbf{K}}{M}$, as is the energy $\dfrac{\hbar^2K^2}{2M}$. The most general state of the atom can now be written in the form,

$$|\psi\rangle = \sum_i \int d^3\mathbf{K}\, c_i(\mathbf{K})|i\rangle \otimes |\mathbf{K}\rangle = \sum_i \int d^3\mathbf{K}\, c_i(\mathbf{K})|i,\mathbf{K}\rangle. \tag{8.3}$$

Note that this state does not necessarily factorize into a tensor product of an internal and an external state.

8.1.2 Momentum conservation

To begin with, we shall use the approach discussed in Chapter 2, where the electromagnetic field is a classical quantity. In the electric dipole approximation, the interaction term in the Hamiltonian of the system is given by (2.75). When the external degrees of freedom of the atom are given a quantum theoretical treatment, the interaction Hamiltonian is

$$\hat{H}_{\mathrm{I}}(t) = -\hat{\mathbf{D}}\cdot\mathbf{E}(\hat{\mathbf{r}},t), \tag{8.4}$$

where the operator $\hat{\mathbf{r}}$ replaces the classical position vector $\mathbf{r}$. If the electromagnetic field is a plane wave of wavevector $\mathbf{k}$, this Hamiltonian can be written more explicitly in the form,

$$\hat{H}_{\mathrm{I}}(t) = -\frac{1}{2}\hat{\mathbf{D}} \cdot \mathbf{E}_0(\mathrm{e}^{\mathrm{i}\mathbf{k}\cdot\hat{\mathbf{r}}-\mathrm{i}\omega t} + \mathrm{e}^{-\mathrm{i}\mathbf{k}\cdot\hat{\mathbf{r}}+\mathrm{i}\omega t}). \tag{8.5}$$

We shall use the perturbative method of Chapter 2 to calculate the probability of transitions between stationary states of the atom induced by such a Hamiltonian, taking the initial state to be $|\psi_i\rangle = |a, \mathbf{K}\rangle$ and the final state $|\psi_f\rangle = |b, \mathbf{K}'\rangle$, where $|a\rangle$ and $|b\rangle$ are two stationary internal states of the atom. If the final state has higher energy than the initial state (absorption process), only the term in $\mathrm{e}^{-\mathrm{i}\omega t}$ in (8.5) contributes (see Section 1.2.4), and the transition probability is (see (1.47))

$$\begin{aligned} P_{i\to f}(T) &= \frac{\pi T}{2\hbar}\left|\langle b, \mathbf{K}'|\hat{\mathbf{D}} \cdot \mathbf{E}_0 \mathrm{e}^{\mathrm{i}\mathbf{k}\cdot\hat{\mathbf{r}}}|a, \mathbf{K}\rangle\right|^2 \delta_T(E_f - E_i - \hbar\omega) \\ &= \frac{\pi T}{2\hbar}\left|\langle b|\hat{\mathbf{D}}|a\rangle \cdot \mathbf{E}_0\right|^2 \left|\langle \mathbf{K}'|\mathrm{e}^{\mathrm{i}\mathbf{k}\cdot\hat{\mathbf{r}}}|\mathbf{K}\rangle\right|^2 \delta_T(E_f - E_i - \hbar\omega). \end{aligned} \tag{8.6}$$

Now the operator $\mathrm{e}^{\mathrm{i}\mathbf{k}\cdot\hat{\mathbf{r}}}$ is simply the translation operator for atomic momentum eigenstates,

$$\mathrm{e}^{\mathrm{i}\mathbf{k}\cdot\hat{\mathbf{r}}}|\mathbf{K}\rangle = |\mathbf{K} + \mathbf{k}\rangle. \tag{8.7}$$

Indeed, representing the states by wavefunctions (8.1), Equation (8.7) simply says that

$$\mathrm{e}^{\mathrm{i}\mathbf{k}\cdot\mathbf{r}}\,\mathrm{e}^{\mathrm{i}\mathbf{K}\cdot\mathbf{r}} = \mathrm{e}^{\mathrm{i}(\mathbf{k}+\mathbf{K})\cdot\mathbf{r}}. \tag{8.8}$$

The transition probability thus only differs from zero if

$$\mathbf{K}' = \mathbf{K} + \mathbf{k}. \tag{8.9}$$

We see that *absorption of radiation modifies the momentum* of the atom. This effect allows the manipulation of atoms by light beams. The change in velocity of the atom for an absorption is equal to $\hbar\mathbf{k}/M$. This has magnitude,

$$V_{\mathrm{R}} = \frac{\hbar k}{M} \simeq \frac{\hbar}{M}\frac{\omega_0}{c} = \frac{E_b - E_a}{Mc}, \tag{8.10}$$

called 'the one photon recoil velocity' or, in short, 'the recoil velocity'. It is typically in the $\mathrm{cm\,s^{-1}}$ range for visible light. This is a very small amount compared with the thermal velocities of atoms.

For a stimulated emission process, where E_f is less than E_i, it is the term in $\mathrm{e}^{\mathrm{i}\omega t}$ in (8.5) that contributes, and we then have

$$\mathbf{K}' = \mathbf{K} - \mathbf{k}. \tag{8.11}$$

Relations (8.9) and (8.11) can be reinterpreted by multiplying both sides by $\hbar$, which yields, respectively,

$$\mathbf{P}' = \mathbf{P} + \hbar\mathbf{k} \quad \text{(absorption)} \tag{8.12}$$

and

$$\mathbf{P}' = \mathbf{P} - \hbar\mathbf{k} \quad \text{(stimulated emission)}. \tag{8.13}$$

These relations admit a very simple interpretation in the framework of the quantum model of the interaction between the atom and the radiation (Chapter 6). More precisely, if we consider a mode ℓ of the radiation characterized by a wavevector $\mathbf{k}_\ell$, and a state $|n_\ell\rangle$ containing n_ℓ photons in mode ℓ, the interaction term,

$$\hat{H}_\mathrm{I} = -\hat{D} \cdot \mathcal{E}_\ell^{(1)} \left(\mathrm{i}\hat{a}_\ell \, \mathrm{e}^{\mathrm{i}\mathbf{k}_\ell \cdot \hat{\mathbf{r}}} - \mathrm{i}\hat{a}_\ell^\dagger \, \mathrm{e}^{-\mathrm{i}\mathbf{k}_\ell \cdot \hat{\mathbf{r}}} \right), \tag{8.14}$$

couples the initial state $|a, \mathbf{K}\rangle \otimes |n_\ell\rangle$ to the final state $|b, \mathbf{K}+\mathbf{k}_\ell\rangle \otimes |n_\ell - 1\rangle$, so one photon is absorbed here and the atom gets the momentum $\hbar\mathbf{k}_\ell$ of the photon. Likewise, if we begin with an initial state $|b, \mathbf{K}\rangle \otimes |n_\ell\rangle$, the second term of the Hamiltonian has a non-zero matrix element with the final state $|a, \mathbf{K}-\mathbf{k}_\ell\rangle \otimes |n_\ell + 1\rangle$, so one photon is emitted in this case and the atom suffers a recoil, i.e. a change in momentum, equal and opposite to the momentum of the emitted photon.

Note that this argument remains valid for the initial state $|b, \mathbf{K}\rangle \otimes |0\rangle$, which is coupled to the state $|a, \mathbf{K} - \mathbf{k}_\ell\rangle \otimes |1_\ell\rangle$ by a process of spontaneous emission in mode ℓ. Here, too, the atom suffers a recoil $\hbar\mathbf{k}_\ell$. This particular phenomenon is only conceivable in the fully quantum model of the atom–radiation interaction.

8.1.3 Energy conservation: the Doppler and the recoil shifts

Equation (8.6) shows that, for an absorption process, the transition probability is non-zero only if we have

$$E_f = E_i + \hbar\omega, \tag{8.15}$$

where $E_f(E_i)$ is the total final (respectively, initial) energy of the atom, the sums of its internal and its kinetic energy. Here again, a relation obtained from a semi-classical treatment gets a simple interpretation if we consider the model in which the radiation initially comprises n photons of energy $\hbar\omega$. Equation (8.15) clearly expresses the conservation of the total energy of the atom–radiation system in the absorption process, where one photon disappears. Introducing the total energy of the atom explicitly for the initial and final states, we obtain

$$E_b + \frac{\hbar^2 \mathbf{K}'^2}{2M} = E_a + \frac{\hbar^2 \mathbf{K}^2}{2M} + \hbar\omega. \tag{8.16}$$

Relations (8.15) and (8.9) can be used to determine the frequency ω responsible for the absorption process:

$$\omega = \omega_0 + \mathbf{k} \cdot \frac{\hbar\mathbf{K}}{M} + \frac{\hbar k^2}{2M} = \omega_0 + \mathbf{k} \cdot \mathbf{V} + \frac{\hbar k^2}{2M}, \tag{8.17}$$

where $\omega_0 = (E_b - E_a)/\hbar$ is the Bohr atomic frequency associated with the transition $a \rightarrow b$ and $\mathbf{V}$ is the initial atomic velocity $\mathbf{P}/M$ (there would be a minus sign before the second term for an emission process). We thus observe that the resonance frequency of the absorption (emission) process is not exactly equal to the Bohr frequency ω_0 of the atom when the external degrees of freedom, i.e. the motion, are taken into account:

- The first correction $\mathbf{k} \cdot \mathbf{V}$, is the *Doppler shift term*, already encountered in (2.70) of Chapter 2. It tells us that we must use the rest frame of the atom to determine the resonance condition. For an atom moving at the average thermal velocity at room temperature, and for a transition in the visible range, the Doppler term is of the order of a few hundred MHz.

- The second correction, which arises even when the atom is initially at rest, is called the *recoil shift*. The atom necessarily recoils under the effect of the photon it absorbs, and this imparts a certain amount of kinetic energy that must come from the photon energy. This term is of the order of a few tens of kHz for a transition in the visible frequency range. Although it was long considered negligible, it now plays an important role in accurate measurements associating ultra-stable lasers with ultracold atoms.

8.2 Radiative forces

8.2.1 Closed two-level atom in a quasi-resonant laser wave

We consider an atom moving in a monochromatic light wave described by a classical electromagnetic field of the general form,

$$\mathbf{E}(\mathbf{r},t) = \boldsymbol{\varepsilon}\, E(\mathbf{r},t) = \boldsymbol{\varepsilon}\, E_0(\mathbf{r}) \cos[\omega t + \varphi(\mathbf{r})]. \tag{8.18}$$

We use the semi-classical model for the interaction between non-quantized radiation and a quantized atom (Chapter 2). The light is quasi-resonant with the transition between the ground state $|a\rangle$ of the atom and an excited state $|b\rangle$, i.e. ω is close to $\omega_0 = (E_b - E_a)/\hbar$. The excited level has lifetime Γ^{-1} due to spontaneous emission, which brings it back down to level $|a\rangle$. We shall use the model of a *closed two-level atom* discussed in Section 2.4.5. We thus ignore all the other atomic energy levels, which is justifiable provided that the detuning $\delta = \omega - \omega_0$ is much smaller than the detuning with respect to other atomic transitions, and that spontaneous emission can only cause the atom to transit from $|b\rangle$ to $|a\rangle$.

By laser cooling, atomic velocities can be considerably reduced, and de Broglie wavelengths therefore increased, until it becomes necessary to consider quantum aspects of the overall motion of the atom, whence the atomic centre of mass motion must be described by a wavefunction, as in Section 8.1. The state of the atom is thus described by a state vector $|\psi\rangle$ of the form (8.3), which for a two-level atom is represented by a two-component spinor:

$$|\psi\rangle = \begin{bmatrix} \psi_a(\mathbf{r},t) \\ \psi_b(\mathbf{r},t) \end{bmatrix}, \tag{8.19}$$

where $\psi_a(\mathbf{r},t)$ is the wavefunction of the ground state and $\psi_b(\mathbf{r},t)$ is the wavefunction of the excited state. In this representation, $\mathbf{r}$ is the position of the atom (or rather, its centre of mass). The atom thus looks like a particle with which we may associate a position operator $\hat{\mathbf{r}}$, and its internal complexity (nucleus and electrons) is totally accounted for by the two-dimensional space of internal states. We have made the long-wavelength approximation (see Section 2.3), in which it is assumed that the distance between the nucleus and the electrons is much shorter than the wavelength of the radiation. The dynamics of the atom is governed by the Hamiltonian,

$$\hat{H} = \hat{H}_0 + \hat{H}_\text{I}. \tag{8.20}$$

The operator $\hat{H}_0$ is the Hamiltonian of the free atom:

$$\hat{H}_0 = \begin{pmatrix} 0 & 0 \\ 0 & \hbar\omega_0 \end{pmatrix} + \frac{\hat{\mathbf{P}}^2}{2M}, \tag{8.21}$$

where the matrix, represented in the basis $\{|a\rangle, |b\rangle\}$, acts on the space $\mathcal{E}_\text{int}$ of internal states, while the kinetic energy term acts on the space $\mathcal{E}_\mathbf{r}$ of external states. ($\hat{\mathbf{P}}$ is the atomic momentum operator, conjugate observable to the position $\hat{\mathbf{r}}$ of the atomic centre of mass.) In the long-wavelength approximation, the electric dipole interaction Hamiltonian can be used (see Section 2.2.4):

$$\hat{H}_\text{I} = -\hat{\mathbf{D}} \cdot \mathbf{E}(\hat{\mathbf{r}}, t), \tag{8.22}$$

where $\hat{\mathbf{D}}$ is the electric dipole operator whose components are 2×2 matrices for a two-level atom. For an electric field of polarization $\boldsymbol{\varepsilon}$,

$$\mathbf{E}(\mathbf{r}, t) = \boldsymbol{\varepsilon}\, E(\mathbf{r}, t), \tag{8.23}$$

we only need to consider the component $\hat{D}_\varepsilon$:

$$\hat{D}_\varepsilon = \hat{\mathbf{D}} \cdot \boldsymbol{\varepsilon} = \begin{pmatrix} 0 & d_\varepsilon \\ d_\varepsilon & 0 \end{pmatrix}, \tag{8.24}$$

taking d_ε real to simplify. With this notation, the Hamiltonian $\hat{H}_\text{I}$ is given by

$$\hat{H}_\text{I} = -\hat{D}_\varepsilon\, E(\hat{\mathbf{r}}, t). \tag{8.25}$$

The interaction Hamiltonian (8.25) thus acts on both the internal state and the external state of the atom. The radiation is treated classically, and only the absorption and stimulated emission processes are taken into account. In principle, spontaneous emission is 'omitted' from this description. In fact, we shall see in Section 8.3 that it must be taken into account in order to provide a correct description of the way the light affects the atomic motion.

8.2.2 Localized atomic wave packet and classical limit

In order to define the radiative force acting upon the atom, we consider the classical limit, in which the motion of the atom is described as that of a classical particle, with a well-determined position at any time. Further, we shall assume that this position is the same for the ground state and the excited state. The atomic state is thus given by

$$|\psi\rangle = |\psi_\text{int}\rangle \otimes |\psi_r\rangle = \begin{bmatrix} c_a(t) \\ c_b(t) \end{bmatrix} \psi(\mathbf{r}, t), \tag{8.26}$$

where $\psi(\mathbf{r}, t)$ is a wave packet with very small spatial extent on the scale of the wavelength λ of the light wave. Under these conditions, the classical position $\mathbf{r}_{\text{at}}$ can be taken as the expectation value of the position observable, i.e.

$$\mathbf{r}_{\text{at}} = \langle\psi|\hat{\mathbf{r}}|\psi\rangle = \int d^3 r |\psi(\mathbf{r}, t)|^2 \mathbf{r}. \tag{8.27}$$

We may then apply Ehrenfest's theorem,[4] to find that the classical velocity,

$$\mathbf{v}_{\text{at}} = \frac{\mathrm{d}}{\mathrm{d}t}\mathbf{r}_{\text{at}}, \tag{8.28}$$

is the same as the expected value of the momentum operator $\hat{\mathbf{P}}$ divided by the mass M of the atom. Indeed, recall that

$$\frac{\mathrm{d}}{\mathrm{d}t}\langle\hat{r}_i\rangle = \frac{1}{\mathrm{i}\hbar}\langle[\hat{r}_i, \mathbf{H}]\rangle = \frac{1}{\mathrm{i}\hbar}\left\langle\left[\hat{r}_i, \frac{\hat{\mathbf{P}}^2}{2M}\right]\right\rangle$$

$$= \frac{1}{\mathrm{i}\hbar}\left\langle\left[\hat{r}_i, \hat{P}_i\right]\frac{\hat{P}_i}{M}\right\rangle = \frac{\langle\hat{P}_i\rangle}{M},$$

where i indexes the Cartesian coordinates. It follows that

$$\mathbf{v}_{\text{at}} = \frac{\mathrm{d}}{\mathrm{d}t}\langle\hat{\mathbf{r}}\rangle = \frac{\langle\hat{\mathbf{P}}\rangle}{M}. \tag{8.29}$$

It is important not to confuse the classical limit for the motion of the atom with the long-wavelength approximation. The latter stipulates that the *internal* structure of the atom is smaller than the light wavelength $\lambda = 2\pi c/\omega$, whence the motion of the atom can be characterized by a wavefunction of a single observable $\hat{\mathbf{r}}$ (identified with the position of the nucleus, or the centre of mass). The long-wavelength approximation does not impose any restriction on the wavefunction describing the atomic motion. On the other hand, in the classical motion limit, we assume further that this wavefunction is a wave packet, highly localized on the scale of λ, which means that we may introduce a *classical atomic position*, and hence a classical atomic trajectory. The validity of this approximation is not totally obvious, but a general discussion goes beyond the scope of this book.[5] For our present purposes, we shall simply give a necessary condition for the validity of such a hypothesis.

The assumption of a localized wave packet on the scale of the optical wavelength can be expressed by

$$\Delta r_i < \frac{\lambda}{2\pi}, \tag{8.30}$$

[4] See Section 2.2.2.

[5] See, for example, C. Cohen-Tannoudji, 'Atomic Motion in Laser Light', in *Fundamental Systems in Quantum Optics*, Elsevier (1991) (Cours des Houches, Session LIII 1990).

where Δr_i is the root-mean-square deviation of the coordinate r_i of the atom:

$$\Delta r_i = \sqrt{\langle\psi|(\hat{r}_i - r_{\mathrm{at},i})^2|\psi\rangle} \\ = \sqrt{\int d^3r(r_i - r_{\mathrm{at},i})^2|\psi(\mathbf{r},t)|^2}. \tag{8.31}$$

The factor of $1/2\pi$ in expression (8.30) ensures that all points within the wave packet experience roughly the same field value.

We know that the dispersions in the position and momentum satisfy the Heisenberg relations,

$$\Delta r_i \cdot \Delta P_i \geq \frac{\hbar}{2}. \tag{8.32}$$

Condition (8.30) thus implies that

$$\Delta P_i > \frac{1}{2}\hbar k. \tag{8.33}$$

This brings in the quantity $\hbar k$ which we have already encountered in Section 8.1, i.e. the momentum of a photon of wavevector $k = 2\pi/\lambda$ (Chapter 4). We thus see that the classical approximation for the motion of the atom requires that the dispersion in the velocity of the atom, i.e.

$$\Delta V_i = \frac{\Delta P_i}{M}, \tag{8.34}$$

should be greater than the recoil velocity $V_{\mathrm{R}} = \hbar k/M$. Now most laser cooling mechanisms give residual velocity dispersion in excess of V_{R}, and in this case, a classical description of the atomic motion is therefore acceptable. However, the cooling process does sometimes create a situation where ΔV is of the same order or smaller than V_{R}, and in this case, one must abandon the classical limit approximation to describe the atomic motion. We shall encounter such a situation in Section 8.3.9 and again in Complement 8A.

8.2.3 Radiative forces: general expression

Here we shall still assume the classical limit for the atomic motion, i.e. a localized wave packet, and now apply Ehrenfest's theorem to the momentum operator. This yields

$$\frac{\mathrm{d}}{\mathrm{d}t}\langle\hat{\mathbf{P}}\rangle = \frac{1}{\mathrm{i}\hbar}\langle[\hat{\mathbf{P}},\hat{H}]\rangle = \frac{1}{\mathrm{i}\hbar}\langle[\mathbf{P},\hat{H}_{\mathrm{I}}]\rangle, \tag{8.35}$$

since $\hat{\mathbf{P}}$ commutes with $\hat{H}_0$ (see (8.21)). In $\hat{H}_{\mathrm{I}}$ (see (8.25)), $\hat{\mathbf{P}}$ does not commute with $E(\hat{\mathbf{r}},t)$, and in the $\{\mathbf{r}\}$ representation, we have

$$[\mathbf{P}, E(\hat{\mathbf{r}},t)] = \frac{\hbar}{\mathrm{i}}\nabla\{E(\mathbf{r},t)\}, \tag{8.36}$$

whence,

$$\frac{\mathrm{d}}{\mathrm{d}t}\langle\hat{\mathbf{P}}\rangle = \langle\hat{D}_{\varepsilon}\cdot\nabla\{E(\mathbf{r},t)\}\rangle. \tag{8.37}$$

Using the factorization (8.26), we obtain

$$\frac{\mathrm{d}}{\mathrm{d}t}\langle\hat{\mathbf{P}}\rangle = \langle\psi_{\mathrm{int}}|\hat{D}_{\boldsymbol{\varepsilon}}|\psi_{\mathrm{int}}\rangle \cdot \langle\psi_r|\boldsymbol{\nabla}\{E(\mathbf{r},t)\}|\psi_r\rangle. \tag{8.38}$$

Since the wave packet $\psi(\mathbf{r},t)$ is spatially localized around $\mathbf{r}_{\mathrm{at}}$, the second term is simply,

$$\int d^3r|\psi(\mathbf{r},t)|^2\boldsymbol{\nabla}\{E(\mathbf{r},t)\} = \boldsymbol{\nabla}\{E(\mathbf{r},t)\}_{\mathbf{r}_{\mathrm{at}}}, \tag{8.39}$$

that is, the gradient of $E(\mathbf{r},t)$ taken at $\mathbf{r}_{\mathrm{at}}$. Finally, the dynamical equation for the average value of the momentum is given by

$$\frac{\mathrm{d}}{\mathrm{d}t}\langle\mathbf{P}\rangle = \langle\hat{D}_{\boldsymbol{\varepsilon}}\rangle\boldsymbol{\nabla}\{E(\mathbf{r},t)\}_{\mathbf{r}_{\mathrm{at}}}. \tag{8.40}$$

Given (8.28–8.29), the dynamical equation for the average position $\mathbf{r}_{\mathrm{at}}$ of the classical wave packet is then,

$$M\frac{\mathrm{d}^2}{\mathrm{d}t^2}\mathbf{r}_{\mathrm{at}} = \mathbf{F}, \tag{8.41}$$

with

$$\mathbf{F} = \langle\hat{D}_{\boldsymbol{\varepsilon}}\rangle\boldsymbol{\nabla}\{E(\mathbf{r},t)\}_{\mathbf{r}_{\mathrm{at}}}. \tag{8.42}$$

Equation (8.41) shows that the quantity $\mathbf{F}$ plays the role of a classical force driving the motion of a classical particle of mass M and position $\mathbf{r}_{\mathrm{at}}$.

To obtain the full expression for this force $\mathbf{F}$, we must work out the expected value $\langle\hat{D}_{\boldsymbol{\varepsilon}}\rangle$ of the quantum dipole at the point $\mathbf{r}_{\mathrm{at}}$. This is what we shall now do for the closed two-level atom.

Comment Equation (8.42) for the force is the same as the equation obtained for a classical electric dipole, placed at $\mathbf{r}_{\mathrm{at}}$ in an electric field $\boldsymbol{\varepsilon}\,E(\mathbf{r},t)$, and with component $\mathcal{D}_{\boldsymbol{\varepsilon}} = \langle\hat{D}_{\boldsymbol{\varepsilon}}\rangle$ in the direction $\boldsymbol{\varepsilon}$ of the electric field.

8.2.4 Steady-state radiative forces for a closed two-level atom

We now assume that the internal state of the atom reaches a steady state at the point $\mathbf{r}_{\mathrm{at}}$ under the effect of the field,

$$\mathbf{E}(\mathbf{r},t) = \boldsymbol{\varepsilon}\,\mathcal{E}(\mathbf{r},t) + \boldsymbol{\varepsilon}^*\,\mathcal{E}^*(\mathbf{r},t), \tag{8.43}$$

with

$$\mathcal{E}(\mathbf{r},t) = \frac{E_0(\mathbf{r})}{2}\mathrm{e}^{-\mathrm{i}\varphi(\mathbf{r})}\,\mathrm{e}^{-\mathrm{i}\omega t}. \tag{8.44}$$

As we have already seen in Section 2.4.5, and more precisely in Complement 2C, the atom undergoes forced oscillations at the frequency ω of the field, and we can write

$$\langle\hat{D}_{\boldsymbol{\varepsilon}}\rangle = \varepsilon_0\alpha\,\mathcal{E} + \text{c.c.}, \tag{8.45}$$

where α is the polarizability of the atom. This is a complex number,

$$\alpha = \alpha' + \mathrm{i}\alpha'', \tag{8.46}$$

which in principle depends on ω and E_0, and has dimensions of a length cubed.

Substituting (8.45) into (8.42), we obtain four terms, two of which oscillate at 2ω and give rise to no effects when averaged over time. The two remaining terms do not oscillate at all, and give rise to a force:

$$\begin{aligned}\mathbf{F} &= \varepsilon_0\alpha\,\mathcal{E}\{\nabla\mathcal{E}^*\}_{\mathbf{r}_{\mathrm{at}}} + \varepsilon_0\alpha^*\,\mathcal{E}^*\{\nabla\mathcal{E}\}_{\mathbf{r}_{\mathrm{at}}} \\ &= \varepsilon_0\alpha'\frac{E_0}{2}\{\nabla E_0(\mathbf{r})\}_{\mathbf{r}_{\mathrm{at}}} - \varepsilon_0\alpha''\frac{E_0^2}{2}\{\nabla\varphi(\mathbf{r})\}_{\mathbf{r}_{\mathrm{at}}}.\end{aligned} \tag{8.47}$$

The radiative force thus comprises two contributions. One of these, related to the real part α' of the polarizability (the reactive part of the atomic response), depends on the gradient of the amplitude (or the intensity) of the electromagnetic wave. This is called the *dipole force*, or the *gradient force*. The other contribution, related to the imaginary part α'' of the polarizability (the dissipative part of the atomic response), depends on the gradient of the phase of the wave. This is the *resonance-radiation pressure*.

From (2C.75), we deduce the polarizability of a two-level atom with zero velocity in the case of a quasi-resonant excitation:

$$\alpha = \frac{d^2}{\varepsilon_0\hbar}\frac{\omega_0 - \omega + \mathrm{i}\Gamma/2}{(\omega-\omega_0)^2 + \frac{\Omega_1^2}{2} + \frac{\Gamma^2}{4}}. \tag{8.48}$$

This can be rewritten as the product of the linear polarizability, independent of the wave intensity (proportional to Ω_1^2) by a saturation term $(1+s)^{-1}$ which tends to zero at high intensities:

$$\alpha = \frac{d^2}{\varepsilon_0\hbar}\frac{\omega_0 - \omega + \mathrm{i}\Gamma/2}{(\omega-\omega_0)^2 + \Gamma^2/4}\frac{1}{1+s}. \tag{8.49}$$

The saturation parameter s, which we have already encountered in (2.189) of Chapter 2, is given by

$$s = \frac{\Omega_1^2/2}{(\omega-\omega_0)^2 + \Gamma^2/4} = \frac{I}{I_{\mathrm{sat}}}\frac{1}{1+4\left(\frac{\omega-\omega_0}{\Gamma}\right)^2}, \tag{8.50}$$

with

$$I_{\mathrm{sat}} = \frac{2}{\Gamma^2}\frac{\Omega_1^2}{I}. \tag{8.51}$$

If now the velocity $\mathbf{V}$ of the atom is non-zero, the Doppler effect (see Section 8.1.3) must be taken into account in the expression for the polarizability, by replacing ω by $\omega - \mathbf{k}\cdot\mathbf{V}$ in (8.48), (8.49) and (8.50). This Doppler shift plays a very important role in many effects associated with radiative forces, as for instance in Doppler cooling (Section 8.3.1).

Comments

(i) Equation (8.47) for the force is only valid if the atomic dipole $\langle \hat{\mathbf{D}} \rangle$ is under forced oscillation conditions at the point $\mathbf{r}_{\mathrm{at}}$. When the atom is moving, it often happens that this hypothesis is not satisfied. One must then return to (8.42), and it is generally found that the force depends on the atomic velocity, as we shall see for example in Section 8.3.

(ii) The above calculation of the radiative force contains several classical features and it is wise to be clear about them. For one thing, the electromagnetic field is classical, and we do not take into account the fact that momentum exchanges between atom and radiation occur in quanta $\hbar k$. We shall see later that these effects in fact lead to a statistical scatter in the motions of the various atoms. The quantity $\mathbf{F}$ then arises as an average force in the statistical sense. In addition, we have adopted a classical view of the motion of the atom by describing it with an almost pointlike wave packet. When this last hypothesis cannot be upheld, the above results can nevertheless be used, remembering that they describe the motion of the quantum expectation of the atomic position, i.e. the barycentre of the wave packet. In this case, $\mathbf{F}$ can be considered as an average force in the quantum sense.

8.2.5 Resonance-radiation pressure

As in Section 8.1, we consider a plane travelling wave,

$$\mathcal{E}(\mathbf{r},t) = \frac{E_0}{2} \mathrm{e}^{\mathrm{i}\mathbf{k}\cdot\mathbf{r}}\, \mathrm{e}^{-\mathrm{i}\omega t}, \tag{8.52}$$

with constant amplitude E_0 and wavevector $\mathbf{k}$ (where $|\mathbf{k}| = k = \omega/c$). As can be seen from (8.47), the dipole force is then zero, and the only radiative force is the resonance-radiation pressure:

$$\mathbf{F}_1 = \varepsilon_0 \alpha'' \frac{E_0^2}{2} \nabla(\mathbf{k}\cdot\mathbf{r}) = \varepsilon_0 \alpha'' \frac{E_0^2}{2} \mathbf{k}. \tag{8.53}$$

This force lies along the direction of propagation $\mathbf{k}$ of the wave. Using (8.48) for the polarizability of a closed two-level atom, $\mathbf{F}_1$ becomes

$$\mathbf{F}_1 = \frac{d^2}{2\hbar} \frac{\Gamma}{(\omega-\omega_0)^2 + \frac{\Omega_1^2}{2} + \frac{\Gamma^2}{4}} \frac{E_0^2}{2} \mathbf{k}. \tag{8.54}$$

A simpler expression can be obtained by recalling that the Rabi angular frequency Ω_1 is equal to $-dE_0/\hbar$ (see (2.86)), and then using (8.50):

$$\mathbf{F}_1 = \hbar\mathbf{k} \frac{\Gamma}{2} \frac{\Omega_1^2/2}{(\omega-\omega_0)^2 + \frac{\Omega_1^2}{2} + \frac{\Gamma^2}{4}} = \hbar\mathbf{k}\, \frac{\Gamma}{2}\, \frac{s}{1+s}. \tag{8.55}$$

As the parameter s behaves resonantly according to (8.50), the radiation pressure has a (Lorentzian) resonance around the atomic frequency ω_0. This resonance is narrow, with width of the same order as the width Γ of the atomic line, if s is not very large compared to 1. We thus understand why a laser is needed here, with linewidth less than Γ, in order to observe a significant effect. Otherwise the light power is 'diluted' over a broad frequency band compared with Γ, and the force will be weak.

Equations (8.50) and (8.55) show that, for small values of the saturation parameter ($s \ll 1$), the radiation pressure is proportional to the light intensity. However, at high intensities ($s \gg 1$), this force saturates, and cannot exceed the maximal value,

$$\mathbf{F}_1^{\max} = \hbar\mathbf{k}\frac{\Gamma}{2}. \tag{8.56}$$

The maximal force is in fact very large, as can be observed by calculating the acceleration of an atom when $\mathbf{F}_1^{\max}$ acts on it:

$$\gamma_{\max} = \frac{F_1^{\max}}{M} = \frac{\Gamma}{2}\frac{\hbar k}{M} = \frac{\Gamma}{2}V_{\mathrm{R}}. \tag{8.57}$$

For a typical recoil velocity of $2.9 \times 10^{-2}\,\mathrm{m.s^{-1}}$ and a radiative linewidth Γ equal to $4 \times 10^7\ \mathrm{s^{-1}}$ (as for the sodium resonance line at $0.589\,\mu\mathrm{m}$), the maximal acceleration is

$$\gamma_{\max} \simeq 5.9 \times 10^5\,\mathrm{m.s^{-2}},$$

or 6×10^4 times the acceleration due to gravity at the Earth's surface.

Accelerations of that order are relatively easily reached using a laser. At resonance, a laser intensity equal to I_{sat} (typically a few mW per $\mathrm{cm^2}$) suffices to obtain an acceleration of $\gamma_{\max}/2$. This is enough to stop the atoms in a thermal beam, with typical velocities of $300\,\mathrm{m.s^{-1}}$, over a distance of less than 1 m, provided that the atom can be held in resonance with the laser during the deceleration (which naturally alters the Doppler shift). The success of such experiments at the beginning of the 1980s boosted the field of research on the laser control of atomic motions (see Figure 8.1).

The calculation leading to expression (8.55) for the resonance-radiation pressure is based on a classical model of light, where the photon concept has no place. However, the quantity $\hbar\mathbf{k}$ appears in the formula, and this is the momentum of photons associated with the light wave exerting the radiation pressure. It is thus tempting to interpret the expression for force $\mathbf{F}_1$ in terms of momentum exchanges between photons and atoms. To do this, note that the quantity,

$$\Gamma_{\mathrm{fluo}} = \frac{\Gamma}{2}\frac{s}{1+s}, \tag{8.58}$$

is precisely the number of fluorescence cycles (absorption and spontaneous emission) per second. This is shown using the optical Bloch equations, but it follows much more simply (although less rigorously) from rate equations analogous to those written down in Section 2.6.2. Letting π_a and π_b be the populations (probabilities of occupation) in levels a and b, and using the rates $\Gamma s/2$ for absorption and stimulated emission, and Γ for spontaneous emission, the rate equations describing the evolution of π_a and π_b for the closed two-level system in Figure 8.2 can be written in the form,

$$\frac{\mathrm{d}\pi_b}{\mathrm{d}t} = -\frac{\mathrm{d}\pi_a}{\mathrm{d}t} = \Gamma\frac{s}{2}\pi_a - \Gamma\frac{s}{2}\pi_b - \Gamma\pi_b. \tag{8.59}$$

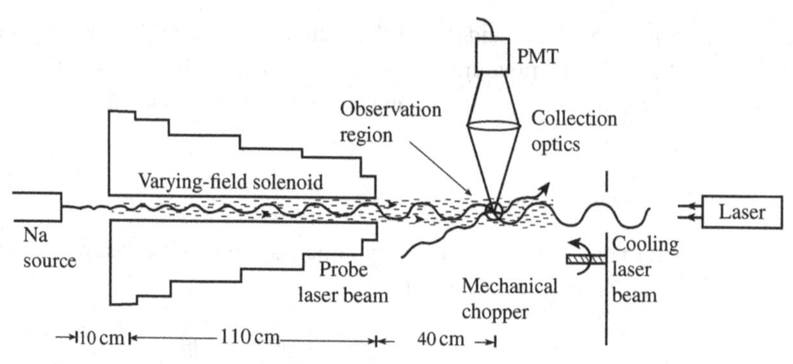

Figure 8.1 Slowing down atoms in a sodium beam by resonance-radiation pressure (from Prodan, Phillips and Metcalf).[6] The atoms are decelerated by a quasi-resonant laser shining in the opposite direction. The solenoid produces a variable magnetic field in its centre, where the atoms propagate under vacuum, and this shifts the atomic resonance ω_0 in a position-dependent way by the Zeeman effect. For a judiciously chosen field, the variation of the atomic resonance exactly balances the change in the Doppler effect due to the deceleration of the atoms. The latter can thus remain in resonance during the whole deceleration process, making it extremely effective, and indeed, even able to completely stop the atoms.

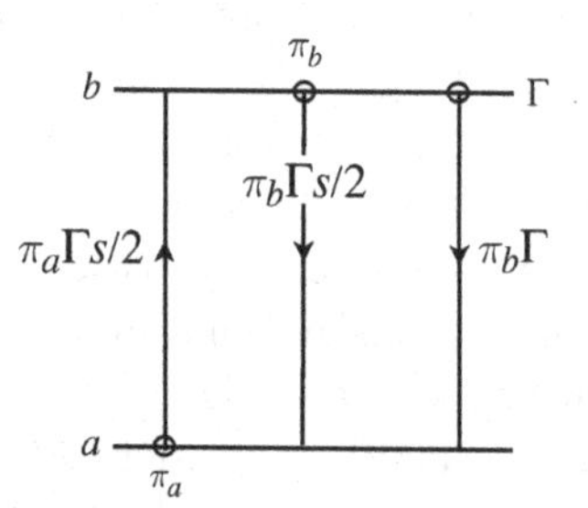

Figure 8.2 The processes of absorption, induced emission and emission of scattered photons contributing to the rate equations describing a closed two-level system $\{a, b\}$. The transition rates are indicated for a wave with saturation parameter s as a function of the populations π_a and π_b of the two states (i.e. the fraction of atoms in a and b, respectively). Level b can only decay to the stable level a, and the sum of the populations is unity, i.e. $\pi_a + \pi_b = 1$.

Since $\pi_a + \pi_b = 1$, the steady-state solution of (8.59) is

$$\pi_b = \frac{1}{2}\frac{s}{s+s}. \tag{8.60}$$

The number of fluorescence cycles per second is equal to the number of photons spontaneously scattered per second:

$$\Gamma_{\text{fluo}} = \Gamma \pi_b = \frac{\Gamma}{2}\frac{s}{1+s}, \tag{8.61}$$

which is just (8.58). In this formula, s is the saturation parameter of (8.50).

[6] J. V. Prodan, W. D. Phillips and H. Metcalf, Laser Production of a Very Slow Monoenergetic Atomic Beam, *Physical Review Letters* **49**, 1149 (1982).

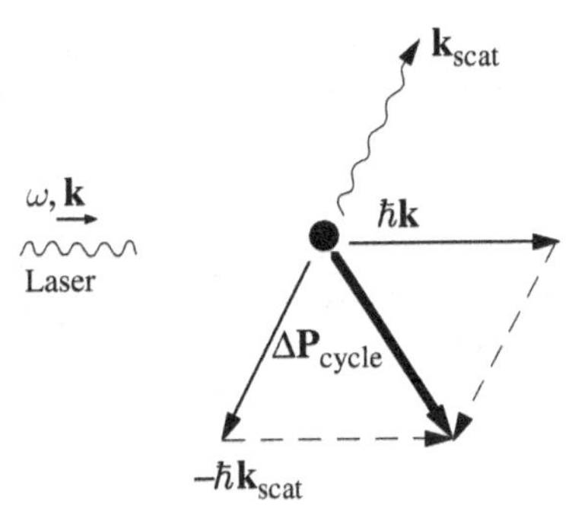

Figure 8.3 Momentum exchanges between the atom and photons in an elementary fluorescence cycle. If this cycle is repeated many times, the recoil $-\hbar\mathbf{k}_{\rm scat}$ due to the spontaneously scattered photons averages out to 0, and the average momentum received by the atom is directed along the wavevector $\mathbf{k}$ of the laser.

Given (8.58), expression (8.55) for the force can be rewritten in the form,

$$\mathbf{F}_1 = \hbar\mathbf{k}\Gamma_{\rm fluo}, \tag{8.62}$$

which is easy to reinterpret by working out the balance of momentum exchanges between the atom and the photons during each fluorescence cycle (see Figure 8.3). Indeed, during one elementary cycle, a laser photon disappears, and its momentum $\hbar\mathbf{k}$ is imparted to the atom, while a scattered photon appears, with momentum $\hbar\mathbf{k}_{\rm scat}$. The change in momentum of the atom during one fluorescence cycle is, therefore,

$$\Delta\mathbf{P}_{\rm cycle} = \hbar\mathbf{k} - \hbar\mathbf{k}_{\rm scat}. \tag{8.63}$$

When the process repeats itself, the term $\hbar\mathbf{k}$ associated with the incident laser photon is the same each cycle, whereas the direction of the scattered photon is random, with a radiation pattern that is symmetric about the coordinate origin. On average, over a great many fluorescence cycles, we thus have,

$$\langle\mathbf{k}_{\rm scat}\rangle = 0, \tag{8.64}$$

whence an average change in the atomic momentum per cycle is given by,

$$\langle\Delta\mathbf{P}_{\rm cycle}\rangle = \hbar\mathbf{k}. \tag{8.65}$$

The average change in the atomic momentum per second is therefore equal to

$$\left\langle\frac{\mathrm{d}}{\mathrm{d}t}\mathbf{P}_{\rm at}\right\rangle = \Gamma_{\rm fluo}\,\hbar\mathbf{k}, \tag{8.66}$$

which is the same as expression (8.62) for the force $\mathbf{F}_1$, obtained using the semi-classical model. As announced earlier, we have thus interpreted it in terms of momentum exchanges between laser photons and atoms.

Comments

(i) We have not considered absorption/stimulated emission cycles. Indeed, in the case of a single plane wave, the total momentum transfer is zero in such a cycle, since,

$$\Delta \mathbf{P}_{\text{sti}} = -\Delta \mathbf{P}_{\text{abs}} = \hbar \mathbf{k}. \tag{8.67}$$

(ii) The scattered photon, re-emitted in the fluorescence process of Figures 8.2 or 8.3, is sometimes called a spontaneous photon. This terminology may be misleading because it might obscure the fact that the spectral properties of the re-emitted light are closely related to those of the laser light. For example, at low saturation ($s < 1$), and for a fixed atom, the scattered light has strictly the same frequency ω as the incident laser light, whereas the light spontaneously emitted by an atom previously raised to level b would have a spectrum of width Γ centred on ω_0. On the other hand, the word 'spontaneous' reminds one that the direction of emission is random. Moreover, it is consistent with the expression $\Gamma\pi_b$ of the rate of scattering (see Equation (8.61) or Figure 8.2). A good terminology, not widely used however, might be 'spontaneously scattered photons'.

8.2.6 Dipole force

Consider now a laser field whose amplitude $E_0(\mathbf{r})$ is non-uniform. For example, we take into account the Gaussian profile,

$$E_0(x, y) = A_0 \, \mathrm{e}^{-\frac{x^2+y^2}{w_0^2}}, \tag{8.68}$$

of a beam propagating in the Oz direction. The field has complex amplitude,

$$\mathcal{E}(\mathbf{r}, t) = \frac{A_0}{2} \, \mathrm{e}^{-\frac{x^2+y^2}{w_0^2}} \, \mathrm{e}^{ikz} \, \mathrm{e}^{-i\omega t}. \tag{8.69}$$

The first term in (8.47) now gives a non-zero contribution, namely, the *dipole force*:

$$\mathbf{F}_2 = \varepsilon_0 \alpha' \frac{E_0}{2} \cdot \boldsymbol{\nabla} E_0 = \varepsilon_0 \frac{\alpha'}{4} \boldsymbol{\nabla}(E_0^2). \tag{8.70}$$

Using expression (8.48) for the polarizability, we obtain

$$\mathbf{F}_2 = \frac{\hbar(\omega_0 - \omega)}{2} \frac{\boldsymbol{\nabla}(\Omega_1^2/2)}{(\omega - \omega_0)^2 + \frac{\Omega_1^2}{2} + \frac{\Gamma^2}{4}}. \tag{8.71}$$

Here we have used the definition $-dE_0(\mathbf{r}) = \hbar\Omega_1(\mathbf{r})$ of the Rabi angular frequency once again.

The dipole force thus varies as the gradient of the light intensity,

$$I = \frac{2\Omega_1^2}{\Gamma^2} I_{\text{sat}}. \tag{8.72}$$

For a negative detuning ($\omega < \omega_0$), the atoms are attracted toward regions of high intensity, while for positive detuning ($\omega > \omega_0$), they are repelled from such regions.

Integrating (8.71), we find that the dipole force $\mathbf{F}_2$ derives from a potential, i.e.

$$\mathbf{F}_2 = -\boldsymbol{\nabla} U_{\text{dip}}(\mathbf{r}), \tag{8.73}$$

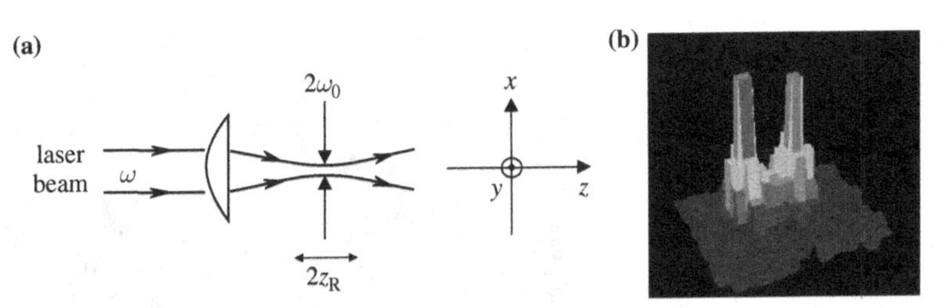

Figure 8.4 **(a) Trapping atoms attracted into the focal region of a negatively detuned laser ($\omega < \omega_0$). The attractive dipole potential can trap atoms provided that they have small enough kinetic energy. (b) Using two beams highly focused at neighbouring points, two single atoms can be trapped separately, and displaced relative to one another, by steering each laser beam independently. Each atom can be observed by the light it scatters when irradiated with resonant light, typically 10^7 photons.s^{-1}, which is easily detected, as can be seen from this figure showing the fluorescence detected on a CCD camera. (Figure courtesy of P. Grangier, Institut d'Optique.)**

with

$$\begin{aligned} U_{\text{dip}}(\mathbf{r}) &= \frac{\hbar(\omega - \omega_0)}{2} \ln\left[1 + \frac{\Omega_1^2(\mathbf{r})/2}{(\omega - \omega_0)^2 + \Gamma^2/4}\right] \\ &= \frac{\hbar(\omega - \omega_0)}{2} \ln\left[1 + s(\mathbf{r})\right]. \end{aligned} \tag{8.74}$$

(The choice of integration constant gives a zero potential outside the laser beam.) This property is very important, because it suggests that it may be possible to trap atoms at the focal point of a laser beam (see Figure 8.4), if the latter is negatively detuned, i.e. if $\omega < \omega_0$. Indeed, at this point, the saturation parameter has a maximum s_{max}, and the potential has a minimum, not only in the x and y directions (potential well of typical width w_0, see (8.69)), but also in the z direction (well of typical width $z_R = \pi w_0^2/\lambda$, where z_R is the Rayleigh length, see Complement 3B). The atoms thus find themselves in a potential well of depth,

$$\Delta U = \frac{\hbar(\omega_0 - \omega)}{2} \ln\left(1 + s_{\text{max}}\right),$$

which increases as the maximal intensity is increased. They will be trapped there provided that their kinetic energy is less than the depth of the well. In practice, with ordinary lasers, it is difficult to produce potential wells able to trap atoms with temperatures much above a few hundred millikelvins, and the atoms first have to be cooled using the methods we shall describe in Section 8.3, in order to be able to trap them. However, with sufficiently cold atoms, such traps are very widely used. Note that it is easy to move the focal point of the laser, and with it the atoms (see Figure 8.4). This device is called an optical tweezer.

The dipole force also plays an important role in atom optics. It underlies devices used to reflect atoms (atom mirrors), to focus them (atom lenses) and to diffract them (atom diffraction gratings). Another important application is the so-called 'optical lattice' scheme, based

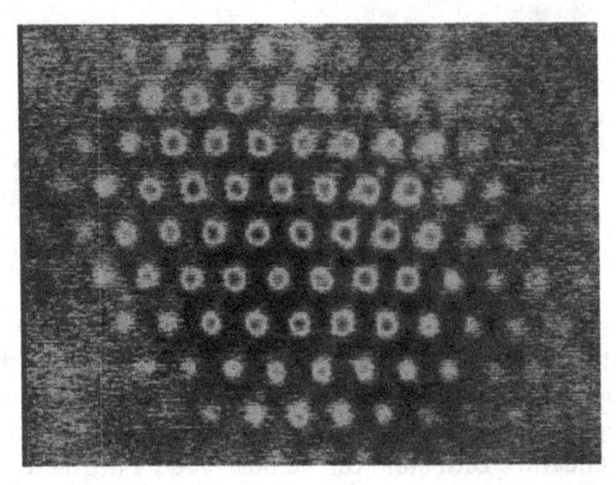

Figure 8.5 Ultracold atoms trapped in a periodic array of microtraps, 10 μm apart, at the antinodes of a standing wave realized with laser beams detuned toward low frequencies ($\omega < \omega_0$). The atoms are observed by absorption of the resonant light. (Image courtesy of D. Boiron and C. Salomon.)

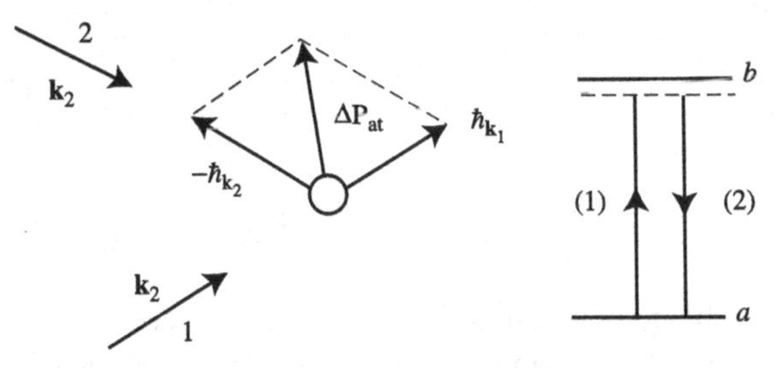

Figure 8.6 Interpretation of the dipole force in terms of the redistribution of photons between wave 1 (absorption) and wave 2 (stimulated emission). During such an elementary cycle, the atom receives an impulse $\hbar(k_1 - k_2)$. Note that the opposite cycle (absorption in wave 2 and stimulated emission in wave 1) communicates the same impulse to the atom but with opposite sign.

on the dipole potential of a 3D standing wave. It realizes a 3D periodic array of microtraps, in which cold atoms may be trapped (see Figure 8.5).

The dipole force has been calculated using the semi-classical approach, where it can be simply interpreted as resulting from the coupling between the electric field of the light and the dipole moment which it induces in the atom. One might wonder whether it could be reinterpreted in terms of momentum exchanges between photons and atoms. The answer is in fact affirmative, but it is not so easy this time. A non-uniform wave decomposes into several plane waves with distinct wavevectors. Consider the simplest case of two plane waves with wavevectors $\mathbf{k}_1$ and $\mathbf{k}_2$, and consider cycles involving the absorption of a photon in one wave, e.g. wave 1, followed by a stimulated emission in the other wave (see Figure 8.6). The result is a force directed along $\mathbf{k}_1 - \mathbf{k}_2$. But we could also consider the opposite process, i.e. absorption in wave 2 followed by a stimulated emission in wave 1, leading to a force of the opposite sign. It can be shown that what favours one process or the other is the relative phase of the two waves 1 and 2 at the point where the atom is located.[7] The photon interpretation is difficult to handle here, in contrast with the picture resulting from the semi-classical model, which rests on the well-known fact that a dipole placed in a non-uniform field suffers a force along the gradient of this field (the sign depending on whether the dipole is in phase with the field, or out of phase).

This is a good illustration of the fact that it is worth having several models for the atom–radiation interaction. The semi-classical approach, in which the classical electromagnetic field interacts with the atomic dipole, provides a simple picture of the dipole force $\mathbf{F}_2$. However, it is the quantum description of light, in terms of photons with a certain momentum, that provides the simplest picture of the resonance-radiation pressure $\mathbf{F}_1$. Apart from its simplicity, we shall see that the latter picture also suggests the existence of a heating process which eventually limits the temperature reduction that can be obtained by Doppler cooling, to be described shortly.

Comments

(i) In principle, in a non-uniform wave like the one shown in Figure 8.4, the two forces $\mathbf{F}_1$ and $\mathbf{F}_2$ will both be at work. It is interesting to identify the parameters that favour one or the other of these forces. To compare expressions (8.55) and (8.71) for $\mathbf{F}_1$ and $\mathbf{F}_2$, respectively, we consider the case of a light intensity with very high gradient, assumed to vary significantly on the scale of the wavelength. In this case $|\boldsymbol{\nabla}|$ is of the order of $\mathbf{k}$, and

$$\frac{|\mathbf{F}_1|}{|\mathbf{F}_2|} \simeq \frac{\Gamma}{|\omega_0 - \omega|}.$$

For low detuning ($|\omega - \omega_0| \ll \Gamma$), it is the resonance-radiation pressure that dominates, while at high values of the detuning, the dipole force is favoured. When we compare the effects of the forces $\mathbf{F}_1$ and $\mathbf{F}_2$, it should generally be borne in mind that $\mathbf{F}_2$ derives from a potential, whereas $\mathbf{F}_1$ does not.

(ii) In the limit of very high values of the detuning and small saturation ($|\delta| \gg \Gamma, \Omega_1$, hence $s \ll 1$), the dipole potential (8.74) takes the value

$$U_{\text{dip}}(\mathbf{r}) \simeq \frac{\hbar\Omega_1^2(\mathbf{r})}{4(\omega - \omega_0)}, \tag{8.75}$$

which coincides with the light-shift in the ground state a under the effect of the laser wave (see Section 2.3.4 of Chapter 2). It is clear how this should be interpreted. Under these conditions, there are very few fluorescence cycles (see Equation (8.58)), and the atom is almost always in the ground state $|a\rangle$, whose light-shift is then almost equal to the dipole potential, which is in fact the average of the light-shift of the ground state and the excited state, with weights proportional to the time spent by the atom in each level. Moreover, if the fluorescence rate is lower than the reciprocal of the duration of the experiment, most of the atoms remain in the ground state for the whole duration of the experiment, and for them the dipole potential is strictly equal to that of the ground state, with no fluctuations.

(iii) The case of closed two-level atoms is important because radiative forces have a strong effect only if they are allowed to act for long enough. If the two-level system $\{a, b\}$ is not closed, spontaneous emission may bring the atom into a level other than a, where the radiative forces will usually be negligible, because the detuning will be very large. Since resonance-radiation

[7] See Footnote 5, this chapter.

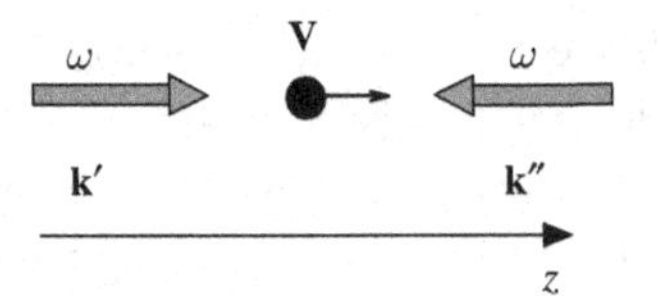

Figure 8.7 Doppler cooling in one dimension. The atom is placed in two counterpropagating waves ($\mathbf{k}'' = -\mathbf{k}'$), of the same frequency ω, lower than the atomic frequency ω_0, and of the same intensity. In the reference frame of the moving atom, the wave that opposes its motion is closer to resonance than the one in the direction of the motion. This is therefore the one that will dominate, and the motion of the atom is damped.

pressure is based on fluorescence cycles involving spontaneously scattered photons, it has a significant effect only for closed two-level systems. On the other hand, dipole force can play an important role even when there is no spontaneous emission, see comment (ii) above, and there is no need to have a closed two-level system to make use of the dipole force.

8.3 Laser cooling and trapping of atoms, optical molasses[8]

8.3.1 Doppler cooling

Consider a closed two-level atom with non-zero velocity $\mathbf{V}$, placed in two counterpropagating laser waves of the same frequency ω, slightly below the atomic frequency ω_0 (see Figure 8.7). We denote the detuning from resonance by

$$\delta = \omega - \omega_0, \tag{8.76}$$

which is thus negative.

These two waves form a stationary wave and, in principle (see (8.47)), one might think that the radiation pressure $\mathbf{F}_1$ would be zero, since the phase is constant, whence the atom would suffer only the dipole force resulting from the spatial intensity variations associated with the nodes and antinodes of the stationary wave. But in fact, for a moving atom, the changes in light intensity are so fast that the internal state never reaches the steady state, and care must be taken in applying the results of Section 8.2. We must use the optical Bloch equations (Complement 2C) to calculate the instantaneous internal state, taking the motion into account, and then deduce the radiative force. It is thereby shown that, for a saturation parameter smaller than unity, and averaging the effects over a half-wavelength, coherence between the waves $\mathbf{k}'$ and $\mathbf{k}''$ can be neglected.[9] The forces due to each plane wave can thus be taken into account separately, whence we must add the resonance-radiation pressures exerted by each wave. We thereby obtain the average radiative force $\mathbf{F}$.

[8] See Footnote 1, this chapter.
[9] See Footnote 5, this chapter.

Let us calculate the radiation pressures exerted by each of these waves, of the same small intensity compared with I_{sat} (Ω_1 small compared with Γ). We have

$$\mathbf{F}' = \hbar\mathbf{k}'\frac{\Gamma}{2}s', \tag{8.77}$$

with

$$s' = \frac{\Omega_1^2/2}{(\delta - kV_z)^2 + \frac{\Gamma^2}{4}} = \frac{s_0}{1 + 4\left(\frac{\delta - kV_z}{\Gamma}\right)^2} \tag{8.78}$$

and

$$s_0 = \frac{2\Omega_1^2}{\Gamma^2} = \frac{I}{I_{sat}}. \tag{8.79}$$

We have taken into account the Doppler effect. V_z is the component of the velocity in the Oz direction and $k = |\mathbf{k}'|$. Likewise, for the second wave ($\mathbf{k}'' = -\mathbf{k}'$),

$$\mathbf{F}'' = \hbar\mathbf{k}''\frac{\Gamma}{2}s'', \tag{8.80}$$

with

$$s'' = \frac{s_0}{1 + 4\left(\frac{\delta + kV_z}{\Gamma}\right)^2}. \tag{8.81}$$

Note that the Doppler effect is of the opposite sign to the Doppler effect for the first wave. The sum of the two forces $\mathbf{F}'$ and $\mathbf{F}''$ has the following component in the Oz direction:

$$F = \hbar k\frac{\Gamma}{2}s_0\left[\frac{1}{1 + \frac{4}{\Gamma^2}(\delta - kV_z)} - \frac{1}{1 + \frac{4}{\Gamma^2}(\delta + kV_z)}\right]. \tag{8.82}$$

Figure 8.8 shows this force as a function of the velocity for a negative detuning ($\delta = -\Gamma/2$). We observe that this force always has opposite sign to the velocity. So whatever direction it moves in, the atom is always slowed down. The result is interpreted by noting that, in the reference frame of the atom, the Doppler effect brings the wave opposed to the motion of the atom closer to resonance, and it is therefore the radiation pressure of this wave that will dominate (see Figure 8.7).

A force opposing the velocity and proportional to it is a friction force. It will damp all the velocities of a whole assembly of atoms, which is thereby cooled. The phenomenon we have just described is called Doppler cooling. Applied to previously decelerated atoms (see, for example, Figure 8.1), Doppler cooling can produce temperatures below the millikelvin, as we shall show in the following sections.

8.3.2 Coefficient of friction and Doppler molasses

A truncated expansion of the total force F_z (8.82) about $V_z = 0$ corresponding to the linear part of the curve of Figure 8.8 around the origin, gives

$$F_z = -M\gamma\, V_z, \tag{8.83}$$

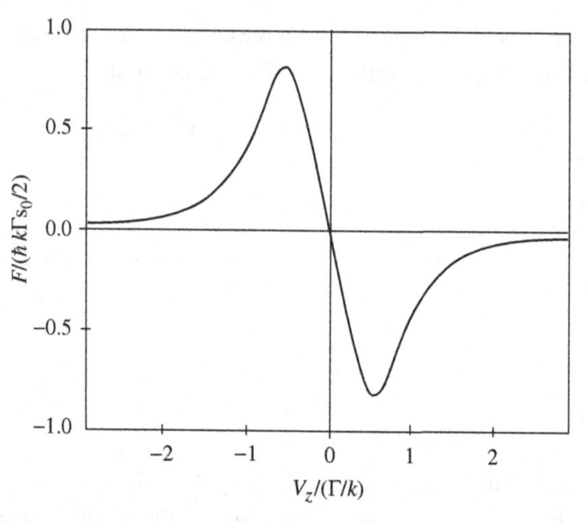

Figure 8.8 **Doppler cooling. The resultant force of the two waves in Figure 8.7 (see (8.82)) is plotted in reduced units ($s_0\hbar k\Gamma/2$), and for $\delta = -\Gamma/2$. The total radiative force always opposes the velocity of the atom (frictional force). An ensemble of atoms with initial velocities distributed over an interval of the order of a few times Γ/k is very efficiently cooled.**

where

$$\gamma = \frac{8\hbar k^2}{M} s_0 \frac{\left(-\frac{\delta}{\Gamma}\right)}{\left(1 + 4\frac{\delta^2}{\Gamma^2}\right)^2}, \tag{8.84}$$

is the coefficient of friction, positive if $\delta < 0$. The quantity γ is the reciprocal of the time required for the velocity to drop by a factor of $1/e$, since (8.83) leads to

$$\frac{dV_z}{dt} = -\gamma V_z, \tag{8.85}$$

a differential equation whose solution is a damped exponential, with time constant γ.

If the atom is now placed in three mutually orthogonal pairs of counterpropagating waves (see Figure 8.9), the three components of the velocity are damped. The atom is slowed down, whatever the direction of its initial velocity. Under low saturation conditions ($s_0 \ll 1$), the above calculation generalizes immediately, and we may write

$$\frac{d\mathbf{V}}{dt} = -\gamma \mathbf{V}, \tag{8.86}$$

where the coefficient γ has the value (8.84) if the six waves have the same intensity and detuning. The coefficient of friction is maximal for $\delta = -\Gamma/2$, and takes the value,

$$\gamma_{\max} = s_0 \frac{\hbar k^2}{M}. \tag{8.87}$$

For alkali metal atoms, $\hbar k^2/M$ is typically of order $10^4\ \mathrm{s}^{-1}$ and, for s_0 of order 0.1, the damping time $1/\gamma_{\max}$ for the velocities is less than a millisecond. It is as though the atoms were caught in an extremely viscous 'medium'. For this reason, this kind of configuration was referred to as 'optical molasses' by S. Chu, the first to observe the effect.

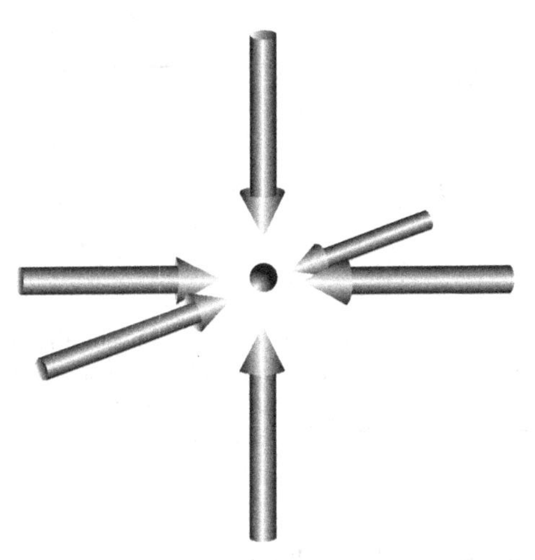

Figure 8.9 Three-dimensional optical molasses. The velocity of an atom, placed in three pairs of counterpropagating waves aligned with three mutually orthogonal axes, is very quickly reduced (in less than one millisecond).

Note that the range over which the frictional force is linear in the velocity is of order Γ/k (see Figure 8.8). More generally, if the velocity is much greater than $-\delta/k$, also called the capture velocity, then the friction is negligible. The capture velocity typically lies in the range 1–10 m.s^{-1}. It is very small compared with thermal atomic velocities at room temperature (several hundred metres per second). However, the atoms in an atomic beam slowed down and stopped by resonance-radiation pressure have residual velocities of a few metres per second. If they are then caught in an optical molasses, these residual velocities are very quickly damped out (in a few milliseconds), and the atomic vapour is cooled extremely efficiently. The average power (in the statistical sense) supplied to each atom by the frictional force is

$$\left[\frac{\mathrm{d}W}{\mathrm{d}t}\right]_{\mathrm{ref}} = \langle \mathbf{F} \cdot \mathbf{V} \rangle = -\gamma M \langle \mathbf{V}^2 \rangle. \tag{8.88}$$

The angle brackets here indicate a statistical average over the ensemble of atoms in the sample. If γ is positive, the average power (8.88) received by the atoms is negative, which means that the atomic sample is cooled. As can be seen from (8.88), the efficiency of cooling drops when $\langle V^2 \rangle$, i.e. the temperature, decreases. Since there is also a heating process independent of the velocity, as will be seen in Section 8.3.6, a limit temperature will be reached (Section 8.3.7).

8.3.3 Magneto-optical trap

The effect of the optical molasses is to drastically reduce the velocities of the atoms, but it acts in a uniform way at every point of the region where the six beams overlap. The atoms are cooled, i.e. their velocities are reduced, but their spatial density remains unchanged. It can be extremely convenient to concentrate the cooled atoms within a very small region of

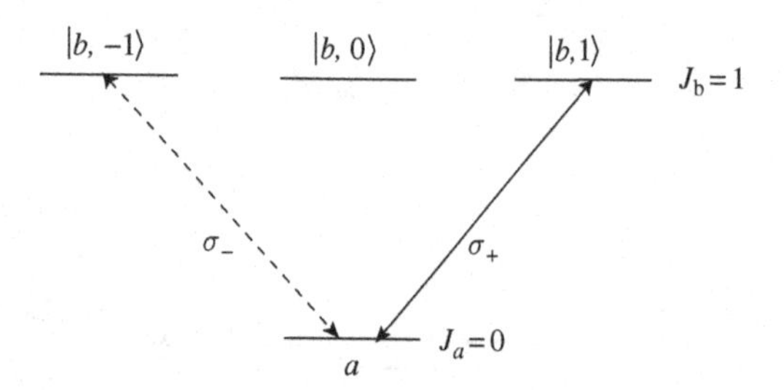

Figure 8.10 $J_a = 0 \leftrightarrow J_b = 1$ system. Light with circular polarization σ_+ interacts solely with the transition $|a\rangle \leftrightarrow |b, 1\rangle$, while light with circular polarization σ_- interacts solely with the transition $|a\rangle \leftrightarrow |b, -1\rangle$. The same axis Oz is used to define the circular polarizations σ_+ and the Zeeman sub-levels $m = 0, \pm 1$.

space. This can be done with the magneto-optical trap, which we shall now describe, and which can be considered as a trapping molasses, or as we will see, a damped trap.

We consider an atom in which the ground state $|a\rangle$ has total angular momentum $J_a = 0$ and the excited state has total angular momentum $J_b = 1$. This level thus comprises three Zeeman sub-levels $\{|b,-1\rangle, |b,0\rangle, |b,+1\rangle\}$. We have seen in Section 2B.1.3 of Complement 2B that a wave with circular polarization σ_+ interacts only with the transition $|a\rangle \leftrightarrow |b,+1\rangle$, while a wave σ_- interacts only with $|a\rangle \leftrightarrow |b,-1\rangle$ (see Figure 8.10).

In the presence of a magnetic field $\mathbf{B} = B\mathbf{e}_z$ in the Oz direction, which is taken as the axis of quantization of the levels $|b,m\rangle$, the Zeeman effect modifies the energy of these levels by an amount $m\hbar\omega_B$, with $\omega_B = g\dfrac{\mu_B}{\hbar}B$, where μ_B is the Bohr magneton ($\mu_B/\hbar$ is equal to roughly 14 GHz.T^{-1}) and the Landé factor g is a number of order unity. The two sub-levels $|b,+1\rangle$ and $|b,-1\rangle$ are shifted in opposite directions.

Now consider the situation shown in Figure 8.11, where the atom interacts with two travelling waves with polarizations σ_+ and σ_- propagating in opposite directions along the z-axis. The radiation pressure exerted by each of these waves is calculated by a similar method to the one used in Section 8.3.1, taking into account the Zeeman shifts of different sign for each of the two polarizations. The detuning $\omega - \omega_0$ thus becomes $\omega - \omega_0 - \omega_B$ for the σ_+ wave, and $\omega - \omega_0 + \omega_B$ for the σ_- wave. Equation (8.82) is therefore simply adapted to the configuration in Figure 8.11 by modifying the Lorentzian denominators to give

$$F = \hbar k \frac{\Gamma}{2} s_0 \left[\frac{1}{1 + \frac{4}{\Gamma^2}(\omega - \omega_0 - \omega_B - kV_z)} - \frac{1}{1 + \frac{4}{\Gamma^2}(\omega - \omega_0 + \omega_B + kV_z)} \right]. \tag{8.89}$$

In fact, we may simply replace kV_z by $kV_z + \omega_B$ in (8.82) to obtain the total force for this polarization and field configuration. In the limit $|kV_z + \omega_B| \ll \sqrt{\delta^2 + \Gamma^2/4}$, the force calculated to first order in kV_z and ω_B is

$$F_z = M\frac{\mathrm{d}V_z}{\mathrm{d}t} = -\gamma\left(V_z + \frac{\omega_B}{k}\right), \tag{8.90}$$

where γ is still given by (8.84).

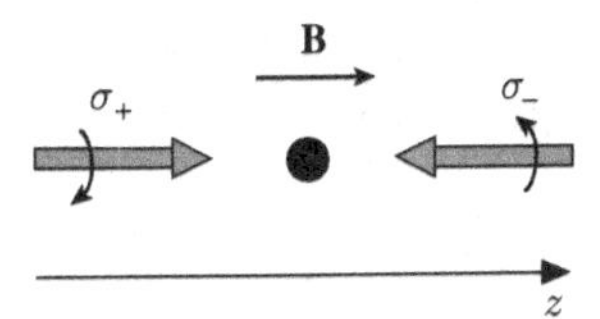

Figure 8.11 Atom subjected to two counterpropagating waves with polarizations σ_+ and σ_-, quasi-resonant with a transition $J = 0 \leftrightarrow J = 1$, in the presence of a magnetic field **B**. We now have a moving molasses which cools atoms around the velocity $V_z = -\omega_B/k$.

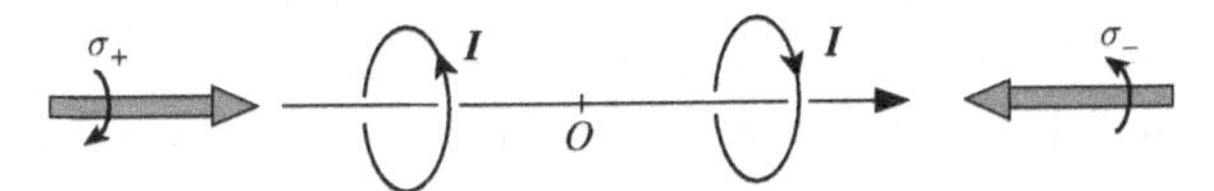

Figure 8.12 One-dimensional magneto-optical trap. The situation here is analogous to the one shown in Figure 8.11, but the two coils now create a magnetic field B_z that varies linearly as a function of z around $z = 0$. As a result, the atoms are cooled and trapped around O.

If $\omega < \omega_0$, there is still a frictional force, but it tends to accumulate the atoms around the velocity $V_z = -\omega_B/k$, rather than around $V_z = 0$. This gives a sort of 'moving molasses', in which all the atoms move at almost the same velocity. As ω_B varies linearly with B, the magnetic field can be used to adjust the velocity of the cooled atoms around a non-zero value.

To obtain a trapping effect, in addition to the friction force, one can use a magnetic field that varies in the z direction, and which is zero at a point where the atoms accumulate. For example, one can use two identical coils in which the currents flow in opposite directions (see Figure 8.12). Half-way between the coils (the point we take as origin), the field is zero, and the velocity $V_z = -\omega_B/k$ around which the atoms are cooled is zero. The atoms thus accumulate around this point. For a quantitative analysis, it suffices to retain the first non-zero term in the expansion of the component B of the field on the z-axis of the field near the origin. The latter varies linearly with z, whence ω_B is proportional to z, and (8.90) becomes

$$M\frac{\mathrm{d}}{\mathrm{d}t}V_z = -\gamma\left[V_z + \frac{g\mu_B}{\hbar k}\frac{\partial B}{\partial z}z\right]. \tag{8.91}$$

If $\gamma > 0$ and $g\dfrac{\partial B}{\partial z} > 0$, the force is attractive both toward $V_z = 0$ in velocity space and toward $z = 0$ in position space. This configuration can thus simultaneously *slow the atoms down* and *group them together in space*, i.e. trap them. Equation (8.91) is in fact the equation for a damped harmonic oscillator. The atoms are both cooled and trapped.

As in the case of the optical molasses, the above argument can be generalized to three dimensions. The magneto-optical trap works with three pairs of waves aligned with the axes Ox, Oy and Oz. In each pair, the waves propagate in opposite directions and have opposite circular polarizations.

The magneto-optical trap is remarkably effective, and has become the basic tool (the 'workhorse') in cold atoms physics. It can even cool and trap atoms directly from a vapour at room temperature, in which the slowest atoms (in the low-velocity tail of the Maxwell–Boltzmann distribution) have velocities in the capture range.

8.3.4 Fluctuations and heating

The argument leading to expression (8.42) for the radiative force is based on a classical model of the electromagnetic field, considered as an electromagnetic wave of perfectly well-defined amplitude and phase. In a quantum description of laser light, the classical field (8.18) can be considered as the quantum expectation value of the field operator, with the radiation in the quasi-classical state describing the laser wave (see Section 5.3.4). The observables associated with the quantum field have fluctuations, for which simple physical pictures can often be devised (see Chapter 5). For example, the intensity fluctuations of a quasi-classical state can be associated with a description in terms of photons with a Poisson distribution.

When we take into account the quantum nature of the field, the radiative force also becomes a quantum observable, and the value for **F** given in (8.42) is its quantum expectation value, taken in the state describing the field and atoms. This force has fluctuations around its expectation value, and the atomic trajectories will be scattered around the average trajectory. This implies in particular that the atomic velocities will be scattered around the average velocity (Figure 8.13). This dispersion in the velocities corresponds to a kinetic energy of agitation W in the frame corresponding to the average velocity:

$$W = \frac{1}{2} M \left\langle |\mathbf{V} - \langle \mathbf{V} \rangle|^2 \right\rangle . \tag{8.92}$$

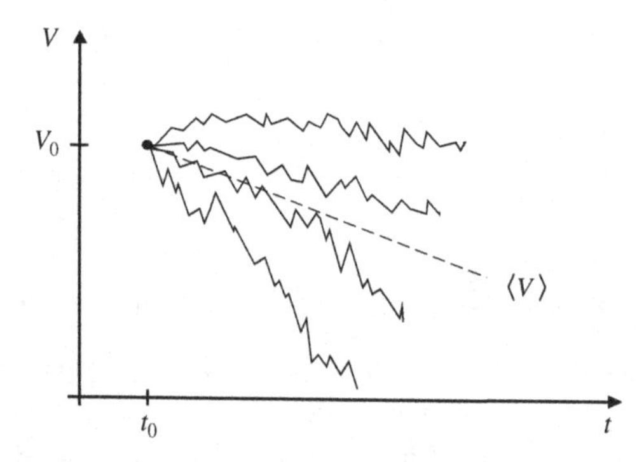

Figure 8.13 Fluctuations (dispersion) in the force, and heating. An ensemble of atoms, each initially with the same velocity V_0, is subjected to a random force, which exhibits statistical scatter in the sense that it is different for each atom. Under the effect of this random force, the trajectories of the various atoms begin to spread out, and the atomic velocities exhibit a dispersion which increases with time, around the average velocity $\langle V \rangle$, which for its part evolves under the effect of the average force. The result is that the sample is heated, the initial temperature being taken as zero in the ideal case considered here.

A temperature can be associated with this kinetic energy of agitation (if the velocity distribution is Gaussian and isotropic around the average velocity). For each direction i, we write

$$\frac{1}{2}k_B T = \frac{1}{2}M\left\langle (V_i - \langle V_i \rangle)^2 \right\rangle = \frac{W}{3}. \tag{8.93}$$

As we shall see below, this temperature tends to increase under the effect of quantum fluctuations, and thus leads to a heating effect which competes with the cooling phenomenon described so far, until an equilibrium temperature is eventually reached. We shall evaluate this heating effect, first in the case of the resonance-radiation pressure, and then for a Doppler molasses.

Comments

(i) In (8.92–8.93), we have deliberately left some ambiguity in the kind of average $\langle\ \rangle$ intended. It is fundamentally a quantum average. However, a quantum average is nothing but an average over a hypothetical statistical ensemble obtained by preparing a large number of systems in a strictly identical way. For an ensemble of atoms placed in a homogeneous laser beam, and without mutual interactions, we do in fact have a large number of identical experiments running in parallel, and we may consider the quantum average to coincide with the statistical average over the atom ensemble. In the following, unless otherwise mentioned, the averages between angle brackets can be taken as denoting either quantum expectations or statistical averages.

(ii) Any fluctuation induces heating. For example, we have mentioned (see comment (ii) in Section 8.2.6) that the dipole potential (8.74) is in fact the average of the potential felt by the atom in its ground state and in its excited state, between which it jumps randomly. The atoms caught in the dipole trap of Figure 8.4(a) are thus heated and end up escaping from the trap. We thus see the interest of using a large detuning from resonance so that the average time interval between two jumps from one state to the other is greater than the duration of the experiment, i.e. several seconds.

8.3.5 Fluctuations of the resonance-radiation pressure

In order to take into account the quantum fluctuations in the resonance-radiation pressure, we refer to the model in Section 8.2.5, in which the resonance radiation pressure force $\mathbf{F}_1$ results from the average momentum transfer between the photons and the atoms during fluorescence cycles. A classical stochastic model can then be used, in which the force is a stochastic process $\mathcal{F}_1(t)$, with (statistical) average equal to $\mathbf{F}_1$:

$$\mathbf{F}_1 = \langle \mathcal{F}_1(t) \rangle. \tag{8.94}$$

We can thus write,

$$\mathcal{F}_1(t) = \mathbf{F}_1 + \mathbf{f}(t), \tag{8.95}$$

where $\mathbf{f}(t)$ is a stochastic process with zero mean. The force $\mathbf{f}(t)$ is a series of impulses applied at random to the atom. Such a force is called a *Langevin force*. It was introduced in the context of Brownian motion, where a large particle is struck randomly by the molecules of a liquid, causing its velocity to change randomly. This then results in diffusion of the particle through space (random walk). We only need to know the statistical properties of the Langevin force $\mathbf{f}(t)$ to deduce its effect on the motion of the Brownian particles. The Langevin model can be applied here by treating the atom as a Brownian particle subjected to momentum transfers from incident and scattered photons during the fluorescence process (see Figure 8.3).

Let us examine the properties of $\mathbf{f}(t)$ in the case considered in Section 8.2.5. The atoms are placed in a travelling plane wave of wavevector $\mathbf{k}$. We further assume that the wave is non-saturating, so that we may neglect the effects of stimulated emission. The average rate of fluorescence cycles is, therefore (see (8.58) with $s \ll 1$),

$$\Gamma_{\text{fluo}} = \frac{\Gamma}{2} s, \tag{8.96}$$

and the average force is equal to (see (8.62))

$$\mathbf{F}_1 = \hbar \mathbf{k} \Gamma_{\text{fluo}}. \tag{8.97}$$

We model $\mathcal{F}_1(t)$ by a sequence of Dirac peaks of weight $\hbar k$, corresponding to a discontinuous variation in the atomic momentum by quanta $\hbar k$. We shall distinguish the effects of absorbed laser photons from those of scattered photons, the latter having arbitrary directions.

The absorbed photons each impart a well-defined impulse $\hbar \mathbf{k}$ at times t_j. The corresponding force can be written

$$\mathcal{F}_{\text{abs}} = \sum_j \hbar \mathbf{k}\, \delta(t - t_j) = \mathbf{F}_1 + \mathbf{f}_{\text{abs}}(t). \tag{8.98}$$

The random process $\mathbf{f}_{\text{abs}}$ is a series of Dirac peaks with zero mean, with variance (average of the squares) $(\hbar k)^2$, and with an average rate Γ_{fluo}.

Re-emitted photons give a contribution $\mathbf{f}_{\text{scat}}$, with zero average, which can be written

$$\mathbf{f}_{\text{scat}} = \sum_j \mathbf{e}_{\text{scat}}(t_j) \hbar k\, \delta(t - t_j). \tag{8.99}$$

Each vector $\mathbf{e}_{\text{scat}}$ has unit length and random direction. For an isotropic distribution, we have

$$\langle \mathbf{e}_{\text{scat}}(t_j) \rangle = 0, \tag{8.100}$$

and along any axis α, the variance of $\mathbf{e}_{\text{scat}}$ is

$$\langle [\mathbf{e}_{\text{scat}}(t_j)]_\alpha^2 \rangle = \frac{1}{3} \langle |\mathbf{e}_{\text{scat}}(t_j)|^2 \rangle = \frac{1}{3}; \quad \alpha = x, y, z. \tag{8.101}$$

Successively scattered photons are emitted in independent directions, whence,

$$\langle \mathbf{e}_{\text{scat}}(t_j) \cdot \mathbf{e}_{\text{scat}}(t_k) \rangle = \delta_{jk}. \tag{8.102}$$

Comment Strictly speaking, for a particular atom, it is not actually true that the successive times t_j are statistically independent. Indeed, when an atom has just emitted a fluorescence photon, a certain time is required before it can emit a second. This phenomenon is known as photon antibunching. In practice, the effect will have negligible consequences on the heating phenomenon we are concerned with here.

8.3.6 Momentum fluctuations and heating for a Doppler molasses

Under the effect of the fluctuations in the radiative forces, an ensemble of atoms undergoes heating (see Figure 8.13). To calculate this, consider an ensemble of atoms with identical initial momenta $\mathbf{p}(t_0)$, and assume to begin with that they are subjected only to the Langevin force $\mathbf{f}_{\text{scat}}(t)$ associated with scattered photons (Equation 8.99). Each atom carries out a *random walk* of step $\hbar k$ in the 3D momentum space. Between t_0 and t, its momentum will vary under the effect of $\Gamma_{\text{fluo}}(t-t_0)$ fluorescence cycles, and will be equal to

$$\mathcal{P}(t) = \mathbf{p}(t_0) + \sum_{j=1}^{\Gamma_{\text{fluo}}(t-t_0)} \hbar k \, \mathbf{e}_{\text{scat}}(t_j), \tag{8.103}$$

where $\mathbf{e}_{\text{scat}}(t_j)$ is an isotropically randomly distributed unit vector (see (8.100) and (8.101)). According to (8.100), we have

$$\langle \mathcal{P}(t) \rangle = \mathbf{p}(t_0). \tag{8.104}$$

The average momentum remains unchanged, which is not surprising since we have ignored the force due to absorbed photons. This corresponds to the case where the average force would be zero in Figure 8.13. Let us now calculate the variance of the momentum at time t:

$$\Delta \mathcal{P}^2(t) = \left\langle [\mathcal{P}(t) - \langle \mathcal{P}(t) \rangle]^2 \right\rangle = \hbar^2 k^2 \left\langle \sum_j \sum_k \mathbf{e}_{\text{scat}}(t_j) \cdot \mathbf{e}_{\text{scat}}(t_k) \right\rangle. \tag{8.105}$$

Using (8.102), we obtain, recalling that there are Γ_{fluo} scattered photons per unit time,

$$\Delta \mathcal{P}^2(t) = \hbar^2 k^2 \sum_j 1 = \hbar^2 k^2 \Gamma_{\text{fluo}} (t - t_0). \tag{8.106}$$

Under the effect of the fluctuations associated with spontaneous emission, *the variance of the momentum increases linearly with time*:

$$\left[\frac{\mathrm{d}}{\mathrm{d}t} \Delta \mathcal{P}^2(t) \right]_{\text{scat}} = \Gamma_{\text{fluo}} \hbar^2 k^2. \tag{8.107}$$

This is a well-known result for a *Brownian diffusion* phenomenon.

We may go further than just finding the variance, and actually write down the probability distribution for the momentum $\mathcal{P}(t)$. Applying the central limit theorem to the random variable $\mathcal{P}(t) - \mathbf{p}(t_0)$, formed by adding a large number of independent contributions, it can be shown that each component $\mathcal{P}_i(t)$ has a Gaussian distribution ρ with variance $\Delta \mathcal{P}_i^2(t)$:

$$\rho[\mathcal{P}_i(t)] = \frac{1}{\sqrt{2\pi} \Delta \mathcal{P}_i(t)} \exp \left\{ - \frac{[\mathcal{P}_i(t) - \langle \mathcal{P}_i \rangle]^2}{2 \Delta \mathcal{P}_i^2(t)} \right\} \tag{8.108}$$

where

$$\Delta\mathcal{P}_i^2 = \frac{1}{3}\Delta\boldsymbol{\mathcal{P}}^2. \tag{8.109}$$

An isotropic Gaussian distribution of momenta can be used to define a temperature T by the relation,

$$\frac{1}{2}k_{\mathrm{B}}T = \frac{\Delta\mathcal{P}_i^2}{2M} = \frac{1}{3}\frac{\Delta\boldsymbol{\mathcal{P}}^2}{2M}, \tag{8.110}$$

where k_{B} is the Boltzmann constant. The fluctuations in the radiative force thus lead to *heating*, since $\Delta\boldsymbol{\mathcal{P}}^2$, and hence also the temperature, increase linearly with time (see (8.107)).

In a similar way, the fluctuations in the force (8.98) due to absorption of the laser photons lead to heating in the component $\mathcal{P}_{\mathbf{k}}$ of the momentum parallel to the direction $\mathbf{k}$ of the laser beam. It can be shown by similar reasoning to the above that this heating is, for each laser beam,

$$\left[\frac{\mathrm{d}}{\mathrm{d}t}\Delta\mathcal{P}_{\mathbf{k}}^2(t)\right]_{\mathrm{abs}} = \Gamma_{\mathrm{fluo}}\hbar^2k^2. \tag{8.111}$$

The various heating terms correspond to independent random processes, so the variances are additive. Moreover, if we have several *non-saturating* travelling plane waves, the heating terms of the various waves are also additive in variance. Hence, for a Doppler molasses comprising three mutually orthogonal counterpropagating pairs of waves, each giving rise to $\mathcal{N}$ fluorescence cycles per second, we will have

$$\left[\frac{\mathrm{d}}{\mathrm{d}t}\Delta\mathcal{P}_i^2(t)\right]_{\mathrm{scat}} = 2\Gamma_{\mathrm{fluo}}\hbar^2k^2 \tag{8.112}$$

and

$$\left[\frac{\mathrm{d}}{\mathrm{d}t}\Delta\mathcal{P}_i^2(t)\right]_{\mathrm{abs}} = 2\Gamma_{\mathrm{fluo}}\hbar^2k^2. \tag{8.113}$$

Note that the term in (8.107) has been multiplied by six, and we have used (8.109), because for each laser beam we admit that the scattering is isotropic. On the other hand, the term in (8.111) is only multiplied by two, because the heating term (8.111) due to absorption only affects the direction of the laser beam. Moreover, we have taken the same value for the fluorescence rate Γ_{fluo} of each wave, ignoring the detuning due to the Doppler effect in this calculation. The rate Γ_{fluo} arising in (8.112) and (8.113) is therefore, in the case $s_0 \ll 1$,

$$\Gamma_{\mathrm{fluo}}(V=0) = \frac{\Gamma}{2}\frac{s_0}{1+4\frac{\delta^2}{\Gamma^2}}. \tag{8.114}$$

If now, instead of an ensemble of atoms which all have the same initial momentum, we consider a statistical ensemble with initial dispersion $\Delta\mathcal{P}_i^2(0)$, it can be shown that, under the effect of fluctuations, the variance will grow at the rates calculated above, and that the distribution at time t is Gaussian as in (8.108), with variance equal to the sum of the initial variance and the variances due to the dynamical evolution.

8.3.7 Equilibrium temperature for a Doppler molasses

The calculation in Section 8.3.6 only took into account the fluctuations in the resonance-radiation pressure. To find the evolution of an assembly of atoms, we must of course include the effects of the average force. Let us do this in the very important case of an assembly of atoms immersed in a 3D Doppler molasses formed by six non-saturating waves of intensity I and with negative detuning δ, where

$$s_0 = \frac{I}{I_{\text{sat}}} = 2\frac{\Omega_1^2}{\Gamma^2}. \tag{8.115}$$

Each atom is subjected to the friction force,

$$\frac{\mathrm{d}\mathcal{P}}{\mathrm{d}t} = \mathbf{F} = -\gamma \mathcal{P}(t), \tag{8.116}$$

where γ is given by (8.84). Under the effect of this friction alone, the mean value of $\mathcal{P}(t)$ tends exponentially to zero, with time constant γ^{-1}. Regarding the variance of $\mathcal{P}$, it falls off twice as fast, and for each axis we have

$$\left[\frac{\mathrm{d}}{\mathrm{d}t}\Delta\mathcal{P}_i^2\right]_{\text{friction}} = -2\gamma\,\Delta\mathcal{P}_i^2. \tag{8.117}$$

This is the cooling term we have already encountered.

Now the heating effect discussed in Section 8.3.6 will come into play to oppose the cooling, and we shall assume that the two terms can be added as independent effects. For each component i of the velocity, we then have

$$\begin{aligned}\frac{\mathrm{d}}{\mathrm{d}t}\Delta\mathcal{P}_i^2 &= \left[\frac{\mathrm{d}}{\mathrm{d}t}\Delta\mathcal{P}_i^2\right]_{\text{friction}} + \left[\frac{\mathrm{d}}{\mathrm{d}t}\Delta\mathcal{P}_i^2\right]_{\text{scat}} + \left[\frac{\mathrm{d}}{\mathrm{d}t}(\Delta\mathcal{P}_i^2\right]_{\text{abs}} \\ &= -2\gamma\,\Delta\mathcal{P}_i^2 + 4\Gamma_{\text{fluo}}\,\hbar^2k^2.\end{aligned} \tag{8.118}$$

Using (8.84) and (8.114), we obtain the steady-state value,

$$\Delta\mathcal{P}_i^2 = \hbar^2k^2\frac{2\Gamma_{\text{fluo}}}{\gamma} = M\frac{\hbar\Gamma}{8}\frac{1+4\delta^2/\Gamma^2}{-\delta/\Gamma}, \tag{8.119}$$

corresponding to a temperature given by

$$k_{\text{B}}T = \frac{\hbar\Gamma}{8}\frac{1+4\delta^2/\Gamma^2}{-\delta/\Gamma}. \tag{8.120}$$

This low-intensity calculation thus predicts that the resulting temperature should not depend on the intensity of the waves forming the molasses. However, it does depend on the detuning, and it goes through a minimum at $\delta = -\frac{\Gamma}{2}$:

$$T_{\text{Dop}}^{\text{min}} = \frac{\hbar\Gamma}{2k_{\text{B}}}. \tag{8.121}$$

We observe that *the values of this temperature can be remarkably low* (less than a millikelvin, e.g. 240 μK for sodium). Temperatures in this range were indeed observed in the first Doppler molasses experiments by S. Chu and his team in 1985. These spectacular results did much to stimulate research in laser cooling of atoms.

Comments

(i) The relation (8.120) can be written in the form found by Einstein when he investigated Brownian motion:

$$k_B T = M\frac{D}{\gamma}. \tag{8.122}$$

In this equation γ is the coefficient of friction of the Brownian particle due to the viscosity of the liquid, while D is the diffusion coefficient of the velocity $\mathbf{V}$, defined by,

$$2D = \gamma \Delta V^2, \tag{8.123}$$

where ΔV^2 is the mean squared velocity jump at each impact of the liquid molecules on the Brownian particle, and γ is the rate of these velocity jumps.

This relation brings out the deep relationship between *dissipation* (friction) and *fluctuations* (responsible for Brownian diffusion). On the microscopic level, the two phenomena have the same origin, namely, impacts from the liquid molecules (or from the photons in the case of laser cooling).

(ii) The very low temperature obtained in an optical molasses can be used to *confine* atoms in the dipole trap described in Section 8.2.6 (see Figure 8.4). To avoid mutual perturbations between the confinement and cooling phenomena, the different effects can be applied alternately: first the atoms are cooled with a Doppler molasses, then the lasers are shut down and the dipole confinement is applied, then the whole process is repeated at a sufficiently high frequency relative to the timescale of motion of the atoms to ensure that the effects reach their average separately.

8.3.8 Going under the Doppler temperature and Sisyphus cooling

For alkali-metal atoms like sodium, caesium and rubidium, a detailed experimental study of the Doppler temperature as a function of the detuning shows that, in certain circumstances, the temperature of the atoms does not obey (8.120), and temperatures can be reached that are one or two orders of magnitude below the 'Doppler limit' (8.121)! In fact, at low intensity and for high detuning, the temperature varies proportionally to $I/|\delta|$. It decreases when the detuning $|\delta|$ is increased toward low frequencies, and when the intensity is reduced, whereas (8.120) predicts that the resulting temperature should be independent of the intensity, and increase in proportion to $-\delta$. This remarkable phenomenon, discovered by W. Phillips and coworkers, was independently interpreted by S. Chu, and also by C. Cohen-Tannoudji and J. Dalibard. It was christened the 'Sisyphus effect'.

Without going into all the details of the explanation for this subtle effect, let us nevertheless outline the main features. The first point to understand is that, in an optical molasses, there is interference between the various waves which causes spatial modulation of the dipole potential. For high detuning, this is equal to the light-shift of the ground state,

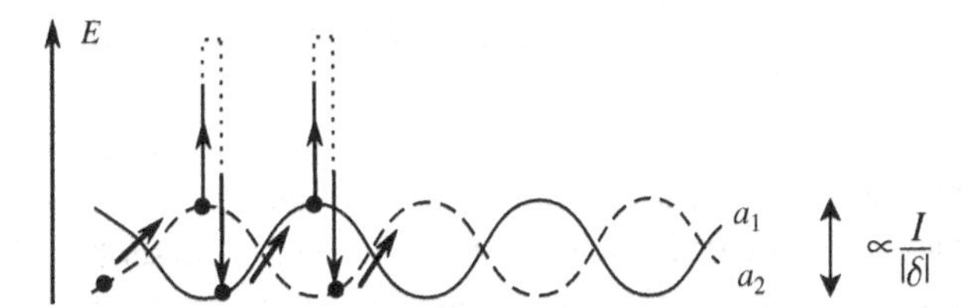

Figure 8.14 Sisyphus cooling. Under the effect of various interfering light waves, the light-shifts of the sub-levels a_1 and a_2 of the atomic ground state are modulated, and the dipole potential felt by the atom depends on the sub-level a_1 or a_2 in which it is located. Furthermore, at the top of the potential hills, the moving atom is subjected preferentially to optical pumping that brings it into a sub-level with a potential minimum at this location. The atom thus gradually loses its kinetic energy during its motion.

proportional to $I/|\delta|$ (see (8.75) in Section 8.2.6). The atom thus moves in a spatially modulated potential, with 'hills' and 'valleys', corresponding to the modulation of the light-shift of this ground state. If the ground state is degenerate, it may happen that the light-shifts of the various sub-levels are in fact different, and the dipole potential felt by the atom depends on the sub-level of the ground state in which it lies. In the Sisyphus configuration, there are two sub-levels, and the hills of one sub-level correspond to the valleys of the other (see Figure 8.14).

When several sub-levels of the ground state interact with the laser beams, we must take into account optical pumping phenomena, able to transfer the atoms from one sub-level to the other (see Complement 2B). In the Sisyphus configuration, the optical pumping preferentially transfers the moving atoms from the top of a hill of one sub-level to the bottom of the valley of another sub-level (see Figure 8.14). The atom is thus made to climb a potential hill, which slows it down, then to transfer without change of velocity to the bottom of a new hill, which it must subsequently climb, slowing it down still further. The mechanism only stops when the atom has kinetic energy less than or equal to the height of the hills, which varies as $I/|\delta|$. In this way, we can understand the resulting dependence of the temperature on the intensity and the detuning.

This decrease in the final temperature with the intensity clearly cannot continue as far as $I = 0$. Both experiment and theory show that the mean squared velocity does not go below a small multiple of the recoil velocity V_R given in (8.10). This result can be interpreted by recalling that each scattered photon causes an uncontrollable recoil of velocity V_R. It is thus easy to understand that, in a cooling mechanism that involves fluorescence cycles, the velocities cannot be controlled to better than a few times V_R. We thus find *the natural limit for radiative cooling*, namely, the *recoil temperature*,

$$T_R = \frac{MV_R^2}{k_B}. \tag{8.124}$$

The recoil temperature is lower than the microkelvin for most alkali metals. Experiment shows that Sisyphus cooling can lead to temperatures of the order of $10T_R$, i.e. a few microkelvins.

8.3.9 Cooling below the recoil temperature

In fact, the recoil temperature (8.124) does not represent the ultimate limit, and several mechanisms have been invented to concentrate a significant fraction of the atoms within a range of velocities below V_R. Complement 8A describes one of these processes in detail, namely, velocity-selective coherent population trapping, which effectively leads to sub-recoil cooling.

One can get round the recoil limit if the cooling does not appeal to a friction force, as in the case of Doppler or Sisyphus cooling. In sub-recoil cooling mechanisms, the atom does a random walk in velocity space, under the effect of momentum exchanges with laser photons during the fluorescence processes, but the fluorescence rate vanishes when the velocity is strictly zero (see Figure 8.15(a)). If during its random walk an atom happens to reach exactly zero velocity, it ceases to interact with the lasers, and its velocity changes no further. In this way, atoms accumulate at $V = 0$ (Figure 8.15(b)) through an optical pumping process like the one discussed in Complement 2B, except that it occurs in the velocity space rather than in the space of internal states of the atom.

The velocity is a continuous variable, and the accumulation will occur over a range of finite width ΔV. If ΔV is less than the recoil velocity, then we have genuine sub-recoil cooling. In the case examined in Complement 8A, the fluorescence rate varies quadratically with the velocity:

$$\Gamma_{\text{fluo}}(V) = \Gamma \frac{V^2}{V_0^2}. \tag{8.125}$$

This rate does indeed vanish at $V = 0$, and remains very small over a finite range of velocities around $V = 0$. Quantitative analysis shows that the situation here is radically different

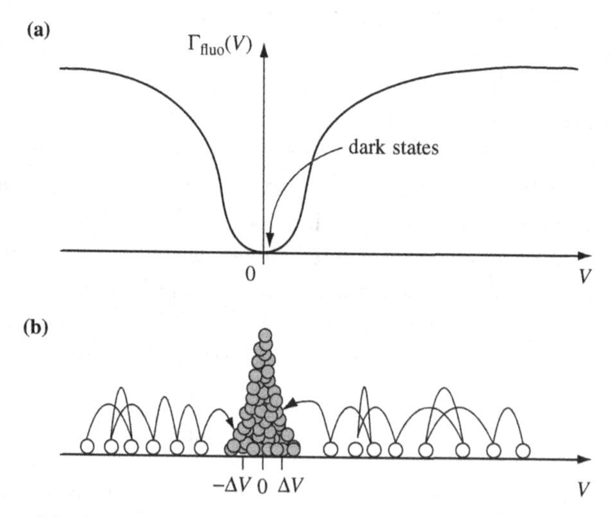

Figure 8.15 Sub-recoil cooling. **(a)** The atom exchanges photons with the radiation, at a rate Γ_{fluo} which goes to zero when $V = 0$. **(b)** The atoms do a random walk in velocity space and accumulate near $V = 0$, where the random walk stops. (Figure courtesy of François Bardou.)

to what happens with the usual cooling processes leading to a steady-state situation and a limiting temperature (see Section 8.3.7). Indeed, if we wish to specify the width ΔV of the range of velocities in which the atoms will accumulate, we are compelled to introduce the duration θ of the interaction between the atoms and the lasers, and to express the fact that an atom is in the trapped velocity band if its velocity is low enough to ensure that the probability $\theta\Gamma_{\text{fluo}}(V)$ of a fluorescence cycle during the time θ remains less than 1. Then (8.125) implies that

$$\Delta V = \frac{V_0}{\sqrt{\Gamma\theta}}, \tag{8.126}$$

which means that the range ΔV gets narrow as the interaction time θ gets longer. We thus reach a steady-state situation, and cooling continues indefinitely. In practice, perturbations related to technical limits (finite widths of laser lines, residual magnetic fields) will eventually lay down a limit to this form of cooling, but temperatures in the nanokelvin range have been achieved.

8.4 Gaseous Bose-Einstein condensates and atom lasers

8.4.1 Bose-Einstein condensation

When a particle ensemble is cooled down sufficiently, certain collective effects can occur, *which even exist for an ideal gas* (without interaction in the usual sense of the term): these are *quantum statistical effects*,[10] which arise when the atomic wave functions overlap, i.e. when their typical dimensions are greater than or equal to their average separation $n^{-1/3}$, where n is the atomic density in position space. Let us express this condition for atoms in thermal equilibrium at temperature T. The Maxwell–Boltzmann distribution of their momenta is a 3D Gaussian with mean squared deviation,

$$\Delta p = \Delta p_x = \Delta p_y = \Delta p_z = \sqrt{M\, k_B T}. \tag{8.127}$$

A wave packet can be associated with it, with typical size $\hbar/\Delta p$, according to Heisenberg relations. Actually, the size is, by convention, taken equal to the thermal de Broglie wavelength Λ_T, which is defined by

$$\Lambda_T = \sqrt{\frac{2\pi\hbar^2}{Mk_B T}}. \tag{8.128}$$

The condition for quantum statistical effects to appear is then,

$$\Lambda_T \gtrsim n^{-1/3}, \tag{8.129}$$

or

$$n\,\Lambda_T^3 \gtrsim 1. \tag{8.130}$$

[10] For example, see C. Kittel, *Elementary Statistical Physics*, Dover (1986).

We introduce the density in the 6D phase space $\{\mathbf{r}, \mathbf{p}\}$, i.e.

$$\nu(\mathbf{r}, \mathbf{p}) = \frac{\mathrm{d}^6 N_{\mathrm{at}}}{\mathrm{d}^3 r\, \mathrm{d}^3 p} = \frac{n(\mathbf{r})}{(2\pi \Delta p)^{3/2}}\, \mathrm{e}^{-p^2/2\Delta p^2}. \tag{8.131}$$

For a uniform spatial density n, the maximal density of the thermal gas in the phase space is

$$\nu(\mathbf{r}, \mathbf{p} = 0) = \frac{n}{(2\pi M\, k_{\mathrm{B}} T)^{3/2}} = h^3\, n \Lambda_T^3. \tag{8.132}$$

Condition (8.130) for the appearance of quantum statistical effects is therefore that the density in phase space should be of the order of one atom per elementary cell of volume h^3. The dimensionless quantity $n\,\Lambda_T^3$ is called the degeneracy parameter, and condition (8.130) characterizes the quantum degeneracy regime. Note that, for a gas at standard temperature and pressure, $n\,\Lambda_T^3$ is much less than unity.

In a famous but highly speculative paper written in 1924, Einstein predicted the existence of a quantum phase transition known today as Bose–Einstein condensation, for the precise value,

$$n\,\Lambda_T^3 = 2.61. \tag{8.133}$$

When $n\,\Lambda_T^3$ has this value, the atomic gas at thermodynamic equilibrium is *saturated*, where the term 'saturation' is used by analogy with a saturated vapour in a liquid–vapour equilibrium. Any excess atoms then 'condense' into the quantum ground state of the trap in which the atoms have been confined.

This quantum phase transition is remarkable because, in contrast to typical phase transitions like vapour–liquid or liquid–solid transitions, it is not due to the effects of attractive interactions prevailing over thermal agitation below some critical temperature. Here, condensation results from a pure quantum interference effect (involving several particles), which favours situations where several indistinguishable bosonic particles are in the same quantum state. More precisely, following a collision between two bosons, the probability of ending up in a state already occupied by n bosons is $n + 1$ times greater than if this state were initially empty. We have encountered an analogous property in Chapter 6 (see Section 6.3.3) in the case of photon emission (recall that photons are bosons). The probability of (stimulated) emission in a mode containing n photons is $n + 1$ times greater than the probability of (spontaneous) emission in a mode that is initially empty (this property underlies the laser effect).

Pursuing the above analogy, we may compare Bose–Einstein condensation, when the temperature goes below the critical temperature, with the laser transition when the gain goes above the threshold value. This comparison is interesting, as we shall see later, but it should be remembered that there is an important difference: a Bose–Einstein condensate is a priori a system in thermodynamic equilibrium with a non-condensed fraction at a finite temperature, whereas a laser beam is a zero-temperature system, described by a pure quantum state, if we ignore spontaneous emission.

8.4.2 Obtaining dilute atomic Bose–Einstein condensates. Laser cooling and evaporative cooling

The Einstein criterion (8.133) suggests that it would be possible to reach the condensation threshold either by lowering the temperature (see expression (8.128) for Λ_T), or by increasing the density, or both. In fact, to satisfy criterion (8.133) at the lowest temperatures accessible by standard cryogenic techniques (a fraction of a kelvin, or a few millikelvins), this would require densities of the order of the densities of liquids or solids, i.e. $n \sim 10^{28}\ \mathrm{m}^{-3}$, where electromagnetic interactions (chemical bonds, van der Waals forces...) are responsible for ordinary phase transitions to a solid or liquid phase, and this would mask any quantum statistical effects. But there is one exception, namely helium, which is inert, i.e. interactions between neighbours are extremely weak. Furthermore, it is well known that, at a temperature of around 2 K, liquid helium goes into a superfluid state. After several decades of debate during the first half of the twentieth century, a general consensus was reached regarding the role of Bose–Einstein condensation in the appearance of superfluidity, characterized by a complete absence of viscosity. But in the superfluid helium liquid, ordinary interactions remain strong and make the physics of this system very complex, a long way from the ideal condensate envisaged by Einstein.

To observe Bose–Einstein condensation unmasked by ordinary interactions, the range of which does not exceed a nanometre for chemical bonds and a fraction of a micrometre for van der Waals interactions, the answer is to work in very dilute conditions, at a density below $10^{18}\ \mathrm{m}^{-3}$. The critical temperature given by the Einstein criterion (8.133) is then in the microkelvin range, well below temperatures accessible to standard cryogenic techniques. However, by combining laser cooling and trapping with a new method, forced evaporative cooling, Bose–Einstein condensation was finally achieved under dilute conditions in 1995.

The remarkable progress in experimental techniques which led to this result, rewarded by the Nobel Prize in 2001,[11] is quite clear when we consider what is happening to the phase-space density (see Figure 8.16). We begin with an atomic vapour containing atoms such as rubidium, of density $10^{18}\ \mathrm{m}^{-3}$ and at a few degrees Celsius. At this temperature, the de Broglie wavelength is of the order of 10^{-11} m, so the degeneracy parameter $n\,\Lambda_T^3$ is of the order of 10^{-15}. After slowing down, by means of a laser, an atomic beam produced from this vapour, the real space density n has been reduced by six orders of magnitude, but velocities have been reduced by a factor of more than 100, and Λ_T^3 has increased by more than six orders of magnitude, leading to an increase in the degeneracy parameter, which is then equal to 10^{-13}. Now cooling by laser molasses, combined with magneto-optical trapping (see Section 8.3) yields a spectacular gain, taking the degeneracy parameter to a value of the order of 10^{-6}. Thanks to laser cooling and trapping methods, more than ten orders of magnitude have been gained from the phase-space density. But experience shows that it is very difficult to go beyond this point, because laser cooling

[11] See Footnote 2, this chapter.

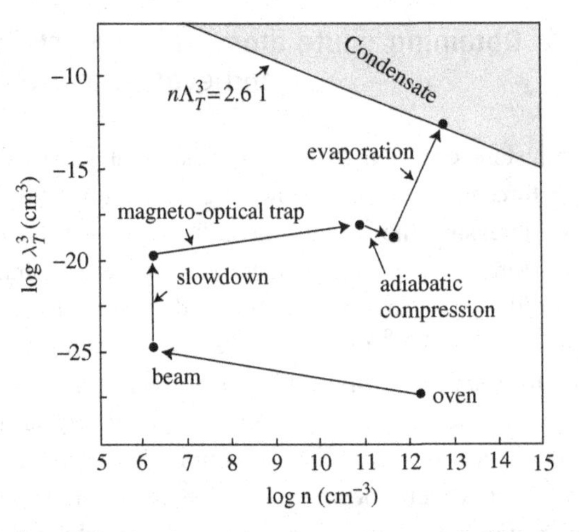

Figure 8.16 Path in the space $(n\Lambda_T^3)$ leading from the atomic vapour in the oven that produces the atomic beam to the final Bose–Einstein condensate. (Straight lines of slope −1 correspond to a constant degeneracy parameter and the bold line corresponds to the condensation threshold given by (8.133).) The final density is of the same order as the initial density, but the temperature has gone down from 500 K to 500 nK, and the density in velocity space (and hence the density in phase space) has gone up by 14 orders of magnitude. This spectacular gain was obtained by laser cooling followed by evaporative cooling.

methods come up against the problem of heating due to reabsorption of scattered photons when the density exceeds 10^{17} m^{-3}. (The scattering cross-section at resonance is of the order of the square of the optical wavelength, typically around 10^{-12} m^2. For a detuning $|\delta| = 10\,\Gamma$, the mean free path of photons in an atomic vapour at this density is then equal to 10^{-3} m.)

In order to remove the last six orders of magnitude required to reach the critical value $n\Lambda_T^3 = 2.61$, one appeals to *evaporative cooling*. This begins by transferring the atoms to a conservative trap, i.e. they are subjected to a static potential with an absolute minimum, where they are sheltered from any heating by re-scattered photons. This can be a magnetic trap, but that method can only be used for paramagnetic atoms with a permanent magnetic moment. It can also be a dipole laser trap (see Figure 8.4), using a highly detuned and very powerful laser, e.g. a CO_2 laser of wavelength 10 μm and power a few tens of watts, focused on a few μm (see Section 8.2.6).

The idea of evaporative cooling is to eliminate the most energetic trapped atoms, e.g. by reducing the depth of the trap, in order to reduce the average energy of the remaining atoms. Typically, atoms of energy greater than $6\,k_BT$ are eliminated. This is higher than the average energy of the atoms, which is $3\,k_BT$ at thermal equilibrium in a harmonic trap. The sample is then allowed to rethermalize under the effect of elastic collisions, which yields a new Maxwell–Boltzmann equilibrium distribution at a lower temperature. We then have a cooler sample, but also one that is denser, since the atoms tend to concentrate at the bottom of the trap when the temperature goes down. The gain in phase space is thus considerable. One subsequently reduces again the depth of the trap, and obtains a

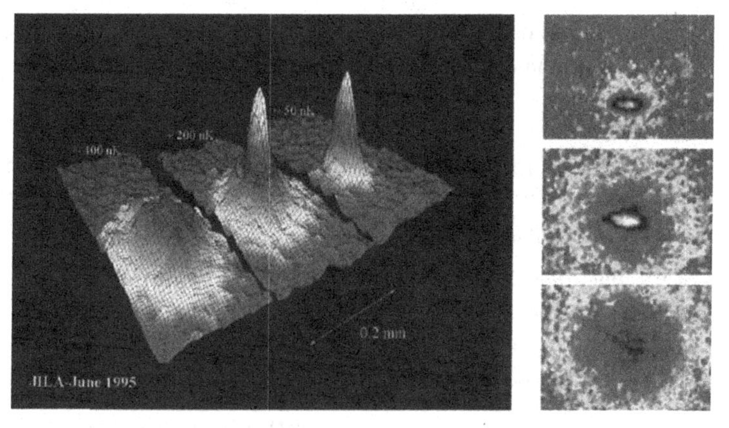

Figure 8.17 **Observing Bose–Einstein condensation. *Left*: Two-dimensional atomic velocity distribution, after integrating along the line of sight, at different temperatures. *Right*: Regions of constant density in velocity space. Condensation manifests itself through the appearance of a double structure, corresponding to the condensed phase in equilibrium with a thermal cloud. The central elliptical structure represents the velocity distribution of the Bose–Einstein condensate. The isotropic halo surrounding it represents the velocity distribution of the cloud of non-condensed atoms at thermal equilibrium. (From E. Cornell and C. Wieman.[12])**

new evaporation, and a new gain in the phase space at the centre of the trap. By pursuing this forced evaporation process, condensation is observed at a temperature of a few hundred nanokelvins, in agreement with (8.133) when expressed for the density at the centre of the trap.

In order to actually *observe* Bose–Einstein condensation, the method of choice is to determine the velocity distribution of the trapped atoms. To do this, the trap is suddenly switched off, and the atomic cloud allowed to spread out in free flight over a time τ long enough for it to become much bigger than its initial size. The spatial density $n(\mathbf{r}, \tau)$ at this time then reflects the initial velocity distribution $\rho(\mathbf{V}, 0)$, and we have

$$n(\mathbf{r}, \tau) = \rho\left(\frac{\mathbf{r}}{\tau}, 0\right). \tag{8.134}$$

It is relatively easy to measure $n(\mathbf{r}, \tau)$ by making an absorption image of the atomic cloud, illuminated by a laser at resonance. The resulting absorption profile is then proportional to the integral of the density $n(\mathbf{r}, \tau)$ along the line of sight (column density). The 3D distribution is then retrieved from the symmetry properties of the trap.

Figure 8.17 shows the atomic velocity profiles as a function of temperature. At a temperature above the critical temperature (160 nK), we observe that the velocity profile is well represented by a Gaussian distribution, fitted to a Maxwell–Boltzmann distribution to obtain the temperature. At lower temperatures, a double structure appears, with a narrow component and a broad component. The wings of the broad component are still very well fitted by a Gaussian, which provides the temperature. The narrow component in the

[12] See Footnote 2, this chapter.

centre is isolated by subtracting the broad component. It corresponds to the atoms in the Bose–Einstein condensate, which we shall describe in more detail below.

8.4.3 Ideal Bose-Einstein condensate and atomic wavefunction

As mentioned above, the appearance of a Bose–Einstein condensate is reminiscent of what we find when we cross the laser threshold. In Chapters 3 and 5, we stressed that a key property of lasers is the presence of a very large number of photons in the same mode of the electromagnetic field, i.e. the mode of the cavity in which the laser effect is produced. This mode is described by a complex classical electromagnetic field,

$$\mathcal{E}(\mathbf{r},t) = \alpha_p \, u_p(\mathbf{r}) \, \mathrm{e}^{-\mathrm{i}\omega_p t}, \tag{8.135}$$

taken as scalar to simplify (we may assume that the polarization is uniform). The function $u_p(\mathbf{r})$, which has the physical dimensions of an electric field, characterizes the spatial structure of the mode (see Complement 3B), and we normalize it in such a way that the corresponding electromagnetic energy is that of one photon:

$$\hbar\omega_p = 2\varepsilon_0 \int d^3r |u_p(\mathbf{r})|^2. \tag{8.136}$$

Clearly, the squared modulus of the dimensionless complex number α_p can be interpreted as the average number of photons $\langle N_p \rangle$ in the mode, and we have

$$\mathcal{E}(\mathbf{r},t) = \sqrt{\langle N_p \rangle} \, \mathrm{e}^{-\mathrm{i}\varphi} u_p(\mathbf{r}) \, \mathrm{e}^{-\mathrm{i}\omega_p t}. \tag{8.137}$$

As suggested in Section 5.6, the expression (8.137) can to a certain extent be interpreted as the wavefunction of $\langle N_p \rangle$ photons all in the same state described by the wavefunction $\mathrm{e}^{-\mathrm{i}\varphi} u_p(\mathbf{r}) \mathrm{e}^{-\mathrm{i}\omega_p t}$. This description is incomplete, however, because it contains no information about the fluctuations in the number of photons, characteristic of a quasi-classical state (see Section 5.3.4). But if the average number of photons $\langle N_p \rangle$ is very large, the fluctuations (of order $\langle N_p \rangle^{1/2}$) are relatively small. The interpretation of (8.137) as the wavefunction of $\langle N_p \rangle$ identical photons then allows us to write the probability of detecting a photon in the volume $\mathrm{d}^3\mathbf{r}$ around the point $\mathbf{r}$ in the suggestive form,

$$\mathrm{d}P = |\mathcal{E}(\mathbf{r},t)|^2 = \langle N_p \rangle |u_p(\mathbf{r})|^2 \, \mathrm{d}^3\mathbf{r}, \tag{8.138}$$

(we have assumed a quantum efficiency of 1 for the detector). The quantity $|\mathcal{E}(\mathbf{r},t)|^2$ can thus be considered as the average density of photons at point $\mathbf{r}$.

In the same way, we can describe the N atoms of an ideal (non-interacting) Bose–Einstein condensate by the N-particle wave function,[13]

$$\Psi = \sqrt{N} \, \mathrm{e}^{-\mathrm{i}\varphi} u_0(\mathbf{r}) \, \mathrm{e}^{-\mathrm{i}\omega t}, \tag{8.139}$$

representing N atoms in the same state, described by the stationary wave function $u_0(\mathbf{r}) \exp(-\mathrm{i}\omega t)$ of an atom in the ground state of the trap in which the atoms have been

[13] See, for example, Chapter 2 of L. Pitaevskii and S. Stringari, *Bose–Einstein Condensation*, Oxford University Press (2003), or C. J. Pethick and H. Smith, *Bose–Einstein Condensation in Dilute Gases*, Cambridge University Press (2008).

condensed. As in the case of a laser beam, this formalism is less rigorous than the genuine Fock space wavefunction describing N bosons, but if N is big enough, and for zero or weak interactions between the atoms, it is an excellent approximation, providing a simple way of describing many phenomena which we shall exemplify below. The complex function Ψ is sometimes called the order parameter of the Bose–Einstein condensate.

A delicate point in the formalism (8.139) is the appearance of a phase φ, which is not in fact determined by any equation. It arises as a result of spontaneous gauge symmetry breaking, which fixes φ at an arbitrary but well-determined value for each specific realization of the condensate. Note that this feature arises in the same manner for the phase of the laser in (8.137). This phase can be considered to have an a-priori arbitrary value, but well-determined for a particular working laser.

Comment For both the laser and the non-interacting Bose–Einstein condensate, rigorous use of the N-body quantum formalism (second quantization) shows that the heuristic hypothesis of spontaneous symmetry breaking is not strictly necessary. In this formalism, the phase arises whenever we make a measurement whose result depends on the phase.[14]

8.4.4 Observing the wavefunction of the Bose–Einstein condensate

For a Bose–Einstein condensate in a harmonic trap, with cylindrical symmetry in the Oz direction, function $u_0(\mathbf{r})$ is the ground state of a particle of mass M (the mass of the atom) immersed in the potential,

$$V(x, y, z) = \frac{1}{2} M\,\omega_\perp^2 (x^2 + y^2) + \frac{1}{2} M\,\omega_z^2\, z^2. \tag{8.140}$$

It has the form,

$$u_0(x, y, z) = \frac{1}{(2\pi^2\, a_\perp^2\, a_z)^{1/2}} \exp\left(-\frac{x^2 + y^2}{4a_\perp^2}\right) \exp\left(-\frac{z^2}{4a_z^2}\right). \tag{8.141}$$

The half-widths at $e^{-1/2}$ of $|u_0|^2$ ('standard' widths), corresponding to the root-mean-square deviations in the position, are given by,

$$a_\perp = \sqrt{\frac{\hbar}{M\,\omega_\perp}} \quad \text{and} \quad a_z = \sqrt{\frac{\hbar}{M\,\omega_z}}. \tag{8.142}$$

In the common case where $\omega_\perp$ and ω_z have very different values, the condensate is stretched out or flattened in certain directions, i.e. if $\omega_\perp \gg \omega_z$, it has a 'cigar' shape, while if $\omega_\perp \ll \omega_z$, it looks like a 'pancake' perpendicular to the Oz-axis. These shapes are characterized by the 'aspect ratio' $a_z/a_\perp = (\omega_\perp/\omega_z)^{1/2}$. In practice the dimensions of the condensate are usually too small (in the micrometre range for rubidium with a trapping

[14] See, for example, Chapter 13 of C. J. Pethick and H. Smith, *Bose–Einstein Condensation in Dilute Gases*, Cambridge University Press (2nd edition 2008). For photons, see K. Mølmer, Optical Coherence, a Convenient Fiction, *Physical Review A* **55**, 3195 (1997).

frequency $\omega/2\pi$ of 100 Hz) to be able to observe the shape of the condensate directly. However, if the velocity distribution is observed by the time-of-flight method discussed in Section 8.4.2, the characteristics of the wavefunction $u_0(\mathbf{r})$, or to be precise, its Fourier transform, can be determined. Indeed, the velocity distribution is the squared modulus of the spatial Fourier transform of the position space wavefunction (8.141). It is therefore a Gaussian, and its mean squared deviations are related to $a_\perp$ and a_z by the Heisenberg relations,

$$\Delta V_\perp = \frac{1}{2}\sqrt{\frac{\hbar\omega_\perp}{M}}, \quad \Delta V_z = \frac{1}{2}\sqrt{\frac{\hbar\omega_z}{M}}. \tag{8.143}$$

The corresponding velocities (5×10^{-4}m.s^{-1} in the above case) give an observable spreading after ballistic flight for several tens of milliseconds. We thus obtain an inverted anisotropic shape as compared with the condensate itself, i.e. a pancake if $\omega_\perp \gg \omega_z$, and a cigar shape otherwise.

The anisotropy of the velocity distribution in a Bose–Einstein condensate differs strikingly from the isotropic profile observed for a cloud at thermal equilibrium at temperature T. Indeed, in this case, the velocity distribution is Gaussian and the standard width is the same for any axis we care to choose:

$$\Delta V_i = \sqrt{\frac{k_B T}{M}}. \tag{8.144}$$

This quantity is clearly independent of the trap stiffness, characterized by ω_i.

Figure 8.17 shows the double structure in the velocity distribution of a Bose–Einstein condensate in thermodynamic equilibrium with a non-condensed atomic cloud. By fitting a Gaussian to the wings of the profile, which contains only non-condensed atoms, the mean squared velocity is obtained directly, and from this the temperature via (8.144). The observation of the double structure, with an anisotropic central profile that appears inverted relative to the trap, was a convincing indication for E. Cornell and C. Wieman that they really had obtained a gaseous Bose–Einstein condensate in their 1995 experiment.

8.4.5 Dilute Bose–Einstein condensate with interactions

If the condensate contains several thousand atoms, interactions between them, usually repulsive, will modify the above description. In the mean field approximation, where each atom feels a time-independent potential resulting from the average of all the interactions with the other atoms, the wavefunction $u_0(\mathbf{r})$ (see (8.139)) obeys a nonlinear Schrödinger equation, also known as the Gross–Pitaevski equation:

$$-\frac{\hbar^2}{2M}\Delta u_0(\mathbf{r}) + V(\mathbf{r})u_0(\mathbf{r}) + Ng|u_0(\mathbf{r})|^2 u_0(\mathbf{r}) = \mu\, u_0(\mathbf{r}). \tag{8.145}$$

Without the third term, this equation is the time-independent Schrödinger equation of a particle of energy μ in the potential $V(\mathbf{r})$, whose minimum is zero. The third term describes the interaction felt by an atom. It involves the atomic density at point $\mathbf{r}$, i.e. $N|u_0(\mathbf{r})|^2$, where N is the total number of atoms in the condensate. The constant,

$$g = \frac{4\pi\hbar^2}{M}a, \tag{8.146}$$

characterizes the interaction between ultracold atoms by a single parameter, the diffusion length a (e.g. $a \simeq 5 \times 10^{-9}$ m for rubidium 87). The energy μ associated with the wavefunction $u_0(\mathbf{r})$ is the chemical potential, i.e. the change in energy of the system when an atom is added.

A simple approximate solution of (8.145) can be found when the repulsive interaction is strong enough to ensure that the ground state of the system is much more extended than the ground state of the harmonic oscillator. The first term in (8.145) involving spatial derivatives can then be neglected (Thomas–Fermi approximation), and the atomic density is given by

$$Ng|u_0(\mathbf{r})|^2 = \mu - V(\mathbf{r}), \tag{8.147}$$

if $V(\mathbf{r}) \leq \mu$, while $u_0(\mathbf{r}) = 0$ if $V(\mathbf{r}) > \mu$. For a harmonic potential (8.140), the density profile is an inverted paraboloid vanishing on an ellipsoidal surface with semi-axes (Thomas–Fermi radii) given by

$$R_{x,y} = \frac{1}{\omega_\perp}\left(\frac{2\mu}{M}\right)^{1/2} \quad \text{and} \quad R_z = \frac{1}{\omega_z}\left(\frac{2\mu}{M}\right)^{1/2}. \tag{8.148}$$

Once again, we note that the condensate is anisotropic, but the anisotropy is more marked here (aspect ratio $\omega_\perp/\omega_z$) than for the non-interacting case (aspect ratio $(\omega_\perp/\omega_z)^{1/2}$). This anisotropy can be observed in the velocity distribution obtained by time-of-flight. These properties have been abundantly confirmed by experiments that have also provided tests for many theoretical predictions regarding ideal (non-interacting) or dilute (weakly interacting) condensates, and in particular regarding their elementary excitations (oscillations, vortices).

8.4.6 Coherence properties of a Bose–Einstein condensate and interference between two Bose–Einstein condensates

A Bose–Einstein condensate, which contains many atoms described by the same wavefunction, is analogous to the mode of a laser cavity containing many photons described by the same classical electromagnetic field. We thus expect luminosity and coherence properties reminiscent of the laser, and interference experiments between condensates have indeed revealed such analogies in a spectacular way (see Figure 8.18).

In the experiment leading to Figure 8.18, two condensates initially trapped at the same height are released at time $t = 0$. The two condensates are allowed to expand ballistically, and when they overlap, the atomic density is observed by absorption of resonant laser light (see Figure 8.18). To account for the observed fringes, we assume that we have two non-interacting condensates located initially in harmonic traps at $-\mathbf{d}/2$ and $\mathbf{d}/2$, taken to be isotropic to simplify the calculation. The wavefunctions are Gaussian and we write the total wavefunction in the form,

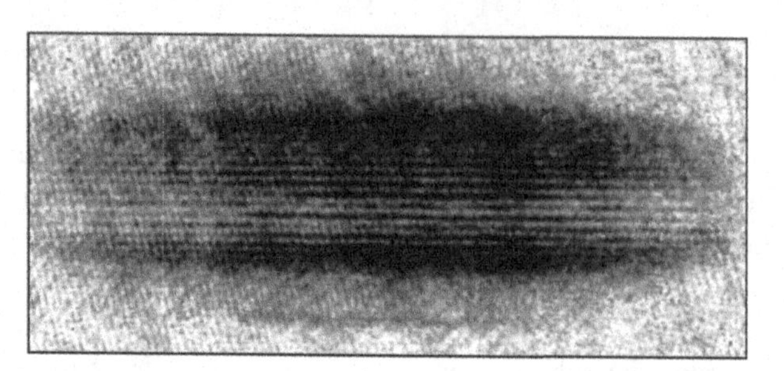

Figure 8.18 **Interference between two gaseous Bose–Einstein condensates (from W. Ketterle).[15] Observation of interference fringes in the region of superposition of the two ballistically expanding condensates shows that each condensate contains a large number of atoms (typically 10^6) described by the same wavefunction.**

$$\Psi(\mathbf{r}, t=0) = \sqrt{N_1}\, \mathrm{e}^{\mathrm{i}\phi_1} u_0\left(\mathbf{r} + \frac{\mathbf{d}}{2}\right) + \sqrt{N_2}\, \mathrm{e}^{\mathrm{i}\phi_2} u_0\left(\mathbf{r} - \frac{\mathbf{d}}{2}\right), \tag{8.149}$$

with

$$u(\mathbf{r}) = \frac{1}{(2\pi a_0^2)^{3/4}} \exp\left(-\frac{r^2}{4a_0^2}\right). \tag{8.150}$$

When the harmonic traps are switched off, each Gaussian spreads in a freely falling frame, as adopted to simplify the treatment. The expansion of a one-particle Gaussian wave packet is obtained by solving the Schrödinger equation,[16] and this yields

$$u(\mathbf{r}, t) = \frac{\mathrm{e}^{\mathrm{i}\delta_t}}{(2\pi a_t^2)^{3/4}} \exp\left\{-\frac{r^2(1 - \mathrm{i}\hbar t/2Ma_0^2)}{4a_t^2}\right\}, \tag{8.151}$$

with

$$a_t^2 = a_0^2 + \left(\frac{\hbar t}{2Ma_0}\right)^2 \tag{8.152}$$

and

$$\tan \delta_t = -\frac{\hbar t}{2Ma_0^2}. \tag{8.153}$$

Under asymptotic conditions $\hbar t \gg 2Ma_0^2$, where the wave packet is several times bigger than it was initially, the phase δ_t is equal to $\delta_\infty = \frac{\pi}{2}$, whereas the size of the wave packet grows linearly with time.

[15] See Footnote 2, this chapter.
[16] See, for example, CDL, complement G_{I}.

The wavefunction describing the superposition of two expanding condensates at time t is

$$\Psi(\mathbf{r},t) = \sqrt{N_1}\,\mathrm{e}^{\mathrm{i}\phi_1}u_1(\mathbf{r},t) + \sqrt{N_2}\,\mathrm{e}^{\mathrm{i}\phi_2}u_2(\mathbf{r},t). \tag{8.154}$$

Absorption imaging yields the atomic density,

$$\begin{aligned} n(\mathbf{r},t) = & |\Psi(\mathbf{r},t)|^2 = N_1|u_1(\mathbf{r},t)|^2 + N_2|u_2(\mathbf{r},t)|^2 \\ & + 2\sqrt{N_1N_2}\,\mathrm{Re}\left\{\mathrm{e}^{\mathrm{i}(\phi_1-\phi_2)}u_1(\mathbf{r},t)u_2^*(\mathbf{r},t)\right\}. \end{aligned} \tag{8.155}$$

The last term differs from zero if we wait long enough for the two wave packets to overlap. This is an interference term. It is calculated using (8.151):

$$u_1(\mathbf{r},t)u_2^*(\mathbf{r},t) = \frac{1}{(2\pi a_t^2)^{3/2}}\exp\left(-\frac{r^2+d^2/4}{2a_t^2}\right)\exp\left(\frac{\mathrm{i}\hbar t}{4Ma_0^2a_t^2}\mathbf{r}\cdot\mathbf{d}\right). \tag{8.156}$$

In the neighbourhood of $\mathbf{r} = 0$, half-way between the initial positions of the condensates, and neglecting r^2 in comparison with $d^2/4$, the density in (8.155) becomes,

$$n(\mathbf{r},t) = n(0,t)\left[1 + \frac{2\sqrt{N_1N_2}}{N_1+N_2}\cos\left(\phi_1 - \phi_2 + \frac{\hbar t}{4Ma_0^2a_t^2}\mathbf{r}\cdot\mathbf{d}\right)\right]. \tag{8.157}$$

At a given time t, the density $n(\mathbf{r},t)$ is modulated along the axis $\mathbf{d}$ joining the initial positions of the condensates. The fringe contrast $2\sqrt{N_1N_2}\big/(N_1+N_2)$ is equal to 1 if $N_1 = N_2$. The fringe separation is

$$i = \frac{8\pi Ma_0^2a_t^2}{\hbar td} \simeq \frac{2\pi\hbar t}{Md}, \tag{8.158}$$

where we have used the asymptotic form of (8.152). It is equal to half the de Broglie wavelength $\lambda_{\mathrm{dB}} = h/MV$ for particles of velocity $V = d/2t$, which have moved from the traps to the observation area at distance $d/2$ from the traps during expansion time t. We can thus give a direct-wave interpretation of these fringes. Note that the surfaces generated by the fringes are plane, rather than hyperboloids. The reader can check that the wave interpretation does indeed give this result, by attributing the wavevectors $\mathbf{k}_\pm = (\mathbf{r}\pm\mathbf{d}/2)\,M/\hbar t$ to the two waves at $\mathbf{r}$ arriving from $-\mathbf{d}/2$ and $+\mathbf{d}/2$, respectively, and propagating for time t.

The term $\phi_1 - \phi_2$ in (8.148) determines the positions of the fringes. Insofar as ϕ_1 and ϕ_2 are random quantities which are different for each run of the experiment, one expects the positions of the fringes to change from one experiment to the next. This is indeed what is observed. However, if the two interfering condensates come from the same condensate which has been cut in two (by causing a potential barrier to grow in the middle of the initial condensate), it is found that the fringes are always in the same place, as predicted by (8.148) when we set $\phi_1 = \phi_2$.

The observations described here, first made by W. Ketterle and collaborators, and in particular the observation of fringes for a single sample, provide convincing confirmation of the validity of the representation of the Bose–Einstein condensate by a macroscopic wavefunction describing a large number of atoms all described by the same wavefunction.

8.4.7 Atom lasers

Apart from the trapped condensate, analogous to the mode of a laser cavity containing a large number of photons, genuine atom lasers can be produced, i.e. beams of free atoms all in the same quantum state. To do this, the trapped condensate is coupled with a wavefunction describing freely propagating atoms, just as the output mirror of a laser couples the cavity mode with a free beam. For example, this can be done by using a radio-frequency field that can make the atoms transit from a state where they are trapped in a non-uniform magnetic field to a state of zero magnetic quantum number ($m = 0$), insensitive to the trap (see Figure 8.19(a)). The transition occurs at a well-defined height z_0, after which the atoms fall freely under the effect of gravity, and they are described by a wavefunction,

$$\psi_{E_0}(z,t) = \Theta_{E_0}(z)\,\mathrm{e}^{-\mathrm{i}\frac{E_0}{\hbar}t}, \tag{8.159}$$

where $\Theta_{E_0}(z)$ is the solution of the time-independent Schrödinger equation,

$$-\frac{\hbar^2}{2M}\frac{\mathrm{d}^2\Theta}{\mathrm{d}z^2} - Mgz\Theta = E_0\Theta. \tag{8.160}$$

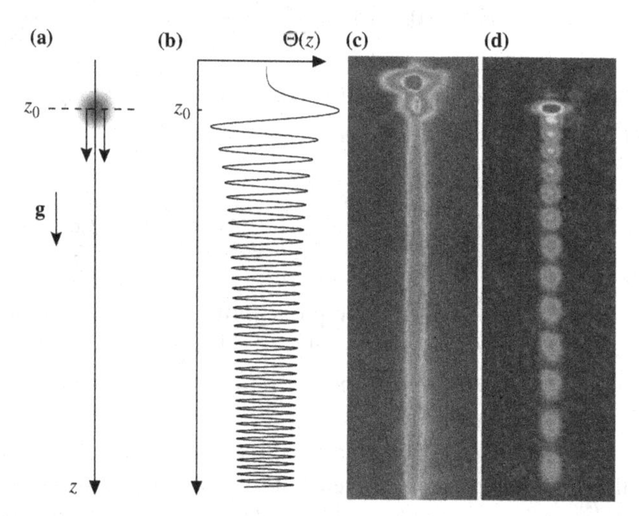

Figure 8.19 **Atom laser. (a) Atoms of a trapped Bose–Einstein condensate are subjected to a radio-frequency field which causes them to transit to a non-trapped state at height z_0. They subsequently fall under the effect of gravity. (b) Real part of the wavefunction describing an atom released at zero velocity at height z_0. The imaginary part has the same shape, oscillating in quadrature (see (8.163)). All of the atoms in the atom laser are described by this wavefunction. (c) Absorption image of an atom laser (width 0.1 mm, height 5 mm).[17] (d) Absorption image of the beat between two atom lasers falling from two slightly different heights z_0' and z_0'', hence with two slightly different energies.[18]**

[17] I. Bloch, T. W. Hänsch and T. Esslinger, Atom Laser with a cw Output Coupler, *Physical Review Letters* **82**, 3008 (1999).

[18] I. Bloch, T. W. Hänsch and T. Esslinger, Measurement of the Spatial Coherence of a Trapped Bose Gas at the Phase Transition, *Nature* **403**, 166 (2000).

(Note that the z-axis is pointing downwards, the gravitational potential energy being $-Mgz$.) The solution is known, and given in terms of the Airy function $\mathrm{Ai}(u)$ in the form,

$$\Theta_{E_0}(z) = C\,\mathrm{Ai}\left(\frac{z - z_0}{\ell}\right), \tag{8.161}$$

where $\ell = (\hbar^2/2M^2 g)^{1/3}$ ($\ell \approx 300$ nm for rubidium of atomic mass 87) and

$$E_0 = -mg\,z_0. \tag{8.162}$$

The parameter z_0 can be interpreted as the turning point of the classical trajectory with total energy E_0 (the kinetic energy vanishes at z_0). The Airy function is a special function that can be found in published tables or specialized software, and whose real part is shown in Figure 8.19(b). The function $\Theta_{E_0}(z)$ extends slightly above z_0 into the classically forbidden region, but it decreases exponentially there. On the other hand, it extends to infinity downwards, and for $z - z_0 \gg \ell$, it has the asymptotic form,

$$\Theta_{E_0}(z) \approx \left(\frac{z - z_0}{\ell}\right)^{-1/4} \exp\left\{\mathrm{i}\frac{2}{3}\left|\frac{z - z_0}{\ell}\right|^{3/2}\right\}. \tag{8.163}$$

The form of (8.163) shows that the de Broglie wavelength of the atom decreases as z increases, which is easy to understand since the atom accelerates during free fall. More precisely, the local wavevector, associated with the wavefunction (8.163), is given at height z by

$$K_{\mathrm{at}} = \frac{\mathrm{d}}{\mathrm{d}z}\left[\frac{2}{3}\left(\frac{z - z_0}{\ell}\right)^{3/2}\right] = \frac{(z - z_0)^{1/2}}{\ell^{3/2}}. \tag{8.164}$$

Replacing ℓ by its value, we obtain

$$K_{\mathrm{at}} = \frac{M}{\hbar}\sqrt{2g(z - z_0)} = \frac{M\,V_{\mathrm{at}}(z)}{\hbar}, \tag{8.165}$$

where $V_{\mathrm{at}}(z)$ is the velocity of a classical particle released at zero velocity from a height z_0. We obtain the expected relationship, i.e. $MV_{\mathrm{at}} = \hbar K_{\mathrm{at}}$, between the de Broglie wavelength $2\pi/K_{\mathrm{at}}$ and the velocity at height z.

Comment The atom laser setup discussed here is clearly one of finite duration, since the laser will stop when all the atoms of the condensate have been extracted. In this sense, it is reminiscent of the first photon laser, the ruby laser, where the initial population inversion is exhausted during the laser emission. A challenge is to find arrangements leading to continuous-wave atom lasers.

We may now interpret the spectacular experiment displaying the beat between two atom lasers extracted from the same Bose–Einstein condensate at two different heights z_0' and z_0'', shown in Figure 8.19(d). The resulting wavefunction is

$$\psi(z,t) = \frac{1}{\sqrt{2}}\left[\psi_{E_0'}(z,t) + \psi_{E_0''}(z,t)\right]. \tag{8.166}$$

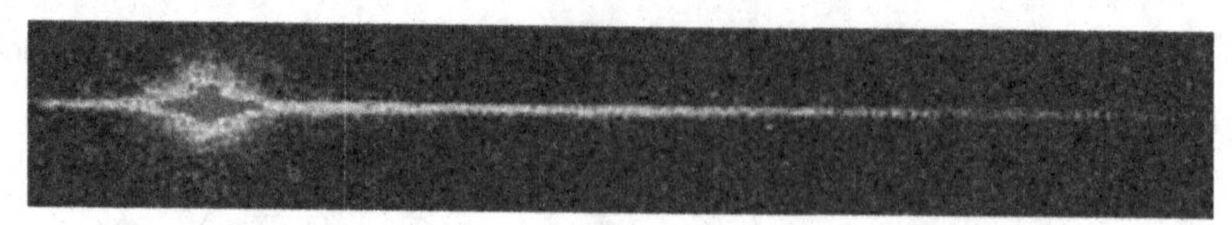

Figure 8.20 **Guided atom laser. The Bose–Einstein condensate is coupled to the modes of a horizontal matter wave guide, i.e. a horizontal laser beam detuned below the resonance. We thus obtain an atom laser in which the atoms have constant velocities. (Figure courtesy of W. Guérin and P. Bouyer, Institut d'Optique.)**

By absorption of resonant light, the atomic density $|\psi(z,t)|^2$ can be measured at height z and time t. Using the asymptotic forms (8.163) of $\psi_{E'_0}$ and $\psi_{E''_0}$, and carrying out a truncated expansion in powers of $z'_0 - z_0$ and $z''_0 - z_0$, where $z_0 = (z'_0 + z''_0)/2$, we obtain, neglecting corrections to the prefactors,

$$|\psi(z,t)|^2 \propto \left(\frac{z-z_0}{\ell}\right)^{1/2}\left(1+\cos\left\{(z''_0 - z'_0)\left[\frac{(z-z_0)^{1/2}}{\ell^{3/2}} + \frac{Mg}{\hbar}t\right]\right\}\right). \quad (8.167)$$

At a given point, a beat is observed at frequency $M\,g(z''_0 - z'_0)/\hbar = (E'_0 - E''_0)/\hbar$, as one would expect for two wavefunctions of different energies, i.e. E'_0 and E''_0. Regarding the signal as a function of z at a given time, shown in Figure 8.19(d), it exhibits a modulation with a period that increases with z. Indeed, the beat period at height z is

$$(\Delta z)_b = \frac{2\pi}{\mathrm{d}\phi/\mathrm{d}z} = 2\frac{\ell^{3/2}}{z''_0 - z'_0}(z-z_0)^{1/2}, \quad (8.168)$$

where $\phi(z,t)$ is the argument of the cosine in (8.167), which can also be written,

$$\phi(z,t) = \frac{z''_0 - z'_0}{\ell^{3/2}}\left[(z-z_0)^{1/2} + \left(\frac{g}{2}\right)^{1/2} t\right]. \quad (8.169)$$

This shows that the beat pattern 'falls' under the effect of gravity, while expanding.

Equation (8.165) shows the fast decrease in the de Broglie wavelength of the atoms accelerated during free fall. This phenomenon can be avoided by coupling the atom laser with a horizontal matter wave guide, made for example using a horizontal laser beam with a small diameter and with a frequency below the atomic resonance. Under the effect of the dipole force (see Section 8.2.6), the atoms are subjected to a confining potential in the plane perpendicular to the horizontal axis of the laser beam. If the guiding laser is intense enough, the confining potential will be able to balance the weight. Figure 8.20 shows such a horizontal guided laser. Matter wave guides can also be made using magnetic fields produced by electrical microcircuits fabricated with the latest nanotechnology. Genuine atomic chips can be obtained, which should facilitate practical applications of ultracold quantum gases.[19]

[19] R. Folman *et al.*, Microscopic Atom Optics: from Wires to an Atom Chip, *Advances in Atomic, Molecular and Optical Physics* **48**, 263 (2002).

8.4.8 Conclusion. From photon optics to atom optics and beyond

In this chapter, we have seen how laser control of atomic motion can produce colder and colder assemblies of atoms, i.e. with ever narrower velocity dispersion. In the analogy between photon optics and atom optics,[20] these samples are compared with photons obtained from very narrow spectral lines: the wavelength dispersion is very small, but these are nevertheless incoherent sources. This does not prohibit accurate interferometric measurements, as the history of optics has shown, since many highly accurate measurements were based on interference effects long before the invention of the laser, e.g. the Michelson experiment to investigate the isotropy of the velocity of light. Indeed, interference effects can be obtained with incoherent light, provided that suitable interferometers are used. Likewise, cold atom samples can be used in atom interferometers to carry out very precise measurements of inertial or gravitational effects. The atoms – massive objects – are very sensitive to these effects. Potential applications of atom interferometry range from tests of general relativity to inertial navigation systems, or survey of the Earth's sub-surface through tiny variations in the gravitational field.

We may pursue the analogy between photon and atom optics and consider the joint probability of detecting two atoms at two points and at two times. In previous chapters, we stressed the importance of photon correlation measurements, which have inspired the development of modern quantum optics. By studying atom correlations, one can also develop a genuine atom quantum optics, studying analogous effects to those that brought about photon quantum optics. However, in the case of atoms, there is an extra degree of freedom, namely the possibility of working either with bosons or with fermions, a choice which leads to radically different behaviours. For example, the Hanbury-Brown–Twiss effect, or photon bunching, observed in 1956 in light emitted by a thermal source, has its counterpart in atom optics: in a gas of bosonic atoms in thermal equilibrium, the joint probability of observing two atoms very close to one another is greater than it would be in the case where we ignore their quantum statistics. This fully quantum effect reflects bosonic stimulation, i.e. the increased probability of finding two bosons in the same elementary cell of phase space. In contrast, when the atoms are fermions, the probability of finding two atoms very close to one another is zero, which reflects the 'Pauli exclusion principle' (see Figure 8.21).

Since its beginnings as a pure product of fundamental research, a subject of perfectly academic investigation for the physicist, the photon laser has rapidly become a practical tool with many applications (see the complements to Chapter 3). We may wonder whether atom lasers are likely to play the same role in atom optics, for example, in the form of atom interferometers. In fact, there is a major difference between photons and atoms, when their density increases: in general, atoms interact together. This is described by the last term on the left-hand side of the Gross–Pitaevski equation (8.145). This term is in fact strictly

[20] P. Meystre, *Atom Optics*, Springer (2001).

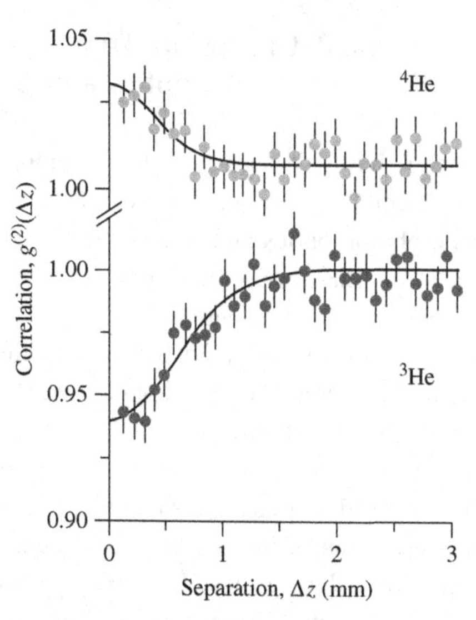

Figure 8.21 **Atomic Hanbury-Brown–Twiss effect. Correlation function for the positions of helium-4 atoms (bosons) and helium-3 atoms (fermions) in an ultracold thermal gas.[21] The bosons tend to group together, while the fermions avoid one another. This is a purely quantum statistical effect.**

equivalent to the nonlinear term arising for light in Kerr media (see Complement 7B), but while in photon optics this term only plays a significant role for very strong light intensities, it is omnipresent in atom optics, and it can lead to phase differences that are difficult to control, a major problem for some applications of atom interferometry. On the other hand, as in photon optics, nonlinear effects give rise to quite spectacular phenomena. For example, the collision of three Bose–Einstein condensates causes the appearance of a fourth wave of matter, an analogous effect to 'four-wave mixing' in nonlinear photon optics (see Figure 8.23).

In nonlinear atom optics, interactions offer completely novel possibilities, because the phenomenon known as Feschbach resonance allows one to vary these interactions from zero to very high values. It then becomes possible to create entangled atomic states, which could be used for quantum information processing (see Complement 5E).

Using interacting ultracold atoms, it is also possible to carry out experimental studies of situations analogous to those existing in solid-state physics, where a large number of particles, e.g. electrons in a solid, are in a 'highly correlated' quantum state. For example, the ultracold atoms can be placed in a 3D periodic dipole potential produced by laser standing waves (see Section 8.2.6), in order to study their quantum transport properties. The parameters of the system can be varied in such a way as to reconstruct the standard

[21] T. Jeltes *et al.*, Comparison of the Hanbury-Brown–Twiss Effect for Bosons and Fermions, *Nature* **445**, 402 (2007).

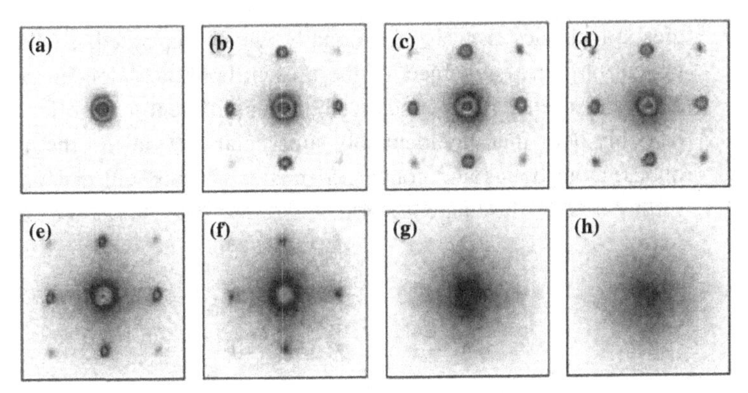

Figure 8.22 Mott insulator–conductor transition. When ultracold atoms are placed in a 3D periodic potential produced by standing light waves (see Figure 8.5), certain phenomena can be observed directly for which only indirect evidence exists for electrons in solids, e.g. the Mott insulator–conductor transition. The figure shows the atomic velocity distribution for an ensemble of ultracold atoms with repulsive interactions, placed in a periodic potential of increasing amplitude. (**b**) and (**d**) show very narrow peaks, associated with the modulation of a wavefunction that is coherent over the whole sample (perfect conductor). When the amplitude of the potential is large enough, the atoms are localized in wells, with exactly one atom per well, and coherence is lost ((**g**) and (**h**)). The system becomes a Mott insulator.

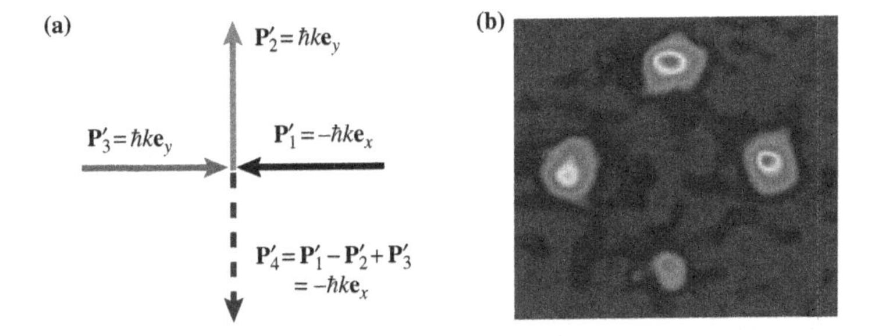

Figure 8.23 Four-wave mixing of Bose–Einstein condensates.[22] (**a**) Three condensates collide, and a fourth appears as a result of the nonlinear term in the Gross–Pitaevskii equation. (**b**) Atomic density map some time after the collision. The three incident condensates can be seen pursuing their motion, losing a few atoms in the process, while the condensate produced in the collision moves towards the bottom of the figure. This is analogous to four-wave mixing in nonlinear photon optics (see Figure 7B.5 of Section 7B.4.1).

problems of condensed matter physics, such as the insulator–conductor transition[23] (see Figure 8.22). These systems are genuine quantum simulators, providing a way of exploring phenomena that are extraordinarily difficult to handle theoretically, e.g. high-temperature superconductivity. In short, we are witnessing a remarkably fertile encounter between two

[22] L. Deng *et al.*, Four Wave Mixing with Matter Waves, *Nature* **398**, 218 (1999).

[23] M. Greiner, O. Mandel, T. Esslinger, T. W. Hänsch and I. Bloch, Quantum Phase Transition from a Superfluid to a Mott Insulator in a Gas of Ultracold Atoms, *Nature* **415**, 39 (2002).

areas of physics that are traditionally widely separated, namely atomic, molecular and optical physics, the subject of the present book, and condensed matter physics. And, as always in science, the boundaries between different fields offer a wealth of prospects. We can only hope that this inevitably superficial overview of the interface between quantum optics, atom optics and condensed matter physics will encourage the reader to venture further into this fascinating world.

8A Complement 8A Cooling to sub-recoil temperatures by velocity-selective coherent population trapping

In Section 8.3.9, we mentioned that the recoil temperature was not the ultimate barrier to cooling. Several mechanisms have been conceived and implemented to concentrate the atomic velocity distribution in a range of values of width less than V_R. In contrast to Doppler or Sisyphus cooling, these mechanisms do not appeal to a friction force, but rather to an optical pumping process (see Complement 2B), able to accumulate the atoms in this narrow range of velocities. We now present one such mechanism called 'velocity-selective coherent population trapping', considering only the case where the motion occurs in one space direction (the Oz-axis). Note, however, that the method generalizes to two or three dimensions.

8A.1 Coherent population trapping

Coherent population trapping is a phenomenon that occurs when an atom has a system of three energy levels in the Λ configuration shown in Figure 8A.1, where each leg of the Λ interacts with a laser L_- or L_+. This was discussed in detail in Section 2D.2 of Complement 2D. Here we consider the case where the two ground states g_+ and g_-, have exactly the same energy, and the two lasers have frequencies ω_+ and ω_-, very close to the resonance frequency ω_0. In this situation, the selectivity of the interactions is obtained via the polarization selection rules (see Complement 2B): if g_-, g_+ and e have magnetic quantum numbers $m = -1, +1$ and 0, respectively, the laser wave L_- is right-circularly polarized (σ_+ polarization), while the laser wave L_+ is left-circularly polarized (σ_- polarization). As long as the frequencies ω_+ and ω_- of the two lasers differ slightly, the atom is observed to emit fluorescence light, i.e. spontaneous photons associated with transitions $e \to g_-$ or $e \to g_+$. But if the two frequencies become strictly equal, the fluorescence is found to cease. The curve representing the fluorescence rate as a function of ω_+ and ω_- in Figure 8A.1(b) exhibits an inverted resonance, sometimes called a dark resonance. One remarkable feature of this resonance is that it is much narrower than the width Γ of the level e.

To understand the disappearance of fluorescence when $\omega_+ = \omega_-$, we use the semi-classical model of Section 2.3, where the laser waves are described by monochromatic fields of equal frequency ω, and phases φ_+ and φ_- at the location of the atom.

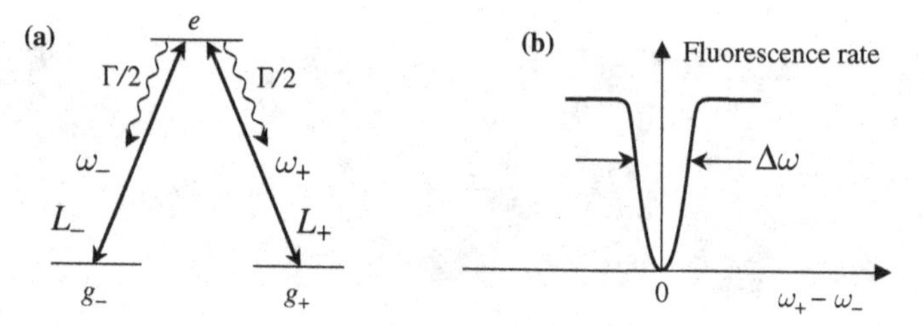

Figure 8A.1 Dark resonance. (a) Three-level atom interacting selectively with a wave of frequency ω_- for the transition $g_- \leftrightarrow e$ and with a wave of frequency ω_+ for the transition $g_+ \leftrightarrow e$. Levels g_- and g_+ have the same energy. From level e (lifetime Γ^{-1}), the atom can decay to g_- or g_+ by emitting fluorescence light. (b) Fluorescence ceases when the frequencies ω_+ and ω_- are exactly equal. The width $\Delta\omega$ of the dark resonance can be much smaller than Γ.

In this setup, the interaction Hamiltonian is

$$\begin{aligned}\hat{H}_\mathrm{I} &= \hbar\Omega_1 \cos(\omega t + \varphi_-)\Big(|e\rangle\langle g_-| + |g_-\rangle\langle e|\Big) \\ &\quad + \hbar\Omega_1 \cos(\omega t + \varphi_+)\Big(|e\rangle\langle g_+| + |g_+\rangle\langle e|\Big).\end{aligned} \tag{8A.1}$$

The couplings with the lasers, characterized by the Rabi angular frequency Ω_1, are assumed equal.

Since the two states $|g_+\rangle$ and $|g_-\rangle$ have the same energy, any combination of them, e.g.

$$|\psi\rangle = \lambda|g_-\rangle + \mu|g_+\rangle, \tag{8A.2}$$

will be an eigenstate of the atomic Hamiltonian, and the perturbative method of Section 1.2.4 can be used to calculate the probability of transition between state $|\psi\rangle$ and state $|e\rangle$. More precisely, the results of Section 1.2.4 relating to the case of a sinusoidal excitation of the form (1.42) can be generalized. We make the resonance approximation, i.e. we keep only the terms $\exp\{-\mathrm{i}\omega t\}\,|e\rangle\,\langle g|$ and conjugate terms in $\hat{H}_\mathrm{I}$, and set

$$\left[\hat{H}_\mathrm{I}\right]_\mathrm{res} = \hat{W}\,\mathrm{e}^{-\mathrm{i}\omega t} + \text{h.c.}, \tag{8A.3}$$

with

$$\hat{W} = \frac{\hbar\Omega_1}{2}\left(\mathrm{e}^{-\mathrm{i}\varphi_-}|e\rangle\,\langle g_-| + \mathrm{e}^{-\mathrm{i}\varphi_+}|e\rangle\,\langle g_+|\right). \tag{8A.4}$$

The probability of a transition from $|\psi\rangle$ to $|e\rangle$ is then proportional to the squared modulus of the matrix element $\langle e|\hat{W}|\psi\rangle$, which is given for the state (8A.2) by

$$W_{e,\psi} = \frac{\hbar\Omega_1}{2}\left(\mathrm{e}^{-\mathrm{i}\varphi_-}\langle g_-|\psi\rangle + \mathrm{e}^{-\mathrm{i}\varphi_+}\langle g_+|\psi\rangle\right) = \frac{\hbar\Omega_1}{2}\left(\lambda\,\mathrm{e}^{-\mathrm{i}\varphi_-} + \mu\,\mathrm{e}^{-\mathrm{i}\varphi_+}\right). \tag{8A.5}$$

Clearly, state (8A.2), such that $\lambda\,\mathrm{e}^{-\mathrm{i}\varphi_-} = -\mu\,\mathrm{e}^{-\mathrm{i}\varphi_+} = 1/\sqrt{2}$, which is given by

$$|\psi_\mathrm{NC}\rangle = \frac{1}{\sqrt{2}}\left(\mathrm{e}^{\mathrm{i}\varphi_-}|g_-\rangle - \mathrm{e}^{\mathrm{i}\varphi_+}|g_+\rangle\right), \tag{8A.6}$$

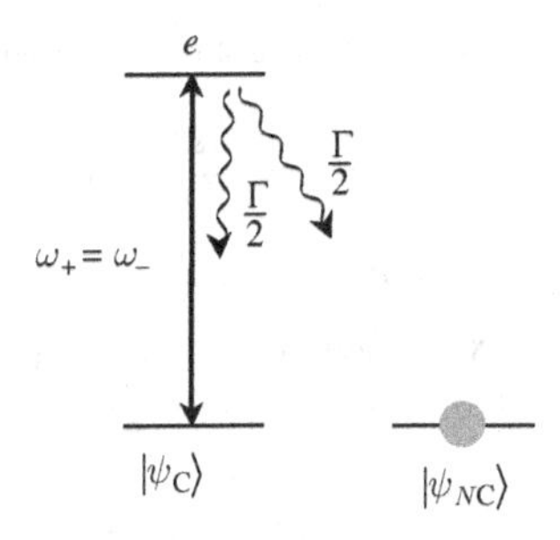

Figure 8A.2 **Coherent population trapping. If $\omega_+ = \omega_-$ and the two states $|g_+\rangle$ and $|g_-\rangle$ have the same energy, the setup of Figure 8A.1 can be transformed into one with a coupled state $|\psi_C\rangle$ and a non-coupled state $|\psi_{NC}\rangle$. By optical pumping, all the atoms are eventually brought into the non-coupled state and fluorescence will cease.**

is not coupled to $|e\rangle$ by the lasers, since the matrix element (8A.5) is then zero. In this non-coupled state, the atom no longer fluoresces. Such a non-coupled state is also called a dark state. In contrast, the state orthogonal to $|\psi_{NC}\rangle$, i.e.

$$|\psi_C\rangle = \frac{1}{\sqrt{2}}\left(e^{i\varphi_-}|g_-\rangle + e^{i\varphi_+}|g_+\rangle\right), \tag{8A.7}$$

is coupled to $|e\rangle$ by the lasers, with the matrix element (8A.5) given by $\hbar\Omega_1/\sqrt{2}$. In this state, the atom does fluoresce.

To understand the total absence of fluorescence in the steady-state regime, we represent the three-level system in the basis $\{|e\rangle, |\psi_{NC}\rangle, |\psi_C\rangle\}$ (see Figure 8A.2). Only the state $|\psi_C\rangle$ is coupled to the state $|e\rangle$ by the lasers. However, the state $|e\rangle$ can decay by spontaneous emission to either $|\psi_{NC}\rangle$ or $|\psi_C\rangle$. So if the atom is in $|\psi_{NC}\rangle$, it will remain there indefinitely, despite the presence of the resonant lasers. But if the atom is in the state $|\psi_C\rangle$, it will go through a fluorescence cycle which has one chance in two of ending up in $|\psi_{NC}\rangle$, whereupon the fluorescence will cease. If on the other hand the atom falls into $|\psi_C\rangle$, a new cycle will occur, once again with probability 1/2 of ending up in $|\psi_{NC}\rangle$. Clearly, after a certain number of cycles (around two on average), the fluorescence will stop. This is an optical pumping process (see Complement 2B), which soon brings all the atoms into the state $|\psi_{NC}\rangle$ (after a time equal to a small multiple of Γ^{-1}). We have already discussed this phenomenon, known as coherent population trapping, in Section 2D.2.1.

When ω_+ and ω_- are different, there is no coherent population trapping, because the state $|\psi_{NC}\rangle$ of (8A.7) remains coupled to e by the lasers. To see this, we write down the interaction Hamiltonian,

$$\begin{aligned}\hat{H}_I &= \hbar\Omega_1\cos(\omega_- t + \varphi_-)(|e\rangle\langle g_-| + |g_-\rangle\langle e|)\\ &+ \hbar\Omega_1\cos(\omega_+ t + \varphi_+)(|e\rangle\langle g_+| + |g_+\rangle\langle e|),\end{aligned} \tag{8A.8}$$

and take $\omega_\pm = \omega \pm \frac{\Delta\omega}{2}$. Keeping the definition (8A.3), the operator $\hat{W}$ can now be written,

$$\hat{W} = \frac{\hbar\Omega_1}{2}\left(e^{-i\left(\varphi_- - \frac{\Delta\omega}{2}t\right)}|e\rangle\langle g_-| + e^{-i\left(\varphi_+ + \frac{\Delta\omega}{2}t\right)}|e\rangle\langle g_+|\right), \tag{8A.9}$$

and the matrix element for the coupling between the state (8A.2) and $|e\rangle$ is

$$W_{e,\phi}(\Delta\omega) = \frac{\hbar\Omega_1}{2}\left[\lambda e^{-i(\varphi_- - \frac{\Delta\omega}{2}t)} + \mu e^{-i(\varphi_+ + \frac{\Delta\omega}{2}t)}\right]. \tag{8A.10}$$

As it is time-dependent, there is no longer any state that remains uncoupled at all times. For example, the matrix element W for the coupling between $|\psi_{NC}\rangle$ in (8A.6) and the excited state, under the effect of the lasers, is given by

$$\hat{W}_{e,\phi_{NC}}(\Delta\omega) = \frac{\hbar\Omega_1}{2\sqrt{2}}\left(e^{i\frac{\Delta\omega}{2}t} - e^{-i\frac{\Delta\omega}{2}t}\right). \tag{8A.11}$$

At time $t = 0$, this matrix element is zero and the state $|\psi_{NC}\rangle$ is a non-coupled state. But at time $t = 2\pi/\Delta\omega$, the matrix element is $W_{e,\psi_{NC}} = -i\hbar\Omega_1/\sqrt{2}$, and the state $|\psi_{NC}\rangle$ is then coupled to $|e\rangle$, to which it therefore has a certain probability of transition. If this happens, it will emit a fluorescence photon.

The above argument shows that the fluorescence will resume all the more quickly as the frequency difference increases. We thus understand why the resonance curve of Figure 8A.1(b) is narrower when the experiment lasts longer. We shall return to this point.

8A.2 Velocity-selective coherent population trapping and sub-recoil cooling

Consider the situation in which the two laser beams L_- and L_+, with polarizations σ_+ and σ_-, respectively, propagate in opposite directions along the Oz-axis (see Figure 8A.3).

If the atom is in the state $|\psi_{NC}\rangle$ and if its velocity V in the Oz direction is zero, there will be no photon exchange and the atom will remain at zero velocity. But if the atom has a velocity V that differs from zero, the two frequencies $\omega_- = \omega - kV$ and $\omega_+ = \omega + kV$ felt by the atom in its frame will be different and there will be no dark state. The atom therefore goes through fluorescence cycles, during which a photon is absorbed either from L_- or from L_+, then rescattered in an arbitrary direction. With each fluorescence cycle, the velocity of the atom is altered by a random amount between $-2V_R$ and $+2V_R$. The velocity V thus does a random walk along the V-axis, which may bring it to the point $V = 0$ (see Figure 8.15(b)). During the last jump which takes it to $V = 0$, there is one chance in two that the atom will go into the state $|\psi_{NC}\rangle$, whereupon it will cease to interact with the light and its velocity V will remain at zero. In all other cases, it resumes its random walk, but later it may fall into the state $|\psi_{NC}\rangle$ at $V = 0$. Gradually, a larger and larger fraction of

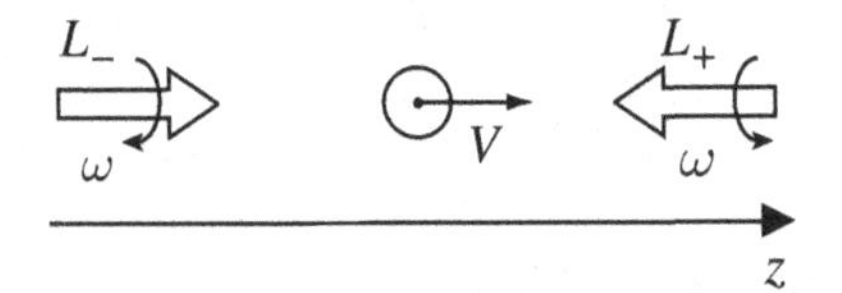

Figure 8A.3 Sub-recoil cooling by velocity-selective coherent population trapping.

the atoms will gather around $V = 0$, in a range of velocities whose width ΔV decreases indefinitely with the duration θ of the interaction with the lasers.

More precisely, in Section 8A.4 (see (8A.26)), we will calculate the fluorescence rate of an atom falling into $|\psi_{NC}\rangle$ at a velocity V slightly different from zero:

$$\Gamma_{NC}(V) = \frac{2k^2V^2}{\Omega_1^2}\Gamma. \tag{8A.12}$$

For an experiment lasting a time θ, the half-width ΔV of the velocity peak is given by the condition,

$$\Gamma_{NC}(\Delta V)\cdot\theta \approx 1, \tag{8A.13}$$

since an atom of velocity greater than ΔV will go through at least one fluorescence cycle in less than θ. It follows that

$$\Delta V = \frac{\Omega_1}{k}\frac{1}{\sqrt{2\Gamma\theta}}. \tag{8A.14}$$

As this shows, there is nothing to stop ΔV going below the recoil velocity V_R. To give a specific example, for $\Gamma = 10^7\,\mathrm{s}^{-1}$, $\Omega_1 = 0,3\Gamma$, and $k \simeq 7 \times 10^6\,\mathrm{m}^{-1}$, i.e. the case of metastable helium, θ only has to exceed a few milliseconds for ΔV to go below the centimetre per second, below the recoil velocity, which is $V_R = 9.2\,\mathrm{cm.s}^{-1}$ for helium absorbing or emitting one photon at $1.08\,\mu\mathrm{m}$. Using this mechanism, widths as low as $V_R/20$ have in fact been observed.

Cooling by velocity-selective coherent population trapping is radically different from cooling by the Doppler or Sisyphus mechanisms, because it does not use a friction force that reduces the velocities of all the atoms. Here we have a scattering process which gets each atom doing a random walk in the velocity space. When, by pure chance, this random walk happens to pass through the zero velocity, the scattering may stop, and it is this that causes a large number of atoms to accumulate around $V = 0$ as time goes by.

Another specific feature of cooling by velocity-selective coherent population trapping is that there is no steady state at finite temperature. Here the width ΔV of the peak (8A.14) in which the atoms accumulate goes on decreasing indefinitely as time goes by, and there is in principle no limit to this kind of cooling, provided that the coherence between the two components of $|\psi_{NC}\rangle$ and the two lasers can be maintained, which raises ever greater experimental difficulties as θ increases (see Section 8A.5). The absence of a steady state corresponds to the fact that this is a non-ergodic random process, because there is no characteristic time, no matter how long, over which a time average referring to the evolution of a single atom would be equivalent to an ensemble average over a large number of atoms. This non-ergodic feature is in turn related to the fact that the dwell time of atoms in non-coupled states obeys anomalous statistics, studied in particular by the mathematician Paul Lévy, characterized by broad distributions with divergent moments. A great many features of sub-recoil cooling have indeed been explained with reference to such statistics,[1] just as normal Gaussian statistics can help to explain the characteristics of Doppler cooling.

[1] F. Bardou, J.-P. Bouchaud, A. Aspect and C. Cohen-Tannoudji, *Lévy Statistics and Laser Cooling. When rare events bring atoms to rest*, Cambridge University Press (2001).

Comments

(i) There is another sub-recoil cooling mechanism, based on a Raman process. It involves inhomogeneous scattering in the velocity space, allowing the atoms to accumulate around $V = 0$.

(ii) A cooling process is characterized not only by achieving a narrow velocity distribution, but also by the fact that the density in the velocity space increases during cooling. (If this is not the case, we speak of filtering, rather than cooling.) This feature has been checked both experimentally and theoretically for the sub-recoil cooling mechanism discussed above.

(iii) The process presented above for just one space dimension can be generalized to two or three. Since the probability of reaching $V = 0$ by chance decreases as the dimension increases, it is then essential that there be a frictional force orientating the diffusion of the atoms toward $V = 0$.

(iv) Since it causes the atoms to accumulate in a small region of velocity space, cooling by velocity-selective coherent population trapping is reminiscent of the famous 'Maxwell demon', which was supposed to accumulate the particles of a gas in one of the two halves of a box. In both cases, the drop in entropy of the sample is associated with an increase in the entropy of the rest of the system that is at least as large. Here it is the entropy of the radiation that increases, as the photons are scattered out of the laser beam into initially empty modes.

8A.3 Quantum description of the atomic motion

In Section 8.2.2, we pointed out that the classical description of atomic motion is not valid if the velocity distribution of the atomic wave packet is narrower than the recoil velocity V_R (condition (8.33)). Now sub-recoil cooling leads to a distribution of width ΔV that is indeed less than V_R. We should therefore reconsider the problem with a quantum description of the centre of mass of the atom (see Section 8.1.1). To do this, we take the basis $\{|p\rangle\}$ of plane waves of momentum p in the Oz direction. An atomic state will thus be written $|i,p\rangle$, where i denotes the internal state and p the atomic momentum. With this notation, we can reconsider the setup in Figure 8A.1(a), introducing the family $\mathcal{F}(p)$ of the three states coupled by the lasers: $\{|e,p\rangle; |g_-,p-\hbar k\rangle; |g_+,p+\hbar k\rangle\}$. We have taken into account the conservation of momentum in the interaction processes with the two counterpropagating lasers, which impart momenta $+\hbar k$ and $-\hbar k$ to the atoms during the absorption processes, and the opposite momenta during stimulated emission processes. Figure 8A.4 depicts the family $\mathcal{F}(p)$.

Each family $\mathcal{F}(p)$ is closed with respect to the interaction Hamiltonian, which has the form,

$$\begin{aligned}\hat{H}_{\mathrm{I}} =& \frac{\hbar\Omega_1}{2}\mathrm{e}^{-\mathrm{i}\omega t}\,\mathrm{e}^{-\mathrm{i}\varphi_-}\,|e,p\rangle\langle g_-,p-\hbar k| + \mathrm{h.c.}\\ &+ \frac{\hbar\Omega_1}{2}\mathrm{e}^{-\mathrm{i}\omega t}\,\mathrm{e}^{-\mathrm{i}\varphi_+}\,|e,p\rangle\langle g_+,p+\hbar k| + \mathrm{h.c.},\end{aligned} \tag{8A.15}$$

keeping only the resonant terms.

A change of basis can be made within the family $\mathcal{F}(p)$, introducing the coupled and non-coupled states:

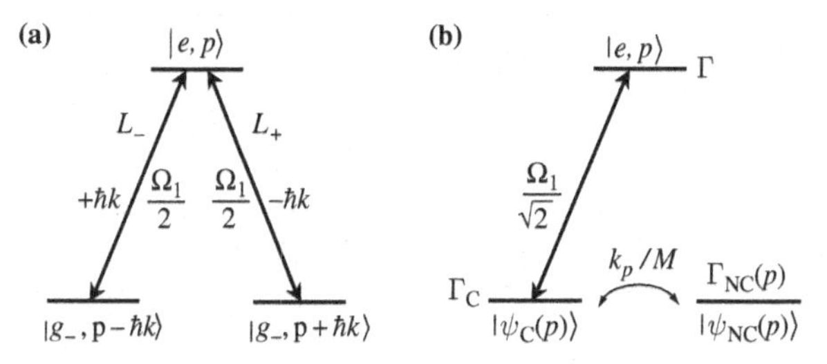

Figure 8A.4 Family $\mathcal{F}(p)$ of three states coupled by the two lasers L_- and L_+, propagating in the positive and negative z directions, respectively.

$$|\psi_{\mathrm{C}}(p)\rangle = \frac{1}{\sqrt{2}}\left(\mathrm{e}^{\mathrm{i}\varphi_+}|g_-,p-\hbar k\rangle + \mathrm{e}^{\mathrm{i}\varphi_-}|g_+,p+\hbar k\rangle\right) \tag{8A.16}$$

$$|\psi_{\mathrm{NC}}(p)\rangle = \frac{1}{\sqrt{2}}\left(\mathrm{e}^{\mathrm{i}\varphi_+}|g_-,p-\hbar k\rangle - \mathrm{e}^{\mathrm{i}\varphi_-}|g_+,p+\hbar k\rangle\right). \tag{8A.17}$$

Once again, $|\psi_{\mathrm{NC}}\rangle$ is not coupled to $|e,p\rangle$ by the lasers, due to destructive interference between the two terms appearing in the matrix element:

$$\begin{aligned}\langle e,p|\hat{H}_{\mathrm{I}}|\psi_{\mathrm{NC}}(p)\rangle =& \frac{\hbar\Omega_1}{2\sqrt{2}}\,\mathrm{e}^{-\mathrm{i}\omega t}\langle e,p|\Big[\mathrm{e}^{-\mathrm{i}\varphi_-}|e,p\rangle\langle g_-,p-\hbar k|\mathrm{e}^{\mathrm{i}\varphi_-}|g_-,p-\hbar k\rangle \\ &- \mathrm{e}^{-\mathrm{i}\varphi_+}|e,p\rangle\langle g_+,p+\hbar k|\mathrm{e}^{\mathrm{i}\varphi_+}|g_+,p+\hbar k\rangle\Big] = 0.\end{aligned} \tag{8A.18}$$

In contrast, $|\psi_{\mathrm{C}}\rangle$ is coupled to e:

$$\langle e,p|\hat{H}_{\mathrm{I}}|\psi_{\mathrm{C}}(p)\rangle = \mathrm{e}^{-\mathrm{i}\omega t}\frac{\hbar\Omega_1}{\sqrt{2}}. \tag{8A.19}$$

These couplings have been indicated in Figure 8A.4(b).

Since the centre of mass motion has been quantized, the Hamiltonian for the atom without the lasers is now,

$$\hat{H}_{\mathrm{At}} = \hat{H}_{\mathrm{int}} + \frac{\hat{p}^2}{2M}, \tag{8A.20}$$

where $\hat{H}_{\mathrm{int}} = \hbar\omega_0|e\rangle\langle e|$ describes the quantization of the internal degrees of freedom (taking $E_{g_+} = 0 = E_{g_-}$ and $\hbar\omega_0 = E_e - E_{g_+}$), and $\hat{p}^2/2M$ is the kinetic energy associated with the motion in the Oz direction. We then observe that $|g_-,p-\hbar k\rangle$ and $|g_+,p+\hbar k\rangle$ are eigenstates of $\hat{H}_{\mathrm{At}}$, but that their energies $(p\pm\hbar k)^2/2M$ are different. We thus conclude that $|\psi_{\mathrm{NC}}(p)\rangle$, composed of two eigenstates of $\hat{H}_{\mathrm{At}}$ corresponding to different eigenvalues, cannot be an eigenstate of $\hat{H}_{\mathrm{At}}$. It is not a stationary state. The above argument no longer applies in the case where $p=0$. Indeed, the state $|\psi_{\mathrm{NC}}(p=0)\rangle$ is the only stationary state of the total Hamiltonian $\hat{H}_{\mathrm{At}} + \hat{H}_{\mathrm{I}}$. Quite generally,

$$\langle\psi_{\mathrm{C}}(p)|\hat{H}_{\mathrm{At}}|\psi_{\mathrm{NC}}(p)\rangle = \frac{1}{4M}\left[(p-\hbar k)^2 - (p+\hbar k)^2\right] = -\hbar k\frac{p}{M}. \tag{8A.21}$$

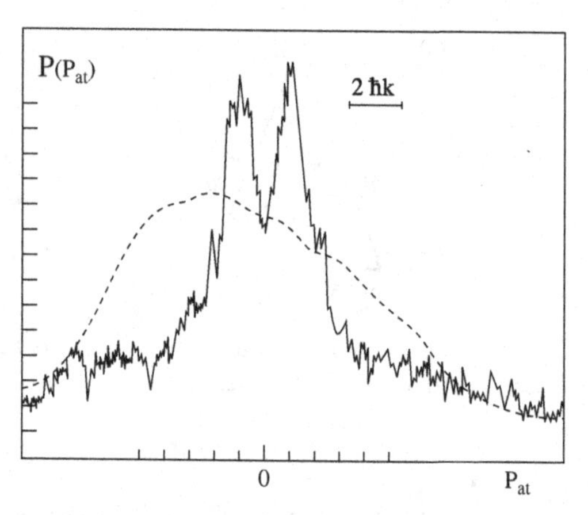

Figure 8A.5 Atomic velocity distribution after cooling by velocity-selective coherent population trapping. (Figure taken from A. Aspect *et al.*)[2] The two peaks, $2V_R$ apart, have greater heights than they did initially (dotted curve), which indicates that this really is a cooling mechanism, not merely a filtering in velocity space.

So if p differs from zero, an atom in the state $|\psi_{NC}(p)\rangle$ evolves towards the state $|\psi_C(p)\rangle$, which is coupled to $|e, p\rangle$ by the lasers (see Figure 8A.4(b)). Only the state $|\psi_{NC}(p = 0)\rangle$ is a non-coupled stationary state.

The formalism we have just set up leads back to the cooling process via velocity-selective coherent population trapping discussed in Section 8A.2. As long as the atom is in a state belonging to a family $\mathcal{F}(p)$ where p differs from zero, it is coupled to $|e, p\rangle$ and can emit a fluorescence photon in a random direction, causing it to move into a new family $\mathcal{F}(p')$, where p' lies between $p - 2\hbar k$ and $p + 2\hbar k$. In this way, the atom may end up in the family $\mathcal{F}(p = 0)$, with one chance in two of falling into the state $|\psi_{NC}(p = 0)\rangle$. In this case, it will remain there indefinitely. A significant fraction of the atoms will thus accumulate in the state $|\psi_{NC}(p)\rangle$, with p very close to 0. So what will be observed if we measure the atomic velocity distribution in the Oz direction? For atoms in

$$|\psi_{NC}(0)\rangle = \frac{1}{\sqrt{2}}\Big(|g_-, -\hbar k\rangle - |g_+, +\hbar k\rangle\Big), \tag{8A.22}$$

we may find either $-\hbar k/M = -V_R$ or $+\hbar k/M = -V_R$. There will therefore be two peaks, at $\pm V_R$. More generally, for states $|\psi_{NC}(p)\rangle$ with $|p| < \hbar k$, two distinct peaks will be observed, of half-width less than V_R, centred on $\pm V_R$. Figure 8A.5 shows an example of a signal obtained by C. Cohen-Tannoudji and coworkers. The width of the peaks is clearly less than V_R. Moreover, the density within the peaks is greater than the initial density, which indicates that this is indeed a cooling mechanism.

[2] A. Aspect, E. Arimondo, R. Kaiser, N. Vansteenkiste and C. Cohen-Tannoudji, Laser Cooling below the One-Photon Recueil Energy by Velocity-Selective Coherent Population Trapping, *Physical Review Letters* **61**, 826 (1988).

8A.4 Fluorescence rate of a state $|\psi_{NC}(p)\rangle$

As mentioned above, the width Δp of the peaks is expected to decrease as the time θ of the interaction with the lasers gets longer, since the fluorescence rate $\Gamma_{NC}(p)$ from $|\psi_{NC}(p)\rangle$ decreases with p. The formalism set up in Section 8A.3 obtains an expression for $\Gamma_{NC}(p)$ in the perturbative limit $kp/M \ll \Omega_1 \ll \Gamma$. As suggested by Figure 8A.4(b), we shall treat the coupling of $|\psi_C(p)\rangle$ to $|e\rangle$, under the effect of the lasers, separately from the coupling of $|\psi_{NC}(p)\rangle$ to $|\psi_C(p)\rangle$ due to the Doppler effect.

We begin by calculating the exit rate Γ_C from $|\psi_{NC}(p)\rangle$. It is equal to the fluorescence rate from $|\psi_C\rangle$ under the effect of the coupling to the state $|e\rangle$ with lifetime Γ^{-1} by the matrix element (8A.19). At resonance, the saturation parameter (2.189) is given by

$$s = \frac{(\Omega_1/\sqrt{2})^2}{\Gamma^2/4} = 2\frac{\Omega_1^2}{\Gamma^2}, \tag{8A.23}$$

and the population of the state $|e\rangle$ is equal to $s/2$, when s is much less than unity (see (2.161)). We deduce the fluorescence rate to be

$$\Gamma_C = \Gamma\frac{s}{2} = 2\frac{\Omega_1^2}{\Gamma}. \tag{8A.24}$$

We can then treat the state $|\psi_C\rangle$ as having a lifetime Γ_C^{-1}, and calculate the exit rate from $|\psi_{NC}(p)\rangle$ under the effect of the coupling to $|\psi_C(p)\rangle$ by an analogous expression, using the matrix element (8A.21). Replacing $\Omega_1/2$ by kp/M, the saturation parameter is

$$s' = 8\frac{k^2p^2}{M^2\Gamma_C^2}, \tag{8A.25}$$

whence the exit rate from $|\psi_{NC}(p)\rangle$ is finally given by

$$\Gamma_{NC}(p) = \Gamma_C\frac{s'}{2} = 2\frac{k^2p^2}{M^2\Omega_1^2}\Gamma. \tag{8A.26}$$

This is the expression used in Section 8A.2 to evaluate the width of the peak of the cooled atoms after an interaction time θ with the lasers.

8A.5 Practical limits. The fragility of coherence

If we want to obtain widths ΔV as small as possible (which correspond to very low temperatures, in the nanokelvin range), $|\psi_{NC}(0)\rangle$ must be kept decoupled from $|e\rangle$ for as long a time θ as possible, despite the presence of the lasers. Now this decoupling results from destructive interference between two quantum amplitudes, and any phenomenon perturbing the relative phase between these two amplitudes will destroy the interference, thereby allowing a direct transfer from $|\psi_{NC}(0)\rangle$ to $|e\rangle$. For example, if the phases of the lasers change, the state $\psi_{NC}(p)$ of (8A.17) will become coupled to e, as can be seen in (8A.18).

Indeed, in that expression, the term $e^{-i\varphi_-}$ refers to the laser phase, while the term $e^{i\varphi_-}$ refers to the state $\psi_{NC}(p)$, and if the laser phase changes there is no longer cancelling of the two terms. More precisely, it is readily seen that the important point is to have the phase difference between the two lasers constant. In fact, the relative phase between the two lasers L_+ and L_- always fluctuates a little, and after a time τ_C, it will have drifted by an amount that can exceed one radian. This fixes a first maximal limit τ_C for the time θ over which the process is expected to be effective. There may also be relative fluctuations in the energies of the levels $|g_-\rangle$ and $|g_+\rangle$, e.g. under the effect of an unwanted magnetic field in the case where $|g_+\rangle$ and $|g_-\rangle$ are magnetic sub-levels with different quantum numbers m_+ and m_-. In practice, it has been possible to control these technical sources of decoherence for a few tens of microseconds, thereby obtaining widths below $V_R/20$, associated with temperatures below 10 nanokelvins.

These extremely low temperatures have only been obtained with samples of very low density. Otherwise, there is a certain probability that the photons scattered by atoms with $p \neq 0$ will interact with an atom already in the state $|\psi_{NC}(0)\rangle$. Here is a limit that will be difficult to get around if we try to increase the density in the phase space $\{\mathbf{r}, \mathbf{p}\}$, i.e. in both velocity space and real space at the same time, whence the difficulty in achieving Bose–Einstein condensation with this technique.

Index

Printed in the United States
By Bookmasters